Geometric Properties of Area Elements

Material Property Relations

Poisson's ratio

$$\nu = -\frac{\epsilon_{\text{lat}}}{\epsilon_{\text{long}}}$$

Generalized Hooke's Law

$$\epsilon_x = \frac{1}{E}\left[\sigma_x - \nu(\sigma_y + \sigma_z)\right]$$

$$\epsilon_y = \frac{1}{E}\left[\sigma_y - \nu(\sigma_z + \sigma_x)\right]$$

$$\epsilon_z = \frac{1}{E}\left[\sigma_z - \nu(\sigma_x + \sigma_y)\right]$$

$$\gamma_{xy} = \frac{1}{G}\tau_{xy}, \; \gamma_{yz} = \frac{1}{G}\tau_{yz}, \; \gamma_{zx} = \frac{1}{G}\tau_{zx}$$

where

$$G = \frac{E}{2(1+\nu)}$$

Relations Between w, V, M

$$\frac{dV}{dx} = -w(x), \qquad \frac{dM}{dx} = V$$

Elastic Curve

$$\frac{1}{\rho} = \frac{M}{EI}$$

$$EI\,\frac{d^4v}{dx^4} = -w(x)$$

$$EI\,\frac{d^3v}{dx^3} = V(x)$$

$$EI\,\frac{d^2v}{dx^2} = M(x)$$

Buckling

Critical axial load

$$P_{\text{cr}} = \frac{\pi^2 EI}{(KL)^2}$$

Critical stress

$$\sigma_{\text{cr}} = \frac{\pi^2 EI}{(KL/r)^2}, \qquad r = \sqrt{I/A}$$

Secant formula

$$\sigma_{\text{max}} = \frac{P}{A}\left[1 + \frac{ec}{r^2}\sec\left(\frac{L}{2r}\sqrt{\frac{P}{EI}}\right)\right]$$

Energy Methods

Conservation of energy

$$U_e = U_i$$

Strain energy

$$U_i = \frac{N^2 L}{2AE} \qquad \text{constant axial load}$$

$$U_i = \int_0^L \frac{M^2 dx}{EI} \qquad \text{bending moment}$$

$$U_i = \int_0^L \frac{f_s V^2 dx}{2GA} \qquad \text{transverse shear}$$

$$U_i = \frac{T^2 L}{2GJ} \qquad \text{constant torsional moment}$$

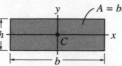

$A = bh$

$I_x = \frac{1}{12}bh^3$

$I_y = \frac{1}{12}hb^3$

Rectangular area

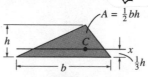

$A = \frac{1}{2}bh$

$I_x = \frac{1}{36}bh^3$

Triangular area

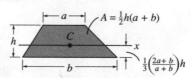

$A = \frac{1}{2}h(a+b)$

$\frac{1}{3}\left(\frac{2a+b}{a+b}\right)h$

Trapezoidal area

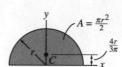

$A = \frac{\pi r^2}{2}$

$\frac{4r}{3\pi}$

$I_x = \frac{1}{8}\pi r^4$

$I_y = \frac{1}{8}\pi r^4$

Semicircular area

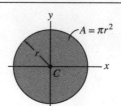

$A = \pi r^2$

$I_x = \frac{1}{4}\pi r^4$

$I_y = \frac{1}{4}\pi r^4$

Circular area

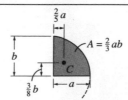

$\frac{2}{5}a$

$A = \frac{2}{3}ab$

$\frac{3}{8}b$

Semiparabolic area

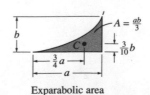

$A = \frac{ab}{3}$

$\frac{3}{10}b$

$\frac{3}{4}a$

Exparabolic area

Mechanics of Materials

R. C. HIBBELER

Macmillan Publishing Company
New York

Collier Macmillan Canada
Toronto

Maxwell Macmillan International
New York   Oxford   Singapore   Sydney

Editor: David Johnstone
Production Supervisor: Dora Rizzuto
Production Manager: Sandra E. Moore
Cover Designer: Blake Logan

This book was set in 10/12 Times Roman by York Graphic Services Inc,
printed by R.R. Donnelly & Sons Company, and bound by R.R. Donnelly & Sons Company
The cover was printed by Phoenix Color Comp.

Macmillan Publishing Company
866 Third Avenue, New York, New York 10022

Collier Macmillan Canada, Inc.
1200 Eglinton Avenue East
Suite 200
Don Mills, Ontario, M3C 3N1

LIBRARY OF CONGRESS CATALOGING-IN-PUBLICATION DATA

Hibbeler, R. C.
    Mechanics of materials / Russell C. Hibbeler.
        p.   cm.
    ISBN 0-02-354451-1
    1. Strength of materials.   2. Structural analysis (Engineering)
I. Title.
TA405.M516   1991   90-6539
620.1'12—dc20   CIP
Printing: 1 2 3 4 5 6 7 8   Year:1 2 3 4 5 6 7 8 9 0

Preface

This book is intended to provide the student with a clear and thorough presentation of both the theory and application of the fundamental principles of mechanics of materials. Understanding is based on first explaining the physical behavior of materials under load and then modeling this behavior to develop the theory. The development emphasizes the importance of satisfying equilibrium, compatibility of deformation, and material behavior requirements. Many analysis and design applications are presented for basic mechanical elements and structural members often encountered in engineering practice.

Organization and Approach. In order to aid both the instructor and the student, the contents of each chapter are organized into well-defined sections. Selected groups of sections contain an explanation of specific topics, followed by illustrative example problems and a set of homework problems. The topics within each section are often placed in subgroups denoted by boldface titles. The purpose of this is to present a structured method for introducing each new definition or concept and to make the book convenient for later reference and review.

As in my other textbooks, a ''procedure for analysis'' is used throughout the book, providing the student with a logical and orderly method to follow when applying the theory. The example problems are then solved using this outlined method in order to clarify its numerical application. It is to be understood, however, that once the relevant principles have been mastered and enough confidence and judgment have been acquired, the student can then develop his or her own procedures for solving problems. In most cases, it is felt that the first step in any procedure should be to draw a diagram. In doing so, the student forms the habit of tabulating the necessary data while focusing on the physical aspects of the problem and its associated geometry. If this step is correctly performed, applying the relevant equations becomes somewhat methodical, since the data can be taken directly from the diagram.

The subject matter is organized into 14 chapters. Chapter 1 begins with a review of the important concepts of statics, followed by a formal definition of both normal and shear stress, and a discussion of normal stress in axially loaded members and average shear stress caused by direct shear. In Chapter 2 normal and shear strain are defined, and in Chapter 3 a discussion of some of the important mechanical properties of materials is given. Separate treatments of axial load, torsion, and bending are presented in Chapters 4, 5, and 6, respectively. In each of these chapters, both linear-elastic and plastic behavior of the material is considered. Also, topics related to stress concentrations and residual stress are included. Transverse shear is discussed in Chapter 7, along with a discussion of thin-walled tubes, shear flow, and the shear center. Chapter 8 provides a partial review of the material covered in the previous chapters, in which the state of stress resulting from combined loadings is discussed. In Chapter 9 the concepts for transforming multiaxial states of stress are presented. In a similar manner, Chapter 10 discusses the methods for strain transformation, including the application of various theories of failure. Chapter 11 provides a means for a further summary and review of previous material by covering design applications of beams and shafts. In Chapter 12 various methods for computing deflections of beams and shafts are covered. Also included is a discussion for finding the reactions on these members if they are statically indeterminate. Chapter 13 provides a discussion of column buckling, and lastly, in Chapter 14 the problem of impact and the application of various energy methods for computing deflections are considered.

Sections of the book that contain more advanced material are indicated by a star (★). Time permitting, some of these topics may be included in the course. Furthermore, this material provides a suitable reference for basic principles when it is covered in other courses, and it can be used as a basis for assigning special projects.

Alternative Method of Coverage. Some instructors prefer to cover stress and strain transformations *first,* before discussing specific applications of axial load, torsion, bending, and shear. One possible method for doing this would be first to cover stress and its transformation, Chapter 1 and Chapter 9, followed by strain and its transformation, Chapter 2 and the first part of Chapter 10. The discussion and example problems in these later chapters have been styled so that this is possible. Chapters 3 through 8 can then be covered with no loss in continuity.

Problems. Numerous problems in the book depict realistic situations encountered in engineering practice. It is hoped that this realism will both stimulate the student's interest in the subject and provide a means for developing the skill to reduce any such problem from its physical description to a model or symbolic representation to which the principles may be applied.

Throughout the text there is an approximate balance of problems using either SI or FPS units. Furthermore, in any set, an attempt has been made to arrange the problems in order of increasing difficulty. The answers to all but every fourth problem are listed in the back of the book. To alert the user to a problem without a reported answer, an asterisk (*) is placed before the problem number. Answers are reported to three significant figures, even though the data for material properties may be known with less accuracy. Although this might appear to be poor practice, it is done simply to be consistent and to allow the student a better chance to validate his or her solution. All the problems and their solutions have been independently checked for accuracy.

Computer Tutorial. Included with this book is a tutorial disk that contains several hundred review questions and problems on the major topics discussed in the book. It has been designed to run on an IBM or compatible computer having color graphics capability. The problems on this disk are very elementary and can be solved without the use of a pocket calculator. Many of my students have found it useful for self testing their understanding of the concepts and as a means for building their confidence prior to solving problems or taking examinations.

Acknowledgments. Preparation of the manuscript for this book has taken many years, during which time it has undergone several reviews and rewritings. In this regard, many of my colleagues in the teaching profession have reviewed successive revisions of the manuscript and I owe them a personal debt of gratitude. Their encouragement and willingness to provide constructive criticism are very much appreciated. In particular, Professor Edward C. Ting of Purdue University was especially helpful, since he reviewed the manuscript several times. Also, Professors Edward E. Hornsey, University of Missouri–Rolla; Charles O. Harris, Michigan State University (formerly); and Will Liddell, Jr., Auburn University–Montgomery, assisted in reviewing and/or checking many of the problems. Many others helped in the development and review of this work. They include: Professors Nicholas J. Altiero, Michigan State University; J. R. Barber, University of Michigan; Peter B. Cooper, Kansas State University; George Costello, University of Illinois at Urbana-Champaign; J. Edmund Fitzgerald, Georgia Institute of Technology; Reese J. Goodwin, Brigham Young University; Frederick M. Graham, Iowa State University; Serope Kalpakjian, Illinois Institute of Technology; John C. McWhorter, Mississippi State University; George J. Murnen, University of Toledo; Thomas M. Murray, USAF Academy; Mario Paz, University of Louisvile; Edward W. Price, North Dakota State University; John D. Ray, Memphis State University; C. E. Taylor, University of Florida; Jan. J. Tuma, Arizona State University; and Han-Chin Wu, University of Iowa. Furthermore, the computer tutorial was reviewed by Gordon Baston, Clarkson University; Royce Beckett, Auburn University; Anthony Chang, Stevens Institute of Technology; Reese J. Goodwin, Brigham Young University; and Stephen Bechtel, Ohio State University.

I would also like to thank all my students who have used the manuscript and the computer tutorial as well as their revisions, and made comments to improve their contents. A particular note of thanks goes to one of my former graduate students, Kai Beng Yap, who has been a great help to me in this regard. A special note of gratitude also goes to my editors and the staff at Macmillan, who have all been very supportive in allowing me to have a more creative and artistic license in the design and execution of this book. It has been a pleasure to work with all of them. Finally, appreciation goes to my wife, Conny, who has been a source of encouragement and has helped with the details of preparing the manuscript and its revisions.

Russell Charles Hibbeler

Contents

4

Axial Load   111

5

Torsion   171

6

Bending   247

7

Transverse Shear 325

8

Combined Loadings 373

9

Stress Transformation 399

10

Strain Transformation 449

11

Design of Beams and Shafts   499

12

Deflections of Beams and Shafts   557

13

Buckling of Columns   631

14

Energy Methods 685

A

Geometric Properties of an Area 753

B

Properties of Wide-Flange Sections 770

Answers 773

Index 789

Mechanics of Materials

1 Stress

In this chapter we will review some of the important principles of statics and show how they are used to determine the internal resultant loadings in a body. Afterwards the concepts of normal and shear stress will be introduced, and specific applications of the analysis and design of members subjected to an axial load or direct shear will be discussed.

1.1 Introduction

Mechanics of materials is a branch of mechanics that develops relationships between the *external* loads applied to a deformable body and the intensity of *internal* forces acting within the body. This subject is also concerned with computing the *deformations* of the body, and it provides a study of the body's *stability* when the body is subjected to external forces.

In the design of any structure or machine, it is *first* necessary to use the principles of statics to determine the forces acting both on and within its various members. Furthermore, the size of the members, their deflection, and their stability depend not only on these internal loadings, but also on the nature of the material from which the members are made. As a result, an accurate determination and fundamental understanding of *material behavior* is of vital importance to the development of the necessary equations used in mechanics of materials. Here we will not be interested in the specific details of experimental methods. Instead, we will state the experimental results and explain how they are used. Realize that many formulas and rules for design, as defined in engineering codes, are based on the fundamentals of mechanics of materials, and for this reason an understanding of the principles of this subject is very important.

Historical Development. The origin of mechanics of materials dates back to the beginning of the seventeenth century, at which time Galileo performed experiments to study the effects of loads on rods and beams made of various materials. For a proper understanding, however, it was necessary to have accurate experimental descriptions of a material's mechanical properties. Methods for doing this were remarkably improved at the beginning of the eighteenth century. At that time both experimental and theoretical study in this subject was undertaken in France by such notables as Saint-Venant, Poisson, Lamé, and Navier. Because their efforts were based on material-body applications of mechanics, they called this study "strength of materials." Currently, however, it is usually referred to as "mechanics of deformable bodies" or simply "mechanics of materials."

Over the years, after many of the fundamental problems of mechanics of materials had been solved, it became necessary to use advanced mathematical and computer techniques to solve more complex problems. As a result, this subject expanded into other subjects of advanced mechanics such as the *theory of elasticity* and the *theory of plasticity*. Research in these fields is ongoing, not only to meet the demands for solving advanced engineering problems, but to justify further the use and limitations upon which the fundamental theory of mechanics of materials is based.

1.2 Equilibrium of a Deformable Body

Since statics plays an important role in both the development and application of mechanics of materials, it is necessary to have a good grasp of its fundamentals. For this reason we will review some of the main principles of statics that will be used throughout the text.

External Loads. A body can be subjected to several different types of external loads; however, any one of these can be classified as either a surface force or a body force.

Surface Forces. As the name implies, *surface forces* are caused by the direct contact of one body with the surface of another. In all cases these forces are distributed over the *area* of contact between the bodies, Fig. 1–1a. In particular, if this area is small in comparison with the total surface area of the body, then the surface force may be *idealized* as a single *concentrated force*, which is applied to a *point* on the body, Fig. 1–1a. For example, this might be done to represent the effect of the ground on the wheels of a bicycle when studying the loading on the bicycle. Also, if the surface loading is applied along a narrow area, the loading may be *idealized* as a *linear distributed load, w(s)*. Here the loading is measured as having an intensity of force/length along the area and is represented graphically by a series of arrows along the line *s*, Fig. 1–1a. The loading along the length of a beam is a typical example

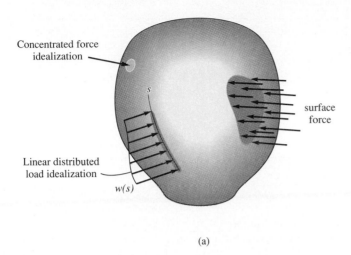

Concentrated force idealization

s

surface force

Linear distributed load idealization

w(s)

(a)

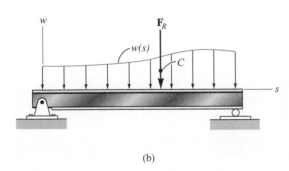

w

F$_R$

w(s)

C

s

(b)

Fig. 1–1

of where this idealization is often applied, Fig. 1–1*b*. The *resultant force F$_R$* of *w(s)* is equivalent to the *area under the distributed loading curve*, and this resultant acts through the *centroid C* or geometric center of this area.

Body Force.   A *body force* occurs when one body exerts a force on another body without direct physical contact between the bodies. Examples include the effects caused by the earth's gravitation or its electromagnetic field. Although body forces affect each of the particles composing the body, they are normally represented by a single concentrated force acting on the body. In the case of gravitation, this force is called the *weight* of the body and acts through the body's center of gravity.

Table 1–1

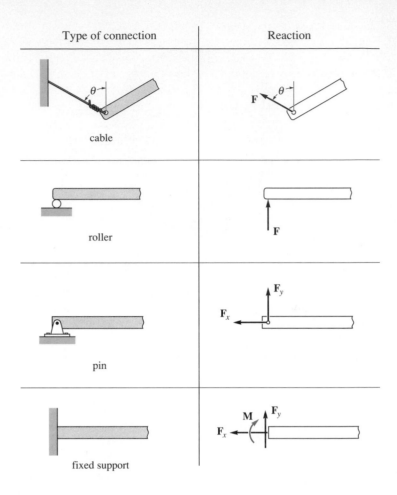

Type of connection	Reaction
cable	
roller	
pin	
fixed support	

Support Reactions. The surface forces that develop at the supports or points of support between bodies are called *reactions*. For two-dimensional problems, i.e., bodies subjected to coplanar force systems, the supports most commonly encountered are shown in Table 1–1. Note the symbol used to represent each support and the type of reactions it exerts on its contacting member. One possible way to determine a type of support reaction is to imagine the attached member as being translated or rotated in a particular direction. If the support *prevents translation* in a given direction, then a *force* is developed on the member in that direction. Likewise, if *rotation* is prevented, a *couple moment* is exerted on the member. For example, a roller support only prevents translation in the contact direction, perpendicular or normal to the surface. Hence, the roller exerts a normal force $\mathbf{F}$ on the member at the point of contact. Since the member is free to rotate about the roller, a couple moment cannot be developed by the roller on the member at the point of contact.

The concentrated forces and couple moment shown in Table 1–1 actually represent the *resultants* of surface forces that exist between each support and its contacting member. Although it is these resultants that are determined in practice, it is generally not important to determine the actual load distribution, since the surface area over which it acts is considerably smaller than the total surface area of the contacting member.

Equations of Equilibrium.

Equilibrium of a body requires both a *balance of forces*, to prevent the body from translating or moving along a straight or curved path, and a *balance of moments*, to prevent the body from rotating. These conditions can be expressed mathematically by the two vector equations

$$\Sigma \mathbf{F} = \mathbf{0}$$
$$\Sigma \mathbf{M}_O = \mathbf{0}$$

(1–1)

Here, $\Sigma \mathbf{F}$ represents the sum of all the forces acting on the body, and $\Sigma \mathbf{M}_O$ is the sum of the moments of all the forces about any point O either on or off the body. If an x, y, z coordinate system is established with the origin at point O, the force and moment vectors can be resolved into components along the coordinate axes and the above two equations can be written in scalar form as six equations, namely,

$$\Sigma F_x = 0 \qquad \Sigma F_y = 0 \qquad \Sigma F_z = 0$$
$$\Sigma M_x = 0 \qquad \Sigma M_y = 0 \qquad \Sigma M_z = 0$$

(1–2)

Often in engineering practice the loading on a body can be represented as a system of *coplanar forces*. If this is the case, and the forces lie in the x–y plane, then the conditions for equilibrium of the body can be specified by only three scalar equilibrium equations; that is,

$$\Sigma F_x = 0$$
$$\Sigma F_y = 0$$
$$\Sigma M_O = 0$$

(1–3)

In this case, if point O is the origin of coordinates, the moments will be directed along the z axis, which is perpendicular to the plane that contains the forces.

Successful application of the equations of equilibrium requires complete specification of all the known and unknown forces that act *on* the body. *The best way to account for these forces is to draw the body's free-body diagram.* Obviously, if the free-body diagram is drawn correctly, the effects of all the applied forces and couple moments can be accounted for when the equations of equilibrium are written.

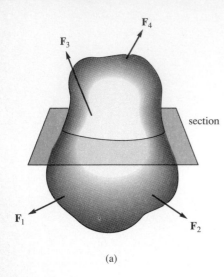

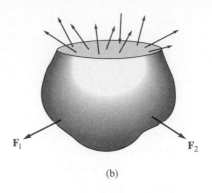

(b)

(a)

Fig. 1–2(a, b)

Internal Loadings. One of the most important applications of statics in the analysis of problems involving mechanics of materials is to be able to determine the resultant force and moment acting *within* a body, which are necessary to hold the body together when the body is subjected to external loads. For example, consider the body shown in Fig. 1–2a, which is held in equilibrium by the four external forces.* In order to obtain the *internal loadings* acting on a specific region within the body, it is necessary to use the *method of sections*. This requires that an imaginary section or "cut" be made through the region where the internal loadings are to be determined. The two parts of the body are then separated, and a free-body diagram of one of the parts is drawn. If we consider the section shown in Fig. 1–2a, then the resulting free-body diagram of the bottom part of the body is shown in Fig. 1–2b. Here it can be seen that there is actually a *distribution* of internal force acting on the "exposed" area of the section. These forces represent the effects of the material of the top part of the body acting on the adjacent material of the bottom part. Although this exact distribution may be *unknown*, we can use statics to determine the *resultant internal force and moment*, $\mathbf{F}_R$ and $\mathbf{M}_{R_O}$, *this distribution exerts at a specific point O on the sectioned area*, Fig. 1–2c.

Since the entire body is in equilibrium, then each part of the body is also in equilibrium. Consequently, $\mathbf{F}_R$ and $\mathbf{M}_{R_O}$ can be determined by applying Eqs. 1–1 to any one of the two parts of the sectioned body. When doing so, note that $\mathbf{F}_R$ is at point O, although its computed value will *not* depend on the location of this point. On the other hand, $\mathbf{M}_{R_O}$ does depend on this location, since the moment arms must extend from O to the line of action of each force on the free-body diagram. It will be shown in later portions of the text that point O is most often chosen at the *centroid* of the sectioned area, and so we will always choose this location for O, unless otherwise stated. Also, if a member is long and slender, as in the case of a rod or beam, the section to be considered is generally taken *perpendicular* to the longitudinal axis of the member. This section is referred to as the *cross section*.

---

*The body's weight is not shown, since it is assumed to be quite small, and therefore negligible compared with the other loads.

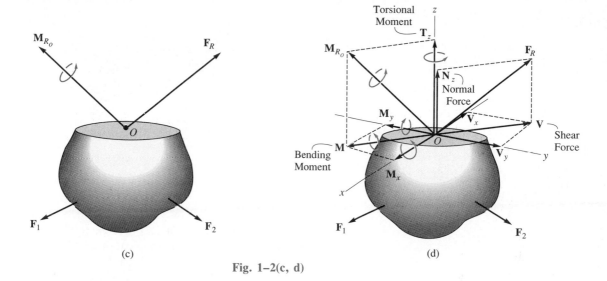

Fig. 1–2(c, d)

Later in this text we will show how to relate the resultant internal force and moment to the *distribution of force* on the sectioned area, Fig. 1–2*b*, and thereby develop equations that can be used for analysis and design of the body. To do this, however, the components of $\mathbf{F}_R$ and $\mathbf{M}_{R_O}$, acting both normal or perpendicular to the sectioned area and within the plane of the area, must be considered. If we establish x, y, z axes with origin at point O, as shown in Fig. 1–2*d*, then $\mathbf{F}_R$ and $\mathbf{M}_{R_O}$ can each be resolved into three components. Four different types of loadings can then be defined as follows:

$\mathbf{N}_z$ is called the *normal force*, since it acts perpendicular to the area. This force is developed when the external loads tend to push or pull on the two segments of the body.

$\mathbf{V}$ is called the *shear force*, and it can be determined from its two components using vector addition, $\mathbf{V} = \mathbf{V}_x + \mathbf{V}_y$. The shear force lies in the plane of the area and is developed when the external loads tend to cause the two segments of the body to slide over one another.

$\mathbf{T}_z$ is called the *torsional moment* or *torque*. It is developed when the external loads tend to twist one segment of the body with respect to the other.

$\mathbf{M}$ is called the *bending moment*. It is determined from the vector addition of its two components, $\mathbf{M} = \mathbf{M}_x + \mathbf{M}_y$. The bending moment is caused by the external loads that tend to bend the body about an axis lying within the plane of the area.

In this text, note that representation of a moment or torque is shown in three dimensions as a vector with an associated curl, Fig. 1–2*d*. By the *right-hand rule*, the thumb gives the arrowhead sense of the vector and the fingers or curl indicate the tendency for rotation (twist or bending).

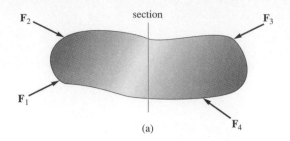

(a)

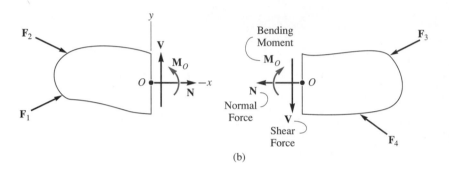

(b)

Fig. 1–3

Each of the six unknown x, y, z components of force and moment shown in Fig. 1–2d can be determined directly from the six equations of equilibrium, that is, Eqs. 1–2, applied to either segment of the body. If, however, the body is subjected to a *coplanar system of forces*, then only normal-force, shear, and bending-moment components will exist at the section. To show this, consider the body in Fig. 1–3a. If it is in equilibrium, then the internal resultant components, acting at the indicated section, can be determined by first "cutting" the body into two parts, as shown in Fig. 1–3b, and then applying the equations of equilibrium to one of the parts. Here the internal resultants consist of the normal force N, shear force V, and bending moment M_O. These loadings must be equal in magnitude and opposite in direction on each of the sectioned parts (Newton's third law). Furthermore, the magnitude of each unknown is determined by applying the three equations of equilibrium to either one of these parts, Eqs. 1–3. If we use the x, y, z coordinate axes, with origin established at point O, as shown on the left segment, then a direct solution for N can be obtained by applying $\Sigma F_x = 0$, and V can be obtained directly from $\Sigma F_y = 0$. Finally, the bending moment M_O can be determined by summing moments about point O (the z axis), $\Sigma M_O = 0$, in order to eliminate the moments caused by the unknowns N and V.

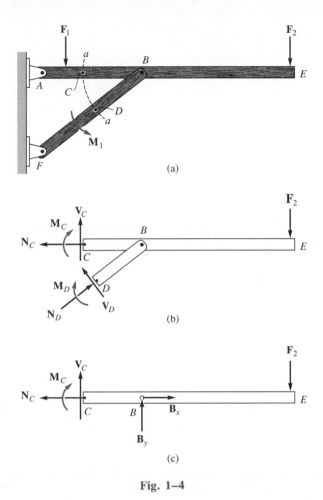

Fig. 1–4

If the body represents one of the members or parts of a structure or machine, then the structure or machine may have to be disassembled in order to determine the support reactions on the member *before* computing the internal loadings. For example, consider the frame shown in Fig. 1–4a. If section *a–a* passes through the frame at C and D, the resulting free-body diagram of the right portion of the section is shown in Fig. 1–4b. At each cross-sectioned member there are *three* unknown internal resultants. As a result, we cannot apply the *three* available equations of equilibrium to obtain these *six unknowns*. Instead, to solve this problem we must *first* disassemble the frame and determine the reactions at the pin connections by applying the equations of equilibrium to each member. Then each member may be sectioned at its appropriate point and the equilibrium equations applied to determine the internal resultants. In this regard, the free-body diagram of segment CBE, Fig. 1–4c, can be used to determine the three internal loadings at C, provided the pin reactions $\mathbf{B}_x$ and $\mathbf{B}_y$ are known.

PROCEDURE FOR ANALYSIS

The method of sections is used to determine the *internal* loadings at a point located on the section of a body. These resultants are statically equivalent to the forces that are *distributed* over the material at the sectioned area. If the body is static, that is, at rest or moving with constant velocity, the resultants must be in equilibrium with the external loadings acting on either one of the sectioned segments of the body.

We will now present a procedure that can be used for applying the method of sections to determine the internal resultant normal force, shear force, bending moment, and torsional moment at a specific location in a body.

Support Reactions.   Before the body is sectioned, it may first be necessary to determine either its support reactions or the reactions at its connections. This is done by drawing the body's free-body diagram, establishing a coordinate system, and then applying the equations of equilibrium to the body.

Free-Body Diagram.   Keep all external distributed loadings, couple moments, torques, and forces acting on the body in their *exact locations*, then pass an imaginary section through the body at the point where the internal resultant loadings are to be determined. If the body represents a member of a structure or mechanical device, this section is often taken *perpendicular* to the axis of the member. Draw a free-body diagram of one of the "cut" segments and indicate the unknown resultants **N, V, M,** and **T** at the section. In most cases, these resultants are placed at the point representing the geometric center or *centroid* of the sectioned area. In particular, if the member is subjected to a *coplanar* system of forces, only **N, V,** and **M** act at the centroid.

Establish the x, y, z coordinate axes at the centroid and show the resultant components acting along the axes.

Equations of Equilibrium.   Apply the equations of equilibrium to obtain the unknown resultants. Moments should be summed at the section, about the axes where the resultants act. Doing this eliminates the unknown forces **N** and **V** and allows a direct solution for **M** (and **T**). If the solution of the equilibrium equations yields a negative value for a resultant, the assumed *directional sense* of the resultant is *opposite* to that shown on the free-body diagram.

The following examples illustrate this procedure numerically and also provide a *review* of some of the important principles of statics. *Since statics plays such an important role in the analysis of problems in mechanics of materials, it is highly recommended that one solves as many problems as possible of those that follow these examples.*

Example 1–1

Determine the resultant internal loadings acting on the cross section at C of the machine shaft shown in Fig. 1–5a. The shaft is supported by bearings at A and B, which exert only vertical forces on the shaft.

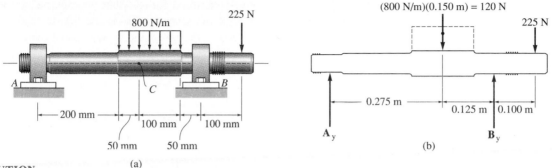

(a)

(b)

SOLUTION

Support Reactions. A free-body diagram of the entire shaft is shown in Fig. 1–5b. Note that the distributed loading has been replaced by its resultant, which has a magnitude equal to the area under the loading curve (rectangle) and acts through the centroid of this area.

$\zeta + \ \Sigma M_A = 0;\ \ \ -(120\ \text{N})(0.275\ \text{m}) + B_y(0.400\ \text{m})$
$$- (225\ \text{N})(0.500\ \text{m}) = 0 \quad B_y = 363.75\ \text{N}$$

$+\uparrow \ \Sigma F_y = 0; \qquad A_y - 120\ \text{N} + 363.75\ \text{N} - 225\ \text{N} = 0$
$$A_y = -18.75\ \text{N}$$

The negative sign for A_y indicates that $\mathbf{A}_y$ acts in the *opposite sense* to that shown in Fig. 1–5b.

Free-Body Diagrams. Passing an imaginary section perpendicular to the axis of the shaft through C yields the free-body diagrams of segments AC and CB shown in Fig. 1–5c. It is important when constructing these diagrams to keep the distributed loading *exactly* where it is until *after* the section is made. Only then should this loading be replaced by a single resultant force. Also, notice that $\mathbf{N}_C$, $\mathbf{V}_C$, and $\mathbf{M}_C$ have an equal magnitude but opposite direction on each segment—Newton's third law.

Equations of Equilibrium. Applying the equations of equilibrium to segment AC, we have

$\xrightarrow{+} \ \Sigma F_x = 0; \qquad\qquad\qquad N_C = 0 \qquad\qquad Ans.$

$+\uparrow \ \Sigma F_y = 0; \qquad\qquad -18.75\ \text{N} - 40\ \text{N} - V_C = 0$
$$V_C = -58.8\ \text{N} \qquad\qquad Ans.$$

$\zeta + \ \Sigma M_C = 0; \ \ M_C + 40\ \text{N}(0.025\ \text{m}) + 18.75\ \text{N}(0.250\ \text{m}) = 0$
$$M_C = -5.69\ \text{N} \cdot \text{m} \qquad\qquad Ans.$$

What do the negative signs for V_C and M_C indicate? As an exercise, try to obtain the same results using segment CB of the shaft.

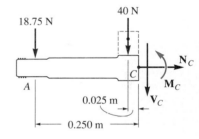

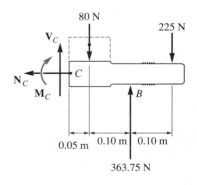

(c)

Fig. 1–5

Example 1–2

The hoist in Fig. 1–6a consists of the beam *AB* and attached pulleys, the cable, and the motor. Determine the resultant internal loadings acting on the cross section at *C* if the motor is lifting the 500-lb load *W* with constant velocity. Neglect the weight of the pulleys and beam.

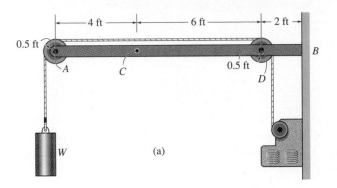

(a)

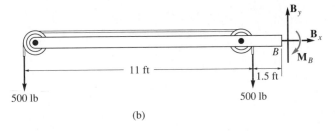

(b)

Fig. 1–6

SOLUTION I

Since the load is hoisted at constant velocity, the equations of equilibrium apply.

Support Reactions.   The free-body diagram of the beam *including* the pulleys and a portion of the cable is shown in Fig. 1–6b. The fixed wall at *B* exerts three reactions on the beam. Also, the cable is subjected to a constant tension force of 500 lb at its ends. Applying the equations of equilibrium yields

$$\xrightarrow{+} \ \Sigma F_x = 0; \qquad\qquad\qquad\qquad B_x = 0$$

$$+\uparrow \ \Sigma F_y = 0; \qquad B_y - 500 \text{ lb} - 500 \text{ lb} = 0 \qquad B_y = 1000 \text{ lb}$$

$$\downarrow^+ \ \Sigma M_B = 0; \qquad -M_B + 500 \text{ lb}(1.5 \text{ ft}) + 500(12.5 \text{ ft}) = 0$$

$$M_B = 7000 \text{ lb} \cdot \text{ft}$$

Free-Body Diagram. Using these results, the free-body diagrams of the pulley at *D* and beam segment *CB* are shown in Fig. 1–6c. Note that the 500-lb force components of the beam on the center of the pulley must be equal and opposite to the 500-lb force components of the pulley on the beam—Newton's third law.

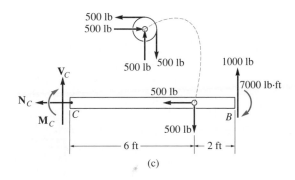

(c)

Equations of Equilibrium. For segment *CB* we have

$\xrightarrow{+} \Sigma F_x = 0;$ $-N_C - 500 \text{ lb} = 0$ $N_C = -500 \text{ lb}$ *Ans.*

$+\uparrow \Sigma F_y = 0;$ $V_C - 500 \text{ lb} + 1000 \text{ lb} = 0$ $V_C = -500 \text{ lb}$ *Ans.*

$\zeta + \Sigma M_C = 0;$ $-M_C - 500 \text{ lb}(6 \text{ ft}) + 1000 \text{ lb}(8 \text{ ft}) - 7000 \text{ lb} \cdot \text{ft} = 0$

$$M_C = -2000 \text{ lb} \cdot \text{ft} \qquad \textit{Ans.}$$

The negative signs indicate that each of these loadings acts in the opposite direction to that shown in Fig. 1–6c.

SOLUTION **II**

This problem can be solved in a more *direct manner* by sectioning the cable and the beam at *C* and then considering the entire left segment.

Free-Body Diagram. See Fig. 1–6d.

Equations of Equilibrium

$\xrightarrow{+} \Sigma F_x = 0;$ $500 \text{ lb} + N_C = 0$ $N_C = -500 \text{ lb}$ *Ans.*

$+\uparrow \Sigma F_y = 0;$ $-500 \text{ lb} - V_C = 0$ $V_C = -500 \text{ lb}$ *Ans.*

$\zeta + \Sigma M_C = 0;$ $500 \text{ lb}(4.5 \text{ ft}) - 500 \text{ lb}(0.5 \text{ ft}) + M_C = 0$

$$M_C = -2000 \text{ lb} \cdot \text{ft} \qquad \textit{Ans.}$$

Try obtaining these same results by removing the pulley at *A* from the beam and showing the 500-lb force components of the pulley acting on the beam segment *AC*.

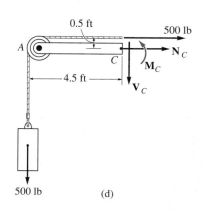

(d)

Example 1–3

Determine the resultant internal loadings acting on the cross section at *G* of the wooden beam shown in Fig. 1–7*a*. Assume the joints at *A*, *B*, *C*, *D*, and *E* are pin-connected.

Fig. 1–7

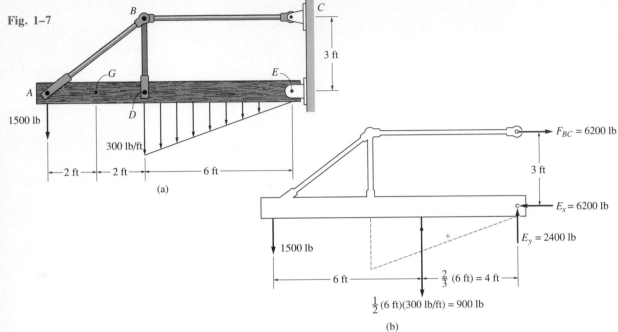

(a)

(b)

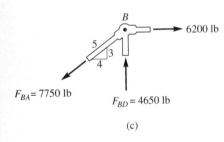

(c)

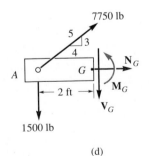

(d)

SOLUTION

Support Reactions. A free-body diagram of the *entire* structure is shown in Fig. 1–7*b*. Verify the computed reactions at *E* and *C*. In particular, note that *BC* is a *two-force member*, so the reaction at *C* must be horizontal. Why?

Also, since *BA* and *BD* are two-force members, the free-body diagram of joint *B* is shown in Fig. 1–7*c*. Again, verify the magnitudes of the computed forces $\mathbf{F}_{BA}$ and $\mathbf{F}_{BD}$.

Free-Body Diagrams. Using the computed reactions at *A*, *D*, and *E*, the left section of the beam is shown in Fig. 1–7*d*.

Equations of Equilibrium. Applying the equations of equilibrium to segment *AG*, since it involves the least number of forces, we have

$\xrightarrow{+} \Sigma F_x = 0;$ $7750 \text{ lb}(\frac{4}{5}) + N_G = 0$ $N_G = -6200 \text{ lb}$ *Ans.*

$+\uparrow \Sigma F_y = 0;$ $-1500 \text{ lb} + 7750 \text{ lb}(\frac{3}{5}) - V_G = 0$

$V_G = 3150 \text{ lb}$ *Ans.*

$\zeta+ \Sigma M_G = 0;$ $M_G - (7750 \text{ lb})(\frac{3}{5})(2 \text{ ft}) + 1500 \text{ lb}(2 \text{ ft}) = 0$

$M_G = 6300 \text{ lb} \cdot \text{ft}$ *Ans.*

As an exercise, compute these same results using segment *GE*.

Example 1–4

Determine the resultant internal loadings acting on the cross section at *B* of the pipe shown in Fig. 1–8*a*. The pipe has a mass of 2 kg/m and is subjected to both a vertical force of 50 N and a couple moment of 70 N · m at its end *A*. It is fixed-connected to the wall at *C*.

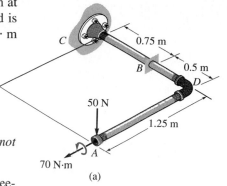

(a)

SOLUTION

The problem can be solved by considering segment *AB*, which does *not* involve the support reactions at *C*.

Free-Body Diagram. The *x, y, z* axes are established at *B* and the free-body diagram of segment *AB* is shown in Fig. 1–8*b*. The resultant force and moment components at the section are assumed to act in the positive coordinate directions and to pass through the *centroid* of the cross-sectional area at *B*. The weight of each segment of pipe is calculated as follows:

$$W_{BD} = (2 \text{ kg/m})(0.5 \text{ m})(9.81 \text{ N/kg}) = 9.81 \text{ N}$$
$$W_{AD} = (2 \text{ kg/m})(1.25 \text{ m})(9.81 \text{ N/kg}) = 24.52 \text{ N}$$

These forces act through the center of gravity of each segment.

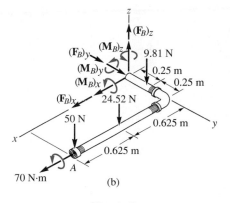

(b)

Fig. 1–8

Equations of Equilibrium. Applying the six scalar equations of equilibrium, we have*

$$\Sigma F_x = 0; \qquad\qquad (F_B)_x = 0 \qquad\qquad Ans.$$
$$\Sigma F_y = 0; \qquad\qquad (F_B)_y = 0 \qquad\qquad Ans.$$
$$\Sigma F_z = 0; \quad (F_B)_z - 9.81 \text{ N} - 24.52 \text{ N} - 50 \text{ N} = 0$$
$$\qquad\qquad\qquad (F_B)_z = 84.3 \text{ N} \qquad\qquad Ans.$$
$$\Sigma(M_B)_x = 0; \quad (M_B)_x + 70 \text{ N} \cdot \text{m} - 50 \text{ N}(0.5 \text{ m}) - 24.52 \text{ N}(0.5 \text{ m})$$
$$\qquad\qquad\qquad\qquad - 9.81 \text{ N}(0.25 \text{ m}) = 0$$
$$\qquad\qquad\qquad (M_B)_x = -30.3 \text{ N} \cdot \text{m} \qquad\qquad Ans.$$
$$\Sigma(M_B)_y = 0; \quad (M_B)_y + 24.52 \text{ N}(0.625 \text{ m}) + 50 \text{ N}(1.25 \text{ m}) = 0$$
$$\qquad\qquad\qquad (M_B)_y = -77.8 \text{ N} \cdot \text{m} \qquad\qquad Ans.$$
$$\Sigma(M_B)_z = 0; \qquad\qquad (M_B)_z = 0 \qquad\qquad Ans.$$

What do the negative signs for $(M_B)_x$ and $(M_B)_y$ indicate? Note that the normal force $N_B = (F_B)_y = 0$, whereas the shear force is $V_B = \sqrt{(0)^2 + (84.3)^2} = 84.3$ N. Also, the torsional moment is $T_B = (M_B)_y = 77.8$ N · m and the bending moment is $M_B = \sqrt{(30.3)^2 + (0)^2} = 30.3$ N · m.

---

*The *magnitude* of each moment about an axis is equal to the magnitude of each force times the perpendicular distance from the axis to the line of action of the force. The *direction* of each moment is determined using the right-hand rule, with positive moments directed along the positive coordinate axes.

15

PROBLEMS

1–1. Determine the resultant internal normal force acting on the cross section through point A in each column. In (a), segment BC weighs 180 lb/ft and segment CD weighs 250 lb/ft. In (b), the column has a mass of 200 kg/m.

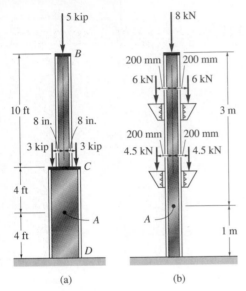

Prob. 1–1

1–2. Determine the resultant internal torque acting on the cross sections through points C and D on each shaft. The support bearings at A and B in (a) allow free turning of the shaft.

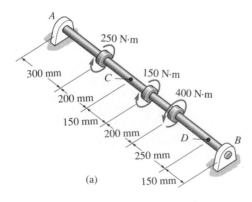

Prob. 1–2a

Prob. 1–2b

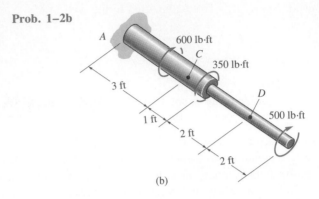

(b)

1–3. Determine the resultant internal normal and shear force in the member at (a) section a–a and (b) section b–b, each of which passes through point A. The 500-lb load is applied along the centroidal axis of the member.

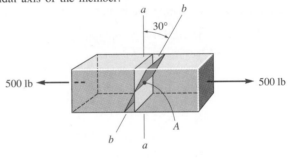

Prob. 1–3

*1–4. The beam supports the distributed load shown. Determine the resultant internal loadings on the cross section through point C. Assume the reactions at the supports A and B are vertical.

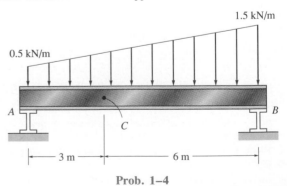

Prob. 1–4

1–5. Determine the resultant internal loadings on the cross section through point *D* of member *AB*.

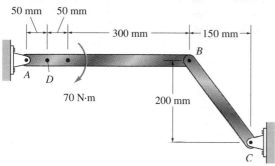

Prob. 1–5

1–6. Determine the resultant internal loadings on the cross sections located through points *D* and *E* of the frame.

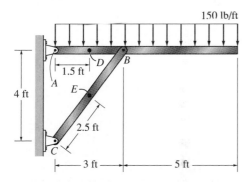

Prob. 1–6

1–7. Determine the resultant internal loadings on the cross sections located through points *D* and *E* of the frame.

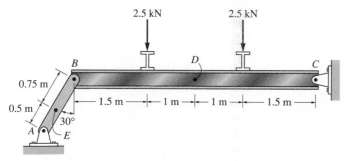

Prob. 1–7

*1–8. The force *F* = 80 lb acts on the gear tooth. Determine the resultant internal loadings at the root of the tooth, i.e., through point *A*, which is located at the centroid of section *a–a*.

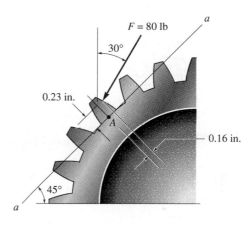

Prob. 1–8

1–9. Determine the resultant internal loadings on the sections through points *C* and *D* of the pliers. There is a pin at *A*, and the jaws at *B* are smooth.

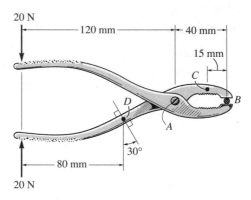

Prob. 1–9

1–10. Determine the resultant internal loadings acting on (*a*) section *a–a* and (*b*) section *b–b*. Each section is located through the centroid, point *C*.

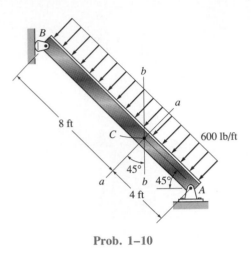

Prob. 1–10

1–11. The serving tray *T* used on an airplane is supported on *each side* by an arm. The tray is pin-connected to the arm at *A* and at *B* there is a smooth pin. (The pin can move within the slot in the arms to permit folding the tray against the front passenger seat when not in use.) Determine the resultant internal loadings in the arm on the cross section through point *C* when the tray arm supports the loads shown.

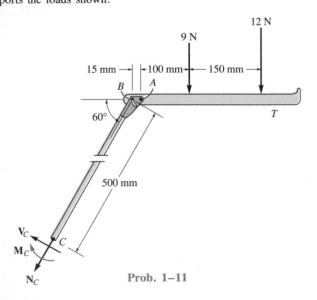

Prob. 1–11

***1–12.** The beam *AB* is pin-supported at *A* and supported by a cable *BC*. A *separate* cable *CG* is used to hold up the frame. If *AB* weighs 120 lb/ft and the column *FC* has a weight of 180 lb/ft, determine the resultant internal loadings acting on cross sections located at points *D* and *E*. Neglect the thickness of both the beam and column in the calculation.

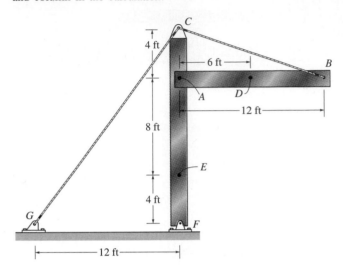

Prob. 1–12

1–13. Determine the resultant internal loadings acting on the cross section through point *C* in the beam. The load *D* has a mass of 300 kg and is being hoisted by the motor *M* with constant velocity.

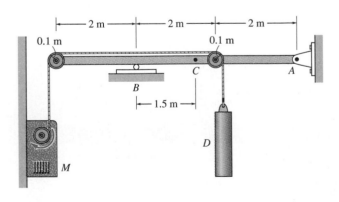

Prob. 1–13

1–14. The 800-lb load is being hoisted at a constant speed using the motor M, which has a weight of 90 lb. Determine the resultant internal loadings acting on the cross section through point B in the beam. The beam has a weight of 40 lb/ft and is fixed to the wall at A.

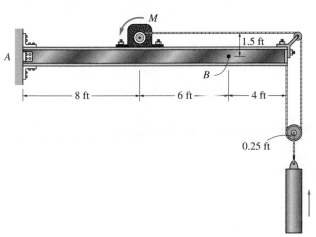

Prob. 1–14

1–15. The pipe has a mass of 12 kg/m. If it is fixed to the wall at A, determine the resultant internal loadings acting on the cross section located at B. Neglect the weight of the wrench CD.

*1–16. Determine the resultant internal loadings acting on the cross section through point B of the signpost. The post is fixed to the ground and a uniform pressure of $p = 7$ lb/ft^2 acts perpendicular to the face of the sign.

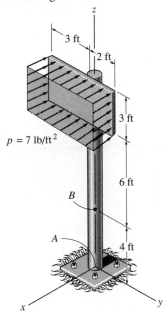

Prob. 1–16

1–17. The shaft is supported at its ends by two bearings A and B and is subjected to the forces applied to the pulleys fixed to the shaft. Determine the resultant internal loadings acting on the cross section located at point C. The 300-N forces act in the $-z$ direction and the 500-N forces act in the $+x$ direction. The journal bearings at A and B exert only x and z components of force on the shaft.

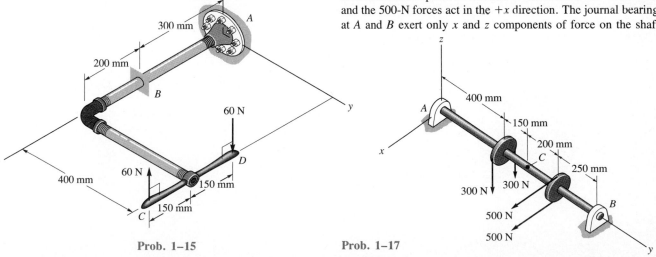

Prob. 1–15   Prob. 1–17

1–18. A hand crank that is used in a press has the dimensions shown. Determine the resultant internal loadings acting on the cross section at A if a vertical force of 50 lb is applied to the handle as shown. Assume the crank is fixed to the shaft at B.

*1–20. A differential element taken from a curved bar is shown in the figure. Show that $dN/d\theta = V$, $dV/d\theta = -N$, $dM/d\theta = -T$, and $dT/d\theta = M$.

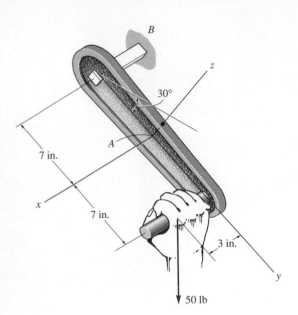

Prob. 1–18

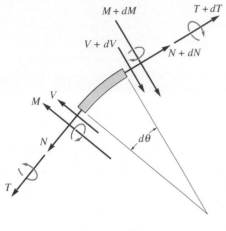

Prob. 1–20

1–19. The curved rod AD of radius r has a weight per length of w. If it lies in the horizontal plane, determine the resultant internal loadings acting on the cross section through point B. *Hint*: The distance from the centroid C of segment AB to point O is $CO = 0.9745r$.

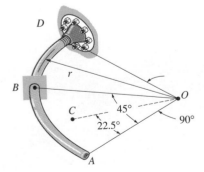

Prob. 1–19

1.3 Stress

In Sec. 1.2 we showed how to determine the internal resultant force and moment acting at a specified point on the sectioned area of a body, Fig. 1–9a. It was stated that these two loadings represent the resultant effects of the actual *distribution of force* acting over the sectioned area, Fig. 1–9b. Obtaining this *distribution* of internal loading, however, is one of the major problems in mechanics of materials.

To solve this problem it will be necessary to study how the body deforms under load, since each internal force distribution will *deform* the material in a *unique* way. In later chapters we will show how measurements of material deformation can be related to the force distribution. Before this can be done, however, it is first necessary to develop a means for describing each of the internal forces at all *points* on the sectioned area. To do this we will establish the concept of stress.

Notice that if the sectioned area is subdivided into small areas, such as the one ΔA shown shaded in Fig. 1–9b, then, on each of these areas, the force distribution will become more *uniform* as the area gets smaller and smaller. As we reduce the area in this manner, however, we must make two assumptions regarding the properties of the material. We will consider the material to be *continuous*, that is, to consist of a continuum or uniform distribution of matter having no voids, rather than being composed of a finite number of distinct atoms or molecules. Furthermore, the material must be *cohesive*, meaning that all portions of it are connected together, rather than having breaks, cracks, or separations. Now, as the subdivided area ΔA of this continuous-cohesive material is reduced to one of *infinitesimal size*, the distribution of force acting over the entire sectioned area will consist of an *infinite number* of forces, each acting at a specific point on the area. A typical finite yet very small force $\Delta \mathbf{F}$, acting on its associated area ΔA, is shown in Fig. 1–9c. This force, like all the others, will have a unique direction, but for further discussion we will replace it by two of its *components*, namely, $\Delta \mathbf{F}_n$ and $\Delta \mathbf{F}_t$, which are taken normal and tangent to the area, respectively. As the area ΔA becomes smaller and smaller, and approaches zero, so do the force $\Delta \mathbf{F}$ and its components; however, the quotient of the force and area will, in general, approach a finite limit. This quotient is called *stress*, and it describes the *intensity of the internal force* on a *specific plane* passing through a point.

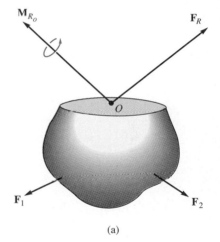

(a)

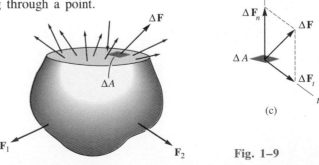

(b)

(c)

Fig. 1–9

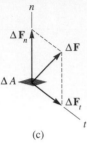

Fig. 1–9c

Normal Stress. The *intensity* of force, or force per unit area, acting normal to ΔA is defined as the *normal stress*, σ (sigma). Mathematically it can be expressed as

$$\sigma = \lim_{\Delta A \to 0} \frac{\Delta F_n}{\Delta A} \qquad (1\text{--}4)$$

If the normal force or stress "pulls" on the area element ΔA as shown in Fig. 1–9c, it is referred to as *tensile stress*, whereas if it "pushes" on ΔA it is called *compressive stress*.

Shear Stress. Likewise, the intensity of force, or force per unit area, acting tangent to ΔA is called the *shear stress*, τ (tau). This component is expressed mathematically as

$$\tau = \lim_{\Delta A \to 0} \frac{\Delta F_t}{\Delta A} \qquad (1\text{--}5)$$

In Fig. 1–9c, note that the orientation of the area ΔA completely specifies the direction of $\Delta \mathbf{F}_n$, which is always perpendicular to the area. On the other hand, each shear force $\Delta \mathbf{F}_t$ can act in an infinite number of directions within the plane of the area. Provided, however, the direction of $\Delta \mathbf{F}$ is known, then the direction of $\Delta \mathbf{F}_t$ can be established as shown in the figure.

Cartesian Stress Components. To specify further the direction of the shear stress, we will resolve it into rectangular components, and to do this we will make reference to x, y, z coordinate axes, oriented as shown in Fig. 1–10a. Here the element of area $\Delta A = \Delta x \, \Delta y$ and the three Cartesian components of $\Delta \mathbf{F}$ are shown in Fig. 1–10b. We can now express the normal-stress component as

$$\sigma_z = \lim_{\Delta A \to 0} \frac{\Delta F_z}{\Delta A}$$

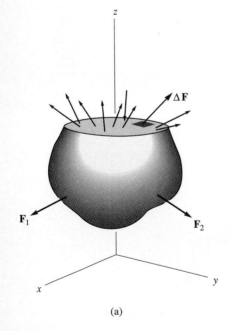

(a)

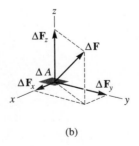

(b)

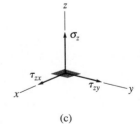

(c)

Fig. 1–10

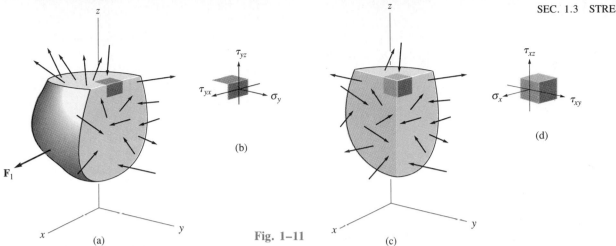

Fig. 1–11

and the two shear-stress components as

$$\tau_{zx} = \lim_{\Delta A \to 0} \frac{\Delta F_x}{\Delta A}$$

$$\tau_{zy} = \lim_{\Delta A \to 0} \frac{\Delta F_y}{\Delta A}$$

The subscript notation z in σ_z is used to reference the *direction* of the outward normal line, which specifies the orientation of the area ΔA. Two subscripts are used for the shear-stress components, τ_{zx} and τ_{zy}. The z specifies the orientation of the area, and x and y refer to the direction lines for the shear stresses.

To summarize these concepts, the intensity of the internal force at a *point* in a body *must* be described on an *area having a specified orientation*. This intensity can then be measured using three components of stress acting on the area. The normal component acts normal or perpendicular to the area, and the shear components act within the plane of the area. These three stress components are shown graphically in Fig. 1–10c.

Now consider passing another imaginary section through the body parallel to the x–z plane and intersecting the front side of the element shown in Fig. 1–10a. The resulting free-body diagram is shown in Fig. 1–11a. Resolving the force acting on the area $\Delta A = \Delta x \, \Delta z$ into its rectangular components, and then determining the intensity of these force components, leads to the normal-stress and shear-stress components shown in Fig. 1–11b. Using the same notation as before, the subscript y in σ_y, τ_{yx}, and τ_{yz} refers to the direction of the normal line associated with the orientation of the area, and x and z in τ_{yx} and τ_{yz} refer to the corresponding direction lines for the shear stress. Lastly, one more section of the body parallel to the y–z plane, as shown in Fig. 1–11c, gives rise to normal stress σ_x and shear stresses τ_{xy} and τ_{xz}, Fig. 1–11d. If we continue in this manner, using corresponding parallel planes, we can "cut out" a cubic volume element of material that represents the *state of stress* acting at the point of its location in the body, Fig. 1–12.

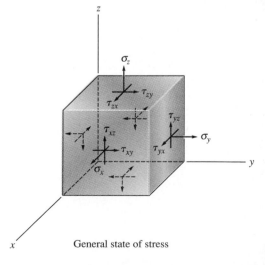

General state of stress

Fig. 1–12

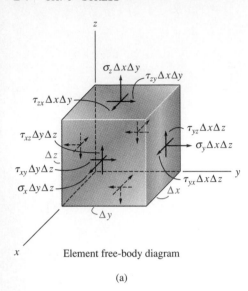

Element free-body diagram

(a)

Fig. 1–13

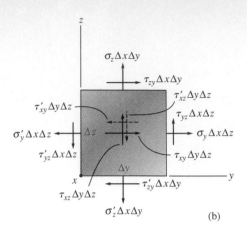

(b)

Equilibrium Requirements. Although each of the six faces of the element in Fig. 1–12 will have three components of stress acting on it, if the stress at the point is *constant*, some of these stress components can be related by satisfying both force and moment equilibrium for the element. To show the relationships between the components we will consider a free-body diagram of the element, Fig. 1–13*a*. This element has a volume of $\Delta V = \Delta x\, \Delta y\, \Delta z$, and in accordance with Eqs. 1–4 and 1–5, the forces acting on each face are determined from the product of the average stress times the area of the face. For simplicity, we have not labeled the "dashed" forces acting on the "hidden" sides of the element. Instead, to view, and thereby label, some of these forces, the element is shown from a front view in Fig. 1–13*b*. Here it should be noted that the force components on the "hidden" sides of the element are designated with stresses having primes, and these forces are shown in the opposite direction to their counterparts acting on the opposite faces of the element.

If we now consider force equilibrium in the y direction, we have

$$\xrightarrow{+}\ \Sigma F_y = 0;\quad \sigma_y\ \underset{\substack{\text{stress}\ \text{area}}}{\underbrace{\Delta x\ \Delta z}} - \sigma_y'\ \Delta x\ \Delta z + \tau_{zy}\ \Delta x\ \Delta y - \tau_{zy}'\ \Delta x\ \Delta y + \tau_{xy}\ \Delta y\ \Delta z$$
$$- \tau_{xy}'\ \Delta y\ \Delta z = 0$$

force

Letting the side $\Delta x \to 0$, we get $\tau_{xy}\ \Delta y\ \Delta z - \tau_{xy}'\ \Delta y\ \Delta z = 0$, which requires $\tau_{xy} = \tau_{xy}'$. In a similar manner, if instead $\Delta y \to 0$, then it is necessary that $\sigma_y = \sigma_y'$; and lastly, if only $\Delta z \to 0$, then $\tau_{zy} = \tau_{zy}'$. In other words, as Δx, Δy, Δz approach zero each stress component acting in the y direction must be equal in magnitude but opposite in direction to its counterpart acting on the opposite face of the element. We must also satisfy force equilibrium in the x and z directions using the same type of analysis. When we are finished, we may again conclude that corresponding normal and shear stress components acting on opposite sides of the element must be equal in magnitude but opposite in direction.

In order to satisfy moment equilibrium, a further restriction has to be placed on the shear-stress components. To show this, consider summing moments about the x axis. From Fig. 1–13*b*, since $\sigma_y = \sigma_y'$, $\sigma_z = \sigma_z'$, $\tau_{xy} = \tau_{xy}'$, $\tau_{xz} = \tau_{xz}'$, the forces these components create are in equal but opposite pairs, and so their moments about the x axis will also cancel. Therefore, excluding these components, we have

$$\zeta + \Sigma M_x = 0; \qquad (\tau_{yz} \; \Delta x \; \Delta z) \; \Delta y - (\tau_{zy} \; \Delta x \; \Delta y) \; \Delta z = 0$$

where τ_{yz} and τ_{zy} are labeled stress, $\Delta x \Delta z$ and $\Delta x \Delta y$ are labeled area, the products are labeled force, Δy and Δz are labeled arm, and force times arm is labeled moment.

If we divide through by $\Delta x \; \Delta y \; \Delta z$, then it is necessary that

$$\tau_{yz} = \tau_{zy}$$

In a similar manner, we can show that moment equilibrium about the y and z axes requires

$$\tau_{xz} = \tau_{zx}$$

and

$$\tau_{xy} = \tau_{yx}$$

Thus, we may conclude that the above pairs of shear stresses on adjacent faces of the element must have equal magnitude and be directed either toward or away from the corners of the element, as in the manner shown in Fig. 1–13. This is sometimes referred to as the *complementary property* of shear.

To summarize the above analysis, the *state of stress* at a point is characterized by *six* independent stress components, namely, three normal stresses σ_x, σ_y, σ_z and three shear stresses τ_{xy}, τ_{yz}, τ_{xz}. Realize that these six components depend only on the *orientation* of the element, since the distribution of force acting on a sectioned plane through the body depends on the orientation of the sectioned plane. In other words, there is a unique set of six stress components describing the state of stress for each particular orientation of the element at the point. In Chapter 8 we will show that if these stress components are known on an element having a specified orientation, it will then be possible to determine the stress components on an element having some other orientation at the point.

Units. In the International Standard or SI system, the magnitudes of both normal and shear stress components are specified in the basic units of newtons per square meter (N/m^2). This unit, called a *pascal* ($1 \; Pa = 1 \; N/m^2$) is rather small, and in engineering work prefixes such as kilo- (10^3), symbolized by k, mega- (10^6), symbolized by M, or giga- (10^9), symbolized by G, are used to represent larger, more realistic values of stress.* Likewise, in the U.S. Customary or Foot-Pound-Second system of units, engineers usually express stress in pounds per square inch (psi) or kilopounds per square inch (ksi), where 1 kilopound (kip) = 1000 lb.

---

* Sometimes stress is expressed in units of N/mm^2, where $1 \; mm = 10^{-3} \; m$. However, in the SI system, prefixes are not allowed in the denominator of a fraction and therefore it is better to use the equivalent $1 \; N/mm^2 = 1 \; MN/m^2 = 1 \; MPa$.

1.4 Average Normal Stress in an Axially Loaded Bar

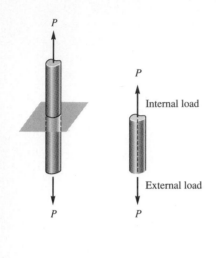

P

P

(a)

P

Internal load

External load

P

(b)

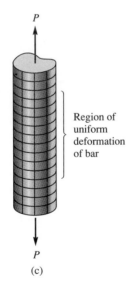

P

Region of uniform deformation of bar

P

(c)

Fig. 1–14

Frequently structural or mechanical members are made long and slender. Also, they are subjected to axial loads that are usually applied to the ends of the member. Truss members, hangers, and bolts are typical examples. In this section we will determine the average stress distribution in an axially loaded bar, such as the one having the general form shown in Fig. 1–14*a*. For the analysis the internal loading on the cross section will be considered. This section defines the *cross-sectional area* of the bar, and since all such cross sections are the same, the bar is referred to as being *prismatic*. If we neglect the weight of the bar and section it as indicated, Fig. 1–14*b*, then, for equilibrium of the segment, the internal resultant force acting on the cross-sectional area must be equal in magnitude, opposite in direction, and collinear to the force acting at the end of the bar.

Assumptions. Before we determine the average stress distribution acting over the bar's cross section, it is first necessary to make some simplifying assumptions concerning the material description and the specific application of the load. In this regard, the analysis will depend on the bar remaining straight both before and after the load is applied, and also, the cross section should remain flat or plane during the deformation, that is, during the time the bar changes its volume and shape. If this occurs, then horizontal and vertical grid lines inscribed on the bar will *deform uniformly* when the bar is subjected to the load, Fig. 1–14*c*. Here we will not consider regions of the bar near its ends, where application of the external loads can cause localized distortions. Instead we will focus only on the stress distribution within the bar's midsection.

As will be shown below, uniform deformation occurs provided **P** is applied along the *centroidal axis* of the cross section. Also, uniform deformation of the bar will occur if the material is assumed to be homogeneous and isotropic. *Homogeneous material* has the same physical and mechanical properties throughout its volume, and *isotropic material* has these same properties in all directions. Many engineering materials may be approximated as being both homogeneous and isotropic as assumed here. Steel, for example, contains thousands of randomly oriented crystals in each cubic millimeter of its volume, and since most problems involving this material have a physical size that is much larger than a single crystal, the above assumption regarding its material composition is quite realistic. It should be mentioned, however, that steel can be made anisotropic by cold-rolling, i.e., rolling or forging it at subcritical temperatures. *Anisotropic materials* have different properties in different directions, and although this is the case, if the anisotropy is in a specified direction throughout the material, it will also deform uniformly when subjected to an axial load. For example, timber, due to its grains or fibers of wood, is an example of an engineering material that is homogeneous and anisotropic in a specified direction and is therefore suited for the foregoing analysis.

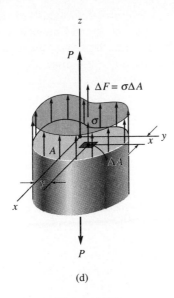

(d)

Fig. 1–14(d)

Average Normal Stress Distribution.   Provided the bar is subjected to a constant uniform deformation as noted, it is reasonable to assume further that this deformation is caused by a *constant* normal stress σ, which is then uniformly distributed over the bar's cross-sectional area, Fig. 1–14d. Since each area ΔA on the cross section is subjected to a force $\Delta F = \sigma \Delta A$, then the sum of these forces acting over the entire cross-sectional area must be equivalent to the internal force resultant **P** at the section. If we let $\Delta A \rightarrow dA$ and therefore $\Delta F \rightarrow dF$, then, since σ is *constant*, we have

$$F_{R_z} = \Sigma F_z; \qquad \int dF = \int_A \sigma \, dA$$

$$P = \sigma A$$

or

$$\boxed{\sigma = \frac{P}{A}} \qquad\qquad (1\text{–}6)$$

Here

σ = average normal stress at any point on the cross-sectional area

P = internal resultant normal force, which is applied through the *centroid* of the cross-sectional area. P is determined using the method of sections and the equations of equilibrium

A = cross-sectional area of the bar

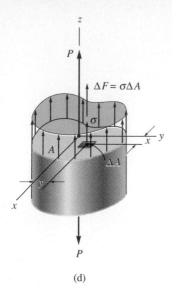

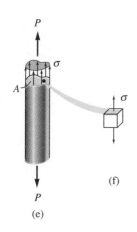

Fig. 1-14(d, e, f)

In order to maintain the uniform deformation of the bar, the stress distribution cannot produce a resultant internal moment on the cross section. To show that this is actually the case, the sum of the moments about the centroidal x and y axes must be *equal to zero*, since P creates zero moment about these axes, Fig. 1–14d. We have

$$(M_R)_x = \Sigma M_x; \qquad 0 = \int_A y \, dF = \int_A y\sigma \, dA = \sigma \int_A y \, dA$$

$$(M_R)_y = \Sigma M_y; \qquad 0 = \int_A x \, dF = \int_A x\sigma \, dA = \sigma \int_A x \, dA$$

These equations are satisfied, since by definition of the centroid,[*] $\int y \, dA = 0$ and $\int x \, dA = 0$.

In summary, then, an axial force **P,** acting on the ends of a homogeneous straight prismatic bar and passing through the centroid of the bar's cross-sectional area, will cause a uniform normal stress distribution over the cross-sectional area, Fig. 1–14e. This stress has a *magnitude* of $\sigma = P/A$ and an arrowhead *sense* of direction that is the *same* as that of the internal resultant force **P,** since all the normal stresses on the cross section develop this resultant. At a specific point in the bar it should therefore be evident that only a normal stress exists on a volume element of material located at the point. In other words, the stress components in Fig. 1–12 are all zero except σ_z, Fig. 1–14f. Under these conditions, the material is said to be subjected to a state of *uniaxial stress*.

Returning to the stress distribution shown in Fig. 1–14e, notice that *graphically* the *magnitude* of the internal resultant force **P** is *equivalent* to the *volume* under the stress diagram; that is, $P = \sigma A$ (volume = height × base). Furthermore, as a consequence of the balance of moments, this resultant *passes through the centroid of this volume*.

The above analysis applies as well to *short members* subjected to a compressive force, Fig. 1–15a. Here the cross section is subjected to a uniaxial compressive stress, Fig. 1–15b, and so is a volume element of material located at a point within the member, Fig. 1–15c. Note that a restriction on the length of the member is necessary, since if the member is long and slender and **P** is large, the member may become unstable and buckle.

Although we have developed this analysis for *prismatic* bars, this assumption can be relaxed somewhat to include bars that have a *slight taper*. For example, it can be shown, using the more exact analysis of the theory of elasticity, that for a tapered bar of rectangular cross section, for which the angle between two adjacent sides is 15°, the average normal stress, as calculated by Eq. 1–6, is only 2.2% less than its value found from the theory of elasticity.

*The centroid is defined in Appendix A.

PROCEDURE FOR ANALYSIS

It is important to be aware that the equation $\sigma = P/A$ gives the *average* normal stress on the sectional area of a member when the section is subjected to a resultant normal force. For axially loaded members, application of the equation requires the following steps.

Internal Loading.   Section the member *perpendicular* to its longitudinal axis at the point where the stress is to be determined and use the necessary free-body diagram and equation of force equilibrium to obtain the internal axial force **P** at the section.

Average Normal Stress.   Determine the member's cross-sectional area at the section and compute the average normal stress $\sigma = P/A$. It is suggested that σ be shown acting on a small volume element of the material located at a point on the section where the stress is calculated. To do this, first draw σ on the face of the element coincident with the sectioned area A. This is indicated as the shaded face in Figs. 1–14*f* and 1–15*c*. Here σ acts in the *same direction* as **P**. For force equilibrium, the normal stress σ acting on the opposite face of the element can then be drawn in its appropriate direction.

The following examples numerically illustrate applications of this procedure.

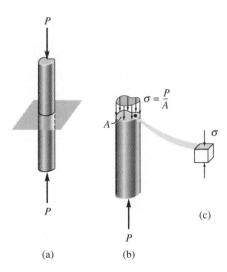

Fig. 1–15

■ Example 1–5

The bar in Fig. 1–16a has a constant width of 35 mm and a thickness of 10 mm. Determine the largest normal stress in the bar when it is subjected to the loading shown.

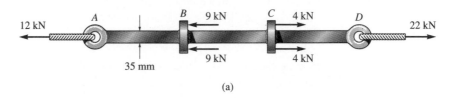

(a)

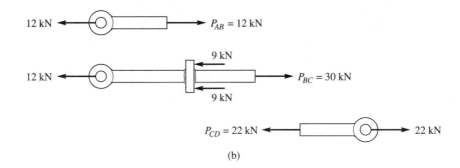

(b)

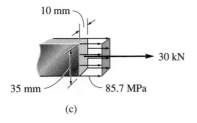

(c)

Fig. 1–16

SOLUTION

Internal Loading.   By inspection, the internal axial forces in regions *AB*, *BC*, and *CD* are all constant yet have different magnitudes. Using the method of sections, these loadings are computed in Fig. 1–16b. By comparison, the largest loading is in region *BC*, where P_{BC} = 30 kN. Since the cross-sectional area of the bar is constant, the largest normal stress also occurs within this region of the bar.

Average Normal Stress.   Applying Eq. 1–6, we have

$$\sigma_{BC} = \frac{P_{BC}}{A} = \frac{30(10^3) \text{ N}}{(0.035 \text{ m})(0.010 \text{ m})} = 85.7 \text{ MPa} \qquad Ans.$$

The stress distribution acting on an arbitrary cross section of the bar within region *BC* is shown in Fig. 1–16c. Graphically the *volume* (or "block") represented by this distribution of stress is equivalent to the load of 30 kN; that is, 30 kN = (85.7 MPa)(35 mm)(10 mm).

Example 1–6

The 80-kg lamp is supported by two rods AB and BC as shown in Fig. 1–17a. If AB has a diameter of 10 mm and BC has a diameter of 8 mm, determine which rod is subjected to the greater normal stress.

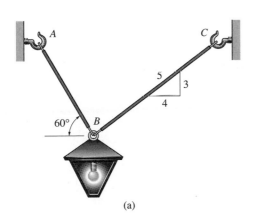

(a)

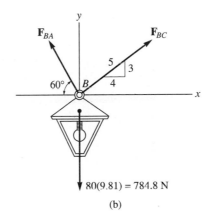

(b)

SOLUTION

Internal Loading. We must first determine the axial force in each rod. A free-body diagram of the lamp is shown in Fig. 1–17b. Applying the equations of force equilibrium yields

$$\xrightarrow{+} \Sigma F_x = 0; \qquad F_{BC}\left(\tfrac{4}{5}\right) - F_{BA} \cos 60° = 0$$

$$+\uparrow \Sigma F_y = 0; \qquad F_{BC}\left(\tfrac{3}{5}\right) + F_{BA} \sin 60° - 784.8 \text{ N} = 0$$

$$F_{BC} = 395.2 \text{ N}, \qquad F_{BA} = 632.4 \text{ N}$$

By Newton's third law of action, equal but opposite reaction, these forces also represent the internal axial forces developed within the rods.

Average Tensile Stress. Applying Eq. 1–6, we have

$$\sigma_{BC} = \frac{F_{BC}}{A_{BC}} = \frac{395.2 \text{ N}}{\pi(0.004 \text{ m})^2} = 7.86 \text{ MPa}$$

$$\sigma_{BA} = \frac{F_{BA}}{A_{BA}} = \frac{632.4 \text{ N}}{\pi(0.005 \text{ m})^2} = 8.05 \text{ MPa} \qquad Ans.$$

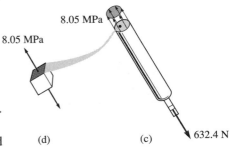

(d)   (c)

The average normal stress distribution acting over a cross section of rod AB is shown in Fig. 1–17c, and at a point on this cross section, an element of material is stressed as shown in Fig. 1–17d. The shaded face of this element represents part of the sectioned area shown shaded in Fig. 1–17c.

Fig. 1–17

Example 1–7

The casting shown in Fig. 1–18a is made of steel having a specific weight of $\gamma_{st} = 490$ lb/ft³. Determine the average compressive stress acting at points A and B.

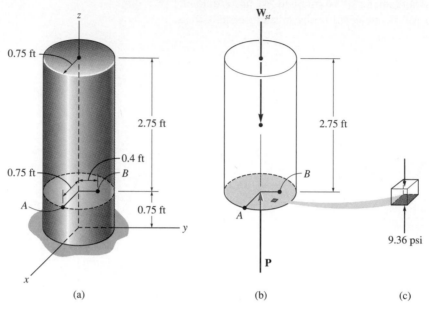

Fig. 1–18
(a)　　　　　　　　　　(b)　　　　　　　　(c)

SOLUTION

Internal Loading. A free-body diagram of the top segment of the casting where the section passes through points A and B is shown in Fig. 1–18b. The weight of this segment is determined from $W_{st} = \gamma_{st}V_{st}$. Thus the internal axial force P at the section is

$$+\uparrow \; \Sigma F_z = 0; \qquad\qquad P - W_{st} = 0$$

$$P - (490 \text{ lb/ft}^3)(2.75 \text{ ft})\pi(0.75 \text{ ft})^2 = 0$$

$$P = 2381 \text{ lb}$$

Average Compressive Stress. The cross-sectional area at the section is $A = \pi(0.75 \text{ ft})^2$, and so the compressive stress becomes

$$\sigma = \frac{P}{A} = \frac{2381 \text{ lb}}{\pi(0.75 \text{ ft})^2}$$

$$= 1347.5 \text{ lb/ft}^2 = 1347.5 \text{ lb/ft}^2(1 \text{ ft}^2/144 \text{ in}^2)$$

$$= 9.36 \text{ psi} \qquad\qquad\qquad\qquad\qquad\qquad\qquad Ans.$$

The stress shown on the volume element of material in Fig. 1–18c is representative of the conditions at either point A or B. Notice that this stress acts *upward* on the bottom or shaded face of the element since this face forms part of the bottom surface area of the cut section, and furthermore, the resultant internal force **P** is pushing upward on this section.

Example 1–8

Member AC shown in Fig. 1–19a is subjected to a vertical force of 3 kN. Determine the position x of this force so that the compressive stress at C is equal to the tensile stress in the tie rod AB. The rod has a cross-sectional area of 400 mm^2 and the contact area at C is 650 mm^2.

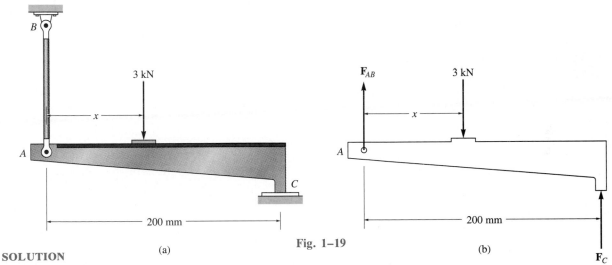

Fig. 1–19

(a)   (b)

SOLUTION

Internal Loading. The free-body diagram for member AC is shown in Fig. 1–19b. There are three unknowns, namely, F_{AB}, F_C, and x. To solve this problem we will work in units of newtons and millimeters.

$$+\uparrow \ \Sigma F_y = 0; \qquad\qquad F_{AB} + F_C - 3000 \text{ N} = 0 \qquad\qquad (1)$$

$$\downarrow+ \ \Sigma M_A = 0; \qquad\qquad -3000 \text{ N}(x) + F_C (200 \text{ mm}) = 0 \qquad\qquad (2)$$

Average Normal Stress. A necessary third equation can be written that requires the tensile stress in the bar AB and the compressive stress at C to be equivalent; i.e.,

$$\sigma = \frac{F_{AB}}{400 \text{ mm}^2} = \frac{F_C}{650 \text{ mm}^2}$$

$$F_C = 1.625 F_{AB}$$

Substituting this into Eq. 1, solving for F_{AB}, then solving for F_C, we obtain

$$F_{AB} = 1143 \text{ N}$$

$$F_C = 1857 \text{ N}$$

The position of the applied load is determined from Eq. 2.

$$x = 124 \text{ mm} \qquad\qquad\qquad Ans.$$

Note that $0 < x < 200$ mm, as expected.

1.5 Average Shear Stress

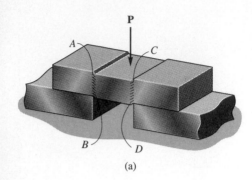

(a)

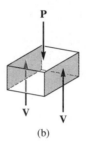

(b)

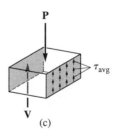

(c)

Fig. 1–20

As discussed in Sec. 1.3, shear stress has been defined as the stress component that acts *in the plane* of the sectioned area. In order to show how this stress can develop, we will consider the effect of applying a force **P** to the bar in Fig. 1–20a. If the supports are considered rigid, and **P** is large enough, it will cause the material of the bar to deform and fail along the planes identified by *AB* and *CD*. A free-body diagram of the unsupported center segment of the bar, Fig. 1–20b, indicates that the shear force $V = P/2$ must be applied at each section to hold the segment in equilibrium. The *average shear stress* distributed over each sectioned area that develops this shear force is defined by

$$\tau_{avg} = \frac{V}{A} \qquad (1\text{–}7)$$

Here

τ_{avg} = average shear stress at the section, which is assumed to be the *same* at each point located on the section

V = internal resultant shear force at the section determined from the equations of equilibrium

A = area at the section

The distribution of average shear stress is shown acting over the right-hand section in Fig. 1–20c. Notice that τ_{avg} is in the *same direction* as **V**, since the shear stress must create forces that contribute to the internal resultant force **V** at the section.

The loading case discussed in Fig. 1–20 is an example of *simple* or *direct shear,* since the shear is caused by the *direct action* of the applied load **P**. This type of shear often occurs in various types of simple connections that use bolts, pins, welding material, etc. In all these cases, however, application of Eq. 1–7 is only approximate. A more precise investigation of the shear-stress distribution over the critical section often reveals that much larger shear stresses occur in the material than those predicted by Eq. 1–7. Although this may be the case, application of Eq. 1–7 may be acceptable for many problems in engineering design and analysis. For example, engineering codes allow the use of Eq. 1–7 when considering design sizes for fasteners such as bolts and for obtaining the bonding strength of joints subjected to shear loadings. In this regard, two types of shear frequently occur in practice, which deserve separate treatment.

Single Shear. The steel and wood joints shown in Fig. 1–21*a* and 1–21*b*, respectively, are examples of *single-shear connections* and are often referred to as *lap joints*. Here we will assume that the members are thin and that the nut *A* in Fig. 1–21*a* is not tightened to any great extent so friction between the members can be neglected. Passing a section between the members yields the free-body diagrams shown in Fig. 1–21*c* and 1–21*d*. Since the members are thin, we can neglect the moment created by the force **P**. Hence the cross-sectional area of the bolt in Fig. 1–21*c* and the bonding surface between the members in Fig. 1–21*d* are subjected only to a *single shear force* $V = P$. This force is used in Eq. 1–7 to determine the average shear stress acting on the section.

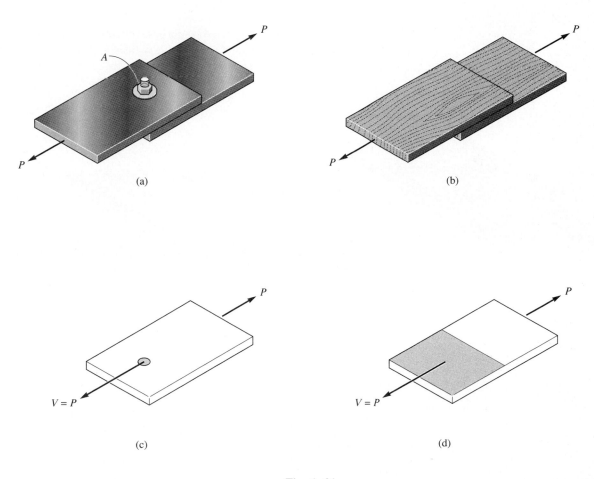

Fig. 1–21

Double Shear. When the joint is constructed as shown in Fig. 1–22*a* or 1–22*b*, *two shear surfaces* must be considered. These types of connections are often called *double lap joints*. If we pass a section between each of the members, the free-body diagrams of the center member are shown in Fig. 1–22*c* and 1–22*d*. Here we have a condition of *double shear*. Consequently, $V = P/2$ acts on *each* sectioned area and this shear, like that in Fig. 1–20*b*, must be considered when applying Eq. 1–7.

Consider now a volume element of material taken at a point located on the surface of any sectioned area on which the average shear stress acts, Fig. 1–23. It was shown in Sec. 1.3 that force and moment equilibrium requires τ_{avg}, acting on the top face of the element, to be accompanied by shear stress acting on three other faces. As shown, all four shear stresses must have equal magnitude and be directed either toward or away from each other at opposite edges of the element. Under these conditions, the material is subjected to *pure shear*.

Although we have considered here a case of simple shear as caused by the *direct* action of a load, in later chapters we will show that shear stress can also arise *indirectly* due to the action of a bending or torsional moment.

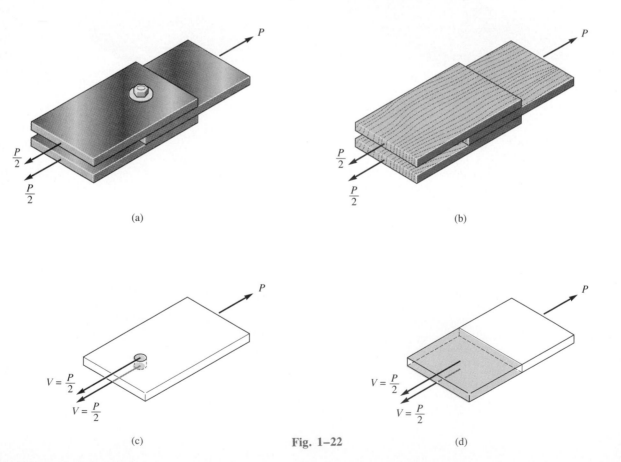

(a)

(b)

(c)

Fig. 1–22

(d)

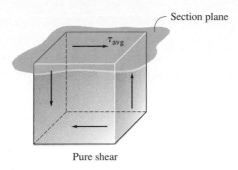

Pure shear

Fig. 1–23

PROCEDURE FOR ANALYSIS

Equation 1–7, $\tau_{avg} = V/A$, is used to compute only the *average shear stress* in the material. Its use in engineering is generally restricted to the design and analysis of members or parts that are thin or small, so that bending can be neglected. Application requires the following steps.

Internal Shear.   Section the member at the point where the average shear stress is to be determined. Draw the necessary free-body diagram and compute the internal shear force **V** acting at the section.

Average Shear Stress.   Determine the sectioned area A, and compute the average shear stress $\tau_{avg} = V/A$. It is suggested that τ_{avg} be shown on a small volume element of material located at a point on the section where it is computed. To do this, first draw τ_{avg} on the face of the element, coincident with the sectioned area A. This shear stress acts in the same direction as **V**. The shear stresses acting on the three adjacent planes can then be drawn in their appropriate directions following the scheme shown in Fig. 1–23.

Example 1–9

The bar shown in Fig. 1–24a has a square cross section for which the depth and thickness are 40 mm. If an axial force of 800 N is applied along the centroidal axis of the bar's cross-sectional area, determine the average normal stress and average shear stress acting on the material along (a) section plane a–a and (b) section plane b–b.

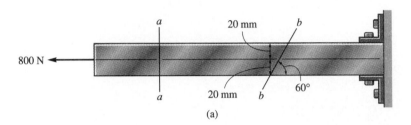

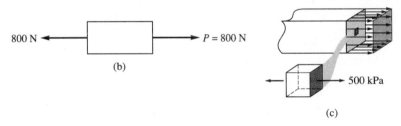

(c)

Fig. 1–24

SOLUTION
Part (a)

Internal Loading. The bar is sectioned, Fig. 1–24b, and the internal resultant loading consists only of an axial force for which $P = 800$ N.

Average Stress. The average normal stress is determined from Eq. 1–6.

$$\sigma = \frac{800 \text{ N}}{(0.04 \text{ m})(0.04 \text{ m})} = 500 \text{ kPa} \qquad \textit{Ans.}$$

No shear stress exists on the section, since the shear force at the section is zero.

$$\tau_{\text{avg}} = 0 \qquad \textit{Ans.}$$

The distribution of average normal stress over the cross section and the stress on a shaded element located on the cross section are shown in Fig. 1–24c.

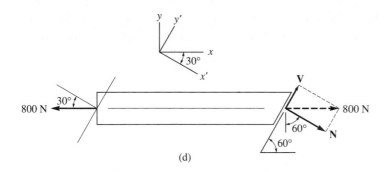

(d)

Part (b)

Internal Loading.   If the bar is sectioned along *b–b*, the free-body diagram shown in Fig. 1–24*d* results. Here both a normal force (**N**) and shear force (**V**) act on the sectioned area. Using *x, y* axes, we require

$$\xrightarrow{+}\ \Sigma F_x = 0; \quad -800\text{ N} + N \sin 60° + V \cos 60° = 0$$

$$+\uparrow\ \Sigma F_y = 0; \quad\quad\quad V \sin 60° - N \cos 60° = 0$$

or, more directly, using *x′, y′* axes,

$$+\searrow\ \Sigma F_{x'} = 0; \quad\quad N - 800\text{ N} \cos 30° = 0$$

$$+\nearrow\ \Sigma F_{y'} = 0; \quad\quad V - 800\text{ N} \sin 30° = 0$$

Solving either set of equations,

$$N = 692.8\text{ N}$$

$$V = 400\text{ N}$$

Average Stresses.   In this case the sectioned area has a thickness and depth of 40 mm and 40 mm/sin 60° = 46.19 mm, respectively, Fig. 1–24*a*. Thus the average normal stress is

$$\sigma = \frac{N}{A} = \frac{692.8\text{ N}}{(0.04\text{ m})(0.04619\text{ m})} = 375\text{ kPa} \quad\quad\quad Ans.$$

and the average shear stress is

$$\tau_{\text{avg}} = \frac{V}{A} = \frac{400\text{ N}}{(0.04\text{ m})(0.04619\text{ m})} = 217\text{ kPa} \quad\quad\quad Ans.$$

The stress distribution and a shaded element of material located at a point on the sectioned area are shown in Fig. 1–24*e*. Comparing this element with that of Fig. 1–24*c*, it can be seen that the stress components at the point depend on the *orientation of the element*. As discussed in Sec. 1.3, any one of these elements can be used to describe the state of stress at the point.

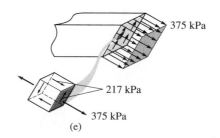

(e)

Example 1-10

The wooden strut shown in Fig. 1–25a is suspended from a 10-mm-diameter steel rod, which is fastened to the wall. If the strut supports a vertical load of 5 kN, compute the average shear stress in the rod at the wall and along the two shaded planes of the strut, one of which is indicated as *abcd*.

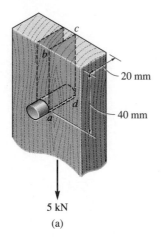

5 kN

(a)

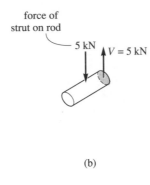

(b)

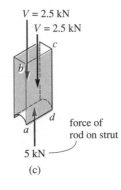

(c)

Fig. 1–25

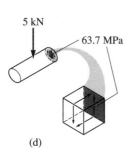

(d)

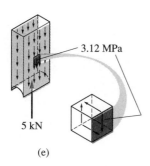

(e)

SOLUTION

Internal Shear. As shown on the free-body diagram in Fig. 1–25b, the rod resists a shear force of 5 kN where it is fastened to the wall. A free-body diagram of the sectioned segment of the strut that is in contact with the rod is shown in Fig. 1–25c. Here the shear force acting along each shaded plane is 2.5 kN.

Average Shear Stress. For the rod,

$$\tau_{avg} = \frac{V}{A} = \frac{5000 \text{ N}}{\pi(0.005 \text{ m})^2} = 63.7 \text{ MPa} \qquad \textit{Ans.}$$

For the strut,

$$\tau_{avg} = \frac{V}{A} = \frac{2500 \text{ N}}{(0.04 \text{ m})(0.02 \text{ m})} = 3.12 \text{ MPa} \qquad \textit{Ans.}$$

The average-shear-stress distribution on the sectioned rod and strut segment is shown in Fig. 1–25d and 1–25e. Also shown with these figures is a typical volume element of the material taken at a point located on the surface of each section. Note carefully how the shear stress must act on each shaded face of these elements and then on the other faces of the elements.

Example 1–11

The inclined member in Fig. 1–26a is subjected to a compressive force of 600 lb. Determine the average compressive stress along the areas of contact defined by *AB* and *BC*, and the average shear stress along the horizontal plane defined by *EDB*.

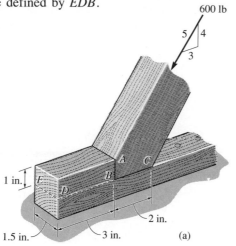

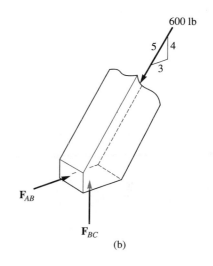

Fig. 1–26

(a)

(b)

SOLUTION

Internal Loadings. The free-body diagram of the inclined member is shown in Fig. 1–26b. The compressive forces acting on the areas of contact are

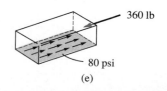

(c)

$$\overset{+}{\rightarrow} \Sigma F_x = 0; \qquad F_{AB} - 600 \text{ lb}(\tfrac{3}{5}) = 0 \qquad F_{AB} = 360 \text{ lb}$$

$$+\uparrow \Sigma F_y = 0; \qquad F_{BC} - 600 \text{ lb}(\tfrac{4}{5}) = 0 \qquad F_{BC} = 480 \text{ lb}$$

Also, from the free-body diagram of the top segment of the bottom member, Fig. 1–26c, the shear force acting on the sectioned horizontal plane *EDB* is

$$\overset{+}{\rightarrow} \Sigma F_x = 0; \qquad V = 360 \text{ lb}$$

Average Stress. The average compressive stresses along the horizontal and vertical planes of the inclined member are

$$\sigma_{AB} = \frac{360 \text{ lb}}{(1 \text{ in.})(1.5 \text{ in.})} = 240 \text{ psi} \qquad Ans.$$

$$\sigma_{BC} = \frac{480 \text{ lb}}{(2 \text{ in.})(1.5 \text{ in.})} = 160 \text{ psi} \qquad Ans.$$

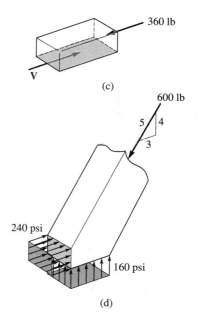

(d)

These stress distributions are shown in Fig. 1–26d.

The average shear stress acting on the horizontal plane defined by *EDB* is

$$\tau_{avg} = \frac{360 \text{ lb}}{(3 \text{ in.})(1.5 \text{ in.})} = 80 \text{ psi} \qquad Ans.$$

This stress is shown distributed over the sectioned area in Fig. 1–26e.

(e)

41

PROBLEMS

1–21. The column is subjected to an axial force of 8 kN at its top. If the cross-sectional area has the dimensions shown in the figure, determine the normal stress acting at section *a–a*. Show this distribution of stress acting over the area's cross section.

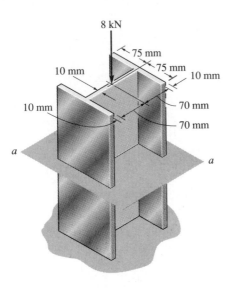

Prob. 1–21

1–22. The yoke-and-rod connection is subjected to a tensile force of 5 kN. Determine the normal stress in each rod and the average shear stress in the pin *A* between the members.

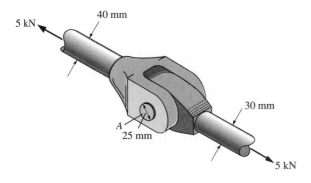

Prob. 1–22

1–23. The beam is supported by a pin at *A* and a short link *BC*. Determine the average shear stress developed in the pins at *A*, *B*, and *C*. All pins are in double shear as shown, and each has a diameter of 18 mm.

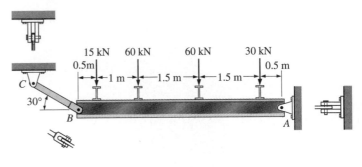

Prob. 1–23

*1–24. The 50-lb lamp is supported by three steel rods connected together by a ring at *A*. Determine which rod is subjected to the greatest normal stress and compute its value. The diameter of each rod is given in the figure.

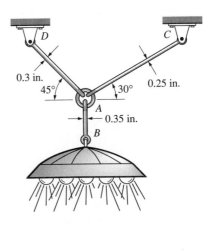

Prob 1–24

1–25. The pedestal has a triangular cross section as shown. If it is subjected to a compressive force of 500 lb, specify the x and y coordinates for the location of point $P(x, y)$, where the load must be applied on the cross section, so that the bearing stress is uniform. Compute the bearing stress and sketch the distribution of the bearing stress acting on the cross section at a location removed from the point of load application.

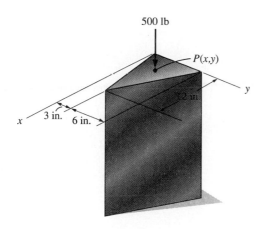

Prob. 1–25

1–26. The two steel members are joined together using a 60° scarf weld. Determine the normal stress and average shear stress resisted in the plane of the weld.

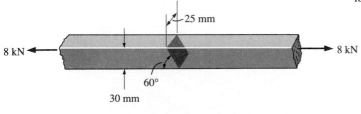

Prob. 1–26

1–27. The board is subjected to a tensile force of 85 lb. Determine the average normal and average shear stress developed in the wood fibers that are orientated along section a–a at 15° with the axis of the board.

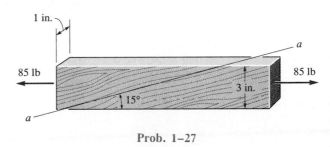

Prob. 1–27

*1–28. A tension specimen having a cross-sectional area A is subjected to an axial force **P.** Determine the maximum average shear stress in the specimen and indicate the orientation θ of a section on which it occurs.

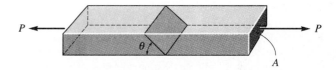

Prob. 1–28

1–29. The bars of the truss each have a cross-sectional area of 1.25 in^2. Determine the normal stress in each member due to the loading shown. State whether the stress is tensile or compressive.

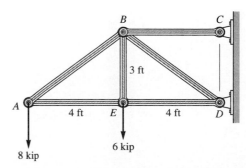

Prob. 1–29

1–30.   The thrust bearing is subjected to the loads shown. Determine the average normal stress developed on the cross sections at B, C, and D. Sketch the results on a volume element located at each section.

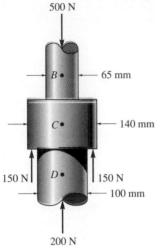

Prob. 1–30

1–31.   The column is made of concrete having a density of 2.30 Mg/m³. At its top B it is subjected to an axial compressive force of 15 kN. Determine the compressive stress in the column as a function of the distance z measured from its base. *Note:* The result will be useful only for finding the average compressive stress at a section removed from the ends of the column, because of localized deformation at the ends.

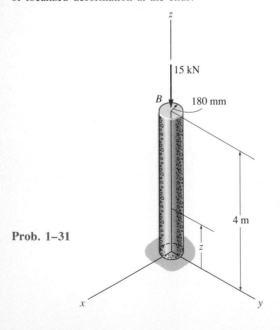

Prob. 1–31

*1–32.   The block in the shape of a frustum of a cone is made of concrete having a specific weight of 150 lb/ft³. Determine the bearing stress acting in the column at its midheight, $z = 4$ ft. *Hint:* The volume of a cone of radius r and height h is $V = \frac{1}{3}\pi r^2 h$.

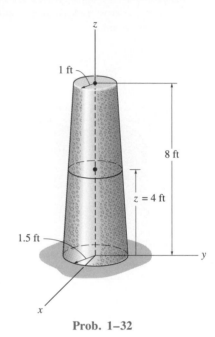

Prob. 1–32

1–33.   Rods AB and BC have diameters of 4 mm and 6 mm, respectively. If a vertical load of 8 kN is applied to the ring at B, determine the angle θ of rod BC so that the normal stress in each rod is equivalent.

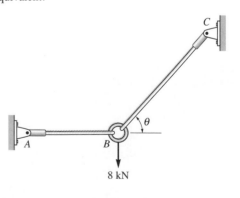

Prob. 1–33

1–34. The long bolt passes through the 30-mm-thick plate. If the force in the bolt shank is 8 kN, determine the normal stress in the shank, the average shear stress along the cylindrical area of the plate defined by the section lines a, and the average shear stress in the bolt head along the cylindrical area defined by the section lines b.

*1–36. The lever is held to the fixed shaft using a tapered pin AB, which has a mean diameter of 6 mm. If a couple is applied to the lever, determine the average shear stress in the pin between the pin and lever.

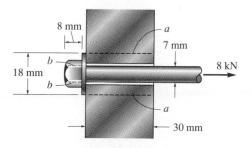

Prob. 1–34

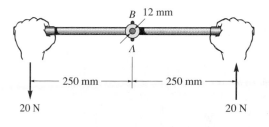

Prob. 1–36

1–35. The pulley is held fixed to the 20-mm-diameter shaft using a key that fits within a groove cut into the pulley and shaft. If the suspended load has a mass of 50 kg, determine the average shear stress in the key along section a–a. The key is 5 mm by 5 mm square and 12 mm long.

1–37. The uniform beam is supported by two rods AB and CD that have cross-sectional areas of 12 mm^2 and 8 mm^2, respectively. Determine the position d of the 6-kN load so that the normal stress in each rod is the same.

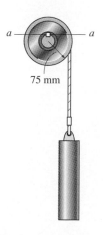

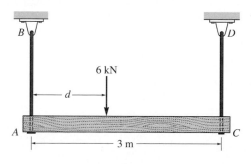

Prob. 1–37

Prob. 1–35

1–38. The two-member frame is subjected to the loading shown. Determine the normal stress and the average shear stress acting at sections *a–a* and *b–b*. Member *CB* has a square cross section of 2 in. on each side.

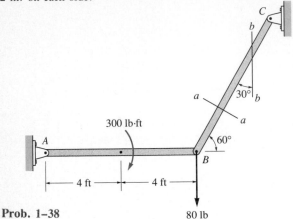

Prob. 1–38

300 lb·ft

80 lb

1–39. The two-member frame is subjected to the distributed loading shown. Determine the normal stress and average shear stress acting at sections *a–a* and *b–b*. Member *CB* has a square cross section of 35 mm on each side. Take $w = 8$ kN/m.

*1–40. The two-member frame is subjected to the distributed loading shown. Determine the intensity w of the largest uniform loading that can be applied to the frame without causing either the normal stress or the average shear stress at section *b–b* to exceed $\sigma = 15$ MPa and $\tau = 16$ MPa, respectively. Member *CB* has a square cross-section of 35 mm on each side.

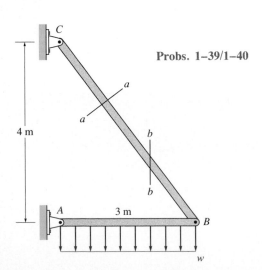

Probs. 1–39/1–40

1–41. The pier is made of material having a specific weight γ. If it has a square cross section, determine its width w as a function of z so that the normal stress in the pier remains constant. The pier supports a constant load **P** at its top where its width is w_1.

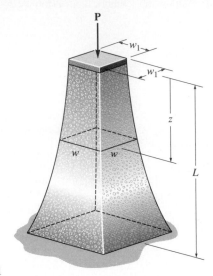

Prob. 1–41

1–42. The pedestal supports a load **P** at its center. If the material has a mass density ρ, determine the radial dimension r as a function of z so that the normal stress in the pedestal remains constant. The cross section is circular.

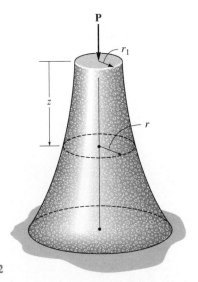

Prob. 1–42

1.6 Allowable Stress

An engineer in charge of the *design* of a structural member or mechanical element must restrict the stress in the material to a level that will be safe. Furthermore, a structure or machine that is currently in use may, on occasion, have to be *analyzed* to see what additional loadings its members or parts can support. So again it becomes necessary to perform the calculations using a safe or allowable stress.

To ensure safety, it is necessary to choose an allowable stress that restricts the applied load to one that is less than the load the member can fully support. There are several reasons for this. For example, the load for which the member is designed may be different from actual loadings placed on it. The intended measurements of a structure or machine may not be exact due to errors in fabrication or in the assembly of its component parts. Unknown vibrations, impact, or accidental loadings can occur that may not be accounted for in the design. Atmospheric corrosion, decay, or weathering tend to cause materials to deteriorate during service. And lastly, some materials, such as wood, concrete, or fiber-reinforced composites, can show high variability in mechanical properties.

One method of specifying the allowable load for the design or analysis of a member is to use a number called the factor of safety. The *factor of safety* (F.S.) is a ratio of a theoretical maximum load that can be carried by the member until it fails in a particular manner (P_{fail}) divided by an allowable load (P_{allow}), which has been determined from experience or experiments to be safe under similar conditions of loading and geometry. Stated mathematically,

$$\text{F.S.} = \frac{P_{\text{fail}}}{P_{\text{allow}}} \tag{1-8}$$

If the load applied to the member is *linearly related* to the stress developed within the member, as in the case of using $\sigma = P/A$ and $\tau_{\text{avg}} = V/A$, then we can express the factor of safety as a ratio of the failure stress σ_{fail} (or τ_{fail}) to the allowable stress σ_{allow} (or τ_{allow});* that is,

$$\text{F.S.} = \frac{\sigma_{\text{fail}}}{\sigma_{\text{allow}}} \tag{1-9}$$

or

$$\text{F.S.} = \frac{\tau_{\text{fail}}}{\tau_{\text{allow}}} \tag{1-10}$$

*In some cases, such as columns, the applied load is *not* linearly related to stress and therefore only Eq. 1-8 can be used to compute the factor of safety. See Chapter 13.

In any of these equations, the factor of safety is generally chosen to be *greater* than 1 in order to avoid the potential for failure. Specific values depend on the types of materials to be used and the intended purpose of the structure or machine. For example, the F.S. used in the design of aircraft or space-vehicle components may be as low as 1 in order to reduce the weight of the vehicle. On the other hand, in the case of a nuclear power plant, the factor of safety for some of its components may be as high as 3 since there may be uncertainties in loading or material behavior. In general, however, factors of safety and therefore the allowable loads or stresses for both structural and mechanical elements have become well standardized, since their design uncertainties have been reasonably evaluated. Their values, which can be found in design codes and engineering handbooks, are intended to form a balance of ensuring public and environmental safety and providing for a reasonable economic solution to design.

1.7 Design of Simple Connections

By making simplifying assumptions regarding the behavior of the material, the equations $\sigma = P/A$ and $\tau_{avg} = V/A$ can often be used to analyze or design a simple connection or a mechanical element. In particular, if a member is subjected to a *normal force* at a section, its required area at the section is determined from

$$A = \frac{P}{\sigma_{allow}} \tag{1-11}$$

On the other hand, if the section is subjected to a *shear force*, then the required area at the section is

$$A = \frac{V}{\tau_{allow}} \tag{1-12}$$

As discussed in Sec. 1.6, the allowable stress used in each of these equations is determined either by applying a factor of safety to a specified normal or shear stress or by finding these stresses directly from an appropriate design code.

We will now discuss four common types of problems for which the above equations can be used.

Cross-Sectional Area of a Tension Member. The cross-sectional area of a prismatic member subjected to a tension force can be determined using Eq. 1–11, provided the force has a line of action that passes through the centroid of the cross section. For example, consider the "eye bar" shown in Fig. 1–27a. At the intermediate section a–a, the stress distribution is uniform over the cross section and the shaded area A is determined from Eq. 1–11, as shown in Fig. 1–27b.

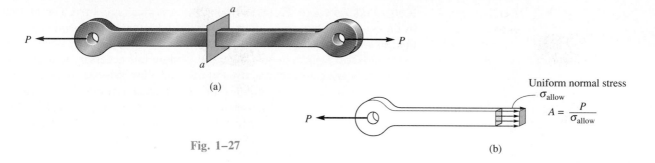

$$A = \frac{P}{\sigma_{allow}}$$

Uniform normal stress
σ_{allow}

Fig. 1–27

(a)

(b)

Cross-Sectional Area of a Connector Subjected to Shear.

Often bolts or pins are used to connect plates, boards, or several members together. When these connectors are subjected to shear, Eq. 1–12 can be used to determine their cross-sectional area. For example, consider the lap joint shown in Fig. 1–28a. If the bolt is loose or the clamping force of the bolt is unknown, it is safe to assume that any frictional force between the plates is negligible. As a result, the free-body diagram for a section passing between the plates and through the bolt is shown in Fig. 1–28b. The bolt is subjected to a resultant internal shear force of $V = P$ at its cross section. Assuming that the shear stress causing this force is *uniformly distributed* over the cross section, the bolt's cross-sectional area A is determined as shown in Fig. 1–28c.

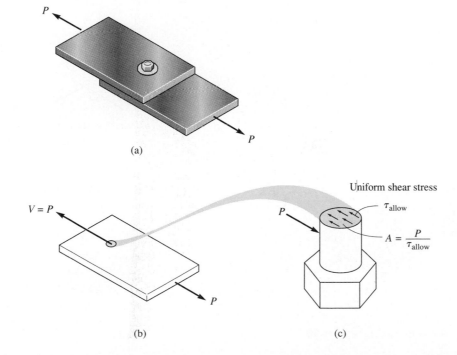

(a)

$V = P$

Uniform shear stress
τ_{allow}

$$A = \frac{P}{\tau_{allow}}$$

Fig. 1–28

(b)

(c)

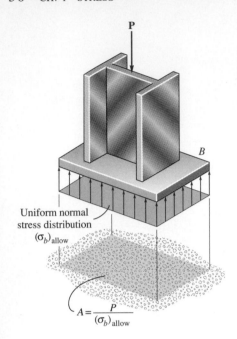

Required Area to Resist Bearing. A normal stress that is produced by the compression of one surface against another is called a *bearing stress*. If this stress becomes large enough, it may crush or locally deform one or both of the surfaces. Hence, in order to prevent failure it is necessary to determine the proper bearing area for the material using an allowable bearing stress. For example, the area A of the column base plate B shown in Fig. 1–29 is determined from the allowable bearing stress of the concrete, $(\sigma_b)_{allow}$, using Eq. 1–11. This assumes, of course, that the bearing stress is uniformly distributed between the plate and the concrete as shown in the figure.

Engineers often use Eq. 1–11 to determine the thickness of plates connected together by bolts or pins. The lap joint in Fig. 1–30a is an example of such a situation. Friction between the plates caused by the unknown clamping force of the bolt can be neglected, so the load **P** is transmitted from one plate to the next only by the *bearing* of the bolt shank against the edge of the hole in each plate. The material strength of the bolt will generally be greater than that of the plate, and therefore crushing of the plate, not the bolt, can occur. The actual stress distribution that the curved surface of the bolt shank exerts on the plate is very difficult to determine. To simplify the design, however, engineering codes often specify the use of the *projected area of contact* between the bolt and plate. As shown in Fig. 1–30b, this area is equal to the product of the plate thickness t and the diameter of the bolt, d_b, not the diameter of the hole. By assuming that the bearing stress is uniformly distributed over this area, it is possible to calculate t using Eq. 1–11, Fig. 1–30b.

Uniform normal stress distribution $(\sigma_b)_{allow}$

$$A = \frac{P}{(\sigma_b)_{allow}}$$

Fig. 1–29

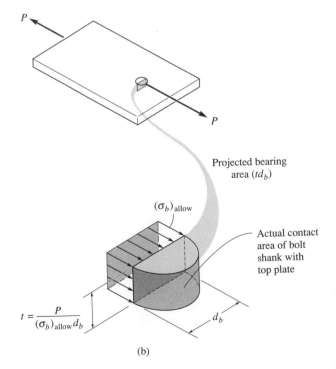

Projected bearing area (td_b)

$(\sigma_b)_{allow}$

Actual contact area of bolt shank with top plate

$$t = \frac{P}{(\sigma_b)_{allow} d_b}$$

d_b

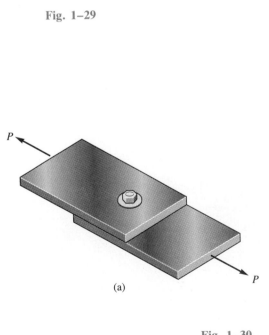

(a)

Fig. 1–30

(b)

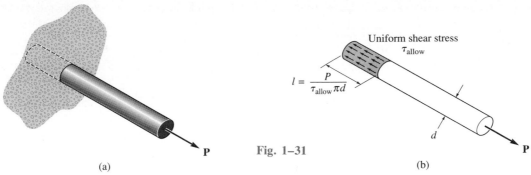

(a)   Fig. 1–31   (b)

Required Area to Resist Shear Caused by Axial Load. Occasionally rods or other members will be supported in such a way that shear stress can be developed in the member even though the member may be subjected to an axial load. An example of this situation would be a steel rod whose end is encased in concrete and loaded as shown in Fig. 1–31a. A free-body diagram of the rod, Fig. 1–31b, shows that *shear stress* acts over the area of contact of the rod with the concrete. This area is $(\pi d)l$, where d is the rod's diameter and l is the length of embedment. Although the actual shear-stress distribution along the rod would be difficult to determine, if we assume it is *uniform,* we can use Eq. 1–12 to calculate l, provided we know d and τ_{allow}, Fig. 1–31b.

PROCEDURE FOR ANALYSIS

The above four cases represent just a few of the many applications of Eqs. 1–11 and 1–12 in engineering design or analysis. Whenever these equations are applied, however, it is important to be aware that the stress distribution is assumed to be *uniformly distributed* over the section. When solving problems, a careful consideration should first be made as to the section over which the stress is to be determined. Once this section is determined, the member must then be designed to have a sufficient area at the section to resist the stress that acts on it. To determine this area, application requires the following steps.

Internal Loading.   Section the member through the area and draw a free-body diagram of a segment of the member. The internal resultant force at the section is then determined using the equations of equilibrium.

Required Area.   Provided the allowable stress is known or can be determined, the required area needed to sustain the load at the section is then computed from Eq. 1–11 or Eq. 1–12. Although this area will then be sufficient at the section considered, it may be necessary to repeat the above process and compute the areas at other sections where the internal loading may be different.

The following examples numerically illustrate the above concepts.

Example 1–12

The two members are pinned together at B as shown in Fig. 1–32a. Top views of the pin connections at A and B are also given in the figure. If the pins have an allowable shear stress of $\tau_{allow} = 12.5$ ksi and the allowable tensile stress of rod CB is $(\sigma_t)_{allow} = 16.2$ ksi, determine to the nearest $\frac{1}{16}$ in. the smallest diameter of pins A and B and the diameter of rod CB necessary to support the load.

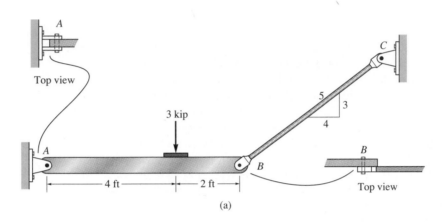

(a)

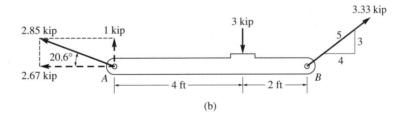

(b)

Fig. 1–32

SOLUTION

Recognizing CB to be a two-force member, the free-body diagram of member AB along with the computed reactions at A and B is shown in Fig. 1–32b. As an exercise, verify the computations and notice that the *resultant force* at A must be used for the design of pin A, since this is the force the pin resists. Why?

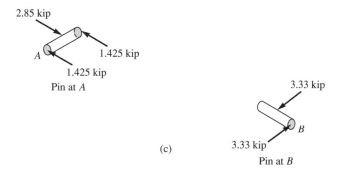

(c)

Diameter of the Pins.   From Fig. 1–32a and the free-body diagrams shown in Fig. 1–32c, pin A is subjected to *double shear,* whereas pin B is subjected to *single shear.* We can compute the required diameter of each pin using Eq. 1–12; that is,

$$A_A = \frac{V_A}{\tau_{\text{allow}}} = \frac{1.425 \text{ kip}}{12.5 \text{ kip/in}^2} = 0.114 \text{ in}^2 = \pi\left(\frac{d_A{}^2}{4}\right) \qquad d_A = 0.381 \text{ in.}$$

$$A_B = \frac{V_B}{\tau_{\text{allow}}} = \frac{3.33 \text{ kip}}{12.5 \text{ kip/in}^2} = 0.267 \text{ in}^2 = \pi\left(\frac{d_B{}^2}{4}\right) \qquad d_B = 0.583 \text{ in.}$$

Although these values represent the *smallest* allowable pin diameters, a *fabricated* or available pin size will have to be chosen. We will choose a size larger to the nearest $\frac{1}{16}$ in. as required.

$$d_A = \tfrac{7}{16} \text{ in.} = 0.438 \text{ in.} \qquad\qquad \textit{Ans.}$$
$$d_B = \tfrac{5}{8} \text{ in.} = 0.625 \text{ in.} \qquad\qquad \textit{Ans.}$$

Diameter of Rod.   The required diameter of the rod throughout its midsection is determined from Eq. 1–11; that is,

$$A_{BC} = \frac{P}{(\sigma_t)_{\text{allow}}} = \frac{3.33 \text{ kip}}{16.2 \text{ kip/in}^2} = \left(0.206 \text{ in}^2\right) = \pi\left(\frac{d_{BC}{}^2}{4}\right)$$

$$d_{BC} = 0.512 \text{ in.}$$

We will choose

$$d_{BC} = \tfrac{9}{16} \text{ in.} = 0.563 \text{ in.} \qquad\qquad \textit{Ans.}$$

Example 1–13

The control arm is subjected to the loading shown in Fig. 1–33a. Determine to the nearest $\frac{1}{4}$ in. the required diameter of the steel pin at C if the allowable shear stress for the steel is $\tau_{\text{allow}} = 8$ ksi. Note in the figure that the pin is subjected to double shear.

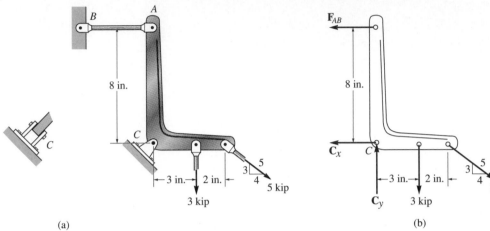

(a)

(b)

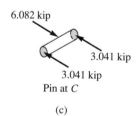

6.082 kip

3.041 kip

3.041 kip

Pin at C

(c)

Fig. 1–33

SOLUTION

Internal Shear Force. A free-body diagram of the arm is shown in Fig. 1–33b. For equilibrium we have

$$\curvearrowright^+ \ \Sigma M_C = 0; \qquad F_{AB}(8 \text{ in.}) - 3 \text{ kip}(3 \text{ in.}) - 5 \text{ kip}(\tfrac{3}{5})(5 \text{ in.}) = 0$$

$$F_{AB} = 3 \text{ kip}$$

$$\xrightarrow{+} \ \Sigma F_x = 0; \qquad -3 \text{ kip} - C_x + 5 \text{ kip}(\tfrac{4}{5}) = 0 \qquad C_x = 1 \text{ kip}$$

$$+\uparrow \ \Sigma F_y = 0; \qquad C_y - 3 \text{ kip} - 5 \text{ kip}(\tfrac{3}{5}) = 0 \qquad C_y = 6 \text{ kip}$$

The resultant force at C is therefore

$$F_C = \sqrt{(1 \text{ kip})^2 + (6 \text{ kip})^2} = 6.082 \text{ kip}$$

Since the pin is subjected to double shear, a shear force of 3.041 kip is resisted over its cross-sectional area between the arm and each supporting leaf for the pin, Fig. 1–33c.

Required Area. Applying Eq. 1–12, we have

$$A = \frac{V}{\tau_{\text{allow}}} = \frac{3.041 \text{ kip}}{8 \text{ kip/in}^2} = 0.3802 \text{ in}^2$$

$$\pi\left(\frac{d}{2}\right)^2 = 0.3802 \text{ in}^2$$

$$d = 0.696 \text{ in.}$$

Use a pin having a diameter of

$$d = \frac{3}{4} \text{ in.} = 0.750 \text{ in.} \qquad\qquad Ans.$$

Example 1–14

The suspender rod is supported at its end by a fixed-connected circular disk as shown in Fig. 1–34*a*. If the rod passes through a 40-mm-diameter hole, determine the minimum required diameter of the rod and the minimum thickness of the disk needed to support the 20-kN load. The allowable normal stress for the rod is $\sigma_{allow} = 60$ MPa, and the allowable shear stress for the plate is $\tau_{allow} = 35$ MPa.

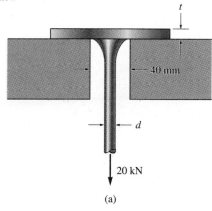

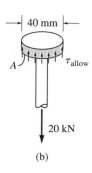

(a)

Fig. 1–34

(b)

SOLUTION

Diameter of Rod. By inspection, the axial force in the rod is 20 kN. Thus the required cross-sectional area is

$$A = \frac{P}{\sigma_{allow}} = \frac{20(10^3) \text{ N}}{60(10^6) \text{ N/m}^2} = 0.333(10^{-3}) \text{ m}^2$$

So that

$$A = \pi \frac{d^2}{4} = 0.333(10^{-3}) \text{ m}^2$$

$$d = 0.0206 \text{ m} = 20.6 \text{ mm} \qquad\qquad \textit{Ans.}$$

Thickness of Plate. As shown on the free-body diagram of the core section of the disk, Fig. 1–34*b*, the material at the sectioned area must resist *shear stress* to prevent movement of the disk through the hole. If this shear stress is *assumed* to be distributed uniformly over the sectioned area, then, since $V = 20$ kN, we have

$$A = \frac{V}{\tau_{allow}} = \frac{20(10^3) \text{ N}}{35(10^6) \text{ N/m}^2} = 0.571(10^{-3}) \text{ m}^2$$

Since the sectioned area $A = 2\pi(0.02 \text{ m})(t)$, the required thickness of the plate is

$$t = \frac{0.571(10^{-3}) \text{ m}^2}{2\pi(0.02 \text{ m})} = 4.55(10^{-3}) \text{ m} = 4.55 \text{ mm} \qquad \textit{Ans.}$$

Example 1–15

An axial load on the shaft shown in Fig. 1–35a is resisted by the collar at C, which is attached to the shaft and located on the right side of the bearing at B. Determine the largest value of P for the two axial forces so that the stress in the collar does not exceed an allowable bearing stress of $(\sigma_b)_{allow} = 75$ MPa and the normal stress in the shaft does not exceed an allowable tensile stress of $(\sigma_t)_{allow} = 55$ MPa.

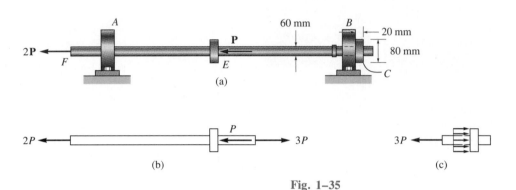

Fig. 1–35

SOLUTION

To solve the problem we will determine P for each possible failure condition. Then we will choose the *smallest* value. Why?

Axial Stress.   Using the method of sections, the axial load within region FE of the shaft is 2P, whereas the *largest* axial load occurs in region EC, Fig. 1–35b. Since the cross-sectional area of the entire shaft is constant, region EC will be subjected to the maximum normal stress. We have

$$55(10^6) \text{ N/m}^2 = \frac{3P}{\pi(0.03 \text{ m})^2}$$

$$P = 51.8 \text{ kN}$$

Bearing Stress.   As shown on the free-body diagram in Fig. 1–35c, the collar at C must resist the load of 3P, which acts over a bearing area of $A_b = [\pi(0.04 \text{ m})^2 - \pi(0.03 \text{ m})^2] = 2.20(10^{-3}) \text{ m}^2$. Thus,

$$75(10^6) \text{ N/m}^2 = \frac{3P}{2.20(10^{-3}) \text{ m}^2}$$

$$P = 55.0 \text{ kN}$$

By comparison, the largest load that can be applied to the shaft is $P = 51.8$ kN, since any load larger than this will cause the allowable tensile stress in the shaft to be exceeded.

Example 1–16

The rigid bar AB shown in Fig. 1–36a is supported by a steel rod AC having a diameter of 20 mm and an aluminum block having a cross-sectional area of 1800 mm². The 18-mm-diameter pins at A and C are subjected to *single shear*. If the failure stress for the steel and aluminum is $(\sigma_{st})_{fail} = 680$ MPa and $(\sigma_{al})_{fail} = 70$ MPa, respectively, and the failure shear stress for each pin is $\tau_{fail} = 900$ MPa, determine the largest load P that can be applied to the bar. Apply a factor of safety of F.S. = 2.0.

SOLUTION

Using Eqs. 1–9 and 1–10, the allowable stresses are

$$(\sigma_{st})_{allow} = \frac{(\sigma_{st})_{fail}}{F.S.} = \frac{680 \text{ MPa}}{2} = 340 \text{ MPa}$$

$$(\sigma_{al})_{allow} = \frac{(\sigma_{al})_{fail}}{F.S.} = \frac{70 \text{ MPa}}{2} = 35 \text{ MPa}$$

$$\tau_{allow} = \frac{\tau_{fail}}{F.S.} = \frac{900 \text{ MPa}}{2} = 450 \text{ MPa}$$

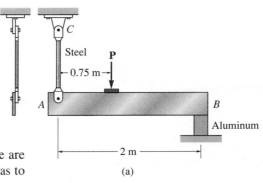

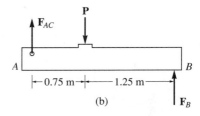

(a)

The free-body diagram for the bar is shown in Fig. 1–36b. There are three unknowns. Here we will apply the equations of equilibrium so as to express F_{AC} and F_B in terms of the applied load P. We have

$$\zeta^+ \ \Sigma M_B = 0; \qquad P(1.25 \text{ m}) - F_{AC}(2 \text{ m}) = 0 \qquad (1)$$
$$\zeta^+ \ \Sigma M_A = 0; \qquad F_B(2 \text{ m}) - P(0.75 \text{ m}) = 0 \qquad (2)$$

We will now determine each value of P that creates the allowable stress in the rod, block, and pins, respectively. In particular, note that the pins at A and C are each subjected to a shear F_{AC}, since single shear exists, Fig. 1–36a.

(b)

Rod AC. This requires

$$F_{AC} = (\sigma_{st})_{allow}(A_{AC}) = 340(10^6) \text{ N/m}^2[\pi(0.01 \text{ m})^2] = 106.8 \text{ kN}$$

Fig. 1–36

Using Eq. 1,

$$P = \frac{(106.8 \text{ kN})(2 \text{ m})}{1.25 \text{ m}} = 171 \text{ kN}$$

Block B. In this case,

$$F_B = (\sigma_{al})_{allow} A_B = 35(10^6) \text{ N/m}^2[1800 \text{ mm}^2(10^{-6}) \text{ m/mm}^2] = 63.0 \text{ kN}$$

Using Eq. 2,

$$P = \frac{(63.0 \text{ kN})(2 \text{ m})}{0.75 \text{ m}} = 168 \text{ kN}$$

Pin A or C. Here

$$V = F_{AC} = \tau_{allow} A = 450(10^6) \text{ N/m}^2[\pi(0.009 \text{ m})^2] = 114.5 \text{ kN}$$

From Eq. 1,

$$P = \frac{114.5 \text{ kN}(2 \text{ m})}{1.25 \text{ m}} = 183 \text{ kN}$$

By comparison, when P reaches its *smallest value* (168 kN), it develops the allowable normal stress in the aluminum block. Hence,

$$P = 168 \text{ kN} \qquad\qquad\qquad Ans.$$

PROBLEMS

1–43.   The steel lap joint is held together by two bolts. If the allowable shear stress for the bolts is $(\tau_{allow})_b = 12$ ksi, determine to the nearest $\frac{1}{4}$ in. the required diameter for each bolt.

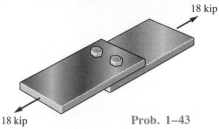

18 kip

18 kip   Prob. 1–43

*1–44.   The joint is fastened together using two bolts. Determine the required diameter of the bolts if the allowable shear stress for the bolts is $\tau_{allow} = 110$ MPa.

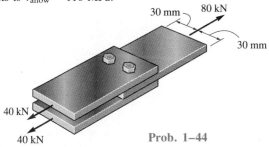

30 mm   80 kN

30 mm

40 kN

40 kN   Prob. 1–44

1–45.   The two aluminum rods support the horizontal force of 18 kN. Determine their required diameters if the allowable tensile stress for the aluminum is $\sigma_{allow} = 150$ MPa.

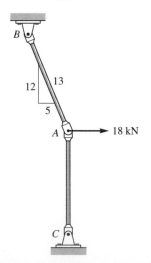

B

12   13

5

A   18 kN

C

Prob. 1–45

1–46.   Member B is subjected to a compressive force of 800 lb. If A and B are both made of wood and are $\frac{3}{8}$ in. thick, determine to the nearest $\frac{1}{4}$ in. the smallest dimension h of the support so that the shear stress does not exceed $\tau_{allow} = 600$ psi.

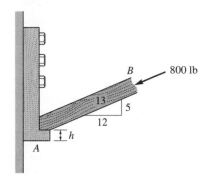

B   800 lb

13

5

12

h

A

Prob. 1–46

1–47.   If the allowable bearing stress for the material under the supports at A and B is $(\sigma_b)_{allow} = 400$ psi, determine the size of *square* bearing plates A' and B' required to support the load. Dimension the plates to the nearest $\frac{1}{2}$ in. The reactions at the supports are vertical.

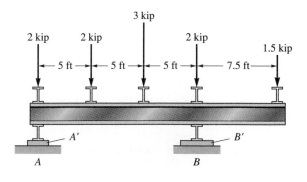

3 kip

2 kip   2 kip   2 kip

1.5 kip

5 ft   5 ft   5 ft   7.5 ft

A'   B'

A   B

Prob. 1–47

*1–48. The clevis C and pin are made of steel having an allowable normal stress of $\sigma_{\text{allow}} = 21$ ksi and an allowable shear stress of $\tau_{\text{allow}} = 12$ ksi. Determine to the nearest $\frac{1}{4}$ in. the diameters of the rod, d_r, and the pin, d_p needed to support the load. Notice that the screw at the end of the rod has ''upset'' threads, so the rod will *not* fail in this region.

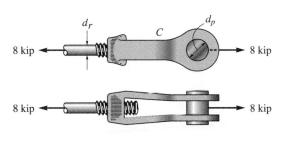

Prob. 1–48

1–49. The column has a cross-sectional area of $12(10^3)$ mm^2. It is subjected to an axial force of 50 kN. If the base plate to which the column is attached has a length of 250 mm, determine its width d so that the average bearing stress under the plate at the ground is one-third of the average compressive stress in the column. Sketch the stress distributions acting over the column's cross-sectional area and at the bottom of the base plate.

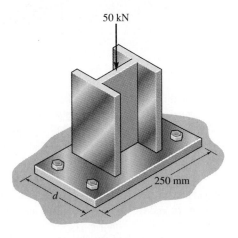

Prob. 1–49

1–50. Determine the required cross-sectional area of member BC and the diameter of the pins at A and B if the allowable normal stress is $\sigma_{\text{allow}} = 3$ ksi and the allowable shear stress is $\tau_{\text{allow}} = 4$ ksi.

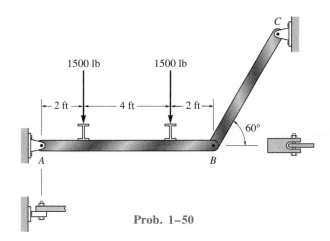

Prob. 1–50

1–51. The two steel cables AB and AC are used to support the chain. If both cables have an allowable tensile stress of $\sigma_{\text{allow}} = 170$ MPa, and cable AB has a diameter of 8 mm and AC has a diameter of 5 mm, determine the greatest force P that can be applied to the chain before one of the cables fails.

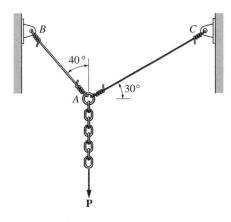

Prob. 1–51

*1–52. The rod BC is made of steel having an allowable tensile stress of $\sigma_{allow} = 155$ MPa. Determine its smallest diameter so that it can support the load shown. The beam is assumed to be pin-connected at A.

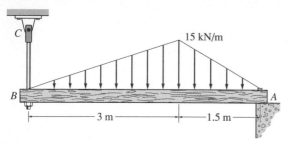

15 kN/m

3 m

1.5 m

Prob. 1–52

1–53. The tension member is fastened together using *two* bolts, one acting on each side of the member as shown. Each bolt has a diameter of 0.3 in. Determine the maximum load P that can be applied to the member if the allowable average shear stress for the bolts is $\tau_{allow} = 12$ ksi and the allowable normal stress is $\sigma_{allow} = 20$ ksi.

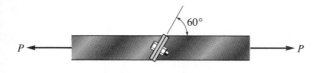

60°

P

P

Prob. 1–53

1–54. The compound wooden beam is connected together by a bolt at B. Assuming that the connections at A, B, C, and D exert only vertical forces on the beam, determine the required diameter of the bolt and the required outer diameter of the washers if the allowable tensile stress for the bolt is $(\sigma_t)_{allow} = 150$ MPa and the allowable bearing stress for the wood is $(\sigma_b)_{allow} = 28$ MPa. Assume that the hole in the washers is the same size as the bolt diameter.

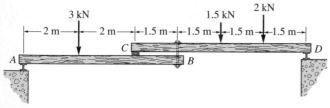

3 kN

1.5 kN

2 kN

2 m — 2 m — 1.5 m — 1.5 m — 1.5 m — 1.5 m

A   C   B   D

Prob. 1–54

1–55. The connection is made using a bolt and nut and two washers. If the allowable bearing stress for the boards is $(\sigma_b)_{allow} = 2$ ksi, and the allowable tensile stress for the bolt shank S is $(\sigma_t)_{allow} = 18$ ksi, determine the maximum allowable tension in the bolt shank. The bolt shank has a diameter of 0.31 in., and the washers have an outer diameter of 0.75 in. and inner diameter (hole) of 0.50 in.

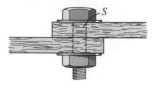

S

Prob. 1–55

*1–56. The bar is held in equilibrium by the pin supports at A and B. Note that the support at A has a single leaf and therefore involves single shear in the pin, and the support at B has a double leaf and therefore involves double shear. The allowable shear stress for both pins is $\tau_{allow} = 150$ MPa. If a force of 3.5 kN is suspended from the bar, determine its maximum allowable position x from B. Pins A and B each have a diameter of 8 mm. Neglect any axial force in the bar.

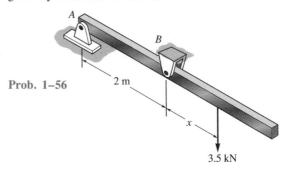

A

B

Prob. 1–56

2 m

x

3.5 kN

1–57. The assembly is used to support the distributed loading of $w = 500$ lb/ft. Determine the factor of safety with respect to yielding for the steel rod BC and the pins at B and C if the yield stress for the steel in tension is $\sigma_y = 36$ ksi and in shear $\tau_y = 18$ ksi. The rod has a diameter of 0.4 in., and the pins each have a diameter of 0.30 in.

1–58. If the allowable shear stress for each of the 0.3-in.-diameter steel pins at A, B, and C is $\tau_{allow} = 12.5$ ksi, and the allowable normal stress for the 0.40-in.-diameter rod is $\sigma_{allow} = 22$ ksi, determine the largest intensity w of the uniform distributed load that can be suspended from the beam.

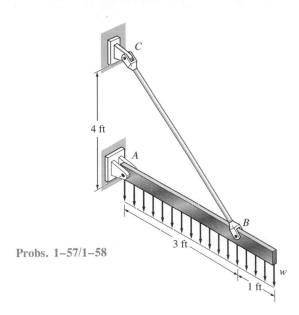

Probs. 1–57/1–58

1–61. The thrust bearing consists of a circular collar A fixed to the shaft B. Determine the maximum axial force P that can be applied to the shaft so that it does not cause the shear stress along a cylindrical surface a or b to exceed an allowable shear stress of $\tau_{\text{allow}} = 170$ MPa.

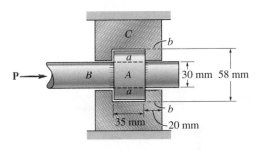

Prob. 1–61

1–59. The hanger assembly is used to support a distributed loading of $w = 0.8$ kip/ft. Determine the average shear stress in the 0.40-in.-diameter bolt at A and the average tensile stress in rod AB, which has a diameter of 0.5 in. If the yield shear stress for the bolt is $\tau_y = 25$ ksi, and the yield tensile stress for the rod is $\sigma_y = 38$ ksi, determine the factor of safety in each case.

*1–60. Determine the intensity w of the maximum distributed load that can be supported by the hanger assembly so that an allowable shear stress of $\tau_{\text{allow}} = 13.5$ ksi is not exceeded in the 0.40-in.-diameter bolts at A and B, and an allowable tensile stress of $\sigma_{\text{allow}} = 22$ ksi is not exceeded in the 0.5-in.-diameter rod AB.

1–62. The pin joint is used to connect the two steel rods. If the material has an allowable shear stress of $\tau_{\text{allow}} = 155$ MPa, determine the maximum force P that can be applied to the rods. Investigate the shear in member A along planes a, member B along planes b, and the pin along planes c.

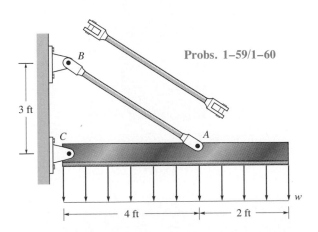

Probs. 1–59/1–60

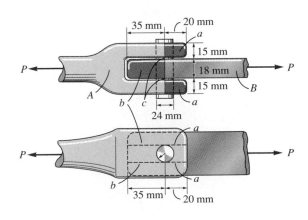

Prob. 1–62

REVIEW PROBLEMS

1–63. The truss is made from three pin-connected members having the cross-sectional areas shown in the figure. Determine the normal stress developed in each member when the truss is subjected to the load shown. State whether the stress is tensile or compressive.

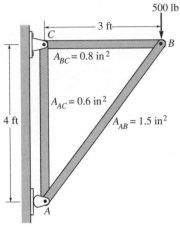

Prob. 1–63

1–65. The member ABC is supported by a pin at A and a rocker at C. Determine the resultant internal loadings acting on the cross sections located through points D and E.

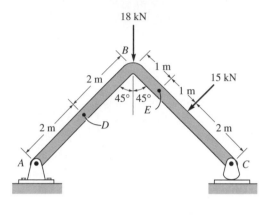

Prob. 1–65

1–66. The plastic block is subjected to an axial compressive force of 600 N. Assuming that the caps at the top and bottom distribute the load uniformly throughout the block, determine the normal and average shear stress acting along section a–a.

*1–64. A force of 8 kN is applied at the *center* of the wooden post. If the post is mounted at the corner of its base plate B, can the bearing stress that the base plate exerts on the slab S be assumed uniformly distributed? Why or why not? What is the average compressive stress in the wooden post?

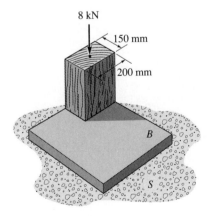

Prob. 1–64

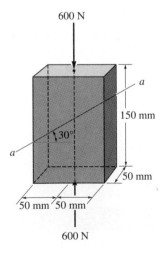

Prob. 1–66

1–67.   The built-up shaft consists of a pipe *AB* and solid rod *BC*. The pipe has an inner diameter of 20 mm and outer diameter of 28 mm. The rod has a diameter of 12 mm. Determine the normal stress at points *D* and *E* and represent the stress on a volume element located at each of these points.

1–69.   The beam *AB* is fixed to the wall and has a uniform weight of 80 lb/ft. If the trolley supports a load of 1500 lb, determine the resultant internal loadings on the cross sections through points *C* and *D*.

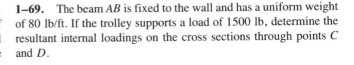

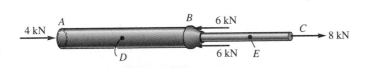

Prob. 1–67

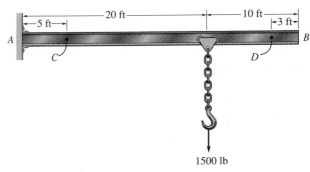

Prob. 1–69

***1–68.**   The frame supports a force of 125 lb. The pins at *B* and *C* each have a diameter of 0.25 in. If these pins are subjected to *double shear*, compute the average shear stress in each pin.

1–70.   If the force on the punch is 20 N, determine the average shear stress developed in the aluminum plate along a cylindrical section taken through the plate just under the punch. The load is not great enough to punch a hole in the plate.

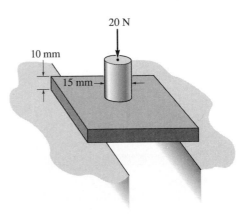

Prob. 1–68

Prob. 1–70

1–71. The beam supports a distributed load and a couple moment. Determine the resultant internal loadings on the cross sections located just to the left and just to the right of the roller support at B.

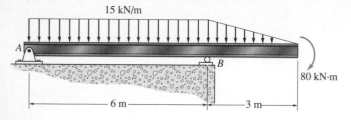

15 kN/m

A

B

80 kN·m

6 m

3 m

Prob. 1–71

1–74. The curved rod AD of radius r has a weight per length of w. If it lies in the vertical plane, determine the resultant internal loadings acting on the cross section through point B. *Hint:* The distance from the centroid C of segment AB to point O is $OC = [2r \sin(\theta/2)]/\theta$.

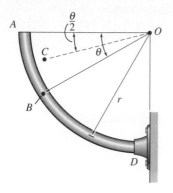

A

$\frac{\theta}{2}$

C

θ

O

B

r

D

Prob. 1–74

*1–72. The two members used in the construction of an aircraft fuselage are joined together using a 30° fish-mouth weld. Determine the normal and average shear stress on the plane of each weld.

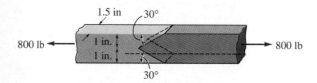

1.5 in   30°

800 lb

1 in.

1 in.

800 lb

30°

Prob. 1–72

1–73. The lever is attached to the shaft A using a key that has a width of 8 mm and length of 25 mm. If the shaft is fixed and a vertical force of 50 N is applied perpendicular to the handle, determine the average shear stress developed along section a–a of the key.

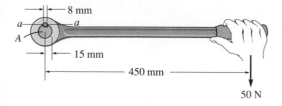

8 mm

a

a

A

15 mm

450 mm

50 N

Prob. 1–73

2   Strain

In engineering the deformation of a body is specified using the concept of normal and shear strain. In this chapter we will define these quantities and show how they can be determined for various types of problems.

2.1 Deformation

Whenever a force is applied to a body, it will tend to change the body's shape and size. These changes are referred to as *deformation*, and they may be either highly visible or practically unnoticable, without the use of equipment to make precise measurements. For example, a rubber band will undergo a very large deformation when stretched. On the other hand, only slight deformations of structural members occur when a building is occupied with people walking about. Deformation of a body can also occur when the temperature of the body is changed. A typical example is the thermal expansion or contraction of a roof caused by the weather.

   Displacement is a vector quantity that is used to measure the movement of a particle or point from one position to the next. Hence, if a body is classified as being deformable, then its adjacent particles can be displaced relative to one another when forces are applied to the body. On the other hand, if the body is rigid, then no relative displacement can occur between the particles.

   Consider the body shown in Fig. 2–1, which is made from a continuous and cohesive material and is shown in the initial undeformed state. The three particles *A, B,* and *C* are located in the body at points measured from a fixed coordinate system. When a loading causes the body to deform, and thus move into its final position, the particles are displaced to points *A', B',* and *C',* Fig.

2–1. For example, the displacement of point A is indicated by the vector $\mathbf{u}(A)$. Because of the deformation, the once straight lines AB and AC become curves $A'B'$ and $A'C'$; and as a result, the length of AB and AC and the angle θ will be different from the curved length of $A'B'$ and $A'C'$ and the angle θ'. In other words, the difference between the lengths and relative orientations of the two lines in the body is a consequence of the displacements caused by the *deformation*. Measurements of the deformation must therefore account for *both* the changes made in the length of line segments inscribed in the body *and* the changes made in the angle between them.

In the general sense, the deformation of a body will not be uniform throughout its volume, and so the change in geometry of a line segment within the body may vary along its length. For example, one portion of the line may elongate, whereas another portion may contract. As shorter and shorter line segments are considered, however, they remain straighter after the deformation, and so to study deformational changes in a more uniform manner, we will consider the lines to be very short and located in the neighborhood of a point. In doing so, realize that any line segment located at one point in the body will change by a different amount from one located at some other point. Furthermore, these changes will also depend on the orientation of the line segment at the point. For example, a line segment may elongate if it is orientated in one direction, whereas it may contract if it is orientated in another direction.

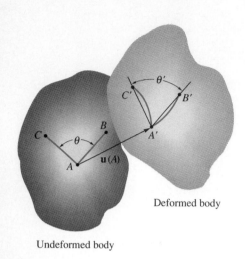

Undeformed body

Deformed body

Fig. 2–1

2.2 Strain

In order to describe the deformation by changes in length of line segments and the changes in the angles between them, we will develop the concept of strain. Measurements of strain are actually made by experiments, and once the strains are obtained, it will be shown later in the text how they may be related to the applied loads, or stresses, acting within the body.

Normal Strain. The elongation or contraction of a line segment per unit of length is referred to as *normal strain*. To develop a formalized definition of normal strain, consider the line AB, which is contained within the undeformed body shown in Fig. 2–2a. This line lies along the n axis and has an original length of Δs. During deformation, points A and B are displaced to A' and B', and the line becomes a curve having a length of $\Delta s'$, Fig. 2–2b. The change in length of the line is therefore $\Delta s' - \Delta s$. If we define the *average normal strain* using the symbol ϵ_{avg} (epsilon), then

$$\epsilon_{\text{avg}} = \frac{\Delta s' - \Delta s}{\Delta s} \qquad (2\text{–}1)$$

As point B is chosen closer and closer to point A, the length of the line becomes shorter and shorter, such that $\Delta s \to 0$. Also, this causes B' to ap-

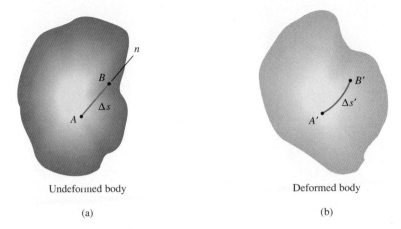

Undeformed body   Deformed body

(a)   (b)

Fig. 2–2

proach A', such that $\Delta s' \to 0$. Consequently, in the limit the normal strain at *point A* and in the *direction* of n is

$$\epsilon = \lim_{B \to A \text{ along } n} \frac{\Delta s' - \Delta s}{\Delta s} \qquad (2\text{–}2)$$

If the normal strain is known, we can use this equation to obtain the approximate final length of a *short* line segment after it is deformed. We have

$$\Delta s' \approx (1 + \epsilon)\, \Delta s \qquad (2\text{–}3)$$

Hence, when ϵ is positive the initial line Δs will elongate, whereas if ϵ is negative the line contracts.

Units. Notice that normal strain is a *dimensionless quantity,* since it is a ratio of two lengths. Although this is the case, it is common practice to state it in terms of a ratio of length units. If the SI system is used, then the basic units will be meters/meter (m/m). Ordinarily, for most engineering applications ϵ will be very small, so measurements of strain are in micrometers per meter (μm/m), where 1 μm $= 10^{-6}$ m. In the Foot-Pound-Second system, strain can be stated in units of inches per inch (in./in.). Sometimes for experimental work, strain is expressed as a percent, e.g., 0.001 m/m $= 0.1\%$. As an example, a normal strain of $480(10^{-6})$ can be reported as $480(10^{-6})$ in./in., 480 μm/m, or 0.0480%. Also, one can state this answer as simply 480 μ (480 "micros").

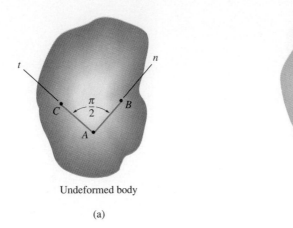

Undeformed body

(a)

Deformed body

(b)

Fig. 2–3

Shear Strain. The change in angle that occurs between two line segments that were originally *perpendicular* to one another is referred to as *shear strain*. This angle is denoted by γ (gamma) and is measured in radians (rad). To show how it is developed, consider the line segments AB and AC originating from the same point A in a body, and directed along the perpendicular n and t axes, Fig. 2–3a. After deformation, the ends of the lines are displaced, and the lines themselves become curves, such that the angle between them at A is θ', Fig. 2–3b. Hence we define the shear strain at point A that is associated with the n and t axes as

$$\gamma_{nt} = \frac{\pi}{2} - \lim_{\substack{B \to A \text{ along } n \\ C \to A \text{ along } t}} \theta' \qquad (2\text{–}4)$$

Notice that if θ' is smaller than $\pi/2$ the shear strain is positive, whereas if θ' is larger than $\pi/2$ the shear strain is negative.

Cartesian Strain Components. Using the above definitions of normal and shear strain, we will now show how they can be used to describe the deformation of the body shown in Fig. 2–4a. To do so, imagine the body to be subdivided into small elements such as the one shown in Fig. 2–4b. This element is rectangular, has undeformed dimensions Δx, Δy, and Δz, and is located in the neighborhood of a point in the body, Fig. 2–4a. Assuming that the element's dimensions are very small, the deformed shape of the element, shown in Fig. 2–4c, will be a parallelepiped, since very small line segments will remain approximately straight after the body is deformed. In order to achieve this deformed shape, we may first consider how the normal strain changes the lengths of the sides of the rectangular element, and then how the shear strain changes the angles of each side. Hence, using Eq. 2–3, in reference to the lines Δx, Δy, and Δz, the approximate lengths of the sides of the parallelepiped are

$$(1 + \epsilon_x)\,\Delta x \qquad (1 + \epsilon_y)\,\Delta y \qquad (1 + \epsilon_z)\,\Delta z$$

And the approximate angles between the sides, again originally defined by the sides Δx, Δy, and Δz, are

$$\frac{\pi}{2} - \gamma_{xy} \qquad \frac{\pi}{2} - \gamma_{yz} \qquad \frac{\pi}{2} - \gamma_{zx}$$

In particular, notice that the *normal strains* cause a change in *volume* of the rectangular element, whereas the *shear strains* cause a change in its *shape*. Of course, both of these effects occur simultaneously during the deformation. It might also be added that because the other elements of the body deform, the element under consideration will not only deform but will also both translate and rotate, since it is being displaced by the deformations of all the other elements. For example, imagine the element located near the end of a fishing pole while the pole deflects when supporting a fish. These *rigid-body displacements,* however, will be neglected here, since we are interested only in the relative displacements of the element due to its deformation. Specifically, it is these *deformation displacements* that will allow us later to determine the stress in the body.

In summary, then, the *state of strain* at a point in a body requires specifying three normal strains, ϵ_x, ϵ_y, ϵ_z, and three shear strains, γ_{xy}, γ_{yz}, γ_{zx}. These strains completely describe the deformation of a rectangular volume element of material located at the point and orientated so that its sides are originally parallel to the x, y, z axes. Once these strains are defined at all points in the body, the deformed shape of the body can then be described. It should also be added that by knowing the state of strain at a point, defined by its six components, it is possible to determine the strain components on an element orientated at the point in any other direction. This is discussed in Chapter 10.

Small Strain Analysis.

Most engineering design involves applications for which only small deformations are allowed. For example, almost all structures and machines appear to be rigid, and the deformations that occur during use are hardly noticeable. Furthermore, even if the deflection of a member such as a thin plate or slender rod may appear to be large, the material from which it is made may only be subjected to very small deformations. In this text, therefore, we will assume that the deformations that take place within a body are almost infinitesimal, so that the *normal strains* occurring within the material are *very small* compared to 1, that is, $\epsilon \ll 1$. This assumption based on the magnitude of the strain has wide practical application in engineering, and its application is often referred to as a *small strain analysis.*

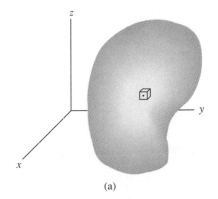

(a)

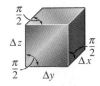

Undeformed element

(b)

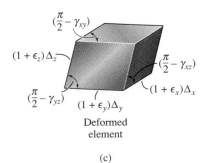

Deformed element

(c)

Fig. 2–4

Whenever such an analysis is used, since $\epsilon \ll 1$, it is permissible to *neglect* the product of either two strains or deformation displacements, or any term involving strain or deformation displacement that is raised to a power greater than 1. Such "first-order" approximations may be used to simplify calculations involving strain and deformation. For example, if Δ is a *very small number*, then we can use the binomial theorem to expand terms having the form $(1 + \Delta)^n$, where n is any exponent. This theorem, in general form, is

$$(1 + \Delta)^n = 1 + n\,\Delta + \frac{n(n-1)}{2!}\Delta^2 + \ldots$$

So that, as an approximation, we can neglect the higher order terms which then yields

$$(1 + \Delta)^n \approx 1 + n\Delta \tag{2-5}$$

For some typical applications, take $n = 3$, $n = \frac{1}{2}$, or $n = -1$, in which case we have

$$(1 + \Delta)^3 \approx 1 + 3\Delta$$
$$\sqrt{1 + \Delta} \approx 1 + \tfrac{1}{2}\Delta$$
$$\frac{1}{1 + \Delta} \approx 1 - \Delta$$

If Δ and Δ' are two small numbers, then as an approximation of the product

$$(\Delta + 1)(\Delta' + 1) = \Delta\Delta' + \Delta + \Delta' + 1$$
$$\approx \Delta + \Delta' + 1$$

Lastly, if the sine or cosine of a *very small angle* $\Delta\theta$ is to be found, then by the power series expansions of these functions, we can make the following approximations:

$$\sin \Delta\theta = \Delta\theta - \frac{\Delta\theta^3}{3!} + \frac{\Delta\theta^5}{5!} - \ldots$$
$$\sin \Delta\theta \approx \Delta\theta \tag{2-6}$$

and

$$\cos \Delta\theta = 1 - \frac{\Delta\theta^2}{2!} + \frac{\Delta\theta^4}{4!} - \ldots$$
$$\cos \Delta\theta \approx 1 \tag{2-7}$$

Also, note that

$$\tan \Delta\theta = \frac{\sin \Delta\theta}{\cos \Delta\theta} \approx \frac{\Delta\theta}{1} = \Delta\theta$$

The following examples illustrate numerical applications of some of these approximations in the calculation of normal and shear strain.

Example 2–1

The slender rod shown in Fig. 2–5 is subjected to an increase of temperature along its axis, which creates a normal strain in the rod of $\epsilon_z = 40(10^{-3})z^{1/2}$, where z is given in meters. Determine (a) the displacement of the end B of the rod due to the temperature increase, and (b) the average normal strain in the rod.

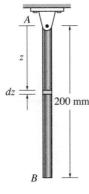

Fig. 2–5

SOLUTION

Part (a). Since the normal strain is reported at each point along the rod, a differential segment dz, located at position z, Fig. 2–5, has a deformed length that can be determined from Eq. 2–3; that is,

$$dz' = [1 + 40(10^{-3})z^{1/2}]\, dz$$

The sum total of these segments along the axis yields the *deformed length* of the rod, i.e.,

$$L' = \int_0^{0.2\ \text{m}} [1 + 40(10^{-3})z^{1/2}]\, dz$$

$$= z + 40(10^{-3})(\tfrac{2}{3}\, z^{3/2}) \Big|_0^{0.2\ \text{m}}$$

$$= 0.20239\ \text{m}$$

The displacement of the end of the rod is therefore

$$\Delta_B = 0.20239\ \text{m} - 0.2\ \text{m} = 0.00239\ \text{m} = 2.39\ \text{mm}\ \downarrow \qquad \textit{Ans.}$$

Part (b). The average normal strain in the rod is determined from Eq. 2–1, which assumes that the rod or "line segment" has an original length of 200 mm and a change in length of 2.39 mm. Hence,

$$\epsilon_{\text{avg}} = \frac{2.39\ \text{mm}}{200\ \text{mm}} = 0.0119\ \text{mm/mm} \qquad \textit{Ans.}$$

Example 2–2

A force acting on the grip of the lever arm shown in Fig. 2–6a causes the arm to rotate clockwise through an angle of $\theta = 0.002$ rad. Determine the average normal strain developed in the wire BC.

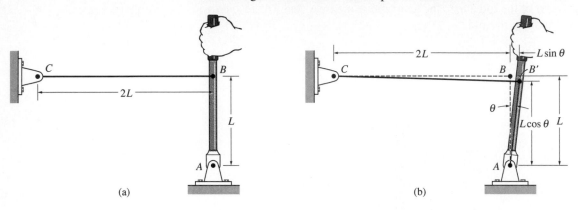

Fig. 2–6

SOLUTION I

The final length of the wire, CB', can be computed from the displacement diagram shown in Fig. 2–6b. We have

$$CB' = \sqrt{(2L + L \sin \theta)^2 + (L - L \cos \theta)^2}$$
$$= L\sqrt{(2 + \sin \theta)^2 + (1 - \cos \theta)^2}$$
$$= L\sqrt{4 + 4 \sin \theta + \sin^2 \theta + 1 - 2 \cos \theta + \cos^2 \theta}$$

Since $\sin^2 \theta + \cos^2 \theta = 1$, then

$$CB' = L\sqrt{6 + 4 \sin \theta - 2 \cos \theta}$$

For small angle θ, $\sin \theta \approx \theta$ and $\cos \theta \approx 1$, so that

$$CB' \approx 2L\sqrt{1 + \theta} = 2L(1 + \theta)^{1/2}$$

Finally, using Eq. 2–5 with $n = 1/2$, we have

$$CB' \approx 2L\left(1 + \tfrac{1}{2}\theta\right) = 2L(1 + 0.001) = 2.002L$$

The average normal strain in the wire is therefore

$$\epsilon_{\text{avg}} = \frac{CB' - CB}{CB} = \frac{2.002L - 2L}{2L} = 0.001 \qquad Ans.$$

SOLUTION II

Since θ is small, the stretch in wire CB, Fig. 2–6b, is $L \sin \theta \approx L\theta$, using Eq. 2–6. The average normal strain in the wire is then

$$\epsilon_{\text{avg}} = \frac{L\theta}{2L} = \frac{\theta}{2} = \frac{0.002}{2} = 0.001 \qquad Ans.$$

Example 2-3

The plate is deformed into the dashed shape shown in Fig. 2–7a. If in this deformed shape horizontal lines on the plate remain horizontal, determine (a) the average normal strain along the side AB, and (b) the average shear strain in the plate relative to the x and y axes.

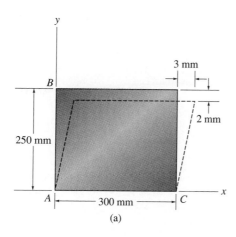

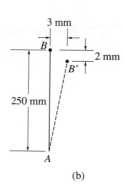

(a)
(b)

Fig. 2-7

SOLUTION

Part (a).   Line AB, coincident with the y axis, becomes line AB' after deformation, as shown in Fig. 2–7b. The length of this line is

$$AB' = \sqrt{(250 - 2)^2 + (3)^2} = 248.018 \text{ mm}$$

The average normal strain for AB is therefore

$$(\epsilon_{AB})_{\text{avg}} = \frac{AB' - AB}{AB} = \frac{248.018 \text{ mm} - 250 \text{ mm}}{250 \text{ mm}}$$

$$= -7.93(10^{-3}) \text{ mm/mm} \qquad \qquad \textit{Ans.}$$

The negative sign indicates the strain causes a contraction of AB.

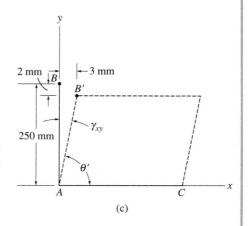

Part (b).   As noted in Fig. 2–7c, the once 90° angle BAC between the sides of the plate, referenced from the x, y axes, changes to θ' due to the displacement of B to B'. Since $\gamma_{xy} = \pi/2 - \theta'$, then γ_{xy} is the angle shown in the figure. Thus,

$$\gamma_{xy} = \tan^{-1}\left(\frac{3 \text{ mm}}{250 \text{ mm} - 2 \text{ mm}}\right) = 0.0121 \text{ rad} \qquad \textit{Ans.}$$

(c)

Example 2-4

The plate shown in Fig. 2–8a is held in the rigid horizontal guides at its top and bottom, AD and BC. If its right side CD is given a uniform horizontal displacement of 2 mm, determine (a) the average normal strain along the diagonal AC, and (b) the shear strain at E relative to the x, y axes.

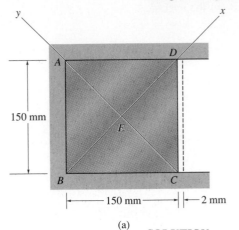

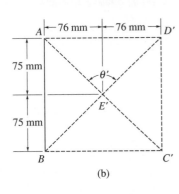

(b)

Fig. 2–8   (a)

SOLUTION

Part (a). When the plate is deformed, the diagonal AC becomes AC', Fig. 2–8b. The length of diagonals AC and AC' can be computed from the Pythagorean theorem. We have

$$AC = \sqrt{(0.150)^2 + (0.150)^2} = 0.21213 \text{ m}$$
$$AC' = \sqrt{(0.150)^2 + (0.152)^2} = 0.21355 \text{ m}$$

Therefore the average normal strain along the diagonal is

$$(\epsilon_{AC})_{\text{avg}} = \frac{AC' - AC}{AC} = \frac{0.21355 \text{ m} - 0.21213 \text{ m}}{0.21213 \text{ m}}$$
$$= 0.00669 \text{ mm/mm} \qquad\qquad Ans.$$

Part (b). To find the shear strain at E relative to the x and y axes, it is first necessary to find the angle θ', which specifies the angle between these axes after deformation, Fig. 2–8b. We have

$$\tan\left(\frac{\theta'}{2}\right) = \frac{76 \text{ mm}}{75 \text{ mm}}$$

$$\theta' = 90.759° = \frac{\pi}{180°}(90.759°) = 1.58404 \text{ rad}$$

Applying Eq. 2–4, the shear strain at E is therefore

$$\gamma_{xy} = \frac{\pi}{2} - 1.58404 \text{ rad} = -0.0132 \text{ rad} \qquad\qquad Ans.$$

According to the sign convention, the *negative sign* indicates that the angle θ' is *greater than* 90°.

PROBLEMS

2–1. A thin strip of rubber has an unstretched length of 15 in. If it is stretched around a pipe having an outer diameter of 5 in., determine the average normal strain in the strip.

2–2. An air filled rubber ball has a diameter of 6 in. If the air pressure within it is increased until the ball's diameter becomes 7 in., determine the average normal strain in the rubber.

2–3. The rigid beam is supported by a pin at A and wires BD and CE. If the load P on the beam causes the end C to be displaced 10 mm downward, determine the strain developed in wires CE and BD.

2–5. The wire AB is unstretched when $\theta = 45°$. If a load is applied to the bar AC, which causes $\theta = 47°$, determine the normal strain in the wire.

2–6. If a load applied to bar AC causes point A to be displaced to the right by an amount ΔL, determine the normal strain in the wire AB. Originally, $\theta = 45°$.

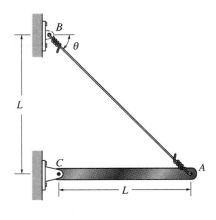

Probs. 2–5/2–6

Prob. 2–3

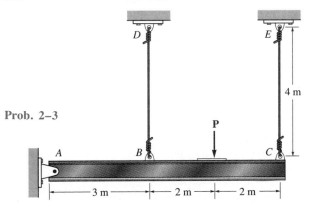

*2–4. Due to its weight, the rod is subjected to a normal strain that varies along its length such that $\epsilon = kz$, where k is a constant. Determine the displacement ΔL of its end B when it is suspended as shown.

2–7. The rectangular membrane has an unstretched length L_1 and width L_2. If the sides are increased by a small amount ΔL_1 and ΔL_2, determine the normal strain along the diagonal AB.

Prob. 2–4

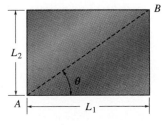

Prob. 2–7

*2–8.   The two wires are connected together at A. If the force **P** causes point A to be displaced horizontally 2 mm, determine the normal strain developed in each wire.

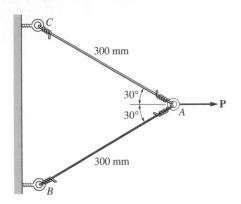

Prob. 2–8

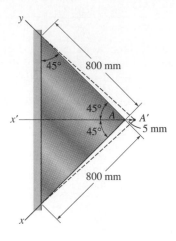

Probs. 2–10/2–11/2–12

2–9.   A thin wire, lying along the x axis, is strained such that each point on the wire is displaced $\Delta x = kx^2$ along the x axis. What is the normal strain at any point P along the wire?

Prob. 2–9

2–10.   The triangular plate is fixed at its base and its apex A is given a horizontal displacement of 5 mm. Determine the shear strain γ_{xy} at A.

2–11.   The triangular plate is fixed at its base, and its apex A is given a horizontal displacement of 5 mm. Determine the average normal strain ϵ_x along the x axis.

*2–12.   The triangular plate is fixed at its base, and its apex A is given a horizontal displacement of 5 mm. Determine the average normal strain $\epsilon_{x'}$ along the x' axis.

2–13.   The corners of the square plate are given the displacements indicated. Determine the average normal strains ϵ_x and ϵ_y along the x and y axes, and the shear strain γ_{xy} at point A.

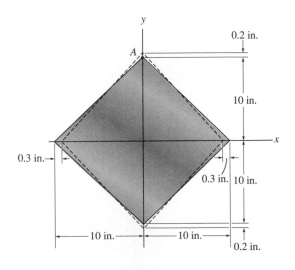

Prob. 2–13

2–14. The rectangular plate is subjected to the deformation shown by the dashed line. Determine the shear strains γ_{xy} and $\gamma_{x'y'}$ developed at point A.

***2–16.** The nonuniform loading causes a normal strain in the shaft that can be expressed as $\epsilon_x = kx^2$, where k is a constant. Determine the displacement of the end B. Also, what is the average strain in the rod?

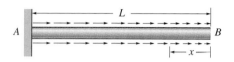

Prob. 2–16

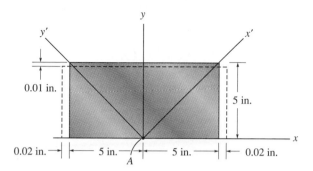

Prob. 2–14

2–17. If the normal strain is defined in reference to the final length, that is,

$$\epsilon_n' = \lim_{P \to P'} \left(\frac{\Delta s' - \Delta s}{\Delta s'} \right)$$

2–15. Bar ABC is originally in a horizontal position. If loads cause the end A to be displaced downwards $\Delta_A = 0.002$ in. and the bar rotates $\theta = 0.2°$, determine the strains in the rods AD, BE, and CF.

instead of in reference to the original length, Eq. 2–2, show that the difference in these strains is represented as a second-order term, namely, $\epsilon_n - \epsilon_n' = \epsilon_n \epsilon_n'$.

2–18. The rectangular plate is subjected to the deformation shown by the dashed line. Determine the average shear strain γ_{xy} of the plate.

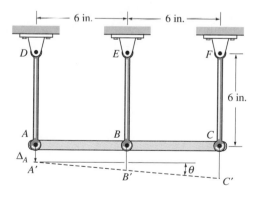

Prob. 2–15

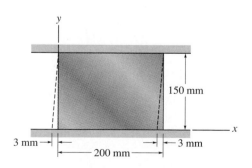

Prob. 2–18

2–19. The rectangular plate is distorted into a parallelogram as shown by the dashed lines. If the diagonal AB' is 50.03 mm, determine the shear strain γ_{xy}.

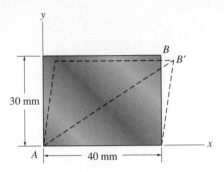

Prob. 2–19

2–21. The fiber AB has a length L and orientation θ. If its ends A and B undergo very small displacements u_A and v_B, respectively, determine the normal strain in the fiber when it is in position $A'B'$.

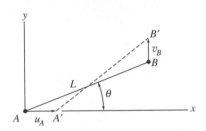

Prob. 2–21

*2–20. The piece of rubber is originally rectangular. Determine the average shear strain γ_{xy} if the corners B and D are subjected to the displacements that cause the rubber to distort as shown by the dashed lines.

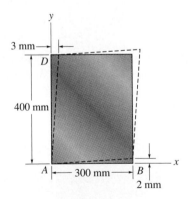

Prob. 2–20

3 Mechanical Properties of Materials

Having discussed the basic concepts of stress and strain, we will in this chapter show how stress can be related to strain by using experimental methods to determine the stress–strain diagram for a specific material. The properties of this diagram will then be discussed for materials that are commonly used in engineering. Also, other mechanical properties and tests that are related to the development of mechanics of materials will be discussed.

3.1 The Tension and Compression Test

The strength of a material depends on its ability to sustain a load without undue deformation or failure. This property is inherent in the material itself and must be determined by *experiment*. As a result, several types of tests have been developed to evaluate a material's strength under loads that are static, cyclic, extended in duration, or impulsive. Over the years, each of these tests has become standardized so that the results obtained from different laboratories can be compared. In the United States the American Society for Testing and Materials (ASTM) has published guidelines for performing such tests and provides limits for which the use of a particular material is considered acceptable.

One of the most important tests to perform is the *tension or compression test*. Although many important mechanical properties of a material can be determined from this test, it is used primarily to determine the relationship between the average normal stress and normal strain in many engineering materials consisting of metal, ceramics, polymers, and composites. To perform this test a specimen of the material is made into a "standard" shape and

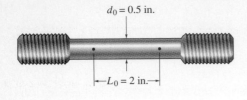

$d_0 = 0.5$ in.

$L_0 = 2$ in.

Fig. 3–1

size. Before testing, two small punch marks are made along the specimen's length. These marks are located away from both ends of the specimen because the stress distribution at the ends is somewhat complex due to gripping at the connections where the load is applied. Measurements are taken of both the specimen's initial cross-sectional area, A_0, and the *gauge-length* distance L_0 between the punch marks. For example, when a steel specimen is used in a tension test it generally has an initial diameter of $d_0 = 0.5$ in. and a gauge length of $L_0 = 2$ in., Fig. 3–1. In order to apply an axial load with no bending of the specimen, the ends are usually seated into ball-and-socket joints. A testing machine like the one shown in Fig. 3–2 is then used to stretch the specimen at a very slow, constant rate until it reaches the breaking point.

At frequent intervals during the test, data is recorded of the applied load P, as read on the dial of the machine. Also, the elongation $\Delta L = L - L_0$ between the punch marks on the specimen may be measured using either a caliper or a mechanical or optical device called an *extensometer*. This value of ΔL is then used to determine the average normal strain in the specimen. Sometimes, however, this measurement is not taken, since it is also possible to read the strain *directly* by using an *electrical-resistance strain gauge*, which looks like the one shown in Fig. 3–3. The operation of this gauge is based on the change in electrical resistance of a very thin wire or piece of metal foil under strain. Essentially the gauge is cemented to the specimen in a specified direction. If the cement is very strong in comparison to the gauge, then the gauge is in effect an integral part of the specimen, so that when the specimen is strained in the direction of the gauge, the wire and specimen will experience the same strain. By measuring the electrical resistance of the wire, the gauge may be calibrated to read values of normal strain directly.

Fig. 3–2

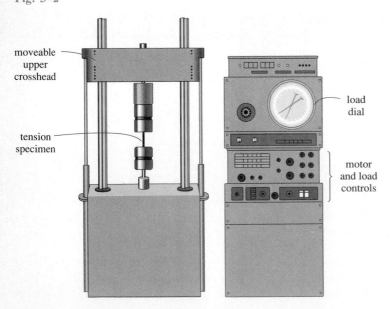

moveable upper crosshead

load dial

tension specimen

motor and load controls

Electrical–resistance strain gauge

Fig. 3–3

3.2 The Stress–Strain Diagram

From the data of a tension or compression test, it is possible to compute the stress and corresponding strain in the specimen and then plot the results. The resulting curve is called the *stress–strain diagram,* and there are two ways in which it is normally described.

Conventional Stress and Strain Diagram. Using the recorded data, we can determine the *nominal* or *engineering stress* by dividing the applied load P by the specimen's *original* cross-sectional area A_0. This calculation assumes that the stress is *constant* over the cross section and throughout the region between the gauge points. We have

$$\sigma = \frac{P}{A_0} \qquad (3-1)$$

Likewise, the *nominal* or *engineering strain* is found directly from the strain gauge reading, or by dividing the change in the specimen's gauge length, ΔL, by the specimen's original gauge length L_0. Here the strain is assumed to be constant throughout the region between the gauge points. Thus,

$$\epsilon = \frac{\Delta L}{L_0} \qquad (3-2)$$

If the corresponding values of σ and ϵ are plotted as a graph, for which the ordinate is the stress and the abscissa is the strain, the resulting curve is called a *conventional stress–strain diagram.* This diagram is very important in engineering since it provides the means for obtaining data about a material's tensile (or compressive) strength *without* regard for the material's physical size or shape. Realize, however, that no two stress–strain diagrams for a particular material will be *exactly* the same, since the results depend on such variables as the material's composition, microscopic imperfections, the way it is manufactured, the rate of loading, and the temperature during the time of the test.

We will now discuss the characteristics of the conventional stress–strain curve as it pertains to *steel,* a commonly used material for making both structural members and mechanical elements. Using the method described above, the characteristic stress–strain diagram for a steel specimen is shown in Fig. 3–4. From this curve we can identify four different ways in which the material behaves, depending on the amount of strain induced in the material.

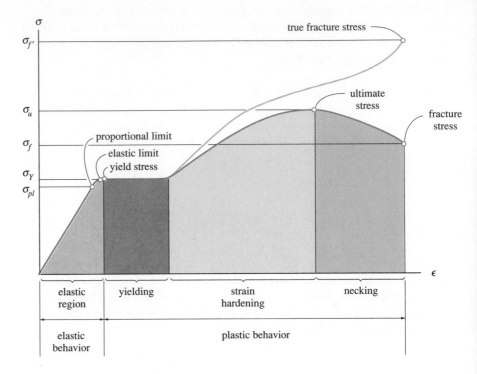

Conventional and true stress-strain diagrams for ductile material (steel)
(not to scale)

Fig. 3–4

Elastic Behavior. The specimen is said to respond *elastically* if it returns to its *original shape* or length when the load acting on it is removed. This elastic behavior occurs when the strains in the specimen are within the lightly shaded region shown in Fig. 3–4. It can be seen that the curve is actually a *straight line* throughout most of this region, so the stress is *proportional* to the strain. In other words, the material is said to be *linear elastic*. The upper stress limit to this linear relationship is called the *proportional limit, σ_{pl}*. If the stress slightly exceeds the proportional limit, the material may still respond elastically; however, the curve tends to flatten out causing a greater increment of strain for a corresponding increment of stress. This continues until the stress reaches the *elastic limit*. To determine this point for any specimen, one must apply and then release an increased loading until a permanent deformation is detected in the specimen. Normally for steel, however, the elastic limit is seldom determined, since it is very close to the proportional limit and therefore rather difficult to detect.

Yielding. A slight increase in stress above the elastic limit will result in a breakdown of the material and cause it to *deform permanently*. This behavior is called *yielding,* and it is indicated by the dark-shaded region of the curve, Fig. 3–4. The stress that causes yielding is called the *yield stress* or *yield point,* σ_Y, and the deformation that occurs is called *plastic deformation*. Unlike elastic loading, a load that causes yielding of the material will permanently change the properties of the material. Although not shown in Fig. 3–4, for low-carbon steels or those that are hot-rolled, the yield point is often distinguished by two values. The *upper yield point* occurs first, followed by a sudden decrease in load-carrying capacity to a *lower yield point*. Once the lower yield point is reached, however, then as shown in Fig. 3–4, the specimen will continue to elongate with *no increase in load*. Realize that Fig. 3–4 is not drawn to scale. If it was, the induced strains due to yielding would be about 10 to 40 times greater than those produced up to the elastic limit. When the material is in this state, it is often referred to as being *perfectly plastic*.

Strain Hardening. When yielding has ended, a further load can be applied to the specimen, resulting in a curve that rises continuously but becomes flatter until it reaches a maximum stress referred to as the *ultimate stress, σ_u*. The rise in the curve in this manner is called *strain hardening,* and it is identified in Fig. 3–4 as the region in light-shaded color. Throughout the test, while the specimen is elongating, its cross-sectional area will decrease. This decrease in area is fairly *uniform* over the specimen's entire gauge length, even up to the strain corresponding to the ultimate stress.

Necking. At the ultimate stress, the cross-sectional area begins to decrease in a *localized* region of the specimen, instead of over its entire length. As a result, a constriction or "neck" gradually tends to form in this region as the specimen elongates further, Fig. 3–5*a*. Since the cross-sectional area within the region is continually decreasing, the smaller area can only carry an ever-decreasing load. Hence the stress–strain diagram tends to curve downward until the specimen breaks at the *fracture stress σ_f*. This region of the curve due to necking is indicated in dark color in Fig. 3–4. The result due to failure of a specimen is shown in Fig. 3–5*b*.

True Stress–Strain Diagram.

Instead of always using the *original* cross-sectional area and specimen length to calculate the (engineering) stress and strain, we could have used the *actual* cross-sectional area and specimen length at the *instant* the load is measured. The values of stress and strain computed from these measurements are called *true stress* and *true strain,* and a plot of their values is called the *true stress–strain diagram*. When this diagram is plotted it has a form shown by the dashed curve in Fig. 3–4. Note that both the conventional and true σ–ϵ diagrams are practically coincident when the strain is small. The differences between the diagrams begin to appear in the strain-hardening range, where the magnitude of strain becomes more significant. In particular, notice the large divergence within the necking

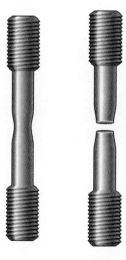

Failure of a
ductile material

(a)   (b)

Fig. 3–5

region. Here it can be seen from the conventional σ–ϵ diagram that the specimen *actually* supports a *decreasing load,* since A_0 is constant when calculating engineering stress, $\sigma = P/A_0$. However, from the true σ–ϵ diagram, the actual area A within the necking region is always decreasing until fracture, $\sigma_{f'}$, and so the material actually sustains *increasing stress,* since $\sigma = P/A$.

Although the true and conventional stress–strain diagrams are different, most engineering design is done within the elastic range. Provided the material is "stiff," like most metals, the strain up to the elastic limit will remain small and the error in using the engineering values of σ and ϵ is very small (about 0.1%) compared with their true values. This is one of the primary reasons for using conventional stress–strain diagrams.

The above concepts can be summarized with reference to Fig. 3–6, which shows an actual conventional stress–strain diagram for a mild steel specimen. In order to enhance the details, the elastic region of the curve has been shown in color to an exaggerated strain scale, also shown in color. Tracing the behavior, the proportional limit is reached at $\sigma_{pl} = 35$ ksi (241 MPa), where $\epsilon_{pl} = 0.0012$ in./in. This is followed by an upper yield point of $(\sigma_Y)_u = 38$ ksi (262 MPa), then suddenly a lower yield point of $(\sigma_Y)_l = 36$ ksi (248 MPa). The end of yielding occurs at a strain of $\epsilon_1 = 0.03$ in./in., which is 25 times greater than the strain at the proportional limit! Continuing, the specimen is strain-hardened until it reaches the ultimate stress of $\sigma_u = 63$ ksi (434 MPa), then it begins to neck down until failure occurs, $\sigma_f = 47$ ksi (324 MPa). By comparison, the strain at failure, $\epsilon_f = 0.380$ in./in., is 317 times greater than ϵ_{pl}!

Fig. 3–6

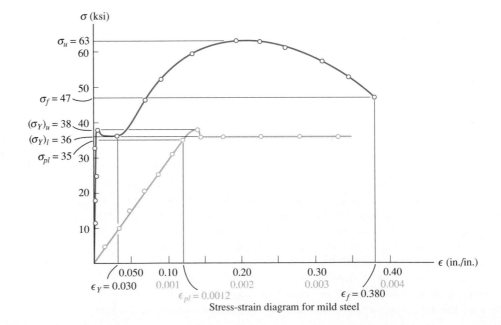

Stress-strain diagram for mild steel

3.3 Stress–Strain Behavior of Ductile and Brittle Materials

Materials can be classified as either being ductile or brittle, depending on their stress–strain characteristics. Each will now be given separate treatment.

Ductile Materials. Any material that can be subjected to large strains before it ruptures is called a *ductile material*. Mild steel, as discussed above, is a typical example. Engineers often choose ductile materials for design because these materials are capable of absorbing shock or energy, and if they become overloaded, they will usually exhibit large deformation before failing.

One way to specify the ductility of a material is to report its percent elongation or percent reduction in area at the time of fracture. The *percent elongation* is the specimen's fracture strain expressed as a percent. Thus, if the specimen's original gauge-mark length is L_0 and its length at fracture is L_f, then

$$\text{Percent elongation} = \frac{L_f - L_0}{L_0} (100\%) \qquad (3\text{–}3)$$

As seen in Fig. 3–6, since $\epsilon_f = 0.380$, this value would be 38% for a mild steel specimen.

The *percent reduction in area* is another way to specify ductility. It is defined within the region of necking as follows:

$$\text{Percent reduction of area} = \frac{A_0 - A_f}{A_0} (100\%) \qquad (3\text{–}4)$$

Here A_0 is the specimen's original cross-sectional area and A_f is the area at fracture. Mild steel has a typical value of 60%.

Besides steel, other metals such as brass, molybdenum, and zinc may also exhibit ductile stress–strain characteristics similar to steel, whereby they undergo elastic stress–strain behavior, yielding at constant stress, strain hardening, and finally necking until rupture. In most metals, however, constant yielding will *not occur* beyond the elastic range. One metal for which this is the case is aluminum. Actually, this metal often does not have a well-defined *yield point,* and consequently it is standard practice to define a *yield strength* for aluminum using a graphical procedure called the *offset method*. Normally a 0.2% strain (0.002 in./in.) is chosen, and from this point on the ϵ axis, a line parallel to the initial straight-line portion of the stress–strain diagram is drawn. The point where this line intersects the curve defines the yield strength. An example of the construction for determining the yield strength for an aluminum alloy is shown in Fig. 3–7. From the graph, the yield strength is $\sigma_{YS} = 51$ ksi (352 MPa).

Realize that the yield strength is not a physical property of the material, since it is a stress that caused a *specified* permanent strain in the material. In this text, however, we will assume that the yield strength, yield point, elastic

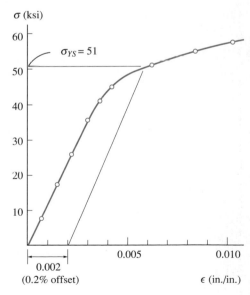

Yield strength for aluminum alloy

Fig. 3–7

limit, and proportional limit all *coincide* unless otherwise stated. For example, an exception would be natural rubber, which in fact does not even have a proportional limit, since stress and strain are *not* linearly related, Fig. 3–8. Instead, this material, which is known as a polymer, exhibits *nonlinear elastic behavior*.

Wood is a material that is often moderately ductile, and as a result it is usually designed to respond only to elastic loadings. The strength characteristics of wood vary greatly from one species to another, and for each species they depend on the moisture content, age, and the size and arrangement of knots in the wood. Since wood is a fibrous material, its tensile or compressive characteristics will differ greatly when it is loaded either parallel or perpendicular to its grain. Specifically, wood splits easily when it is loaded in tension perpendicular to its grain, and consequently tensile loads are almost always intended to be applied parallel to the grain of wood members.

Brittle Materials.   Materials that exhibit little or no yielding before failure are referred to as *brittle materials*. Gray cast iron is an example, having a stress–strain diagram in tension as shown by portion *AB* of the curve in Fig. 3–9. Here fracture at $\sigma_f = 22$ ksi (152 MPa) takes place initially at an imperfection or microscopic crack and then spreads rapidly across the specimen, causing complete fracture. As a result of this type of failure, brittle materials do not have a well-defined tensile fracture stress, since the appearance of cracks in a specimen is quite random. Instead the average fracture stress from a set of observed tests is generally reported. A typical failed specimen is shown in Fig. 3–10a.

Fig. 3–8

Fig. 3–9

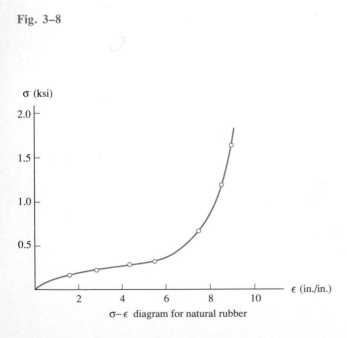

σ–ϵ diagram for natural rubber

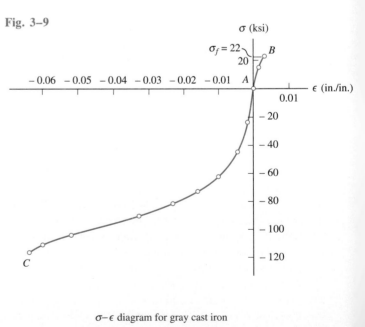

σ–ϵ diagram for gray cast iron

Compared with their behavior in tension, brittle materials, such as gray cast iron, exhibit a much higher resistance to axial compression, as evidenced by portion *AC* of the curve in Fig. 3–9. For this case any cracks or imperfections in the specimen tend to close-up, and as the load increases the material will generally bulge out or become barrel-shaped as the strains become larger, Fig. 3–10*b*.

Like gray cast iron, concrete is also classified as a brittle material, and it has a low strength capacity in tension. The characteristics of its stress–strain diagram depend primarily on the mix of concrete (water, sand, gravel, and cement) and the time and temperature of curing. A typical example of a ''complete'' stress–strain diagram for concrete is given in Fig. 3–11. By inspection, its maximum compressive strength is almost 12.5 times greater than its tensile strength, $(\sigma_c)_{max} = 5$ ksi (34.5 MPa) versus $(\sigma_t)_{max} = 0.40$ ksi (2.76 MPa). For this reason, concrete is almost always reinforced with steel bars or rods whenever it is designed to support tensile loads.

It can generally be stated that most materials exhibit both ductile and brittle behavior. For example, steel has brittle behavior when it contains a high carbon content, and it is ductile when the carbon content is reduced. Also, at low temperatures materials become harder and more brittle, whereas when the temperature rises they become softer and more ductile.

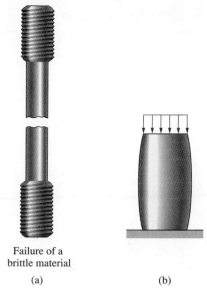

Failure of a brittle material

(a)   (b)

Fig. 3–10

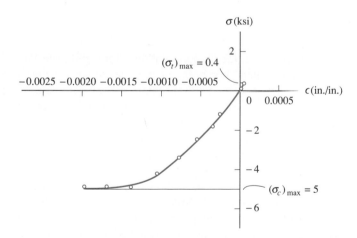

σ- ϵ diagram for typical concrete mix

Fig. 3–11

3.4 Hooke's Law

As noted in the previous section, the stress–strain diagrams for most engineering materials exhibit a *linear relationship* between stress and strain within the elastic region. Consequently, an increase in stress causes a proportionate increase in strain. This fact was discovered by Robert Hooke in 1676 using springs and is known as *Hooke's law*. It may be expressed mathematically as

$$\sigma = E\epsilon \tag{3–5}$$

Here E represents the constant of proportionality, which is called the *modulus of elasticity* or *Young's modulus,* after Thomas Young, who published an account of it in 1807.

Equation 3–5 actually represents the equation of the *initial straight-lined portion* of the stress–strain diagram up to the proportional limit. Furthermore, the modulus of elasticity represents the *slope* of this line. Since strain is dimensionless, from Eq. 3–5, E will have units of stress, such as psi, ksi, or pascals. As an example of its calculation, consider the stress–strain diagram for steel shown in Fig. 3–6. Here $\sigma_{pl} = 35$ ksi and $\epsilon_{pl} = 0.0012$ in./in., so that

$$E = \frac{\sigma_{pl}}{\epsilon_{pl}} = \frac{35 \text{ ksi}}{0.0012 \text{ in./in.}} = 29(10^3) \text{ ksi}$$

As shown in Fig. 3–12, the proportional limit for a particular type of steel depends on its alloy content; however, most grades of steel, from the softest rolled steel to the hardest tool steel, have about the same modulus of elasticity, generally accepted to be $E_{st} = 29(10^3)$ ksi or 200 GPa. Common values of E for other engineering materials are often tabulated in engineering code and reference books. It should be noted that the modulus of elasticity is a mechanical property that indicates the *stiffness* of a material. Materials that are very stiff, such as steel, have large values of E [$E_{st} = 29(10^3)$ ksi or 200 GPa], whereas spongy materials such as vulcanized rubber may have low values [$E_r = 0.10(10^3)$ ksi or 0.70 MPa].

The modulus of elasticity is one of the most important mechanical properties used in the development of equations presented in this text. It must always be remembered, though, that E can be used only if a material has *linear-elastic* behavior. Also, if the stress in the material is *greater* than the proportional limit, the stress–strain diagram ceases to be a straight line and Eq. 3–5 is no longer valid.

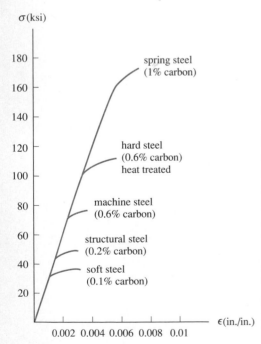

Fig. 3–12

Strain Hardening. If a specimen of ductile material, such as steel, is loaded into the *plastic range* and then unloaded, *elastic strain is recovered* as the material returns to its equilibrium state. The *plastic strain remains,* however, and as a result the material is subjected to a *permanent set*. For example, a wire when bent (plastically) will spring back a bit (elastically) when the load is removed; however, it will not fully return to its original position. This behavior can be illustrated on the stress–strain diagram shown in Fig. 3–13*a*. Here the specimen is first loaded beyond its yield point *A* to the point *A′*. Since interatomic forces have to be overcome to elongate the specimen *elastically,* then these same forces pull the atoms back together when the load is removed, Fig. 3–13*a*. Consequently, the modulus of elasticity, *E*, is the same, and therefore the slope of line *O′A′* has the same slope as line *OA*.

If the load is reapplied, the atoms in the material will again be displaced until yielding occurs at or near the stress *A′*, and the stress–strain diagram continues along the same path as before, Fig. 3–13*b*. It should be noted, however, that this new stress–strain diagram, defined by *O′A′B*, now has a *higher* yield point (*A′*), a consequence of strain-hardening. In other words, the material now has a *greater elastic region;* however, it has *less ductility* than when it was in its original state.

It should be mentioned, however, that in the true sense some heat or *energy* may be *lost* as the specimen is unloaded from *A′* and then again loaded to this same stress. As a result, slight curves in the paths *A′* to *O′* and *O′* to *A′* will occur during a carefully measured cycle of loading. This is shown by the dashed curves in Fig. 3–13*b*. The area between these curves represents lost energy and is called *mechanical hysteresis*. It becomes an important consideration when selecting materials to serve as dampers for vibrating structural or mechanical equipment, although its effects will not be considered in this text.

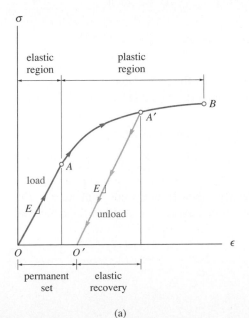

(a)

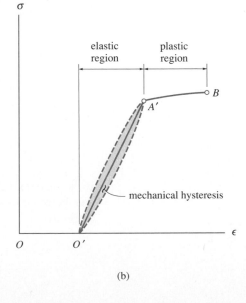

(b)

Fig. 3–13

3.5 Strain Energy

As a material is deformed by an external loading, the material tends to store energy *internally* throughout its volume. Since this energy is related to the strains in the material, it is referred to as *strain energy*. In this section we will formulate the strain energy stored in a volume element of material when the energy is caused by a uniaxial stress. The results will then be used to identify two important material properties that are often used to specify a material's ability to *absorb* energy. A further discussion of strain energy concepts, as they relate to more complicated states of stress, will be given in Secs. 10.7 and 14.1.

When a tension-test specimen is subjected to an axial load, a volume element of the material is subjected to uniaxial stress as shown in Fig. 3–14. This stress develops a force $\Delta F = \sigma \, \Delta A = \sigma(\Delta x \, \Delta y)$ on the top and bottom faces of the element *after* the element undergoes a vertical displacement $\epsilon \, \Delta z$. The work of this force will now be formulated.

By definition, *work* is determined by the product of the force and displacement in the direction of the force. Since the force ΔF is increased uniformly from zero to its final magnitude ΔF when the displacement $\epsilon \, \Delta z$ is attained, the work done on the element by the force is equal to the *average* force magnitude $(\Delta F/2)$ times the displacement $\epsilon \, \Delta z$. This "external work" is equivalent to the internal work or strain energy stored in the element— assuming that no energy is lost in the form of heat. Consequently, the strain energy is

$$\Delta U = \left(\frac{1}{2} \, \Delta F\right) \epsilon \, \Delta z = \left(\frac{1}{2}\sigma \, \Delta x \, \Delta y\right) \epsilon \, \Delta z$$

Since the volume of the element is $\Delta V = \Delta x \, \Delta y \, \Delta z$, then

$$\Delta U = \frac{1}{2}\sigma \epsilon \, \Delta V$$

It is sometimes convenient to formulate the strain energy per unit volume of material. This is called the *strain-energy density,* and from the above equation it can be expressed as

$$u = \frac{\Delta U}{\Delta V} = \frac{1}{2}\sigma \epsilon \qquad (3\text{–}6)$$

If the material behavior is linear-elastic, then Hooke's law applies, $\sigma = E\epsilon$, and therefore we can express the strain-energy density in terms of the uniaxial stress as

$$u = \frac{1}{2}\frac{\sigma^2}{E} \qquad (3\text{–}7)$$

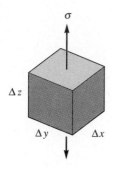

Fig. 3–14

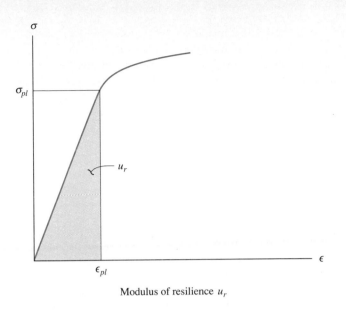

Modulus of resilience u_r

(a)

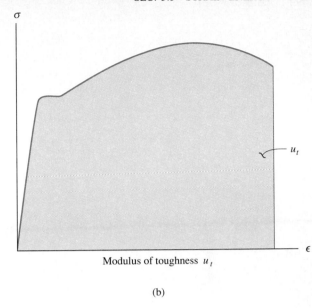

Modulus of toughness u_t

(b)

Fig. 3–15

Modulus of Resilience.

In particular, when the stress σ reaches the proportional limit, the strain-energy density, as calculated by Eq. 3–6 or 3–7, is referred to as the *modulus of resilience,* i.e.,

$$u_r = \frac{1}{2}\sigma_{pl}\epsilon_{pl} = \frac{1}{2}\frac{\sigma_{pl}^2}{E} \qquad (3\text{--}8)$$

From the elastic region of the stress–strain diagram, Fig. 3–15a, we can see that u_r is equivalent to the shaded *triangular area* under the diagram. Physically a material's resilience represents the ability of the material to absorb energy without any permanent damage to the material.

Modulus of Toughness.

Another important property of a material is the *modulus of toughness, u_t.* This quantity represents the *entire area* under the stress–strain diagram, Fig. 3–15b, and therefore it indicates the strain-energy density of the material just before it fractures. By comparing various values of u_t, one can determine the toughness of a particular material. This consideration becomes important when designing members that may be accidentally overloaded. Materials with a high modulus of toughness will distort greatly due to an overloading; however, they may be preferable to those with a low value, since materials having a low u_t may suddenly fracture without warning of an approaching failure. Alloying metals can also change their resilience and toughness. For example, by changing the percentage of carbon in steel, the resulting stress–strain diagrams in Fig. 3–16 indicate how the degrees of resilience and toughness can be changed among the three alloys.

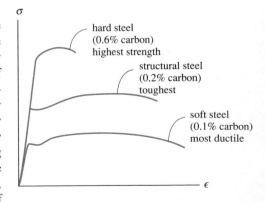

Fig. 3–16

Example 3–1

A tension test for a steel alloy results in the stress–strain diagram shown in Fig. 3–17. Calculate the modulus of elasticity and the yield strength based on a 0.2% offset. Identify on the graph the ultimate stress, and the fracture stress.

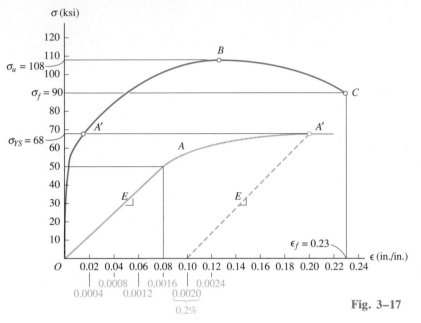

Fig. 3–17

SOLUTION

Modulus of Elasticity. We must calculate the *slope* of the initial straight-line portion of the graph. Using the magnified scale shown in color in Fig. 3–17, this line extends from point O to point A. A point on this line has coordinates of approximately (50 ksi, 0.0016 in./in.). Therefore,

$$E = \frac{50 \text{ ksi}}{0.0016 \text{ in./in.}} = 31.25(10^3) \text{ ksi} \qquad \textit{Ans.}$$

Note that the equation of the line OA is thus $\sigma = 31.25(10^3)\epsilon$.

Yield Strength. For a 0.2% offset, we begin at a strain of 0.2% or 0.002 in./in. and graphically extend a (dashed) line parallel to OA until it intersects the σ–ϵ curve at A', Fig. 3–17. The yield strength is approximately

$$\sigma_{YS} = 68 \text{ ksi} \qquad \textit{Ans.}$$

Ultimate Stress. This is defined by the peak of the σ–ϵ graph, point B in Fig. 3–17.

$$\sigma_u = 108 \text{ ksi} \qquad \textit{Ans.}$$

Fracture Stress. When the specimen is strained to its maximum of $\epsilon_f = 0.23$ in./in., it fractures at point C, Fig. 3–17. Thus,

$$\sigma_f = 90 \text{ ksi} \qquad \textit{Ans.}$$

Example 3–2

The stress–strain diagram for an aluminum alloy that is used to make parts for an aircraft is shown in Fig. 3–18. If a specimen of this material is stressed to 600 MPa, determine the permanent strain that remains in the specimen when the load is released. Also, compute the modulus of resilience both before and after the load application.

SOLUTION

Permanent Strain. When the specimen is subjected to a stress of 600 MPa, it strain-hardens until point B is reached on the σ–ϵ diagram, Fig. 3–18. The strain at this point is approximately 0.023 mm/mm. When the load is released, the material behaves by following the straight line BC, which is parallel to line OA. Since both lines have the same slope, the strain at point C can be determined analytically. The slope of line OA is the modulus of elasticity, i.e.,

$$E = \frac{\sigma_{pl}}{\epsilon_{pl}} = \frac{450 \text{ MPa}}{0.006 \text{ mm/mm}} = 75.0 \text{ GPa}$$

From triangle CBD in Fig. 3–18, we require

$$E = \frac{BD}{CD} = \frac{600(10^6) \text{ Pa}}{CD} = 75.0(10^9) \text{ Pa}$$

$$CD = 0.008 \text{ mm/mm}$$

This strain represents the amount of *recovered elastic strain*. The permanent strain, ϵ_{OC}, Fig. 3–18, is thus

$$\epsilon_{OC} = 0.023 \text{ mm/mm} - 0.00800 \text{ mm/mm}$$

$$= 0.0150 \text{ mm/mm} \qquad \textit{Ans.}$$

Note: If gauge marks on the specimen were originally 50 mm apart, then after the load is *released* these marks will be 50 mm + (0.0150) (50 mm) = 50.75 mm apart.

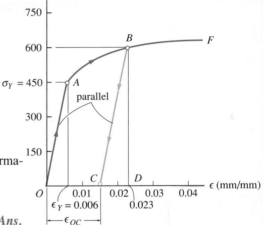

Fig. 3–18

Modulus of Resilience. Applying Eq. 3–8, we have[*]

$$(u_r)_{\text{initial}} = \frac{1}{2}\sigma_{pl}\epsilon_{pl} = \frac{1}{2}(450 \text{ MPa}) (0.006 \text{ mm/mm})$$

$$= 1.35 \text{ MJ/m}^3 \qquad \textit{Ans.}$$

$$(u_r)_{\text{final}} = \frac{1}{2}\sigma_{pl}\epsilon_{pl} = \frac{1}{2}(600 \text{ MPa}) (0.008 \text{ mm/mm})$$

$$= 2.40 \text{ MJ/m}^3 \qquad \textit{Ans.}$$

The effect of strain-hardening the material has caused an increase in the modulus of resilience as noted by comparing the answers; however, note that the modulus of toughness for the material has decreased since the area under the original curve, $OABF$, is larger than the area under curve CBF.

[*]Work in the SI system of units is measured in joules, where $1 \text{ J} = 1 \text{ N} \cdot \text{m}$.

Example 3–3

An aluminum rod shown in Fig. 3–19a has a circular cross section and is subjected to an axial load of 10 kN. If a portion of the stress–strain diagram for the material is shown in Fig. 3–19b, determine the approximate elongation of the rod when the load is applied. If the load is removed, does the rod return to its original length? Take $E_{al} = 70$ GPa.

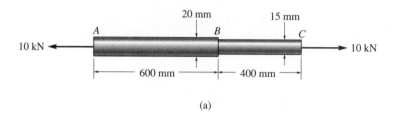

(a)

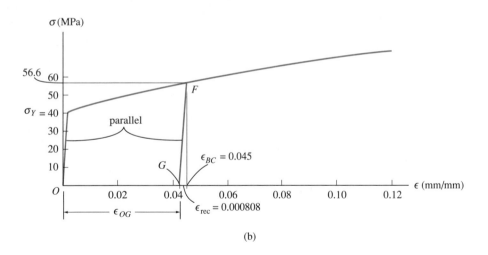

(b)

Fig. 3–19

SOLUTION

For the analysis we will neglect the *localized deformations* at the point of load application and where the rod's cross-sectional area suddenly changes. (These effects will be discussed in Secs. 4.1 and 4.6.) Throughout the midsection of each segment the normal stress and deformation are uniform.

In order to study the deformation of the rod, we must obtain the strain. This is done by first calculating the stress, then using the stress–strain diagram to obtain the strain. The normal stress within each segment is

$$\sigma_{AB} = \frac{P}{A} = \frac{10(10^3) \text{ N}}{\pi (0.01 \text{ m})^2} = 31.8 \text{ MPa}$$

$$\sigma_{BC} = \frac{P}{A} = \frac{10(10^3) \text{ N}}{\pi (0.0075 \text{ m})^2} = 56.6 \text{ MPa}$$

From the stress–strain diagram, the material in region AB is strained *elastically* since $\sigma_Y = 40$ MPa > 31.8 MPa. Using Hooke's law,

$$\epsilon_{AB} = \frac{\sigma_{AB}}{E_{al}} = \frac{31.8(10^6) \text{ Pa}}{70(10^9) \text{ Pa}} = 0.0004543 \text{ mm/mm}$$

The material within region BC is strained plastically, since $\sigma_Y = 40$ MPa < 56.6 MPa. From the graph, for $\sigma_{BC} = 56.6$ MPa,

$$\epsilon_{BC} \approx 0.0450 \text{ mm/mm}$$

The approximate elongation of the rod is therefore

$$\Delta = \Sigma\epsilon L = 0.0004543(600 \text{ mm}) + 0.045(400 \text{ mm})$$
$$= 18.3 \text{ mm} \qquad\qquad Ans.$$

When the 10-kN load is removed, segment AB of the rod will be restored to its original length. Why? On the other hand, the material in segment BC will recover elastically along line FG, Fig. 3–19b. Since the slope of FG is E_{al}, the elastic strain recovery is

$$\epsilon_{\text{rec}} = \frac{\sigma_{BC}}{E_{al}} = \frac{56.6(10^6) \text{ Pa}}{70(10^9) \text{ Pa}} = 0.000808 \text{ mm/mm}$$

The remaining plastic strain in segment BC is then

$$\epsilon_{OG} = 0.0450 - 0.000808 = 0.0442 \text{ mm/mm}$$

Therefore, when the load is removed the rod remains elongated by an amount

$$\Delta' = \epsilon_{OG} L_{BC} = 0.0442(400 \text{ mm}) = 17.7 \text{ mm} \qquad Ans.$$

PROBLEMS

3–1. Data taken from a stress–strain test for a ceramic is given in the table. The curve is linear between the origin and the first point. Plot the curve, and determine the modulus of elasticity and the modulus of resilience.

σ (MPa)	ϵ (mm/mm)
0	0
229	0.0008
314	0.0012
341	0.0016
355	0.0020
368	0.0024

Prob. 3–1

3–2. Data taken from a stress–strain test is given in the table. The curve is linear between the origin and the first point. Plot the curve, and determine the modulus of elasticity and the modulus of resilience.

σ (ksi)	ϵ (in./in.)
0	0
32.0	0.0016
33.5	0.0018
40.0	0.0030
41.2	0.0050

Prob. 3–2

3–3. A concrete cylinder having a diameter of 6.00 in. and gauge length of 12 in. is tested in compression. The results of the test are reported in the table as load versus contraction. Draw the stress–strain diagram using scales of 1 in. = 0.5 ksi and 1 in. = $0.2(10^{-3})$ in./in. From the diagram, determine approximately the modulus of elasticity.

Load (kip)	Contraction (in.)
0	0
5.0	0.0006
9.5	0.0012
16.5	0.0020
20.5	0.0026
25.5	0.0034
30.0	0.0040
34.5	0.0045
38.5	0.0050
46.5	0.0062
50.0	0.0070
53.0	0.0075

Prob. 3–3

*3–4. A tension test was performed on a steel specimen having an original diameter of 0.503 in. and a gauge length of 2.00 in. The data is listed in the table. Plot the stress–strain diagram and determine approximately the modulus of elasticity, the ultimate stress, and the rupture stress. Use a scale of 1 in. = 15 ksi and 1 in. = 0.05 in./in. Redraw the linear-elastic region, using the same stress scale but a strain scale of 1 in. = 0.001 in.

3–5. A tension test was performed on a steel specimen having an original diameter of 0.503 in. and gauge length of 2.00 in. Using the data listed in the table, plot the stress–strain diagram and determine approximately the modulus of toughness.

Load (kip)	Elongation (in.)
0	0
2.50	0.0009
6.50	0.0025
8.50	0.0040
9.20	0.0065
9.80	0.0098
12.0	0.0400
14.0	0.1200
14.5	0.2500
14.0	0.3500
13.2	0.4700

Prob. 3–4/3–5

3–6. A tension test was performed on a steel specimen having an original diameter of 0.503 in. and gauge length of 2.00 in. The data is listed in the table below. Plot the stress–strain diagram and determine approximately the modulus of elasticity, the yield stress, the ultimate stress, and the rupture stress. Use a scale of 1 in. = 20 ksi and 1 in. = 0.05 in./in. Redraw the elastic region, using the same stress scale but a strain scale of 1 in. = 0.001 in./in.

Load (kip)	Elongation (in.)
0	0
1.50	0.0005
4.60	0.0015
8.00	0.0025
11.00	0.0035
11.80	0.0050
11.80	0.0080
12.00	0.0200
16.60	0.0400
20.00	0.1000
21.50	0.2800
19.50	0.4000
18.50	0.4600

Prob. 3–6

3–7. The stress–strain diagram for a steel bar is shown in the figure. Determine approximately the modulus of elasticity, the proportional limit, the ultimate stress, and the modulus of resilience. If the bar is loaded until it is stressed to 72 ksi, determine the amount of elastic recovery and the permanent set or strain in the bar when it is unloaded.

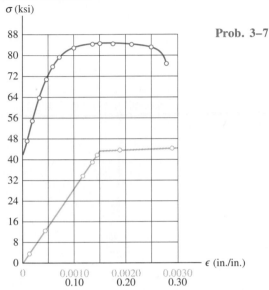

Prob. 3–7

3–9. The change in weight of an airplane is determined from reading the strain gauge A mounted in the plane's aluminum wheel strut. *Before* the plane is loaded, the strain-gauge reading in a strut is $\epsilon_1 = 0.00100$ in./in., whereas after loading $\epsilon_2 = 0.00243$ in./in. Determine the change in the force on the strut if the cross-sectional area of the strut is 3.5 in². $E_{al} = 10(10^3)$ ksi.

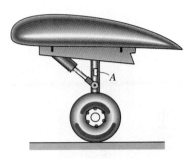

Prob. 3–9

*3–8. The stress–strain diagram for a steel specimen having an original diameter of 0.5 in. and a gauge length of 2 in. is given in the figure. Determine approximately the modulus of elasticity for the material, the load on the specimen that causes yielding, and the ultimate load the specimen will support.

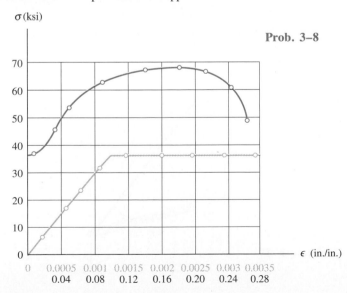

Prob. 3–8

3–10. The pole is supported by a pin at C and a steel guy wire AB. If the wire has a diameter of 0.2 in., determine how much it stretches when a horizontal force of 2.5 kip acts on the pole. $E_{st} = 29(10^3)$ ksi.

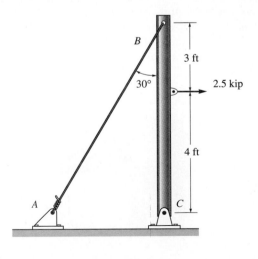

Prob. 3–10

3–11. The 8-mm-diameter bolt is made from an aluminum alloy. It fits through a magnesium sleeve that has an inner diameter of 12 mm and an outer diameter of 20 mm. If the original lengths of the bolt and sleeve are 80 mm and 50 mm, respectively, determine the strains in the sleeve and the bolt if the nut on the bolt is tightened so that the tension in the bolt is 8 kN. Assume the material at A is rigid. $E_{al} = 70$ GPa, $E_{mg} = 45$ GPa.

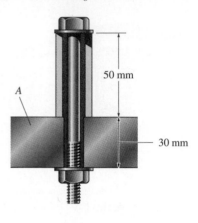

50 mm

A

30 mm

Prob. 3–11

*3–12. The head H is connected to the cylinder of a compressor using six steel bolts. If the clamping force in each bolt is 800 lb, determine the normal strain in the bolts. Each bolt has a diameter of $\frac{3}{16}$ in. If $\sigma_y = 40$ ksi and $E_{st} = 29(10^3)$ ksi, what is the strain in each bolt when the nut is unscrewed so that the clamping force is released?

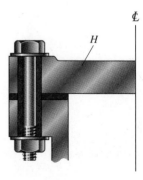

¢

H

Prob. 3–12

3–13. The bar DB is rigid and is originally held in the horizontal position when the weight W is supported from C. If the weight causes the end B to be displaced downward 0.025 in., determine the strain in wires DE and BC. Also, if the wires are made of steel and have a cross-sectional area of 0.002 in^2, determine the weight W. $E_{st} = 29(10^3)$ ksi.

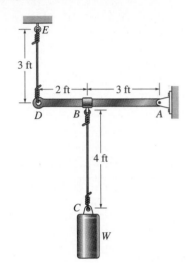

E

3 ft

2 ft   3 ft

D B A

4 ft

C

W

Prob. 3–13

3–14. A structural member in a nuclear reactor is made from a zirconium alloy. If an axial load of 4 kip is to be supported by the member, determine its required cross-sectional area. Use a factor of safety of 3 with respect to yielding. What is the load on the member if it is 3 ft long and its elongation is 0.02 in.? $E_{zr} = 14(10^3)$ ksi, $\sigma_Y = 57.5$ ksi. The material has elastic behavior.

3–15. The steel wire AB has a cross-sectional area of 10 mm^2 and is unstretched when $\theta = 45.0°$. Determine the applied load P needed to cause $\theta = 44.9°$. $E_{st} = 200$ GPa.

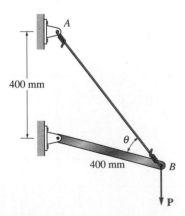

A

400 mm

400 mm

θ

B

P

Prob. 3–15

3.6 Poisson's Ratio

When a deformable body is subjected to an axial tensile force, not only does it elongate but it also contracts laterally. For example, if a rubber band is stretched, it can be noted that both the thickness and width of the band are decreased. Likewise, a compressive force acting on a body causes it to contract in the direction of the force and yet its sides expand laterally. These two cases are illustrated in Fig. 3–20 for a bar having an original radius r and length L.

When a load **P** is applied to the bar, it changes the bar's length by an amount Δ and its radius by an amount δ (delta). Strains in the longitudinal or axial direction and in the lateral or radial direction are, respectively,

$$\epsilon_{long} = \frac{\Delta}{L} \quad \text{and} \quad \epsilon_{lat} = \frac{\delta}{r}$$

In the early 1800s, the French scientist S. D. Poisson realized that within the *elastic range* the *ratio* of these strains is a *constant,* since the deformations Δ and δ are proportional. This constant is referred to as *Poisson's ratio,* ν (nu), and it has a numerical value that is unique for a particular material that is both *homogeneous and isotropic.* Stated mathematically it is

$$\nu = -\frac{\epsilon_{lat}}{\epsilon_{long}} \tag{3–9}$$

The negative sign is used here since *longitudinal elongation* (positive strain) causes *lateral contraction* (negative strain), and vice versa. Notice that this lateral strain is the *same* in all lateral (or radial) directions. Furthermore, this strain is caused only by the axial or longitudinal force; i.e., no force or stress acts in a lateral direction in order to strain the material in this direction.

Poisson's ratio is seen to be *dimensionless,* and for most nonporous solids it has a value that is generally between $\frac{1}{4}$ and $\frac{1}{3}$. In particular, an ideal material having no lateral movement when it is stretched or compressed will have $\nu = 0$. Furthermore, it will be shown in Sec. 10.6 that the *maximum* possible value for Poisson's ratio is $\frac{1}{2}$. Therefore $0 \leq \nu \leq \frac{1}{2}$.

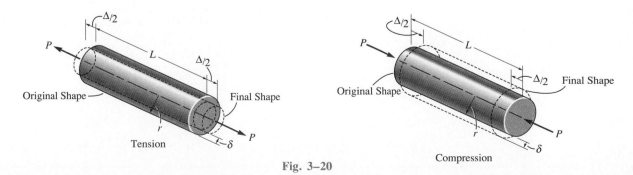

Tension

Compression

Fig. 3–20

Example 3–4

A steel bar has the dimensions shown in Fig. 3–21. If an axial force of $P = 80$ kN is applied to the bar, determine the change in its length and the change in the dimensions of its cross section after applying the load. Take $E_{st} = 200$ GPa and $\nu_{st} = 0.3$. The material behaves elastically.

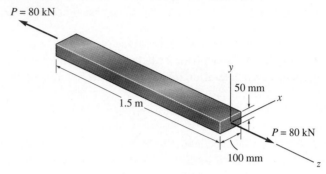

Fig. 3–21

SOLUTION

The normal stress in the bar is

$$\sigma_z = \frac{P}{A} = \frac{80(10^3) \text{ N}}{(0.1 \text{ m}) (0.05 \text{ m})} = 16.0(10^6) \text{ Pa}$$

Thus the strain in the z direction is

$$\epsilon_z = \frac{\sigma_z}{E_{st}} = \frac{16.0(10^6) \text{ Pa}}{200(10^9) \text{ Pa}} = 80(10^{-6}) \text{ mm/mm}$$

The axial elongation of the bar is therefore

$$\Delta L_z = \epsilon_z L_z = [80(10^{-6})] \ (1.5 \text{ m}) = 120 \ \mu\text{m} \qquad \textit{Ans.}$$

Using Eq. 3–9, the contraction strains in *both* the x and y directions are

$$\epsilon_x = \epsilon_y = -\nu_{st}\epsilon_z = -0.30[80(10^{-6})] = -24 \ \mu\text{m/m}$$

Thus the changes in the dimensions of the cross section are

$$\Delta L_x = -\epsilon_x L_x = -[24(10^{-6})] \ (0.1 \text{ m}) = -2.40 \ \mu\text{m} \qquad \textit{Ans.}$$

$$\Delta L_y = -\epsilon_y L_y = -[24(10^{-6})] \ (0.05 \text{ m}) = -1.20 \ \mu\text{m} \qquad \textit{Ans.}$$

3.7 The Shear Stress–Strain Diagram

In Sec. 1.3 it was shown that when an element of material is subjected to *pure shear,* equilibrium requires that equal shear stresses must be developed on four faces of the element. These stresses must be directed toward or away from diagonally opposite corners of the element as shown on the side view in Fig. 3–22a. Furthermore, if the material is *homogeneous* and *isotropic,* then the shear stress will distort the element uniformly as shown in Fig. 3–22b. As mentioned in Sec. 2.2, the shear strain γ_{xy} measures the angular distortion of the element relative to the sides originally along the x and y axes.

The behavior of a material subjected to pure shear can be studied in a laboratory by using specimens in the shape of thin tubes and subjecting them to a torsional loading. If measurements are made of the applied torque and the resulting angle of twist, then by the methods to be explained in Chapter 5, the data can be converted into shear stress and shear strain, and a shear stress–strain diagram plotted. An example of such a diagram for a ductile material is shown in Fig. 3–23. Like the tension test, this material when subjected to shear will exhibit linear-elastic behavior and it will have a defined *proportional limit* τ_{pl}. Also, strain hardening will occur until an *ultimate shear stress* τ_u is reached. And finally, the material will begin to lose its shear strength until it reaches a point where it fractures, τ_f.

For most engineering materials, like the one just described, the elastic behavior is *linear,* and so Hooke's law for shear can be written as

$$\boxed{\tau = G\gamma} \tag{3-10}$$

Here G is called the *shear modulus of elasticity* or the *modulus of rigidity.* Its value can be measured as the slope of the line on the τ–γ diagram, that is, $G = \tau_{pl}/\gamma_{pl}$, Fig. 3–23. Notice that the units of measurement for G will be the *same* as those for E (Pa or psi), since γ is measured in radians, a dimensionless quantity.

It will be shown in Sec. 10.6 that the three material constants, E, ν, and G are actually *related* by the equation

$$\boxed{G = \frac{E}{2(1 + \nu)}} \tag{3-11}$$

Provided E and G are known, the value of ν can be determined from this equation rather than through experimental measurement. For example, for mild steel, $E_{st} = 29(10^3)$ ksi and $G_{st} = 11.2(10^3)$ ksi, so that, from Eq. 3–11, $\nu_{st} = 0.29$.

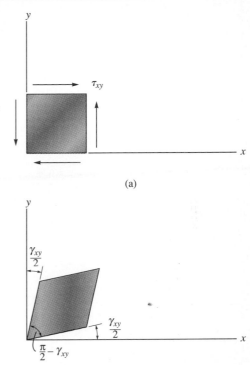

(a)

(b)

Fig. 3–22

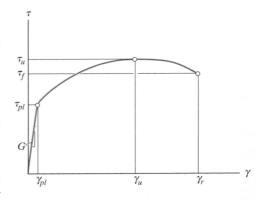

Fig. 3–23

Example 3–5

A specimen of titanium alloy is tested in torsion and the shear stress–strain diagram is shown in Fig. 3–24a. Determine the shear modulus G, the proportional limit, and the ultimate shear stress. Also, determine the maximum distance d that the top of a block of this material, shown in Fig. 3–24b, would be displaced horizontally if the material behaves elastically when acted upon by a shear force **V**. What is the magnitude of **V** necessary to cause this displacement?

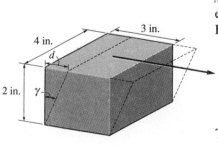

(a)

SOLUTION

Shear Modulus. This value represents the slope of the straight-line portion OA of the τ–γ diagram. The coordinates of point A are (52 ksi, 0.008 rad). Thus,

$$G = \frac{52 \text{ ksi}}{0.008 \text{ rad}} = 6500 \text{ ksi} \qquad \textit{Ans.}$$

The equation of line OA is therefore $\tau = 6500\gamma$, which is Hooke's law for shear.

Proportional Limit. By inspection, the graph ceases to be linear at point A. Thus,

$$\tau_{pl} = 52 \text{ ksi} \qquad \textit{Ans.}$$

Ultimate Stress. This value represents the maximum shear stress, point B. From the graph,

$$\tau_u = 73 \text{ ksi} \qquad \textit{Ans.}$$

Maximum Elastic Displacement and Shear Force. Since the maximum elastic shear strain is 0.008 rad, a very small angle, the top of the block in Fig. 3–24b will be displaced horizontally:

$$\tan(0.008 \text{ rad}) \approx 0.008 \text{ rad} = \frac{d}{2 \text{ in.}}$$

$$d = 0.016 \text{ in.} \qquad \textit{Ans.}$$

The corresponding *average* shear stress in the block is $\tau_{pl} = 52$ ksi. Thus, the shear force V needed to cause displacement is

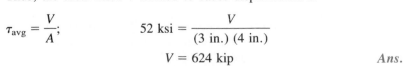

$$\tau_{\text{avg}} = \frac{V}{A}; \qquad 52 \text{ ksi} = \frac{V}{(3 \text{ in.}) (4 \text{ in.})}$$

$$V = 624 \text{ kip} \qquad \textit{Ans.}$$

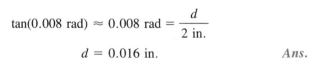

(b)

Fig. 3–24

Example 3-6

An aluminum specimen shown in Fig. 3–25 has a diameter of $d_0 = 25$ mm and a gauge length of $L_0 = 250$ mm. If a force of 165 kN elongates the gauge length 1.20 mm, determine the modulus of elasticity. Also, determine by how much the force causes the diameter of the specimen to contract. Take $G_{al} = 26$ GPa and $\sigma_Y = 440$ MPa.

SOLUTION

Modulus of Elasticity.   The normal stress in the specimen is

$$\sigma = \frac{P}{A} = \frac{165(10^3) \text{ N}}{(\pi/4)(0.025 \text{ m})^2} = 336.1 \text{ MPa}$$

and the normal strain is

$$\epsilon = \frac{\Delta L}{L} = \frac{1.20 \text{ mm}}{250 \text{ mm}} = 0.00480 \text{ mm/mm}$$

Since $\sigma < \sigma_y = 440$ MPa, the material behaves elastically. The modulus of elasticity is

$$E_{al} = \frac{\sigma}{\epsilon} = \frac{336.1(10^6) \text{ Pa}}{0.00480} = 70.0 \text{ GPa} \qquad \textit{Ans.}$$

Contraction of Diameter.   First we will compute Poisson's ratio for the material using Eq. 3–11.

$$G = \frac{E}{2(1 + \nu)}$$

$$26 \text{ GPa} = \frac{70 \text{ GPa}}{2(1 + \nu)}$$

$$\nu = 0.346$$

Since $\epsilon_{long} = 0.00480$, then by Eq. 3–9,

$$\nu = -\frac{\epsilon_{lat}}{\epsilon_{long}}$$

$$0.346 = -\frac{\epsilon_{lat}}{0.00480}$$

$$\epsilon_{lat} = -0.00166 \text{ mm/mm}$$

The contraction of the diameter is therefore

$$\Delta d = (0.00166)(25 \text{ mm})$$

$$= 0.0415 \text{ mm} \qquad \textit{Ans.}$$

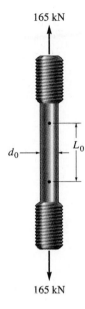

165 kN

165 kN

d_0

L_0

Fig. 3–25

3.8 Failure of Materials Due to Creep and Fatigue

The mechanical properties of a material have up to this point been discussed only for a static or slowly applied load at constant temperature. In some cases, however, a member may have to be used in an environment for which loadings must be sustained over long periods of time at elevated temperatures, or in other cases, the loading may be repeated or cycled. We will not consider these effects in this book, although we will briefly mention how one determines a material's strength for these conditions since they are given special treatment in design.

Creep. When a material has to support a load for a very long period of time, it may continue to deform until a sudden fracture occurs or its usefulness is impaired. This time-dependent permanent deformation is known as *creep*. Normally creep is considered when metals and ceramics are used for structural members or mechanical parts that are subjected to high temperature. For some materials, however, such as polymers and composite materials—including wood or concrete—temperature is *not* an important factor, and yet creep can occur. As a typical example, consider the fact that a rubber band will not return to its original shape after being released from a stretched position in which it was held for a very long period of time. In the general sense, therefore, both *stress and temperature* play a significant role in the *rate* of creep.

For practical purposes, when creep becomes important, a material is usually designed to resist a specified creep strain for a given period of time. In this regard, an important mechanical property that is used for the design of members subjected to creep is *creep strength*. This value represents the highest initial stress the material can withstand during a specified time without causing a given amount of creep strain. The creep strength will vary with temperature, and for design, a given temperature, duration of loading, and allowable creep strain must all be specified. For example, a creep strain of 0.1% per year has been suggested for steel in bolts and piping, and 0.25% per year for lead sheathing on cables.

Several methods exist for determining an allowable creep strength for a particular material. One of the simplest involves testing several specimens simultaneously at a constant temperature, but with each subjected to a *different* axial stress. By measuring the length of time needed to produce either an allowable strain or the rupture strain for each specimen, a curve of stress versus time can be established. Normally these tests are run for 1000 hr. An example of the results for stainless steel at a temperature of 1200°F is shown in Fig. 3–26. This material has a yield strength of 40 ksi (276 MPa) at room temperature (0.2% offset), and the prescribed creep strain is 1%. From the graph, the creep strength at 1000 hr is determined to be approximately $\sigma_c = 20$ ksi (138 MPa).

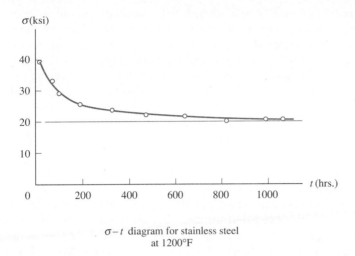

$\sigma - t$ diagram for stainless steel
at 1200°F

Fig. 3–26

In general, the creep strength will *decrease* for *higher temperatures* or for *higher applied stresses*. For longer periods of time, extrapolations from the curves must be made. To do this usually requires a certain amount of experience with creep behavior, and some supplementary knowledge as to the creep properties of the material to be used. Once the material's creep strength has been determined, however, a factor of safety is applied to obtain an appropriate allowable stress for design.

Fatigue. When a metal is subjected to repeated *cycles* of stress or strain, it causes its structure to break down, ultimately leading to fracture. This behavior is called *fatigue,* and it is usually responsible for a large percentage of failures in connecting rods and crankshafts of engines; steam or gas turbine blades; connections or supports for bridges, railroad wheels, and axles; and other parts subjected to cyclic loading. In all these cases, fracture will occur at a stress that is *less* than the material's yield stress.

The nature of this failure apparently results from the fact that there are microscopic regions, usually on the surface of the member, where the localized stress becomes *much greater* than the average stress acting over the cross section. As this higher stress is cycled, it leads to the formation of minute cracks. Occurrence of these cracks causes a further increase of stress at their tips or boundaries, which in turn causes a further extension of the cracks into the material as the stress continues to be cycled. Eventually the cross-sectional area of the member is reduced to the point where the load can no longer be sustained, and as a result sudden fracture occurs. The material, even though known to be ductile, behaves as if it were *brittle*.

In order to specify a safe strength for a metallic material under repeated loading, it is necessary to determine a limit below which no evidence of failure can be detected after applying a load for a specified number of cycles. This limiting stress is called the *endurance* or *fatigue limit*. Using a testing machine for this purpose, a series of specimens are each subjected to a specified stress and cycled to failure. The results are plotted as a graph representing the stress S (or σ) as the ordinate and the number of cycles-to-failure N as the abscissa. This graph is called an *S–N diagram* or *stress–cycle diagram,* and most often the values of N are plotted on a logarithmic scale since they are generally quite large.

Examples of *S–N* diagrams for two common engineering metals are shown in Fig. 3–27. The endurance limit is that stress for which the *S–N* graph becomes horizontal or asymptotic. As noted, it has a well-defined value of $(S_{el})_{st} = 27$ ksi (186 MPa) for steel. For aluminum, however, the endurance limit is not well defined, and so it is normally specified as the stress having a limit of 500 million cycles, $(S_{el})_{al} = 19$ ksi (131 MPa). Typical values of endurance limits for various engineering materials are usually reported in handbooks. Once a particular value is obtained, it is assumed that for any stress below this value the fatigue life is infinite, and therefore the number of cycles to failure is no longer given consideration.

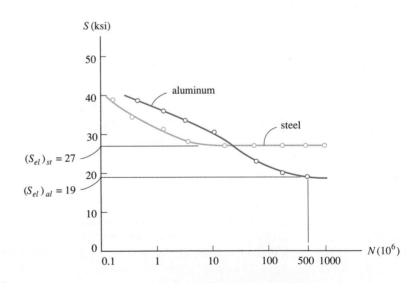

S– N diagram for steel and aluminum alloys
(N axis has a logarithmic scale)

Fig. 3–27

PROBLEMS

*3–16.   The acrylic plastic rod is 200 mm long and 15 mm in diameter. If an axial load of 300 N is applied to it, determine the change in its length and the change in diameter. $E_p = 2.70$ GPa, $\nu_p = 0.4$.

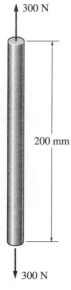

300 N

200 mm

300 N

Prob. 3–16

3–18.   The elastic portion of the stress–strain diagram for a steel alloy is shown in the figure. The specimen from which it was obtained had an original diameter of 13 mm and a gauge length of 50 mm. When the applied load on the specimen is 50 kN, the diameter is 12.99265 mm. Determine Poisson's ratio for the material.

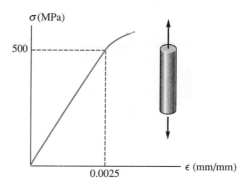

σ (MPa)

500

ϵ (mm/mm)

0.0025

Prob. 3–18

3–17.   The short cylindrical block of aluminum, having an original diameter of 0.5 in. and a length of 1.5 in., is placed in the smooth jaws of a vice and squeezed until the axial load applied is 800 lb. Determine (a) the decrease in its length and (b) its new diameter. $E_{al} = 10(10^3)$ ksi, $\nu_{al} = 0.33$.

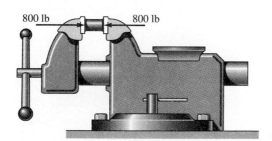

800 lb   800 lb

Prob. 3–17

3–19.   The rubber block is subjected to an elongation of 0.03 in. along the x axis, and its vertical faces are given a tilt so that $\theta = 89.3°$. Determine the strains ϵ_x, ϵ_y, and γ_{xy} at each point in the block. Take $\nu_r = 0.5$.

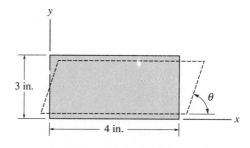

y

3 in.

θ

4 in.

x

Prob. 3–19

*3–20.   The plug has a diameter of 30 mm and fits within a rigid sleeve having an inner diameter of 32 mm. Both the plug and the sleeve are 50 mm long. Determine the axial pressure p that must be applied to the top of the plug to cause it to contact the sides of the sleeve. Also, how far must the plug be compressed downward in order to do this? The plug is made from a material for which $E = 5$ MPa, $\nu = 0.45$.

Prob. 3–20

3–21.   The support consists of three rigid plates, which are connected together using two symmetrically placed rubber pads. If a vertical force of 50 N is applied to plate A, determine the approximate vertical displacement of this plate due to shear strains in the rubber. Each pad has cross-sectional dimensions of 30 mm and 20 mm. $G_r = 0.20$ MPa.

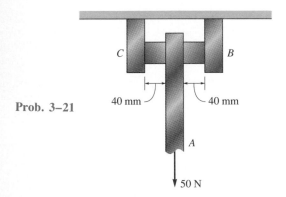

Prob. 3–21

3–22.   The stone has a mass of 800 kg and center of gravity at G. It rests on a pad at A and a roller at B. The pad is fixed to the ground and has a compressed height of 30 mm, a width of 140 mm, and a length of 150 mm. If the coefficient of static friction between the pad and the stone is $\mu_s = 0.8$, determine the approximate horizontal displacement of the stone, caused by the shear strains in the pad, before the stone begins to slip. Assume the normal force at A acts 1.5 m from G as shown. The pad is made made from a material having $E = 4$ MPa and $\nu = 0.35$. Assume the normal force at A acts 1.5 m from G as shown.

Prob. 3–22

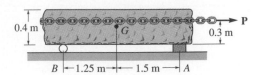

3–23.   A shear spring is made from two blocks of rubber, each having a height h, width b, and thickness a. The blocks are bonded to three plates as shown. If the plates are rigid and the shear modulus of the rubber is G, determine the displacement of point A if a vertical load P is applied at this point. Assume that the displacement is small so that $\delta = a \tan \gamma \approx a\gamma$.

Prob. 3–23

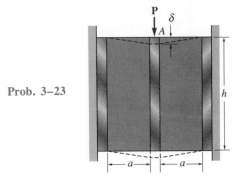

*3–24.   A shear spring is made by bonding the rubber annulus to a rigid fixed ring and a plug. When an axial load $\mathbf{P}$ is placed on the plug, show that the slope at point y in the rubber is $dy/dr = -\tan \gamma = -\tan(P/2\pi hGr)$. For small angles we can write $dy/dr = -P/(2\pi hGr)$. Integrate this expression and evaluate the constant of integration using the condition that $y = 0$ at $r = r_o$. From the result compute the deflection $y = \delta$ of the plug.

Prob. 3–24

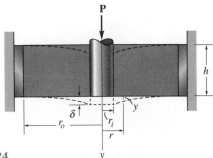

REVIEW PROBLEMS

3–25. The aluminum block has a rectangular cross section and is subjected to an axial compressive force of 8 kip. If the 1.5-in. side changed its length to 1.500132 in., determine Poisson's ratio and the new length of the 2-in. side. $E_{al} = 10(10^3)$ ksi.

Prob. 3–25

3–27. The rigid beam rests in the horizontal position on two aluminum cylinders having the *unloaded* lengths shown. If each cylinder has a diameter of 30 mm, determine the placement x of the applied 80-kN load so that the beam remains horizontal. What is the new diameter of cylinder A after the load is applied? $E_{al} = 70$ GPa, $\nu_{al} = 0.33$.

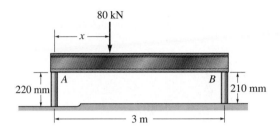

Prob. 3–27

3–26. The steel wires AB and AC support the 200-kg mass. If the allowable axial stress for the wires is $\sigma_{allow} = 130$ MPa, determine the required diameter of each wire. Also, what is the new length of wire AB after the load is applied? Take the unstretched length of AB to be 750 mm. $E_{st} = 200$ GPa.

*3–28. The shear stress–strain diagram for a steel alloy is shown in the figure. If a bolt having a diameter of 0.25 in. is made from this material and used in the lap joint, determine the force P required to cause the material to yield.

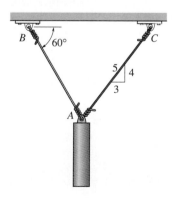

Prob. 3–26

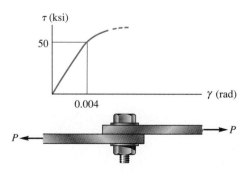

Prob. 3–28

3–29. While undergoing a tension test, a copper-alloy specimen having a gauge length of 2 in. is subjected to a strain of 0.40 in./in. when the stress is 70 ksi. If $\sigma_Y = 45$ ksi when $\epsilon_y = 0.0025$ in./in., determine the distance between the gauge points when the load is released.

3–30. The elastic portion of the tension stress–strain diagram for an aluminum specimen is shown in the figure. The specimen used for the test has a gauge length of 2 in. and a diameter of 0.5 in. When the applied load is 9 kip, the new diameter of the specimen is 0.49935 in. Compute the shear modulus G_{al} for the aluminum.

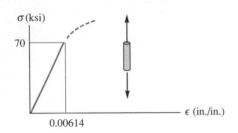

Prob. 3–30

3–31. A bar having a length of 5 in. and cross-sectional area of 0.7 in^2 is subjected to an axial force of 8000 lb. If the bar stretches 0.002 in., determine the modulus of elasticity of the material. The material has linear-elastic behavior.

Prob. 3–31

*3–32. An 8-mm-diameter brass rod has a modulus of elasticity of $E_{br} = 100$ GPa. If it is 3 m long and subjected to an axial load of 2 kN, determine its elongation. What is its elongation under the same load if its diameter is 6 mm?

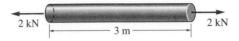

Prob. 3–32

4 Axial Load

In Chapter 1 we developed the method for finding the normal stress in axially loaded members. In this chapter we will discuss how to determine the deformation of these members, and we will also develop a method for finding the support reactions when these reactions cannot be determined strictly from the equations of equilibrium. An analysis of the effects of thermal stress, stress concentrations, inelastic deformations, and residual stress will also be discussed.

4.1 Saint-Venant's Principle

In the previous chapters we have developed the concept of *stress* as a means of measuring the force distribution within a body and *strain* as a means of measuring a body's deformation. We have also shown that the mathematical relationship between stress and strain depends on the type of material from which the body is made. In particular, if the stress creates a linear elastic response from the material, then Hooke's law applies and there is a proportional relationship between stress and strain. Furthermore, for this case, since stress can be related to the load and strain can be related to displacement, there must also be a proportional relationship between the applied load and the displacement of points in the body.

For example, consider the manner in which a rectangular bar will deform elastically when the bar is subjected to a force **P** applied along its centroidal axis, Fig. 4–1a. Here the bar is fixed-connected at one end, with the force applied through a hole at its other end. Due to the loading, the bar deforms as indicated by the distortions of the once horizontal and vertical grid lines

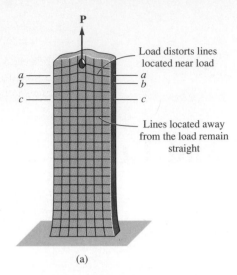

(a)

Fig. 4–1

drawn on the bar. Notice the *localized deformation* that occurs at each end. This effect tends to *decrease* as measurements are taken farther and farther away from the ends. Furthermore, the deformations even out and become uniform throughout the midsection of the bar.

Since the deformation is related to stress within the bar, we can state that stress will be distributed more uniformly throughout the cross-sectional area if the section is taken farther and farther from the point where the external load is applied. To show this, consider a profile of the variation of the stress distribution acting at sections *a–a, b–b,* and *c–c,* each of which is shown in Fig. 4–1*b*. In each of these three cases, force equilibrium requires the magnitude of the *resultant* force developed by the stress distribution to be equal to P. Furthermore, *moment equilibrium* requires each stress distribution to be *symmetrical* over the cross section, and it is for this reason that **P** is applied through the centroid of the cross section.

By comparison, the stress distribution *almost* reaches a uniform value at section *c–c,* which is sufficiently removed from the end. In other words, section *c–c* is far enough away from the application of **P** so that the localized deformation caused by **P** *vanishes*. The minimum distance from the bar's end where this occurs can be determined using a mathematical analysis based on the theory of elasticity. However, as a *general rule,* which applies as well to many other cases of loading and member geometry, we can consider this distance to be at least equal to the *largest dimension* of the loaded cross section. Hence, for the bar in Fig. 4–1*b*, section *c–c* should be located at a distance at least equal to the width (not the thickness) of the bar.* This rule is based on *experimental observation of material behavior,* and only in special

---

*When section *c–c* is so located, the theory of elasticity predicts the maximum stress to be $\sigma_{\max} = 1.02\sigma_{\text{avg}}$.

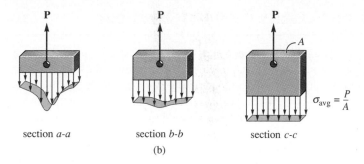

section *a-a*   section *b-b*   section *c-c*

(b)

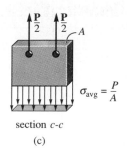

section *c-c*

(c)

cases, like the one discussed here, has it been validated mathematically. It should be noted, however, that this rule does not apply to every type of member and loading case. For example, members made from thin-walled elements, and subjected to loadings that cause large deflections, may create localized stresses and deformations that have an influence a considerable distance away from the point of application of loading.

At the support, in Fig. 4–1*a* notice how the bar is prevented from decreasing its width, which should occur due to the bar's lateral elongation—a consequence of the "Poisson effect," discussed in Sec. 3–6. By the same arguments given above, however, we could demonstrate that the stress distribution at the support will also even out and become uniform over the cross section at a short distance from the support; and furthermore, the magnitude of the resultant force created by this stress distribution must also equal P.

The fact that stress and deformation behave in this manner is referred to as *Saint-Venant's principle,* since it was first noticed by the French scientist Barré de Saint-Venant. Essentially it states that the stress and strain produced at points in a body sufficiently removed from the region of load application will be the *same* as the stress and strain produced by any applied loadings that have the same statically equivalent resultant and are applied to the body within the same region. In other words, by application of Saint-Venant's principle, we can assume that the *same* uniform stress distribution at section *c–c* in Fig. 4–1*b* will occur if the load $\mathbf{P}$ is replaced by *any other statically equivalent loading*. For example, if two symmetrically applied forces $P/2$ act on the bar, Fig. 4–1*c*, the stress distribution at section *c–c*, which is sufficiently removed from the localized effects of these loads, will be uniform and therefore equivalent to $\sigma_{\text{avg}} = P/A$ as before.

To summarize, then, we do not have to consider the somewhat complex stress distributions that may actually develop at points of load application, or at supports, when studying the stress distribution in a body at sections *sufficiently removed* from the points of load application. Saint-Venant's principle claims that the localized effects caused by any load acting on the body will dissipate or smooth out within regions that are sufficiently removed from the location of the load. Furthermore, the resulting stress distribution at these regions will be the *same* as that caused by any other statically equivalent load applied to the body within the same localized area.

4.2 Elastic Deformation of an Axially Loaded Member

Using Hooke's law and the definitions of stress and strain, we will now develop an equation that can be used to determine the elastic deformation of a member subjected to axial loads. To generalize the development, consider the bar shown in Fig. 4–2a, which has a cross-sectional area that *gradually* varies along its length L. The bar is subjected to concentrated loads at its ends and a variable external load distributed along its length. This distributed load could, for example, represent the weight of a vertical bar, or friction forces acting on the bar's surface. Here we wish to find the *relative displacement* Δ of one end of the bar with respect to the other end as caused by this loading. In the following analysis we will neglect the localized deformations that occur at points of concentrated loading and where the cross section suddenly changes. As noted in Sec. 4.1, these effects occur within small regions of the bar's length and will therefore have only a slight effect on the final result. For the most part, the bar will deform uniformly, so the normal stress will be uniformly distributed over the cross section.

Using the method of sections, a differential element (or wafer) of length dx and area $A(x)$ is isolated from the bar at the arbitrary position x along the bar's length. The free-body diagram of this element is shown in Fig. 4–2b. The resultant internal axial force is represented as $P(x)$, since the external loading will cause it to vary along the length of the bar. This load, $P(x)$, will deform the element into the shape indicated by the dashed outline, and therefore the displacement of one end of the element with respect to the other end is δx. The stress and strain in the element are

$$\sigma = \frac{P(x)}{A(x)} \qquad \text{and} \qquad \epsilon = \frac{\delta x}{dx}$$

Provided these quantities do not exceed the proportional limit, we can relate them using Hooke's law; i.e.,

$$\sigma = E\epsilon$$

$$\frac{P(x)}{A(x)} = E\left(\frac{\delta x}{dx}\right)$$

$$\delta x = \frac{P(x)\, dx}{A(x)\, E}$$

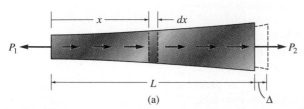

(a)

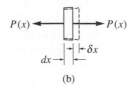

(b)

Fig. 4–2

For the entire length L of the bar, we must integrate this expression to find the required end displacement. This yields

$$\Delta = \int_0^L \frac{P(x)\ dx}{A(x)\ E}$$

(4–1)

where

Δ = displacement of one point on the bar relative to another point

L = distance between the points

$P(x)$ = internal axial force at the section, located a distance x from one end

$A(x)$ = cross-sectional area of the bar, expressed as a function of x

E = modulus of elasticity for the material

Constant Load and Cross-Sectional Area.

In many cases the bar will have a constant cross-sectional area. Furthermore, if a constant external force is applied at each end, Fig. 4–3a, then the internal force P throughout the length of the bar is also constant, Fig. 4–3b. As a result, Eq. 4–1 can be integrated to yield

$$\Delta = \frac{PL}{AE}$$

(4–2)

Since Hooke's law was used in the development of the above two equations, it is important that the material exhibits linear-elastic behavior, and the applied load does not cause yielding of the material. Furthermore, the material must be homogeneous, since E is constant.

If the bar is subjected to several different axial forces, or the cross-sectional area or modulus of elasticity changes abruptly from one region of the bar to the next, the above equation can be applied to each *segment* of the bar where these quantities are all *constant*. The displacement of one end of the bar with respect to the other is then found from the *vector addition* of the end displacements of each segment. For this general case,

$$\Delta = \sum \frac{PL}{AE}$$

(4–3)

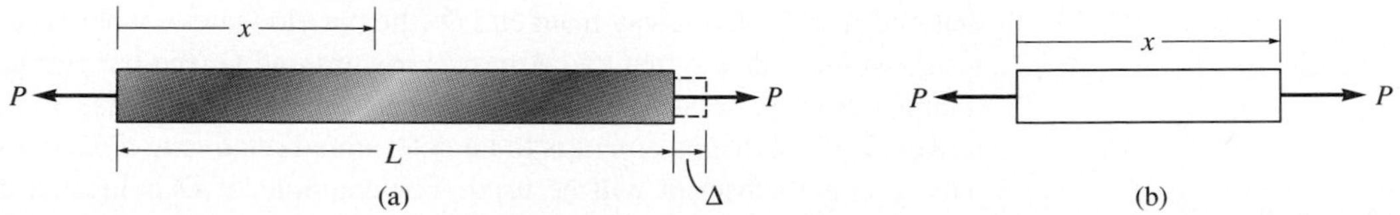

Fig. 4–3

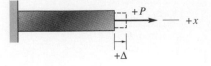

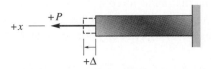

Positive sign convention for P and Δ

Fig. 4–4

Sign Convention. In order to apply Eq. 4–3, we must develop a sign convention for the internal axial force and the displacement of one end of the bar with respect to the other end. To do so, we will consider both the force and displacement to be positive if they cause tension and elongation, respectively; whereas a negative force and displacement will cause compression and contraction, respectively, Fig. 4–4.

For example, consider the bar shown in Fig. 4–5a. The *internal axial forces "P,"* computed by the method of sections for each segment, are $P_{AB} = +5$ kN, $P_{BC} = -3$ kN, $P_{CD} = -7$ kN, Fig. 4–5b. Applying Eq. 4–3 to obtain the displacement of end A relative to end D, we have

$$\Delta_{A/D} = \sum \frac{PL}{AE} = \frac{(5 \text{ kN}) L_{AB}}{AE} + \frac{(-3 \text{ kN}) L_{BC}}{AE} + \frac{(-7 \text{ kN}) L_{CD}}{AE}$$

If the other data are substituted and a positive answer is computed, it means that end A will move away from end D (the bar elongates), whereas a negative result would indicate that end A moves toward end D (the bar shortens). The double subscript notation is used to indicate this relative displacement ($\Delta_{A/D}$); however, if the displacement is to be determined relative to a *fixed point*, then only a single subscript will be used. For example, if D is located at a *fixed* support, then the computed displacement will be denoted as simply Δ_A.

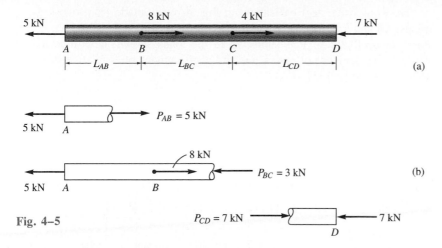

Fig. 4–5

PROCEDURE FOR ANALYSIS

The relative displacement between two points *A* and *B* on an axially loaded member can be determined by applying Eq. 4–1 (or Eq. 4–2). Since Hooke's law has been used in the development of these equations, it is important that the external loads do not cause yielding of the material and that the material is homogeneous and behaves in a linear-elastic manner. Application of the necessary equation requires the following steps.

Internal Force.   To obtain the internal axial force *P,* one should use the method of sections and the equation of force equilibrium applied along the axis of the member. If this force varies along the member's length, a section should be made at the arbitrary location *x* from one end of the member and the force represented as a function of *x, P(x)*. If several *constant external forces* act on the member, the internal force in each *segment* of the member, between any two external forces, must then be determined. For any segment, an internal *tensile force* is *positive* and an internal *compressive force* is *negative*.

Displacement.   When the member's cross-sectional area varies along its axis, the area must be expressed as a function of its position *x, A(x)*. On the other hand, if the cross-sectional area, the modulus of elasticity, or the internal loading suddenly changes between points *A* and *B,* then Eq. 4–2 should be applied to each segment for which these qualities are constant (Eq. 4–3). When substituting the data into Eqs. 4–1 through 4–3, be sure to account for the proper sign for *P,* as discussed above, and use a consistent set of units. For any segment, if the computed result is a *positive* numerical quantity, it indicates *elongation;* if it is *negative,* it indicates a *contraction*.

The following examples illustrate numerical applications of this method.

Example 4–1

The composite steel bar shown in Fig. 4–6a is made from two segments, AB and BD, having cross-sectional areas of $A_{AB} = 1$ in^2 and $A_{BD} = 2$ in^2, respectively. If it is subjected to the loads shown, determine the vertical displacement of end A and the displacement of B relative to C. Take $E_{st} = 29(10^3)$ ksi.

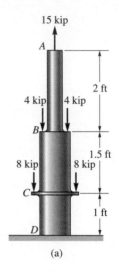

(a)

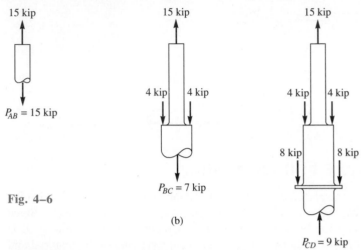

Fig. 4–6

(b)

SOLUTION

Internal Force. Due to the application of the external loadings, the *internal axial forces* in regions AB, BC, and CD will all be *different*. These forces are obtained by applying the method of sections and the equation of vertical force equilibrium as shown in Fig. 4–6b.

Displacement. Using the sign convention, i.e., the internal tensile forces are positive and the compressive forces are negative, the vertical displacement of A relative to the *fixed* support (D) is

$$\Delta_A = \sum \frac{PL}{AE} = \frac{[+15 \text{ kip}](2 \text{ ft})(12 \text{ in./ft})}{(1 \text{ in}^2)[29(10^3) \text{ kip/in}^2]} + \frac{[+7 \text{ kip}](1.5 \text{ ft})(12 \text{ in./ft})}{(2 \text{ in}^2)[29(10^3) \text{ kip/in}^2]}$$

$$+ \frac{[-9 \text{ kip}](1 \text{ ft})(12 \text{ in./ft})}{(2 \text{ in}^2)[29(10^3) \text{ kip/in}^2]}$$

$$= +0.0127 \text{ in.} \qquad\qquad Ans.$$

Since the result is *positive*, the bar *elongates* and so the displacement at A is upward.

Applying Eq. 4–2 between points B and C, we can obtain the relative displacement of B with respect to C; that is,

$$\Delta_{B/C} = \frac{P_{BC}L_{BC}}{A_{BC}E} = \frac{[+7 \text{ kip}](1.5 \text{ ft})(12 \text{ in./ft})}{(2 \text{ in}^2)[29(10^3) \text{ kip/in}^2]} = +0.00217 \text{ in.} \qquad Ans.$$

Here point B moves away from point C, since the segment elongates.

Example 4-2

The assembly shown in Fig. 4–7a consists of an aluminum tube AB having a cross-sectional area of 400 mm². A steel rod having a diameter of 10 mm is attached to a rigid collar and passes through the tube. If a tensile load of 80 kN is applied to the rod, determine the displacement of the end C of the rod. Take E_{st} = 200 GPa, E_{al} = 70 GPa.

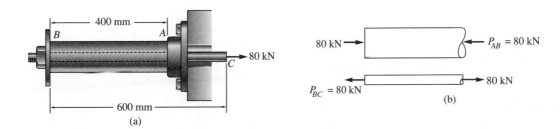

(a)

(b)

Fig. 4–7

SOLUTION

Internal Force.   The free-body diagram of the tube and rod, Fig. 4–7b, shows that the rod is subjected to a tension of 80 kN and the tube is subjected to a compression of 80 kN.

Displacement.   Using Eq. 4–2, we will first determine the displacement of end C with respect to end B. Working in units of newtons and meters, we have

$$\Delta_{C/B} = \frac{PL}{AE} = \frac{[+80(10^3)\ \text{N}](0.6\ \text{m})}{(\pi/4)(0.01\ \text{m})^2[200(10^9)\ \text{N/m}^2]} = +0.003056\ \text{m} \rightarrow$$

The positive sign indicates that end C moves *to the right* relative to end B, since the bar elongates.

Now, applying Eq. 4–2 to the tube, in order to determine the displacement of end B with respect to the *fixed* end A, yields

$$\Delta_B = \frac{PL}{AE} = \frac{[-80(10^3)\ \text{N}](0.4\ \text{m})}{400\ \text{mm}^2((10^{-6})\ \text{m}^2/\text{mm}^2)[70(10^9)\ \text{N/m}^2]}$$

$$= -0.001143\ \text{m} = 0.001143\ \text{m} \rightarrow$$

Here the negative sign indicates that the tube shortens and so B moves to the *right* relative to A.

Since both computed displacements are to the right, the resultant displacement of C relative to A is therefore

$$(\overset{+}{\rightarrow}) \qquad \Delta_C = \Delta_B + \Delta_{C/B} = 0.001143\ \text{m} + 0.003056\ \text{m}$$

$$= 0.00420\ \text{m} = 4.20\ \text{mm} \rightarrow \qquad\qquad \textit{Ans.}$$

Example 4–3

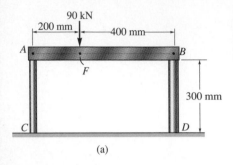

(a)

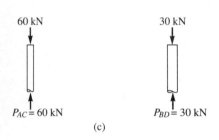

(b)

(c)

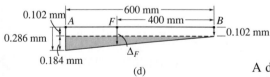

(d)

Fig. 4–8

A *rigid* beam AB rests on the two short posts shown in Fig. 4–8a. AC is made of steel and has a diameter of 20 mm, and BD is made of aluminum and has a diameter of 40 mm. Determine the displacement of point F on AB if a vertical load of 90 kN is applied over this point. Take $E_{st} = 200$ GPa, $E_{al} = 70$ GPa.

SOLUTION

Internal Force. The compressive forces acting at the top of each post are determined from the equilibrium of member AB, Fig. 4–8b. These forces are equal to the internal forces in each post, Fig. 4–8c.

Displacement. Applying Eq. 4–2 to compute the displacement of the top of each post, we have

Post AC:

$$\Delta_A = \frac{P_{AC}L_{AC}}{A_{AC}E_{st}} = \frac{[-60(10^3) \text{ N}](0.300 \text{ m})}{\pi(0.010 \text{ m})^2[200(10^9) \text{ N/m}^2]} = -286(10^{-6}) \text{ m}$$

$$= 0.286 \text{ mm} \downarrow$$

Post BD:

$$\Delta_B = \frac{P_{BD}L_{BD}}{A_{BD}E_{al}} = \frac{[-30(10^3) \text{ N}](0.300 \text{ m})}{\pi(0.020 \text{ m})^2[70(10^9) \text{ N/m}^2]} = -102(10^{-6}) \text{ m}$$

$$= 0.102 \text{ mm} \downarrow$$

A diagram showing the centerline displacements at points A, B, and F on the beam is shown in Fig. 4–8d. By proportion of the shaded triangle, the displacement of point F is therefore

$$\Delta_F = 0.102 \text{ mm} + (0.184 \text{ mm})\left(\frac{400 \text{ mm}}{600 \text{ mm}}\right) = 0.225 \text{ mm} \downarrow \quad \textit{Ans.}$$

Example 4–4

A member is made from a material that has a specific weight γ and modulus of elasticity E. If it is formed into a *cone* having the dimensions shown in Fig. 4–9a, determine how far its end is displaced due to gravity when it is suspended in the vertical position.

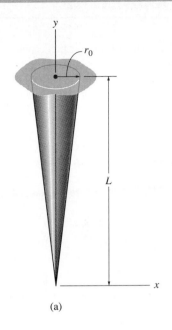

SOLUTION

Internal Force. The internal axial force varies along the member since it is dependent on the weight $W(y)$ of a segment of the member below any section, Fig. 4–9b. Hence, to calculate the displacement, we must use Eq. 4–1. At the section located a distance y from its bottom end, the radius x of the cone is determined by proportion; i.e.,

$$\frac{x}{y} = \frac{r_0}{L}; \qquad x = \frac{r_0}{L}y$$

The volume of a cone having a base of radius x and height y is

$$V = \frac{\pi}{3}yx^2 = \frac{\pi r_0^2}{3L^2}y^3$$

Since $W = \gamma V$, the internal force at the section becomes

$$+\uparrow \ \Sigma F_y = 0; \qquad P(y) = \frac{\gamma \pi r_0^2}{3L^2}y^3$$

Displacement. The area of the cross section is also a function of position y, Fig. 4–9b. We have

$$A(y) = \pi x^2 = \frac{\pi r_0^2}{L^2}y^2$$

Applying Eq. 4–1 between the limits of $y = 0$ and $y = L$ yields

$$\Delta = \int_0^L \frac{P(y)\ dy}{A(y)\ E} = \int_0^L \frac{[(\gamma \pi r_0^2/3L^2)\ y^3]\ dy}{[(\pi r_0^2/L^2)\ y^2]\ E}$$

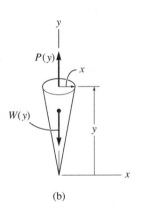

$$= \frac{\gamma}{3E} \int_0^L y\ dy$$

Fig. 4–9

$$= \frac{\gamma L^2}{6E} \qquad\qquad\qquad \textit{Ans.}$$

As a partial check of this result, notice how the units of the terms, when canceled, give the deflection in units of length as expected.

PROBLEMS

4–1. The composite shaft, consisting of aluminum, copper, and steel sections, is subjected to the loading shown. Determine the displacement of end A with respect to end D and the normal stress in each section. The cross-sectional area and moduli of elasticity for each section are shown in the figure. Neglect the size of the collars at B and C.

4–2. Determine the displacement of B with respect to C of the composite shaft in Prob. 4–1.

*4–4. The steel column is used to support the symmetric loads from the two floors of a building. Determine the vertical displacement of its top A if the column has a cross-sectional area of 23.4 in^2. $E_{st} = 29(10^3)$ ksi.

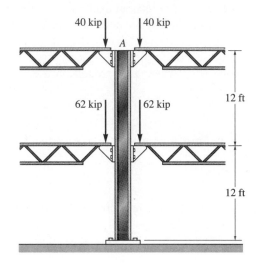

Prob. 4–4

Aluminum	Copper	Steel
$E_{al} = 10(10^3)$ ksi	$E_{cu} = 18(10^3)$ ksi	$E_{st} = 29(10^3)$ ksi
$A_{AB} = 0.09$ in^2	$A_{BC} = 0.12$ in^2	$A_{CD} = 0.06$ in^2

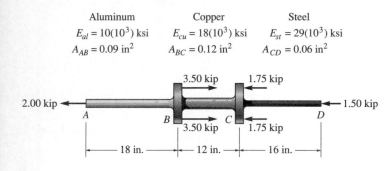

Probs. 4–1/4–2

4–3. The assembly consists of a steel rod CB and an aluminum rod BA, each having a diameter of 12 mm. If the rod is subjected to the axial loadings at A and at the coupling B, determine the displacement of the coupling B and the end A. The unstretched length of each segment is shown in the figure. Neglect the size of the connections at B and C, and assume that they are rigid. $E_{st} =$ 200 GPa, $E_{al} = 70$ GPa.

4–5. The copper shaft is subjected to the axial loads show.1. Determine the displacement of end A with respect to end D if the diameters of each segment are $d_{AB} = 0.75$ in., $d_{BC} = 1$ in., and $d_{CD} = 0.5$ in. Take $E_{cu} = 18(10^3)$ ksi.

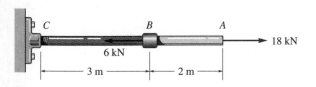

Prob. 4–3

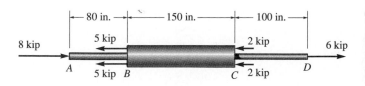

Prob. 4–5

4–6. The steel rod is subjected to the loading shown. If the cross-sectional area of the rod is 60 mm^2 and E_{st} = 200 GPa, determine the displacement of B and A. Neglect the size of the couplings at B, C, and D.

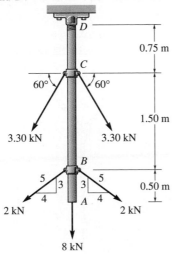

Prob. 4–6

*4–8.** The 15-mm-diameter steel shaft AC is supported by a rigid collar, which is fixed to the shaft at B. If it is subjected to an axial load of 80 kN at its end, determine the uniform pressure distribution p on the collar required for equilibrium. Also, what is the elongation of segment BC and segment BA? E_{st} = 200 GPa.

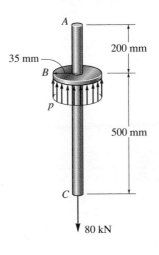

Prob. 4–8

4–7. The assembly consists of a 30-mm-diameter aluminum bar ABC with fixed collar at B and a 10-mm-diameter steel rod CD. Determine the displacement of point D when the assembly is loaded as shown. Neglect the size of the collar at B and the connection at C. E_{st} = 200 GPa, E_{al} = 70 GPa.

Prob. 4–7

4–9. The truss is made of three steel members, each having a cross-sectional area of 400 mm^2. If E_{st} = 200 GPa, determine the horizontal displacement of the roller at C when the truss supports the loads shown.

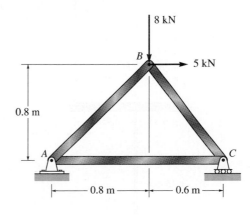

Prob. 4–9

4–10. The assembly consists of two steel suspender rods AC and BD attached to the 100-lb uniform rigid beam AB. Determine the position x for the 300-lb loading so that the beam remains in a horizontal position both before and after the load is applied. Each rod has a diameter of 0.5 in. $E_{st} = 29(10^3)$ ksi.

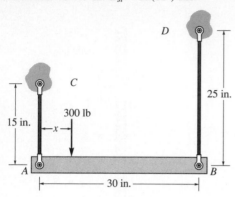

Prob. 4–10

4–11. The assembly consists of three titanium rods and a rigid bar AC. The cross-sectional area of each rod is given in the figure. If a vertical force of $P = 20$ kN is applied to the ring F, determine the vertical displacement of point F. $E_{ti} = 350$ GPa.

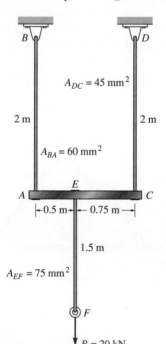

Prob. 4–11

*4–12. The linkage is made of three pin-connected steel members, each having a cross-sectional area of 0.730 in². If a vertical force of 50 kip is applied to the end B of member AB, determine the vertical displacement of point B. $E_{st} = 29(10^3)$ ksi.

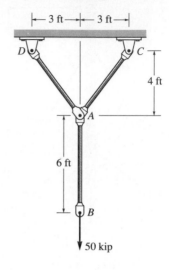

Prob. 4–12

4–13. The bar has a length L and cross-sectional area A. Determine its elongation due to both the force **P** and its own weight. The material has a specific weight γ (weight/volume) and a modulus of elasticity E.

Prob. 4–13

4–14.   The steel drill shaft of an oil well extends 12,000 ft into the ground. Assuming that the pipe used to drill the well is suspended freely from the derrick at A, determine the maximum stress in the pipe and the elongation of its end D with respect to the fixed end at A. The shaft consists of three different sizes of pipe, AB, BC, and CD, each having the length, weight per unit length, and cross-sectional area indicated. $E_{st} = 29(10^3)$ ksi. *Hint:* Use the results of Prob. 4-13.

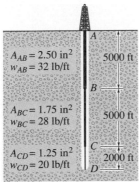

$A_{AB} = 2.50$ in^2
$w_{AB} = 32$ lb/ft
5000 ft

$A_{BC} = 1.75$ in^2
$w_{BC} = 28$ lb/ft
5000 ft

$A_{CD} = 1.25$ in^2
$w_{CD} = 20$ lb/ft
2000 ft

Prob. 4–14

4–15.   The segments of pipe and couplings used for drilling an oil well 12,000 ft deep are made of steel weighing 25 lb/ft. They have an outside diameter of 5.60 in. and an inside diameter of 4.70 in. In order to prevent buckling or sidesway of the pipe due to its own weight, it is partially supported at its top by the drawworks of the rig. If this force is $P = 298.7$ kip, determine the force **F** of the drill bit on the ground and the elongation of the pipe for this condition. $E_{st} = 29(10^3)$ ksi.

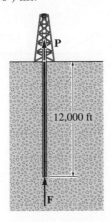

12,000 ft

Prob. 4–15

*__*4–16.__   The pipe is stuck in the ground so that when it is pulled upward the frictional force along its length varies linearly from zero at B to f_{max} (force/length) at C. Determine the initial force P required to pull the pipe out and the pipe's associated elongation just before it starts to slip. The pipe has a length L, cross-sectional area A, and the material from which it is made has a modulus of elasticity E.

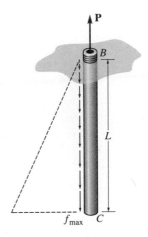

f_{max}

Prob. 4–16

4–17.   Determine the relative displacement of one end of the slightly tapered shaft with respect to the other end when it is subjected to the axial load P.

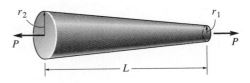

Prob. 4–17

4–18. The rod has a slight taper and length L. It is suspended from the ceiling and supports its own weight and a load **P** at its end. Determine the displacement of its end due to both of these loads. The material has a specific weight γ (weight/volume) and a modulus of elasticity E.

***4–20.** Show that the relative displacement of one end of the tapered plate with respect to the other end when it is subjected to an axial load P is $\Delta = \{Ph/[tE(d_2 - d_1)]\}\ln(d_2/d_1)$.

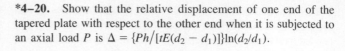

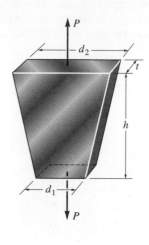

Prob. 4–20

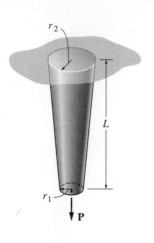

Prob. 4–18

4–19. Determine the elongation of the tapered steel shaft when it is subjected to an axial force of 18 kip. $E_{st} = 29(10^3)$ ksi. *Hint:* Use the result of Prob. 4-17.

4–21. Determine the elongation of the aluminum strap when it is subjected to an axial force of 30 kN. $E_{al} = 70$ GPa. *Hint:* Use the result of Prob. 4-20.

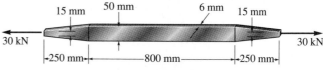

Prob. 4–21

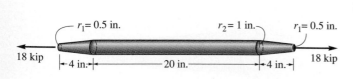

Prob. 4–19

4.3 Principle of Superposition

The principle of superposition is often used to determine the stress or displacement at a point in a member when the member is subjected to a complicated loading. By subdividing the loading into components, the *principle of superposition* states that the resultant stress or displacement at the point can be determined by first finding the stress or displacement caused by each component load acting *separately* on the member; then the resultant stress or displacement can be determined by adding the contributions caused by each of the components.

The following two conditions must be valid if the principle of superposition is to be applied.

1. *The loading must be linearly related to the stress or displacement that is to be determined.* For example, the equations $\sigma = P/A$ and $\Delta = PL/AE$ involve a linear relationship between P and σ or Δ.

2. *The loading must not significantly change the original geometry or configuration of the member.* If significant changes do occur, the direction and location of the applied forces and their moment arms will change, and consequently, application of the equilibrium equations will yield different results. For example, consider the slender rod shown in Fig. 4–10a, which is subjected to the load **P.** In Fig. 4–10b, **P** is replaced by two of its components, $\mathbf{P} = \mathbf{P}_1 + \mathbf{P}_2$. If **P** causes the rod to deflect a large amount, as shown, the moment of the load about its support, Pd, will *not* equal the sum of the moments of its component loads, $Pd \neq P_1 d_1 + P_2 d_2$, because $d_1 \neq d_2 \neq d$.

Most of the equations involving load, stress, and displacement that are developed in this text consist of linear relationships between these quantities. Also, members or bodies that are to be considered will be such that the loading will produce deformations that are so small that the change in position and direction of the loading will be insignificant and can be neglected. One exception to this rule, however, will be discussed in Chapter 13. It consists of a column that carries an axial load that is equivalent to the critical or buckling load. It will be shown that when this load increases only slightly, it will cause the column to have a large lateral deflection, even if the material remains linear-elastic. These deflections, associated with the components of any axial load, *cannot* be superimposed.

Fig. 4–10

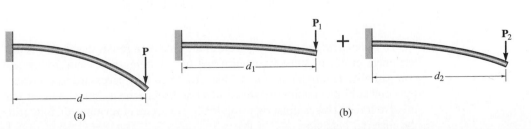

(a)   (b)

4.4 Statically Indeterminate Axially Loaded Member

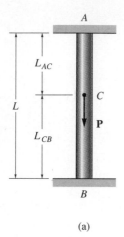

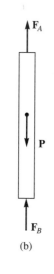

(a)

(b)

Fig. 4–11

When a bar is fixed-supported at one end and is subjected to an axial force, the force equilibrium equation applied along the axis of the bar is *sufficient* to find the reaction at the fixed support. A problem such as this, where the reactions can be determined strictly from the equations of equilibrium, is called *statically determinate*. If the bar is fixed at *both ends,* however, as in Fig. 4–11a, then two unknown reactions occur, Fig. 4–11b, and the force equilibrium equation becomes

$$+\uparrow \ \Sigma F = 0; \qquad\qquad F_B + F_A - P = 0$$

In this case the bar in Fig. 4–11a is called *statically indeterminate,* since the equilibrium equation(s) are not sufficient to determine the reactions.

In order to establish an additional equation needed for solution, it is necessary to consider the geometry of the deformation. Specifically, an equation that specifies the conditions for displacement is referred to as a *compatibility* or *kinematic condition.* For the bar in Fig. 4–11a, a suitable compatibility condition would require the relative displacement of one end of the bar with respect to the other end to be equal to zero, since the end supports are fixed. Hence, we can write

$$\Delta_{A/B} = 0$$

This equation can be expressed in terms of the applied loads by using a *load-displacement relationship,* which depends on the material behavior. For example, if linear-elastic behavior occurs, Eq. 4–2 can be used. Realizing that the internal force in segment AC is $+F_A$, and in segment CB the internal force is $-F_B$, the compatibility equation can be written as

$$\frac{F_A L_{AC}}{AE} - \frac{F_B L_{CB}}{AE} = 0$$

Assuming that AE is constant, we can solve the above two equations for the reactions, which gives

$$F_A = P\left(\frac{L_{CB}}{L}\right) \quad \text{and} \quad F_B = P\left(\frac{L_{AC}}{L}\right)$$

To summarize the above procedure, the unknown forces in statically indeterminate problems must be determined by satisfying both equilibrium and compatibility requirements for the bar. To do so it is necessary to transform the compatibility conditions, which involve displacement, into equations that involve force. This is done by using load-displacement relations that describe the material behavior.

Superposition of Forces. For some types of problems it may be easier to write the compatibility equation using the superposition of the forces acting on the free-body diagram. This method of solution is often referred to as the *flexibility* or *force method of analysis*. To show how it is applied, consider again the bar in Fig. 4–11a. In order to write the necessary equation of compatibility, we will first choose any one of the two supports as "redundant" and temporarily remove its effect on the bar. The word *redundant*, as used here, indicates that the support is not needed to hold the bar in stable equilibrium, so that when it is removed, the bar becomes statically determinate. Here we will choose the support at B as redundant. By using the principle of superposition, the bar having its original loading on it, Fig. 4–11c, is then equivalent to the bar subjected only to the external load **P,** Fig. 4–11d, plus the bar subjected only to the redundant load F_B, Fig.4–11e. Although not shown, notice that the reaction at the support A satisfies force equilibrium, $F_A = P - F_B$, as determined either from Fig. 4–11b or from the sum of the reactions at A in Fig. 4–11d and 4–11e.

If the load **P** causes B to be displaced *downward* by an amount Δ_B, Fig. 4–11d, the reaction F_B must be capable of displacing the end B of the bar *upward* by an amount δ_B, Fig. 4–11e, such that no displacement occurs at B, Fig. 4–11e, when the two loadings are superimposed.

$$(+\downarrow) \qquad\qquad 0 = \Delta_B - \delta_B$$

This equation represents the compatibility equation for displacements at point B, for which we have assumed that displacements are positive downward.

Applying the load–displacement relationship, we have $\Delta_B = PL_{AC}/AE$ and $\delta_B = F_B L/AE$. Consequently,

$$0 = \frac{PL_{AC}}{AE} - \frac{F_B L}{AE}$$

$$F_B = P\left(\frac{L_{AC}}{L}\right)$$

From the free-body diagram of the bar, Fig. 4–11b, the reaction at A can now be determined from the equation of equilibrium,

$$+\uparrow \ \Sigma F_y = 0; \qquad\qquad P\left(\frac{L_{AC}}{L}\right) + F_A - P = 0$$

Since $L_{CB} = L - L_{AC}$, then

$$F_A = P\left(\frac{L_{CB}}{L}\right)$$

These results are the same as those obtained previously, except that here we have applied the condition of compatibility and then the equilibrium condition to obtain the solution. Also note that the principle of superposition can be used here since the displacement and the load are linearly related ($\Delta = PL/AE$), which assumes, of course, that the material behaves in a linear-elastic manner.

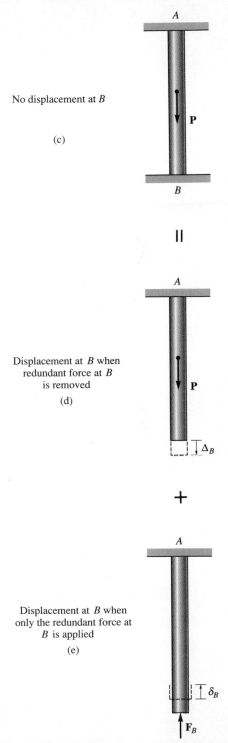

No displacement at B

(c)

Displacement at B when redundant force at B is removed

(d)

Displacement at B when only the redundant force at B is applied

(e)

PROCEDURE FOR ANALYSIS

Either one of the above two methods can be used to determine the unknown axial forces in a member that is statically indeterminate. Both methods require that the conditions of equilibrium and compatibility be satisfied.

To solve a particular problem it is important *first* to draw a *free-body diagram* of the member in order to identify all the forces that act on it. The sequence for applying the necessary equilibrium and compatibility equations is arbitrary. If it seems difficult to establish the compatibility equation, it is suggested that it be obtained using a superposition of forces. This method requires choosing one of the supports as *redundant*. The known displacement at the redundant support, which may be zero, is then equated to the displacement at this support caused by the external loads acting on the member, *excluding* the redundant support reaction, plus (vectorially) the displacement at the support caused *only* by the redundant reaction acting on the member. These two displacements are then expressed in terms of the loadings by using the load–displacement relationship $\Delta = PL/AE$. Once established, the compatibility equation can then be solved for the magnitude of the force at the redundant support. Any other unknown forces are then determined from the equilibrium equations.

If any of the unknown force magnitudes has a negative numerical value, it indicates that this force acts in the opposite sense of direction of that indicated on the free-body diagram.

The following examples numerically illustrate both methods of solution.

Example 4–5

The steel rod shown in Fig. 4–12a has a diameter of 5 mm. It is attached to the fixed wall at A, and before it is loaded there is a gap between the wall at B' and the rod of 1 mm. Determine the reactions at A and B' if the rod is subjected to an axial force of $P = 20$ kN as shown. Neglect the size of the collar at C. Take $E_{st} = 200$ GPa.

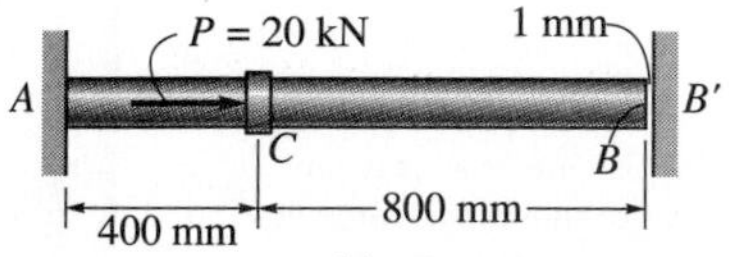

SOLUTION I
Equilibrium.   As shown on the free-body diagram, Fig. 4–12b, we will assume that the force P is large enough to cause the rod's end B to contact the wall at B'. The problem is statically indeterminate since there are two unknowns and only one equation of equilibrium.

Equilibrium of the rod requires

$$\xrightarrow{+} \Sigma F_x = 0; \qquad -F_A - F_B + 20(10^3) \text{ N} = 0 \qquad (1)$$

Compatibility.   The compatibility condition for the rod is

$$\Delta_{B/A} = 0.001 \text{ m}$$

$P = 20$ kN   1 mm

A —————————— B'

C B

400 mm   800 mm

(a)

Fig. 4–12

This displacement can be expressed in terms of the unknown reactions by using the load–displacement relationship, Eq. 4–2, applied to segments AC and CB, Fig. 4–12c. Working in units of newtons and meters, we have

$$\Delta_{B/A} = 0.001 \text{ m} = \frac{F_A L_{AC}}{AE} - \frac{F_B L_{CB}}{AE}$$

$$0.001 \text{ m} = \frac{F_A(0.4 \text{ m})}{\pi(0.0025 \text{ m})^2[200(10^9) \text{ N/m}^2]}$$

$$- \frac{F_B(0.8 \text{ m})}{\pi(0.0025 \text{ m})^2[200(10^9) \text{ N/m}^2]}$$

or

$$F_A(0.4 \text{ m}) - F_B(0.8 \text{ m}) = 3927.0 \text{ N} \cdot \text{m} \qquad (2)$$

Solving Eqs. 1 and 2 yields

$$F_A = 16.6 \text{ kN} \qquad F_B = 3.40 \text{ kN} \qquad Ans.$$

Since the answer for F_B is *positive*, indeed the end B contacts the wall at B' as originally assumed.*

SOLUTION II

Compatibility. This problem can also be solved using the superposition of forces. Here we will consider the support at B' as redundant. Using the principle of superposition, Fig. 4–12d, we have

$$(\overset{+}{\rightarrow}) \qquad\qquad 0.001 \text{ m} = \Delta_B - \delta_B \qquad (3)$$

The deflections Δ_B and δ_B are determined from Eq. 4–2. Working in units of newtons and meters, we have

$$\Delta_B = \frac{PL_{AC}}{AE} = \frac{[20(10^3) \text{ N}](0.4 \text{ m})}{\pi(0.0025 \text{ m})^2[200(10^9) \text{ N/m}^2]} = 0.002037 \text{ m}$$

$$\delta_B = \frac{F_B L_{AB}}{AE} = \frac{F_B(1.20 \text{ m})}{\pi(0.0025 \text{ m})^2[200(10^9) \text{ N/m}^2]} = 0.3056(10^{-6})F_B$$

Substituting into Eq. 3, we get

$$0.001 \text{ m} = 0.002037 \text{ m} - 0.3056(10^{-6})F_B$$

$$F_B = 3.40(10^3) \text{ N} = 3.40 \text{ kN} \qquad Ans.$$

Equilibrium. From the free-body diagram, Fig. 4–12b, the reaction at A is thus

$$\overset{+}{\rightarrow} \Sigma F_x = 0; \qquad -F_A + 20 \text{ kN} - 3.40 \text{ kN} = 0$$

$$F_A = 16.6 \text{ kN} \qquad Ans.$$

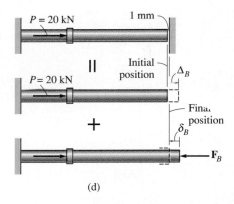

(b)

(c)

(d)

---

*If F_B were a negative quantity, the problem would be statically determinate, so that $F_B = 0$ and $F_A = 20 \text{ kN}$.

Example 4–6

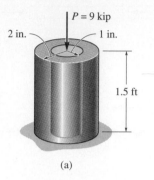

(a)

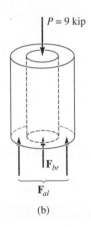

(b)

(c)

Fig. 4–13

The aluminum post shown in Fig. 4–13a is reinforced with a brass core. If it supports a resultant axial compressive load of $P = 9$ kip, determine the average normal stress in the aluminum and the brass. Take $E_{al} = 10(10^3)$ ksi and $E_{br} = 15(10^3)$ ksi.

SOLUTION

Equilibrium. The free-body diagram of the post is shown in Fig. 4–13b. Here the resultant axial force at the base is represented by the unknown components carried by the aluminum, $\mathbf{F}_{al}$, and brass, $\mathbf{F}_{br}$. The problem is statically indeterminate. Why?

Vertical force equilibrium requires

$$+\downarrow \ \Sigma F_y = 0; \qquad 9 \text{ kip} - F_{al} - F_{br} = 0 \qquad (1)$$

Compatibility. In order to satisfy compatibility requirements, the displacement at the top of the post for both the aluminum and brass must be the *same;* i.e.,

$$\Delta_{al} = \Delta_{br}$$

Using the load–displacement relationships,

$$\frac{F_{al}L}{A_{al}E_{al}} = \frac{F_{br}L}{A_{br}E_{br}}$$

$$F_{al} = F_{br}\left(\frac{A_{al}}{A_{br}}\right)\left(\frac{E_{al}}{E_{br}}\right)$$

$$F_{al} = F_{br}\left[\frac{\pi[(2 \text{ in.})^2 - (1 \text{ in.})^2]}{\pi(1 \text{ in.})^2}\right]\left[\frac{10(10^3) \text{ ksi}}{15(10^3) \text{ ksi}}\right]$$

$$F_{al} = 2F_{br} \qquad (2)$$

Solving Eqs. 1 and 2 simultaneously yields

$$F_{al} = 6 \text{ kip} \qquad F_{br} = 3 \text{ kip}$$

The average normal stress in the aluminum and brass is therefore

$$\sigma_{al} = \frac{6 \text{ kip}}{\pi[(2 \text{ in.})^2 - (1 \text{ in.})^2]} = 0.637 \text{ ksi} \qquad \textit{Ans.}$$

$$\sigma_{br} = \frac{3 \text{ kip}}{\pi(1 \text{ in.})^2} = 0.955 \text{ ksi} \qquad \textit{Ans.}$$

The stress distributions are shown on the top section of the post in Fig. 4–13c.

Example 4–7

The three steel bars shown in Fig. 4–14a are pin-connected to a *rigid* member. If the applied load on the member is 15 kN, determine the force developed in each bar. Bars *AB* and *EF* each have a cross-sectional area of 25 mm², and bar *CD* has a cross-sectional area of 15 mm². Take E_{st} = 200 GPa.

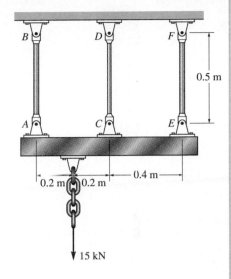

(a)

SOLUTION

Equilibrium.   The free-body diagram of the rigid member is shown in Fig. 4–14b. This problem is statically indeterminate since there are three unknowns and only two available equilibrium equations. These equations are

$$+\uparrow \ \Sigma F_y = 0; \qquad F_A + F_C + F_E - 15 \text{ kN} = 0 \qquad (1)$$
$$\downdownarrows + \ \Sigma M_C = 0; \qquad -F_A(0.4 \text{ m}) + 15 \text{ kN}(0.2 \text{ m}) + F_E(0.4 \text{ m}) = 0 \qquad (2)$$

Compatibility.   Due to the displacements at the ends of each bar, line *ACE* shown in Fig. 4–14c will move to the inclined position $A'C'E'$. From this position, the displacements of points *A, C,* and *E* can be related by proportional triangles. Thus the compatibility equation for these displacements is

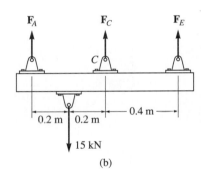

$$\frac{\Delta_A - \Delta_E}{0.8 \text{ m}} = \frac{\Delta_C - \Delta_E}{0.4 \text{ m}}$$

$$\Delta_C = \frac{1}{2}\Delta_A + \frac{1}{2}\Delta_E$$

Using the load–displacement relationship, Eq. 4–2, we have

$$\frac{F_C L}{(15 \text{ mm}^2)E_{st}} = \frac{1}{2}\left[\frac{F_A L}{(25 \text{ mm}^2)E_{st}}\right] + \frac{1}{2}\left[\frac{F_E L}{(25 \text{ mm}^2)E_{st}}\right]$$
$$F_C = 0.3F_A + 0.3F_E \qquad (3)$$

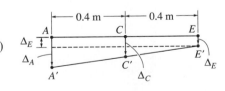

(c)

Solving Eqs. 1–3 simultaneously yields

$$F_A = 9.52 \text{ kN} \qquad\qquad Ans.$$
$$F_C = 3.46 \text{ kN} \qquad\qquad Ans.$$
$$F_E = 2.02 \text{ kN} \qquad\qquad Ans.$$

Fig. 4–14

Example 4–8

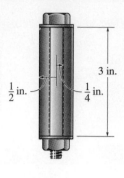

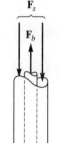

(a)

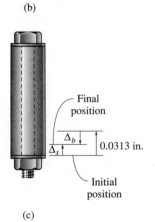

F_s

F_b

(b)

Final position

Δ_b

Δ_s

0.0313 in.

Initial position

(c)

Fig. 4–15

The bolt shown in Fig. 4–15a is made of an aluminum alloy and is tightened so it compresses a cylindrical spool made from a magnesium alloy. The spool has an outer radius of $\frac{1}{2}$ in. and it is assumed that both the inner radius of the spool and the radius of the bolt are $\frac{1}{4}$ in. The washers at the top and bottom of the spool are considered to be rigid and have a negligible thickness. Initially the nut is hand-tightened slightly; then, using a wrench, the nut is further tightened one-half turn. If the bolt has 16 threads per inch, determine the stress in the bolt. $E_{al} = 10(10^3)$ ksi, $E_{mg} = 6.5(10^3)$ ksi, $(\sigma_Y)_{al} = 70$ ksi, $(\sigma_Y)_{mg} = 40$ ksi.

SOLUTION

Equilibrium. The free-body diagram of a section of the bolt and the spool, Fig. 4–15b, is considered in order to relate the force in the bolt F_b to that in the spool, F_s. Equilibrium requires

$$+\uparrow \ \Sigma F_y = 0; \qquad\qquad F_b - F_s = 0 \qquad\qquad (1)$$

The problem is statically indeterminate since there are two unknowns in the equation.

Compatibility. Tightening the nut one-half turn advances it a distance of $(\frac{1}{2})(\frac{1}{16}$ in.$) = 0.0313$ in. along the bolt. This causes the bolt to *elongate* Δ_b and the spool to *shorten* Δ_s, Fig. 4–15c. As shown, compatibility requires

$$(+\uparrow) \qquad\qquad \Delta_s = 0.0313 \text{ in.} - \Delta_b$$

Assuming that the material has linear-elastic behavior, the load–displacement relationship is given by Eq. 4–2. Thus,

$$\frac{F_s(3 \text{ in.})}{\pi[(0.5 \text{ in.})^2 - (0.25 \text{ in.})^2][6.5(10^3) \text{ ksi}]} = 0.0313 \text{ in.} - \frac{F_b(3 \text{ in.})}{\pi(0.25 \text{ in.})^2[10(10^3) \text{ ksi}]}$$

$$0.8205F_s = 32.725 - 1.60F_b \qquad\qquad (2)$$

Solving Eqs. 1 and 2 simultaneously, we get

$$F_b = F_s = 13.52 \text{ kip}$$

The stresses in the bolt and spool are therefore

$$\sigma_b = \frac{F_b}{A_b} = \frac{13.52 \text{ kip}}{\pi(0.25 \text{ in.})^2} = 68.9 \text{ ksi} \qquad\qquad Ans.$$

$$\sigma_s = \frac{F_s}{A_s} = \frac{13.52 \text{ kip}}{\pi[(0.5 \text{ in.})^2 - (0.25 \text{ in.})^2]} = 23.0 \text{ ksi}$$

These stresses are less than the reported yield stress for each material, $(\sigma_Y)_{al} = 70$ ksi and $(\sigma_Y)_{mg} = 40$ ksi, and therefore this "elastic" analysis is valid.

PROBLEMS

4–22. The steel pipe is filled with concrete and subjected to a compressive force of 80 kN. Determine the stress in the concrete and the steel due to this loading. The pipe has an outer diameter of 80 mm and an inner diameter of 70 mm. $E_{st} = 200$ GPa, $E_c = 24$ GPa.

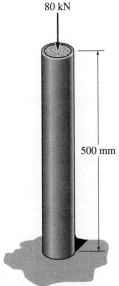

80 kN

500 mm

Prob. 4–22

4–23. The concrete column is reinforced using four steel reinforcing rods, each having a diameter of 18 mm. Determine the stress in the concrete and the steel if the column is subjected to an axial load of 800 kN. $E_{st} = 200$ GPa, $E_c = 25$ GPa.

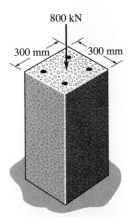

800 kN

300 mm   300 mm

Prob. 4–23

*4–24. The steel column, having a cross-sectional area of 15 in^2, is encased in concrete as shown. If an axial force of 50 kip is applied to the column, determine the compressive stress in the concrete and in the steel. How far does the column shorten? It has an original length of 6 ft. $E_{st} = 29(10^3)$ ksi, $E_c = 3.6(10^3)$ ksi.

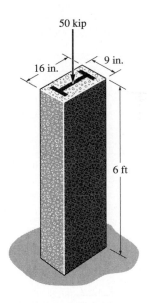

50 kip

16 in.   9 in.

6 ft

Prob. 4–24

4–25. The column is constructed from concrete and six steel reinforcing rods. If it is subjected to an axial force of 18 kip, determine the stress in the concrete and in each rod. Each rod has a diameter of 0.5 in. $E_{st} = 29(10^3)$ ksi, $E_c = 3.5(10^3)$ ksi.

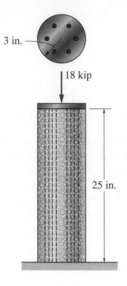

Prob. 4–25

4–26. The steel pipe has an outer radius of 20 mm and an inner radius of 15 mm. If it fits snugly between the fixed walls before it is loaded, determine the reaction at the walls when it is subjected to the load shown. $E_{st} = 200$ GPa.

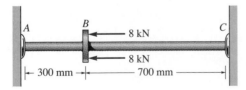

Prob. 4–26

4–27. The composite bar consists of a 20-mm-diameter steel segment AB and 50-mm-diameter brass end segments DA and CB. Determine the normal stress in each segment due to the applied 250-kN load. $E_{br} = 100$ GPa, $E_{st} = 200$ GPa.

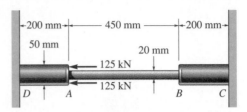

Prob. 4–27

*4–28. The steel post A is surrounded by a brass tube B. Both rest on the rigid surface. If a force of 5 kip is applied to the rigid cap, determine the normal stress developed in the post and the tube. $E_{st} = 29(10^3)$ ksi, $E_{br} = 15(10^3)$ ksi.

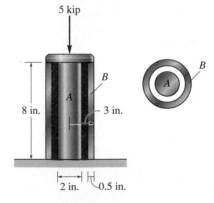

Prob. 4–28

4–29. The load of 2800 lb is to be supported by the two essentially vertical wires. If originally wire *AB* is 60 in. long and wire *AC* is 40 in. long, determine the force developed in each wire after the load is suspended. Each wire has a cross-sectional area of 0.02 in². E_{st} = 29(10³) ksi.

4–31. The assembly consists of a steel bolt and a brass tube. If the nut is drawn up snug against the tube so that L = 75 mm, then turned an additional amount so that it advances 2 mm on the bolt, determine the force in the bolt and the tube. The bolt has a diameter of 7 mm and the tube has a cross-sectional area of 100 mm². E_{st} = 200 GPa, E_{br} = 100 GPa.

Prob. 4–31

Prob. 4–29

4–30. The steel bolt *AB* has a diameter of 20 mm and passes through a steel sleeve that has an inner diameter of 40 mm and an outer diameter of 50 mm. The bolt and sleeve are secured to the rigid brackets as shown. If the bolt length is 150 mm and the sleeve length is 120 mm, determine the tension in the bolt when a force of 20 kN is applied to the brackets. E_{st} = 200 GPa.

*****4–32.** Two steel pipes, each having a cross-sectional area of 0.32 in², are screwed together using a union at *B* as shown. Originally the assembly is adjusted so that no load is on the pipe. If the union is then tightened so that its screw, having a lead of 0.15 in., undergoes two full turns, determine the normal stress developed in the pipe. Assume that the union at *B* and couplings at *A* and *C* are rigid. Neglect the size of the union. E_{st} = 29(10³) ksi. *Note:* The lead would cause the pipe, when *unloaded*, to shorten 0.15 in. when the union is rotated one revolution.

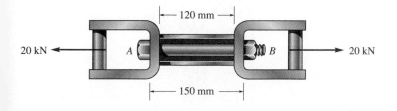

Prob. 4–30

Prob. 4–32

4–33. The rigid member is held in the position shown by three steel tie rods. Assuming that each rod has an unstretched length of 0.75 m and a cross-sectional area of 125 mm², determine the forces in the rods if a turnbuckle on rod *EF* undergoes one full turn. The lead of the screw is 1.5 mm. Neglect the size of the turnbuckle and assume that it is rigid. *Note:* The lead would cause the rod, when *unloaded,* to shorten 1.5 mm when the turnbuckle is rotated one revolution. E_{st} = 200 GPa.

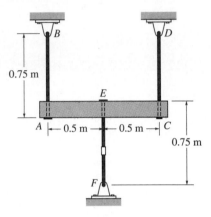

Prob. 4–33

4–35. The three bars are pinned together and subjected to the force **P.** If each bar has the same length *L,* cross-sectional area *A,* and modulus of elasticity *E,* determine the force in each bar.

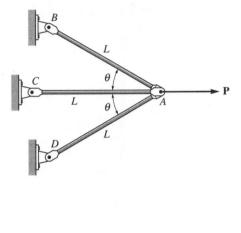

Prob. 4–35

4–34. Two steel wires are used to hoist the 650-lb engine. Originally, *AB* is 32 in. long and *A'B'* is 32.008 in. long. Determine the force supported by each wire when the engine is suspended from them. Each wire has a cross-sectional area of 0.01 in². E_{st} = 29(10³) ksi.

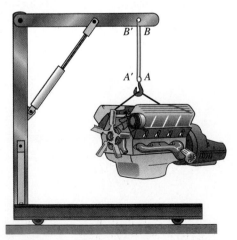

Prob. 4–34

*4–36. The three steel wires each have a diameter of 2 mm and unloaded lengths of L_{CA} = 1.60 m and $L_{AB} = L_{AD}$ = 2.00 m. Determine the force in each wire after the 150-kg mass is suspended from the ring at *A.* E_{st} = 200 GPa.

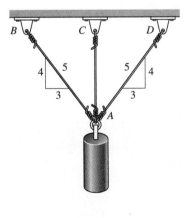

Prob. 4–36

4–37. The three suspender bars are made of the same material and have equal cross-sectional areas A. Determine the stress in each bar if the rigid beam ACE is subjected to the force **P**.

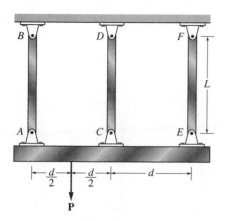

Prob. 4–37

4–38. The rigid block is supported by a symmetrical arrangement of bars of equal area A and length L. Bars AB and CD have a modulus of elasticity E_1 and bars EF and GH have a modulus of elasticity E_2. Determine the stress in each bar if a couple moment $\mathbf{M}_0$ is applied to the block.

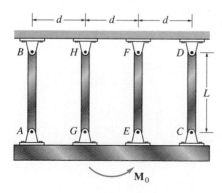

Prob. 4–38

4–39. The bracket is held to the wall using three steel bolts at B, C, and D. Each bolt has a diameter of 0.5 in. and an unstretched length of 4 in. If a force of 600 lb is placed on the bracket as shown, determine the force developed in each bolt. For the calculation, assume that the bolts carry no shear; rather, the vertical force of 600 lb is supported by the toe at A. Also, assume that the wall and bracket are rigid. $E_{st} = 29(10^3)$ ksi. A greatly exaggerated deformation of the bolts is shown.

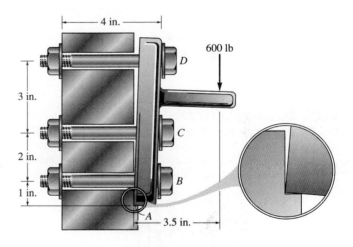

Prob. 4–39

*4–40. The rigid beam is supported by two posts, each having a width d and thickness d and a length L. If the modulus of elasticity for material A is E_A, and for material B it is E_B, determine the distance x for placement of the force **P** so that the beam remains horizontal.

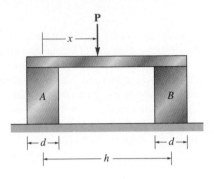

Prob. 4–40

4–41. The center post B of the assembly has an original length of 124.7 mm, whereas posts A and C have a length of 125 mm. If the caps on the top and bottom can be considered rigid, determine the axial stress in each post. The posts are made of aluminum and have a cross-sectional area of 400 mm². E_{al} = 70 GPa.

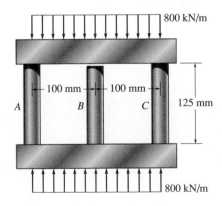

Prob. 4–41

4–42. The distributed loading is supported by the three suspender bars. AB and EF are made from aluminum and CD is made from steel. If each bar has a cross-sectional area of 450 mm², determine the maximum intensity w of the distributed loading so that an allowable stress of $(\sigma_{allow})_{st}$ = 180 MPa in the steel, and $(\sigma_{allow})_{al}$ = 94 MPa in the aluminum is not exceeded. E_{st} = 200 GPa, E_{al} = 70 GPa.

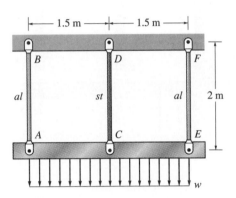

Prob. 4–42

4–43. The beam is pinned at A and supported by two aluminum rods, each having a diameter of 1 in. and a modulus of elasticity $E_{al} = 10(10^3)$ ksi. If the beam is assumed to be rigid and initially horizontal, determine the displacement of the end B when the force of 5 kip is applied.

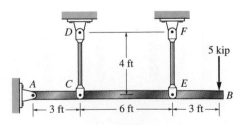

Prob. 4–43

*4–44.   The horizontal beam is assumed to be rigid and supports the distributed load shown. Determine the vertical reactions at the supports. Each support consists of a wooden post having a diameter of 120 mm and an unloaded (original) length of 1.40 m. Take $E_w = 12$ MPa.

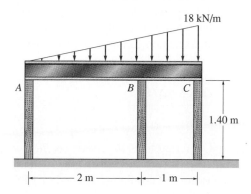

Prob. 4–44

4–45.   The rigid link is supported by a pin at A and two steel wires, each having an unstretched length of 12 in. and cross-sectional area of 0.0125 in^2. Determine the force developed in the wires when the link supports the vertical load of 350 lb. $E_{st} = 29(10^3)$ ksi.

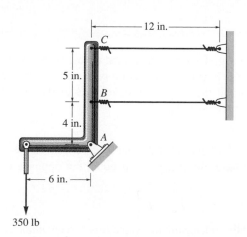

Prob. 4–45

4–46.   The rigid bar is supported by the two short wooden posts and a spring. If each of the posts has an unloaded length of 500 mm and a cross-sectional area of 800 mm^2, and the spring has a stiffness of $k = 1.8$ MN/m and an unstretched length of 520 mm, determine the force in each post after the load is applied to the bar. $E_w = 11$ GPa.

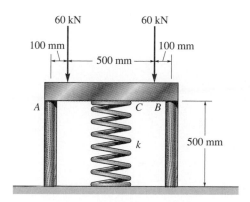

Prob. 4–46

4–47.   The spring has an unstretched length of 250 mm and stiffness $k = 400$ kN/m. If it is compressed and placed over the 200-mm-long portion of the aluminum bar AB and released, determine the force that the bar exerts on the wall at A. Before loading there is a gap of 0.1 mm between the bar and the wall at B. The bar is fixed to the wall at A. $E_{al} = 70$ GPa.

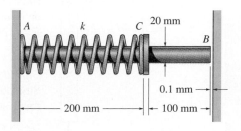

Prob. 4–47

4.5 Thermal Stress

A change in temperature can cause a material to change its dimensions. If the temperature increases, generally a material expands, whereas if the temperature decreases, the material will contract. Ordinarily this expansion or contraction is *linearly* related to the temperature increase or decrease that occurs. If this is the case, and the material is homogeneous and isotropic, it has been found from experiment that the deformation can be calculated using the formula

$$\Delta L = \alpha \Delta T L \qquad\qquad (4\text{--}4)$$

where

α = a property of the material, referred to as the *linear coefficient of thermal expansion*. The units measure strain per degree of temperature. They are $1/°F$ (Fahrenheit) in the Foot-Pound-Second or FPS system, and $1/°C$ (Celsius) or $1/°K$ (Kelvin) in the SI system

ΔT = the change in temperature

L = the original length of the member

ΔL = the change in length of the member

    The change in length of a *statically determinate* member can readily be computed using Eq. 4–4, since the member is free to expand or contract when it undergoes a temperature change. However, in a *statically indeterminate* member these thermal displacements can be constrained by the supports, producing *thermal stresses* that must be considered in design.

    Computations of these thermal stresses can be made using the procedure for analysis outlined in the previous section. The following examples illustrate some applications.

Example 4–9

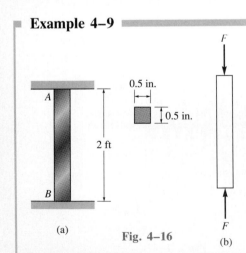

(a)

Fig. 4–16

(b)

The steel bar shown in Fig. 4–16 is constrained to just fit between two fixed supports when $T_1 = 60°F$. If the temperature is raised to $T_2 = 120°F$, determine the average normal thermal stress developed in the bar. Take $E_{st} = 29(10^3)$ ksi and $\alpha_{st} = 6.5(10^{-6})/°F$.

SOLUTION

Equilibrium.   The free-body diagram of the bar is shown in Fig. 4–16b. Since there is no external load, the force at A is equal but opposite to the force acting at B; that is,

$$+\uparrow\ \Sigma F_y = 0; \qquad\qquad F_A = F_B = F$$

    The problem is statically indeterminate, since this force cannot be determined from equilibrium.

Example 4–9 (*Continued*)

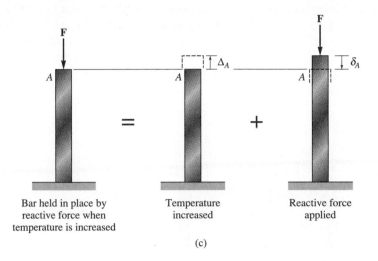

Bar held in place by
reactive force when
temperature is increased

Temperature
increased

Reactive force
applied

(c)

Compatibility.   Using the principle of superposition, Fig. 4–16c, the redundant support at A is removed, and the thermal displacement Δ_A at A
occurs. The force $\mathbf{F}$, developed at the redundant, A, then pushes the bar δ_A
back to its original position; i.e., the compatibility condition at A becomes

$$(+\uparrow) \qquad\qquad 0 = \Delta_A - \delta_A$$

Applying the thermal and load–displacement relationships, Eq. 4–4
and Eq. 4–2, we have

$$0 = \alpha \Delta T L - \frac{FL}{AE}$$

Thus,

$$F = \alpha \Delta T A E$$
$$= [6.5(10^{-6})/°F](120°F - 60°F)(0.5 \text{ in.})^2[29(10^3)] \text{ kip/in}^2$$
$$= 2.83 \text{ kip}$$

From the magnitude of $\mathbf{F}$ it should be apparent that changes in temperature can cause large reaction forces in statically indeterminate members.

Since $\mathbf{F}$ also represents the internal axial force within the bar, the
average normal compressive (thermal) stress is thus

$$\sigma = \frac{F}{A} = \frac{2.83 \text{ kip}}{(0.5 \text{ in.})^2} = 11.3 \text{ ksi} \qquad\qquad \textit{Ans.}$$

Example 4-10

The rigid bar shown in Fig. 4–17a is fixed to the top of the three posts made of steel and aluminum. The posts each have a length of 250 mm when no load is applied to the bar and the temperature is $T_1 = 20°C$. Determine the force supported by each post if the bar is subjected to a uniform distributed load of 150 kN/m and the temperature is raised to $T_2 = 80°C$. The diameter of each post and its material properties are listed in the figure.

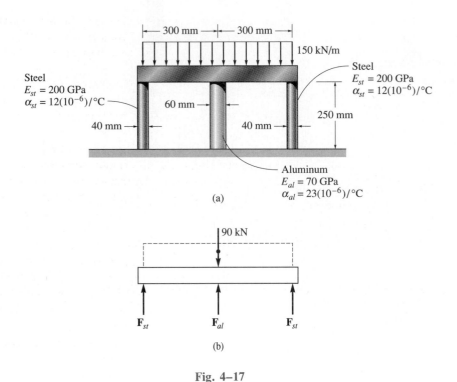

(a)

(b)

Fig. 4–17

SOLUTION

Equilibrium. The free-body diagram of the bar is shown in Fig. 4–17b. Moment equilibrium about the bar's center requires the forces in the steel posts to be equal. Summing forces on the free-body diagram, we have

$$+\uparrow \ \Sigma F_y = 0; \qquad 2F_{st} + F_{al} - 90(10^3) \ \text{N} = 0 \qquad (1)$$

Compatibility. Due to load, geometry, and material symmetry, the top of each post is displaced by an equal amount. Hence, the compatibility equation is

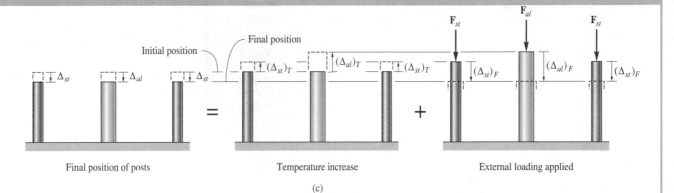

Final position of posts   Temperature increase   External loading applied

(c)

$(+\downarrow)$ $\Delta_{st} = \Delta_{al}$ (2)

We will now consider the steel and aluminum posts separately and apply the principle of superposition, Fig. 4–17c. Here the rigid bar will be treated as the redundant. Hence, the actual displacement at the top of each post, or the displacement of the redundant, is equal to the displacement of the top of each post caused by temperature, plus the displacement of the top of each post caused by the axial force that the redundant creates in each post. Thus, for a steel and aluminum post we have

$(+\downarrow)$ $\Delta_{st} = -(\Delta_{st})_T + (\Delta_{st})_F$

$(+\downarrow)$ $\Delta_{al} = -(\Delta_{al})_T + (\Delta_{al})_F$

Applying Eq. 2 gives

$$-(\Delta_{st})_T + (\Delta_{st})_F = -(\Delta_{al})_T + (\Delta_{al})_F$$

Using Eqs. 4–2 and 4–4, we get

$$-[12(10^{-6})/°C](80°C - 20°C)(0.250 \text{ m}) + \frac{F_{st}(0.250 \text{ m})}{\pi(0.020 \text{ m})^2[200(10^9) \text{ N/m}^2]}$$

$$= -[23(10^{-6})/°C](80°C - 20°C)(0.250 \text{ m}) + \frac{F_{al}(0.250 \text{ m})}{\pi(0.03 \text{ m})^2[70(10^9) \text{ N/m}^2]}$$

$$F_{st} = 1.270F_{al} - 165.9(10^3) \qquad (3)$$

To be *consistent*, all numerical data has been expressed in terms of newtons, meters, and degrees Celsius. Solving Eqs. 1 and 3 simultaneously yields

$$F_{st} = -14.6 \text{ kN} \qquad\qquad Ans.$$

$$F_{al} = 119 \text{ kN} \qquad\qquad Ans.$$

The negative value for F_{st} indicates that this force acts opposite to that shown in Fig. 4–17b. In other words, the steel posts are in tension and the aluminum post is in compression.

Example 4–11

An aluminum tube having a cross-sectional area of 600 mm^2 is used as a sleeve for a steel bolt having a cross-sectional area of 400 mm^2, Fig. 4–18a. When the temperature is $T_1 = 15°C$, the nut holds the assembly in a snug position such that the axial force in the bolt is negligible. If the temperature increases to $T_2 = 80°C$, determine the average normal stress in the bolt and sleeve. Take $E_{st} = 200$ GPa, $\alpha_{st} = 12(10^{-6})/°C$, and $E_{al} = 70$ GPa, $\alpha_{al} = 23(10^{-6})/°C$.

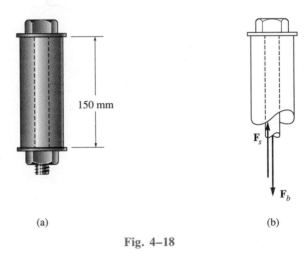

(a)   (b)

Fig. 4–18

SOLUTION

Equilibrium. A free-body diagram of a sectioned segment of the assembly is shown in Fig. 4–18b. The forces F_b and F_s are produced since the bolt and sleeve have different coefficients of thermal expansion and therefore will expand by different amounts when the temperature is increased. The problem is statically indeterminate since these forces cannot be determined from equilibrium. However, it is required that

$$+\uparrow \ \Sigma F_y = 0; \qquad\qquad F_s = F_b \qquad\qquad (1)$$

Compatibility. Using the principle of superposition, Fig. 4–18c, the nut and washer (redundant) are removed, and the temperature increase causes the bolt and sleeve to expand Δ_b and Δ_s, respectively. The redundant forces F_b and F_s return these thermal displacements to the final position. Hence, the compatibility condition becomes

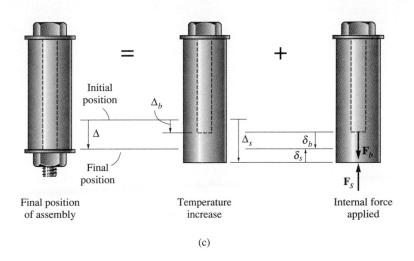

Final position   Temperature   Internal force
of assembly   increase   applied

(c)

$(+\downarrow)$ $\qquad\qquad\qquad \Delta = \Delta_b + \delta_b = \Delta_s - \delta_s$

Applying Eqs. 4–2 and 4–4, we have

$[12(10^{-6})/°\text{C}](80°\text{C} - 15°\text{C})(0.150 \text{ m})$

$$+ \frac{F_b(0.150 \text{ m})}{(400 \text{ mm}^2)(10^{-6} \text{ m}^2/\text{mm}^2)[200(10^9) \text{ N/m}^2]}$$

$= [23(10^{-6})/°\text{C}](80°\text{C} - 15°\text{C})(0.150 \text{ m})$

$$- \frac{F_s(0.150 \text{ m})}{600 \text{ mm}^2(10^{-6} \text{ m}^2/\text{mm}^2)[70(10^9) \text{ N/m}^2]}$$

Using Eq. 1 and solving gives

$$F_s = F_b = 19.67 \text{ kN}$$

The average normal stress in the bolt and sleeve is therefore

$$\sigma_b = \frac{19.67 \text{ kN}}{400 \text{ mm}^2(10^{-6} \text{ m}^2/\text{mm}^2)} = 49.2 \text{ MPa} \qquad \textit{Ans.}$$

$$\sigma_s = \frac{19.67 \text{ kN}}{600 \text{ mm}^2(10^{-6} \text{ m}^2/\text{mm}^2)} = 32.8 \text{ MPa} \qquad \textit{Ans.}$$

Since linear elastic material behavior was assumed in this analysis, the calculated stresses should be checked to make sure that they do not exceed the proportional limits for the material.

PROBLEMS

*4–48. The 40-ft-long steel rails on a train track are laid with a small gap between them to allow for thermal expansion. Determine the required gap δ so that the rails just touch one another when the temperature is increased from $T_1 = -30°F$ to $T_2 = 70°F$. Using this gap, what would be the axial force in the rails if the temperature were to rise to $T_3 = 100°F$? The cross-sectional area of each rail is 4.25 in². $E_{st} = 29(10^3)$ ksi, $\alpha_{st} = 6.5(10^{-6})/°F$.

Prob. 4–48

4–49. A 6-ft-long steam pipe is connected directly to two turbines A and B as shown. The pipe has an outer radius of 1.5 in. and a wall thickness of 0.25 in. The connection was made at $T_1 = 85°F$. If the turbines' points of attachment are assumed rigid, determine the force the pipe exerts on the turbines when the steam and thus the pipe reach a temperature of $T_2 = 250°F$. $E_{st} = 29(10^3)$ ksi, $\alpha_{st} = 6.5(10^{-6})/°F$.

4–50. A 6-ft-long steam pipe is connected directly to two turbines A and B as shown. The pipe has an outer diameter of 3 in. and a wall thickness of 0.25 in. The connection was made at $T_1 = 85°F$. If the turbines' points of attachment are assumed to have a stiffness of $k = 80(10^3)$ kip/in., determine the force the pipe exerts on the turbines when the steam and thus the pipe reach a temperature of $T_2 = 250°F$. $E_{st} = 29(10^3)$ ksi, $\alpha_{st} = 6.5(10^{-6})/°F$.

Probs. 4–49/4–50

4.51. Three bars each made of different materials are connected together and placed between two walls when the temperature is $T_1 = 12°C$. Determine the force exerted on the (rigid) supports when the temperature becomes $T_2 = 18°C$. The material properties and cross-sectional area of each bar are given in the figure.

Steel	Brass	Copper
$E_{st} = 200$ GPa	$E_{br} = 100$ GPa	$E_{cu} = 120$ GPa
$\alpha_{st} = 12(10^{-6})/°C$	$\alpha_{br} = 21(10^{-6})/°C$	$\alpha_{cu} = 17(10^{-6})/°C$

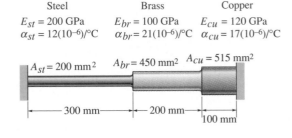

Prob. 4–51

*4–52. The two circular rod segments, one of aluminum and the other of copper, are fixed to the rigid walls such that there is a gap of 0.008 in. between them when $T_1 = 60°F$. What larger temperature T_2 is required in order to just close the gap? Each rod has a diameter of 1.25 in. $\alpha_{al} = 13(10^{-6})/°F$, $E_{al} = 10(10^3)$ ksi, $\alpha_{cu} = 9.4(10^{-6})/°F$, $E_{cu} = 18(10^3)$ ksi. Determine the stress in each rod if $T_2 = 200°F$.

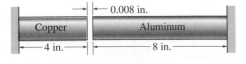

Prob. 4–52

4–53. A 0.25-in.-diameter steel rivet having a temperature of 1500°F is secured between two plates such that at this temperature it is 2 in. long and exerts a clamping force of 250 lb between the plates. Determine the approximate clamping force between the plates when the rivet cools to 70°F. For the calculation, assume that the heads of the rivet and the plates are rigid. Take $\alpha_{st} = 8(10^{-6})/°F$, $E_{st} = 29(10^3)$ ksi. Is the result a conservative estimate of the actual answer? Why or why not?

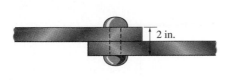

Prob. 4–53

4–54. The center rod CD of the assembly is heated from $T_1 = 30°C$ to $T_2 = 180°C$ using electrical resistance heating. At the lower temperature T_1 the gap between C and the rigid bar is 0.7 mm. Determine the force in rods AB and EF caused by the increase in temperature. Rods AB and EF are made of steel, and each has a cross-sectional area of 125 mm². CD is made of aluminum and has a cross-sectional area of 375 mm². $E_{st} = 200$ GPa, $E_{al} = 70$ GPa, and $\alpha_{al} = 23(10^{-6})/°C$.

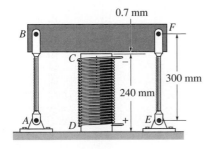

Prob. 4–54

4–55. The center rod CD of the assembly is heated from $T_1 = 30°C$ to $T_2 = 180°C$ using electrical resistance heating. Also, the two end rods AB and EF are heated from $T_1 = 30°C$ to $T_2 = 50°C$. At the lower temperature T_1 the gap between C and the rigid bar is 0.7 mm. Determine the force in rods AB and EF caused by the increase in temperature. Rods AB and EF are made of steel, and each has a cross-sectional area of 125 mm². CD is made of aluminum and has a cross-sectional area of 375 mm². $E_{st} = 200$ GPa, $E_{al} = 70$ GPa, $\alpha_{st} = 12(10^{-6})/°C$, and $\alpha_{al} = 23(10^{-6})/°C$.

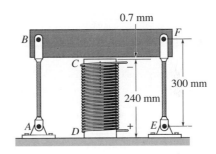

Prob. 4–55

*4–56. The aluminum bar has a diameter of 0.6 in. and is attached to the rigid supports at A and B when $T_1 = 80°F$. If the temperature becomes $T_2 = -10°F$, and an axial force of 16 lb is applied to the rigid collar as shown, determine the reactions at A and B. $E_{al} = 10(10^3)$ ksi, $\alpha_{al} = 13(10^{-6})/°F$.

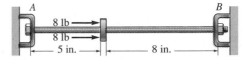

Prob. 4–56

4–57. The steel bolt has a diameter of 7 mm and fits through an aluminum sleeve as shown. The sleeve has an inner diameter of 8 mm and an outer diameter of 10 mm. The nut at A is adjusted so that it just presses up against the sleeve. If the assembly is originally at a temperature of $T_1 = 20°C$ and then is heated to a temperature of $T_2 = 100°C$, determine the stress developed in the bolt and the sleeve. $E_{st} = 200$ GPa, $E_{al} = 70$ GPa, $\alpha_{st} = 14(10^{-6})/°C$, $\alpha_{al} = 23(10^{-6})/°C$.

4–59. The bar has a cross-sectional area A, length L, modulus of elasticity E, and coefficient of thermal expansion α. The temperature of the bar changes uniformly from an original temperature of T_A to T_B so that at any point x along the bar $T = T_A + x (T_B - T_A)/L$. Determine the force it exerts on the rigid walls. Initially no axial force is in the bar.

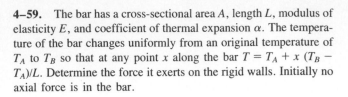

Prob. 4–57

Prob. 4–59

4–58. The three bars are made of steel and form a pin-connected truss. If the truss is constructed when $T_1 = 30°F$, determine the force in each bar when $T_2 = 100°F$. Each bar has a cross-sectional area of 2 in². $E_{st} = 29(10^5)$ ksi, $\alpha_{st} = 6.5(10^{-6})/°F$.

*4–60. The assembly consists of two bars AB and CD of the same material having a modulus of elasticity E_1 and coefficient of thermal expansion α_1, and a bar EF having a modulus of elasticity E_2 and coefficient of thermal expansion α_2. All the bars have the same length L and cross-sectional area A. If the rigid block is originally horizontal, at temperature T_1, determine the angle it makes with the horizontal when the temperature is increased to T_2.

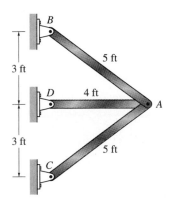

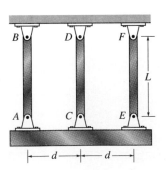

Prob. 4–58

Prob. 4–60

4.6 Stress Concentrations

In Sec. 4.1 it was pointed out that when an axial force is applied to a member it creates a complex stress distribution within a localized region of the point of load application. Such typical stress distributions are shown in Fig. 4–1. Not only do complex stress distributions arise just under a concentrated loading, however, they also arise at sections where the member's cross-sectional area changes. For example, consider the bar in Fig. 4–19a, which is subjected to an axial force P. Here it can be seen that the once horizontal and vertical grid lines deflect into an irregular pattern around the hole centered in the bar. The maximum normal stress in the bar occurs on section a–a, which is taken through the bar's *smallest* cross-sectional area. Provided the material behaves in a linear-elastic manner, the stress distribution acting on this section can be determined either from a mathematical analysis using the theory of elasticity, or experimentally by measuring the strain normal to section a–a and then computing the stress using Hooke's law, $\sigma = E\epsilon$. Regardless of the method used, the general shape of the stress distribution will be like that shown in Fig. 4–19b. In a similar manner, if the bar has a reduction in its cross section, achieved using shoulder fillets, Fig. 4–20a, then again the maximum normal stress in the bar will occur at the *smallest* cross-sectional area, section a–a. Here the stress distribution is shown in Fig. 4–20b.

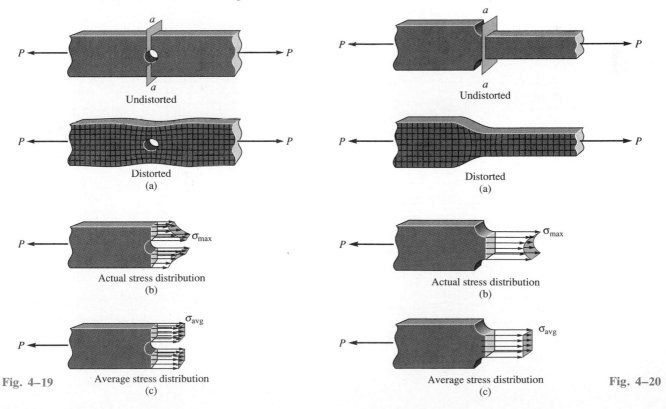

Undistorted

Distorted
(a)

Actual stress distribution
(b)

Average stress distribution
(c)

Fig. 4–19

Undistorted

Distorted
(a)

Actual stress distribution
(b)

Average stress distribution
(c)

Fig. 4–20

In both of the above cases, *force equilibrium* requires the magnitude of the *resultant force* developed by the stress distribution to be equal to *P*. In other words, from Eq. 1–6,

$$P = \int_A \sigma \, dA \qquad (4-5)$$

As stated in Sec. 1.4, this integral *graphically* represents the *volume* under each of the stress-distribution diagrams shown in Fig. 4–19b or 4–20b. Furthermore, moment equilibrium requires each stress distribution to be symmetrical over the cross section, so that **P** must pass through the *centroid* of each *volume*.

In engineering practice, though, the actual stress distribution does *not* have to be determined. Instead, only the *maximum stress* at these sections must be known, and the member is then designed to resist this stress when the axial load **P** is applied. In cases where a member's cross-sectional area changes, such as those discussed above, specific values of the maximum normal stress at the critical section can be determined by experimental methods or by advanced mathematical techniques using the theory of elasticity. The results of these investigations are usually reported in graphical form using a *stress-concentration factor K*. We define *K* as a ratio of the maximum stress to the average stress acting at the smallest cross section; i.e.,

$$K = \frac{\sigma_{\max}}{\sigma_{\text{avg}}} \qquad (4-6)$$

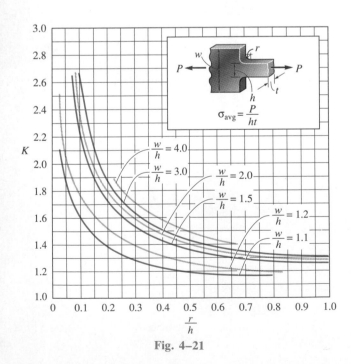

Fig. 4–21

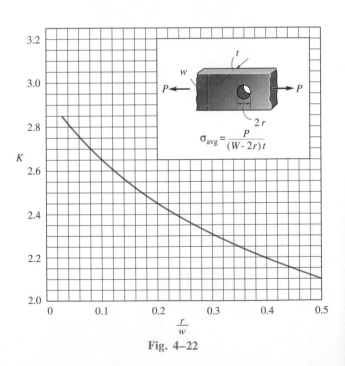

Fig. 4–22

Provided K is known, and the average normal stress has been calculated from $\sigma_{avg} = P/A$, where A is the *smallest* cross-sectional area, Figs. 4–19c and 4–20c, then from the above equation the maximum stress at the cross section is $\sigma_{max} = K(P/A)$.

Specific values of K are generally reported in graphical form in handbooks related to stress analysis. Examples of these graphs for the bars shown in Figs. 4–19a and 4–20a are given in Figs. 4–21 and 4–22, respectively.* In particular, note that K is independent of the bar's material properties; rather, it *depends* only on the bar's *geometry* and the type of discontinuity. As the size r of the discontinuity is *decreased*, the stress concentration is increased. For example, if a bar requires a change in cross section, theoretically it has been determined that a sharp corner, Fig. 4–23a, produces a stress-concentration factor greater than 3. In other words, the maximum stress will be three times greater than the average stress on the smallest cross section. However, this can be reduced to, say, 1.5 by introducing a fillet, Fig. 4–23b. And a still further reduction can be made by means of small grooves or holes placed at the transition, Fig. 4–23c and 4–23d. In all of these cases the designs help to reduce the rigidity of the material surrounding the corners, so that both the strain and the stress are more evenly spread throughout the bar.

The stress-concentration factors given in Figs. 4–21 and 4–22 were determined on the basis of a static loading, with the assumption that the stress in the material does not exceed the proportional limit. If the material is *very brittle*, the proportional limit may be at the rupture stress, and so for this material, failure will begin *at* the point of stress concentration when the proportional limit is reached. Essentially what happens is that a crack begins to form at this point and a higher stress concentration will develop at the *tip* of this crack. This, in turn, causes the crack to progress over the cross section, resulting in sudden fracture. For this reason, it very important to use stress-concentration factors in design when using brittle materials. On the other hand, if the material is ductile and subjected to a static load, designers usually neglect using stress-concentration factors since any stress that exceeds the proportional limit will not result in a crack. Instead, the material will have reserve strength due to yielding and strain-hardening. In the next section we will discuss the effects caused by this phenomenon.

Stress concentrations are also responsible for many failures of structural members or mechanical elements subjected to *fatigue loadings*. For these cases, a stress concentration will cause the material to crack if the stress exceeds the material's endurance limit, whether or not the material is ductile or brittle. What happens is that the material *localized* at the tip of the crack remains in a *brittle state*, and so the crack continues to grow, leading to a progressive fracture. Consequently, engineers involved in the design of such members must seek ways to limit the amount of damage that can be caused by fatigue.

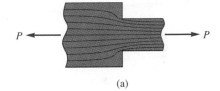

(a)

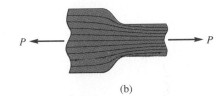

(b)

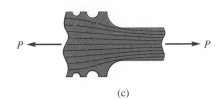

(c)

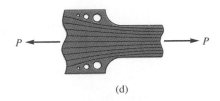

(d)

*See Lipson, C. and Juvinall, R.C., *Handbook of Stress and Strength,* Macmillan, 1963.

Fig. 4–23

Example 4–12

A steel bar has the dimensions shown in Fig. 4–24. If the allowable stress is $\sigma_{allow} = 16.2$ ksi, determine the largest axial force P that the bar can carry.

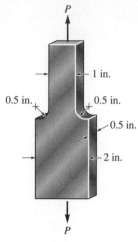

P

1 in.

0.5 in.   0.5 in.

0.5 in.

2 in.

P

Fig. 4–24

SOLUTION

Because there is a shoulder fillet, the stress-concentration factor can be determined using the graph in Fig. 4–21. Calculating the necessary geometric parameters yields

$$\frac{r}{h} = \frac{0.5 \text{ in.}}{1 \text{ in.}} = 0.50$$

$$\frac{w}{h} = \frac{2 \text{ in.}}{1 \text{ in.}} = 2$$

Thus, from the graph,

$$K = 1.4$$

Computing the average normal stress at the *smallest* cross section, we have

$$\sigma_{avg} = \frac{P}{(1 \text{ in.})(0.5 \text{ in.})} = 2P$$

Applying Eq. 4–6 with $\sigma_{allow} = \sigma_{max}$ yields

$$\sigma_{allow} = K\sigma_{avg}$$
$$16.2 \text{ ksi} = 1.4(2P)$$
$$P = 5.79 \text{ kip} \qquad\qquad \textit{Ans.}$$

Example 4–13

The steel strap shown in Fig. 4–25 is subjected to an axial load of 80 kN. Determine the maximum normal stress developed in the strap, and the displacement of one end of the strap with respect to the other end. The steel has a yield stress of $\sigma_Y = 700$ MPa, and $E_{st} = 200$ GPa.

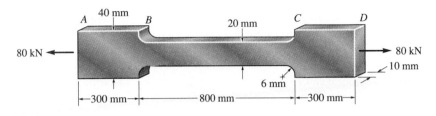

Fig. 4–25

SOLUTION

Maximum Normal Stress.   By inspection, the maximum normal stress occurs at the smaller cross section, where the shoulder fillet begins at B or C. The stress-concentration factor is determined from Fig. 4–21. We require

$$\frac{r}{h} = \frac{6 \text{ mm}}{20 \text{ mm}} = 0.3; \qquad \frac{w}{h} = \frac{40 \text{ mm}}{20 \text{ mm}} = 2$$

Thus, $K = 1.6$.

The maximum stress is therefore

$$\sigma_{max} = K \frac{P}{A} = 1.6 \frac{80(10^3) \text{ N}}{(0.02 \text{ m})(0.01 \text{ m})} = 640 \text{ MPa} \qquad \textit{Ans.}$$

Notice that the material remains elastic, since 640 MPa $< \sigma_Y = 700$ MPa.

Displacement.   Here we will neglect the localized deformations surrounding the applied load and at the sudden change in cross section of the shoulder fillet (Saint-Venant's principle). Applying Eq. 4–3, we have

$$\Delta_{A/D} = \sum \frac{PL}{AE} = 2\left\{ \frac{80(10^3) \text{ N}(0.3 \text{ m})}{(0.04 \text{ m})(0.01 \text{ m})[200(10^9) \text{ N/m}^2]} \right\}$$

$$+ \left\{ \frac{80(10^3) \text{ N}(0.8 \text{ m})}{(0.02 \text{ m})(0.01 \text{ m})[200(10^9) \text{ N/m}^2]} \right\}$$

$$\Delta_{A/D} = 2.20 \text{ mm} \qquad \textit{Ans.}$$

4.7 Inelastic Axial Deformation

Up to this point we have only considered loadings that cause the material of a member to behave elastically. Sometimes, however, a member may be designed so that the loading causes the material to yield and thereby permanently deform. Such members are often made from a highly ductile metal such as annealed low-carbon steel having a stress–strain diagram that can be *modeled* as shown in Fig. 4–26b. A material that exhibits this behavior is referred to as being *elastic-perfectly plastic or elastoplastic*.

To illustrate physically how such a material behaves, consider the bar in Fig. 4–26a, which is subjected to the axial load **P**. If the load causes an *elastic stress* $\sigma = \sigma_1$ to be developed in the bar, then applying Eq. 4–5, *equilibrium* requires $P = \int \sigma_1 \, dA = \sigma_1 A$. Furthermore, the stress σ_1 causes the bar to strain ϵ_1 as indicated on the stress–strain diagram, Fig. 4–26b. If P is now increased to P_p such that it causes yielding of the material, that is, $\sigma = \sigma_Y$, then again $P_p = \int \sigma_Y \, dA = \sigma_Y A$. The load P_p is called the *plastic load* since it represents the maximum load that can be supported by an elastoplastic material. For this case, the strains are *not* uniquely defined. Instead, at the instant σ_Y is attained, the bar is *first* subjected to the yield strain ϵ_Y, Fig. 4–26b, after which the bar will *continue to yield* (or elongate) such that the strains ϵ_2, then ϵ_3, etc., are generated. Since our "model" of the material exhibits perfectly plastic material behavior, this elongation will continue indefinitely with no increase in load. In reality, however, the material will, after some yielding, actually begin to strain-harden so that the extra strength it attains will *stop* any further straining. As a result, any design based on this behavior will be safe, since strain-hardening provides the potential for the material to support an *additional* load if necessary.

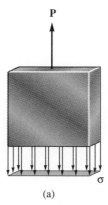

(a)

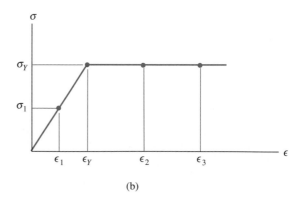

(b)

Fig. 4–26

Consider now the case of a bar having a hole through it as shown in Fig. 4–27a. As the magnitude of **P** is increased, a stress concentration occurs in the material near the hole, along section a–a. The stress here will reach a maximum value of, say, $\sigma_{max} = \sigma_1$ and have a corresponding *elastic strain* of ϵ_1, Fig. 4–27b. The stresses and corresponding strains at other points along the cross section will be smaller, as indicated by the stress distribution shown in Fig. 4–27c. As expected, equilibrium requires $P = \int \sigma \, dA$. In other words, P is geometrically equivalent to the "volume" contained within the stress distribution. If the load is now increased to P', so that $\sigma_{max} = \sigma_Y$, then the material will begin to yield outward from the hole, until the equilibrium condition $P' = \int \sigma \, dA$ is satisfied, Fig. 4–27d. As shown, this produces a stress distribution that has a geometrically *greater* "volume" than that shown in Fig. 4–27c. An even further increase in load will cause the material over the *entire cross section* to yield eventually, until *no greater load* can be sustained by the bar. This *plastic load* P_p is shown in Fig. 4–27e. It can be calculated from the equilibrium condition

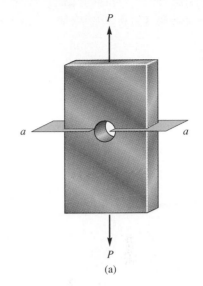

(a)

$$P_p = \int_A \sigma_Y \, dA = \sigma_Y A \qquad (4\text{--}7)$$

Here σ_Y is the yield stress and A is the bar's cross-sectional area at section a–a.

The following examples illustrate numerically how these concepts apply to other types of problems for which the material has elastoplastic behavior.

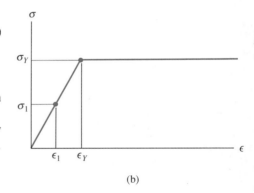

(b)

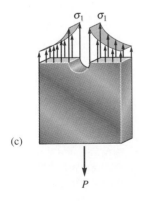

(c)

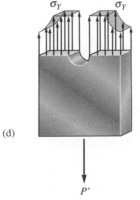

(d)

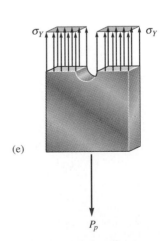

(e)

Fig. 4–27

Example 4–14

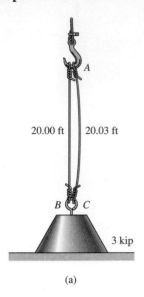

20.00 ft   20.03 ft

3 kip

(a)

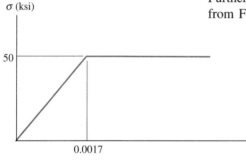

σ (ksi)

50

0.0017

ϵ (in./in.)

(b)

T_{AB} T_{AC}

3 kip

(c)

Fig. 4–28

Two steel wires are used to lift the weight of 3 kip, Fig. 4–28a. Wire AB has an unstretched length of 20.00 ft and wire AC has an unstretched length of 20.03 ft. If each wire has a cross-sectional area of 0.05 in², and the steel can be considered elastic-perfectly plastic as shown by the σ–ϵ graph in Fig. 4–28b, determine the force in each wire and its elongation.

SOLUTION

By inspection, wire AB begins to carry the weight when the hook is lifted. However, if this wire stretches more than 0.03 ft, the load is then carried by both wires. For this to occur the strain in wire AB must be

$$\epsilon_{AB} = \frac{0.03 \text{ ft}}{20 \text{ ft}} = 0.0015$$

which is less than the maximum elastic strain, $\epsilon_Y = 0.0017$, Fig. 4–28b. Furthermore, the stress in wire AB when this happens can be determined from Fig. 4–28b by proportion; i.e.,

$$\frac{0.0017}{50 \text{ ksi}} = \frac{0.0015}{\sigma_{AB}}$$

$$\sigma_{AB} = 44.12 \text{ ksi}$$

As a result, the force in the wire is thus

$$F_{AB} = \sigma_{AB} A = (42.12 \text{ ksi})(0.05 \text{ in}^2) = 2.20 \text{ kip}$$

Since the weight to be supported is 3 kip, we can conclude that both wires must be used for support.

Once the weight is supported, the stress in the wires depends on the corresponding strain. There are three possibilities, namely, the strains in both wires are elastic, wire AB is plastically strained while wire AC is elastically strained, or both wires are plastically strained. We will begin by assuming that both wires remain *elastic*. Investigation of the free-body diagram of the suspended weight, Fig. 4–28c, indicates that the problem is statically indeterminate. The equation of equilibrium is

$$+\uparrow \ \Sigma F_y = 0; \qquad T_{AB} + T_{AC} - 3 \text{ kip} = 0 \qquad (1)$$

Since AC is 0.03 ft longer than AB, then from Fig. 4–28d, compatibility of displacement of the ends B and C requires that

$$\Delta_{AB} = 0.03 \text{ ft} + \Delta_{AC} \qquad (2)$$

The modulus of elasticity, Fig. 4–28b, is $E_{st} = 50 \text{ ksi}/0.0017 = 29.4(10^3)$ ksi. Since this is an elastic analysis, the load–displacement relationship is $\Delta = PL/AE$, and therefore

$$\frac{T_{AB}(20.00 \text{ ft})(12 \text{ in./ft})}{(0.05 \text{ in}^2)[29.4(10^3) \text{ ksi}]} = 0.03 \text{ ft}(12 \text{ in./ft}) + \frac{T_{AC}(20.03 \text{ ft})(12 \text{ in./ft})}{(0.05 \text{ in}^2)[29.4(10^3) \text{ ksi}]}$$

$$20.00 T_{AB} = 44.11 + 20.03 T_{AC} \qquad (3)$$

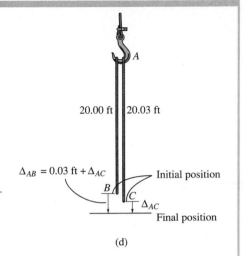

(d)

Solving Eqs. 1 and 3, we have

$$T_{AB} = 2.60 \text{ kip}$$
$$T_{AC} = 0.400 \text{ kip}$$

The stress in wire AB is thus

$$\sigma_{AB} = \frac{2.60 \text{ kip}}{0.05 \text{ in}^2} = 52.0 \text{ ksi}$$

This stress is greater than the maximum elastic stress allowed ($\sigma_Y = 50$ ksi), and therefore wire AB becomes plastically strained and supports its maximum load of

$$T_{AB} = 50 \text{ ksi} \ (0.05 \text{ in}^2) = 2.50 \text{ kip} \qquad \textit{Ans.}$$

From Eq. 1,

$$T_{AC} = 0.500 \text{ kip} \qquad \textit{Ans.}$$

Note that wire AC remains elastic since the stress in the wire is $\sigma_{AC} = 0.500 \text{ kip}/0.05 \text{ in}^2 = 10 \text{ ksi} < 50$ ksi. The corresponding elastic strain is determined by proportion, Fig. 4–28b; i.e.,

$$\frac{\epsilon_{AC}}{10 \text{ ksi}} = \frac{0.0017}{50 \text{ ksi}}$$

$$\epsilon_{AC} = 0.000340$$

The elongation of AC is thus

$$\Delta_{AC} = (0.000340)(20.03 \text{ ft}) = 0.00681 \text{ ft} \qquad \textit{Ans.}$$

Applying Eq. 2, the elongation of AB is then

$$\Delta_{AB} = 0.03 \text{ ft} + 0.00681 \text{ ft} = 0.0368 \text{ ft} \qquad \textit{Ans.}$$

Example 4–15

The bar in Fig. 4–29a is made of steel that can be assumed to be elastic-perfectly plastic, with $\sigma_Y = 250$ MPa. Determine (a) the maximum value of the applied load P that can be applied without causing the steel to yield, and (b) the maximum value of P that the bar can support. Sketch the stress distribution at the critical section for each case.

SOLUTION

Part (a). When the material behaves *elastically*, we must use a stress-concentration factor determined from Fig. 4–21 that is unique for the bar's geometry. Here

(a)

$$\frac{r}{h} = \frac{4 \text{ mm}}{(40 \text{ mm} - 8 \text{ mm})} = 0.125$$

$$\frac{w}{h} = \frac{40 \text{ mm}}{40 \text{ mm} - 8 \text{ mm}} = 1.25$$

so that $K \approx 1.75$.

The maximum load, without causing yielding, occurs when $\sigma_{max} = \sigma_Y$. The average normal stress is $\sigma_{avg} = P/A$. Using Eq. 4–6, we have

$$\sigma_{max} = K\sigma_{avg}; \qquad \sigma_Y = K\left(\frac{P_Y}{A}\right)$$

$$250(10^6) \text{ Pa} = 1.75\left[\frac{P_Y}{(0.002 \text{ m})(0.032 \text{ m})}\right]$$

$$P_Y = 9.14 \text{ kN} \qquad \qquad Ans.$$

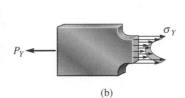

(b)

This load has been calculated using the *smallest* cross section. The resulting stress distribution is shown in Fig. 4–29b. For equilibrium, the "volume" contained within this distribution must equal 9.14 kN.

Part (b). The maximum load sustained by the bar requires *all the material* at the smallest cross section to yield. Therefore, as P is increased to the *plastic load P_p*, it gradually changes the stress distribution from the elastic state shown in Fig. 4–29b to the plastic state shown in Fig. 4–29c. We require

$$\sigma_Y = \frac{P_p}{A}$$

$$250(10^6) \text{ Pa} = \frac{P_p}{(0.002 \text{ m})(0.032 \text{ m})}$$

$$P_p = 16.0 \text{ kN} \qquad \qquad Ans.$$

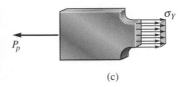

(c)

Fig. 4–29

Here P_p equals the "volume" contained within the stress distribution, which in this case is $P_p = \sigma_Y A$.

4.8 Residual Stress

If an axially loaded member or group of such members forms a statically indeterminate system that can support both tensile and compressive loads, then excessive external loadings, which cause yielding of the material, will create *residual stresses* in the members when the loads are removed. The reason for this has to do with the elastic recovery of the material that occurs during unloading. For example, consider a prismatic member made from an elastoplastic material having the stress–strain diagram *OAB*, as shown in Fig. 4–30. If an axial load produces a stress σ_Y in the material and a corresponding plastic strain ϵ_C, then when the load is *removed*, the material will respond elastically and follow the line *CD* in order to recover some of the plastic strain. A full recovery to zero stress at point O' will be possible only if the member is statically determinate, since the support reactions for the member must be zero when the load is removed. Under these circumstances the member will be permanently deformed so that the permanent set or strain in the member is $\epsilon_{O'}$. If the member is *statically indeterminate*, however, removal of the external load will cause the support forces to respond to the elastic recovery *CD*. Since these forces will constrain the member from full recovery, they will induce *residual stresses* in the member.

To solve a problem of this kind, the complete cycle of loading and then unloading of the member can be considered as the *superposition* of a positive load (loading) on a negative load (unloading). The loading, *O* to *C*, results in a plastic stress distribution, whereas the unloading, along *CD*, results only in an elastic stress distribution. Superposition requires the loads to cancel; however, the stress distributions will not cancel, and so residual stresses will remain.

The following example illustrates these concepts numerically.

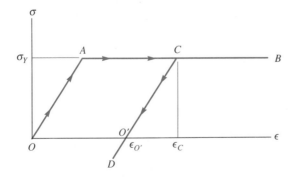

Fig. 4–30

Example 4–16

The rod shown in Fig. 4–31a has a radius of 5 mm and is made from an elastic-perfectly plastic material for which $\sigma_Y = 420$ MPa, $E = 70$ GPa, Fig. 4–31b. If a force of $P = 60$ kN is applied to the rod and then removed, determine the residual stress in the rod and the permanent displacement of the collar at C.

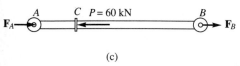

(a)

(b)

SOLUTION

The free-body diagram of the rod is shown in Fig. 4–31c. By inspection, the rod is statically indeterminate. Application of the load **P** will cause one of three possibilities, namely, both segments AC and CB remain elastic, AC is plastic while CB is elastic, or both AC and CB are plastic.*

An *elastic analysis*, similar to that discussed in Sec. 4.4, will produce $F_A = 45$ kN and $F_B = 15$ kN at the supports. However, this results in a stress of

$$\sigma_{AC} = \frac{45 \text{ kN}}{\pi(0.005 \text{ m})^2} = 573 \text{ Mpa (compression)} > \sigma_Y = 420 \text{ MPa}$$

in segment AC, and

$$\sigma_{CB} = \frac{15 \text{ kN}}{\pi(0.005 \text{ m})^2} = 191 \text{ MPa (tension)}$$

in segment CB. Since the material in segment AC will yield, we will assume that AC becomes plastic, while CB remains elastic.

For this case, the maximum possible force developed in AC is

$$(F_A)_Y = \sigma_Y A = 420(10^3) \text{ kN/m}^2[\pi(0.005 \text{ m})^2]$$
$$= 33.0 \text{ kN}$$

and from the equilibrium of the rod, Fig. 4–31c,

$$F_B = 60 \text{ kN} - 33.0 \text{ kN} = 27.0 \text{ kN}$$

The stress in each segment of the rod is therefore

$$\sigma_{AC} = \sigma_Y = 420 \text{ MPa (compression)}$$

$$\sigma_{CB} = \frac{27.0 \text{ kN}}{\pi(0.005 \text{ m})^2} = 344 \text{ MPa (tension)} < 420 \text{ MPa (OK)}$$

(c)

Fig. 4–31

Residual Stress. In order to obtain the residual stress, it is also necessary to know the strain in each segment due to the loading. Since CB responds elastically,

$$\Delta_C = \frac{F_B L_{CB}}{AE} = \frac{[27.0 \text{ kN}](0.300 \text{ m})}{\pi(0.005 \text{ m})^2[70(10^6) \text{ kN/m}^2]} = 0.001473 \text{ m}$$

Thus,

$$\epsilon_{CB} = \frac{\Delta_C}{L_{CB}} = \frac{0.001473 \text{ m}}{0.300 \text{ m}} = +0.004911$$

Also, since Δ_C is known, the strain in AC is

$$\epsilon_{AC} = \frac{\Delta_C}{L_{AC}} = -\frac{0.001473 \text{ m}}{0.100 \text{ m}} = -0.01473$$

Therefore, when **P** is *applied*, the stress–strain behavior for the material in segment CB moves from O to A', Fig. 4–31b, and the stress–strain behavior for the material in segment AC moves from O to B'. If the load **P** is applied in the reverse direction, in other words, the load is removed, then a reverse force of $F_A = 45$ kN and $F_B = 15$ kN must be applied to each segment. This causes the material to *behave elastically* in both segments. As calculated previously, these forces produce stresses $\sigma_{AC} = 573$ MPa (tension) and $\sigma_{CB} = 191$ MPa (compression), and as a result the residual stress in each member is

$$(\sigma_{AC})_r = -420 \text{ MPa} + 573 \text{ MPa} = 153 \text{ MPa} \qquad \textit{Ans.}$$

$$(\sigma_{CB})_r = 344 \text{ MPa} - 191 \text{ MPa} = 153 \text{ MPa} \qquad \textit{Ans.}$$

This tensile stress is the *same* for both segments, which is to be expected. Also note that the stress–strain behavior for segment AC moves from B' to D' in Fig. 4–31b, while the stress–strain behavior for the material in segment CB moves from A' to C'.

Permanent Displacement. From Fig. 4–31b, the *residual strain* in CB is

$$\epsilon'_{CB} = \frac{\sigma}{E} = \frac{153(10^6) \text{ Pa}}{70(10^9) \text{ Pa}} = 0.002185$$

so that the permanent displacement of C is

$$\Delta_C = \epsilon'_{CB} L_{CB} = 0.002185(300 \text{ mm}) = 0.655 \text{ mm} \leftarrow \qquad \textit{Ans.}$$

We can also obtain this result by computing the residual strain ϵ'_{AC} in AC, Fig. 4–31b. Since line $B'D'$ has a slope of E, then

$$\Delta\epsilon_{AC} = \frac{\Delta\sigma}{E} = \frac{(420 + 153)10^6 \text{ Pa}}{70(10^9) \text{ Pa}} = 0.008185$$

Therefore

$$\epsilon'_{AC} = \epsilon_{AC} + \Delta\epsilon_{AC} = -0.01473 + 0.008185 = -0.00655$$

Finally,

$$\Delta_C = \epsilon'_{AC} L_{AC} = 0.00655(100 \text{ mm}) = 0.655 \text{ mm} \leftarrow \qquad \textit{Ans.}$$

*The possibility of CB becoming plastic before AC will not occur, because when point C deforms, the *strain* in AC (since it is shorter) will always be larger than the strain in CB.

PROBLEMS

4–61. The steel bar has the dimensions shown in the figure. Determine the maximum axial force P that can be applied so as not to exceed an allowable tensile stress of $\sigma_{\text{allow}} = 150$ MPa.

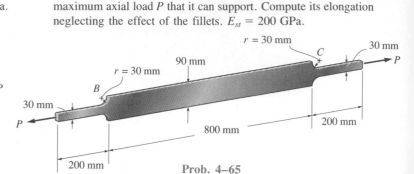

Prob. 4–61

4–62. Determine the maximum normal stress developed in the bar when it is subjected to a tension of $P = 8$ kN.

4–63. If the allowable normal stress for the bar is $\sigma_{\text{allow}} = 120$ MPa, determine the maximum axial force P that can be applied to the bar.

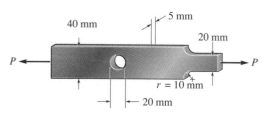

Probs. 4–62/4–63

*4–64. Determine the maximum axial force P that can be applied to the bar. The bar is made from steel and has an allowable stress of $\sigma_{\text{allow}} = 21$ ksi.

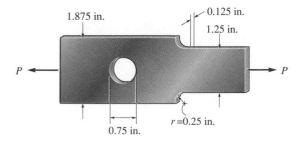

Prob. 4–64

4–65. The steel plate has a thickness of 12 mm. If there are shoulder fillets at B and C, and $\sigma_{\text{allow}} = 150$ MPa, determine the maximum axial load P that it can support. Compute its elongation neglecting the effect of the fillets. $E_{st} = 200$ GPa.

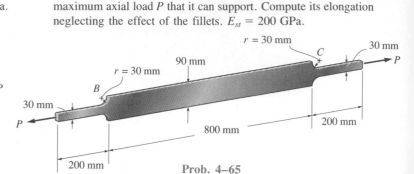

Prob. 4–65

4–66. The member is to be made from a steel plate that is 0.25 in. thick. If a 0.5-in. hole is drilled through its center, determine the approximate width w of the plate so that it can support an axial force of 800 lb. The allowable stress is $\sigma_{\text{allow}} = 22$ ksi.

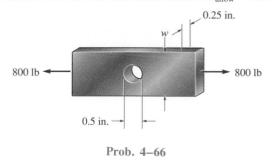

Prob. 4–66

4–67. The resulting stress distribution along the section AB for the bar is shown in the figure. From this distribution, determine the approximate resultant axial force P applied to the bar. Also, what is the stress-concentration factor for this geometry?

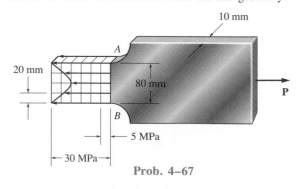

Prob. 4–67

164

*4–68. The resulting stress distribution along the section *AB* for the bar is shown in the figure. From this distribution, determine the approximate resultant axial force *P* applied to the bar. Also, what is the stress-concentration factor for this geometry?

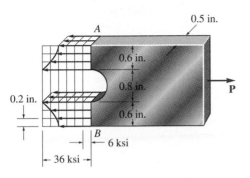

0.5 in.

A

0.6 in.

0.8 in.

P

0.2 in.

0.6 in.

B

6 ksi

36 ksi

Prob. 4–68

4–69. The wire *BC* has a diameter of 0.125 in. and the material has the stress–strain characteristics shown in the figure. Determine the vertical displacement of the handle at *D* if the pull at the grip is slowly increased and reaches a magnitude of (*a*) *P* = 450 lb, (*b*) *P* = 600 lb.

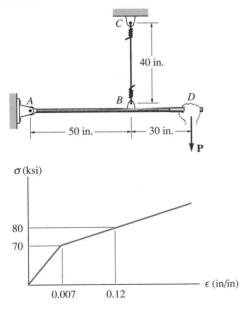

C

40 in.

A   *B*   *D*

50 in.   30 in.

P

σ (ksi)

80

70

0.007   0.12

ϵ (in/in)

Prob. 4–69

4–70. The bar has a cross-sectional area of 0.5 in² and is made from a material that has a stress–strain diagram that can be approximated by the two line segments shown. Determine the elongation of the bar due to the applied loading.

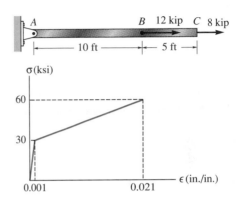

A   *B* 12 kip *C* 8 kip

10 ft   5 ft

σ (ksi)

60

30

0.001   0.021

ϵ (in./in.)

Prob. 4–70

4–71. The weight is suspended from steel and aluminum wires, each having the same initial length of 3 m and cross-sectional area of 4 mm². If the materials can be assumed to be elastic-perfectly plastic, with $(\sigma_Y)_{st}$ = 120 MPa and $(\sigma_Y)_{al}$ = 70 MPa, determine the force in each wire if the weight is (*a*) 600 N and (*b*) 720 N. E_{al} = 70 GPa, E_{st} = 200 GPa.

F_{al} F_{st}

Aluminum   Steel

Prob. 4–71

*4–72. The 130-kip weight is slowly set on the top of a post made of aluminum with a steel core. If both materials can be considered elastic-perfectly plastic such that $(\sigma_Y)_{st} = 36$ ksi and $(\sigma_Y)_{al} = 3$ ksi, determine the stress in each material. $E_{st} = 29(10^3)$ ksi, $E_{al} = 10(10^3)$ ksi.

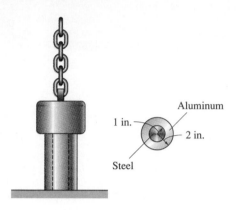

Prob. 4–72

4–73. The rigid beam is supported by the three posts A, B, and C of equal length. Posts A and C have a diameter of 75 mm and are made of aluminum, for which $E_{al} = 70$ GPa and $(\sigma_Y)_{al} = 20$ MPa. Post B has a diameter of 20 mm and is made of brass, for which $E_{br} = 100$ GPa and $(\sigma_Y)_{br} = 590$ MPa. Determine the smallest magnitude of **P** so that (a) only posts A and C yield and (b) all the posts yield.

4–74. The rigid beam is supported by the three posts A, B, and C. Posts A and C have a diameter of 60 mm and are made of aluminum, for which $E_{al} = 70$ GPa and $(\sigma_Y)_{al} = 20$ MPa. Post B is made of brass, for which $E_{br} = 100$ GPa and $(\sigma_Y)_{br} = 590$ MPa. If $P = 130$ kN, determine the largest diameter of post B so that all the posts yield at the same time.

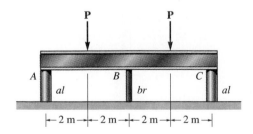

Probs. 4–73/4–74

4–75. The distributed loading is applied to the rigid beam, which is supported by the three bars. Each bar has a cross-sectional area of 1.25 in^2 and is made from a material having a stress–strain diagram that can be approximated by the two line segments shown. If a load of $w = 25$ kip/ft is applied to the bar, determine the stress in each bar and the vertical displacement of the rigid beam.

*4–76. The distributed loading is applied to the rigid beam, which is supported by the three bars. Each bar has a cross-sectional area of 0.75 in^2 and is made from a material having a stress–strain diagram that can be approximated by the two line segments shown. Determine the intensity of the distributed loading w needed to cause the beam to be displaced downward 1.5 in.

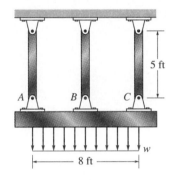

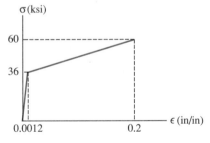

Probs. 4–75/4–76

4–77. The rigid bar is supported by a pin at A and two steel wires, each having a diameter of 4 mm. If the yield stress for the wires is $\sigma_Y = 530$ MPa, determine the smallest intensity of the distributed load w that can be placed on the beam and will cause (a) one of the wires to yield and (b) both wires to yield. $E_{st} = 200$ GPa. For the calculation, assume that the steel is elastic-perfectly plastic.

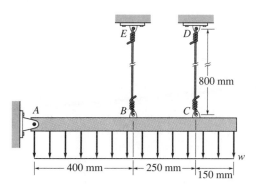

Prob. 4–77

4–78. A material has a stress–strain diagram that can be described by the curve $\sigma = c\epsilon^{1/2}$. Determine the deflection Δ of the end of a rod made from this material if it has a length L, cross-sectional area A, and a specific weight γ.

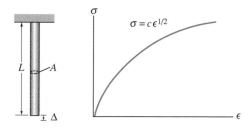

Prob. 4–78

4–79. The force **P** is applied to the bar, which is composed of an elastic-perfectly plastic material. Construct a graph to show how the force in each section AB and BC (ordinate) varies as P (abscissa) is increased. The bar has cross-sectional areas of 1 in^2 in region AB and 4 in^2 in region BC, and $\sigma_Y = 30$ ksi.

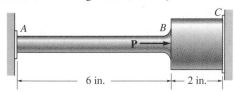

Prob. 4–79

*4–80. The three bars are pinned together and subjected to the load **P**. If each bar has a cross-sectional area A, length L, and is made from an elastic-plastic material, for which the yield stress is σ_Y, determine the largest load (ultimate load) that can be supported by the bars, i.e., the load P that causes all the bars to yield. Also, what is the horizontal displacement of point A when the load reaches its ultimate value? The modulus of elasticity is E.

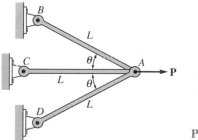

Prob. 4–80

4–81. The bar has a cross section that is 1 in^2. If a force of $P = 45$ kip is applied at B and then removed, determine the residual stress in each part of the bar. $\sigma_Y = 30$ ksi.

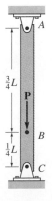

Prob. 4–81

REVIEW PROBLEMS

4–82. A steel surveyor's tape is to be used to measure the length of a line. The tape has a rectangular cross section of 0.05 in. by 0.2 in. and a length of 100 ft when $T_1 = 60°F$ and the tension or pull on the tape is 20 lb. Determine the true length of the line if the tape shows the reading to be 463.25 ft when used with a pull of 35 lb at $T_2 = 90°F$. The ground on which it is placed is flat. $\alpha_{st} = 9.6(10^{-6})/°F$, $E_{st} = 29(10^3)$ ksi.

Prob. 4–82

4–83. The brass plug is force-fitted into the rigid casting. The uniform normal bearing pressure on the plug is estimated to be 15 MPa. If the coefficient of static friction between the plug and casting is $\mu_s = 0.3$, determine the axial force P needed to pull the plug out. Also, calculate the displacement of end B relative to end A just before the plug starts to slip out. $E_{br} = 98$ GPa.

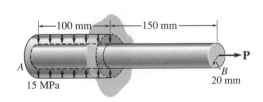

Prob. 4–83

*4–84. The rigid link is supported by a pin at A, a steel wire BC having an unstretched length of 200 mm and cross-sectional area of 22.5 mm², and a short aluminum block having an unloaded length of 50 mm and cross-sectional area of 40 mm². If the link is subjected to the vertical load shown, determine the normal stress in the wire and the block. $E_{st} = 200$ GPa, $E_{al} = 70$ GPa.

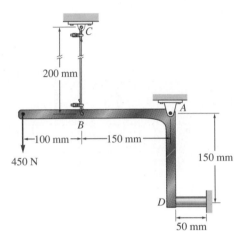

Prob. 4–84

4–85. The rigid beam is supported at its ends by steel tie rods. The rods have diameters $d_{AB} = 0.5$ in. and $d_{CD} = 0.3$ in. If the allowable stress for the steel is $\sigma_{allow} = 16.2$ ksi, determine the maximum value of P and the position x of this force on the beam so the beam remains in the horizontal position when it is loaded. $E_{st} = 29(10^3)$ ksi.

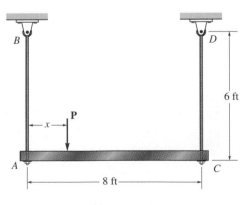

Prob. 4–85

4–86. The 0.4-in.-diameter steel bolt is used to hold the (rigid) assembly together. Determine the clamping force that must be provided by the bolt when $T_1 = 80°F$, so that the clamping force it exerts when $T_2 = 180°F$ is 200 lb. $E_{st} = 29(10^3)$ ksi, $\alpha_{st} = 6(10^{-6})/°F$.

*4–88. The joint is made from three steel plates that are bonded together at their seams. Determine the displacement of end A with respect to end B when the joint is subjected to the axial loads shown. Each plate has a thickness of 5 mm. $E_{st} = 200$ GPa.

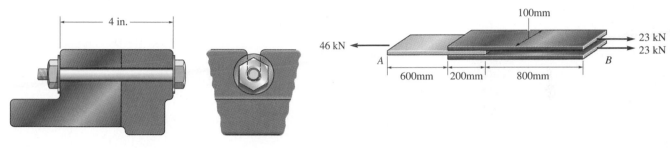

Prob. 4–88

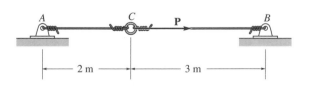

Prob. 4–86

4–89. The steel bar has the original dimensions shown in the figure. If it is subjected to an axial loading of 50 kN, determine the change in its length and its new cross-sectional dimensions at section a–a. $E_{st} = 200$ GPa, $\nu_{st} = 0.29$.

4–87. Two steel wires, each having a cross-sectional area of 2 mm², are tied to a ring at C, and then stretched and tied between the two pins A and B. The initial tension in the wires is 50 N. If a horizontal force $\mathbf{P}$ is applied to the ring, determine the force in each wire if $P = 20$ N. What is the smallest force P that must be applied to the ring to reduce the force in wire CB to zero? Take $\sigma_Y = 300$ MPa. $E_{st} = 210(10^3)$ GPa.

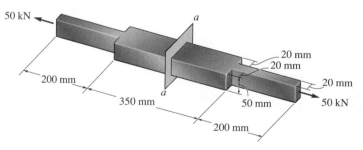

Prob. 4–87

Prob. 4–89

4–90. The 10-mm-diameter shank of the steel bolt has a bronze sleeve bonded to it. The outer diameter of this sleeve is 20 mm. If the bolt is subjected to a tensile force of $P = 20$ kN, determine the stress in the steel and the bronze. $E_{st} = 200$ GPa, $E_{br} = 100$ GPa.

4–91. The 10-mm-diameter shank of the steel bolt has a bronze sleeve bonded to it. The outer diameter of this sleeve is 20 mm. If the yield stress for the steel is $(\sigma_Y)_{st} = 640$ MPa, and for the bronze $(\sigma_Y)_{br} = 520$ MPa, determine the magnitude of the largest elastic load P that can be applied to the assembly. $E_{st} = 200$ GPa, $E_{br} = 100$ GPa.

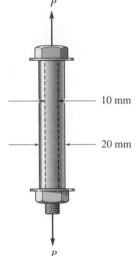

10 mm

20 mm

Probs. 4–90/4–91

4–93. The steel bolt has a diameter of 7 mm and fits through an aluminum sleeve as shown. The sleeve has an inner diameter of 8 mm and an outer diameter of 10 mm. The nut at A is adjusted so that it just presses up against the sleeve. If it is then tightened one-half turn, determine the force in the bolt and the sleeve. The single-threaded screw on the bolt has a lead of 1.5 mm. $E_{st} = 200$ GPa, $E_{al} = 70$ GPa. *Note*: The lead represents the distance the nut advances along the bolt for one complete turn of the nut.

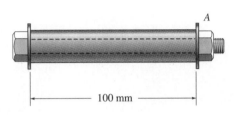

A

100 mm

Prob. 4–93

*4–92.** The rigid beam is supported by three steel wires, each having a length of 4 ft. The cross-sectional area of AB and EF is 0.015 in^2, and the cross-sectional area of CD is 0.006 in^2. If $\sigma_Y = 36$ ksi, determine the largest distributed load w that can be supported by the beam before any of the wires begin to yield. $E_{st} = 29(10^3)$ ksi. If the steel is assumed to be elastic-perfectly plastic, determine how far the beam is displaced downward just before all the wires begin to yield.

4–94. The observation cage C has a weight of 250 kip and, through a system of gears, travels upward at constant velocity along the steel column, which has a height of 200 ft. The column has an outer diameter of 3 ft and is made from steel plate having a thickness of 0.25 in. Neglect the weight of the column, and determine the normal stress in the column at its base, B, as a function of the cage's position y. Also, determine the relative displacement of end A with respect to end B as a function of y. $E_{st} = 29(10^3)$ ksi.

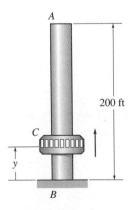

A

200 ft

C

y

B

Prob. 4–94

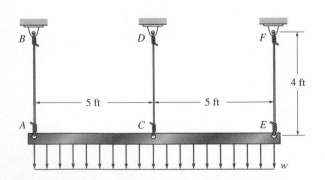

B

D

F

4 ft

5 ft

5 ft

A

C

E

w

Prob. 4–92

5 Torsion

In this chapter we will discuss the effects of applying a torsional loading to a long straight member such as a shaft or tube. Initially we will consider the member to have a circular cross section. We will show how to determine both the stress distribution within the member and the angle of twist, when the material behaves in a linear-elastic manner and also when it is inelastic. Statically indeterminate analysis of shafts and tubes will also be discussed, along with special topics that include those members having noncircular cross sections. Stress concentrations and residual stress caused by torsional loadings will also be given special consideration.

5.1 Torsional Deformation of a Circular Shaft

Torque is a moment that tends to twist a member about its longitudinal axis. Its effect is of primary concern in the design of axles or drive shafts used in vehicles and machinery. In this section we will discuss the type of deformation that occurs when a torque is applied to a circular shaft or tube made of a homogeneous material. Later in the chapter we will then use these concepts to develop equations that give the stress distribution and rotation or twist that can be developed in shafts and tubes that transmit torsional loadings.

We can illustrate physically what happens when a torque is applied to a circular shaft by considering the shaft to be made of a highly deformable material such as rubber, Fig. 5–1a. When the torque is applied, the circles and longitudinal grid lines originally marked on the shaft tend to distort into the pattern shown in Fig. 5–1b. By inspection, twisting causes the circles to *remain circles,* and each longitudinal grid line deforms into a helix that inter-

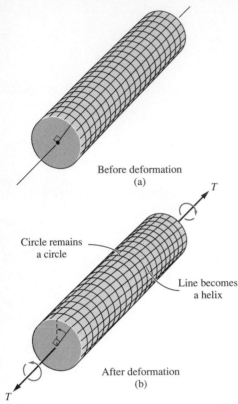

Before deformation
(a)

Circle remains
a circle

Line becomes
a helix

After deformation
(b)

Fig. 5–1

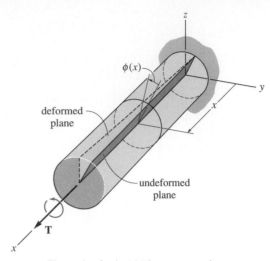

The angle of twist $\phi(x)$ increases as x increases.

Fig. 5–2

sects the circles at equal angles. Also, the cross sections at the *ends* of the shaft remain flat—that is, they do not warp or bulge in or out—and radial lines on these ends remain straight during the deformation. From these observations we will make the assumptions that *every cross section of the shaft remains plane,* and second *radial lines* on these cross sections *remain straight and rotate through the same angle* during deformation. Furthermore, we can assume that if the angle of rotation is *small,* the *length of the shaft* and *its radius* will *remain unchanged.* Thus, if the shaft is fixed at one end as shown in Fig. 5–2 and a torque is applied to its other end, the shaded plane will distort into a skewed form as shown. Here it is seen that a radial line located on the cross section at a distance x from the fixed end of the shaft will rotate through an angle $\phi(x)$. The angle $\phi(x)$, so defined, is called the *angle of twist.* It depends on the position x and will vary along the shaft as shown.

In order to understand how this distortion strains the material, we will now isolate a small element located at a radial distance ρ (rho) from the axis of the shaft, Fig. 5–3*a*. Notice that the front and back faces of the element have sides that are bounded by radial lines, located at x and $x + \Delta x$, respectively. Due to the deformation as noted in Fig. 5–2, each of these faces will undergo a rotation. The one at x by $\phi(x)$, and the one at $x + \Delta x$ by $\phi(x) + \Delta\phi$, Fig. 5–3*b*. As a result, the *difference* in these rotations, $\Delta\phi$, causes the element to be subjected to a *shear strain.* To show this, note that before deformation the angle between the edges AB and AC is 90°; after deformation, however, the edges of the element are AD and AC and the angle between them is θ'. From the definition of shear strain, Eq. 2–4, we have

$$\gamma = \frac{\pi}{2} - \lim_{\substack{C \to A \text{ along } CA \\ D \to A \text{ along } BA}} \theta'$$

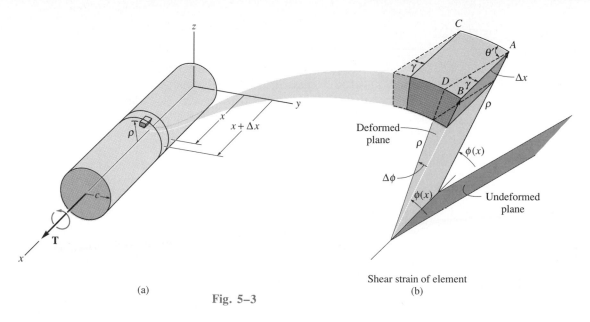

(a)

Fig. 5–3

Shear strain of element
(b)

This angle, γ, is indicated in Fig. 5–3b. It can be related to the length Δx of the element and the difference in the angle of rotation, $\Delta\phi$, between the shaded faces. If we let $\Delta x \to dx$ and $\Delta\phi \to d\phi$, then from Fig. 5–3b we have

$$BD = \rho \, d\phi = dx \, \gamma$$

Therefore,

$$\gamma = \rho \, \frac{d\phi}{dx} \qquad (5\text{–}1)$$

Since dx and $d\phi$ are the *same* for *all elements* located at points within the cross section at x, then $d\phi/dx$ is constant and Eq. 5–1 states that the magnitude of the shear strain for any of these elements varies only with its radial distance ρ from the axis of the shaft. In other words, *the shear strain within the shaft varies linearly along any radial line, from zero at the axis of the shaft to a maximum at its outer boundary.* Assuming that the outer radius of the shaft is c, the shear strain for typical elements located within the shaft at ρ, and at its surface $\rho = c$, is shown in Fig. 5–4. From Eq. 5–1, since $d\phi/dx = \gamma/\rho = \gamma_{max}/c$, then

$$\gamma = \left(\frac{\rho}{c}\right) \gamma_{max} \qquad (5\text{–}2)$$

The results obtained here are also valid for circular tubes. They depend only on the assumptions regarding the deformations mentioned above. Using a more complete analysis of the shaft, it is also possible to show that these assumptions require all other components of both normal and shear strain to be zero when the shaft is subjected to a torque.

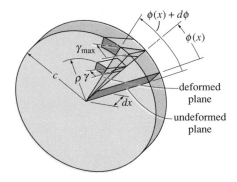

The shear strain for the material increases linearly with ρ, i.e., $\gamma = (\rho/c)\gamma_{max}$.

Fig. 5–4

5.2 The Torsion Formula

If a shaft is subjected to an external torque, then for equilibrium an internal torque **T** must also be developed within the shaft. At any given section along the shaft this internal torque is the resultant of the moment produced by the distribution of shear stress acting over the shaft's cross-sectional area, Fig. 5–5. In this section we will develop an equation that relates this shear-stress distribution to the resultant internal torque resisted at the section of a circular shaft or tube.

If the material is linear-elastic, then Hooke's law applies, $\tau = G\,\gamma$, and consequently a *linear variation in shear strain,* as noted in Sec. 5.1, leads to a corresponding *linear variation in shear stress* along any radial line on the cross section. Hence, like the shear-strain variation, for a solid shaft, τ will vary from zero at the shaft's longitudinal axis to a maximum value, τ_{max}, at its outer surface. This variation is shown in Fig. 5–5 on the front faces of a selected number of elements, located at an intermediate radial position ρ and at the outer radius c. Due to the proportionality of triangles, or by using Hooke's law ($\tau = G\,\gamma$) and Eq. 5–2 [$\gamma = (\rho/c)\gamma_{max}$], we can write

$$\tau = \left(\frac{\rho}{c}\right)\tau_{max} \tag{5–3}$$

This equation expresses the shear-stress distribution as a *function* of the radial position ρ of the element; in other words, it defines the stress distribution in terms of the geometry of the shaft. Using it, we will now apply the condition that requires the torque produced by the stress distribution over the entire cross section to be equivalent to the resultant internal torque T at the section,

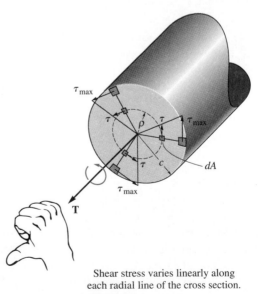

Shear stress varies linearly along each radial line of the cross section.

Fig. 5–5

which holds the shaft in equilibrium, Fig. 5–5. Specifically, each element of area dA, located at ρ, is subjected to a force of $dF = \tau\, dA$. The torque produced by this force is $dT = \rho(\tau\, dA)$. Using Eq. 5–3, we therefore have for the entire cross section

$$T = \int_A \rho(\tau\, dA) = \int_A \rho\left(\frac{\rho}{c}\right)\tau_{max}\, dA \qquad (5\text{–}4)$$

Since τ_{max}/c is constant,

$$T = \frac{\tau_{max}}{c} \int_A \rho^2\, dA \qquad (5\text{–}5)$$

The integral in this equation depends only on the geometry of the shaft. It represents the *polar moment of inertia* of the shaft's cross-sectional area computed about the shaft's longitudinal axis. We will symbolize its value as J, and therefore the above equation can be written in a more compact form, namely,

$$\boxed{\tau_{max} = \frac{Tc}{J}} \qquad (5\text{–}6)$$

where

τ_{max} = the maximum shear stress in the shaft, which occurs at the outer surface

T = the resultant internal torque acting at the cross section. Its value is determined from the method of sections and the equation of moment equilibrium applied about the shaft's longitudinal axis

J = the polar moment of inertia of the cross-sectional area

c = the outer radius of the shaft

Using Eqs. 5–3 and 5–6, the shear stress at the intermediate distance ρ can be determined from a similar equation:

$$\boxed{\tau = \frac{T\rho}{J}} \qquad (5\text{–}7)$$

Either of the above two equations is often referred to as the *torsion formula*. Recall that it is used only if the shaft is circular and the material is homogeneous and behaves in a linear-elastic manner, since the derivation is based on the fact that the shear stress is proportional to the shear strain, and thus both vary linearly along every radial line on the cross section.

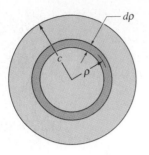

Fig. 5–6

Solid Shaft. If the shaft has a solid circular cross section, the polar moment of inertia J can be determined using an area element in the form of a *differential ring* or annulus having a thickness $d\rho$ and circumference $2\pi\rho$, Fig. 5–6. For this ring, $dA = 2\pi\rho\, d\rho$, so

$$J = \int_A \rho^2\, dA = \int_0^c \rho^2\,(2\pi\rho\, d\rho) = 2\pi \int_0^c \rho^3\, d\rho = 2\pi\left(\frac{1}{4}\right)\rho^4\,\bigg|_0^c$$

$$\boxed{J = \frac{\pi}{2}c^4} \qquad (5\text{–}8)$$

Note that J is always positive. Common units used for its measurement are mm^4 or in^4. It is a *geometric property* of the circular area to be used in *both* Eqs. 5–6 and 5–7.

As noted in the derivation, Eq. 5–7 gives specific values of the shear stress at any point located on the shaft's cross-sectional area. In particular, if the shear stress is plotted along radial line segments, it is represented by a linear distribution as shown in Fig. 5–7a. Notice that each value of τ is directed on the cross section so that the force it creates contributes a *counterclockwise torque* about the center of the shaft. This is necessary since the total torque developed by the entire stress distribution must yield the *counterclockwise* resultant torque **T** at the section, Eq. 5–4.

In Sec. 1.3 it was shown that any isolated *volume element* of material that is subjected to a shear stress on one of its faces must, by reason of both force and moment equilibrium, also develop equal shear stress on three of its adjacent faces. Consequently, typical elements, when isolated from the shaft in Fig. 5–7a, must be subjected to shear stresses directed as shown. As explained above, the shear stress shown on the *shaded* face of each element creates a force that contributes to the resultant torque **T** at the section; whereas, the three other shear stresses acting on the element are necessary for equilibrium of the element.

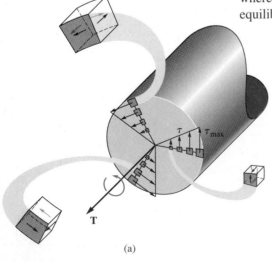

T

(a)

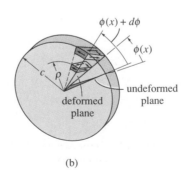

$\phi(x) + d\phi$

$\phi(x)$

undeformed plane

deformed plane

(b)

Fig. 5–7

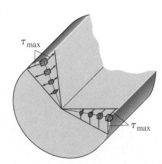

τ_{max}

τ_{max}

Shear stress varies linearly along each radial line of the cross section.

(c)

Perhaps another way of seeing this result is to isolate a differential disk of the shaft having a thickness dx, Fig. 5–7b. The deformation and associated shear stresses acting on two shaded elements, one located at the outer surface and the other at some intermediate radial point, ρ, are shown. Therefore, *not only does the internal torque* **T** *develop a linear distribution of shear stress along each radial line in the plane of the cross-sectional area, but an associated shear-stress distribution must also be developed along an axial plane as noted in Fig. 5–7c.* It is interesting to note that because of this axial distribution of shear stress, shafts made from wood tend to *split* along the axial plane when subjected to excessive torque, Fig. 5–8. This is because wood is an anisotropic material. Its shear resistance parallel to its grains or fibers, directed along the axis of the shaft, is much less than its resistance perpendicular to the fibers, directed in the plane of the cross section.

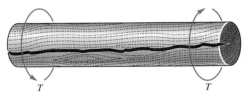

Failure of a wooden shaft due to torsion

Fig. 5–8

Tubular Shaft. If a shaft has a tubular cross section, with inner radius c_i and outer radius c_o, then from Eq. 5–8 we can determine its polar moment of inertia by subtracting J for a shaft of radius c_i from that computed for a shaft of radius c_o. The result is

$$J = \frac{\pi}{2}(c_o^4 - c_i^4) \qquad (5\text{–}9)$$

Like the solid shaft, the shear stress distributed over the tube's cross-sectional area varies linearly along any radial line, Fig. 5–9a. Furthermore, the shear stress varies along an axial plane in this same manner, Fig. 5–9b. Examples of the shear stress acting on typical volume elements are shown in Fig. 5–9a.

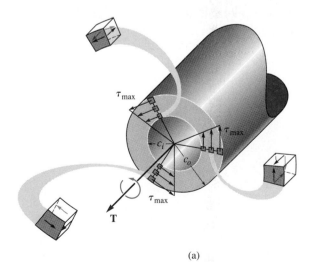

(a)

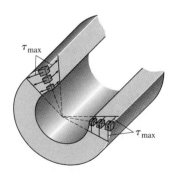

Shear stress varies linearly along each radial line of the cross section.

(b)

Fig. 5–9

PROCEDURE FOR ANALYSIS

The torsion formula can be used to find the shear-stress distribution in a solid shaft or tube having a *circular* cross section that is made of *homogeneous* material and has *linear-elastic* behavior. Saint-Venant's principle requires that this equation be applied at points located a sufficient distance away from any cross-section discontinuities or where an external torque acts. In order to apply the equation, the following procedure is suggested.

Internal Loading. Section the shaft perpendicular to its axis at the point where the shear stress is to be determined, and use the necessary free-body diagram and equations of equilibrium to obtain the internal torque at the section.

Section Property. Compute the polar moment of inertia of the cross-sectional area. For a solid section of radius c, $J = \pi c^4/2$, and for a tube of outer radius c_o and inner radius c_i, $J = \pi(c_o^2 - c_i^2)/2$.

Shear Stress. Specify the radial distance ρ, measured from the center of the cross section to the point where the shear stress is to be computed. Then apply the torsion formula $\tau = T\rho/J$, or if the maximum shear stress is to be computed use $\tau_{\max} = Tc/J$. When substituting the data, make sure to use a consistent set of units.

The shear stress acts on the cross section in a direction that is always perpendicular to ρ. The force it creates must contribute a torque about the axis of the shaft that is in the *same direction* as the internal resultant torque **T** acting on the section. Once this direction is established, a volume element of the material, located at the point where τ is computed, can be isolated, and the direction of τ acting on the remaining three faces of the element can be shown.

The following examples illustrate numerical applications of this procedure.

Example 5–1

The stress distribution in a solid shaft has been plotted along three arbitrary radial lines as shown in Fig. 5–10a. Determine the resultant internal torque at the section (a) using the torsion formula, (b) by finding the resultant of the stress distribution using basic principles.

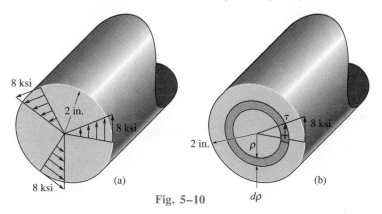

Fig. 5–10

SOLUTION

Part (a).   The polar moment of inertia for the cross-sectional area is

$$J = \frac{\pi}{2}(2 \text{ in.})^4 = 25.13 \text{ in}^4$$

Applying the torsion formula, with $\tau_{max} = 8$ ksi, Fig. 5–10a, we have

$$\tau_{max} = \frac{Tc}{J}; \qquad 8 \text{ kip/in}^2 = \frac{T(2 \text{ in.})}{(25.13 \text{ in}^4)}$$
$$T = 101 \text{ kip} \cdot \text{in.} \qquad \qquad Ans.$$

Part (b).   The shear stress τ acting at the arbitrary distance ρ from the center of the cross section is determined by proportional triangles, Fig. 5–10b. We have

$$\frac{\tau}{\rho} = \frac{8 \text{ ksi}}{2 \text{ in.}}$$
$$\tau = 4\rho$$

This stress acts on all portions of the differential ring element that has an area $dA = 2\pi\rho\, d\rho$. Since the force created by τ is $dF = \tau\, dA$, the torque is

$$dT = \rho\, dF = \rho(\tau dA) = \rho(4\rho)2\pi\rho\, d\rho = 8\pi\rho^3\, d\rho$$

For the entire area over which τ acts, we require

$$T = \int_0^2 8\pi\rho^3\, d\rho = 8\pi\left(\frac{1}{4}\rho^4\right)\Big|_0^2 = 101 \text{ kip} \cdot \text{in.} \qquad Ans.$$

Example 5–2

(a)

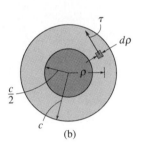

(b)

Fig. 5–11

The *solid* shaft of radius c is subjected to a torque $\mathbf{T}$, Fig. 5–11a. Determine the fraction of $\mathbf{T}$ that is resisted by the material contained within the lighter-shaded region of the shaft, which has an inner radius of $c/2$ and outer radius c.

SOLUTION

The stress in the shaft varies linearly, such that $\tau = (\rho/c)\tau_{max}$, Eq. 5–3. Therefore, the torque dT' on the ring (area) located within the lighter-shaded region, Fig. 5–11b, is

$$dT' = \rho(\tau\, dA) = \rho(\rho/c)\tau_{max}(2\pi\rho\, d\rho)$$

For the entire lighter-shaded area the torque is

$$T' = \frac{2\pi\tau_{max}}{c} \int_{c/2}^{c} \rho^3\, d\rho$$

$$= \frac{2\pi\tau_{max}}{c} \frac{1}{4}\rho^4 \bigg|_{c/2}^{c}$$

So that

$$T' = \frac{15\pi}{32}\tau_{max}c^3 \tag{1}$$

This torque T' can be expressed in terms of the applied torque T by first using the torsion formula to determine the maximum stress in the shaft. We have

$$\tau_{max} = \frac{Tc}{J} = \frac{Tc}{(\pi/2)c^4}$$

or

$$\tau_{max} = \frac{2T}{\pi c^3}$$

Substituting this into Eq. 1 yields

$$T' = \frac{15}{16}T$$

Here approximately 94% of the torque is resisted by the lighter-shaded region, and the remaining 6% of T (or $\frac{1}{16}$) is resisted by the inner "core" of the shaft, $\rho = 0$ to $\rho = c/2$. As a result, the material located at the *outer region* of the shaft is highly effective in resisting torque, which justifies the use of tubular shafts as an efficient means for transmitting torque and thereby saving material.

Example 5–3

The shaft shown in Fig. 5–12a is supported by two bearings and is subjected to three torques. Determine the shear stress developed at points A and B, located at section a–a of the shaft, Fig. 5–12b.

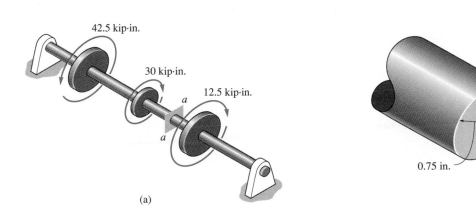

42.5 kip·in.

30 kip·in.

12.5 kip·in.

a

a

(a)

18.9 ksi

z

y

B

3.77 ksi

0.75 in.   0.15 in. x

(b)

SOLUTION

Internal Torque.   The bearing reactions on the shaft are zero, provided the shaft's weight is neglected. Furthermore, the applied torques satisfy moment equilibrium about the shaft's axis.

The internal torque at section a–a will be determined from the free-body diagram of the left segment, Fig. 5–12c. We have

$$\Sigma M_x = 0; \quad 42.5 \text{ kip} \cdot \text{in.} - 30 \text{ kip} \cdot \text{in.} - T = 0 \qquad T = 12.5 \text{ kip} \cdot \text{in.}$$

Section Property.   The polar moment of inertia for the shaft is

$$J = \frac{\pi}{2}(0.75 \text{ in.})^4 = 0.497 \text{ in}^4$$

Shear Stress.   Since point A is at $\rho = c = 0.75$ in.,

$$\tau_A = \frac{Tc}{J} = \frac{(12.5 \text{ kip} \cdot \text{in.})(0.75 \text{ in.})}{(0.497 \text{ in}^4)} = 18.9 \text{ ksi} \qquad \textit{Ans.}$$

Likewise for point B, at $\rho = 0.15$ in., we have

$$\tau_B = \frac{T\rho}{J} = \frac{(12.5 \text{ kip} \cdot \text{in.})(0.15 \text{ in.})}{(0.497 \text{ in}^4)} = 3.77 \text{ ksi} \qquad \textit{Ans.}$$

The directions of these stresses on each element at A and B, Fig. 5–12b, are established from the direction of the resultant internal torque $\mathbf{T}$, shown in Fig. 5–12c. Note carefully how the shear stress acts on the planes of each of these elements.

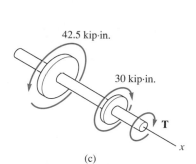

42.5 kip·in.

30 kip·in.

$\mathbf{T}$

x

(c)

Fig. 5–12

Example 5–4

The pipe shown in Fig. 5–13a has an inner diameter of 80 mm and an outer diameter of 100 mm. If its end is tightened against the support at A using a torque wrench at B, determine the shear stress developed in the material at the inner and outer walls along the central portion of the pipe when the 80-N forces are applied to the wrench.

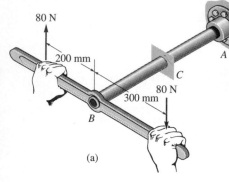

(a)

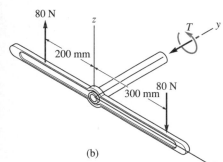

(b)

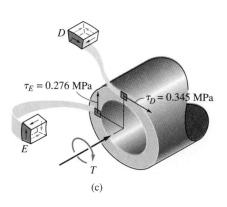

Fig. 5–13

SOLUTION

Internal Torque.   A section is taken at an intermediate location C along the pipe's axis, Fig. 5–13b. The only unknown at the section is the internal torque **T**. Force equilibrium and moment equilibrium about the x and z axes are satisfied. We require

$$\Sigma M_y = 0; \qquad 80 \text{ N}(0.3 \text{ m}) + 80 \text{ N}(0.2 \text{ m}) - T = 0$$

$$T = 40 \text{ N} \cdot \text{m}$$

Section Property.   The polar moment of inertia for the pipe's cross-sectional area is

$$J = \frac{\pi}{2}[(0.05 \text{ m})^4 - (0.04 \text{ m})^4] = 5.80(10^{-6}) \text{ m}^4$$

Shear Stress.   For any point lying on the outside surface of the pipe, $\rho = c_o = 0.05$ m, we have

$$\tau_o = \frac{Tc_o}{J} = \frac{40 \text{ N} \cdot \text{m}(0.05 \text{ m})}{5.80(10^{-6}) \text{ m}^4} = 0.345 \text{ MPa} \qquad Ans.$$

And for any point located on the inside surface, $\rho = c_i = 0.04$ m, so

$$\tau_i = \frac{Tc_i}{J} = \frac{40 \text{ N} \cdot \text{m}(0.04 \text{ m})}{5.80(10^{-6}) \text{ m}^4} = 0.276 \text{ MPa} \qquad Ans.$$

To show how these stresses act at representative points D and E on the cross-sectional area, we will first view the cross section from the front of segment CA of the pipe, Fig. 5–13a. On this section, Fig. 5–13c, the resultant internal torque is equal but opposite to that shown in Fig. 5–13b. The shear stresses at D and E contribute to this torque and therefore act on the shaded faces of the elements in the directions shown. As a consequence, notice how the shear-stress components act on the other three faces. Furthermore, since the top face of D and the inner face of E are on stress-free regions taken from the pipe's inner and outer walls, no shear stress can exist on these faces or on the other corresponding faces of the elements.

5.3 Power Transmission

Shafts and tubes having circular cross sections are often used to transmit power developed by a machine. When used for this purpose they are subjected to torques that depend on the power generated by the machine and the angular speed of the shaft. *Power* is defined as the work performed per unit of time. The work transmitted by a rotating shaft equals the torque applied times the angle of rotation. Therefore, if during an instant of time dt an applied torque $\mathbf{T}$ will cause the shaft to rotate $d\theta$, then the instantaneous power is

$$P = \frac{T\,d\theta}{dt}$$

Since the shaft's angular velocity $\omega = d\theta/dt$, we can also express the power as

$$\boxed{P = T\omega} \tag{5–10}$$

In the SI system, power is expressed in *watts* when torque is measured in newton-meters (N · m) and ω is in radians per second (rad/s) (1 W = 1 N · m/s). In the Foot-Pound-Second or FPS system, the basic units of power are foot-pounds per second (ft · lb/s); however, horsepower (hp) is often used in engineering practice, where

$$1 \text{ hp} = 550 \text{ ft} \cdot \text{lb/s}$$

For machinery, the *frequency* of a shaft's rotation, f, is often reported. This is a measure of the number of revolutions or cycles the shaft makes per second and is expressed in hertz (1 Hz = 1 cycle/s). Since 1 cycle = 2π rad, then $\omega = 2\pi f$ and the above equation for power becomes

$$\boxed{P = 2\pi f T} \tag{5–11}$$

Shaft Design. When the power transmitted by a shaft and its frequency of rotation are known, the torque developed in the shaft can be determined from Eq. 5–11, that is, $T = P/2\pi f$. Knowing T and the allowable shear stress for the material, τ_{allow}, we can determine the size of the shaft's cross section using the torsion formula, provided the material behavior is linear-elastic. Specifically, the design or geometric parameter J/c becomes

$$\frac{J}{c} = \frac{T}{\tau_{\text{allow}}} \tag{5–12}$$

For a *solid shaft*, $J = (\pi/2)c^4$, and thus, upon substitution, a *unique value* for the shaft's radius c can be determined. If the shaft is *tubular*, so that $J = (\pi/2)(c_o^4 - c_i^4)$, design permits a wide range of possibilities for the solution since an *arbitrary choice* can be made for either c_o or c_i and the other radius is determined from Eq. 5–12.

The following examples illustrate numerical applications of the above equations.

Example 5–5

A solid steel shaft AB shown in Fig. 5–14 is to be used to transmit 5 hp from the motor M to which it is attached. If the shaft rotates at $\omega = 175$ rpm and the steel has an allowable shear stress of $\tau_{\text{allow}} = 14.5$ ksi, determine the required diameter of the shaft to the nearest $\frac{1}{8}$ in.

Fig. 5–14

SOLUTION

The torque on the shaft is determined from Eq. 5–10, that is, $P = T\omega$. Expressing P in foot-pounds per second and ω in radians/second, we have

$$P = 5 \text{ hp}\left(\frac{550 \text{ ft} \cdot \text{lb/s}}{1 \text{ hp}}\right) = 2750 \text{ ft} \cdot \text{lb/s}$$

$$\omega = \frac{175 \text{ rev}}{\text{min}}\left(\frac{2\pi \text{ rad}}{1 \text{ rev}}\right)\left(\frac{1 \text{ min}}{60 \text{ s}}\right) = 18.33 \text{ rad/s}$$

Thus,

$$P = T\omega; \qquad 2750 \text{ ft} \cdot \text{lb/s} = T(18.33 \text{ rad/s})$$
$$T = 150.1 \text{ ft} \cdot \text{lb}$$

Applying Eq. 5–12, we have

$$\frac{J}{c} = \frac{\pi}{2}\frac{c^4}{c} = \frac{T}{\tau_{\text{allow}}}$$
$$c = \left(\frac{2T}{\pi\tau_{\text{allow}}}\right)^{1/3} = \left(\frac{2(150.1 \text{ ft} \cdot \text{lb})(12 \text{ in./ft})}{\pi(14{,}500 \text{ lb/in}^2)}\right)^{1/3}$$
$$c = 0.429 \text{ in.}$$

Since $2c = 0.858$ in., select a shaft having a diameter of

$$d = \frac{7}{8} \text{ in.} = 0.875 \text{ in.} \qquad \textit{Ans.}$$

Example 5–6

A tubular shaft, having an inner diameter of 30 mm and an outer diameter of 42 mm, is to be used to transmit 90 kW of power. Determine the frequency of rotation of the shaft so that the shear stress cannot exceed 50 MPa.

SOLUTION

The maximum torque that can be applied to the shaft is determined from the torsion formula.

$$\tau_{max} = \frac{Tc}{J}$$

$$50(10^6) \text{ N/m}^2 = \frac{T(0.021 \text{ m})}{(\pi/2)[(0.021 \text{ m})^4 - (0.015 \text{ m})^4]}$$

$$T = 538 \text{ N} \cdot \text{m}$$

Applying Eq. 5–11, the frequency of rotation is

$$P = 2\pi f T$$

$$90(10^3) \text{ N} \cdot \text{m/s} = 2\pi f(538 \text{ N} \cdot \text{m})$$

$$f = 26.6 \text{ Hz} \qquad\qquad Ans.$$

PROBLEMS

5–1. A shaft is made from a steel alloy having an allowable shear stress of $\tau_{allow} = 12$ ksi. If the diameter of the shaft is 1.5 in., determine the maximum torque **T** that can be transmitted. What would be the maximum torque **T'** if a 1-in.-diameter hole is bored through the shaft? Sketch the shear-stress distribution along a radial line in each case.

5–2. The tube is subjected to a torque of 600 N · m. Determine the amount of this torque that is resisted by the shaded section. Solve the problem two ways: (*a*) by using the torsion formula; (*b*) by finding the resultant of the shear-stress distribution.

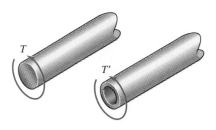

Prob. 5–1

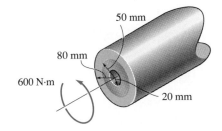

Prob. 5–2

5–3. The solid shaft of radius r is subjected to a torque **T**. Determine the radius r' of the inner core of the shaft that resists one-half of the applied torque ($T/2$). Solve the problem two ways: (*a*) by using the torsion formula, (*b*) by finding the resultant of the shear-stress distribution.

*5–4. The solid shaft of radius r is subjected to a torque **T**. Determine the radius r' of the inner core of the shaft that resists one-quarter of the applied torque ($T/4$). Solve the problem two ways: (*a*) by using the torsion formula, (*b*) by finding the resultant of the shear-stress distribution.

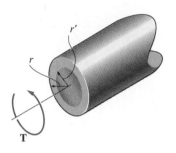

Probs. 5–3/5–4

5–5. The copper pipe has an outer diameter of 40 mm and an inner diameter of 37 mm. If it is tightly secured to the wall at A and three torques are applied to it as shown, determine the maximum shear stress developed in the pipe.

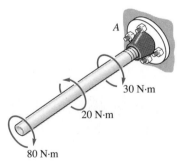

Prob. 5–5

5–6. The solid shaft is fixed to the support at C and subjected to a torque of 950 N · m. Determine the shear stress at points A and B and sketch the shear stress on volume elements located at these points.

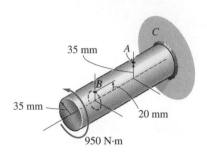

Prob. 5–6

5–7. The copper pipe has an outer diameter of 2.50 in. and an inner diameter of 2.30 in. If it is tightly secured to the wall at C and three torques are applied to it as shown, determine the shear stress developed at points A and B. These points lie on the pipe's outer surface. Sketch the shear stress on volume elements located at A and B.

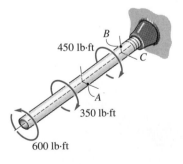

Prob. 5–7

*5–8. The assembly consists of two sections of galvanized steel pipe connected together using a reducing coupling at *B*. The smaller pipe has an outer diameter of 0.75 in. and an inner diameter of 0.68 in., whereas the larger pipe has an outer diameter of 1 in. and an inner diameter of 0.86 in. If the pipe is tightly secured to the wall at *C*, determine the maximum shear stress developed in each section of the pipe when the couple shown is applied to the handles of the wrench.

5–10. The solid shaft has a diameter of 0.75 in. If it is subjected to the torques shown, determine the maximum shear stress developed in regions *BC* and *DE* of the shaft. The bearings at *A* and *F* allow free rotation of the shaft.

5–11. The solid shaft has a diameter of 0.75 in. If it is subjected to the torques shown, determine the maximum shear stress developed in regions *CD* and *EF* of the shaft. The bearings at *A* and *F* allow free rotation of the shaft.

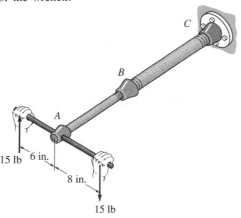

Prob. 5–8

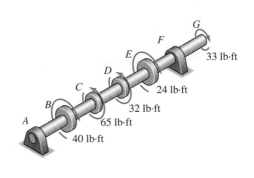

Probs. 5–10/5–11

5–9. The solid 1.25-in.-diameter shaft is used to transmit the torques applied to the gears. If it is supported by smooth bearings at *A* and *B*, which do not resist torque, determine the shear stress developed in the shaft at points *C* and *D*. Indicate the shear stress on volume elements located at these points.

*5–12. The steel shaft is subjected to the torsional loading shown. Determine the shear stress developed at points *A* and *B* and sketch the shear stress on volume elements located at these points. The shaft where *A* and *B* are located has a radius of 60 mm.

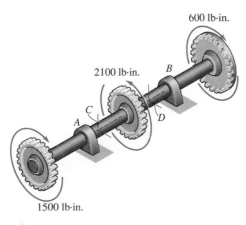

Prob. 5–9

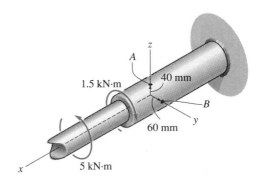

Prob. 5–12

5–13. If the solid shaft to which the valve handle is attached is made of brass and has a diameter of 10 mm, determine the maximum force F that can be applied to the handle just before the material starts to yield. Take $\tau_Y = 235$ MPa.

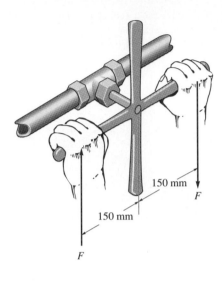

150 mm

150 mm

F

F

Prob. 5–13

5–15. The solid aluminum shaft has a diameter of 50 mm and an allowable shear stress of $\tau_{allow} = 6$ MPa. Determine the largest torques T_1 and T_2 that can be applied to the shaft if it is also subjected to the other torsional loadings. It is required that T_1 and T_2 act in the directions shown.

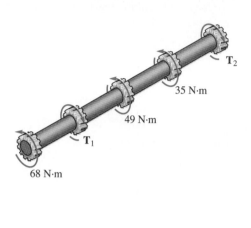

T_2

35 N·m

49 N·m

T_1

68 N·m

Prob. 5–15

5–14. The motor delivers a torque of 50 N · m to the shaft AB. This torque is transmitted to shaft CD using the gears at E and F. Determine the equilibrium torque T' on shaft CD and the maximum shear stress in each shaft. The bearings at B, C, and D allow free rotation of the shafts.

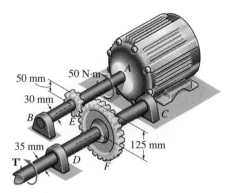

50 mm

30 mm

50 N·m

A

B

E

C

35 mm

125 mm

T'

D

F

Prob. 5–14

***5–16.** The 20-mm-diameter steel shafts are connected together using a brass coupling. If the yield point for the steel is $(\tau_Y)_{st} = 100$ MPa and for the brass $(\tau_Y)_{br} = 250$ MPa, determine the required outer diameter d of the coupling so that the steel and brass begin to yield at the same time when the assembly is subjected to a torque T. Assume that the coupling has an inner diameter of 20 mm.

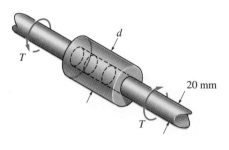

d

T

20 mm

T

Prob. 5–16

5–17. The coupling consists of two disks fixed to separate shafts, each 25 mm in diameter. The shafts are supported on journal bearings that allow free rotation. In order to limit the torque **T** that can be transmitted, a "shear pin" *P* is used to connect the disks together. If this pin can sustain an *average* shear force of 550 N before it fails, determine the maximum constant torque *T* that can be transmitted from one shaft to the other. Also, what is the maximum shear stress in each shaft when the "shear pin" is about to fail?

5–19. The wooden post, which is half buried in the ground, is subjected to a torsional moment of 50 N · m that causes the post to rotate at constant angular velocity. This moment is resisted by a *linear distribution* of torque developed by soil friction, which varies from zero at the ground to t_0 N · m/m at its base. Determine the equilibrium value for t_0 and then calculate the shear stress at points *A* and *B*, which lie on the outer surface of the post.

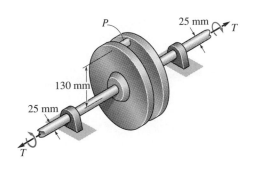

Prob. 5–17

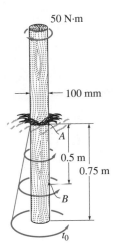

Prob. 5–19

5–18. The glass tube is confined within a rubber stopper, such that when the tube is twisted at constant angular velocity the stopper creates a *constant distribution* of frictional torque along the contacting length *AB* of the tube. If the tube has an inner diameter of 2 mm and an outer diameter of 4 mm, determine the shear stress developed at a point located at its inner and outer walls at a section through point *C*. Show the shear-stress distribution acting along a radial line segment at this section.

*5–20. When drilling a well at constant angular velocity, the bottom end of the drill pipe encounters a torsional resistance T_A. Also, soil along the sides of the pipe creates a distributed frictional torque along its length, varying uniformly from zero at the surface *B* to t_A at *A*. Determine the minimum torque T_B that must be supplied by the drive unit to overcome the resisting torques, and compute the maximum shear stress in the pipe. The pipe has an outer radius r_o and an inner radius r_i.

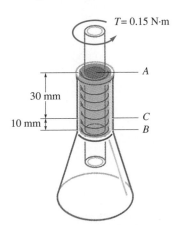

Prob. 5–18

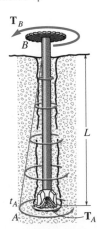

Prob. 5–20

5–21. A cylindrical spring consists of a rubber annulus bonded to a rigid ring and shaft. If the ring is held fixed and a torque **T** is applied to the shaft, determine the maximum shear stress in the rubber.

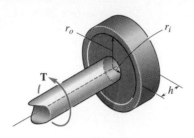

Prob. 5–21

5–22. The solid shaft has a linear taper from r_A at one end to r_B at the other. Derive an equation that gives the maximum shear stress in the shaft at a location x along the shaft's axis.

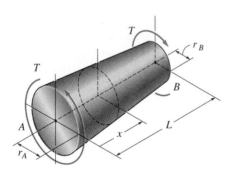

Prob. 5–22

5–23. Determine to the nearest $\frac{1}{8}$ in. the diameter of a shaft that is required to transmit 150 hp at 4000 rev/min. The material has an allowable shear stress of $\tau_{\text{allow}} = 8$ ksi.

*5–24. The drilling pipe on an oil rig is made from steel pipe having an outside diameter of 4.5 in. and a thickness of 0.25 in. If the pipe is turning at 650 rev/min while being powered by a 15-hp motor, determine the maximum shear stress in the pipe.

5–25. The 0.75-in.-diameter shaft in the electric motor develops 0.5 hp and runs at 1740 rev/min. Determine the torque produced and compute the maximum shear stress in the shaft. The shaft is supported by ball bearings at A and B.

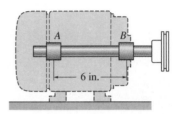

Prob. 5–25

5–26. The 4-hp motor turns the shaft AB at a rate of 1300 rev/min. Determine the maximum shear stress developed in the shaft within region AB, where it has a diameter of 0.5 in., and within region BC, where it has a diameter of 0.65 in.

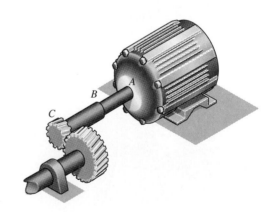

Prob. 5–26

5–27. The drive shaft AB of an automobile is made of a steel tube having an allowable shear stress of $\tau_{allow} = 8$ ksi. If the outer diameter is 2.5 in. and the engine delivers 200 hp to the shaft when it is turning at 1140 rev/min, determine the minimum required thickness of the shaft.

*5–28. The drive shaft AB of an automobile is to be designed as a thin-walled tube. The engine delivers 150 hp when the shaft is turning at 1500 rev/min. Determine the minimum thickness if the outer diameter is 2.5 in. The material has an allowable shear stress of $\tau_{allow} = 7$ ksi.

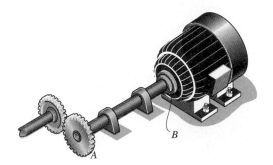

Probs. 5–27/5–28

5–29. The motor delivers 500 hp to the steel shaft AB, which is tubular and has an outer diameter of 2 in. and an inner diameter of 1.84 in. Determine the *smallest* angular velocity at which it can rotate if the allowable shear stress for the material is $\tau_{allow} = 25$ ksi.

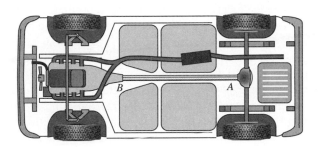

Prob. 5–29

5–30. The motor of a fan delivers 150 W of power to the blade when it is turning at 18 rev/s. Determine the smallest diameter of shaft A that can be used to connect the fan blade to the motor if the allowable shear stress is $\tau_{allow} = 80$ MPa.

Prob. 5–30

5–31. The solid steel shaft AC has a diameter of 25 mm and is supported by smooth bearings at D and E. It is coupled to a motor at C, which delivers 3 kW of power to the shaft while it is turning at 50 rev/s. If gears A and B remove 1 kW and 2 kW, respectively, determine the maximum shear stress developed in the shaft within regions AB and BC. The shaft is free to turn in its support bearings D and E.

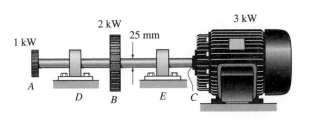

Prob. 5–31

5.4 Angle of Twist

Occasionally the design of a shaft depends on restricting the amount of rotation or twist that may occur when the shaft is subjected to a torque. Furthermore, being able to compute the angle of twist for a shaft is important when analyzing the reactions on statically indeterminate shafts.

In this section we will develop a formula for determining the *angle of twist* ϕ (phi) of one end of a shaft with respect to its other end. The shaft is assumed to have a circular cross section that can gradually vary along its length, Fig. 5–15a, and the material is assumed to be homogeneous and to behave in a linear-elastic manner when the torque is applied. As in the case of an axially loaded bar, we will neglect the localized deformations that occur at points of application of the torques and where the cross section changes abruptly. By Saint-Venant's principle, these effects occur within small regions of the shaft's length and generally have only a slight effect on the final result.

Using the method of sections, a differential disk of thickness dx, located at position x, is isolated from the shaft. Its free-body diagram is shown in Fig. 5–15b. The internal resultant torque is represented as $T(x)$, since the external loading may cause it to vary along the axis of the shaft. Due to $T(x)$ the disk will twist, such that the *relative rotation* of one of its faces with respect to the other face is $d\phi$, Fig. 5–15b. Furthermore, as explained in Sec. 5.1 and shown in Fig. 5–4, an element of material located at an arbitrary radius ρ within the disk will undergo a shear strain γ. The values of γ and $d\phi$ are related by Eq. 5–1, namely,

$$d\phi = \gamma \, \frac{dx}{\rho} \qquad (5\text{--}13)$$

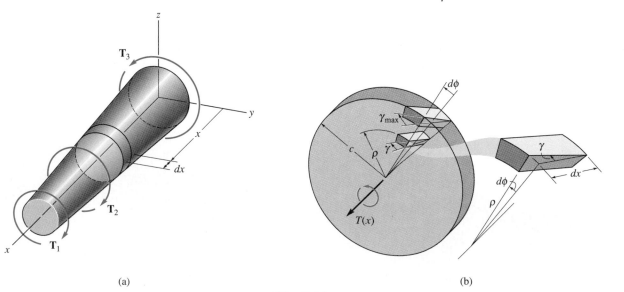

(a)

(b)

Fig. 5–15

Since Hooke's law, $\gamma = \tau/G$, applies, and the shear stress can be expressed in terms of the applied torque using the torsion formula $\tau = T(x)\rho/J(x)$, then $\gamma = T(x)\rho/J(x)G$. Substituting this result into Eq. 5–13, the angle of twist for the disk is

$$d\phi = \frac{T(x)}{J(x)G}\, dx$$

Integrating over the entire length L of the shaft, we obtain the angle of twist for the entire shaft, namely,

$$\phi = \int_0^L \frac{T(x)dx}{J(x)G} \qquad (5\text{–}14)$$

Here

ϕ = the angle of twist of one end of the shaft with respect to the other end, measured in radians

$T(x)$ = the internal torque at the arbitrary position x, found from the method of sections and the equation of moment-equilibrium applied about the shaft's axis

$J(x)$ = the shaft's polar moment of inertia expressed as a function of position x

G = the shear modulus for the material

Constant Torque and Cross-Sectional Area.

Usually in engineering practice the shaft's cross-sectional area and the applied torque are constant along the length of the shaft, Fig. 5–16. If this is the case, the internal torque $T(x) = T$, the polar moment of inertia $J(x) = J$, and Eq. 5–14 can be integrated, which gives

$$\phi = \frac{TL}{JG} \qquad (5\text{–}15)$$

The similarities between the above two equations and those for an axially loaded bar ($\Delta = \int P(x)\,dx/A(x)E$ and $\Delta = PL/AE$) should be noted.

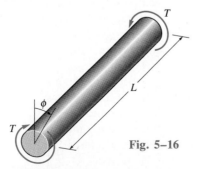

Fig. 5–16

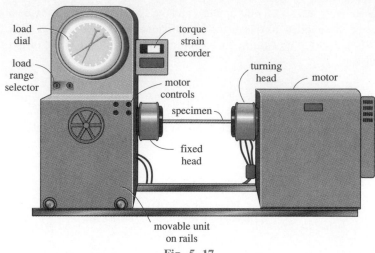

load
dial

load
range
selector

torque
strain
recorder

turning
head

motor

motor
controls

specimen

fixed
head

movable unit
on rails

Fig. 5–17

In particular, Eq. 5–15 is often used to determine the shear modulus G. To do so a specimen of known length and diameter is placed in a torsion testing machine like the one shown in Fig. 5–17. The torque T and angle of twist ϕ are then measured between a gauge length L. Using Eq. 5–15, $G = TL/J\phi$. Usually, to obtain a more reliable value of G, several of these tests are performed and the average value used.

If the shaft is subjected to several different torques, or the cross-sectional area or shear modulus changes abruptly from one region of the shaft to the next, the above equation can be applied to each segment of the shaft where these quantities are all constant. The angle of twist of one end of the shaft with respect to the other is then found from the vector addition of the angles of twist of each segment. For this case,

$$\phi = \sum \frac{TL}{JG} \tag{5–16}$$

Sign Convention.   In order to apply the above equations, we must develop a sign convention for the internal torque and the angle of twist of one end of the shaft with respect to the other end. To do this we will use the right-hand rule, whereby both the torque and angle will be *positive* at the location where they are to be calculated, provided that the *thumb* is directed *outward* from the shaft when the fingers curl to give the tendency for rotation, Fig. 5–18.

To illustrate the use of this sign convention, consider the shaft shown in Fig. 5–19a, which is subjected to four torques. The angle of twist of end A with respect to end D is to be determined. For this problem, three segments of the shaft must be considered, since the internal torque changes at B and C. Using the method of sections, the internal torques are computed for each

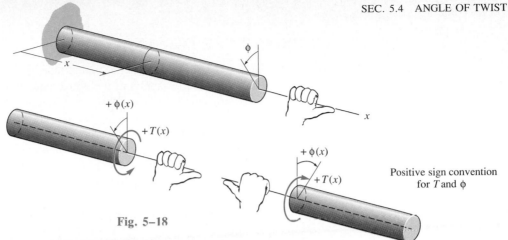

Fig. 5–18

Positive sign convention for T and ϕ

segment, Fig. 5–19b. By the right-hand rule, with positive torques directed away from the *sectioned end* of the shaft, we have $T_{AB} = +80$ N · m, $T_{BC} = -70$ N · m, and $T_{CD} = -10$ N · m. Applying Eq. 5–16, we have

$$\phi_{A/D} = \frac{(+80\text{ N}\cdot\text{m})\,L_{AB}}{JG} + \frac{(-70\text{ N}\cdot\text{m})\,L_{BC}}{JG} + \frac{(-10\text{ N}\cdot\text{m})\,L_{CD}}{JG}$$

If the other data is substituted and the answer is computed as a *positive* quantity, it means that end A will rotate as indicated by the curl of the right-hand fingers when the thumb is directed *away* from the shaft, Fig. 5–19a. The double subscript notation is used to indicate this relative angle of twist ($\phi_{A/D}$); however, if the angle of twist is to be determined relative to a *fixed point,* then only a single subscript is used. For example, if D is located at a fixed support, then the computed angle of twist will be denoted as ϕ_A.

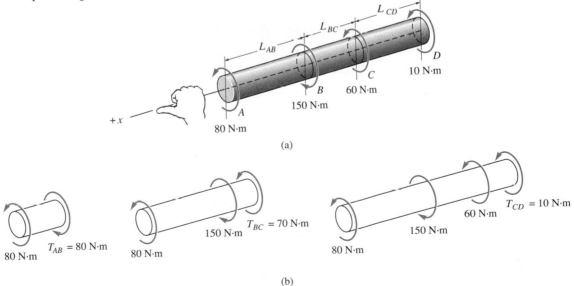

Fig. 5–19

PROCEDURE FOR ANALYSIS

The angle of twist of one end of a shaft or tube with respect to the other end can be determined by applying the above equations. The following is a suggested procedure for doing this.

Internal Torque. The internal torque $T(x)$ can be found at a point on the axis of the shaft by using the method of sections and the equation of moment equilibrium, applied along the shaft's axis. If the torque varies along the shaft's length, a section should be made at the arbitrary position x along the shaft and the torque represented as a function of x. If several constant external torques act on the shaft between its ends, the internal torque in each *segment* of the shaft, between any two external torques, must be determined.

Angle of Twist. When the circular cross-sectional area varies along the shaft's axis, the polar moment of inertia $J(x)$ must be expressed as a function of its position x along the axis. Furthermore, if the polar moment of inertia or the internal torque *suddenly changes* between the ends of the shaft, then Eq. 5–14, $\phi = \int (T(x)/(J(x)G)\,dx$, or Eq. 5–15, $\phi = TL/JG$, must be applied to *each segment* for which J, G, and T are continuous or constant. When substituting the data for the internal torque in each segment, be sure to use a consistent sign convention for the shaft, such as the one discussed above. Also make sure that a consistent set of units is used when substituting numerical data into the equations.

The following examples illustrate application of this procedure.

Example 5–7

The gears attached to the fixed-end steel shaft are subjected to the torques shown in Fig. 5–20a. If the shear modulus of elasticity is $G =$ 80 GPa and the shaft has a diameter of 14 mm, determine the displacement of the tooth P on gear A. The shaft turns freely within the bearing at B.

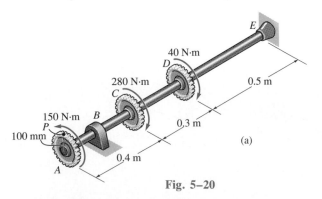

Fig. 5–20

Example 5–7 (*Continued*)

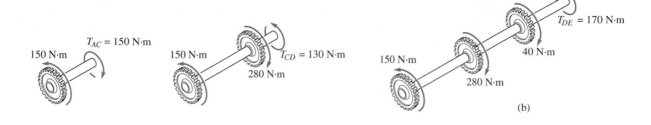

(b)

SOLUTION

Internal Torque.   By inspection, the torques in segments AC, CD, and DE are different yet *constant* throughout each segment. Free-body diagrams of appropriate segments of the shaft along with the calculated internal torques are shown in Fig. 5–20b. Using the right-hand rule and the established sign convention that positive torque is directed away from the sectioned end of the shaft, we have

$$T_{AC} = +150 \text{ N} \cdot \text{m} \qquad T_{CD} = -130 \text{ N} \cdot \text{m} \qquad T_{DE} = -170 \text{ N} \cdot \text{m}$$

Angle of Twist.   The polar moment of inertia for the shaft is

$$J = \frac{\pi}{2}(0.007 \text{ m})^4 = 3.77(10^{-9}) \text{ m}^4$$

Applying Eq. 5–16 to each segment and adding the results algebraically, we have

$$\phi_A = \sum \frac{TL}{JG} = \frac{(+150 \text{ N} \cdot \text{m})(0.4 \text{ m})}{3.77(10^{-9}) \text{ m}^4 [80(10^9) \text{ N/m}^2]}$$

$$+ \frac{(-130 \text{ N} \cdot \text{m})(0.3 \text{ m})}{3.77(10^{-9}) \text{ m}^4 [80(10^9) \text{ N/m}^2]}$$

$$+ \frac{(-170 \text{ N} \cdot \text{m})(0.5 \text{ m})}{3.77(10^{-9}) \text{ m}^4 [80(10^9) \text{ N/m}^2]} = -0.212 \text{ rad}$$

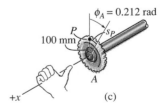

(c)

Since the answer is negative, by the right-hand rule the thumb is directed *toward* the end E of the shaft, and therefore gear A will rotate as shown in Fig. 5–20c.

The displacement of tooth P on gear A is

$$s_P = \phi_A r = (0.212 \text{ rad})(100 \text{ mm}) = 21.2 \text{ mm} \qquad \textit{Ans.}$$

Remember that this analysis is valid only if the shear stress does not exceed the proportional limit of the material.

Example 5–8

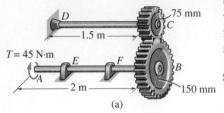

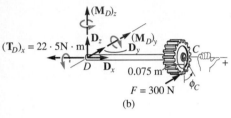

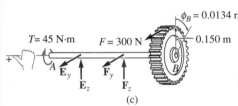

Fig. 5–21

The two solid steel shafts shown in Fig. 5–21a are coupled together using the meshed gears. Determine the angle of twist of end A of shaft AB when the torque $T = 45$ N · m is applied. Take $G = 80$ GPa. Shaft AB is free to rotate within bearings E and F, whereas shaft DC is fixed at D. Each shaft has a diameter of 20 mm.

SOLUTION

Since this problem involves two shafts, we must first calculate the angle of twist of each shaft separately, then add the results vectorially.

Internal Torque.   Free-body diagrams for each shaft are shown in Fig. 5–21b and 5–21c. Summing moments along the x axis of shaft AB yields the tangential reaction between the gears of $F = 45$ N · m/0.15 m = 300 N. Summing moments about the x axis of shaft DC, this force then creates a torque of $(T_D)_x = 300$ N(0.075 m) = 22.5 N · m on shaft DC.

Angle of Twist.   By the positive sign convention, the 22.5-N · m torque on shaft DC rotates gear C in the *positive direction;* i.e., by the right-hand rule the thumb is directed outward from the end of the shaft, Fig. 5–21b. This angle of twist is

$$\phi_C = \frac{TL_{DC}}{JG} = \frac{(+22.5 \text{ N} \cdot \text{m})(1.5 \text{ m})}{(\pi/2)(0.010 \text{ m})^4[80(10^9) \text{ N/m}^2]} = +0.0269 \text{ rad}$$

Since the gears at the end of the shaft are in *mesh,* the rotation ϕ_C causes gear B to rotate ϕ_B, Fig. 5–21c. Thus,

$$\phi_B(0.15 \text{ m}) = (0.0269 \text{ rad})(0.075 \text{ m})$$

$$\phi_B = 0.0134 \text{ rad}$$

Equation 5–15 may now be applied to determine the angle of twist of end A with respect to end B of shaft AB. The 45-N · m torque in the shaft is positive, since the thumb of the right hand is directed *away* from the end of the shaft at A. Therefore,

$$\phi_{A/B} = \frac{T_{AB}L_{AB}}{JG} = \frac{(+45 \text{ N} \cdot \text{m})(2 \text{ m})}{(\pi/2)(0.010 \text{ m})^4[80(10^9) \text{ N/m}^2]} = +0.0716 \text{ rad}$$

Here the positive sign indicates that the rotation, defined by the right-hand thumb at A, is directed to the left, Fig. 5–21c.

The rotation of end A is therefore determined by adding ϕ_B and $\phi_{A/B}$, since both angles are in the *same direction*. We have

$$\phi_A = \phi_B + \phi_{A/B} = 0.0134 \text{ rad} + 0.0716 \text{ rad} = +0.0850 \text{ rad}   \textit{Ans.}$$

This angle of twist is in the same direction as **T**, shown in Fig. 5–21c.

Example 5–9

The 2-in.-diameter cast-iron post shown in Fig. 5–22a is buried 24 in. in soil. If a torque is applied to its top using a rigid wrench, determine the maximum shear stress in the post and the angle of twist at its top. Assume that the torque is about to turn the post, and the soil exerts a uniform torsional resistance of t lb $\cdot$ in./in. along its 24-in. buried length. $G = 5.5(10^3)$ ksi.

SOLUTION

Internal Torque. The internal torque in segment AB of the post is constant. From the free-body diagram, Fig. 5–22b, we have

$$\Sigma M_z = 0; \qquad T_{AB} = 25 \text{ lb}(12 \text{ in.}) = 300 \text{ lb} \cdot \text{in.}$$

The magnitude of the uniform distribution of torque along the buried segment BC can be determined from equilibrium of the entire post, Fig. 5–22c. Here

$$\Sigma M_z = 0; \qquad 25 \text{ lb}(12 \text{ in.}) - t(24 \text{ in.}) = 0$$
$$t = 12.5 \text{ lb} \cdot \text{in./in.}$$

Hence, from a free-body diagram of a section of the post located at the position x within region BC, Fig. 5–22d, we have

$$\Sigma M_z = 0; \qquad T_{BC} - 12.5x = 0$$
$$T_{BC} = 12.5x$$

Maximum Shear Stress. The largest shear stress occurs in region AB, since the torque is largest there and J is constant for the post. Applying the torsion formula, we have

$$\tau_{\max} = \frac{T_{AB}c}{J} = \frac{(300 \text{ lb} \cdot \text{in.})(1 \text{ in.})}{(\pi/2)(1 \text{ in.})^4} = 191 \text{ psi} \qquad \textit{Ans.}$$

Angle of Twist. The angle of twist at the top can be determined relative to the bottom of the post, since it is fixed and yet is about to turn. Both segments AB and BC twist, and so in this case we have

$$\phi_A = \frac{T_{AB}L_{AB}}{JG} + \int_0^{L_{BC}} \frac{T_{BC}\,dx}{JG}$$

$$= \frac{(300 \text{ lb} \cdot \text{in.})(36 \text{ in.})}{JG} + \int_0^{24 \text{ in.}} \frac{12.5x\,dx}{JG}$$

$$= \frac{10{,}800 \text{ lb} \cdot \text{in}^2}{JG} + \frac{12.5[(24)^2/2]\text{lb} \cdot \text{in}^2}{JG}$$

$$= \frac{14{,}400 \text{ lb} \cdot \text{in}^2}{(\pi/2)(1 \text{ in.})^4 5500(10^3) \text{ lb/in}^2} = 0.00167 \text{ rad} \qquad \textit{Ans.}$$

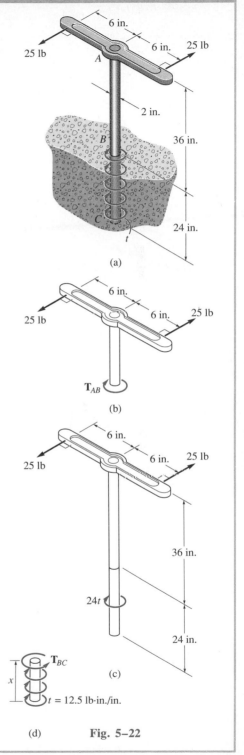

Fig. 5–22

The tapered shaft shown in Fig. 5–23*a* is made of a material having a shear modulus *G*. Determine the angle of twist of its end *B* when subjected to the torque **T**.

SOLUTION

Internal Torque. By inspection or from the free-body diagram of a section located at the arbitrary position *x*, Fig. 5–23*b*, the internal torque is *T*.

Angle of Twist. Here the polar moment of inertia varies along the shaft's axis and therefore we must express it in terms of the coordinate *x*. The radius *c* of the shaft at *x* can be determined in terms of *x* by proportion of the slope of line *AB* in Fig. 5–23*c*. We have

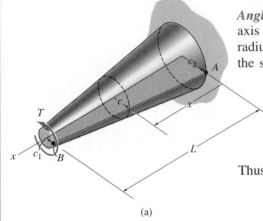

$$\frac{c_2 - c_1}{L} = \frac{c_2 - c}{x}$$

$$c = c_2 - x\left(\frac{c_2 - c_1}{L}\right)$$

Thus, at *x*,

$$J(x) = \frac{\pi}{2}\left[c_2 - x\left(\frac{c_2 - c_1}{L}\right)\right]^4$$

Applying Eq. 5–14, we have

$$\phi = \int_0^L \frac{T\,dx}{\left(\frac{\pi}{2}\right)\left[c_2 - x\left(\frac{c_2 - c_1}{L}\right)\right]^4 G} = \frac{2T}{\pi G}\int_0^L \frac{dx}{\left[c_2 - x\left(\frac{c_2 - c_1}{L}\right)\right]^4}$$

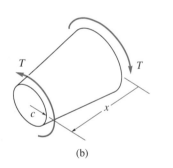

Performing the integration using an integral table, the result becomes

$$\phi = \left(\frac{2T}{\pi G}\right)\frac{1}{3\left(\frac{c_2 - c_1}{L}\right)\left[c_2 - x\left(\frac{c_2 - c_1}{L}\right)\right]^3}\Bigg|_0^L$$

$$= \frac{2T}{\pi G}\left(\frac{L}{3(c_2 - c_1)}\right)\left(\frac{1}{c_1^3} - \frac{1}{c_2^3}\right)$$

Rearranging terms yields

$$\phi = \frac{2TL}{3\pi G}\left(\frac{c_2^2 + c_1 c_2 + c_1^2}{c_1^3 c_2^3}\right) \qquad \textit{Ans.}$$

To partially check this result, note that when $c_1 = c_2 = c$, then

$$\phi = \frac{TL}{[(\pi/2)c^4]G} = \frac{TL}{JG}$$

which is Eq. 5–15.

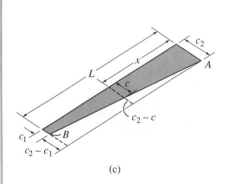

Fig. 5–23

PROBLEMS

*5–32. The 0.75-in.-diameter steel shaft is supported by bearings at A and B. If it is subjected to torques of 35 lb · ft applied to each gear, determine in degrees the angle of twist of end C of the shaft with respect to end B. $G_{st} = 10.8(10^3)$ ksi.

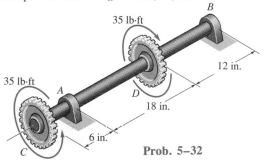

Prob. 5–32

5–33. The aluminum shaft consists of a solid section AB and a tube BC, which is fixed to the wall at C. Determine the angle of twist at its end A when it is subjected to the torsional loading shown. Take $G_{al} = 3.80(10^3)$ ksi. Section AB has a diameter of 1.5 in., and BC has an inner diameter of 1.5 in. and an outer diameter of 2.5 in.

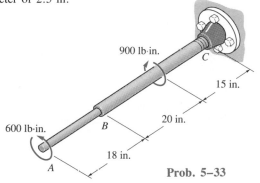

Prob. 5–33

5–34. The solid shaft of radius c is subjected to a torque $\mathbf{T}$ at its ends. Show that the maximum shear strain developed in the shaft is $\gamma_{max} = Tc/JG$. What is the shear strain on an element located at point A, $c/2$ from the center of the shaft? Sketch the strain distortion of this element.

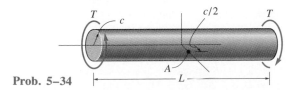

Prob. 5–34

5–35. The splined ends and gears attached to the steel shaft are subjected to the torques shown. Determine the angle of twist of end B with respect to end A. The shaft has a diameter of 40 mm and $G_{st} = 75$ GPa.

*5–36. The splined ends and gears attached to the steel shaft are subjected to the torques shown. Determine the angle of twist of gear C with respect to gear D. The shaft has a diameter of 40 mm and $G_{st} = 75$ GPa.

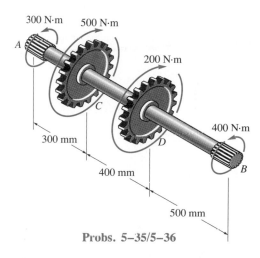

Probs. 5–35/5–36

5–37. The steel axle is made from tubes AB and CD and a solid section BC. It is supported on smooth bearings that allow it to rotate freely. If the gears, fixed to its ends, are subjected to 85-N · m torques, determine the angle of twist of gear A relative to gear D. The tubes have an outer diameter of 30 mm and an inner diameter of 20 mm. The solid section has a diameter of 40 mm. $G_{st} = 75$ GPa.

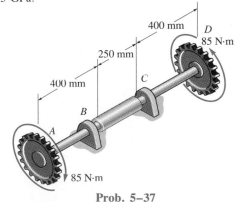

Prob. 5–37

5–38. The engine of the helicopter is delivering 600 hp to the rotor shaft AB when the blade is rotating at 1200 rev/min. Determine to the nearest $\frac{1}{8}$ in. the diameter of the shaft AB if the allowable shear stress is $\tau_{\text{allow}} = 8$ ksi and the vibrations limit the angle of twist of the shaft to 0.05 rad. The shaft is 2 ft long and made from steel. $G_{st} = 11.6(10^3)$ ksi.

5–39. The engine of the helicopter is delivering 600 hp to the rotor shaft AB when the blade is rotating at 1200 rev/min. Determine to the nearest $\frac{1}{8}$ in. the diameter of the shaft AB if the allowable shear stress is $\tau_{\text{allow}} = 10.5$ ksi and the vibrations limit the angle of twist of the shaft to 0.05 rad. The shaft is 2 ft long and made from steel. $G_{st} = 11.6(10^3)$ ksi.

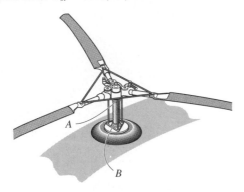

Probs. 5–38/5–39

*5–40. The 8-mm-diameter steel bolt is screwed tightly into a block at A. Determine the couple forces F that should be applied to the wrench so that the maximum shear stress in the bolt becomes 180 MPa. Also, compute the corresponding displacement of each force F needed to cause this stress. Assume that the wrench is rigid. Take $G_{st} = 75$ GPa.

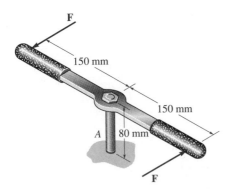

Prob. 5–40

5–41. The rotating flywheel-and-shaft, when brought to a sudden stop at D, begins to oscillate clockwise-counterclockwise such that a point A on the outer edge of the flywheel is displaced through a 6-mm arc. Determine the maximum shear stress developed in the tubular steel shaft due to this oscillation. The shaft has an inner diameter of 24 mm and an outer diameter of 32 mm. The bearings at B and C allow the shaft to rotate freely, whereas the support at D holds the shaft fixed. $G_{st} = 75$ GPa.

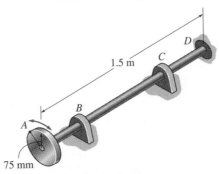

Prob. 5–41

5–42. The steel shaft is 2 m long and has an outer diameter of 40 mm. When it is rotating at 80 rad/s, it transmits 32 kW of power from the engine E to the generator G. Determine the smallest thickness of the shaft if the allowable shear stress is $\tau_{\text{allow}} = 140$ MPa and the shaft is restricted not to twist more than 0.05 rad. $G_{st} = 75$ GPa.

5–43. The solid steel shaft is 3 m long and has a diameter of 50 mm. It is required to transmit 35 kW of power from the engine E to the generator G. Determine the smallest angular velocity the shaft can have if it is restricted not to twist more than 1°. $G_{st} = 75$ GPa.

Probs. 5–42/5–43

*5–44. The motor delivers 40 hp to the solid steel shaft while it rotates at 20 Hz. The shaft is supported on smooth bearings at A and B, which allow free rotation of the shaft. The gears C and D fixed to the shaft remove 25 hp and 15 hp, respectively. Determine the diameter of the shaft to the nearest $\frac{1}{8}$ in. if the allowable shear stress is $\tau_{\text{allow}} = 8$ ksi and the allowable angle of twist of C with respect to D is 0.20°. $G_{st} = 11.0(10^3)$ ksi.

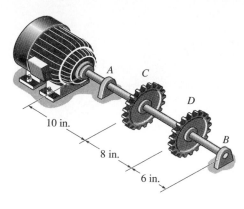

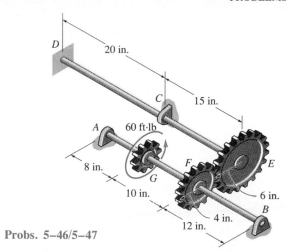

Prob. 5–44

Probs. 5–46/5–47

5–45. The two steel shafts AC and FD are coupled together using the meshed-gear arrangement shown. If F is a fixed support, determine the angle of twist at end A due to the applied torsional loading. Each shaft has a diameter of 30 mm, and the bearings at A, B, and G resist no torque. $G_{st} = 80$ GPa.

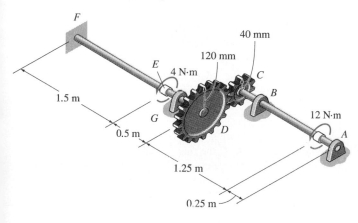

Prob. 5–45

5–46. The two steel shafts each have a diameter of 1 in. and are supported by bearings at A, B, and C, which allow free rotation. If the support at D is fixed, determine the angle of twist of end B when the torque of 60 ft · lb is applied to gear G. $G_{st} = 10.8(10^3)$ ksi.

5–47. The two steel shafts each have a diameter of 1 in. and are supported by bearings at A, B, and C, which allow free rotation. If the support at D is fixed, determine the angle of twist of end A when the torque of 60 ft · lb is applied to gear G. $G_{st} = 10.8(10^3)$ ksi.

*5–48. A shaft is subjected to a torque T. Compare the effectiveness of using the tube shown in the figure with that of a solid section of radius c. To do this, compute the percent increase in torsional stress and angle of twist per unit length for the tube versus the solid section.

Prob. 5–48

5–49. The shaft has a radius c and is subjected to a torque per unit length of t_0, which is distributed uniformly over the shaft's entire length L. If it is fixed at its far end A, determine the angle of twist ϕ of end B.

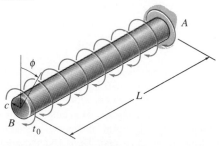

Prob. 5–49

5–50. When drilling a well, the deep end of the drill pipe is assumed to encounter a torsional resistance T_A. Furthermore, soil friction along the sides of the pipe creates a linear distribution of torque per unit length, varying from zero at the surface B to t_0 at A. Determine the necessary torque T_B that must be supplied by the drive unit to turn the pipe. Also, what is the relative angle of twist of one end of the pipe with respect to the other end at the instant the pipe is about to turn? The pipe has an outer radius r_o and an inner radius r_i. The shear modulus is G.

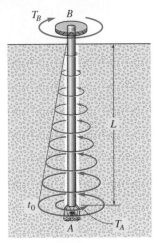

Prob. 5–50

5–51. The tapered shaft has a length L and a radius r at end A and $2r$ at end B. If it is fixed at end B and is subjected to a torque T, determine the angle of twist of end A. The shear modulus is G.

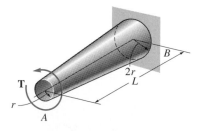

Prob. 5–51

*5–52. The solid steel shaft consists of two tapered ends, AB and CD, and a central portion BC having a constant diameter. It is supported by two smooth bearings, which allow it to rotate freely. If the gears fixed to its ends are subjected to counterbalancing torques of 1700 lb · in., determine the angle of twist of one end of the shaft relative to the other end. *Hint:* Use the results of Prob. 5–51. $G_{st} = 12(10^3)$ ksi.

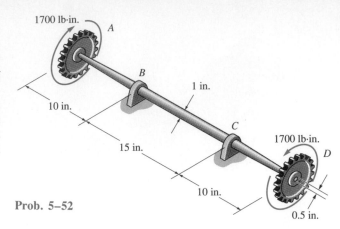

Prob. 5–52

5–53. The contour of the surface of the shaft is defined by the equation $y = e^{ax}$, where a is a constant. If the shaft is subjected to a torque T at its ends, determine the angle of twist of end A with respect to end B. The shear modulus is G.

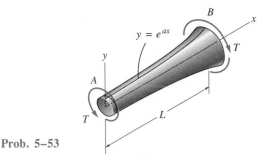

Prob. 5–53

5–54. A cylindrical spring consists of a rubber annulus bonded to a rigid ring and shaft. If the ring is held fixed and a torque T is applied to the rigid shaft, determine the angle of twist of the shaft. The shear modulus of the rubber is G. *Hint:* As shown in the figure, the deformation of the element at radius r can be determined from $r\,d\theta = dr\,\gamma$. Use this expression along with $\tau = T/2\pi r^2 h$, from Prob. 5–21, to obtain the result.

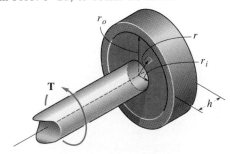

Prob. 5–54

5.5 Statically Indeterminate Torque-Loaded Members

A torsionally loaded shaft may be classified as statically indeterminate if the moment equation of equilibrium, applied about the axis of the shaft, is not adequate to determine the unknown torques acting on the shaft. An example of this situation is shown in Fig. 5–24a. As shown on the free-body diagram, Fig. 5–24b, the reactive torques at the supports A and B are unknown. We require

$$\Sigma M_x = 0; \qquad\qquad T - T_A - T_B = 0$$

Since only one equilibrium equation is relevant and there are two unknowns, this problem is statically indeterminate. In order to obtain a solution we will use the method of analysis discussed in Sec. 4.4.

The necessary condition of compatibility, or the kinematic condition, requires the angle of twist of one end of the shaft with respect to the other end to be equal to zero, since the end supports are fixed. Therefore,

$$\phi_{A/B} = 0$$

In order to write this equation in terms of the unknown torques, we will assume that the material behaves in a linear-elastic manner, so that the load–displacement relationship is expressed by Eq. 5–16. Realizing that the internal torque in segment AC is $+T_A$, and in segment CB the internal torque is $-T_B$, the above compatibility equation can be written as

$$\frac{T_A L_{AC}}{JG} - \frac{T_B L_{BC}}{JG} = 0$$

Here JG is assumed to be constant.

Solving the above two equations for the reactions, realizing that $L = L_{AC} + L_{BC}$, we get

$$T_A = T\left(\frac{L_{BC}}{L}\right)$$

and

$$T_B = T\left(\frac{L_{AC}}{L}\right)$$

To summarize the above procedure, the unknown torques in statically indeterminate problems must be determined by satisfying both equilibrium and compatibility for the shaft. To do so it is necessary to transform the compatibility equation, which involves rotation, into an equation that involves torque. This is done using the load–displacement relation that describes the material behavior.

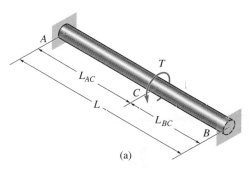

(a)

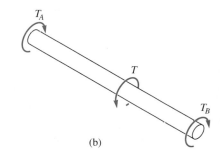

(b)

Fig. 5–24

Superposition of Torques. It is also possible to write the equation of compatibility using the superposition of torques. For example, we will choose the support at B to be redundant. When this support is temporarily removed, the shaft, being fixed at A, becomes statically determinate and stable. Using the principle of superposition, the external torque $\mathbf{T}$ is first applied to the shaft, Fig. 5–24d. This causes the end B of the shaft to undergo an angle of twist ϕ_B. The redundant torque $\mathbf{T}_B$ is then applied to the unloaded shaft, which causes the end B of the shaft to twist $-\phi'_B$, Fig. 5–24e. Since the redundant actually holds the shaft fixed at B, Fig. 5–24c, the compatibility equation at B becomes

$$\phi_B - \phi'_B = 0$$

Applying the load–displacement relationship, we find $\phi_B = TL_{AC}/JG$ and $\phi'_B = T_B L/JG$. Consequently,

$$\frac{TL_{AC}}{JG} - \frac{T_B L}{JG} = 0$$

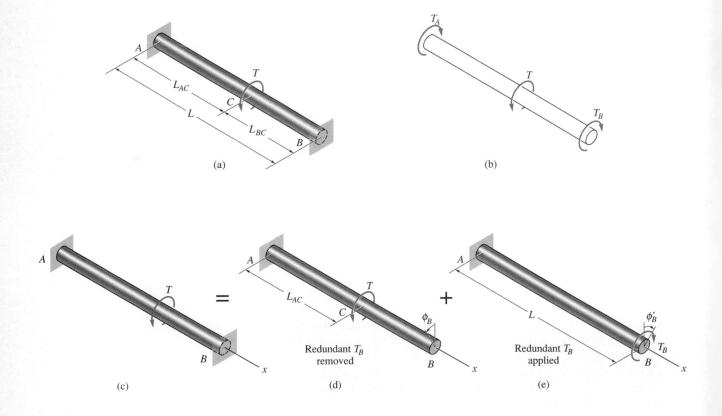

Fig. 5–24

or

$$T_B = T\left(\frac{L_{AC}}{L}\right)$$

From the free-body diagram of the shaft, Fig. 5–24b, the torque at A can now be determined using the moment equation of equilibrium,

$$\Sigma M_x = 0; \qquad\qquad T - T_A - T_B = 0$$

$$T_A = T\left(\frac{L - L_{AC}}{L}\right)$$

Since $L_{BC} = L - L_{AC}$, we have

$$T_A = T\left(\frac{L_{BC}}{L}\right)$$

These results are the same as those obtained previously, except that here we first applied the compatibility condition and then the equilibrium condition to obtain the solution. Also note that the principle of superposition can indeed be used here, since the angles of twist and torque are linearly related, i.e., $\phi = TL/JG$.

PROCEDURE FOR ANALYSIS

Either of the above two methods can be used to determine the unknown torques in a shaft that is statically indeterminate. Each method requires that the conditions of equilibrium and compatibility be satisfied.

In order to solve a particular problem, it is important first to draw a free-body diagram of the shaft to identify all the torques. The sequence for applying the necessary equilibrium and compatibility equations is arbitrary. If it seems difficult to establish the compatibility equation, it is suggested that a superposition of torques be used. This method requires choosing one of the supports as a redundant. This support is then temporarily removed. The known angle of twist at the support (which may be zero) is then expressed as the algebraic addition of the angle of twist at the support caused by the external torques acting on the shaft and the angle of twist at the support caused only by the redundant torque acting on the shaft. These two angles are then expressed in terms of the torques using the load–displacement relationship $\phi = TL/JG$. Once established, the compatibility equation can then be solved for the magnitude of the redundant torque. The other unknown torque is determined from the conditions for equilibrium.

If any of the unknown torque magnitudes has a negative numerical value once it has been determined, then the torque acts in the opposite sense of direction of that indicated on the free-body diagram.

The following examples numerically illustrate both methods of solution.

Example 5–11

The solid steel shaft shown in Fig. 5–25a has a diameter of 20 mm. If it is subjected to the two torques, determine the reactions at the fixed supports A and B. Take $G = 75$ GPa.

Fig. 5–25

(a)

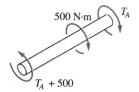

(c)

(b)

SOLUTION I

Equilibrium. By inspection of the free-body diagram, Fig. 5–25b, it is seen that the problem is statically indeterminate since there is only *one* available equation of equilibrium, whereas $\mathbf{T}_A$ and $\mathbf{T}_B$ are unknown. We require

$$\Sigma M_x = 0; \qquad -T_B + 800 - 500 - T_A = 0 \qquad (1)$$

Compatibility. Since the ends of the shaft are fixed, the angle of twist of one end of the shaft with respect to the other must be zero. Hence, the compatibility equation can be written as

$$\phi_{A/B} = 0$$

This condition can be expressed in terms of the unknown torques by using the load–displacement relationship, Eq. 5–15. Here there are three regions of the shaft where the internal torque is constant. On the free-body diagrams in Fig. 5–25c we have shown these internal torques acting on segments of the shaft.* By the sign conventions established in Sec. 5.4, we have

$$\frac{-T_B(0.2 \text{ m})}{JG} + \frac{(T_A + 500 \text{ N} \cdot \text{m})(1.5 \text{ m})}{JG} + \frac{T_A(0.3 \text{ m})}{JG} = 0$$

or

$$1.8T_A - 0.2T_B = -750 \qquad (2)$$

Solving Eqs. 1 and 2 yields

$$T_A = -345 \text{ N} \cdot \text{m} \qquad T_B = 645 \text{ N} \cdot \text{m} \qquad \textit{Ans.}$$

The negative sign indicates that $\mathbf{T}_A$ acts in the opposite direction of that shown in Fig. 5–25b.

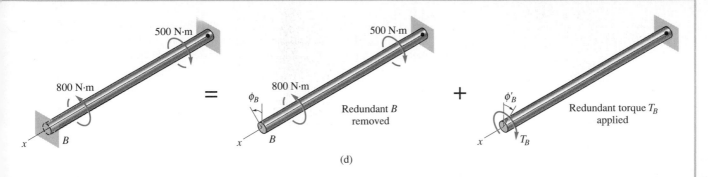

(d)

SOLUTION II

We can also solve this problem using the method of superposition, with the support at B considered as the redundant.

Compatibility. Figure 5–25d shows the shaft loaded with the external torques, causing end B to rotate ϕ_B, plus the shaft loaded only with the redundant torque $\mathbf{T}_B$, which causes a rotation of ϕ_B'. We require

$$0 = \phi_B + \phi_B' \qquad (3)$$

The internal torque in each section of the shaft used for computing ϕ_B is shown on the segments in Fig. 5–25e. Thus,

$$\phi_B = 0 + \frac{(+800 \text{ N} \cdot \text{m})(1.5 \text{ m})}{JG} + \frac{(+300 \text{ N} \cdot \text{m})(0.3 \text{ m})}{JG}$$

$$= \frac{1290}{JG}$$

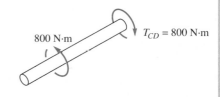

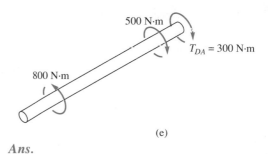

For the shaft subjected only to $\mathbf{T}_B$, Fig. 5–25d, we have

$$\phi_B' = \frac{-T_B (2.0 \text{ m})}{JG}$$

Therefore, Eq. 3 becomes

$$0 = \frac{1290}{JG} - \frac{T_B (2.0)}{JG}$$

$$T_B = 645 \text{ N} \cdot \text{m} \qquad \textit{Ans.}$$

Equilibrium. Using this result and applying moment equilibrium to the shaft in Fig. 5–25b, we obtain the previous result for T_A, i.e.,

$$\Sigma M_x = 0; \quad -645 \text{ N} \cdot \text{m} + 800 \text{ N} \cdot \text{m} - 500 \text{ N} \cdot \text{m} - T_A = 0$$

$$T_A = -345 \text{ N} \cdot \text{m} \qquad \textit{Ans.}$$

*Alternatively, we can use internal loadings of $(T_A - 300)$, $(800 - T_B)$, and $(-T_B + 300)$.

Example 5–12

The shaft shown in Fig. 5–26a is made from a steel tube and brass core. If a torque of $T = 250$ lb · ft is applied at its end, plot the shear-stress distribution along a radial line of its cross-sectional area. Take $G_{st} = 11.4(10^3)$ ksi, $G_{br} = 5.20(10^3)$ ksi.

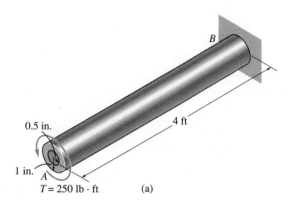

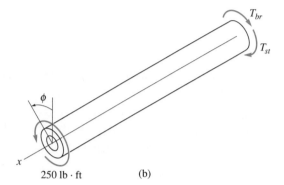

Fig. 5–26

SOLUTION

Equilibrium. A free-body diagram of the shaft is shown in Fig. 5–26b. The reaction at the wall has been represented by the unknown amount of torque resisted by the steel, $\mathbf{T}_{st}$, and by the brass, $\mathbf{T}_{br}$. Working in units of pounds and inches, equilibrium requires

$$-T_{st} - T_{br} + 250 \text{ lb} \cdot \text{ft}(12 \text{ in./ft}) = 0 \qquad (1)$$

Compatibility. We require the angle of twist of end A to be the *same* for both the steel and brass. Thus,

$$\phi = \phi_{st} = \phi_{br}$$

Applying the load–displacement relationship, Eq. 5–15, we have

$$\frac{T_{st}L}{(\pi/2)[(1 \text{ in.})^4 - (0.5 \text{ in.})^4]11.4(10^3) \text{ kip/in}^2} =$$

$$\frac{T_{br}L}{(\pi/2)(0.5 \text{ in.})^4 5.20(10^3) \text{ kip/in}^2}$$

$$T_{st} = 32.88 T_{br} \qquad (2)$$

Solving Eqs. 1 and 2, we get

$$T_{st} = 2911.0 \text{ lb} \cdot \text{in.} = 242.6 \text{ lb} \cdot \text{ft}$$
$$T_{br} = 88.5 \text{ lb} \cdot \text{in.} = 7.38 \text{ lb} \cdot \text{ft}$$

These torques act throughout the entire length of the shaft, since no external torques act at intermediate points along the shaft's axis. The shear stress in the brass core varies from zero at its center to a maximum at the interface where it contacts the steel tube. Using the torsion formula,

$$(\tau_{br})_{max} = \frac{(88.5 \text{ lb} \cdot \text{in.})(0.5 \text{ in.})}{(\pi/2)(0.5 \text{ in.})^4} = 451 \text{ psi}$$

For the steel, the minimum shear stress is also at this interface,

$$(\tau_{st})_{min} = \frac{(2911.0 \text{ lb} \cdot \text{in.})(0.5 \text{ in.})}{(\pi/2)[(1 \text{ in.})^4 - (0.5 \text{ in.})^4]} = 988 \text{ psi}$$

and the maximum shear stress is at the outer surface,

$$(\tau_{st})_{max} = \frac{(2911.0 \text{ lb} \cdot \text{in.})(1 \text{ in.})}{(\pi/2)[(1 \text{ in.})^4 - (0.5 \text{ in.})^4]} = 1977 \text{ psi}$$

The results are plotted in Fig. 5–26c. Note the discontinuity of *shear stress* at the brass and steel interface. This is to be expected, since the materials have different moduli of rigidity; i.e., steel is stiffer than brass ($G_{st} > G_{br}$) and thus it carries more shear stress at the interface. Although the shear stress is discontinuous here, the *shear strain* is not. Rather, the shear strain is the *same* for both the brass and the steel. This can be shown by using Hooke's law, $\gamma = \tau/G$. At the interface, Fig. 5–26d, the shear strain is

$$\gamma = \frac{\tau}{G} = \frac{451 \text{ psi}}{5,200 \text{ psi}} = \frac{988 \text{ psi}}{11,400 \text{ psi}} = 0.0867 \text{ rad}$$

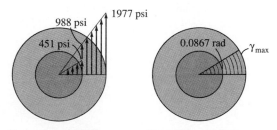

Shear stress distribution

(c)

Shear strain distribution

(d)

PROBLEMS

5–55. The copper pipe has an outer diameter of 1.5 in. and a thickness of 0.125 in. The coupling on it at C is being tightened using a wrench. If the torque developed at A is 125 lb · in., determine the magnitude F of the couple forces. The pipe is fixed-supported at end B. $G_{cu} = 62(10^3)$ ksi.

*5–56. The copper pipe has an outer diameter of 1.5 in. and a thickness of 0.125 in. The coupling on it at C is being tightened using a wrench. If the applied force $F = 30$ lb, determine the maximum shear stress developed in the pipe. The pipe is fixed-supported at end B. $G_{cu} = 62(10^3)$ ksi.

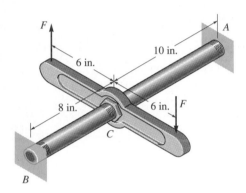

Probs. 5–55/5–56

5–57. A rod is made from two segments: AB is steel and BC is brass. It is fixed at its ends and subjected to a torque of $T = 680$ N · m. If the steel portion has a diameter of 30 mm, determine the required diameter of the brass portion so the reactions at the walls will be the same. $G_{st} = 75$ GPa, $G_{br} = 39$ GPa.

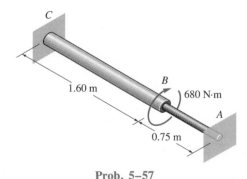

Prob. 5–57

5–58. The composite shaft is made from two segments, one having a diameter $2c$ and the other $4c$. Determine the position x of the applied torque **T** so that the reactive torques at A and B are equal. The material has a shear modulus G.

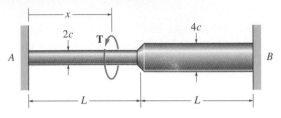

Prob. 5–58

5–59. The steel shaft is made from two segments: Segment AC has a diameter of 0.5 in. and CB has a diameter of 1 in. If it is fixed at its ends A and B and subjected to a torque of 500 lb · ft, determine the maximum shear stress in the shaft. $G_{st} = 10.8(10^3)$ ksi.

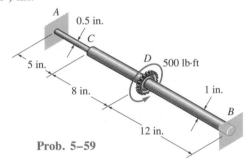

Prob. 5–59

*5–60. The shaft is made from a solid steel section AB and a tubular portion made of steel and having a brass core. If it is fixed to a rigid support at A, and a torque of $T = 50$ lb · ft is applied to it at C, determine the angle of twist that occurs at C and compute the maximum shear stress and shear strain in the brass and steel. Take $G_{st} = 11.5(10^3)$ ksi, $G_{br} = 5.6(10^3)$ ksi.

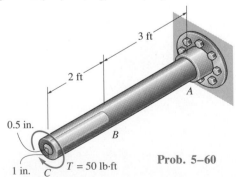

Prob. 5–60

5–61. The aluminum tube has an inner diameter of 35 mm and an outer diameter of 60 mm. A torque of $T = 850$ N · m is applied to its end B. Determine the depth d to which it has to be filled with a steel plug so that the rotation of end B is limited to 0.015 rad. The tube is fixed to the wall at A. $G_{st} = 80$ GPa, $G_{al} = 30$ GPa.

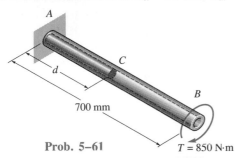

Prob. 5–61

$T = 850$ N·m

5–62. The two steel shafts each have a diameter of 25 mm and are connected together using the gears fixed to their ends. Their other ends are attached to fixed supports at A and B. They are also supported by bearings at C and D, which allow free rotation of the shafts along their axes. If a torque of 500 N · m is applied to the top gear as shown, determine the reactions at A and B. $G_{st} = 80$ GPa.

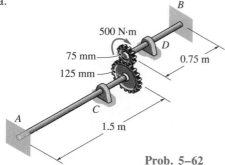

Prob. 5–62

5–63. The motor A develops a torque at gear B of 450 lb · ft, which is applied along the axis of the 2-in.-diameter steel shaft CD. This torque is to be transmitted to the pinion gears at E and F. If these gears are temporarily fixed, determine the maximum shear stress in segments CB and BD of the shaft. Also, what is the angle of twist of each of these segments? The bearings at C and D only exert force reactions on the shaft and do not resist torque. $G_{st} = 12(10^3)$ ksi.

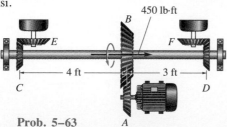

Prob. 5–63

*5–64. The two 3-ft-long aluminum shafts each have a diameter of 1.5 in. and are connected together using the gears fixed to their ends. Their other ends are attached to fixed supports at A and B. They are also supported by bearings at C and D, which allow free rotation of the shafts along their axes. If a torque of 600 lb · ft is applied to the top gear as shown, determine the maximum shear stress in each shaft. $G_{al} = 30$ GPa.

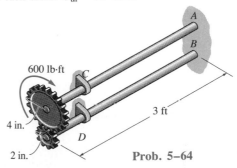

Prob. 5–64

5–65. A portion of the steel shaft is subjected to a constant distributed torsional loading of 200 lb · ft/ft. If the shaft has the dimensions shown, determine the reactions at the fixed supports A and C. Segment AB has a diameter of 1.5 in. and segment BC has a diameter of 0.75 in. $G_{st} = 11.3(10^3)$ ksi.

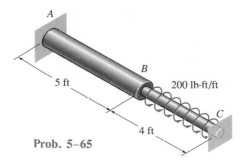

Prob. 5–65

5–66. The shaft of radius c is subjected to a distributed torque t, measured as torque/length of shaft. Determine the reactions at the fixed supports A and B.

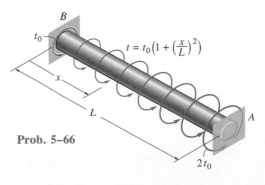

$t = t_0\left(1 + \left(\frac{x}{L}\right)^2\right)$

Prob. 5–66

5.6 Solid Noncircular Shafts

It was demonstrated in Sec. 5.1 that when a torque is applied to a shaft having a circular cross section—that is, one that is axisymmetric—the shear strains vary linearly from zero at its center to a maximum at its outer surface. Furthermore, due to the uniformity of the shear strain at all points on the same radius, the cross section does not deform, but rather remains plane after the shaft has twisted. Shafts that have a noncircular cross section, however, are *not* axisymmetric, and because the shear stress over their cross section is distributed in a very complex manner, their cross sections may *bulge* or *warp* when the shaft is twisted. Evidence of this can be seen from the way grid lines deform on a shaft having a square cross section when the shaft is twisted, Fig. 5–27. As a consequence of this deformation the torsional analysis of *noncircular* shafts becomes considerably complicated and will not be considered in this text.

Using a mathematical analysis based on the theory of elasticity, however, it is possible to determine the shear-stress distribution within a shaft of square cross section. Examples of how this shear stress varies along two radial lines of the shaft are shown in Fig. 5–28a. As stated above, because these shear-stress distributions vary in a complex manner, the shear strains they create will *warp* the cross section as shown in Fig. 5–28b. In particular, notice that the corner points of the shaft will be subjected to zero shear stress and therefore zero shear strain. The reason for this can be shown by considering an element of material located at one of these points, Fig. 5–28c. One would expect the shaded face of this element to be subjected to a shear stress in order to aid in resisting the applied torque **T**. This, however, is *not* the case, since the shear stresses τ and τ', acting on the *outer surface* of the shaft, must be *zero,* which in turn implies that the corresponding shear-stress components τ and τ' on the shaded face must also be equal to zero.

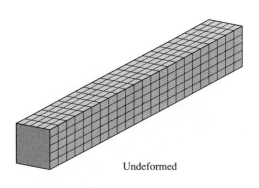

Undeformed

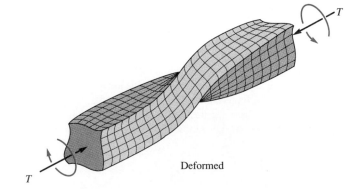

Deformed

Fig. 5–27

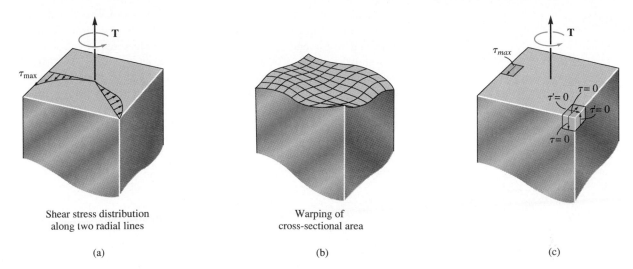

Shear stress distribution
along two radial lines

(a)

Warping of
cross-sectional area

(b)

(c)

Fig. 5–28

The results of the above analysis, along with other results from the theory of elasticity, for shafts having triangular and elliptical cross sections, are reported in Table 5–1. In all cases the *maximum shear stress* occurs at a point on the cross section that is *least distant* from the center axis of the shaft. In Table 5–1 these points are indicated as "dots" on the cross sections. Also given are formulas for the angle of twist of each shaft. By extending these results to a shaft having an *arbitrary* cross section, it can also be shown that a shaft having a *circular* cross section is most efficient, since it is subjected to both a *smaller* maximum shear stress and a *smaller* angle of twist than a corresponding shaft having a noncircular cross section and subjected to the same torque.

Shape of cross-section	τ_{max}	ϕ
Square	$\dfrac{4.81T}{a^3}$	$\dfrac{7.10\,TL}{a^4 G}$
Equilateral triangle	$\dfrac{20\,T}{a^3}$	$\dfrac{46\,TL}{a^4 G}$
Ellipse	$\dfrac{2\,T}{\pi ab^2}$	$\dfrac{(a^2 + b^2)TL}{\pi a^3 b^3 G}$

Tab. 5–1

Example 5–13

The aluminum shaft shown in Fig. 5–29 has a cross-sectional area in the shape of an equilateral triangle. Determine the largest torque **T** that can be applied to the end of the shaft if the allowable shear stress is $\tau_{\text{allow}} = 8$ ksi and the angle of twist at its end is restricted to $\phi_{\text{allow}} = 0.02$ rad. How much torque can be applied to a shaft of circular cross section made from the same amount of aluminum? Take $G_{al} = 4(10^3)$ ksi.

SOLUTION

By inspection, the resultant internal torque at any cross section along the shaft's axis is also **T**. Using the formulas for τ_{max} and ϕ in Table 5–1, we require

$$\tau_{\text{allow}} = \frac{20T}{a^3}; \qquad 8(10^3) \text{ lb/in}^2 = \frac{20T}{(1.5 \text{ in.})^3}$$

$$T = 1350 \text{ lb} \cdot \text{in.}$$

Also,

$$\phi_{\text{allow}} = \frac{46TL}{a^4 G_{al}}; \quad 0.02 \text{ rad} = \frac{46T(4 \text{ ft})(12 \text{ in./ft})}{(1.5 \text{ in.})^4 [4(10^6) \text{ lb/in}^2]}$$

$$T = 183 \text{ lb} \cdot \text{in.} \qquad \qquad \textit{Ans.}$$

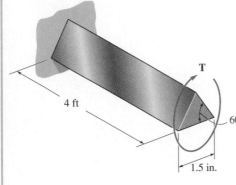

Fig. 5–29

By comparison, the torque is limited due to the angle of twist.

Circular Cross Section. If the same amount of aluminum is to be used in making the same length of shaft having a circular cross section, then the radius of the cross section can be calculated. We have

$$A_{\text{circle}} = A_{\text{triangle}}; \qquad \pi c^2 = \frac{1}{2}(1.5 \text{ in.})(1.5 \sin 60°)$$

$$c = 0.557 \text{ in.}$$

The limitations of stress and angle of twist then require

$$\tau_{\text{allow}} = \frac{Tc}{J}; \qquad 8(10^3) \text{ lb/in}^2 = \frac{T(0.557 \text{ in.})}{(\pi/2)(0.557 \text{ in.})^4}$$

$$T = 2170 \text{ lb} \cdot \text{in.}$$

$$\phi_{\text{allow}} = \frac{TL}{JG_{al}}; \quad 0.02 \text{ rad} = \frac{T(4 \text{ ft})(12 \text{ in./ft})}{(\pi/2)(0.557 \text{ in.})^4 [4(10^6) \text{ lb/in}^2]}$$

$$T = 252 \text{ lb} \cdot \text{in.} \qquad \qquad \textit{Ans.}$$

Again, the angle of twist limits the loading.

Comparing this result (252 lb · in.) with that given above (183 lb · in.), it is seen that a shaft of circular cross section can support 38% more torque than the one having a triangular cross section.

5.7 Thin-Walled Tubes Having Closed Cross Sections

Thin-walled tubes of noncircular shape are often used to construct light-weight frameworks such as those used in aircraft. In some applications, they may be subjected to a torsional loading. In this section we will analyze the effects of applying a torque to a thin-walled tube having a *closed* cross section, that is, a tube that does not have any breaks or slits along its length. Such a tube, having a constant yet arbitrary cross-sectional shape, is shown in Fig. 5–30a. For the analysis we will assume that the walls have a variable thickness t. Since the walls are thin, we will be able to obtain an approximate solution for the shear stress by assuming that this stress is *uniformly distributed* across the thickness of the tube. In other words, we will be able to determine the *average shear stress* in the tube at any given point. Before we do this, however, we will first discuss some preliminary concepts regarding the action of shear stress over the cross section.

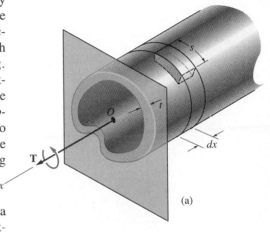

(a)

Shear Flow.   Shown in Fig. 5–30b is a small element of the tube, having a finite length s and differential width dx. At one end the element has a thickness t_A, and at the other end the thickness is t_B. Due to the applied torque **T**, shear stress is developed on the colored face of the element. Specifically, at end A the shear stress is τ_A, and at end B it is τ_B. These stresses can be related by noting that equivalent shear stresses τ_A and τ_B must also act on the longitudinal sides of the element, shown shaded in Fig. 5–30b. Since these sides have *constant* thicknesses t_A and t_B, the forces acting on them are $dF_A = \tau_A(t_A\, dx)$ and $dF_B = \tau_B(t_B\, dx)$. Force equilibrium requires these forces to be of equal magnitude but opposite direction, so that

$$\tau_A t_A = \tau_B t_B$$

This important result states that *the product of the average longitudinal shear stress times the thickness of the tube is the same at each point along the tube's cross-sectional area*. This product is called *shear flow,* q*, and in general terms we can express it as

$$\boxed{q = \tau_{\text{avg}}t} \tag{5–17}$$

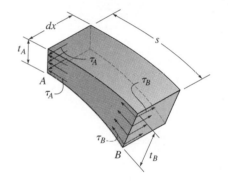

(b)

Since q is constant over the cross section, the *largest* average shear stress will occur where the tube's thickness is the *smallest*.

If a differential element having a thickness t, length ds, and width dx is isolated from the tube, Fig. 5–30c, it is seen that the colored area over which the average shear stress acts is $dA = t\, ds$. Hence, $dF = \tau_{\text{avg}}t\, ds = q\, ds$, or $q = dF/ds$. In other words, *the shear flow, which is constant over the cross-sectional area, measures the force per unit length along the tube's cross-sectional area*.

---

*The terminology "flow" is used since q is analogous to water flowing through an open channel of constant depth and variable width w. Although the water's velocity v at each point along the channel will be different (like τ_{avg}), the flow $q = vw$ will be constant.

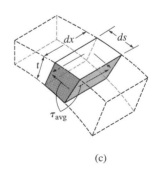

(c)

Fig. 5–30

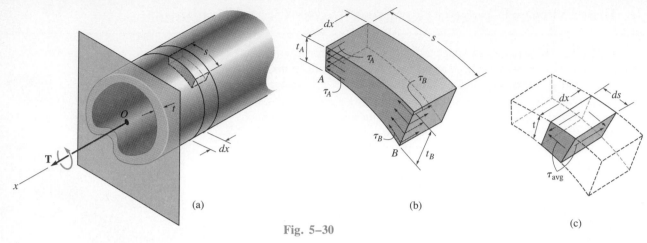

Fig. 5–30

(a)

(b)

(c)

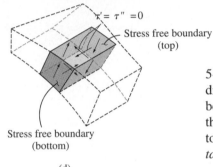

(d)

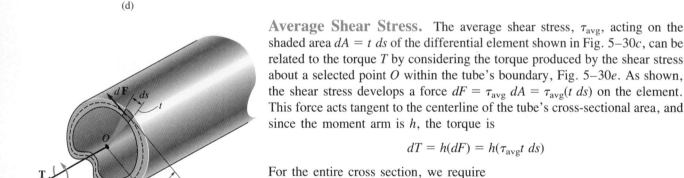

(e)

It is important to note that the shear-stress components shown in Fig. 5–30c are the only ones acting on the tube. Components acting in the other direction, as shown in Fig. 5–30d, cannot exist. This is because the top and bottom faces of the element are at the inner and outer walls of the tube, and these boundaries must be free of stress. Instead, as noted above, the applied torque causes *the shear flow and the average stress to be always directed tangent to the wall of the tube*.

Average Shear Stress. The average shear stress, τ_{avg}, acting on the shaded area $dA = t \, ds$ of the differential element shown in Fig. 5–30c, can be related to the torque T by considering the torque produced by the shear stress about a selected point O within the tube's boundary, Fig. 5–30e. As shown, the shear stress develops a force $dF = \tau_{avg} \, dA = \tau_{avg}(t \, ds)$ on the element. This force acts tangent to the centerline of the tube's cross-sectional area, and since the moment arm is h, the torque is

$$dT = h(dF) = h(\tau_{avg} t \, ds)$$

For the entire cross section, we require

$$T = \oint h \, \tau_{avg} t \, ds$$

Here the "line integral" indicates that integration is performed *around* the entire boundary of the area. Since the shear flow $q = \tau_{avg} t$ is *constant*, these terms together can be factored out of the integral, so that

$$T = \tau_{avg} t \oint h \, ds$$

A graphical simplification can be made for evaluating the integral by noting that the *mean area,* shown by the colored triangle in Fig. 5–30e, is $dA_m = (1/2)h \, ds$. Thus,

$$T = 2\tau_{avg}t \int dA_m = 2\tau_{avg}tA_m$$

Solving for τ_{avg}, we have

$$\boxed{\tau_{avg} = \frac{T}{2tA_m}} \tag{5–18}$$

Here

τ_{avg} = the average shear stress acting over the thickness of the tube

T = the resultant internal torque at the cross section, which is found using the method of sections and the equations of equilibrium

t = the thickness of the tube where τ_{avg} is to be computed

A_m = the mean area enclosed within the boundary of the *centerline* of the tube's thickness. A_m is shown shaded in Fig. 5–30f

(f)

Since $q = \tau_{avg}t$, we can determine the shear flow throughout the cross section using the equation

$$\boxed{q = \frac{T}{2A_m}} \tag{5–19}$$

Angle of Twist. The angle of twist of a thin-walled tube of length L can be determined using energy methods, and the development of the necessary equation is given as a problem later in the text.* If the material behaves in a linear-elastic manner and G is the shear modulus, then this angle ϕ, given in radians, can be expressed as

$$\phi = \frac{TL}{4A_m^2 G} \oint \frac{ds}{t} \tag{5–20}$$

Here the integration must be performed around the entire boundary of the tube's cross-sectional area.

The following examples numerically illustrate application of the above equations.

*See Prob. 14–19.

Example 5–14

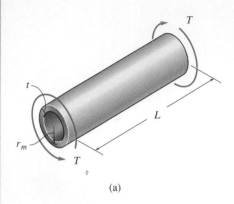

(a)

Compute the average shear stress in a thin-walled tube having a circular cross section of mean radius r_m and thickness t, which is subjected to a torque T, Fig. 5–31a. Also, what is the relative angle of twist if the tube has a length L?

SOLUTION

Average Shear Stress. The mean area for the tube is $A_m = \pi r_m^2$. Applying Eq. 5–18 gives

$$\tau_{avg} = \frac{T}{2tA_m} = \frac{T}{2\pi t r_m^2} \qquad \textit{Ans.}$$

We can check the validity of this result by applying the torsion formula. In this case, using Eq. 5–9, we have

$$J = \frac{\pi}{2}(r_o^4 - r_i^4)$$

$$= \frac{\pi}{2}(r_o^2 + r_i^2)(r_o^2 - r_i^2)$$

$$= \frac{\pi}{2}(r_o^2 + r_i^2)(r_o + r_i)(r_o - r_i)$$

Since $r_m \approx r_o \approx r_i$ and $t = r_o - r_i$, $J = \dfrac{\pi}{2}(2r_m^2)(2r_m)t = 2\pi r_m^3 t$

so that

$$\tau_{avg} = \frac{Tr_m}{J} = \frac{Tr_m}{2\pi r_m^3 t} = \frac{T}{2\pi t r_m^2} \qquad \textit{Ans.}$$

which agrees with the previous result.

The average shear-stress distribution acting throughout the tube's cross section is shown in Fig. 5–31b. Also shown is the shear-stress distribution acting on a radial line as calculated using the torsion formula. Notice how each τ_{avg} acts in a direction such that it contributes to the resultant torque $\mathbf{T}$ at the section. As the tube's thickness decreases, the shear stress throughout the tube becomes more uniform.

Angle of Twist. Applying Eq. 5–20, we have

$$\phi = \frac{TL}{4A_m^2 G} \oint \frac{ds}{t} = \frac{TL}{4(\pi r_m^2)^2 Gt} \oint ds$$

The integral represents the length around the centerline boundary, which is $2\pi r_m$. Substituting, the final result is

$$\phi = \frac{TL}{2\pi r_m^3 Gt} \qquad \textit{Ans.}$$

Show that one obtains this same result using Eq. 5–15.

Actual shear stress
distribution
(torsion formula)

$\tau_{max} \qquad \tau_{avg}$

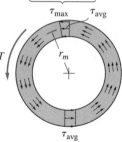

τ_{avg}

Average shear stress
distribution
(thin-wall approximation)

(b)

Fig. 5–31

Example 5–15

The brass tube has a rectangular cross section as shown in Fig. 5–32a. If it is subjected to the two torques, determine the average shear stress in the tube at points A and B. Also, what is the angle of twist of end C? The tube is fixed at E, and $G_{br} = 38$ GPa.

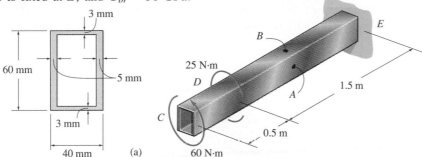

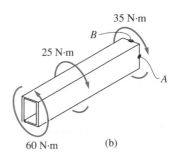

(a)

(b)

SOLUTION

Average Shear Stress. If the tube is sectioned through points A and B, the resulting free-body diagram is shown in Fig. 5–32b. The internal torque is 35 N · m. As shown in Fig. 5–32d, the area A_m is

$$A_m = (0.035 \text{ m})(0.057 \text{ m}) = 0.00200 \text{ m}^2$$

Applying Eq. 5–18 for point A, $t_A = 5$ mm, so that

$$\tau_A = \frac{T}{2tA_m} = \frac{35 \text{ N} \cdot \text{m}}{2(0.005 \text{ m})(0.00200 \text{ m}^2)} = 1.75 \text{ MPa} \qquad \textit{Ans.}$$

And for point B, $t_B = 3$ mm, and therefore

$$\tau_B = \frac{T}{2tA_m} = \frac{35 \text{ N} \cdot \text{m}}{2(0.003 \text{ m})(0.00200 \text{ m}^2)} = 2.92 \text{ MPa} \qquad \textit{Ans.}$$

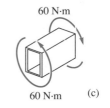

60 N·m

60 N·m   (c)

These results are shown on elements of material located at points A and B, Fig. 5–32e. Note carefully how the 35-N · m torque in Fig. 5–32b creates these stresses on the shaded faces of each element.

Angle of Twist. From the free-body diagrams in Fig. 5–32b and 5–32c, the internal torques in regions DE and CD are 35 N · m and 60 N · m, respectively. Following the sign convention outlined in Sec. 5.4, these torques are both positive. Thus, Eq. 5–20 becomes

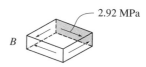

57 mm

35 mm   (d)

$$\phi = \sum \frac{TL}{4A_m^2 G} \oint \frac{ds}{t}$$

$$= \frac{60 \text{ N} \cdot \text{m}(0.5 \text{ m})}{4(0.00200 \text{ m}^2)^2 (38(10^9) \text{ N/m}^2)} \left[2\left(\frac{57 \text{ mm}}{5 \text{ mm}}\right) + 2\left(\frac{35 \text{ mm}}{3 \text{ mm}}\right) \right]$$

$$+ \frac{35 \text{ N} \cdot \text{m}(1.5 \text{ m})}{4(0.00200 \text{ m}^2)^2 (38(10^9) \text{ N/m}^2)} \left[2\left(\frac{57 \text{ mm}}{5 \text{ mm}}\right) + 2\left(\frac{35 \text{ mm}}{3 \text{ mm}}\right) \right]$$

$$= 6.26(10^{-3}) \text{ rad} \qquad \textit{Ans.}$$

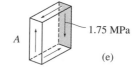

2.92 MPa

1.75 MPa

(e)

Fig. 5–32

Example 5–16

A square aluminum tube has the dimensions shown in Fig. 5–33a. Determine the average shear stress in the tube at point *A* if it is subjected to a torque of 85 lb · ft. Also compute the angle of twist due to this loading. Take $G_{al} = 3.80(10^3)$ ksi.

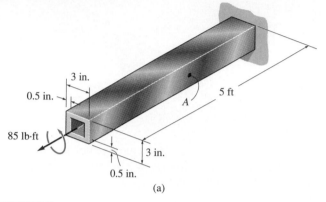

(a)

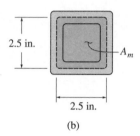

2.5 in.

A_m

2.5 in.

(b)

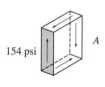

154 psi

A

(c)

Fig. 5–33

SOLUTION

Average Shear Stress.   By inspection, the internal resultant torque at the cross section where point *A* is located is $T = 85$ lb · ft. From Fig. 5–33b, the area A_m, shown shaded, is

$$A_m = (2.5 \text{ in.})(2.5 \text{ in.}) = 6.25 \text{ in}^2$$

Applying Eq. 5–18,

$$\tau_{\text{avg}} = \frac{T}{2tA_m} = \frac{85 \text{ lb} \cdot \text{ft}(12 \text{ in./ft})}{2(0.5 \text{ in.})(6.25 \text{ in}^2)} = 163 \text{ psi} \qquad \textit{Ans.}$$

Since *t* is constant except at the corners, the average shear stress is the same at all points on the cross section. It is shown acting on an element located at point *A* in Fig. 5–33c. Note that τ_{avg} acts upward on the shaded face, since it contributes to the internal resultant torque **T** at the section.

Angle of Twist.   The angle of twist caused by **T** is determined from Eq. 5–20; i.e.,

$$\phi = \frac{TL}{4A_m^2 G} \oint \frac{ds}{t} = \frac{85 \text{ lb} \cdot \text{ft}(12 \text{ in./ft})(5 \text{ ft})(12 \text{ in./ft})}{4(6.25 \text{ in}^2)^2[3.80(10^6) \text{ lb/in}^2]} \oint \frac{ds}{(0.5 \text{ in.})}$$

$$= 0.206(10^{-3}) \text{ in}^{-1} \oint ds$$

Here the integral represents the *length* around the centerline boundary of the tube. Thus,

$$\phi = 0.206(10^{-3}) \text{ in}^{-1}[4(2.5 \text{ in.})] = 2.06(10^{-3}) \text{ rad} \qquad \textit{Ans.}$$

Example 5–17

A thin tube is made from three 5-mm-thick steel plates such that it has a cross section that is triangular as shown in Fig. 5–34a. Determine the maximum torque T to which it can be subjected, if the allowable shear stress is $\tau_{\text{allow}} = 90$ MPa and the tube is restricted to twist no more than $\phi = 2(10^{-3})$ rad. Take $G_{st} = 75$ GPa.

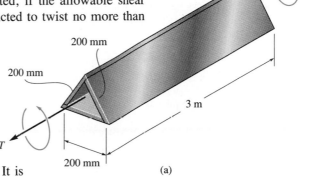

(a)

SOLUTION

The area A_m is shown shaded in Fig. 5–34b. It is

$$A_m = \frac{1}{2}(200 \text{ mm})(200 \text{ mm sin } 60°)$$

$$= 17.32(10^3) \text{ mm}^2(10^{-6} \text{ m}^2/\text{mm}^2) = 17.32(10^{-3}) \text{ m}^2$$

The greatest average shear stress occurs at points where the tube's thickness is smallest, which is along the sides and not at the corners. Applying Eq. 5–18, with $t = 0.005$ m, yields

$$\tau_{\text{avg}} = \frac{T}{2tA_m}; \qquad 90(10^6) \text{ N/m}^2 = \frac{T}{2(0.005 \text{ m})(17.32(10^{-3}) \text{ m}^2)}$$

$$T = 15.6 \text{ kN} \cdot \text{m}$$

Also, from Eq. 5–20, we have

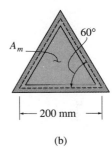

$$\phi = \frac{TL}{4A_m^2 G} \oint \frac{ds}{t}$$

$$0.002 \text{ rad} = \frac{T(3 \text{ m})}{4(17.32(10^{-3}) \text{ m})^2[75(10^9) \text{ N/m}^2]} \oint \frac{ds}{(0.005 \text{ m})}$$

$$300.0 = T \oint ds$$

The integral represents the sum of the dimensions along the three sides; i.e.,

$$300.0 = T[3(0.20 \text{ m})]$$

$$T = 500 \text{ N} \cdot \text{m} \qquad\qquad \textit{Ans.}$$

By comparison, the application of torque is restricted due to the angle of twist.

PROBLEMS

5–67. By what amount is the shaft of circular cross section more efficient than the shaft of elliptical cross section for withstanding torque, when the maximum shear stress is to be the same for both shafts?

*5–68. If $a = 2$ in. and $b = 1.5$ in., determine the maximum shear stress in the circular and elliptical shafts when the applied torque is $T = 16$ lb · ft. By what percentage is the shaft of circular cross section more efficient at withstanding the torque than the shaft of elliptical cross section?

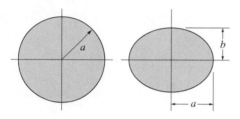

Probs. 5–67/5–68

5–69. Compare the values of the maximum shear stress and the angle of twist developed in steel shafts having square and circular cross sections. Each shaft has a cross-sectional area of 4 in^2, length of 25 in., and is subjected to a torque of 3000 lb · in. $G_{st} = 12(10^3)$ ksi.

Prob. 5–69

5–70. The aluminum rod has a square cross section of 10 mm by 10 mm. If it is 8 m long, determine the torque T that is required to rotate one end relative to the other end by 90°. $G_{al} = 28$ GPa, $(\tau_Y)_{al} = 240$ MPa.

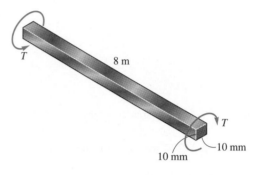

Prob. 5–70

5–71. The shaft is made of a ceramic material having an allowable shear stress of $\tau_{allow} = 7$ ksi and shear modulus of $G_c = 2.8(10^3)$ ksi. If it has a triangular cross section as shown, determine the dimension a of its sides so that it can resist a torque of 15 lb · in. Also, determine the rotation in degrees of its end B.

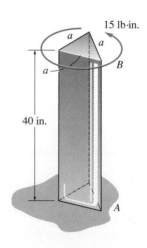

Prob. 5–71

*5–72.   The square shaft is used at the end of a drive cable in order to register the rotation of the cable on a gauge. If it has the dimensions shown and is subjected to a torque of 8 N · m, determine the shear stress in the shaft at point A. Sketch the shear stress on a volume element located at this point.

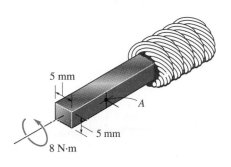

Prob. 5–72

5–73.   The steel shaft is 12 in. long and is screwed into the wall using a wrench. Determine the largest couple force F that can be applied to the shaft without causing the steel to yield. $\tau_Y = 8$ ksi.

5–74.   The steel shaft is 12 in. long and is screwed into the wall using a wrench. Determine the maximum shear stress in the shaft and the amount of displacement that each couple force undergoes if the couple forces have a magnitude of $F = 30$ lb. $G_{st} = 10.8(10^3)$ ksi.

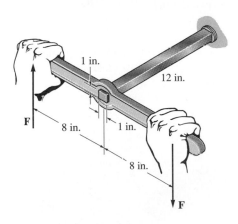

Probs. 5–73/5–74

5–75.   The shaft is made of a plastic and has an elliptical cross-section. If it is subjected to the torsional loading shown, determine the shear stress at point A and show the shear stress on a volume element located at this point. Also, determine the angle of twist ϕ_a at the end B. $G_p = 15$ GPa.

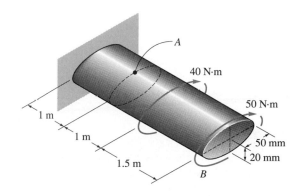

Prob. 5–75

*5–76.   The copper rod has an elliptical cross section. If it is subjected to the five torques shown, determine the maximum shear stress in the rod and the rotation of end A with respect to end B. $G_{cu} = 40$ GPa.

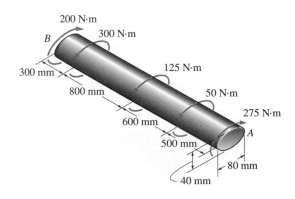

Prob. 5–76

5–77. The material of which each of the three shafts is made has a yield stress of τ_Y and a shear modulus of G. Determine which shaft geometry will resist the largest torque without yielding. What percent of this torque can be carried by the other two shafts? Assume that each shaft is made of the same amount of material so that it has the same cross-sectional area A.

Prob. 5–77

5–78. The aluminum strut is fixed between the two walls at A and B. If it has a 2 in. by 2 in. square cross section, and it is subjected to the torque of 80 lb · ft at C, determine the reactions at the fixed supports. Also, what is the angle of twist at C? $G_{al} = 3.8(10^3)$ ksi.

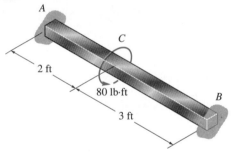

Prob. 5–78

5–79. It is intended to manufacture a circular bar to resist torque; however, the bar is made elliptical in the process of manufacturing, with one dimension smaller than the other by a factor k as shown. Determine the factor by which the maximum shear stress is increased.

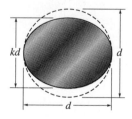

Prob. 5–79

*5–80. Consider a thin-walled tube of mean radius r and thickness t. Show that the maximum shear stress in the tube due to an applied torque T approaches the average shear stress computed from Eq. 5–18 as $r/t \rightarrow \infty$.

Prob. 5–80

5–81. A tube having the dimensions shown is subjected to a torque of $T = 60$ N · m. Neglecting the stress concentrations at its corners, determine the maximum shear stress in the tube. Does the increased thickness from 5 mm to 10 mm in the *vertical walls* have any effect on the result? Why or why not?

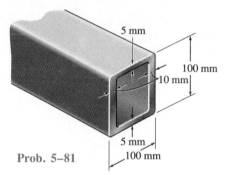

Prob. 5–81

5–82. The steel tube has a thickness of 10 mm. If the allowable shear stress is $\tau_{allow} = 80$ MPa, determine the maximum torque that it can transmit. Also, what is the angle of twist of one end of the tube with respect to the other if the tube is 4 m long? Neglect the stress concentrations at the corners. $G_{st} = 75$ GPa.

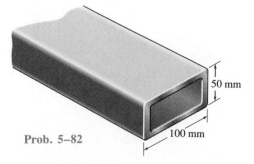

Prob. 5–82

5–83. The aluminum tube has a thickness of 5 mm and the outer cross-sectional dimensions shown. Determine the maximum average shear stress in the tube. If the tube has a length of 5 m, determine the angle of twist. $G_{al} = 28$ GPa.

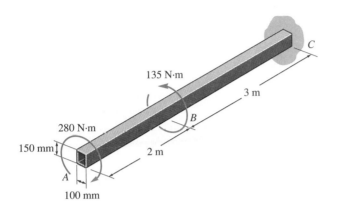

150 mm

100 mm

280 N·m

135 N·m

2 m

3 m

B

C

A

Prob. 5–83

5–85. The steel tube has an elliptical cross section of mean dimensions shown and a constant thickness of $t = 0.2$ in. If the allowable shear stress is $\tau_{allow} = 8$ ksi, and the tube is to resist a torque of $T = 250$ lb · ft, determine the necessary dimension b. The mean area A_m for the ellipse is $\pi b(0.5b)$.

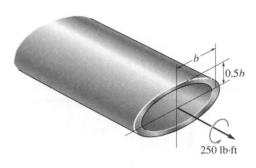

b

$0.5b$

250 lb·ft

Prob. 5–85

*5–84. The steel circular tube is subjected to a torque of 9 kN · m. Determine the shear stress at the mean thickness $\rho = 53$ mm and compute the angle of twist of the tube if it is 3 m long and fixed at its far end. Solve the problem using Eqs. 5–7 and 5–15 and by using Eqs. 5–18 and 5–20. $G_{st} = 80$ GPa.

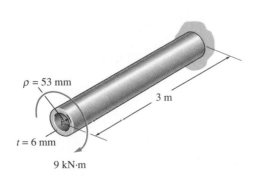

$\rho = 53$ mm

$t = 6$ mm

3 m

9 kN·m

Prob. 5–84

5–86. An aluminum tube is subjected to a torque of 85 N · m. If it has the mean cross-sectional dimensions shown, determine the average shear stress at points A and B and indicate the shear stress on volume elements located at these points. If the tube has a length of 800 mm, determine the angle of twist of one end relative to the other end. $G_{al} = 28$ GPa.

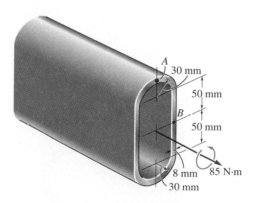

A

30 mm

50 mm

B

50 mm

8 mm

85 N·m

30 mm

Prob. 5–86

5–87. The tube is made from a high-strength steel, having the mean dimensions shown and a thickness of 5 mm. If it is subjected to a torque of $T = 40$ N · m, determine the average shear stress developed at points A and B. Indicate the shear stress on volume elements located at these points.

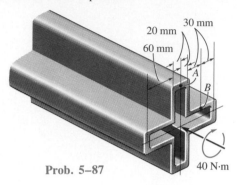

Prob. 5–87

*5–88. The plastic hexagonal tube is subjected to a torque of 150 N · m. Determine the mean dimension a of its sides if the allowable shear stress is $\tau_{allow} = 60$ MPa. Each side has a thickness of $t = 3$ mm.

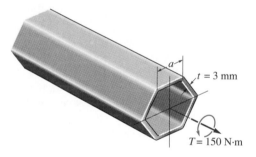

Prob. 5–88

5–89. For a given maximum shear stress, determine the factor by which the torque carrying capacity is increased if the half-circular section is reversed from the dashed-line position to the section shown. The tube is 0.1 in. thick.

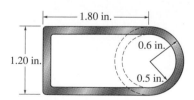

Prob. 5–89

5–90. A torque of 2 kip · in. is applied to the tube shown. If the wall thickness is 0.1 in., determine the maximum shear stress in the tube.

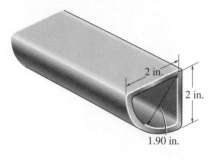

Prob. 5–90

5–91. The tube is made of plastic, is 5 mm thick, and has the mean dimensions shown. Determine the shear stress at points A and B if it is subjected to the torque of $T = 5$ N · m. Show the shear stress on volume elements located at these points.

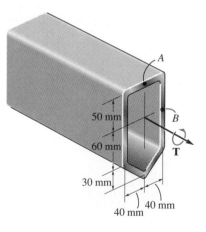

Prob. 5–91

5.8 Stress Concentration

The torsion formula, $\tau_{max} = Tc/J$, can be applied to regions of a shaft having a circular cross section that is constant or tapers slightly. When sudden changes arise in the cross section, both the shear-stress and shear-strain distributions in the shaft become complex and can be obtained only by using experimental methods or possibly by a mathematical analysis based on the theory of elasticity. Three common discontinuities of the cross section that occur in practice are shown in Fig. 5–35. They are at *couplings*, which are used to connect two collinear shafts together, Fig. 5–35*a*, *keyways*, used to connect gears or pulleys to a shaft, Fig. 5–35*b*, and *shoulder fillets*, used to fabricate a single collinear shaft from two shafts having different diameters, Fig. 5–35*c*. In each case the maximum shear stress will occur at the point (dot) indicated on the cross section.

In order to eliminate the necessity for the engineer to perform a complex stress analysis at a shaft discontinuity, the maximum shear stress can be determined for a specified geometry using a *torsional stress-concentration factor, K*. As in the case of axially loaded members, Sec. 4.6, K is usually taken from a graph. An example, for the shoulder-fillet shaft, is shown in Fig. 5–36.* To use this graph, one first computes the geometric ratio D/d to define the appropriate curve, and then once the abscissa r/d is calculated, the value of K is

(a)

(b)

(c)

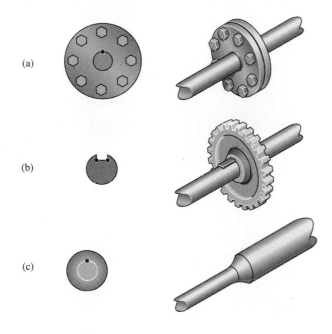

Fig. 5–35

*See Lipson, C. and Juvinall, R.C., *Handbook of Stress and Strength*, Macmillan, 1963.

found along the ordinate. The maximum shear stress is then determined from the equation

$$\tau_{max} = K \frac{Tc}{J}$$ (5–21)

Here the torsion formula is applied to the *smaller* of the two connected shafts, since τ_{max} occurs at the base of the fillet, Fig. 5–35c.

It can be noted from the graph in Fig. 5–36 that an *increase* in fillet radius *r* causes a *decrease* in *K*. Hence the maximum shear stress in the shaft can be reduced by *increasing* the fillet radius. Also, if the diameter of the larger shaft is reduced, the *D/d* ratio will be lower and so the value of *K* and therefore τ_{max} will be lower.

Like the case of axially loaded members, torsional stress concentration factors should *always* be used when designing shafts made from *brittle materials,* or when designing shafts that will be subjected to *fatigue or cyclic torsional loadings*. These types of loadings give rise to the formation of cracks at the stress concentration, and this can often lead to a sudden failure of the shaft. Also realize that if a large *static* torsional loading is applied to a shaft made from *ductile material,* then *inelastic strains* may develop within the shaft. As a result of yielding, the stress distribution will become more *evenly distributed* throughout the shaft, so that the maximum stress that results will be limited at the stress concentration. This phenomenon will be discussed further in the next section.

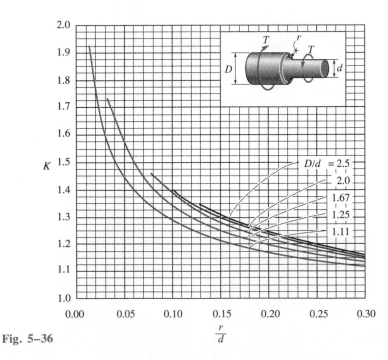

Fig. 5–36

Example 5–18

The stepped shaft shown in Fig. 5–37a is supported by bearings at A and B. Determine the maximum stress in the shaft due to the applied torques. How can this stress be reduced? The fillet at the junction of each shaft has a radius of $r = 6$ mm.

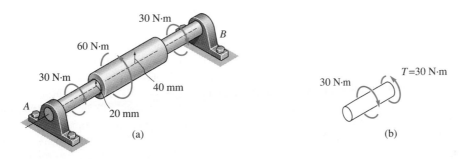

(a)

(b)

SOLUTION

Internal Torque. By inspection, moment equilibrium about the axis of the shaft is satisfied. Since the maximum shear stress occurs at the "rooted" ends of the *smaller*-diameter shafts, the internal torque (30 N · m) can be found there by applying the method of sections, Fig. 5–37b.

Maximum Shear Stress. The stress-concentration factor can be determined by using Fig. 5–36. From the shaft geometry we have

$$\frac{D}{d} = \frac{2(40 \text{ mm})}{2(20 \text{ mm})} = 2$$

$$\frac{r}{d} = \frac{6 \text{ mm}}{2(20 \text{ mm})} = 0.15$$

$\tau_{max} = 3.10$ MPa

Shear–stress distribution predicted by torsion formula

Actual shear–stress distribution caused by stress concentration

(c)

Thus the value of $K = 1.3$ is obtained.
    Applying Eq. 5–21, we have

$$\tau_{max} = K \frac{Tc}{J}; \qquad \tau_{max} = 1.3 \left[\frac{30 \text{ N} \cdot \text{m}(0.020 \text{ m})}{(\pi/2)(0.020 \text{ m})^4} \right] = 3.10 \text{ MPa} \quad Ans.$$

Fig. 5–37

From experimental evidence, the actual stress distribution along a radial line of the cross section at the critical section looks similar to that shown in Fig. 5–37c. Notice how this compares with the linear stress distribution found from the torsion formula.

5.9 Inelastic Torsion

The equations of stress and deformation developed thus far are valid only if the applied torque causes the material to behave in a linear-elastic manner. If the torsional loadings are excessive, however, the material may yield, and consequently a "plastic analysis" must then be used to determine the shear-stress distribution and the angle of twist. To perform this analysis, it is necessary to meet the conditions of both deformation and equilibrium for the shaft.

It was shown in Sec. 5.1 that the shear strains that develop in the material must vary *linearly* from zero at the center of the shaft to a maximum at its outer boundary, Fig. 5–38a. This conclusion was based entirely on geometric considerations and not the material's behavior. Also, the resultant torque at the section must be equivalent to the torque caused by the entire shear-stress distribution about the shaft's axis. This condition can be expressed mathematically by considering the shear stress τ acting on an element of area dA located a distance ρ from the center of the shaft, Fig. 5–38b. The force produced by this stress is $dF = \tau\, dA$, and the torque produced is $dT = \rho\, dF = \rho\tau\, dA$. For the entire shaft we require

$$T = \int_A \rho\tau\, dA \qquad (5\text{–}22)$$

If the area dA over which τ acts can be defined as a *differential ring* having an area of $dA = 2\pi\rho\, d\rho$, Fig. 5–38c, then the above equation can be written as

$$T = 2\pi \int_A \tau\rho^2\, d\rho \qquad (5\text{–}23)$$

These conditions of geometry and loading will now be used to determine the shear-stress distribution in a shaft when the shaft is subjected to three types of torque.

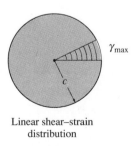

Linear shear–strain distribution

(a)

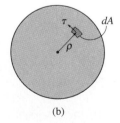

(b)

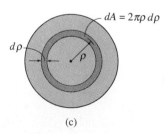

(c)

Fig. 5–38

Maximum Elastic Torque. If the torque produces the maximum *elastic* shear strain γ_Y, at the outer boundary of the shaft, then the shear-strain distribution along a radial line of the shaft will look like that shown in Fig. 5–39*b*. To establish the shear-stress distribution, we must either use Hooke's law or find the corresponding values of shear stress from the material's τ–γ diagram, Fig. 5–39*a*. For example, a shear strain γ_Y produces the shear stress τ_Y at $\rho = c$. Likewise, at $\rho = \rho_1$ in Fig. 5–39*b*, the shear strain is $\gamma_1 = (\rho_1/c)\gamma_Y$. From the τ–γ diagram γ_1 produces τ_1. When these stresses and others like them are plotted at $\rho = c$, $\rho = \rho_1$, etc., the expected *linear* shear-stress distribution in Fig. 5–39*c* results. Since this shear-stress distribution can be described mathematically as $\tau = \tau_Y(\rho/c)$, the maximum elastic torque can be determined from Eq. 5–23; i.e.,

$$T_Y = 2\pi \int_0^c \tau_Y \left(\frac{\rho}{c}\right) \rho^2 \, d\rho$$

or

$$T_Y = \frac{\pi}{2} \tau_Y c^3 \qquad (5\text{–}24)$$

This same result can of course be obtained in a more direct manner using the torsion formula; i.e., $\tau_Y = T_Y c/[(\pi/2)c^4]$. Furthermore, the angle of twist can be determined from Eq. 5–13, namely,

$$d\phi = \gamma \frac{dx}{\rho} \qquad (5\text{–}25)$$

As noted in Sec. 5.4, this equation results in $\phi = TL/JG$, when the shaft is subjected to a constant torque and has a constant cross-sectional area.

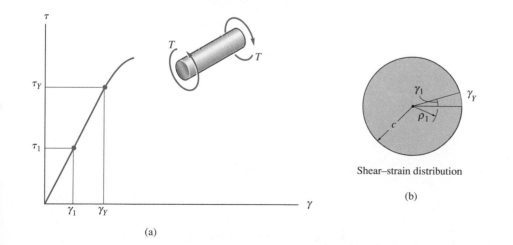

Shear–strain distribution

(b)

Shear–stress distribution

(c)

(a)

Fig. 5–39

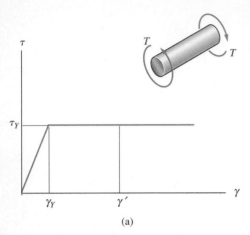

(a)

Elastic-Plastic Torque.   Let us now consider the material in the shaft to exhibit an elastic-perfectly plastic behavior. As shown in Fig. 5–40a, this is characterized by a shear stress–strain diagram for which the material undergoes an increasing amount of shear strain when the shear stress in the material reaches the yield point τ_Y. Thus, as the applied torque increases in magnitude above T_Y, it will begin to cause yielding. First at the outer boundary of the shaft, $\rho = c$, and then, as the shear strain increases to, say, γ', the yielding boundary will progress inward toward the shaft's center, Fig. 5–40b. As shown, this produces an *elastic core,* where, by proportion, the outer radius of the core is $\rho_Y = (\gamma_Y/\gamma')c$. Also, the outer portion of the shaft forms a *plastic annulus* or ring, since the shear strains γ are greater than γ_Y within this region. The corresponding shear-stress distribution along a radial line of the shaft is shown in Fig. 5–40c. It was established by taking successive points on the shear-strain distribution, Fig. 5–40b, and finding the corresponding value of shear stress from the τ–γ diagram, Fig. 5–40a. For example, at $\rho = c$, γ' gives τ_Y, and at $\rho = \rho_Y$, γ_Y also gives τ_Y; etc.

Since τ can now be established as a function of ρ, we can apply Eq. 5–23 to determine the torque. As a general formula for elastic-plastic material behavior, we have

$$
\begin{aligned}
T &= 2\pi \int_0^c \tau\rho^2 \, d\rho \\
&= 2\pi \int_0^{\rho_Y} \left(\tau_Y \frac{\rho}{\rho_Y}\right)\rho^2 \, d\rho + 2\pi \int_{\rho_Y}^c \tau_Y\rho^2 \, d\rho \\
&= \frac{2\pi}{\rho_Y}\tau_Y \int_0^{\rho_Y} \rho^3 \, d\rho + 2\pi\tau_Y \int_{\rho_Y}^c \rho^2 \, d\rho \\
&= \frac{\pi}{2\rho_Y}\tau_Y\rho_Y^4 + \frac{2\pi}{3}\tau_Y(c^3 - \rho_Y^3) \\
&= \frac{\pi\tau_Y}{6}(4c^3 - \rho_Y^3) \qquad\qquad (5\text{–}26)
\end{aligned}
$$

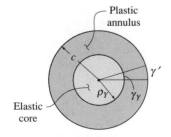

Shear–strain distribution

(b)

Fig. 5–40

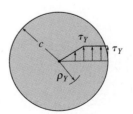

Shear–stress distribution

(c)

Plastic Torque. Further increases in T tend to shrink the radius of the elastic core until all the material will yield, i.e., $\rho_Y \to 0$. The material of the shaft is then subjected to *perfectly plastic behavior* and the shear-stress distribution is constant as shown in Fig. 5–40d. Since then $\tau = \tau_Y$, we can apply Eq. 5–23 to determine the *plastic torque T_p*, which represents the largest possible torque the shaft will support.

$$T_p = 2\pi \int_0^c \tau_Y \rho^2 \, d\rho$$

$$= \frac{2\pi}{3} \tau_Y \, c^3 \tag{5–27}$$

By comparison with the maximum elastic torque T_Y, Eq. 5–24, it can be seen that

$$T_p = \frac{4}{3} T_Y$$

In other words, the plastic torque is 33% greater than the maximum elastic torque.

The angle of twist ϕ for the shear-stress distribution in Fig. 5–40d *cannot* be uniquely defined. This is because $\tau = \tau_Y$ does not correspond to any unique value of shear strain $\gamma \geq \gamma_Y$. As a result, according to our "model" of behavior for the material, Fig. 5–40a, once $\mathbf{T}_p$ is applied, the shaft will continue to deform or twist with no corresponding increase in shear stress.

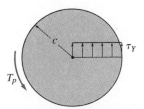

Fully plastic torque

(d)

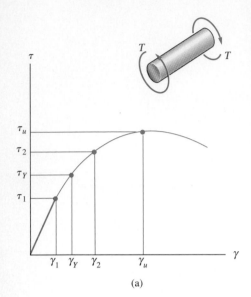

(a)

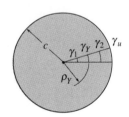

Ultimate shear–strain distribution

(b)

Ultimate Torque. In the general case, most engineering materials will have a shear stress–strain diagram as shown in Fig. 5–41a. Consequently, if T is increased so that the maximum shear strain in the shaft becomes $\gamma = \gamma_u$, Fig. 5–41b, then by proportion γ_Y occurs at $\rho_Y = (\gamma_Y/\gamma_u)c$. Likewise, the shear strains at, say, $\rho = \rho_1$ and $\rho = \rho_2$ can be found by proportion, i.e., $\gamma_1 = (\rho_1/c)\gamma_u$ and $\gamma_2 = (\rho_2/c)\gamma_u$. If corresponding values of τ_1, τ_Y, τ_2, and τ_u are taken from the τ–γ diagram and plotted, we obtain the shear-stress distribution, which acts over a radial line on the cross section, Fig. 5–41c. The torque produced by this stress distribution is called the *ultimate torque*, T_u, since any further increase in shear strain would cause the maximum shear stress at the outer boundary of the shaft to be less than τ_u, and therefore the torque produced by the resulting shear-stress distribution would be *less* than T_u.

The magnitude of $\mathbf{T}_u$ can be determined by "graphically" integrating Eq. 5–23. To do this, the cross-sectional area of the shaft is segmented into a finite number of rings, such as the one shown shaded in Fig. 5–41d. The area of this ring, $\Delta A = 2\pi\rho\,\Delta\rho$, is multiplied by the shear stress τ that acts on it, so that the force $\Delta F = \tau\,\Delta A$ can be determined. The torque created by this force is then $\Delta T = \rho\,\Delta F = \rho(\tau\,\Delta A)$. The addition of all the torques for the entire cross section, as determined in this manner, gives the ultimate torque T_u; that is, Eq. 5–23 becomes $T_u \approx 2\pi\Sigma\tau\rho^2\,\Delta\rho$. On the other hand, if the stress distribution can be expressed as an analytical function, $\tau = f(\rho)$, as in the elastic and plastic torque cases, then the integration of Eq. 5–23 can be carried out directly.

The following examples further illustrate application of the above concepts.

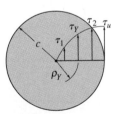

Ultimate shear–stress distribution

(c)

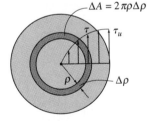

(d)

Fig. 5–41

Example 5–19

The tubular shaft in Fig. 5–42a is made of an aluminum alloy that is assumed to have an elastic-plastic τ–γ diagram as shown. Determine (a) the maximum torque that can be applied to the shaft without causing the material to yield, (b) the maximum torque or plastic torque that can be applied to the shaft. What should the minimum shear strain at the outer radius be in order to develop a plastic torque?

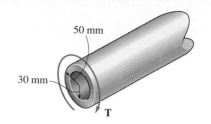

SOLUTION

Maximum Elastic Torque. We require the shear stress at the outer fiber to be 20 MPa. Using the torsion formula, we have

$$\tau_Y = \frac{T_Y c}{J}; \quad 20(10^6)\ \text{N/m}^2 = \frac{T_Y(0.05\ \text{m})}{(\pi/2)[(0.05\ \text{m})^4 - (0.03\ \text{m})^4]}$$

$$T_Y = 3.42\ \text{kN} \cdot \text{m} \qquad\qquad Ans.$$

The shear-stress and shear-strain distributions for this case are shown in Fig. 5–42b. The values at the tube's inner wall are obtained by proportion.

Plastic Torque. The shear-stress distribution in this case is shown in Fig. 5–42c. Application of Eq. 5–23 requires $\tau = \tau_Y$. We have

$$T_p = 2\pi \int_{0.03\ \text{m}}^{0.05\ \text{m}} [20(10^6)\ \text{N/m}^2]\rho^2\ d\rho = 125.66(10^6)\ \frac{1}{3}\rho^3 \Big|_{0.03\ \text{m}}^{0.05\ \text{m}}$$

$$= 4.10\ \text{kN} \cdot \text{m} \qquad\qquad Ans.$$

For this tube T_p represents a 20% increase in torque capacity compared with the elastic torque T_Y.

Outer Radius Shear Strain. The tube becomes fully plastic when the shear strain at the *inner wall* becomes $0.286(10^{-3})$ rad as shown in Fig. 5–42c. Since the shear strain *remains linear* over the cross section, the plastic strain at the outer fibers of the tube in Fig. 5–42c is determined by proportion; i.e.,

$$\frac{\gamma_o}{50\ \text{mm}} = \frac{0.286(10^{-3})\ \text{rad}}{30\ \text{mm}}$$

$$\gamma_o = 0.477(10^{-3})\ \text{rad} \qquad\qquad Ans.$$

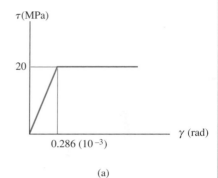

(a)

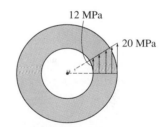

Elastic shear–stress distribution

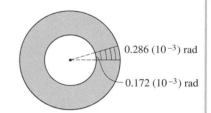

Elastic shear–strain distribution

(b)

Fig. 5–42

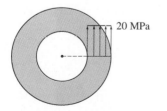

Plastic shear–stress distribution

(c)

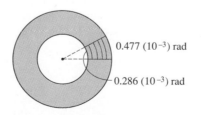

Initial plastic shear–strain distribution

237

Example 5–20

A solid circular shaft has a radius of 20 mm and length of 1.5 m. The material has an elastic-plastic τ–γ diagram as shown in Fig. 5–43a. Determine the torque needed to twist the shaft $\phi = 0.6$ rad.

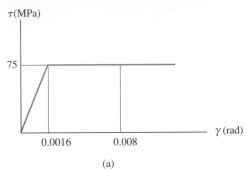

(a)

SOLUTION

To solve the problem we will first obtain the shear-strain distribution, then establish the shear-stress distribution. Once these are known, the applied torque can be determined.

The maximum shear strain occurs at the surface of the shaft, $\rho = c$. Since the angle of twist is $\phi = 0.6$ rad for the entire 1.5-m length of shaft, then using Eq. 5–25, we have

$$\phi = \gamma \frac{L}{\rho}; \qquad\qquad 0.6 = \frac{\gamma_{max}(1.5 \text{ m})}{(0.02 \text{ m})}$$

$$\gamma_{max} = 0.008 \text{ rad}$$

The shear-strain distribution, which always varies linearly, is shown in Fig. 5–43b. Note that yielding of the material occurs since $\gamma_{max} > \gamma_Y = 0.0016$ rad in Fig. 5–43a. The radius of the elastic core, ρ_Y, can be obtained by proportion. From Fig. 5–43b,

$$\frac{\rho_Y}{0.0016} = \frac{0.02 \text{ m}}{0.008}$$

$$\rho_Y = 0.004 \text{ m} = 4 \text{ mm}$$

Based on the shear-strain distribution, the shear-stress distribution, plotted over a radial line segment, is shown in Fig. 5–43c. The torque can now be obtained using Eq. 5–26. Substituting in the numerical data, Fig. 5–43c, yields

$$T = \frac{\pi \tau_Y}{6}(4c^3 - \rho_Y^3)$$

$$= \frac{\pi[75(10^6) \text{ N/m}^2]}{6}[4(0.02 \text{ m})^3 - (0.004 \text{ m})^3]$$

$$= 1.25 \text{ kN} \cdot \text{m} \qquad\qquad\qquad\qquad \textit{Ans.}$$

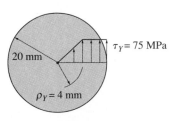

Shear–strain distribution

(b)

Shear–stress distribution

(c)

Fig. 5–43

5.10 Residual Stress

When a shaft is subjected to plastic shear strains caused by torsion, removal of the torque will cause some shear stress to remain in the shaft. This stress is referred to as *residual stress,* and its distribution can be calculated using the principles of superposition and elastic recovery.

Elastic recovery was discussed in Sec. 3.2, and it refers to the fact that whenever a material is plastically strained, some of the strain in the material is recovered when the load is released. For example, if a material is strained to γ_1, shown as point C on the $\tau–\gamma$ curve in Fig. 5–44, the release will cause a reverse shear stress, such that the material behavior will follow the straight-lined segment CD, creating some *elastic recovery* of the shear strain γ_1. This line is parallel to the initial straight-lined portion AB of the $\tau–\gamma$ diagram, and thus both lines have a slope G as indicated.

To illustrate how the residual-stress distribution can be determined in a shaft, we will first consider the shaft to be subjected to a plastic torque $\mathbf{T}_p$. As explained in Sec. 5.9, $\mathbf{T}_p$ creates the shear-stress distribution shown in Fig. 5–45a. We will assume that this distribution is a consequence of straining the material at the shaft's outer boundary to γ_1 in Fig. 5–44. Also, γ_1 is large enough so that the radius of the elastic core is assumed to approach zero, that is, $\gamma_1 \gg \gamma_Y$. If $\mathbf{T}_p$ is removed, the material tends to recover *elastically,* following along line CD in Fig. 5–44, as explained above. Since elastic behavior

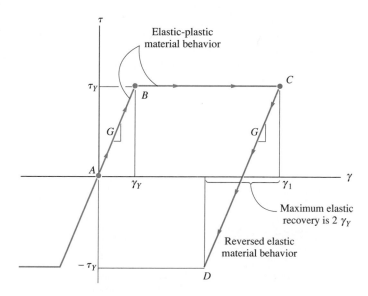

Fig. 5–44

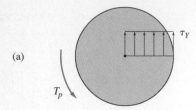

(a)

Plastic torque applied
causing plastic shear–strains
throughout the shaft

+

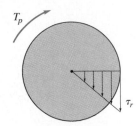

(b)

Plastic torque reversed
causing elastic shear –strains
throughout the shaft

=

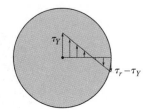

(c)

Residual shear–stress
distribution in shaft

occurs, we can superimpose on the stress distribution in Fig. 5–45*a* a *linear stress distribution* caused by applying the plastic torque $\mathbf{T}_p$ in the *opposite* direction, Fig. 5–45*b*. Here the maximum shear stress τ_r, computed for this stress distribution, is called the *modulus of rupture* for torsion. It is determined from the torsion formula,* which gives

$$\tau_r = \frac{T_p c}{J} = \frac{T_p c}{(\pi/2)c^4}$$

Using Eq. 5–27,

$$\tau_r = \frac{[(2/3)\pi\tau_Y c^3]c}{(\pi/2)c^4} = \frac{4}{3}\tau_Y$$

Note that reversed application of $\mathbf{T}_p$ using the linear shear-stress distribution in Fig. 5–45*b* is possible here, since the maximum recovery for the elastic shear strain is $2\gamma_Y$ as noted in Fig. 5–44. This corresponds to a maximum applied shear stress of $2\tau_Y$, which is greater than the maximum shear stress of $\frac{4}{3}\tau_Y$ computed above. Hence, by superimposing the stress distributions involving application and then removal of the plastic torque, we obtain the residual-shear-stress distribution in the shaft as shown in Fig. 5–45*c*. It should be noted from this diagram that the shear stress at the center of the shaft, shown as τ_Y, must actually be *zero,* since the material along the axis of the shaft is not strained. The reason this is not so is because we assumed that *all* the material of the shaft was strained beyond the proportional limit in order to determine the plastic torque. To be more realistic, an elastic-plastic torque must be considered when modeling the material behavior. Doing so leads to the superposition of the stress distributions shown in Fig. 5–45*d*.

The following example numerically illustrates these principles.

---

*The torsion formula is valid only when the material behaves in a linear-elastic manner; however, the modulus of rupture is so named because it assumes that the material behaves elastically and then suddenly *ruptures* at the proportional limit.

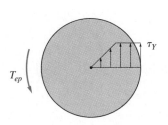

Elastic-plastic torque applied

+

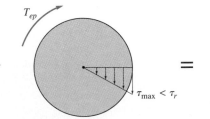

Elastic-plastic torque reversed

(d)

=

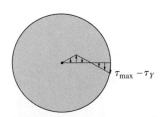

Residual shear–stress
distribution in shaft

Fig. 5–45

Example 5–21

A tube is made from a brass alloy having a length of 5 ft and a cross-sectional area shown in Fig. 5–46a. The material has an elastic-plastic τ–γ diagram, also shown in Fig. 5–46a. Compute the plastic torque $\mathbf{T}_p$. What are the residual-shear-stress distribution and permanent twist of the tube that remain if $\mathbf{T}_p$ is removed *just after* the tube becomes fully plastic?

SOLUTION

Plastic Torque. The plastic torque $\mathbf{T}_p$ will strain the tube such that all the material yields. Hence the stress distribution will appear as shown in Fig. 5–46b. Applying Eq. 5–23, we have

$$T_p = 2\pi \int_{c_i}^{c_o} \tau_Y \rho^2 \, d\rho = \frac{2\pi}{3} \tau_Y (c_o^3 - c_i^3)$$

$$= \frac{2\pi}{3}(12(10^3) \text{ lb/in}^2)[(2 \text{ in.})^3 - (1 \text{ in.})^3] = 175.9 \text{ kip} \cdot \text{in.} \qquad Ans.$$

When the tube just becomes fully plastic, yielding has started at the inner radius, i.e., at $c_i = 1$ in., $\gamma_Y = 0.002$ rad, Fig. 5–46a. The angle of twist that occurs can be determined from Eq. 5–25, which for the entire tube becomes

$$\phi_p = \gamma_Y \frac{L}{c_i} = \frac{(0.002)(5 \text{ ft})(12 \text{ in./ft})}{(1 \text{ in.})} = 0.120 \text{ rad} \; \uparrow$$

When $\mathbf{T}_p$ is *removed,* or in effect reapplied in the opposite direction, then the "fictitious" linear shear-stress distribution shown in Fig. 5–46c must be superimposed on the one shown in Fig. 5–46b. In Fig. 5–46c the maximum shear stress or the modulus of rupture is computed from the torsion formula

$$\tau_r = \frac{T_p c_o}{J} = \frac{(175.9 \text{ kip} \cdot \text{in.})(2 \text{ in.})}{(\pi/2)[(2 \text{ in.})^4 - (1 \text{ in.})^4]} = 14.93 \text{ ksi}$$

Also, at the inner wall of the tube the shear stress is

$$\tau_i = (14.93 \text{ ksi}) \left(\frac{1 \text{ in.}}{2 \text{ in.}} \right) = 7.45 \text{ ksi}$$

From Fig. 5–46a, $G = \tau_Y/\gamma_Y = 12 \text{ ksi}/(0.002 \text{ rad}) = 6000 \text{ ksi}$, so that the corresponding angle of twist ϕ_p' upon removal of $\mathbf{T}_p$ is therefore

$$\phi_p' = \frac{T_p L}{JG} = \frac{(175.9 \text{ kip} \cdot \text{in.})(5 \text{ ft})(12 \text{ in./ft})}{(\pi/2)[(2 \text{ in.})^4 - (1 \text{ in.})^4]6000 \text{ kip/in}^2} = 0.0747 \text{ rad} \; \downarrow$$

The resulting *residual-shear-stress distribution* is therefore shown in Fig. 5–46d. The permanent rotation of the tube after T_p is removed is

$$\phi = 0.120 - 0.0747 = 0.0453 \text{ rad} \; \uparrow \qquad Ans.$$

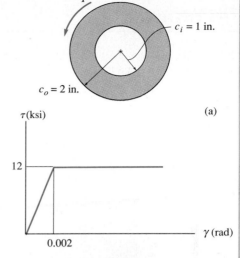

(a)

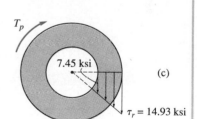

Plastic torque applied

\+

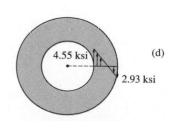

Plastic torque reversed

||

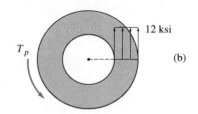

Residual shear–stress distribution
Fig. 5–46

PROBLEMS

*5–92. The assembly is subjected to a torque of 710 lb · in. If the allowable shear stress for the material is $\tau_{allow} = 12$ ksi, determine the radius of the smallest size fillet that can be used to transmit the torque.

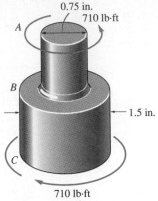

Prob. 5–92

5–93. The shaft is used to transmit 8 hp while turning at 450 rpm. Determine the maximum shear stress in the shaft. The segments are connected together using a fillet weld having a radius of 0.075 in.

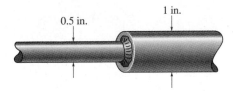

Prob. 5–93

5–94. The built-up shaft is to be designed to rotate at 650 rpm while transmitting 25 kW of power. If the allowable shear stress is $\tau_{allow} = 11.7$ MPa, determine the radius of the smallest fillet that can be used.

5–95. The built-up shaft is designed to rotate at 480 rpm. The diameter of the smaller shaft is $d = 60$ mm, the radius of the fillet weld connecting the shafts is $r = 7.20$ mm, and the allowable shear stress for the material is $\tau_{allow} = 55$ MPa. Determine the maximum power the shaft can transmit.

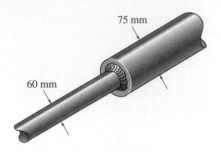

Probs. 5–94/5–95

*5–96. The shaft is fixed to the wall at A and is subjected to the torques shown. Determine the maximum shear stress in the shaft. A fillet, having a radius of 4.5 mm, is used to connect the shafts together at B.

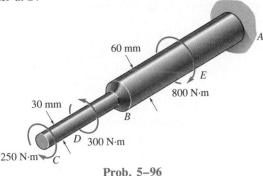

Prob. 5–96

5–97. The steel shaft is made from two segments AB and BC, which are connected together using a fillet weld having a radius of 2.8 mm. If there is a 20-mm-diameter hole bored into segment DC as shown, determine the maximum shear stress developed in the shaft.

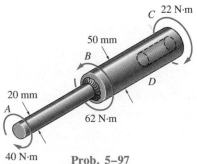

Prob. 5–97

5–98. The steel used for the shaft has an allowable shear stress of $\tau_{allow} = 8$ MPa. If the members are connected together with a fillet weld of radius $r = 2.25$ mm, determine the maximum torque T that can be applied.

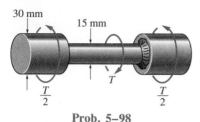

30 mm

15 mm

T

$\dfrac{T}{2}$

$\dfrac{T}{2}$

Prob. 5–98

5–99. Determine the torque needed to twist a short 2-mm-diameter steel wire through several revolutions if it is made from steel assumed to be elastic-plastic and having a yield stress of $\tau_Y = 50$ MPa. Assume that the material becomes fully plastic.

*5–100. A 2-in.-diameter shaft is made from an elastic-plastic material where $\tau_Y = 22$ ksi. If the shaft is 1.5 ft long, determine the torque T that must be applied in order to twist it 3°. $G = 11.0(10^3)$ ksi.

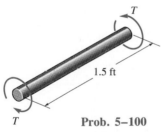

T

1.5 ft

T

Prob. 5–100

5–101. A shaft of radius $c = 0.75$ in. is made from an elastic-plastic material as shown. Determine the torque T that must be applied to its ends so that it has an elastic core of radius $\rho = 0.6$ in. If the shaft is 30 in. long, determine the angle of twist.

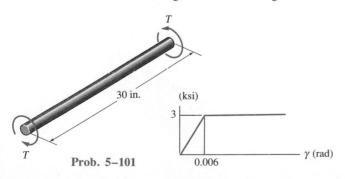

T

30 in.

(ksi)

3

T

Prob. 5–101

0.006

γ (rad)

5–102. A bar having a circular cross section of 3 in. diameter is subjected to a torque of 100 in. · kip. If the material is elastic-plastic, with $\tau_Y = 16$ ksi, determine the radius of the elastic core.

5–103. The solid shaft is made from an elastic-plastic material as shown. Determine the torque T needed to form an elastic core in the shaft having a radius of $\rho_Y = 23$ mm. If the shaft is 2 m long, through what angle does one end of the shaft twist with respect to the other end? When the torque is removed, determine the residual stress distribution in the shaft and the permanent angle of twist.

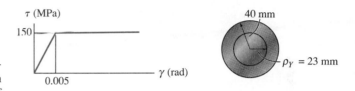

τ (MPa)

150

0.005

γ (rad)

40 mm

$\rho_Y = 23$ mm

Prob. 5–103

*5–104. A solid shaft is subjected to the torque T, which causes the material to yield. If the material is elastic-plastic, show that the torque can be expressed in terms of the angle of twist ϕ of the shaft as $T = \frac{4}{3}T_Y(1 - \phi_Y^3/4\phi^3)$, where T_Y and ϕ_Y are the torque and angle of twist when the material begins to yield.

5–105. A solid shaft has a diameter of 40 mm and length of 1 m. It is made from an elastic-plastic material having a yield stress of $\tau_Y = 100$ MPa. Determine the maximum elastic torque T_Y and the corresponding angle of twist. What is the angle of twist if the torque is increased to $T = 1.2T_Y$? $G = 80$ GPa.

5–106. A tubular shaft has an inside diameter of 20 mm, an outside diameter of 40 mm, and a length of 1 m. It is made from an elastic-plastic material having a yield stress of $\tau_Y = 100$ MPa. Determine the maximum elastic torque T_Y and the corresponding angle of twist. What is the angle of twist if the torque is increased to $T = 1.2T_Y$? $G = 80$ GPa.

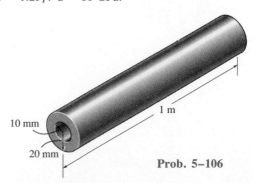

1 m

10 mm

20 mm

Prob. 5–106

5–107. The 2-m-long tube is made from an elastic-plastic material as shown. Determine the applied torque T, which subjects the material at the tube's outer edge to a shearing strain of $\gamma_{max} = 0.008$ rad. What would be the permanent angle of twist of the tube when this torque is removed? Sketch the residual stress distribution in the tube.

45 mm

40 mm

τ (MPa)

240

0.003

γ (rad)

Prob. 5–107

*5–108. The tube has a length of 2 m and is made from an elastic-plastic material as shown. Determine the torque needed to just cause the material to become fully plastic. What is the permanent angle of twist of the tube when this torque is removed?

50 mm

30 mm

τ (MPa)

350

0.007

γ (rad)

Prob. 5–108

5–109. The shaft consists of two sections that are rigidly connected. If the material is elastic-plastic as shown, determine the largest torque T that can be applied to the shaft. Also, draw the shear-stress distribution over a radial line for each section. Neglect the effect of stress concentration.

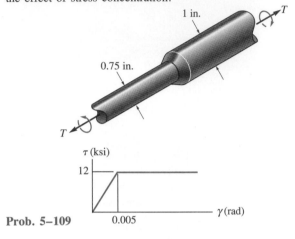

1 in.

0.75 in.

T

τ (ksi)

12

0.005

γ (rad)

Prob. 5–109

5–110. The shear stress-strain diagram for a solid 50-mm-diameter shaft can be approximated as shown in the figure. Determine the torque required to cause a maximum shear stress in the shaft of 125 MPa. If the shaft is 3 m long, what is the corresponding angle of twist?

τ (MPa)

125

50

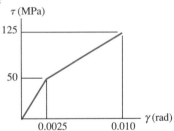

0.0025   0.010 γ (rad)

Prob. 5–110

5–111. A torque is applied to the shaft of radius r. If the material obeys a shear stress–strain relation of $\tau = k\gamma^{1/6}$, where k is a constant, determine the maximum shear stress in the shaft.

r

T

Prob. 5–111

*5–112. The shaft is subjected to a maximum shear strain of 0.0048 rad. Determine the torque applied to the shaft if the material has strain-hardening as shown by the shear stress–strain diagram.

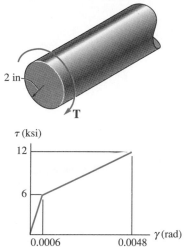

τ (ksi)

12

6

0.0006   0.0048 γ (rad)

Prob. 5–112

5–113. The shaft is made from a strain-hardening material having a τ–γ diagram as shown. Determine the torque T that must be applied to the shaft in order to create an elastic core in the shaft having a radius of $\rho_c = 0.5$ in.

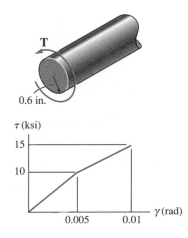

0.6 in.

τ (ksi)

15

10

0.005   0.01 γ (rad)

Prob. 5–113

REVIEW PROBLEMS

5–114. Determine the constant thickness of the rectangular tube if the maximum shear stress is not to exceed 12 ksi when a torque of $T = 20$ kip · in. is applied to the tube. Neglect stress concentrations at the corners. The mean dimensions of the tube are shown.

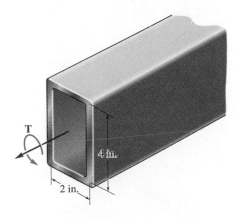

T

4 in.

2 in.

Prob. 5–114

5–115. The brass wire has a triangular cross section, 2 mm on a side. If the yield shear stress for brass is $\tau_Y = 205$ MPa, determine the maximum torque T to which it can be subjected. If this torque is applied to a segment 4 m long, determine the greatest angle of twist of one end of the wire relative to the other end, so as not to cause permanent damage to the wire. $G_{br} = 39$ GPa.

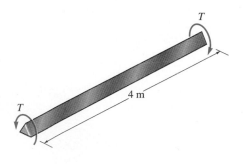

T

4 m

T

Prob. 5–115

*5–116. A steel tube having an outer diameter of 2.5 in. is used to transmit 35 hp when turning at 2700 rev/min. Determine the inner diameter d of the tube to the nearest $\frac{1}{8}$ in. if the allowable shear stress is $\tau_{\text{allow}} = 10$ ksi.

2.5 in.

Prob. 5–116

5–117. The steel shaft has a diameter of 40 mm and is fixed at its ends A and B. If it is subjected to the couple, determine the maximum shear stress in regions AC and CB of the shaft. $G_{st} = 10.8(10^3)$ ksi.

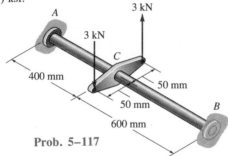

Prob. 5–117

5–118. By what percent is a bar of circular cross section more efficient than a bar of square cross section for resisting torque when the maximum shear stress is to be the same in both bars? Both bars are made from the same amount of material.

5–119. The shaft is made from an elastic-plastic material as shown. Plot the shear-stress distribution acting along a radial line if it is subjected to a torque of $T = 2$ kN $\cdot$ m. What is the residual stress distribution in the shaft when the torque is removed?

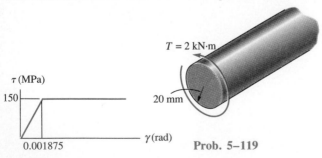

τ (MPa)

150

0.001875

γ (rad)

$T = 2$ kN·m

20 mm

Prob. 5–119

*5–120. The coupling is used to connect the two shafts together. Assuming that the shear stress in the bolts is *uniform*, determine the number of bolts necessary to make the maximum shear stress in the shaft equal to the shear stress in the bolts. Each bolt has a diameter d.

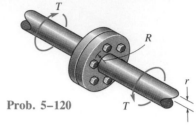

Prob. 5–120

5–121. The inner circle of the tube is eccentric with respect to the outer circle. By what percent is the torsional strength reduced when the eccentricity e is one-fourth of the difference in radii?

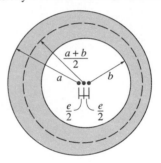

$\frac{a+b}{2}$

a b

$\frac{e}{2}$ $\frac{e}{2}$

Prob. 5–121

5–122. A solid shaft having a diameter of 2 in. is made of elastic-plastic material having a yield stress of $\tau_Y = 16$ ksi and shear modulus of $G = 12(10^3)$ ksi. Determine the torque required to develop an elastic core in the shaft having a diameter of 1 in. Also, what is the plastic torque?

5–123. The steel assembly consists of a tube having an outer radius of 1 in. and a wall thickness of 0.125 in. Using a rigid plate at B, it is connected to the solid 1-in.-diameter shaft AB. Determine the rotation of the tube's end C if a torque of 200 lb $\cdot$ in. is applied to the tube at this end. The end A of the shaft is fixed-supported. $G_{st} = 10.8(10^3)$ ksi.

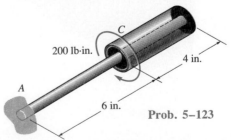

B

C

200 lb·in.

4 in.

A

6 in.

Prob. 5–123

6 Bending

In this chapter we will determine the stress in a member that is subjected to bending. We will begin by assuming the member is straight, has a symmetric cross section, and is made of homogeneous linear-elastic material. Afterwards, we will discuss special cases involving unsymmetric bending and members made of composite materials. Consideration will also be given to curved members, stress concentrations, inelastic bending, and residual stress.

6.1 Bending Deformation of a Straight Member

Members that are slender and support loadings that are applied perpendicular to their longitudinal axis are called *beams*. In this section we will discuss the deformations that occur when a straight prismatic beam, made of a homogeneous material, is subjected to bending. The discussion will be limited to beams having a cross-sectional area that is symmetrical with respect to an axis, and the bending moment is applied about an axis perpendicular to this axis of symmetry as shown in Fig. 6–1. The behavior of members that have unsymmetrical cross sections, or are made from several different materials, is based on similar observations and will be discussed separately in later sections of this chapter.

By using a highly deformable material such as rubber, we can physically illustrate what happens when a straight prismatic member is subjected to a bending moment. Consider, for example, the undeformed bar in Fig. 6–2a, which has a square cross section and is marked with longitudinal and transverse grid lines. When a bending moment is applied, it tends to distort these lines into the pattern shown in Fig. 6–2b. Here it can be seen that the longitu-

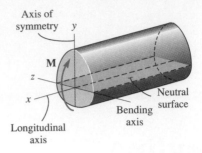

Fig. 6–1

dinal lines become *curved* and the vertical transverse lines *remain straight* and yet undergo a *rotation*.

The behavior of the deformable bar in Fig. 6–2 is the same as that exhibited by the member of arbitrary cross section shown in Fig. 6–1. In both cases, the bending moment causes the material within the bottom portion of the member to stretch and the material within the top portion to compress. Consequently, between these two regions there must be a surface, called the *neutral surface,* in which longitudinal fibers of the material will not undergo a change in length. This surface is located in the x–z plane as shown in Fig. 6–1.

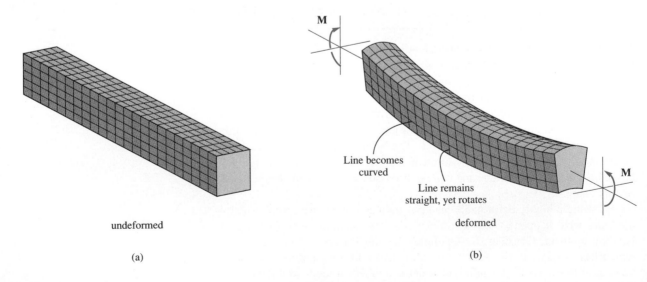

Fig. 6–2

From these observations we will make the following three assumptions regarding the way the stress deforms the material. First, the *longitudinal axis x*, which lies within the neutral surface, does *not* experience any *change in length*. Rather the moment will tend to deform the beam ,so that this line *becomes a curve* that lies in the *x–y* plane of symmetry, Fig. 6–3*b*. Second, all *cross sections* of the beam *remain plane* and perpendicular to the longitudinal axis during the deformation. And third, any *deformation* of the *cross section* within its own plane, as noticed in Fig. 6–2*b*, will be *neglected*. Thus, if the beam is fixed at one end as shown in Fig. 6–3*a*, and a moment is applied to its other end, the neutral surface and longitudinal *x* axis will deform into a curve, while the cross section remains plane, Fig. 6–3*b*. In particular, the *z* axis, lying in the plane of the cross section and about which the cross section rotates, is called the *neutral axis*. Its location will be determined in the next section.

In order to show how this distortion will strain the material in the longitudinal direction, we will isolate a segment of the beam that is located a distance *x* along the beam's length and has an undeformed thickness Δx, Fig. 6–3. This element, taken from the beam in Fig. 6–3, is shown in profile view in the

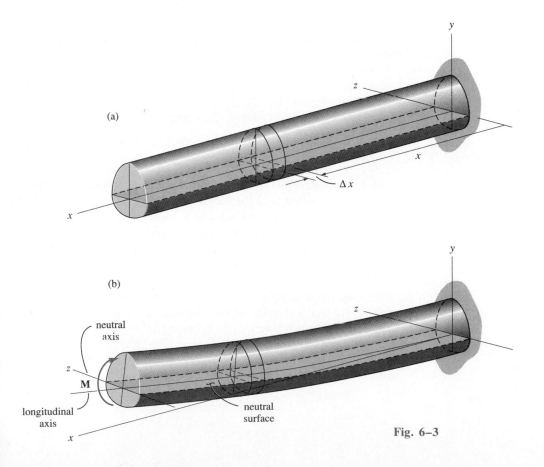

Fig. 6–3

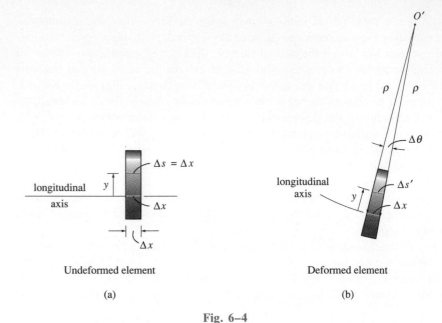

Undeformed element   Deformed element

(a)   (b)

Fig. 6–4

undeformed and deformed positions in Fig. 6–4. Notice that any line segment Δx, located on the neutral surface or along the longitudinal axis, does not change its length, whereas any line segment Δs, located at the arbitrary distance y above the neutral surface, Fig. 6–4a, will contract and become $\Delta s'$ after deformation, Fig. 6–4b. By definition, the normal strain along line segment Δs is determined from Eq. 2–2, namely,

$$\epsilon = \lim_{\Delta s \to 0} \frac{\Delta s' - \Delta s}{\Delta s}$$

We will now represent this strain in terms of the location y of the segment and the radius of curvature ρ of the longitudinal axis of the element. First notice in Fig. 6–4 that the distance y remains *approximately* the same both before and after deformation, since we have assumed that the deformation of the element in the plane of its cross section is not considered to be severe enough to caused any significant change in its dimensions. Before deformation, $\Delta s = \Delta x$, Fig. 6–4a. After deformation Δx has a radius of curvature ρ, with center of curvature at point O', Fig. 6–4b. Since $\Delta \theta$ defines the angle between the cross-sectional sides of the element, $\Delta x = \Delta s = \rho \, \Delta \theta$. In the same manner, the deformed length of Δs becomes $\Delta s' = (\rho - y) \, \Delta \theta$. Substituting into the above equation, we get

$$\epsilon = \lim_{\Delta \theta \to 0} \frac{(\rho - y) \, \Delta \theta - \rho \, \Delta \theta}{\rho \, \Delta \theta}$$

or

$$\epsilon = \frac{-y}{\rho} \qquad (6\text{–}1)$$

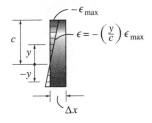

Normal strain distribution

Fig. 6–5

This important result indicates that the longitudinal normal strain of any element within the beam depends on its location y on the cross section and the radius of curvature of the beam's longitudinal axis at the point. In other words, for any specific cross section, the *longitudinal normal strain* will *vary linearly* with y from the neutral axis. In accordance with the sign convention for positive **M** and y shown in Fig. 6–3, Eq. 6–1 states that a contraction $(-\epsilon)$ will occur in fibers located above the neutral axis $(+y)$, whereas elongation $(+\epsilon)$ will occur in fibers located below the axis $(-y)$. This variation in strain over the cross section is shown in Fig. 6–5. Here the maximum strain occurs at the outermost fiber, located a distance c from the neutral axis. Using Eq. 6–1, since $\epsilon_{\max} = c/\rho$, then by division,

$$\frac{\epsilon}{\epsilon_{\max}} = \frac{-y/\rho}{c/\rho}$$

So that

$$\epsilon = -\left(\frac{y}{c}\right)\epsilon_{\max} \tag{6–2}$$

This normal stream depends only on the assumptions made with regard to the *deformation*. Provided, however, that only a moment is applied to each end of the beam, as shown in Fig. 6–1, then it is reasonable to further assume that this moment causes a *normal stress only* in the longitudinal or x direction. All the other components of normal and shear stress are zero, since the beam's surface is free of any other load. It is this uniaxial state of stress that causes the material to have the longitudinal normal strain component ϵ_x, $(\sigma_x = E\epsilon_x)$, defined by Eq. 6–1 or Eq. 6–2. Furthermore, by Poisson's ratio, there must also be associated strain components $\epsilon_y = -\nu\epsilon_x$ and $\epsilon_z = -\nu\epsilon_x$, which deform the plane of the cross-sectional area, although we have neglected these deformations in the development of Eqs. 6–1 and 6–2. Such deformations will, however, cause the *cross-sectional dimensions* to become smaller below the neutral axis and larger above the neutral axis. For example, if the beam has a square cross section, like the one in Fig. 6–2, it will actually deform as shown in Fig. 6–6.

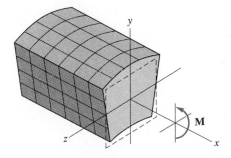

Fig. 6–6

6.2 The Flexure Formula

In this section we will develop an equation that relates the longitudinal stress distribution in a beam to the internal resultant bending moment acting on the beam's cross section. To do this we will assume that the material behaves in a linear-elastic manner so that Hooke's law applies, that is, $\sigma = E\epsilon$. Consequently, a linear variation of *normal strain,* as noted in Sec. 6.1, Fig. 6–7a, must then be the result of a corresponding linear variation in *normal stress* acting over the cross section, Fig. 6–7b. Hence, like the normal strain variation, σ will vary from zero at the member's neutral axis to a maximum value, σ_{max}, a distance c farthest from the neutral axis. A three-dimensional view of this stress distribution is shown in Fig. 6–7c. Due to the applied "positive" moment M, the material located at the very *top* of the beam is subjected to a maximum *compressive* stress, σ_{max}, whereas the material located at the intermediate position y *above* the neutral axis is subjected to the *compressive stress* $\sigma < \sigma_{max}$. Likewise, material *below* the neutral axis $(-y)$ is subjected to *tensile stress,* which varies from zero at the neutral axis to a maximum at the bottom of the member. Because of the proportionality of triangles, or by using Hooke's law, $\sigma = E\epsilon$, and Eq. 6–2, we can write

$$\sigma = -\left(\frac{y}{c}\right)\sigma_{max} \qquad (6\text{–}3)$$

Since σ_{max} and c are constant, this equation expresses the normal stress as a function of the y coordinate; in other words, it represents the stress distribution over the cross-sectional area. The sign convention established here is significant. By the right-hand rule, the moment **M** in Fig. 6–7c is considered positive since it is applied along the *positive z* axis. As a result, positive values of y give negative values for σ, that is, a compressive stress since it acts in the negative x direction. Similarly, negative y values will give positive or tensile values for σ. If a volume element of material is selected at a specific point on the cross section, only these tensile or compressive normal stresses will act on it. For example, the element located at $+y$ is shown in Fig. 6–7d.

We can locate the neutral axis on the cross section by satisfying the condition that the *resultant force* produced by the stress distribution over the cross-sectional area must be equal to *zero.* Noting that the force $dF = \sigma\,dA$ acts on the arbitrary element dA in Fig. 6–7c, then using Eq. 6–3 we require

$$F_R = \Sigma F_x; \qquad 0 = \int_A dF = \int_A \sigma\,dA = \int_A -\left(\frac{y}{c}\right)\sigma_{max}\,dA = \frac{-\sigma_{max}}{c}\int_A y\,dA$$

Since σ_{max}/c is not equal to zero, this equation is satisfied as long as

$$\int_A y\,dA = 0 \qquad (6\text{–}4)$$

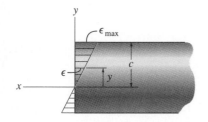

Normal strain variation
(profile view)

(a)

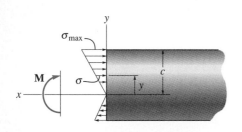

Bending stress variation
(profile view)

(b)

Fig. 6–7

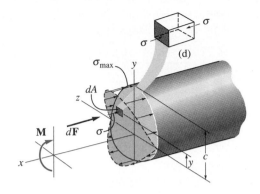

Bending stress variation

(c)

Fig. 6–7

In other words, the first moment of the member's cross-sectional area about the neutral axis must be zero. This condition can only be satisfied if the *neutral axis* is also the horizontal *centroidal axis* for the cross section.* Consequently, once the centroid for the member's cross-sectional area is determined, the location of the neutral axis is known.

In order to relate the stress in the beam to the resultant internal moment M acting at the cross section, we must require this moment to be equal to the moment produced by the stress distribution about the neutral axis. The moment of $d\mathbf{F}$ in Fig. 6–7c about the neutral axis is $dM = y \, dF$. This moment is *positive* since, by the right-hand rule, the thumb is directed along the positive z axis when the fingers are curled with the sense of rotation caused by $d\mathbf{M}$. Since $dF = \sigma \, dA$, using Eq. 6–3, and realizing that the moment is positive when y is positive and σ is negative, we have for the entire cross-section,

$$(M_R)_z = \Sigma M_z; \qquad M = \int_A y \, dF = \int_A y \, (\sigma \, dA) = -\int_A y \left(-\frac{y}{c} \sigma_{max} \right) dA$$

or

$$M = \frac{\sigma_{max}}{c} \int_A y^2 \, dA \qquad\qquad (6\text{–}5)$$

---

*Recall that the location $\bar{y}$ for the centroid of the cross-sectional area is defined from the equation $\bar{y} = \int y \, dA / \int dA$. If $\int y \, dA = 0$, then $\bar{y} = 0$, and so the centroid lies on the reference (neutral) axis. See Appendix A.

Here the integral represents the *moment of inertia* of the beam's cross-sectional area, computed about the neutral axis. We symbolize its value as *I*. Hence, Eq. 6–5 can be solved for σ_{max} and written in general form as

$$\sigma_{max} = \frac{Mc}{I} \qquad (6\text{--}6)$$

Here

σ_{max} = the maximum normal stress in the member, which occurs at a point on the cross-sectional area *farthest away* from the neutral axis

M = the resultant internal moment, determined from the method of sections and the equations of equilibrium, and computed about the neutral axis of the cross section

I = the moment of inertia of the cross-sectional area computed about the neutral axis

c = the perpendicular distance from the neutral axis to a point farthest away from the neutral axis, where σ_{max} acts

Since $\sigma_{max}/c = -\sigma/y$, Eq. 6–3, the normal stress at the intermediate distance *y* can be determined from an equation similar to Eq. 6–6. We have

$$\sigma = -\frac{My}{I} \qquad (6\text{--}7)$$

Note that the negative sign is necessary since it agrees with the established *x, y, z* axes. By the right-hand rule, *M* is positive along the $+z$ axis, *y* is positive upward, and σ therefore must be negative since it acts in the negative *x* direction, Fig. 6–7c.

Either of the above two equations is often referred to as the *flexure formula*. Recall that it is used to determine the normal stress in a straight member, having a cross section that is symmetrical with respect to an axis, and the moment is applied perpendicular to this axis. Furthermore, the material is assumed to be homogeneous and to behave in a linear-elastic manner. Although we have also assumed that the member is prismatic, we can in most cases of engineering design use the flexure formula to determine as well the normal stress in members that have a *slight taper*. For example, using a mathematical analysis based on the theory of elasticity, a member having a rectangular cross section and a taper of 15° on both its top and bottom sides will have an actual maximum normal stress that is about 5.4% *less* than that calculated using the flexure formula.

PROCEDURE FOR ANALYSIS

The flexure formula can be used to find the normal-stress distribution in a *straight prismatic member,* which is made from homogeneous material and has linear-elastic behavior. Saint-Venant's principle requires that this formula be applied at points located a sufficient distance away from any supports, discontinuities in the cross section, or points where any concentrated loading acts. In order to apply the equation, the following procedure is suggested.

Internal Moment.   Section the member perpendicular to its longitudinal axis at the point where the bending or normal stress is to be determined, and use the necessary free-body diagrams and equations of equilibrium to obtain the internal moment M at the section. For this purpose, the centroidal or neutral axis for the cross section must be known, since M *must* be computed about this axis.

Section Property.   Compute the moment of inertia of the cross-sectional area about the neutral axis. Methods used for its computation are discussed in Appendix A, and a table listing values of I for several common shapes is given on the inside back cover.

Normal Stress.   Specify the distance y, measured perpendicular to the neutral axis to the point where the normal stress is to be determined. Then apply the equation $\sigma = My/I$, or if the maximum bending stress is to be computed, $\sigma_{max} = Mc/I$. When substituting the data, make sure to use a consistent set of units.

The stress acts in a direction such that the force $d\mathbf{F}$ it creates at the point contributes a moment about the neutral axis that is in the same direction as the internal moment $\mathbf{M}$, Fig. 6–7c. In this manner the stress distribution acting over the entire cross section can be graphed, or a volume element of the material can be isolated and used to represent graphically the normal stress acting at the point.

The following examples illustrate numerical applications of this procedure.

Example 6–1

A beam has a rectangular cross section and is subjected to a stress distribution as shown in Fig. 6–8a. Determine the internal moment **M** at the section caused by the stress distribution (*a*) using the flexure formula, (*b*) by finding the resultant of the stress distribution using basic principles.

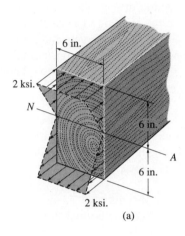

6 in.

2 ksi.

N

6 in.

A

6 in.

2 ksi.

(a)

Fig. 6–8

SOLUTION

Part (a). The flexure formula is $\sigma_{max} = Mc/I$. From Fig. 6–8a, $c = 6$ in. and $\sigma_{max} = 2$ ksi. The neutral axis is defined as line *NA*, because the stress is zero along this line. Since the cross section has a rectangular shape, the moment of inertia for the area about *NA* is determined from the formula for a rectangle given on the inside back cover; i.e.,

$$I = \frac{1}{12}bh^3 = \frac{1}{12}\,(6\text{ in.})(12\text{ in.})^3 = 864\text{ in}^4$$

Therefore,

$$\sigma_{max} = \frac{Mc}{I}; \qquad 2\text{ kip/in}^2 = \frac{M(6\text{ in.})}{864\text{ in}^4}$$

$$M = 288\text{ kip} \cdot \text{in.} = 24\text{ kip} \cdot \text{ft} \qquad\qquad \textit{Ans.}$$

Part (b).   First we will show that the resultant force of the stress distribution is zero. As shown in Fig. 6–8b, the stress acting on the arbitrary element strip $dA = (6 \text{ in.}) \, dy$, located y from the neutral axis, is

$$\sigma = \left(\frac{-y}{6 \text{ in.}}\right)(2 \text{ kip/in}^2)$$

The force created by this stress is $dF = \sigma \, dA$, and thus, for the entire cross section,

$$F_R = \int_A \sigma \, dA = \int_{-6 \text{ in.}}^{6 \text{ in.}} \left[\left(\frac{-y}{6 \text{ in.}}\right)(2 \text{ kip/in}^2)\right](6 \text{ in.}) \, dy$$

$$= (-1 \text{ kip/in}^2) \, y^2 \, \Big|_{-6 \text{ in.}}^{+6 \text{ in.}} = 0$$

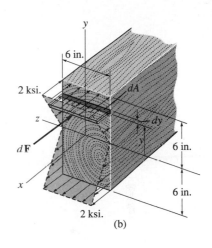

(b)

The resultant moment of the stress distribution about the neutral axis (z axis) must equal M. Since the magnitude of the moment of $d\mathbf{F}$ about this axis is $dM = y \, dF$, and $d\mathbf{M}$ is *always positive*, Fig. 6–8b, then for the entire area,

$$M = \int_A y \, dF = -\int_{-6 \text{ in}}^{6 \text{ in.}} y\left[\left(\frac{-y}{6 \text{ in.}}\right)(2 \text{ kip/in}^2)\right](6 \text{ in.}) \, dy$$

$$= \left(\frac{2}{3} \text{ kip/in}^2\right) y^3 \, \Big|_{-6 \text{ in.}}^{+6 \text{ in.}}$$

$$= 288 \text{ kip} \cdot \text{in.} = 24 \text{ kip} \cdot \text{ft} \qquad \textit{Ans.}$$

The above result can *also* be determined without the need for integration. The resultant force for each of the two *triangular* stress distributions in Fig. 6–8c is graphically equivalent to the *volume* contained within each stress distribution. Thus, each volume is

$$F = \frac{1}{2} \, (6 \text{ in.})(2 \text{ kip/in}^2)(6 \text{ in.}) = 36 \text{ kip}$$

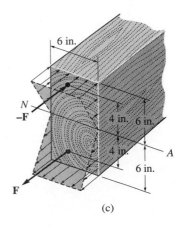

(c)

These forces, which form a couple, act in the same direction as the stresses within each distribution, Fig. 6–8c. Furthermore, they act through the *centroid* of each volume, i.e., $\frac{1}{3}(6 \text{ in.}) = 2 \text{ in.}$ from the top and bottom of the beam. Hence the distance between them is 8 in. as shown. The moment of the couple is therefore

$$M = 36 \text{ kip} (8 \text{ in.}) = 288 \text{ kip} \cdot \text{in.} = 24 \text{ kip} \cdot \text{ft} \qquad \textit{Ans.}$$

Example 6–2

The simply-supported beam in Fig. 6–9a has the cross-sectional area shown in Fig. 6–9b. Determine the bending stress that acts at points B and D, located at section a–a.

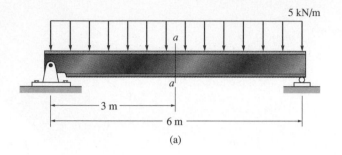

(a)

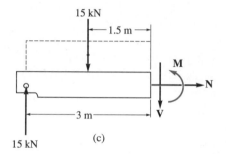

(c)

Fig. 6–9

(b)

SOLUTION

Internal Moment. Due to symmetry of both geometry and loading, the supports each exert a vertical reaction of 15 kN on the beam. Using the method of sections, the free-body diagram of the left segment of section a–a is shown in Fig. 6–9c. The moment is

$$M = 15 \text{ kN } (1.5 \text{ m}) = 22.5 \text{ kN} \cdot \text{m}$$

Section Property. By reasons of symmetry, the centroid C and thus the neutral axis pass through the midheight of the beam, Fig. 6–9b. The area is subdivided into the three parts shown, and the moment of inertia of each part is computed about the neutral axis using the parallel-axis theorem. (See Eq. A-5 of Appendix A.) Choosing to work in meters, we have

$$I = \Sigma(\bar{I} + Ad^2)$$
$$= 2 \left[\frac{1}{12} (0.25)(0.020)^3 + (0.25)(0.020)(0.160)^2 \right]$$
$$+ \left[\frac{1}{12} (0.020)(0.300)^3 \right]$$
$$= 301.3(10^{-6}) \text{ m}^4$$

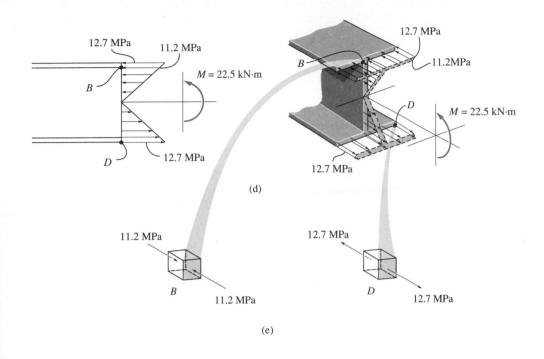

(d)

(e)

Normal Stress. Applying the flexure formula, with $y_B = 150$ mm for σ_B, we have

$$\sigma_B = \frac{My_B}{I}; \qquad \sigma_B = \frac{22.5 \text{ kN} \cdot \text{m}(0.150 \text{ m})}{301.3(10^{-6}) \text{ m}^4} = 11.2 \text{ MPa} \qquad \textit{Ans.}$$

For σ_D, $y_D = c = 170$ mm, so

$$\sigma_D = \frac{Mc}{I}; \qquad \sigma_D = \frac{22.5 \text{ kN} \cdot \text{m}(0.170 \text{ m})}{301.3(10^{-6}) \text{ m}^4} = 12.7 \text{ MPa} \qquad \textit{Ans.}$$

Two-and-three-dimensional views of the stress distribution are shown in Fig. 6–9d. Here the stress at each point on the cross section develops a force that contributes a moment $d\mathbf{M}$ about the neutral axis such that it has the same direction as $\mathbf{M}$. Specifically, the normal stress acting on elements of material located at points B and D is shown in Fig. 6–9e.

Example 6-3

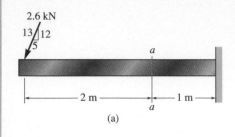

2.6 kN

13/12
/5

a

|← 2 m →|← 1 m →|

a

(a)

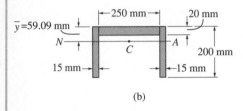

|← 250 mm →| |20 mm|

$\bar{y}$=59.09 mm

N ─┼─ ────── ─┤A ↑

C

200 mm

15 mm →|← →|←15 mm|

(b)

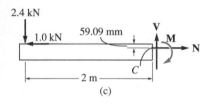

2.4 kN

1.0 kN   59.09 mm   V

────────── ┤ M

N

C

|← 2 m →|

(c)

Fig. 6–10

The beam shown in Fig. 6–10a has a cross-sectional area in the shape of a channel, Fig. 6–10b. Determine the maximum bending stress that occurs in the beam at section a–a.

SOLUTION

Internal Moment. Here the beam's support reactions do not have to be determined. Instead, by the method of sections, the segment to the left of section a–a can be used, Fig. 6–10c. In particular, note that the resultant internal axial force **N** passes through the centroid of the cross section. Also, realize that the resultant internal moment must be computed about the beam's neutral axis at section a–a.

To find the location of the neutral axis, the cross-sectional area is subdivided into three composite parts as shown in Fig. 6–10b. Since the neutral axis passes through the centroid, then using Eq. A–2 of Appendix A, we have

$$\bar{y} = \frac{\Sigma \bar{y}A}{\Sigma A} = \frac{2[100](200)(15) + [10](20)(250)}{2(200)(15) + 20(250)}$$

$$= 59.09 \text{ mm} = 0.05909 \text{ m}$$

This dimension is shown in Fig. 6–10c.

Applying the moment equation of equilibrium about the neutral axis, we have

$$\zeta + \Sigma M_{NA} = 0; \qquad 2.4 \text{ kN}(2 \text{ m}) + 1 \text{ kN}(0.05909 \text{ m}) - M = 0$$

$$M = 4.859 \text{ kN} \cdot \text{m}$$

Section Property. The moment of inertia about the neutral axis is determined using the parallel-axis theorem applied to each of the three composite parts of the cross-sectional area. Working in meters, we have

$$I = \left[\frac{1}{12}(0.250)(0.020)^3 + (0.250)(0.020)(0.05909 - 0.010)^2 \right]$$

$$+ 2 \left[\frac{1}{12}(0.015)(0.200)^3 + (0.015)(0.200)(0.100 - 0.05909)^2 \right]$$

$$= 42.26(10^{-6}) \text{ m}^4$$

Maximum Bending Stress. The maximum bending stress occurs at points farthest away from the neutral axis. This is at the bottom of the beam, $c = 200 \text{ mm} - \bar{y} = 140.9 \text{ mm}$. Thus,

$$\sigma_{max} = \frac{Mc}{I} = \frac{4.859 \text{ kN} \cdot \text{m}(0.1409 \text{ m})}{42.26(10^{-6}) \text{ m}^4} = 16.2 \text{ MPa} \qquad \textit{Ans.}$$

By comparison, the bending stress at the *top* of the beam is

$$\sigma' = \frac{M\bar{y}}{I} = \frac{4.859 \text{ kN}(0.05909 \text{ m})}{42.26(10^{-6}) \text{ m}^4} = 6.79 \text{ MPa}$$

Example 6–4

The member having a rectangular cross section, Fig. 6–11a, is designed to resist a moment of 40 N · m. In order to increase its strength and rigidity, it is proposed that two small ribs be added at its bottom, Fig. 6–11b. Determine the maximum normal stress in the member for both cases.

SOLUTION

Without Ribs. Clearly the neutral axis is at the center of the cross section, so $\bar{y} = c = 15$ mm $= 0.015$ m. Thus,

$$I = \frac{1}{12} (0.06 \text{ m})(0.03 \text{ m})^3 = 0.135(10^{-6}) \text{ m}^4$$

Therefore the maximum normal stress is

$$\sigma_{max} = \frac{Mc}{I} = \frac{(40 \text{ N} \cdot \text{m})(0.015 \text{ m})}{0.135(10^{-6}) \text{ m}^4} = 4.44 \text{ MPa} \qquad Ans.$$

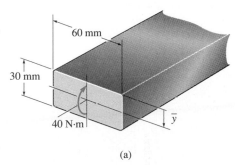

(a)

With Ribs. From Fig. 6–11b, segmenting the area into the large main rectangle and the bottom two rectangles (ribs), the location $\bar{y}$ of the centroid and the neutral axis is determined as follows:

$$\bar{y} = \frac{\Sigma \bar{y} A}{\Sigma A}$$

$$= \frac{[0.015 \text{ m}](0.030 \text{ m})(0.060 \text{ m}) + 2[0.0325 \text{ m}](0.005 \text{ m})(0.010 \text{ m})}{(0.03 \text{ m})(0.060 \text{ m}) + 2(0.005 \text{ m})(0.010 \text{ m})}$$

$$= 0.01592 \text{ m}$$

This value does not represent c. Instead

$$c = 0.035 \text{ m} - 0.01592 \text{ m} = 0.01908 \text{ m}$$

Using the parallel-axis theorem, the moment of inertia about the neutral axis is

$$I = \left[\frac{1}{12} (0.06)(0.03)^3 + (0.06)(0.03)(0.01592 - 0.015)^2 \right]$$

$$+ 2\left[\frac{1}{12} (0.01)(0.005)^3 + (0.01)(0.005)(0.0325 - 0.01592)^2 \right]$$

$$= 0.1642 (10^{-6}) \text{ m}^4$$

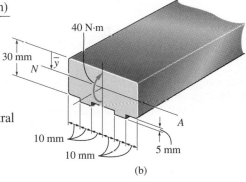

Therefore, the maximum normal stress is

$$\sigma_{max} = \frac{Mc}{I} = \frac{40 \text{ N} \cdot \text{m}(0.01908 \text{ m})}{0.1642 (10^{-6}) \text{ m}^4} = 4.65 \text{ MPa} \qquad Ans.$$

Fig. 6–11

The result indicates that the addition of the ribs to the cross section will *increase* the normal stress rather than decrease it, and for this reason they should be omitted.

PROBLEMS

6–1. A member having the dimensions shown is to be used to resist an internal bending moment of $M = 17$ lb $\cdot$ ft. Determine the maximum stress in the member if the moment is applied (*a*) about the z–z axis, (*b*) about the y–y axis. Sketch the stress distribution for each case.

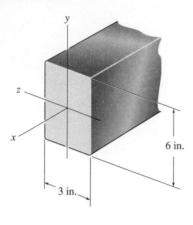

Prob. 6–1

6–2. A beam is constructed from four pieces of wood, glued together as shown. If the internal bending moment is $M = 80$ kip $\cdot$ ft, determine the maximum bending stress in the beam. Sketch a three-dimensional view of the stress distribution acting over the cross section.

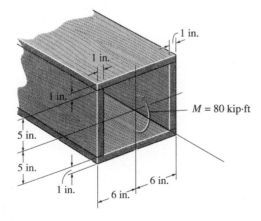

Prob. 6–2

6–3. The tapered casting supports the loading shown. Determine the bending stress at points A and B. The cross section at section a–a is given in the figure.

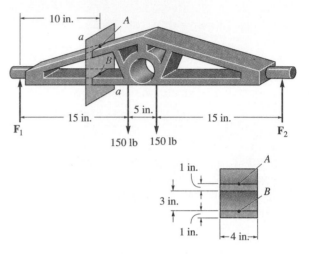

Prob. 6–3

*6–4. The aluminum strut has a cross-sectional area in the form of a cross. If it is subjected to the moment $M = 5$ kN $\cdot$ m, determine the stress acting at points A and B. Also, sketch a three-dimensional view of the stress distribution acting over the entire cross-sectional area.

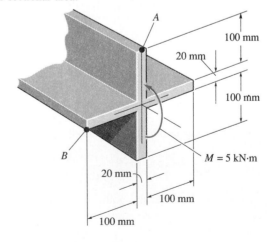

Prob. 6–4

6–5. The chair is supported by an arm that is hinged so it rotates about the vertical axis at A. If the load on the chair is 180 lb and the arm is a hollow tube section having the dimensions shown, determine the maximum bending stress at section a–a.

6–7. The aluminum machine part is subjected to an internal bending moment of $M = 75$ N · m. Determine the stress created at points B and C on the cross section. Sketch the results on a volume element located at each of these points.

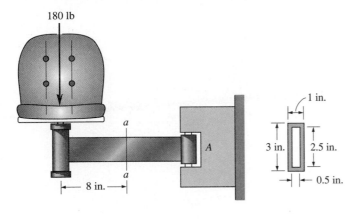

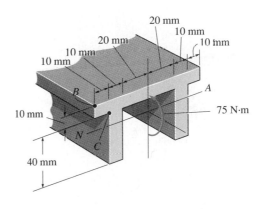

Prob. 6–5

Prob. 6–7

6–6. The beam is made from three boards nailed together as shown. If the moment acting on the cross section is $M = 650$ N · m, determine the resultant force the stress produces on the top board, shown shaded in the figure.

*6–8. A beam has the cross section shown. If it is made of steel that has an allowable stress of $\sigma_{allow} = 24$ ksi, determine the largest internal moment the beam can resist if the moment is applied (*a*) about the z axis, (*b*) about the y axis.

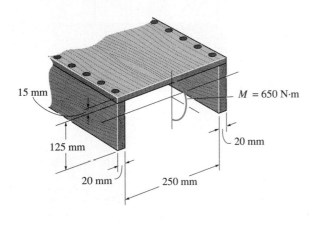

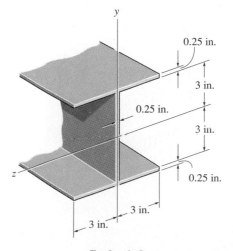

Prob. 6–6

Prob. 6–8

6–9.   A beam is constructed from four pieces of wood, glued together as shown. If the moment acting on the cross section is $M = 450$ N · m, determine the resultant force the stress produces on the top board A and on the side board B.

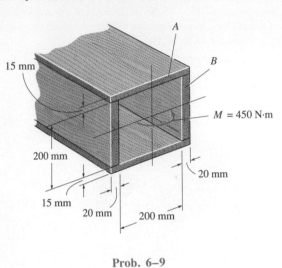

Prob. 6–9

6–10.   The beam is subjected to a moment M. Determine the percent of this moment that is resisted by the stresses acting on both the top and bottom boards, A and B, of the beam.

6–11.   Determine the moment M that should be applied to the beam in order to create a compressive stress at point D of $\sigma_D =$ 30 MPa. Also sketch the stress distribution acting over the cross section and compute the maximum stress developed in the beam.

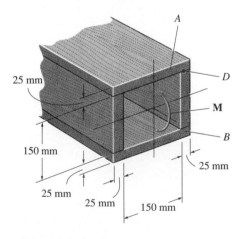

Probs. 6–10/6–11

*6–12.   The simply-supported beam is subjected to the load of 1.5 kN. Determine the maximum bending stress in the beam at a section taken just to the left of its center.

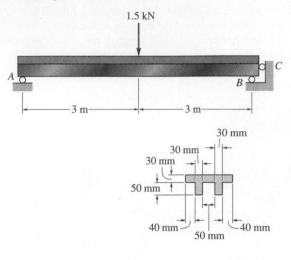

Prob. 6–12

6–13.   Determine the bending stress distribution in the beam at section a–a. Sketch the distribution in three dimensions acting over the cross section.

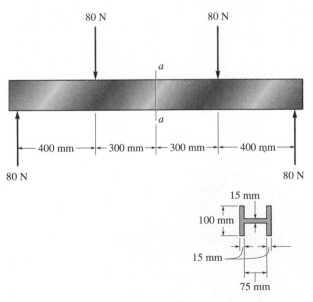

Prob. 6–13

6–14. The steel shaft has a circular cross section with a diameter of 2 in. It is supported on smooth journal bearings A and B, which exert only vertical reactions on the shaft. Determine the maximum bending stress in the shaft at section a–a if it is subjected to the pulley loadings shown.

*6–16.** The beam is constructed from four boards as shown. If it is subjected to an internal bending moment of $M_z = 16$ kip · ft, determine the stress at points A and B. Sketch a three-dimensional view of the stress distribution.

6–17. The beam is constructed from four boards as shown. If it is subjected to an internal bending moment of $M_z = 16$ kip · ft, determine the resultant force the stress produces on the top board C.

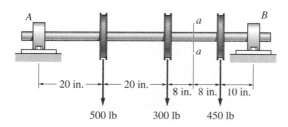

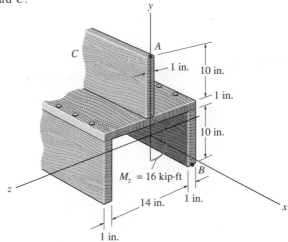

Prob. 6–14

Probs. 6–16/6–17

6–15. The axle of the freight car is subjected to wheel loadings of 20 kip. If it is supported by two journal bearings at C and D, determine the maximum bending stress developed at the center of the axle, where the diameter is 5.5 in.

6–18. The bolster or main supporting girder of a truck body is subjected to the uniform distributed load. Determine the bending stress at points A and B.

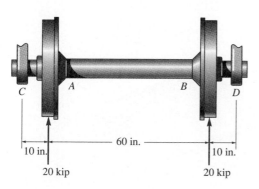

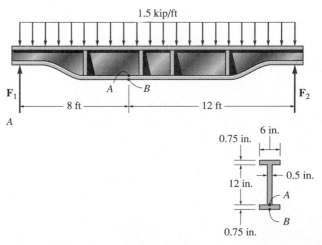

Prob. 6–15

Prob. 6–18

6–19.   The pin is used to connect the three links together. Due to wear, the load is distributed over the top and bottom of the pin as shown on the free-body diagram. If the diameter of the pin is 0.40 in., determine the maximum bending stress on the cross-sectional area at the center section a–a. For the solution it is first necessary to determine the load intensities w_1 and w_2.

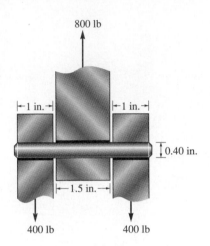

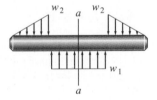

Prob. 6–19

*6–20.   A member has the triangular cross section shown. Determine the largest internal moment M that can be applied to the cross section without exceeding allowable tensile and compressive stresses of $(\sigma_{\text{allow}})_t = 22$ ksi and $(\sigma_{\text{allow}})_c = 15$ ksi, respectively.

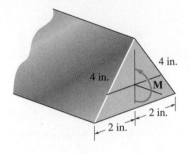

Prob. 6–20

6–21.   A shaft is made of a polymer having an elliptical cross section. If it resists an internal moment of $M = 50$ N · m, determine the maximum bending stress developed in the material (a) using the flexure formula, where $I_z = \frac{1}{4}\pi\,(0.08\text{ m})\,(0.04\text{ m})^3$, ($b$) using integration. Sketch a three-dimensional view of the stress distribution acting over the cross-sectional area.

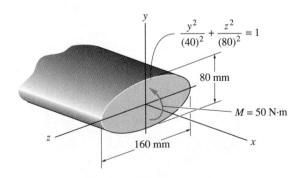

Prob. 6–21

6–22. Determine the magnitude of the maximum loads **P** that can be applied to the beam if the beam is made of a material having an allowable bending stress of $(\sigma_{allow})_c$ = 12 ksi in compression and $(\sigma_{allow})_t$ = 22 ksi in tension. Maximum bending moment in the beam will occur at any section a–a between the loads.

6–23. Determine the magnitude of the maximum loads **P** that can be applied to the beam if the beam is made of a material having an allowable bending stress of $(\sigma_{allow})_c$ = 16 ksi in compression and $(\sigma_{allow})_t$ = 18 ksi in tension. Maximum bending moment in the beam will occur at any section a–a between the loads.

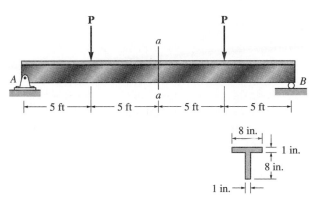

Probs. 6–22/6–23

6–26. The steel beam has the cross-sectional area shown. Determine the largest intensity of distributed load w that it can support so that the bending stress at its center section a–a does not exceed σ_{max} = 22 ksi.

6–27. The steel beam has the cross-sectional area shown. If w = 5 kip/ft, determine the maximum bending stress acting at its center section a–a.

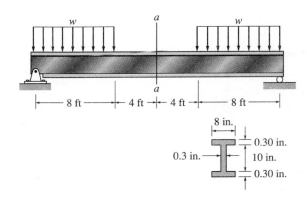

Probs. 6–26/6–27

*6–24. The beam has a rectangular cross section as shown. Determine the largest load P that can be supported on its overhanging ends so that the bending stress at its center section a–a does not exceed σ_{max} = 10 MPa.

6–25. The beam has the rectangular cross section shown. If P = 12 kN, determine the maximum bending stress acting at its center section a–a. Sketch the stress distribution acting over the cross section.

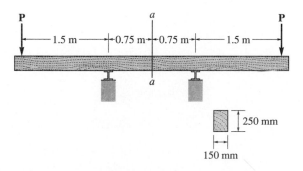

Probs. 6–24/6–25

*6–28. A beam is made of a material that has a modulus of elasticity in compression different from that given for tension. Determine the location c of the neutral axis, and derive an expression for the maximum tensile stress in the beam having the dimensions shown if it is subjected to the bending moment M.

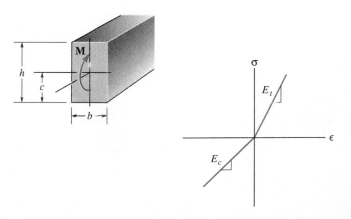

Prob. 6–28

6.3 Unsymmetric Bending

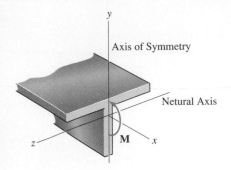

Axis of Symmetry

Netural Axis

M

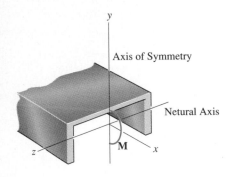

Axis of Symmetry

Netural Axis

M

Fig. 6–12

When developing the flexure formula we imposed a condition that the cross-sectional area be *symmetric* about an axis perpendicular to the neutral axis; furthermore, the resultant internal moment **M** acts along the neutral axis. Such is the case for the "T" or channel sections shown in Fig. 6–12. These conditions, however, are unnecessary, and in this section we will show that the flexure formula can also be applied either to a beam having a cross-sectional area of any shape or to a beam having a resultant internal moment that acts in any direction.

Moment Applied Along Principal Axis. Consider the beam's cross section to have the unsymmetrical shape shown in Fig. 6–13a. We will assume that the beam is both *straight* and *prismatic*. As in Sec. 6.2, the right-handed x, y, z coordinate system is established such that the origin is located at the centroid C of the cross section, and the resultant internal moment **M** acts along the $+z$ axis. We require the stress distribution acting over the entire cross-sectional area to have a zero force resultant, the resultant internal moment about the y axis to be zero, and the resultant internal moment about the z axis to equal **M**.* These three conditions can be expressed mathematically by considering the force acting on the differential element dA located at $(0, y, z)$, Fig. 6–13a. This force is $dF = \sigma \, dA$, and therefore we have

$$F_R = \Sigma F_x; \qquad\qquad 0 = \int_A \sigma \, dA \qquad\qquad (6\text{–}8)$$

$$(M_R)_y = \Sigma M_y; \qquad\qquad 0 = \int_A z\sigma \, dA \qquad\qquad (6\text{–}9)$$

$$(M_R)_z = \Sigma M_z; \qquad\qquad M = \int_A -y\sigma \, dA \qquad\qquad (6\text{–}10)$$

*The condition that moments about the y axis be equal to zero was not considered in Sec. 6.2, since the bending-stress distribution was *symmetric* with respect to the y axis and such a distribution of stress automatically produces zero moment about the y axis. See Fig. 6–7c.

Fig. 6–13

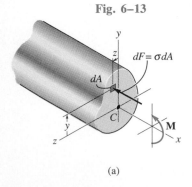

z

$dF = \sigma dA$

dA

C

y

M

x

(a)

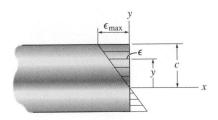

ϵ_{max}

ϵ

c

y

x

Normal Strain Distribution
(Profile View)
(b)

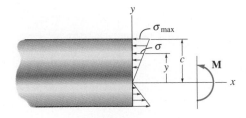

σ_{max}

σ

c

y

M

x

Bending Stress Distribution
(Profile View)
(c)

As shown in Sec. 6.2, Eq. 6–8 is satisfied since the z axis passes through the *centroid* of the cross-sectional area. Also, since the z axis represents the *neutral axis* for the cross section, the normal strain will vary linearly from zero at the neutral axis, to a maximum at a point located the largest y coordinate distance, $y = c$, from the neutral axis, Fig. 6–13b. Provided the material behaves in a linear-elastic manner, the normal-stress distribution over the cross-sectional area is *also* linear, so that $\sigma = -(y/c)\sigma_{max}$, Fig. 6–13c. When this equation is substituted into Eq. 6–10 and integrated, it leads to the flexure formula $\sigma_{max} = Mc/I$. When it is substituted into Eq. 6–9, we get

$$0 = -\frac{\sigma_{max}}{c} \int_A yz \, dA$$

which requires

$$\int_A yz \, dA = 0$$

This integral is called the *product of inertia* for the area. As indicated in Appendix A, it will indeed be zero provided the y and z axes are chosen as *principal axes of inertia* for the area. For an arbitrarily shaped area, like the one shown in Fig. 6–13a, the orientation of the principal axes can always be determined, using either the inertia transformation equations or Mohr's circle of inertia as explained in Appendix A, Sec. A–5. If the area has an axis of symmetry, however, the *principal axes* can easily be established since they will always be oriented *along the axis of symmetry* and *perpendicular* to it.

In summary, then, Eqs. 6–8 through 6–10 will *always* be satisfied *provided* the moment **M** is applied about one of the centroidal principal axes of inertia. For example, consider the members shown in Fig. 6–14, which are subjected to bending. In each of these cases, y and z define the principal axes of inertia for the cross section having the origin located at the area's centroid. In Fig. 6–14a and 6–14b the principal axes are located by symmetry, and in Fig. 6–14c and 6–14d their orientation is determined using the methods of Appendix A. Since **M** is applied about one of the principal axes (z axis), the stress distribution is determined from the flexure formula, $\sigma = My/I_z$, and is shown for each case.

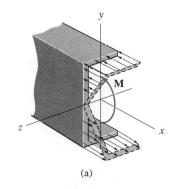

(a)

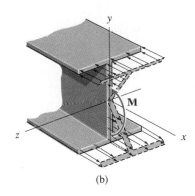

(b)

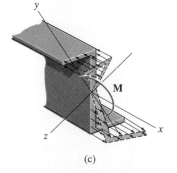

(c)

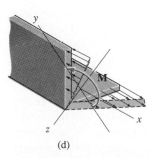

(d)

Fig. 6–14

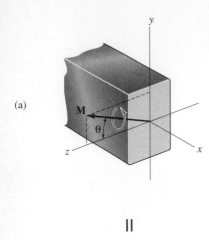

(a)

||

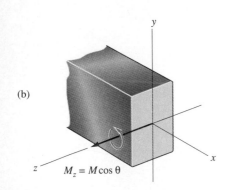

(b)

$M_z = M \cos \theta$

+

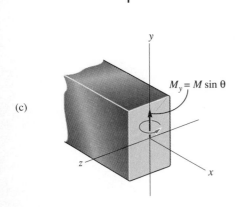

(c)

$M_y = M \sin \theta$

Fig. 6–15

Moment Arbitrarily Applied. Sometimes a member may be loaded such that the resultant internal moment does not act about one of the principal axes of the cross section. When this occurs, the moment should first be resolved into components directed along the principal axes. The flexure formula can then be used to determine the normal stress caused by each moment component. Finally, using the principle of superposition, the resultant normal stress at the point can be determined.

For example, consider the beam to have a rectangular cross section and to be subjected to the moment **M**, Fig. 6–15a. By the right-hand rule, **M** is represented as a vector such that it makes an angle θ with the *principal z* axis. In particular, θ is positive when it is directed from the $+z$ axis towards the $+y$ axis, as shown. Resolving **M** into components along the z and y axes, we have $M_z = M \cos \theta$ and $M_y = M \sin \theta$, respectively. Each of these components is shown separately on the cross section in Fig. 6–15b and 6–15c. The normal-stress distributions that produce **M** and its components **M**$_z$ and **M**$_y$ are shown in Fig. 6–15d, 6–15e, and 6–15f, respectively. Here it is assumed that $(\sigma_x)_{max} > (\sigma_x')_{max}$. By inspection, the maximum tensile and compressive stresses $[(\sigma_x)_{max} + (\sigma_x')_{max}]$ occur at two of the beam's opposite corners, Fig. 6–15d.

Applying the flexure formula to each moment component in Fig. 6–15b and 6–15c, we can express the resultant normal stress at any point on the cross section, Fig. 6–15d, in general terms as

$$\sigma = -\frac{M_z y}{I_z} + \frac{M_y z}{I_y} \qquad (6\text{–}11)$$

Here

σ = the normal stress at the point

y, z = the coordinates of the point measured from x, y, z axes having their origin at the centroid of the cross-sectional area. The x axis is directed outward from the cross-section and the y and z axes represent respectively the principal axes of minimum and maximum moment of inertia for the area

M_y, M_z = the resultant internal moment components directed along the principal y and z axes. They are positive if directed along the $+y$ and $+z$ axes, otherwise they are negative. Or, stated another way, $M_y = M \sin \theta$ and $M_z = M \cos \theta$, where θ is measured positive from the $+z$ axis toward the $+y$ axis

I_y, I_z = the *principal moments of inertia* computed about the z and y axes, respectively. See Appendix A

As noted above, it is *very important* that the x, y, z axes form a right-handed system and that the proper algebraic signs be assigned to the moment components and the coordinates when applying Eq. 6–11. The resulting stress will be *tensile* if it is *positive* and *compressive* if it is *negative*.

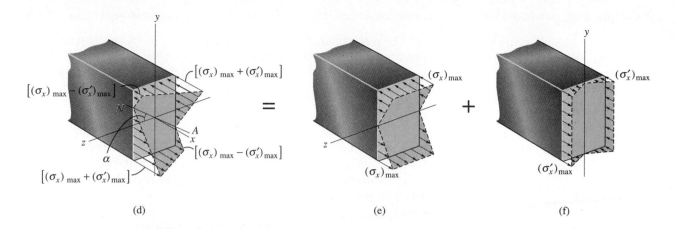

(d)   (e)   (f)

Orientation of the Neutral Axis.

The angle α of the neutral axis in Fig. 6–15d can be determined by applying Eq. 6–11 with $\sigma = 0$, since by definition no normal stress acts on the neutral axis. We have

$$y = \frac{M_y I_z}{M_z I_y} z$$

Since $M_z = M \cos \theta$ and $M_y = M \sin \theta$, Fig. 6–15a, then

$$y = \left(\frac{I_z}{I_y} \tan \theta \right) z \qquad (6\text{–}12)$$

This is the equation of the line that defines the neutral axis for the cross section, Fig. 6–15d. Since the term in parentheses represents the slope of this line (y/z),

$$\boxed{\tan \alpha = \frac{I_z}{I_y} \tan \theta} \qquad (6\text{–}13)$$

Here it can be seen that for *unsymmetrical bending* the angle θ, defining the direction of the moment **M**, Fig. 6–15a, is *not equal* to α, the angle defining the inclination of the neutral axis, Fig. 6–15d, unless $I_z = I_y$. Instead, if the *y axis* is chosen as the principal axis for the *minimum* moment of inertia, and the *z axis* is chosen as the principal axis for the *maximum* moment of inertia, so that $I_y < I_z$, Fig. 6–15a, then from Eq. 6–13 we can conclude that the angle α, which is measured positive from the $+z$ axis toward the $+y$ axis, will lie *between* the line of action of **M** and the *y* axis, Fig. 6–15d.

The following examples numerically illustrate these concepts.

Example 6–5

The rectangular cross section shown in Fig. 6–16a is subjected to a bending moment of $M = 12$ kN · m. Determine the normal stress developed at each corner of the section, and specify the orientation of the neutral axis.

SOLUTION

Internal Moment Components. By inspection it is seen that the y and z axes represent the principal axes of inertia since they are axes of symmetry for the cross section. Here we have established the z *axis* as the principal axis for *maximum* moment of inertia. The moment is resolved into its y and z components, where

$$M_y = -\frac{4}{5}(12 \text{ kN} \cdot \text{m}) = -9.60 \text{ kN} \cdot \text{m}$$

$$M_z = \frac{3}{5}(12 \text{ kN} \cdot \text{m}) = 7.20 \text{ kN} \cdot \text{m}$$

Section Properties. The moments of inertia about the y and z axes are

$$I_y = \frac{1}{12}(0.4 \text{ m})(0.2 \text{ m})^3 = 0.2667(10^{-3}) \text{ m}^4$$

$$I_z = \frac{1}{12}(0.2 \text{ m})(0.4 \text{ m})^3 = 1.067(10^{-3}) \text{ m}^4$$

Bending Stress. Applying Eq. 6–11, that is,

$$\sigma = -\frac{M_z y}{I_z} + \frac{M_y z}{I_y}$$

we have

$$\sigma_B = -\frac{7.20(10^3)(0.2)}{1.067(10^{-3})} + \frac{-9.60(10^3)(-0.1)}{0.2667(10^{-3})} = 2.25 \text{ MPa} \qquad \textit{Ans.}$$

$$\sigma_C = -\frac{7.20(10^3)(0.2)}{1.067(10^{-3})} + \frac{-9.60(0.1)}{0.2667(10^{-3})} = -4.95 \text{ MPa} \qquad \textit{Ans.}$$

$$\sigma_D = -\frac{7.20(10^3)(-0.2)}{1.067(10^{-3})} + \frac{-9.60(0.1)}{0.2667(10^{-3})} = -2.25 \text{ MPa} \qquad \textit{Ans.}$$

$$\sigma_E = -\frac{7.20(10^3)(-0.2)}{1.067(10^{-3})} + \frac{-9.60(-0.1)}{0.2667(10^{-3})} = 4.95 \text{ MPa} \qquad \textit{Ans.}$$

The resultant normal-stress distribution has been sketched using these values, Fig. 6–16b. Since superposition applies, the distribution is linear as shown.

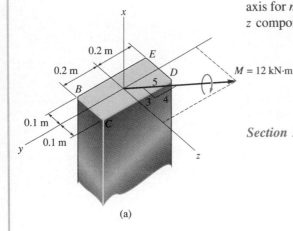

Fig. 6–16

(a)

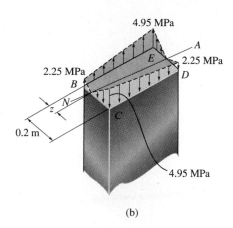

(b)

Orientation of Neutral Axis.   The location z of the neutral axis (*NA*), Fig. 6–16*b*, can be established by proportion. Along the edge *BC*, we require

$$\frac{2.25 \text{ MPa}}{z} = \frac{4.95 \text{ MPa}}{(0.2 \text{ m} - z)}$$

$$0.450 - 2.25z = 4.95z$$

$$z = 0.0625 \text{ m}$$

In the same manner this is also the distance from *D* to the neutral axis in Fig. 6–16*b*.

We can also establish the orientation of the *NA* using Eq. 6–13, which is used to specify the angle α that the axis makes with the z or *maximum* principal axis, Fig. 6–16*c*. According to our sign convention, θ must be measured from the positive z axis, and it is positive if it is directed from the $+z$ axis towards the $+y$ axis, Fig. 6–15*a*. By comparison, in Fig. 6–16*c*, $\theta = -\tan^{-1}\frac{4}{3} = -53.1°$ (or $\theta = +306.9°$). Thus,

$$\tan \alpha = \frac{I_z}{I_y} \tan \theta$$

$$\tan \alpha = \frac{1.067(10^{-3}) \text{ m}^4}{0.2667(10^{-3}) \text{ m}^4} \tan(-53.1°)$$

$$\alpha = -79.4° \qquad\qquad Ans.$$

To verify the previous calculation for z, shown in Fig. 6–16*c*, we can recalculate the same angle α. From the geometry of the figure, we have

$$\tan \alpha = \frac{0.2 \text{ m}}{0.0375 \text{ m}}$$

$$\alpha = 79.4° \qquad\qquad Ans.$$

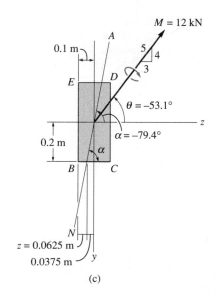

(c)

Example 6–6

A *T*-beam is subjected to the bending moment of 15 kN · m as shown in Fig. 6–17b. Determine the maximum normal stress in the beam and the orientation of the neutral axis.

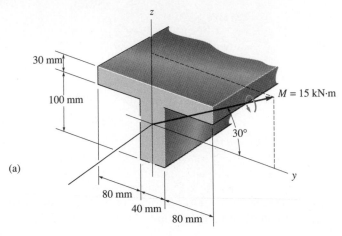

(a)

Fig. 6–17

SOLUTION

Internal Moment Components. The *y* and *z* axes are principal axes of inertia. Why? From Fig. 6–17a, both moment components are positive. We have

$$M_y = (15 \text{ kN} \cdot \text{m}) \cos 30° = 13.0 \text{ kN} \cdot \text{m}$$
$$M_z = (15 \text{ kN} \cdot \text{m}) \sin 30° = 7.50 \text{ kN} \cdot \text{m}$$

Section Properties. With reference to Fig. 6–17b, working in units of meters, we have

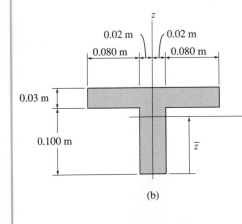

(b)

$$\bar{z} = \frac{\Sigma \tilde{z} A}{\Sigma A} = \frac{[0.05](0.100)(0.04) + [0.115](0.03)(0.200)}{(0.100)(0.04) + (0.03)(0.200)}$$

$$= 0.0890 \text{ m}$$

Using the parallel-axis theorem, Eq. *A–5* of Appendix *A*, the principal moments of inertia are thus

$$I_z = \frac{1}{12}(0.100)(0.04)^3 + \frac{1}{12}(0.03)(0.200)^3 = 20.53(10^{-6}) \text{ m}^4$$

$$I_y = \left[\frac{1}{12}(0.04)(0.100)^3 + (0.100)(0.04)(0.0890 - 0.05)^2 \right]$$
$$+ \left[\frac{1}{12}(0.200)(0.03)^3 + (0.200)(0.03)(0.115 - 0.0890)^2 \right]$$

$$= 13.92(10^{-6}) \text{ m}^4$$

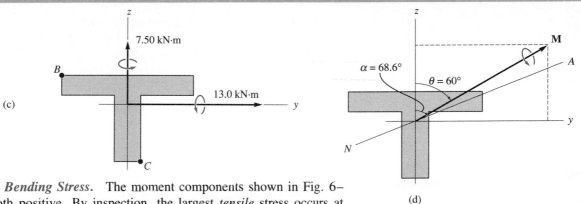

(c)

(d)

Maximum Bending Stress.   The moment components shown in Fig. 6–17c are both positive. By inspection, the largest *tensile* stress occurs at point B, since by superposition both moment components create a tensile stress there. Likewise, the greatest *compressive* stress occurs at point C. Applying Eq. 6–11 to compute these stresses yields

$$\sigma = -\frac{M_z y}{I_z} + \frac{M_y z}{I_y}$$

$$\sigma_B = -\frac{7.50 \text{ kN} \cdot \text{m} \, (-0.100 \text{ m})}{20.53(10^{-6}) \text{ m}^4}$$

$$+ \frac{13.0 \text{ kN} \cdot \text{m} \, (0.130 \text{ m} - 0.0890 \text{ m})}{13.92 \, (10^{-6}) \text{ m}^4}$$

$$= 74.8 \text{ MPa}$$

$$\sigma_C = -\frac{7.50 \text{ kN} \cdot \text{m} \, (0.020 \text{ m})}{20.53(10^{-6}) \text{ m}^4} + \frac{13.0 \text{ kN} \cdot \text{m} \, (-0.0890 \text{ m})}{13.92(10^{-6}) \text{ m}^4}$$

$$= -90.4 \text{ MPa} \hspace{3cm} \textit{Ans.}$$

The largest normal stress is therefore compressive and occurs at point C.

Orientation of Neutral Axis.   When applying Eq. 6–13 it is important to be sure the angles α and θ are defined correctly. As previously stated, y must represent the axis for *minimum* principal moment of inertia, and z must represent the axis for *maximum* principal moment of inertia. These axes are properly positioned here since $I_y < I_z$. Using this setup, θ and α are measured positive from the $+z$ axis toward the $+y$ axis. Hence, from Fig. 6–17a, $\theta = +60°$. Thus,

$$\tan \alpha = \frac{20.53(10^{-6}) \text{ m}^4}{13.92(10^{-6}) \text{ m}^4} \tan 60°$$

$$\alpha = 68.6° \hspace{3cm} \textit{Ans.}$$

The neutral axis is shown in Fig. 6–17d. As expected, it lies between the y axis and the line of action of **M**.

Example 6–7

The Z-section shown in Fig. 6–18a is subjected to the bending moment of $M = 20$ kN $\cdot$ m. Using the methods of Appendix A (see Example A–5 or A–6), the principal axes y and z are oriented as shown, such that they represent the minimum and maximum principal moments of inertia, $I_y = 0.960(10^{-3})$ m^4 and $I_z = 7.54(10^{-3})$ m^4, respectively. Determine the normal stress at point P and the orientation of the neutral axis.

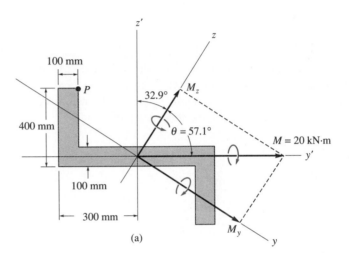

Fig. 6–18

SOLUTION

For use of Eq. 6–13, it is important that the z axis be the principal axis for the *maximum* moment of inertia, which it is because most of the area is located furthest from this axis. Also, notice that the x, y, z and x', y', z' axes are right-handed coordinate systems, with the x and x' axes directed outward from the cross section.

Internal Moment Components. From Fig. 6–18a,

$$M_y = 20 \sin 57.1°$$
$$= 16.79 \text{ kN} \cdot \text{m}$$
$$M_z = 20 \cos 57.1°$$
$$= 10.86 \text{ kN} \cdot \text{m}$$

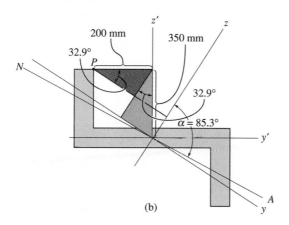

(b)

Bending Stress.   The y and z coordinates of point P must be determined first. Note that the y', z' coordinates of P are $(-200$ mm, 350 mm$)$. Using the colored and shaded triangles from the construction shown in Fig. 6–18b, we have

$$y_P = -350 \sin 32.9° - 200 \cos 32.9° = -358.0 \text{ mm}$$
$$z_P = 350 \cos 32.9° - 200 \sin 32.9° = 185.2 \text{ mm}$$

Applying Eq. 6–11, we have

$$\sigma_P = -\frac{M_z\, y_P}{I_z} + \frac{M_y\, z_P}{I_y}$$

$$= -\frac{(10.86 \text{ kN} \cdot \text{m})(-0.3580 \text{ m})}{7.54(10^{-3}) \text{ m}^4} + \frac{(16.79 \text{ kN} \cdot \text{m})(0.1852 \text{ m})}{0.960(10^{-3}) \text{ m}^4}$$

$$= 3.75 \text{ MPa} \qquad\qquad \textit{Ans.}$$

Orientation of Neutral Axis.   The angle $\theta = 57.1°$ is shown in Fig. 6–18a. Thus,

$$\tan \alpha = \left[\frac{7.54(10^{-3}) \text{ m}^4}{0.960(10^{-3}) \text{ m}^4}\right] \tan 57.1°$$

$$\alpha = 85.3° \qquad\qquad \textit{Ans.}$$

The neutral axis is located as shown in Fig. 6–18b.

PROBLEMS

6–29. The member has a square cross section and is subjected to a resultant internal bending moment of $M = 850$ N · m as shown. Determine the stress at each corner and sketch the stress distribution produced by **M**. Set $\theta = 45°$.

6–30. The member has a square cross section and is subjected to a resultant internal bending moment of $M = 850$ N · m as shown. Determine the stress at each corner and sketch the stress distribution produced by **M**. Set $\theta = 30°$.

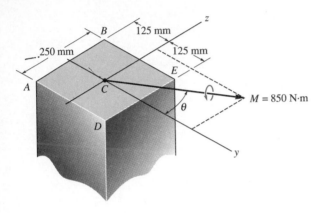

Probs. 6–29/6–30

6–31. The beam has a rectangular cross section. If it is subjected to a bending moment of $M = 3500$ N · m directed as shown, determine the maximum bending stress in the beam and the orientation of the neutral axis.

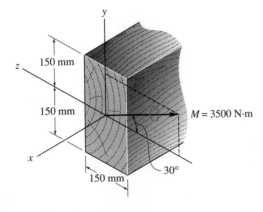

Prob. 6–31

*6–32. The T-beam is subjected to a bending moment of $M = 150$ kip · in. directed as shown. Determine the maximum bending stress in the beam and the orientation of the neutral axis. The location $\bar{y}$ of the centroid, C, must be determined.

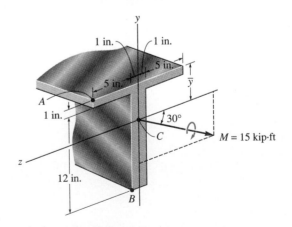

Prob. 6–32

6–33. The resultant internal moment acting on the cross section of the T-beam has a magnitude of $M = 15$ kip · ft and is directed as shown. Determine the bending stress at points A and B. The location $\bar{y}$ of the centroid C must be determined.

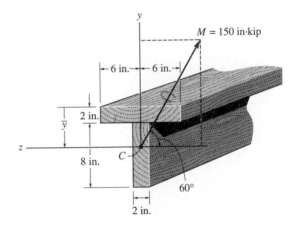

Prob. 6–33

6–34. If the internal moment acting on the cross section of the strut has a magnitude of $M = 800$ N · m and is directed as shown, determine the bending stress at points A and B. The location $\bar{z}$ of the centroid C of the strut's cross-sectional area must be determined. Also, specify the orientation of the neutral axis.

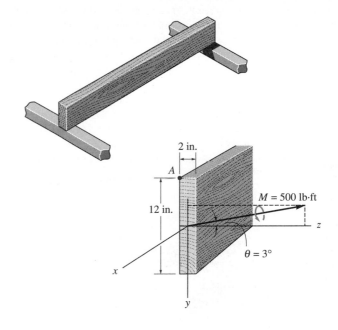

Prob. 6–35

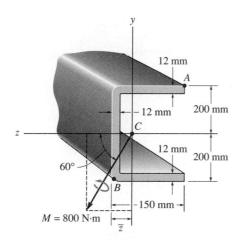

Prob. 6–34

6–35. The board is used as a simply-supported floor joist. If a bending moment of $M = 500$ lb · ft is applied 3° from the z axis, determine the stress developed in the board at the corner A. Compare this stress with that developed by the same moment applied along the z axis ($\theta = 0°$). What is the angle α for the neutral axis when $\theta = 3°$? *Comment:* Normally, floor boards would be nailed to the top of the beam so that $\theta \approx 0°$ and the high stress due to misalignment would not occur.

*6–36. The cantilevered wide-flange steel beam is subjected to the concentrated force **P** at its end. Determine the largest magnitude of this force so that the bending stress developed at A does not exceed $\sigma_{allow} = 180$ MPa.

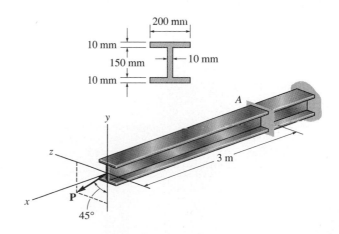

Prob. 6–36

6–37. Consider the general case of a prismatic beam subjected to bending-moment components $\mathbf{M}_y$ and $\mathbf{M}_z$, as shown, when the x, y, z axes pass through the centroid of the cross section. If the material is linear-elastic, the normal stress in the beam is a linear function of position such that $\sigma = a + by + cz$. Using the equilibrium conditions $0 = \int_A \sigma \, dA$, $M_y = \int_A z\sigma \, dA$, $M_z = \int_A -y\sigma \, dA$, determine the constants a, b, and c, and show that the normal stress can be determined from the equation $\sigma = [-(M_zI_y + M_yI_{yz})y + (I_zM_y + M_zI_{yz})z]/(I_yI_z - I_{yz}{}^2)$, where the moments and products of inertia are defined in Appendix A.

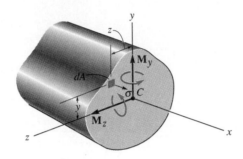

Prob. 6–37

6–38. Using the techniques outlined in Appendix A, Example A–5 or A–6, the "Z" section has principal moments of inertia of $I_y = 0.960(10^{-3})$ m^4 and $I_z = 7.54(10^{-3})$ m^4, computed about the principal axes of inertia y and z, respectively. If the section is subjected to an internal moment of $M = 400$ N · m directed horizontally as shown, determine the stress produced at point A. Solve the problem two ways: (a) using Eq. 6–11 and (b) using the equation developed in Prob. 6–37.

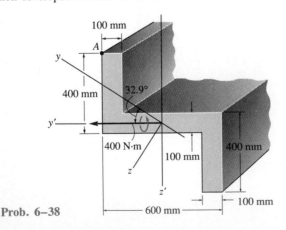

Prob. 6–38

6–39. For the section, $I_{y'} = 31.7(10^{-6})$ m^4, $I_{z'} = 114(10^{-6})$ m^4, $I_{y'z'} = 15.1(10^{-6})$ m^4. Using the techniques outlined in Appendix A, the member's cross-sectional area has principal moments of inertia of $I_y = 29.0(10^{-6})$ m^4 and $I_z = 117(10^{-6})$ m^4, computed about the principal axes of inertia y and z, respectively. If the section is subjected to a moment of $M = 2500$ N · m directed as shown, determine the stress produced at point A, (a) using Eq. 6–11, and (b) using the equation developed in Prob. 6–37.

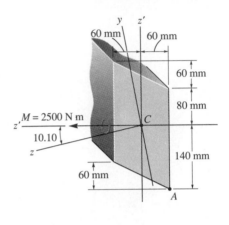

Prob. 6–39

*6–40. The strut has a square cross section a by a and is subjected to the bending moment $\mathbf{M}$ applied at an angle θ as shown. Determine the maximum bending stress in terms of a, M, and θ. What angle θ will give the largest bending stress in the strut? Specify the orientation of the neutral axis for this case.

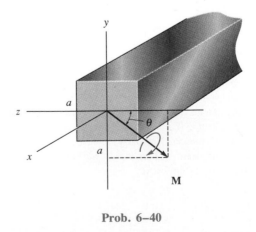

Prob. 6–40

*6.4 Composite Beams

Beams constructed of two or more different materials are referred to as *composite beams*. Examples include those made of wood with straps of steel used at the bottom or top, Fig. 6–19a, or more commonly, concrete beams reinforced with steel rods, Fig. 6–19b. Engineers purposely design beams in this manner in order to develop a more efficient means for carrying applied loads. For example, it has been shown in Sec. 3.3 that concrete is excellent in resisting compressive stress but is very poor in resisting tensile stress. Hence, the steel reinforcing rods shown in Fig. 6–19b have been placed in the tension zone of the beam's cross section so that they resist the tensile stresses that result from the moment **M**.

Since the flexure formula was developed for beams whose material is homogeneous, this formula cannot be applied directly to determine the normal stress in a composite beam. In this section, however, we will develop a method for modifying or "transforming" the beam's cross section into one made of a single material. Once this has been done, the flexure formula can be used for the stress analysis.

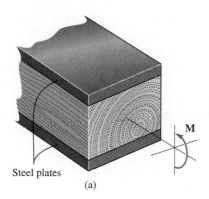

Steel plates

(a)

Fig. 6–19

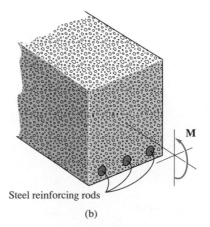

Steel reinforcing rods

(b)

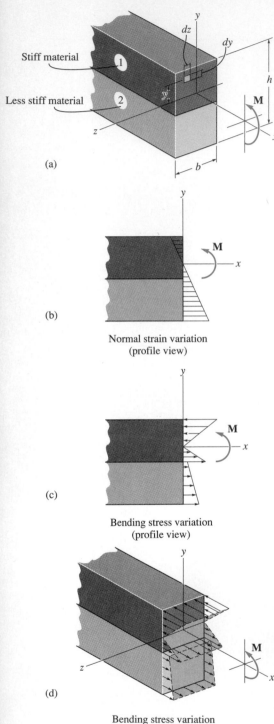

Stiff material

Less stiff material

(a)

(b)

Normal strain variation
(profile view)

(c)

Bending stress variation
(profile view)

(d)

Bending stress variation

Fig. 6–20

To explain how to apply the *transformed-section method,* consider the composite beam to be made of two materials, 1 and 2, which have the cross-sectional areas shown in Fig. 6–20a. If a bending moment is applied to this beam, then, like one that is homogeneous, the total cross-sectional area will *remain plane* after bending, and hence the normal strains will vary linearly from zero at the neutral axis to a maximum in the material located farthest from this axis, Fig. 6–20b. Provided the material has linear-elastic behavior, Hooke's law applies, and at any point the normal stress in material 1 is determined from $\sigma = E_1\epsilon$. Likewise, for material 2 the stress distribution is found from $\sigma = E_2\epsilon$. Obviously, if material 1 is *stiffer* than material 2, e.g., steel versus rubber, most of the load will be carried by material 1, since $E_1 > E_2$. Assuming this to be the case, the stress distribution will look like that shown in Fig. 6–20c or 6–20d. In particular, notice the jump in stress that occurs at the juncture of the materials. Here the *strain* is the *same,* but since the modulus of elasticity or stiffness for the materials suddenly changes, so does the stress. Location of the neutral axis, and determination of the maximum stress in the beam, using this stress distribution, can be based on a trial-and-error procedure. This requires satisfying the conditions that the stress distribution produces a zero resultant force on the cross section and the moment of the stress distribution about the neutral axis must be equal to **M**.

A simpler way to satisfy these two conditions, however, is to transform the beam into one made of a *single material.* For example, if the beam is thought to consist entirely of the less stiff material 2, then the cross section would have to look like that shown in Fig. 6–20e. Here the height h of the beam remains the *same,* since the strain distribution shown in Fig. 6–20b must be preserved. However, the upper portion of the beam must be widened in order to carry a load *equivalent* to that carried by the stiffer material 1 in Fig. 6–20c. The necessary width can be determined by considering the force **dF** acting on an area $dA = dz\,dy$ of the beam in Fig. 6–20a. It is $dF = \sigma\,dA = (E_1\epsilon)dz\,dy$. On the other hand, if the width of a *corresponding element* of height dy in Fig. 6–20e is $n\,dz$, then $dF' = \sigma'\,dA' = (E_2\epsilon)n\,dz\,dy$. Equating these forces, we have

$$E_1\epsilon\,dz\,dy = E_2\epsilon n\,dz\,dy$$

or

$$n = \frac{E_1}{E_2} \qquad (6\text{–}14)$$

This dimensionless number n is called the *transformation factor*. It indicates that the cross section, having a width b on the original beam, Fig. 6–20a, must be increased in width to $b_2 = nb$ in the region where material 1 is being transformed into material 2, Fig. 6–20e. In a similar manner, if the less stiff material 2 is transformed into the stiffer material 1, the cross section will look like that shown in Fig. 6–20f. Here the width of material 2 has been changed to $b_1 = n'b$, where $n' = E_2/E_1$. Note that in this case the transformation factor n' must be *less than one* since $E_1 > E_2$. In other words, we need less of the stiffer material to support a given moment.

Once the beam has been transformed into one having a *single material,* the normal-stress distribution over the transformed cross section will be linear as shown in Fig. 6–20g or 6–20h. Consequently, the centroid (neutral axis) and moment of inertia for the transformed area can be determined and the flexure formula applied in the usual manner to determine the stress at each point on the transformed beam. Realize that the stress in the transformed beam is equivalent to the stress in the *same material* of the actual beam. For the *transformed material,* however, the stress found on the transformed section has to be multiplied by the transformation factor n (or n'), since the area of the transformed material, $dA' = n\, dz\, dy$, is n times the area of actual material $dA = dy\, dz$. That is,

$$dF = \sigma\, dA = \sigma'dA'$$
$$\sigma\, dz\, dy = \sigma'n\, dz\, dy$$
$$\sigma = n\sigma' \qquad (6\text{–}15)$$

The following examples numerically illustrate these concepts.

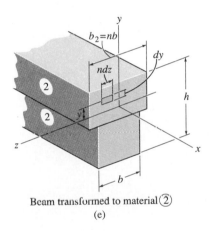

Beam transformed to material ②

(e)

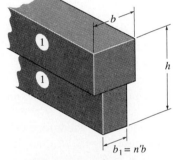

Bending-stress variation for
beam transformed to material ①

(f)

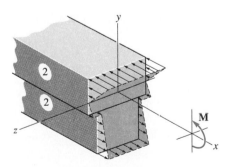

Bending-stress variation for
beam transformed to material ②

(g)

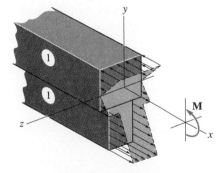

Bending-stress variation for
beam transformed to material ①

(h)

Example 6–8

A composite beam is made of wood and reinforced with a steel strap located on its bottom side. It has the cross-sectional area shown in Fig. 6–21a. If the beam is subjected to a bending moment of $M = 2$ kN $\cdot$ m, determine the normal stress at points B and C. Take $E_w = 12$ GPa and $E_{st} = 200$ GPa.

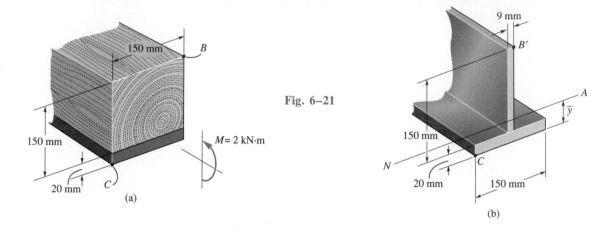

Fig. 6–21

SOLUTION

Section Properties.   Although the choice is arbitrary, here we will transform the section into one made entirely of steel. Since steel has a greater stiffness than wood ($E_{st} > E_w$), the width of the wood must be *reduced* to an equivalent width for steel. Hence n must be less than one. For this to be the case, $n = E_w/E_{st}$, so that

$$b_{st} = nb_w = \frac{12 \text{ GPa}}{200 \text{ GPa}} (150 \text{ mm}) = 9 \text{ mm}$$

The transformed section is shown in Fig. 6–21b.

The location of the centroid (neutral axis), computed from a reference axis located at the *bottom* of the section, is

$$\bar{y} = \frac{\Sigma \tilde{y}A}{\Sigma A} = \frac{[10](20)(150) + [95](9)(150)}{20(150) + 9(150)} = 36.38 \text{ mm}$$

The moment of inertia about the neutral axis is therefore

$$I_{NA} = \left[\frac{1}{12} (150)(20)^3 + (150)(20)(36.38 - 10)^2 \right]$$
$$+ \left[\frac{1}{12} (9)(150)^3 + (9)(150)(95 - 36.38)^2 \right]$$
$$= 9.36(10^6) \text{ mm}^4 (10^{-12} \text{ m /mm})^4 = 9.36(10^{-6}) \text{ m}^4$$

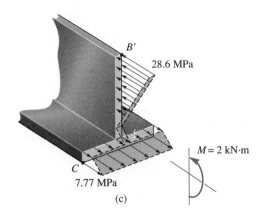

(c)

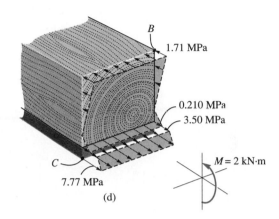

(d)

Normal Stress. Applying the flexure formula, the normal stress at B' and C is

$$\sigma_{B'} = \frac{2 \text{ kN} \cdot \text{m} \ (0.170 \text{ m} - 0.03638 \text{ m})}{9.36(10^{-6}) \text{ m}^4} = 28.6 \text{ MPa}$$

$$\sigma_C = \frac{2 \text{ kN} \cdot \text{m} \ (0.03638 \text{ m})}{9.36(10^{-6}) \text{ m}^4} = 7.77 \text{ MPa} \qquad\qquad Ans.$$

The normal-stress distribution on the transformed (all steel) section is shown in Fig. 6–21c.

The normal stress in the wood, located at B in Fig. 6–21a, is determined from Eq. 6–15; that is,

$$\sigma_B = n\sigma_{B'} = \frac{12 \text{ GPa}}{200 \text{ GPa}}(28.6 \text{ MPa}) = 1.71 \text{ MPa} \qquad Ans.$$

Using these concepts, show that the normal stress in the steel and the wood at the point where they are in contact is $\sigma_{st} = 3.50$ MPa and $\sigma_w = 0.210$ MPa, respectively. The normal-stress distribution in the actual beam is shown in Fig. 6–21d.

Example 6–9

In order to reinforce the steel beam, an oak board is placed between its flanges as shown in Fig. 6–22a. If the allowable normal stress for the steel is $(\sigma_{\text{allow}})_{st} = 24$ ksi, and for the wood $(\sigma_{\text{allow}})_w = 3$ ksi, determine the maximum bending moment the beam can support. $E_{st} = 29(10^3)$ ksi, $E_w = 1.60(10^3)$ ksi. The moment of inertia of the steel beam is $I_z = 20.3$ in^4, and its cross-sectional area is $A = 8.79$ in^2.

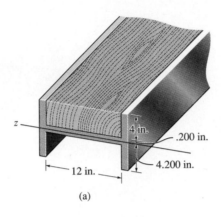

4 in.

.200 in.

4.200 in.

12 in.

(a)

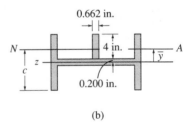

0.662 in.

N

4 in.

z

$\bar{y}$

A

c

0.200 in.

(b)

Fig. 6–22

SOLUTION

Without Board. Here the neutral axis coincides with the z axis. Direct application of the flexure formula to the steel beam yields

$$(\sigma_{\text{allow}})_{st} = \frac{Mc}{I_z}$$

$$24 \text{ kip/in}^2 = \frac{M(4.200 \text{ in.})}{20.3 \text{ in}^4}$$

$$M = 116 \text{ kip} \cdot \text{in.} \qquad \qquad \textit{Ans.}$$

With Board. Since now we have a composite beam, we must transform the section to a single material. It will be easier to transform the wood to an equivalent amount of steel. To do this, $n = E_w/E_{st}$. Why? Thus, the width of an equivalent amount of steel is

$$b_{st} = nb_w = \frac{1.60(10^3) \text{ ksi}}{29(10^3) \text{ ksi}} \, (12 \text{ in.}) = 0.662 \text{ in.}$$

The transformed section is shown in Fig. 6–22b. The neutral axis is at

$$\bar{y} = \frac{\Sigma \bar{y}A}{\Sigma A} = \frac{[0](8.79 \text{ in}^2) + [2.20 \text{ in.}](4 \text{ in.})(0.662 \text{ in.})}{8.79 \text{ in}^2 + 4(0.662 \text{ in}^2)}$$

$$= 0.5093 \text{ in.}$$

And the moment of inertia about the neutral axis is

$$I = [20.3 \text{ in}^4 + (8.79 \text{ in}^2)(0.5093 \text{ in.})^2] +$$

$$\left[\frac{1}{12} (0.662 \text{ in.})(4 \text{ in.})^3 + (0.662 \text{ in.})(4 \text{ in.})(2.200 \text{ in.} - 0.5093 \text{ in.})^2 \right]$$

$$= 33.68 \text{ in}^4$$

The maximum normal stress will occur at the *bottom* of the beam, Fig. 6–22b. Here $c = 4.200 \text{ in.} + 0.5093 \text{ in.} = 4.7093 \text{ in.}$ The maximum moment the steel will support is therefore

$$(\sigma_{\text{allow}})_{st} = \frac{Mc}{I}$$

$$24 \text{ kip/in}^2 = \frac{M(4.7093 \text{ in.})}{33.68 \text{ in}^4}$$

$$M = 172 \text{ kip} \cdot \text{in.}$$

The maximum normal stress in the wood occurs at the top of the beam, Fig. 6–22b. Here $c' = 4.20 \text{ in.} - 0.5093 \text{ in.} = 3.6907 \text{ in.}$ Since $\sigma_w = n\sigma_{st}$, the maximum moment the wood will support is

$$(\sigma_{\text{allow.}})_w = n \frac{M'c'}{I}$$

$$3 \text{ kip/in}^2 = \left[\frac{1.60(10^3) \text{ ksi}}{29(10^3) \text{ ksi}} \right] \frac{M'(3.6907 \text{ in.})}{33.68 \text{ in}^4}$$

$$M' = 496 \text{ kip} \cdot \text{in.}$$

By comparison, the maximum moment is limited by the allowable stress in the steel. Thus,

$$M = 172 \text{ kip} \cdot \text{in.} \qquad \qquad Ans.$$

Note also that by using the board as reinforcement, one provides an additional 48% moment capacity for the beam.

*6.5 Curved Beams

The flexure formula applies to a prismatic member that is *straight,* since it was shown that for a straight member the normal strain varies linearly from the neutral axis. If the member is *curved,* however, this assumption becomes inaccurate, and so we must develop another equation that describes the stress distribution. In this section we will consider the analysis of a *curved beam,* that is, a member that has a curved axis and is subjected to bending. Typical examples include hooks and chain links. In all cases, the members are not slender, but rather have a sharp curve, and their cross-sectional dimensions are large compared with their radius of curvature.

The analysis to be considered assumes that the cross-sectional area is constant and has an axis of symmetry that is perpendicular to the direction of the applied moment **M**, Fig. 6–23*a*. Also, the material is homogeneous and isotropic, and it behaves in a linear-elastic manner when the load is applied. Like the case of a straight beam, we will, for a curved beam, also assume that the *cross sections* of the member *remain plane* after the moment is applied. Furthermore, any distortion of the cross section within its own plane will be neglected.

To perform the analysis, three radii, extending from the center of curvature O' of the member, are identified in Fig. 6–23*a*. They are as follows: $\bar{r}$ references the known location of the *centroid* for the cross-sectional area, R references the yet unspecified location of the *neutral axis,* and r locates the *arbitrary point* or area element dA on the cross section. Notice that the neutral axis lies within the cross section, since the moment **M** creates compression in the beam's top fibers and tension in its bottom fibers, and by definition, the neutral axis is a line of zero stress and strain.

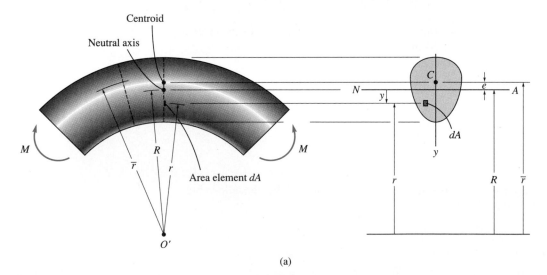

(a)

Fig. 6–23

If we isolate a differential segment of the beam in Fig. 6–23a, the stress tends to deform the material such that each cross section will rotate through an angle $\delta\theta/2$, as shown in Fig. 6–23b. The normal strain ϵ in the arbitrary strip of material located at r will now be determined. This strip has an original length $r\,d\theta$, Fig. 6–23b. Due to the rotations $\delta\theta/2$, however, the strip's total change in length is $\delta\theta(R - r)$. Consequently,

$$\epsilon = \frac{\delta\theta\,(R - r)}{r\,d\theta}$$

Defining $k = \delta\theta/d\theta$, which is constant for any particular element, we have

$$\epsilon = k\!\left(\frac{R - r}{r}\right)$$

Unlike the case of straight beams, here it can be seen that the *normal strain* is a nonlinear function of r, in fact it varies in a *hyperbolic* fashion. This occurs even though the cross section of the beam remains plane after deformation. Since the moment causes the material to behave elastically, Hooke's law applies, and therefore the stress as a function of position is

$$\sigma = Ek\!\left(\frac{R - r}{r}\right) \tag{6--16}$$

This variation is also hyperbolic, and since it has now been established, we can determine the location of the neutral axis and relate the stress distribution to the resultant internal moment M.

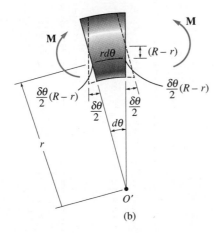

(b)

To obtain the location R of the neutral axis, we require the resultant internal force caused by the stress distribution acting over the cross section to be equal to zero; i.e.,

$$F_R = \Sigma F_x; \qquad \int_A \sigma \, dA = 0$$

$$\int_A Ek \left(\frac{R-r}{r}\right) dA = 0$$

Since Ek and R are constants, we have

$$R \int_A \frac{dA}{r} - \int_A dA = 0$$

Solving for R yields

$$R = \frac{A}{\displaystyle\int_A \frac{dA}{r}} \qquad (6\text{--}17)$$

Here

R = the location of the neutral axis, specified from the center of curvature O' of the member

A = the cross-sectional area of the member

r = the arbitrary position of the area element dA on the cross section, specified from the center of curvature O' of the member

The integral in Eq. 6–17 may be evaluated for various cross-sectional geometries. The results for some common cross sections are listed in Table 6–1.

Shape	Area	$\int_A \frac{dA}{r}$
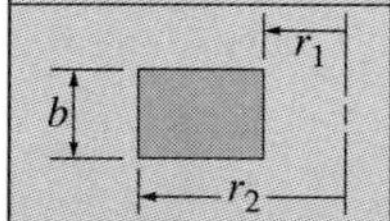	$b(r_2 - r_1)$	$b \ln \frac{r_2}{r_1}$
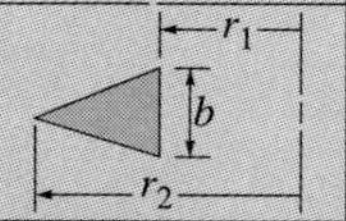	$\frac{b}{2}(r_2 - r_1)$	$\frac{b\,r_2}{(r_2 - r_1)}\left(\ln \frac{r_2}{r_1}\right) - b$
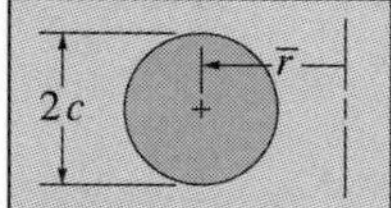	πc^2	$2\pi\left(\bar{r} - \sqrt{\bar{r}^2 - c^2}\right)$
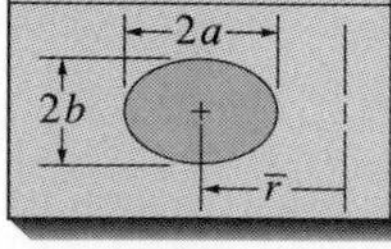	πab	$\frac{2\pi b}{a}\left(\bar{r} - \sqrt{\bar{r}^2 - a^2}\right)$

Table 6–1

In order to relate the stress distribution to the resultant bending moment, we require the resultant internal moment to be equal to the moment of the stress distribution computed about the neutral axis. From Fig. 6–23a, the stress σ, acting on the area element dA and located a distance y from the neutral axis, creates a force of $dF = \sigma \, dA$ on the element and a moment about the neutral axis of $dM = y \, (\sigma \, dA)$. This moment is positive, since by the right-hand rule it is directed in the same direction as **M**. For the entire cross section, we require $M = \int y\sigma \, dA$.

Since $y = R - r$, and σ is defined by Eq. 6–16, we have

$$M = \int_A (R - r)Ek \left(\frac{R - r}{r} \right) dA$$

Expanding, realizing that Ek and R are constants, gives

$$M = Ek \left(R^2 \int_A \frac{dA}{r} - 2R \int_A dA + \int_A r \, dA \right)$$

The first integral is equivalent to A/R as determined from Eq. 6–17, and the second integral is simply the cross-sectional area A. Realizing that the location of the centroid of the cross section is determined from $\bar{r} = \int r \, dA/A$, the third integral can be replaced by $\bar{r}A$. Thus, we can write

$$M = EkA(\bar{r} - R)$$

Solving for Ek in Eq. 6–16, substituting into the above equation, and solving for σ, we have

$$\boxed{\sigma = \frac{M(R - r)}{Ar(\bar{r} - R)}} \qquad (6\text{–}18)$$

Here

σ = the normal stress in the member

M = the internal moment, determined from the method of sections and the equations of equilibrium and computed about the neutral axis for the cross section. This moment is *positive* if it tends to increase the member's radius of curvature, i.e., it tends to straighten out the member

A = the cross-sectional area of the member

R = The distance measured from the center of curvature to the neutral axis, determined from Eq. 6–17

$\bar{r}$ = the distance measured from the center of curvature to the centroid of the cross-sectional area

r = the distance measured from the center of curvature to the point where the stress σ is to be determined

From Fig. 6–23a, $y = R - r$ or $r = R - y$. Also, the constant and usually very small distance $e = \bar{r} - R$. When these results are substituted into Eq. 6–18, we can also write

$$\sigma = \frac{My}{Ae(R - y)}$$

(6–19)

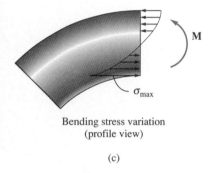

Bending stress variation
(profile view)

(c)

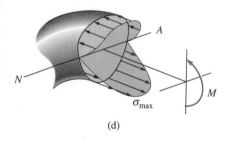

(d)

Fig. 6–23

The above two equations represent two forms of the so-called *curved-beam formula,* which like the flexure formula can be used to determine the normal-stress distribution in a curved member. This distribution is, as previously stated, hyperbolic; an example is shown in Fig. 6–23c and 6–23d. Since the stress acts in the direction of the circumference of the beam, it is sometimes called *circumferential stress.* It should be realized, however, that due to the curvature of the beam, the circumferential stress will create a corresponding component of *radial stress,* so called since this component acts in the radial direction. To show how it is developed, consider the free-body diagram shown in Fig. 6–23e, which is a top segment of the differential element in Fig. 6–23b. Here the radial stress σ_r is necessary since it creates the force $d\mathbf{F}_r$ that is required to balance the components of circumferential forces $d\mathbf{F}$, which act along the line $O'B$.

Sometimes the radial stresses within curved members may become significant, especially if the member is constructed from thin plates and has, for example, the shape of an I-section. Then the radial stress can become as large as the circumferential stress, and consequently the member must be designed to resist both stresses. For most cases, however, these stresses can be neglected, especially if the member's cross section is a *solid section.* For these cases, the curved-beam formula gives results that are in very close agreement with those determined either by experiment or by a mathematical analysis based on the theory of elasticity.

The curved-beam formula is normally used when the curvature of the member is very pronounced, as in the case of hooks or rings. However, if the radius of curvature is greater than five times the depth of the member, the *flexure formula* can normally be used to compute the stress. Specifically, for rectangular sections for which this ratio equals 5, the maximum normal stress, when computed by the flexure formula, will be about 7% *less* than its value when computed by the curved-beam formula. This error is further reduced when the radius of curvature-to-depth ratio is more than 5.*

---

*See, for example, Boresi, A. P., et al., *Advanced Mechanics of Materials,* 3rd ed., p. 333, 1978, John Wiley & Sons, New York.

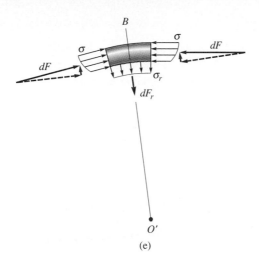

(e)

PROCEDURE FOR ANALYSIS

The curved-beam formula is used to find the normal or circumferential stress distribution in a member having a constant cross section and made of a homogeneous isotropic material that behaves in a linear-elastic manner.

Section Properties. Determine the cross-sectional area A and the location of the centroid, $\bar{r}$, measured from the center of curvature. Compute the location of the neutral axis, R, using Eq. 6–17 or Table 6–1. If the cross-sectional area consists of n ''composite'' parts, compute $\int dA/r$ for *each part*. Then, from Eq. 6–17, for the entire section, $R = \Sigma A/\Sigma \ (\int dA/r)$. In all cases, $R < \bar{r}$.

Normal Stress. The normal stress located at a point r away from the center of curvature is determined from Eq. 6–18. If the distance y to the point is measured from the neutral axis, then compute $e = \bar{r} - R$ and use Eq. 6–19. Since $\bar{r} - R$ generally produces a very *small number,* it is best to calculate $\bar{r}$ and R with sufficient accuracy so that the subtraction leads to a number e having at least three significant figures.

The normal stress σ will always act in a direction such that the force $d\mathbf{F}$ it creates on the cross section contributes a moment about the neutral axis that is in the *same direction* as the moment $\mathbf{M}$, Fig. 6–23c or 6–23d. This can always be checked from the computed value of σ. If the value is positive the stress will be tensile, whereas if the value is negative the stress will be compressive. The stress distribution over the entire cross section can be graphed, or a volume element of the material can be isolated and used to represent the stress acting at the point on the cross section when it has been calculated.

The following examples numerically illustrate this procedure.

Example 6–10

A steel bar having a rectangular cross section is shaped into a circular arc as shown in Fig. 6–24a. If the allowable normal stress is $\sigma_{allow} = 20$ ksi, determine the maximum bending moment M that can be applied to the bar. What would be this moment if the bar was straight?

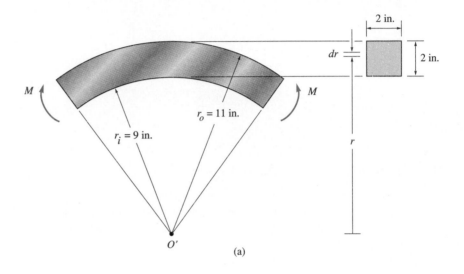

(a)

Fig. 6–24

SOLUTION

Internal Moment. Since M tends to increase the bar's radius of curvature, it is positive.

Section Properties. The location of the neutral axis is determined using Eq. 6–17. From Fig. 6–24a, we have

$$\int_A \frac{dA}{r} = \int_{9\ in.}^{11\ in.} \frac{(2\ in.)\ dr}{r} = (2\ in.)\ \ln r \Big|_{9\ in.}^{11\ in.} = 0.40134\ in.$$

This same result can of course be obtained directly from Table 6–1. Thus,

$$R = \frac{A}{\displaystyle\int_A dA/r} = \frac{(2\ in.)(2\ in.)}{0.40134\ in.} = 9.9666\ in.$$

It is unknown if the normal stress reaches its maximum at the top or bottom of the bar, and so we must compute the moment M for each case separately. Since the normal stress at the bar's top is compressive, $\sigma = -20$ ksi,

$$\sigma = \frac{M(R - r_o)}{Ar_o(\bar{r} - R)}$$

$$-20 \text{ kip/in}^2 = \frac{M(9.9666 \text{ in.} - 11 \text{ in.})}{(2 \text{ in.})(2 \text{ in.})(11 \text{ in.})(10 \text{ in.} - 9.9666 \text{ in.})}$$

$$M = 28.4 \text{ lb} \cdot \text{in.}$$

Likewise, at the bottom of the bar the normal stress will be tensile, so $\sigma = +20$ ksi. Therefore,

$$\sigma = \frac{M(R - r_i)}{Ar_i(\bar{r} - R)}$$

$$20 \text{ kip/in}^2 = \frac{M(9.9666 \text{ in.} - 9 \text{ in.})}{(2 \text{ in.})(2 \text{ in.})(9 \text{ in.})(10 \text{ in.} - 9.9666 \text{ in.})}$$

$$M = 24.9 \text{ kip} \cdot \text{in.} \qquad Ans.$$

By comparison, the maximum moment that can be applied is 24.9 kip · in. and so maximum normal stress occurs at the bottom of the bar. The compressive stress at the top of the bar is then

$$\sigma = \frac{24.9 \text{ kip} \cdot \text{in.} (9.9666 \text{ in.} - 11 \text{ in.})}{(2 \text{ in.})(2 \text{ in.})(11 \text{ in.})(10 \text{ in.} - 9.9666 \text{ in.})}$$

$$= -17.5 \text{ ksi}$$

The stress distribution is shown in Fig. 6–24b. It should be noted that throughout the above calculations R must be calculated to several significant figures to ensure that $(\bar{r} - R)$ is accurate to at least three significant figures.

If the bar was straight, then

$$\sigma = \frac{Mc}{I}$$

$$20 \text{ kip/in}^2 = \frac{M(1 \text{ in.})}{\frac{1}{12}(2 \text{ in.})(2 \text{ in.})^3}$$

$$M = 26.67 \text{ kip} \cdot \text{in.} \qquad Ans.$$

This represents an error of about 7% from the more exact value computed above.

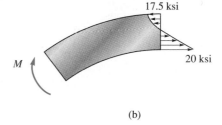

(b)

Example 6-11

The curved bar has a cross-sectional area shown in Fig. 6–25a. If it is subjected to bending moments of 4 kN · m, determine the maximum normal stress developed in the bar.

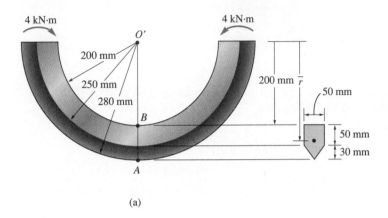

Fig. 6-25

SOLUTION

Internal Moment. Each section of the bar is subjected to the same resultant internal moment of 4 kN · m. Since this moment tends to decrease the bar's radius of curvature, it is negative. Thus $M = -4$ kN · m.

Section Properties. Here we will consider the cross section to be composed of a rectangle and triangle. Calculations will be performed using units of meters. The total cross-sectional area is

$$\Sigma A = (0.05)^2 + \frac{1}{2}(0.05)(0.03) = 3.2500(10^{-3}) \text{ m}^2$$

The location of the centroid is determined with reference to the center of curvature, point O', Fig. 6–25a.

$$\bar{r} = \frac{\Sigma \tilde{r} A}{\Sigma A} = \frac{[0.225](0.05)(0.05) + [0.260]\frac{1}{2}(0.050)(0.030)}{3.250(10^{-3})}$$

$$= 0.23308 \text{ m}$$

We can compute $\int_A dA/r$ for each part using Table 6–1. For the rectangle,

$$\int_A \frac{dA}{r} = 0.05 \ln \frac{0.250}{0.200} = 0.011157 \text{ m}$$

And for the triangle,

$$\int_A \frac{dA}{r} = \frac{(0.05)(0.280)}{(0.280 - 0.250)} \left(\ln \frac{0.280}{0.250} \right) - 0.05 = 0.0028867 \text{ m}$$

Thus the location of the neutral axis is determined from

$$R = \frac{\Sigma A}{\Sigma \int_A dA/r} = \frac{3.2500(10^{-3})}{0.011157 + 0.0028867} = 0.23142 \text{ m}$$

Note that $R < \bar{r}$ as expected. Also, the calculations were performed with sufficient accuracy so that $(\bar{r} - R) = 0.23308 - 0.23142 = 0.00166$ m is now accurate to three significant figures.

Normal Stress. The maximum normal stress occurs either at A or B. Applying the curved-beam formula to calculate the normal stress at B, $r_B = 0.200$ m, we have

$$\sigma_B = \frac{M(R - r_B)}{A r_B(\bar{r} - R)} = \frac{(-4 \text{ kN} \cdot \text{m})(0.23142 \text{ m} - 0.200 \text{ m})}{3.2500(10^{-3}) \text{ m}^2 (0.200 \text{ m})(0.00166 \text{ m})}$$

$$= -116 \text{ MPa}$$

At point A, $r_A = 0.280$ m and the normal stress is

$$\sigma_A = \frac{M(R - r_A)}{A r_A(\bar{r} - R)} = \frac{(-4 \text{ kN} \cdot \text{m})(0.23142 \text{ m} - 0.280 \text{ m})}{3.2500(10^{-3}) \text{ m}^2 (0.280 \text{ m})(0.00166 \text{ m})}$$

$$= 129 \text{ MPa} \qquad\qquad\qquad\qquad\qquad\qquad \textit{Ans.}$$

By comparison, the maximum normal stress is at A. A two-dimensional representation of the stress distribution is shown in Fig. 6–25b.

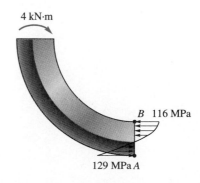

(b)

6.6 Stress Concentrations

The flexure formula, $\sigma_{max} = Mc/I$, can be used to determine the stress distribution within regions of a member where the cross-sectional area is constant or tapers slightly. If the cross section suddenly changes, however, the normal-stress and strain distributions at the section become *nonlinear* and can be obtained only through experiment or, in some cases, by a mathematical analysis using the theory of elasticity. Common discontinuities include members having notches on their surfaces, Fig. 6–26a, holes for passage of fasteners or other items, Fig. 6–26b, or abrupt changes in the outer dimensions of the member's cross section, Fig. 6–26c. The *maximum* normal stress at each of these discontinuities occurs at the section taken through the *smallest* cross-sectional area.

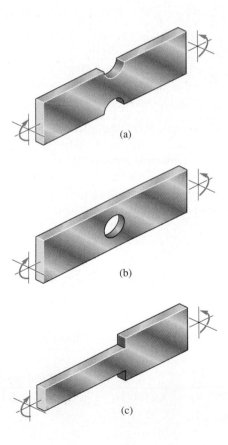

(a)

(b)

(c)

Fig. 6–26

For design, it is generally important to know the maximum normal stress developed at these sections, not the actual stress distribution itself. As in the previous cases of axially loaded bars and torsionally loaded shafts, we can obtain the maximum normal stress due to bending using a stress-concentration factor K. For example, Fig. 6–27 gives values of K for a flat bar that has a change in cross section using shoulder fillets.* To use this graph simply compute the geometric ratios w/h and r/h and then find the corresponding value of K for a particular geometry. Once K is obtained, the maximum bending stress is determined using

$$\sigma_{\max} = K\,\frac{Mc}{I} \qquad\qquad (6\text{–}20)$$

Here the flexure formula is applied to the *smaller* cross-sectional area, since $\sigma_{\max}$ occurs at the base of the fillet, Fig. 6–28. In the same manner, Fig. 6–29 can be used if the discontinuity consists of circular grooves or notches.

Like axial load and torsion, stress concentration for bending should always be considered when designing members made of brittle materials or those that are subjected to fatigue or cyclic loadings. Realize also that stress-concentration factors apply only when the material is subjected to *elastic behavior*. If the applied moment causes yielding of the material, the stress becomes redistributed throughout the member, and the maximum stress that results will be *lower* than that computed using stress-concentration factors. This phenomenon is discussed further in the next section.

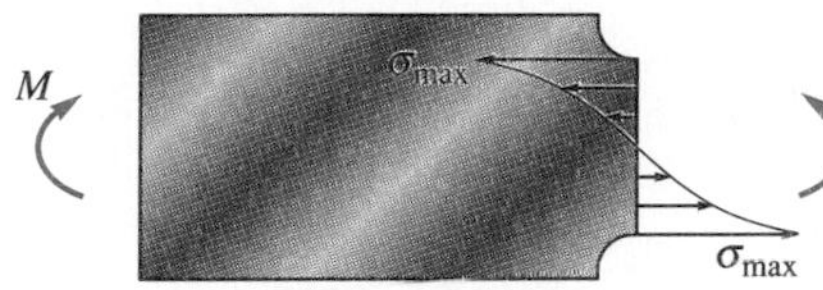

*See C. Lipson and R. C. Juvinall, *Handbook of Stress and Strength*, Macmillan, 1963.

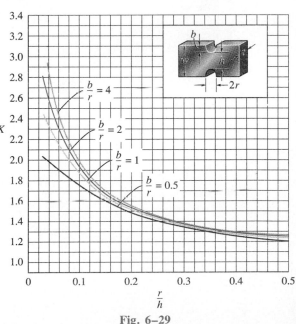

Fig. 6–28

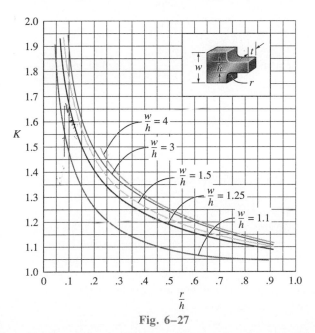

Fig. 6–27

Fig. 6–29

Example 6–12

The transition in the cross-sectional area of the steel bar is achieved using shoulder fillets as shown in Fig. 6–30a. If the bar is subjected to a bending moment of 5 N · m, determine the maximum normal stress developed in the steel. The yield stress is $\sigma_Y = 500$ MPa.

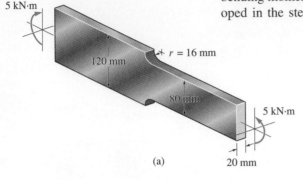

5 kN·m

120 mm

r = 16 mm

80 mm

5 kN·m

(a)

20 mm

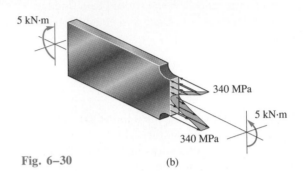

5 kN·m

340 MPa

5 kN·m

340 MPa

Fig. 6–30

(b)

SOLUTION

The moment creates the largest stress in the bar at the base of the fillet, where the cross-sectional area is smallest. The stress-concentration factor can be determined by using Fig. 6–27. From the geometry of the bar, we have $r = 16$ mm, $h = 80$ mm, $w = 120$ mm. Thus,

$$\frac{r}{h} = \frac{16 \text{ mm}}{80 \text{ mm}} = 0.2 \qquad \frac{w}{h} = \frac{120 \text{ mm}}{80 \text{ mm}} = 1.5$$

These values give $K = 1.45$.

Applying Eq. 6–20, we have

$$\sigma_{max} = K\frac{Mc}{I} = (1.45)\frac{(5 \text{ kN} \cdot \text{m})(0.04 \text{ m})}{[\frac{1}{12}(0.020 \text{ m})(0.08 \text{ m})^3]} = 340 \text{ MPa}$$

5 kN·m

234 MPa

5 kN·m

234 MPa

(c)

This result indicates that the steel remains elastic since the stress is below the yield stress (500 MPa).

The normal-stress distribution is nonlinear and is shown in Fig. 6–30b. Realize, however, that by Saint-Venant's principle, Sec. 4.1, these localized stresses smooth out and become linear when one moves (approximately) a distance of 80 mm or more to the right of the transition. In this case, the flexure formula gives $\sigma_{max} = 234$ MPa, Fig. 6–30c. Also note that the choice of a larger-radius fillet will significantly reduce σ_{max}, since as r increases in Fig. 6–27, K will decrease.

PROBLEMS

6–41. The composite beam is made of steel (*A*) bonded to brass (*B*) and has the cross section shown. If it is subjected to a moment of $M = 6.5$ kN · m, determine the maximum stress in the brass and steel. Also, what is the stress in each material at the seam where they are bonded together? $E_{br} = 100$ GPa, $E_{st} = 200$ GPa.

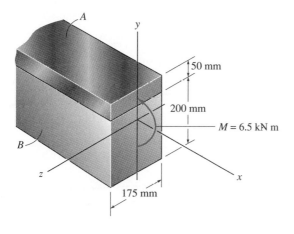

Prob. 6–41

6–42. A wooden beam is reinforced with steel straps at its top and bottom as shown. Determine the maximum stress developed in the wood and steel if the beam is subjected to a bending moment of $M = 5$ kN · m. Sketch the stress distribution acting over the cross section. Take $E_w = 11$ GPa, $E_{st} = 200$ GPa.

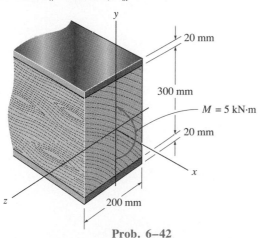

Prob. 6–42

6–43. The composite beam consists of a wood core and two plates of steel. If the allowable bending stress for the wood is $(\sigma_{allow})_w = 20$ MPa, and for the steel $(\sigma_{allow})_{st} = 130$ MPa, determine the maximum moment that can be applied to the beam. $E_w = 11$ GPa, $E_{st} = 200$ GPa.

*6–44. Solve Prob. 6–43 if the moment $M = 5$ kN · m is applied about the *y* axis instead of the *z* axis as shown.

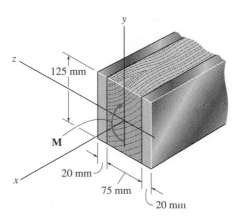

Probs. 6–43/6–44

6–45. The steel channel is used to reinforce the wood beam. Determine the maximum stress in the steel and in the wood if the beam is subjected to a moment of $M = 850$ lb · ft. $E_{st} = 29(10^3)$ ksi, $E_w = 1600$ ksi.

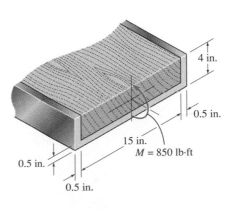

Prob. 6–45

6–46. The member has a brass core bonded to a steel casing. If a couple moment of 8 kN · m is applied at its end, determine the maximum bending stress in the member. E_{br} = 100 GPa, E_{st} = 200 GPa.

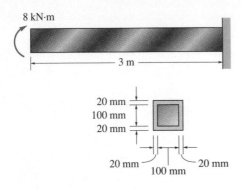

Prob. 6–46

6–47. For the curved beam in Fig. 6–23a, show that when the radius of curvature approaches infinity, the curved-beam formula, Eq. 6–18, reduces to the flexure formula, Eq. 6–6.

*6–48. The curved beam is subjected to a bending moment of M = 85 N · m as shown. Determine the stress at points A and B and show the stress on a volume element located at these points.

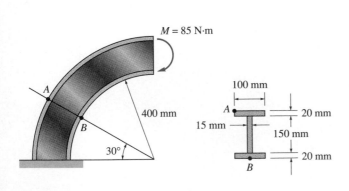

Prob. 6–48

6–49. The fork is used as part of a nosewheel assembly for an airplane. If the maximum wheel reaction at the end of the fork is 840 lb, determine the maximum bending stress in the curved portion of the fork at section a–a. There the cross-sectional area is circular, having a diameter of 2 in.

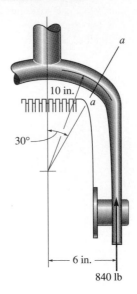

Prob. 6–49

6–50. The curved bar used on a machine has a rectangular cross section. If the bar is subjected to a couple as shown, determine the maximum tensile and compressive stress acting at section a–a. Sketch the stress distribution on the section in three dimensions.

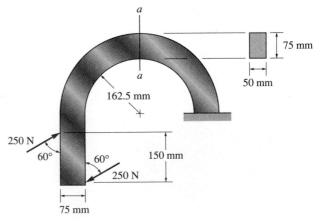

Prob. 6–50

6–51. The circular spring clamp produces a compressive force of 3 N on the plates. Determine the maximum bending stress produced in the spring at A. The spring has a rectangular cross section as shown.

6–53. The steel rod has a circular cross section. If it is gripped at its ends and a couple moment of $M = 12$ lb · in. is developed at each grip, determine the stress acting at points A and B and at the centroid C.

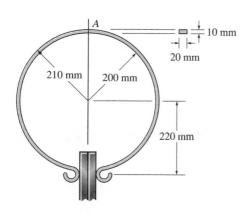

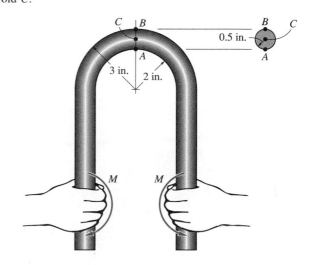

Prob. 6–51

Prob. 6–53

*6–52. Determine the greatest magnitude of the applied moment **M** if the allowable bending stress is $(\sigma_{allow})_c = 50$ MPa in compression and $(\sigma_{allow})_t = 120$ MPa in tension.

6–54. The curved beam is subjected to a bending moment of $M = 40$ lb · ft. Determine the maximum stress in the beam. Also, sketch a two-dimensional view of the stress distribution acting on section a–a.

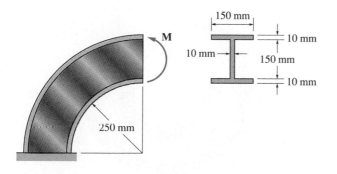

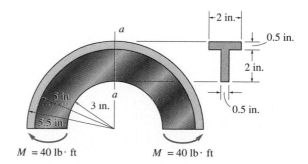

Prob. 6–52

Prob. 6–54

6–55. The curved box member is symmetric and is subjected to a moment of $M = 500$ lb · ft. Determine the stress in the member at points A and B. Show the stress acting on volume elements located at these points.

*6–56. The curved box member is symmetric and is subjected to a moment of $M = 350$ lb · ft. Determine the maximum tensile and compressive stress in the member. Compare these values with those for a straight member having the same cross section and loaded with the same moment.

6–57. The elbow of the pipe has an outer radius of 0.75 in. and an inner radius of 0.63 in. If the assembly is subjected to the moments of $M = 25$ lb · in., determine the maximum stress developed at section a–a.

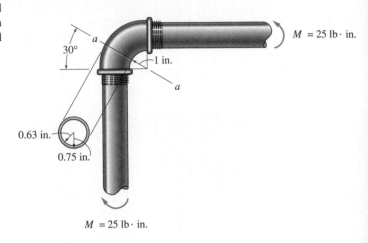

Prob. 6–57

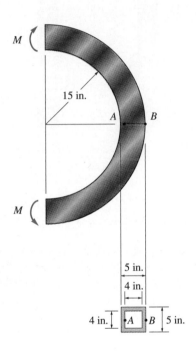

Probs. 6–55/6–56

6–58. The member has an elliptical cross section. If it is subjected to a moment of $M = 50$ N · m, determine the stress at points A and B. Is the stress at point A', which is located on the member near the wall, the same as that at A? Explain.

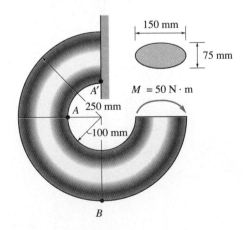

Prob. 6–58

6–59.   The bar is subjected to a moment of $M = 153$ N · m. Determine the smallest radius r of the fillets so that an allowable bending stress of $\sigma_{allow} = 120$ MPa is not exceeded.

6–61.   The bar is subjected to a moment of $M = 15$ N · m. Determine the maximum bending stress in the bar and sketch, approximately, how the stress varies over the critical section.

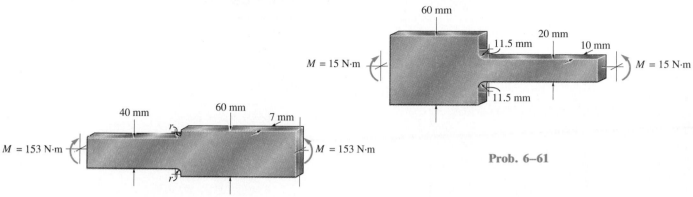

M = 15 N·m

60 mm

11.5 mm

20 mm

10 mm

11.5 mm

M = 15 N·m

Prob. 6–61

40 mm

60 mm

7 mm

r

r

M = 153 N·m

M = 153 N·m

Prob. 6–59

6–62.   Determine the length L of the center portion of the bar so that the maximum bending stress at A, B, and C is the same. The bar has a thickness of 10 mm.

*6–60.   The bar has a thickness of 0.25 in. and is made of a material having an allowable bending stress of $\sigma_{allow} = 18$ ksi. Determine the maximum moment M that can be applied.

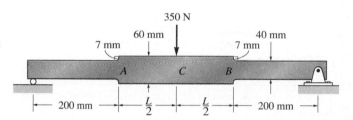

350 N

60 mm

7 mm

40 mm

7 mm

A C B

200 mm

$\dfrac{L}{2}$ $\dfrac{L}{2}$

200 mm

Prob. 6–62

6–63.   Determine the maximum bending stress developed in the bar if it is subjected to the couples shown. The bar has a thickness of 0.25 in.

1 in.

M

4 in.

0.25 in.

M

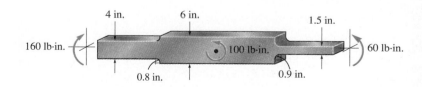

4 in.

6 in.

1.5 in.

160 lb·in.

100 lb·in.

60 lb·in.

0.8 in.

0.9 in.

Prob. 6–60

Prob. 6–63

*6–64. The bar is subjected to four couple moments. If it is in equilibrium, determine the magnitudes of the largest moments **M** and **M'** that can be applied without exceeding an allowable bending stress of $\sigma_{allow} = 22$ ksi.

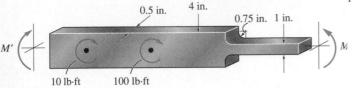

0.5 in.   4 in.   0.75 in.   1 in.

M'

M

10 lb·ft   100 lb·ft

Prob. 6–64

6–65. The stepped bar has a thickness of 12 mm. Determine the maximum moment that can be applied to its ends if it is made of a material having an allowable bending stress of $\sigma_{allow} = 150$ MPa.

6–66. The stepped bar has a thickness of 12 mm. If $M = 20$ N · m, determine the maximum bending stress in the plate. Specify where this stress occurs and sketch the stress distribution over the cross section.

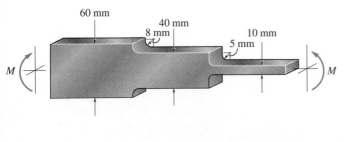

60 mm   40 mm   8 mm   10 mm   5 mm

M   M

Prob. 6–65/6–66

6–67. If the radius of each notch on the plate is $r = 0.5$ in., determine the largest moment that can be applied. The allowable bending stress for the material is $\sigma_{allow} = 18$ ksi.

*6–68. The symmetric notched plate is subjected to bending. If the radius of each notch is $r = 0.5$ in. and the applied moment is $M = 10$ kip · ft, determine the maximum bending stress in the plate.

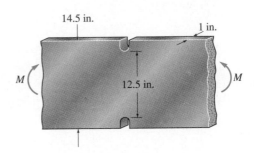

14.5 in.   1 in.

M   M

12.5 in.

Probs. 6–67/6–68

6–69. The notched bar is subjected to two forces **P**. If $r = 10$ mm, determine the largest magnitude of **P** that can be applied without causing the material to yield. The material is steel having a yield point of $\sigma_Y = 680$ MPa.

6–70. The simply-supported notched bar is subjected to the two loads, each having a magnitude of $P = 20$ kN. Determine the maximum bending stress developed in the bar, and sketch the bending-stress distribution acting over the cross section at the center of the bar.

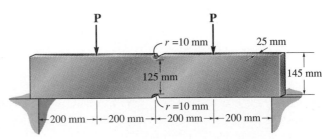

P   P

r = 10 mm   25 mm

125 mm   145 mm

r = 10 mm

200 mm   200 mm   200 mm   200 mm

Probs. 6–69/6–70

*6.7 Inelastic Bending

The equations for determining the normal stress due to bending that have previously been developed are valid only if the material behaves in a linear-elastic manner. If the applied moment causes the material to *yield*, a plastic analysis must then be used to determine the stress distribution. For both elastic and plastic cases, however, realize that for bending of straight members three conditions must be met.

Linear Normal-Strain Distribution. Based on geometric considerations, it was shown in Sec. 6.1 that the normal strains that develop in the material always vary *linearly* from zero at the neutral axis of the cross section to a maximum at the farthest point from the neutral axis.

Resultant Force Equals Zero. Since there is only a resultant internal moment acting on the cross section, the resultant force caused by the stress distribution must be equal to zero. This condition can be expressed mathematically by realizing that the normal stress σ, acting on an element of area dA, creates a force on the area of $dF = \sigma \, dA$, Fig. 6–31. Hence, for the entire cross-sectional area A, we have

$$F_R = \Sigma F_x; \qquad \int_A \sigma \, dA = 0 \qquad (6\text{–}21)$$

This equation provides a means for obtaining the *location of the neutral axis*.

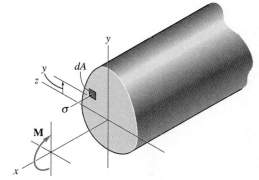

Fig. 6–31

Resultant Moment. The resultant moment at the section must be equivalent to the moment caused by the entire stress distribution about the neutral axis. This condition can be expressed mathematically by considering the moment of the force $dF = \sigma \, dA$ about the neutral axis and then summing the results over the entire cross section, Fig. 6–31. We have,

$$(M_R)_z = \Sigma M_z; \qquad M = \int_A y \, (\sigma \, dA) \qquad (6\text{–}22)$$

These conditions of geometry and loading will now be used to show how to determine the stress distribution in a beam when it is subjected to a resultant internal moment that causes yielding of the material. Throughout the discussion we will assume that the material has a stress–strain diagram that is the *same* in tension as it is in compression. For the sake of simplicity, we will begin by considering the beam to have a cross-sectional area with two axes of symmetry; for example, a rectangle of height h and width b, as shown in Fig. 6–32a. Three cases of loading that are of special interest will be considered.

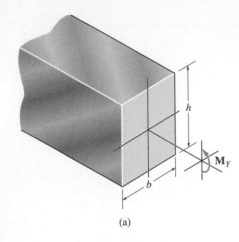

(a)

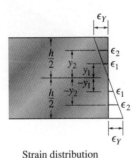

Strain distribution
(profile view)

(b)

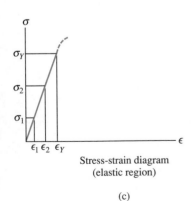

Stress-strain diagram
(elastic region)

(c)

Fig. 6–32

Maximum Elastic Moment. Assume that the applied moment $M = M_Y$ is just sufficient to produce yielding strains in the top and bottom boundaries of the beam as shown in Fig. 6–32b. Since the strain distribution is linear, we can determine the corresponding stress distribution by using the stress–strain diagram, Fig. 6–32c. Here it is seen that the yield strain ϵ_Y causes the yield stress σ_Y, and the intermediate strains ϵ_1 and ϵ_2 cause stresses σ_1 and σ_2, respectively. When these stresses, and others like them, are plotted at the measured points $y = h/2$, $y = y_1$, $y = y_2$, etc., the stress distribution in Fig. 6–32d or 6–32e results. The linearity of the stress is, of course, a consequence of Hooke's law.

Now that the stress distribution has been established, we can check to see if Eq. 6–21 is satisfied. To do so we will first calculate the resultant force for each of the two portions of the stress distribution in Fig. 6–32e. Geometrically this is equivalent to finding the *volumes* under the two triangular blocks. As shown, the top cross section of the member is subjected to compression and the bottom portion is subjected to tension. We have

$$T = C = \frac{1}{2}\left(\frac{h}{2}\,\sigma_Y\right)b = \frac{1}{4}\,bh\sigma_Y$$

Since **T** is equal but opposite to **C**, Eq. 6–21 is satisfied and indeed the neutral axis passes through the centroid of the cross-sectional area.

The maximum elastic moment M_Y is computed from Eq. 6–22, which states that M_Y is equivalent to the moment of the stress distribution about the neutral axis. To apply this equation geometrically, we must compute the moments created by **T** and **C** in Fig. 6–32e about the neutral axis. Since each of the forces acts through the centroid of the volume of its associated triangular stress block, we have

$$M_Y = C\left(\frac{2}{3}\right)\frac{h}{2} + T\left(\frac{2}{3}\right)\frac{h}{2} = 2\left(\frac{1}{4}\,bh\,\sigma_Y\right)\left(\frac{2}{3}\right)\frac{h}{2}$$

$$= \frac{1}{6}\,bh^2\sigma_Y \qquad\qquad (6\text{–}23)$$

This same result can of course be obtained in a more direct manner by using the flexure formula, that is, $\sigma_Y = M_Y(h/2)/[bh^3/12]$, or $M_Y = bh^2\sigma_Y/6$.

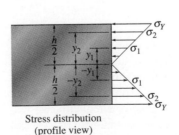

Stress distribution
(profile view)

(d)

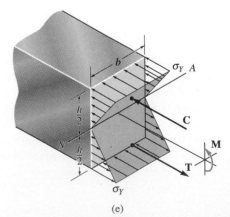

(e)

Plastic Moment. Some materials, such as steel, tend to exhibit elastic-perfectly plastic behavior when the stress in the material exceeds σ_Y. Consider, for example, the member in Fig. 6–33a. If the internal moment $M > M_Y$, the material at the top and bottom of the beam will begin to yield, causing a redistribution of stress over the cross section until the required internal moment M is developed. If the normal-strain distribution so produced is as shown in Fig. 6–33b, the corresponding normal-stress distribution is determined from the stress–strain diagram in the same manner as in the elastic case. Using the stress–strain diagram for the material shown in Fig. 6–33c, the strains ϵ_1, ϵ_Y, ϵ_2, ϵ_3 correspond to stresses σ_1, σ_Y, σ_Y, σ_Y, respectively. When these and other stresses are plotted on the cross section, we obtain the stress distribution shown in Fig. 6–33d or 6–33e. Here the compression and tension stress "blocks" each consist of component rectangular and triangular blocks. Their volumes are

$$T_1 = C_1 = \frac{1}{2}\, y_Y \sigma_Y b$$

$$T_2 = C_2 = \left(\frac{h}{2} - y_Y\right)\sigma_Y b$$

Because of the symmetry, Eq. 6–21 is satisfied and the neutral axis passes through the centroid of the cross section as shown. The applied moment M can be related to the yield stress σ_Y using Eq. 6–22. From Fig. 6–23e, we require

$$M = T_1 \left(\frac{2}{3}y_Y\right) + C_1\left(\frac{2}{3}y_Y\right) + T_2\left[y_Y + \frac{1}{2}\left(\frac{h}{2} - y_Y\right)\right]$$

$$+ C_2\left[y_Y + \frac{1}{2}\left(\frac{h}{2} - y_Y\right)\right]$$

$$= 2\left(\frac{1}{2}y_Y\sigma_Y b\right)\left(\frac{2}{3}y_Y\right) + 2\left[\left(\frac{h}{2} - y_Y\right)\sigma_Y b\right]\left[\frac{1}{2}\left(\frac{h}{2} + y_Y\right)\right]$$

$$= \frac{1}{4}\, bh^2\sigma_Y\left(1 - \frac{4}{3}\frac{y_Y^2}{h^2}\right)$$

Or, using Eq. 6–23,

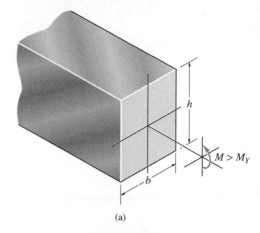

(a)

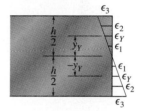

Strain distribution
(profile view)

(b)

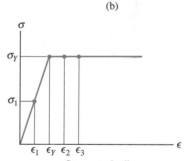

Stress-strain diagram
(elastic-plastic region)

(c)

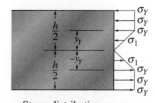

Stress distribution
(profile view)

(d)

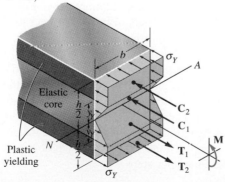

(e)

Fig. 6–33

$$M = \frac{3}{2} M_Y \left(1 - \frac{4}{3} \frac{y_Y^2}{h^2} \right) \qquad (6-24)$$

Inspection of Fig. 6–33e reveals that **M** produces two zones of plastic yielding (shown shaded) and an elastic core (shown colored) in the member. The boundary between them is located a distance $\pm y_Y$ from the neutral axis. As **M** increases in magnitude, y_Y approaches zero. This would render the material entirely plastic and the stress distribution would then look like that shown in Fig. 6–33f. From Eq. 6–24 with $y_Y = 0$, or by computing the moments of the stress "blocks" around the neutral axis, we can write this limiting value as

$$M_p = \frac{1}{4} bh^2 \sigma_Y \qquad (6-25)$$

or, using Eq. 6–23, we have

$$M_p = \frac{3}{2} M_Y \qquad (6-26)$$

This moment is referred to as the *plastic moment*. Its value is unique only for the rectangular section shown in Fig. 6–33f, since the analysis depends on the geometry of the cross section.

Beams used in steel buildings are sometimes designed to resist a plastic moment. When this is the case, codes usually list a design property for a beam called the shape factor. The *shape factor* is defined as a ratio,

$$\boxed{k = \frac{M_p}{M_Y}} \qquad (6-27)$$

This value specifies the additional moment capacity that a beam can support beyond its maximum elastic moment. For example, from Eq. 6–26, a beam having a rectangular cross section has a shape factor of $k = 1.5$. We may therefore conclude that this section will support 50% more bending moment than its maximum elastic moment when it becomes fully plastic.

Ultimate Moment. Consider now the more general case of a beam having a cross section that is symmetrical only with respect to the vertical axis, while the moment is applied about the horizontal axis, Fig. 6–34a. We will assume that the material exhibits strain hardening and that its stress–strain diagrams for tension and compression are different, Fig. 6–34b.

If the moment **M** produces yielding of the beam, difficulty arises in finding *both* the location of the neutral axis and the maximum strain that is produced in the beam. This is because the cross section is unsymmetrical about the horizontal axis and the stress–strain behavior of the material is unsymmetrical in tension and compression. To solve this problem, a trial-and-error procedure requires the following steps:

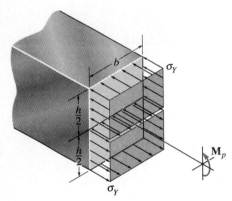

Plastic moment

(f)

Fig. 6–33(f)

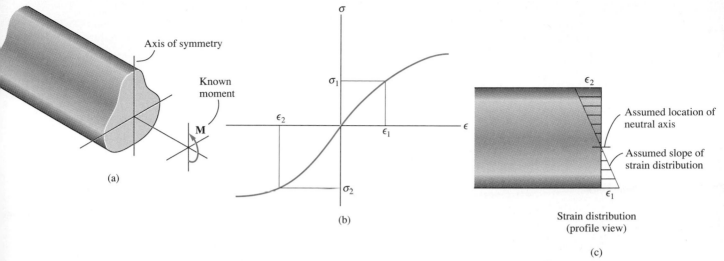

Strain distribution
(profile view)

(c)

1. For a given moment **M**, *assume* the location of the neutral axis and the slope of the ''linear'' strain distribution, Fig. 6–34c.

2. Graphically establish the stress distribution on the member's cross section using the σ–ϵ curve to plot values of stress corresponding to values of strain. The resulting stress distribution, Fig. 6–34d, will then have the same shape as the σ–ϵ curve.

3. Determine the volumes enclosed by the tensile and compressive stress ''blocks.'' (As an approximation, this may require dividing each block into composite regions.) Equation 6–21 requires the volumes of these blocks to be *equal,* since they represent the resultant tensile force **T** and resultant compressive force **C** on the section, Fig. 6–34e. If these forces are unequal, an adjustment as to the *location* of the neutral axis must be made (point of *zero strain*) and the process repeated until Eq. 6–21 ($T = C$) is satisfied.

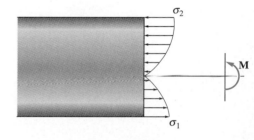

Stress distribution
(profile view)

(d)

4. Once $T = C$, the moments produced by **T** and **C** can be computed about the neutral axis. Here the moment arms for **T** and **C** are measured from the neutral axis to the *centroids of the volumes* defined by the stress distributions, Fig. 6–34e. Equation 6–22 requires $M = Ty' + Cy''$. If this equation is not satisfied, the *slope* of the *strain distribution* must be adjusted and the computations for T and C and the moment must be repeated until close agreement is obtained.

   This calculation process is obviously very tedious, and fortunately it does not occur very often in engineering practice. Most beams are symmetric about two axes, and they are constructed from materials that are assumed to have similar tension-and-compression stress–strain diagrams. Whenever this occurs, the neutral axis will pass through the centroid of the cross section, and the process of relating the stress distribution to the resultant moment is thereby simplified.

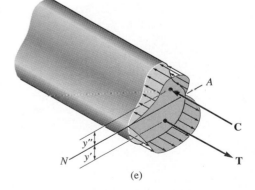

(e)

Fig. 6–34

Example 6–13

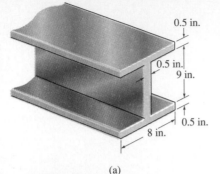

0.5 in.

0.5 in.
9 in.

0.5 in.

8 in.

(a)

The steel wide-flange beam has the dimensions shown in Fig. 6–35a. If it is made of an elastic-plastic material having a tensile and compressive yield stress of $\sigma_Y = 36$ ksi, determine the shape factor for the beam.

SOLUTION

In order to determine the shape factor, it is first necessary to compute the maximum elastic moment M_Y and the plastic moment M_p.

Maximum Elastic Moment. The normal-stress distribution for the maximum elastic moment is shown in Fig. 6–35b. The moment of inertia about the neutral axis is

$$I = \left[\frac{1}{12}(0.5)(9)^3\right] + 2\left[\frac{1}{12}(8)(0.5)^3 + 8(0.5)(4.75)^2\right] = 211.0 \text{ in}^4$$

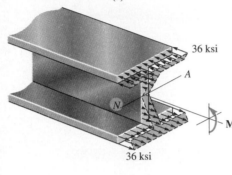

36 ksi

A

N

$\mathbf{M}_Y$

36 ksi

(b)

Applying the flexure formula, we have

$$\sigma_{\max} = \frac{Mc}{I}; \qquad 36 \text{ kip/in}^2 = \frac{M_Y(5 \text{ in.})}{211.0 \text{ in}^4}$$

$$M_Y = 1519.5 \text{ kip} \cdot \text{in.}$$

Plastic Moment. The plastic moment causes the steel over the entire cross section of the beam to yield, so that the normal-stress distribution looks like that shown in Fig. 6–35c. Due to symmetry of the cross-sectional area and since the tension and compression stress–strain diagrams are the same, the neutral axis passes through the centroid of the cross section. In order to compute the plastic moment, the stress distribution is divided into four composite rectangular "blocks," and the force produced by each "block" is equal to the volume of the block. Therefore, we have

$$C_1 = T_1 = 36 \text{ kip/in}^2(0.5 \text{ in.})(4.5 \text{ in.}) = 81 \text{ kip}$$
$$C_2 = T_2 = 36 \text{ kip/in}^2(0.5 \text{ in.})(8 \text{ in.}) = 144 \text{ kip}$$

These forces act through the *centroid* of the volume for each block. Computing the moments of these forces about the neutral axis, we obtain the plastic moment:

$$M_p = 2[(2.25 \text{ in.})(81 \text{ kip})] + 2[(4.75 \text{ in.})(144 \text{ kip})] = 1732.5 \text{ kip} \cdot \text{in.}$$

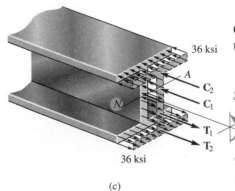

36 ksi

A

C₂
C₁

N

T₁
T₂

$\mathbf{M}_p$

36 ksi

(c)

Fig. 6–35

Shape Factor. Applying Eq. 6–27 gives

$$k = \frac{M_p}{M_Y} = \frac{1732.5}{1519.5} = 1.14 \qquad \textbf{\textit{Ans.}}$$

This value indicates that a wide-flange beam provides a very efficient section for resisting an *elastic moment*. Most of the moment is developed in the flanges, i.e., in the top and bottom segments, whereas the web or vertical segment contributes very little. In this particular case, only 14% additional moment can be supported by the beam beyond that which can be supported elastically.

Example 6–14

A T-beam has the dimensions shown in Fig. 6–36a. If it is made of an elastic-plastic material having a tensile and compressive yield stress of $\sigma_Y = 250$ MPa, determine the plastic moment that can be resisted by the beam.

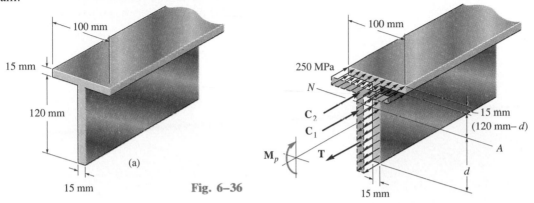

(a)

Fig. 6–36

(b)

SOLUTION

The "plastic" stress distribution acting over the beam's cross-sectional area is shown in Fig. 6–36b. In this case the cross section is not symmetric with respect to a horizontal axis, and consequently, the neutral axis will *not* pass through the centroid of the cross section. To determine the *location* of the neutral axis, d, we require the stress distribution to produce a zero resultant force on the cross section. Assuming that $d \le 120$ mm, we have

$$\int_A \sigma \, dA = 0; \qquad T - C_1 - C_2 = 0$$

$$250(15)(d) - 250(15)(120 - d) - 250(15)(100) = 0$$

$$2d = 220$$

$$d = 110 \text{ mm} < 120 \text{ mm} \qquad \text{OK}$$

Using this result, the forces acting on each segment are

$$T = 250 \text{ MN/m}^2(0.015 \text{ m})(0.110 \text{ m}) = 412.5 \text{ kN}$$

$$C_1 = 250 \text{ MN/m}^2(0.015 \text{ m})(0.010 \text{ m}) = 37.5 \text{ kN}$$

$$C_2 = 250 \text{ MN/m}^2(0.015 \text{ m})(0.100 \text{ m}) = 375 \text{ kN}$$

Hence the resultant plastic moment about the neutral axis is

$$M_p = 412.5 \text{ kN} \left(\frac{0.110 \text{ m}}{2} \right) + 37.5 \text{ kN} \left(\frac{0.01 \text{ m}}{2} \right)$$

$$+ 375 \text{ kN} \left(0.01 \text{ m} + \frac{0.015 \text{ m}}{2} \right)$$

$$M_p = 29.4 \text{ kN} \cdot \text{m} \qquad\qquad\qquad \textit{Ans.}$$

Example 6–15

The beam in Fig. 6-37a is made of an alloy of titanium that has a stress–strain diagram that can in part be approximated by two straight lines. If the material behavior is the *same* in both tension and compression, determine the bending moment that can be applied to the beam that will cause the material at the top and bottom of the beam to be subjected to a strain of 0.050 in./in.

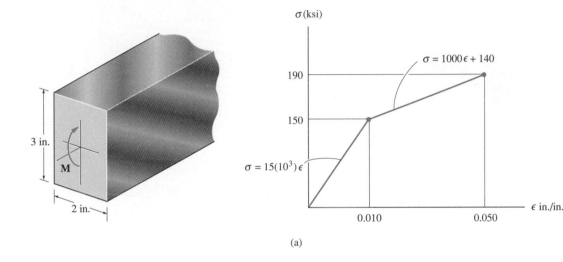

(a)

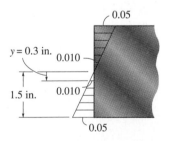

Strain distribution

(b)

Fig. 6–37

SOLUTION I

By inspection of the stress–strain diagram, the material is said to exhibit "elastic-plastic behavior with strain hardening." Since the cross section is symmetric and the tension–compression σ–ϵ diagrams are the same, the neutral axis must pass through the centroid of the cross section. The strain distribution, which is always linear, is shown in Fig. 6–37b. In particular, the point where maximum elastic strain (0.010 in./in.) occurs has been determined by proportion, such that 0.05/1.5 in. = 0.010/y or y = 0.30 in.

The corresponding normal-stress distribution acting over the cross section is shown in Fig. 6–37c. The moment produced by this distribution can be calculated by finding the "volume" of the stress blocks. To do so we will subdivide this distribution into two triangular blocks and a rectangular block in both the tension and compression regions, Fig. 6–37d. Since the beam is 2 in. wide, the resultants and their locations are computed as follows:

$$T_1 = C_1 = \frac{1}{2}(1.2 \text{ in.})(40 \text{ kip/in}^2)(2 \text{ in.}) = 48 \text{ kip}$$

$$y_1 = 0.3 \text{ in.} + \frac{2}{3}(1.2 \text{ in.}) = 1.10 \text{ in.}$$

$$T_2 = C_2 = (1.2 \text{ in.})(150 \text{ kip/in}^2)(2 \text{ in.}) = 360 \text{ kip}$$

$$y_2 = 0.3 \text{ in.} + \frac{1}{2}(1.2 \text{ in.}) = 0.90 \text{ in.}$$

$$T_3 = C_3 = \frac{1}{2}(0.3 \text{ in.})(150 \text{ ksi})(2 \text{ in.}) = 45 \text{ kip}$$

$$y_3 = \frac{2}{3}(0.3 \text{ in.}) = 0.2 \text{ in.}$$

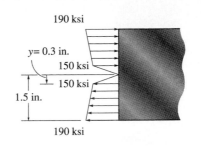

Stress distribution

(c)

The moment produced by this normal-stress distribution about the neutral axis is therefore

$$M = 2 [48 \text{ kip } (1.10 \text{ in.}) + 360 \text{ kip } (0.90 \text{ in.}) + 45 \text{ kip } (0.2 \text{ in.})]$$
$$= 772 \text{ kip} \cdot \text{in.} \qquad\qquad Ans.$$

SOLUTION II

Rather than using the above semigraphical technique, it is also possible to compute the moment analytically. To do this we must express the stress distribution in Fig. 6–37c as a function of position y along the beam. Note that $\sigma = f(\epsilon)$ has been given in Fig. 6–37a. Also, from Fig. 6–37b, the normal strain can be determined as a function of position y by proportional triangles; i.e.,

$$\epsilon = \frac{0.05}{1.5} y \qquad 0 \leqslant y \leqslant 1.5 \text{ in.}$$

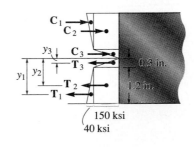

(d)

Substituting this into the σ–ϵ functions shown in Fig. 6–37a gives

$$\sigma = 500y \qquad\qquad 0 \leqslant y \leqslant 0.3 \text{ in.} \qquad (1)$$
$$\sigma = 33.33y + 140 \qquad 0.3 \text{ in.} \leqslant y \leqslant 1.5 \text{ in.} \qquad (2)$$

From Fig. 6–37e, the moment caused by σ acting on the area strip $dA = 2 \, dy$ is

$$dM = y \, (\sigma \, dA) = y\sigma \, (2 \, dy)$$

Using Eqs. 1 and 2, the moment for the entire cross section is thus

$$M = 2 \left[2 \int_0^{0.3} 500y^2 \, dy + 2 \int_{0.3}^{1.5} (33.3y^2 + 140y) \, dy \right]$$

$$= 2[9 + 390 - 13.2]$$

$$= 772 \text{ kip} \cdot \text{in.} \qquad\qquad Ans.$$

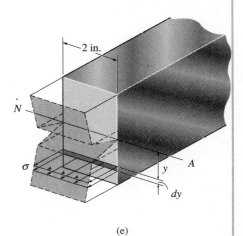

(e)

*6.8 Residual Stress

If a beam is loaded such that it causes the material to yield, then removal of the load will cause *residual stress* to be developed in the beam. Since residual stresses are often important when considering fatigue and other types of mechanical behavior, we will discuss a method used for their computation when a member is subjected to bending.

Like the case for torsion, we can calculate the residual-stress distribution in a member subjected to bending by using the principles of superposition and elastic recovery. To explain how this is done, consider the beam shown in Fig. 6–38a, which has a rectangular cross section and is made of an elastic-plastic material having the same shaped stress–strain diagram in tension as in compression, Fig. 6–38b. Application of the plastic moment M_p causes a stress distribution in the member to be idealized as shown in Fig. 6–38c. From Eq. 6–25, this moment is

$$M_p = \frac{1}{4} bh^2 \sigma_Y$$

If M_p causes the material at the top and bottom of the beam to be strained to ϵ_1 ($>>\epsilon_Y$), as shown by point B on the σ–ϵ curve in Fig. 6–38b, then a release of this moment will cause this material to recover some of this strain elastically by following the dashed path BC. Since this recovery is elastic, we can superimpose on the stress distribution in Fig. 6–38c a linear stress distribution caused by applying the moment in the opposite direction, Fig. 6–38d.

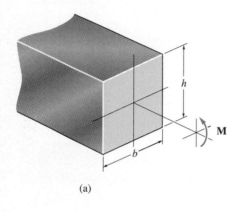

(a)

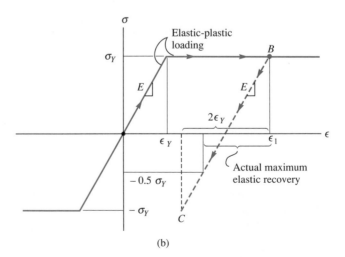

(b)

Fig. 6–38

Here the maximum stress, which is called the *modulus of rupture* for bending, σ_r, can be computed from the flexure formula when the beam is loaded with the plastic moment. We have

$$\sigma_{\max} = \frac{Mc}{I}; \qquad \sigma_r = \frac{M_p\,(\frac{1}{2}\,h)}{(\frac{1}{12}\,bh^3)} = \frac{(\frac{1}{4}\,bh^2\sigma_Y)(\frac{1}{2}\,h)}{(\frac{1}{12}\,bh^3)}$$

$$= 1.5\sigma_Y$$

Note that reversed application of the plastic moment using a linear stress distribution is possible here, since *elastic recovery* of the material at the top and bottom of the beam can have a *maximum recovery strain of* $2\epsilon_Y$ as shown in Fig. 6–38b. This would correspond to a maximum stress of $2\sigma_Y$ at the top and bottom of the beam, which is greater than the *required* stress of $1.5\sigma_Y$ as calculated above, Fig. 6–38d.

The superposition of the plastic moment, Fig. 6–38c, and its removal, Fig. 6–38d, gives the residual-stress distribution shown in Fig. 6–38e. As an exercise, use the component triangular ''blocks'' that represent this stress distribution and show that it results in a zero-force and zero-moment resultant on the member as required.

The following example numerically illustrates application of these principles.

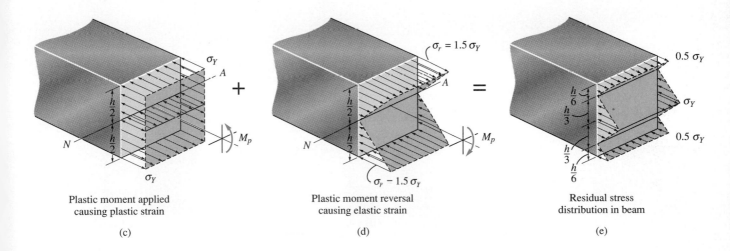

Plastic moment applied
causing plastic strain

(c)

Plastic moment reversal
causing elastic strain

(d)

Residual stress
distribution in beam

(e)

Fig. 6–38

Example 6–16

The steel wide-flange beam shown in Fig. 6–39a is subjected to a fully plastic moment of $\mathbf{M}_p$. If this moment is removed, determine the residual-stress distribution in the beam. The material is elastic-plastic and has a yield stress of $\sigma_Y = 36$ ksi.

SOLUTION

The normal-stress distribution in the beam caused by $\mathbf{M}_p$ is shown in Fig. 6–39b. When $\mathbf{M}_p$ is removed, the material responds elastically. Removal of $\mathbf{M}_p$ requires applying $\mathbf{M}_p$ in its reverse direction and therefore leads to an assumed elastic stress distribution as shown in Fig. 6–39c. The modulus of rupture σ_r is computed from the flexure formula. Using $M_p = 1732.5$ kip · in. and $I = 211.0$ in^4 from Example 6–13, we have

$$\sigma_{\max} = \frac{Mc}{I}; \qquad \sigma_r = \frac{1732.5 \text{ kip} \cdot \text{in. (5 in.)}}{211.0 \text{ in}^4} = 41.1 \text{ ksi}$$

As expected, $\sigma_r < 2\sigma_Y$.

Superposition of the stresses gives the residual-stress distribution shown in Fig. 6–39d. Note that the point of zero normal stress was determined by proportion; i.e., from Fig. 6–39b and 6–39c, we require that

$$\frac{41.1 \text{ ksi}}{5 \text{ in.}} = \frac{36 \text{ ksi}}{y}$$

$$y = 4.37 \text{ in.}$$

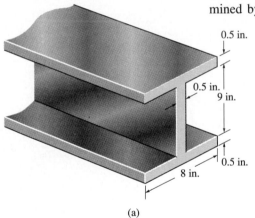

(a)

Fig. 6–39

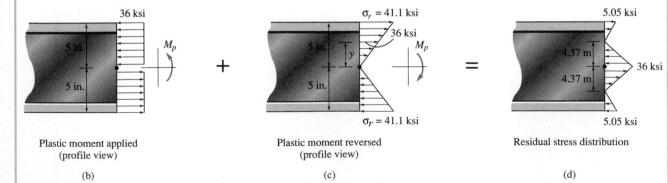

Plastic moment applied
(profile view)

(b)

Plastic moment reversed
(profile view)

(c)

Residual stress distribution

(d)

PROBLEMS

6–71. A bar having a width of 3 in. and height of 2 in. is made of an elastic-plastic material for which $\sigma_Y = 36$ ksi. Determine the moment applied about the horizontal axis that will cause half the bar to yield.

*6–72. Determine the plastic section modulus and the shape factor for the wide-flange beam.

6–73. The beam is made of an elastic-plastic material for which $\sigma_Y = 250$ MPa. Determine the residual stress in the beam at its top and bottom after the plastic moment $\mathbf{M}_p$ is applied and then released.

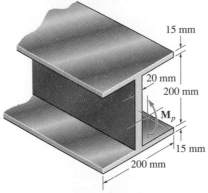

Probs. 6–72/6–73

6–74. Determine the shape factor for the cross section of the H-beam.

6–75. The H-beam is made of an elastic-plastic material for which $\sigma_Y = 250$ MPa. Determine the residual stress in the top and bottom of the beam after the plastic moment $\mathbf{M}_p$ is applied and then released.

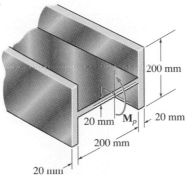

Probs. 6–74/6–75

*6–76. Determine the plastic section modulus and the shape factor for the cross section of the beam.

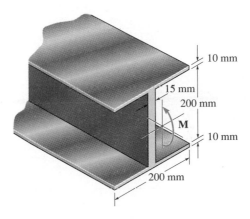

Prob. 6–76

6–77. The T-beam is made of an elastic-plastic material. Determine the maximum elastic moment and the plastic moment that can be applied to the cross section. $\sigma_Y = 36$ ksi.

6–78. Determine the plastic section modulus and the shape factor for the beam.

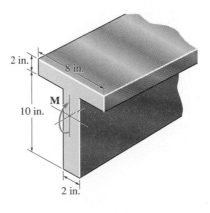

Probs. 6–77/6–78

6–79. The channel strut is made of an elastic-plastic material for which $\sigma_Y = 250$ MPa. Determine the maximum elastic moment and the plastic moment that can be applied to the cross section.

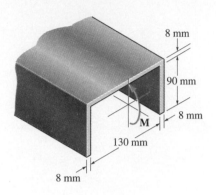

Prob. 6–79

6–81. The rod has a circular cross section. If it is made of an elastic-plastic material, determine the shape factor and the plastic section modulus Z.

Prob. 6–81

*6–80. The member has a square cross section. If it is made of an elastic-plastic material, determine the shape factor and the plastic section modulus Z.

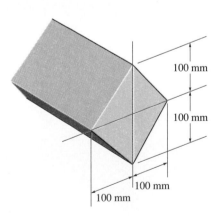

Prob. 6–80

6–82. The beam is made of an elastic-plastic material for which $\sigma_Y = 200$ MPa. If the largest moment in the beam occurs within the center section a–a, determine the magnitude of each force **P** that causes this moment to be (a) the largest elastic moment and (b) the largest plastic moment.

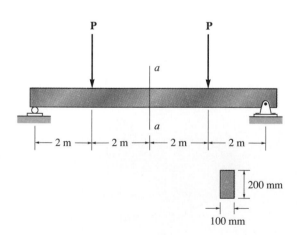

Prob. 6–82

6–83. The beam is made of an elastic-plastic material for which $\sigma_Y = 30$ ksi. If the largest moment in the beam occurs at the center section a–a, determine the intensity of the distributed load w that causes this moment to be (a) the largest elastic moment and (b) the largest plastic moment.

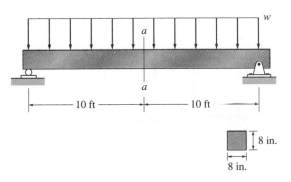

Prob. 6–83

*6–84. The box beam is made of an elastic-plastic material for which $\sigma_Y = 25$ ksi. If the largest moment in the beam occurs at the center section a–a, determine the intensity of the distributed load w_0 that will cause this moment to be (a) the largest elastic moment and (b) the largest plastic moment.

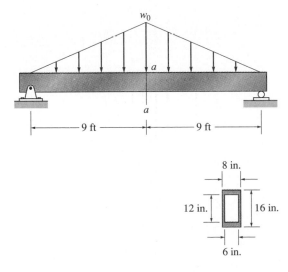

Prob. 6–84

6–85. The bar is made of an aluminum alloy having a stress–strain diagram that can be approximated by the straight-line segments shown. Assuming that this diagram is the same for both tension and compression, determine the moment the bar will support if the maximum strain at the top and bottom fibers of the beam is $\epsilon_{max} = 0.03$.

6–86. The bar is made of an aluminum alloy having a stress–strain diagram that can be approximated by the straight line segments shown. Assuming that this diagram is the same for both tension and compression, determine the moment the bar will support if the maximum strain at the top and bottom fibers of the beam is $\epsilon_{max} = 0.05$.

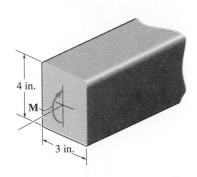

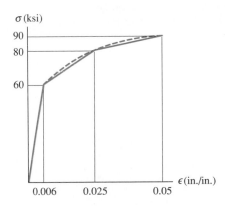

Probs. 6–85/6–86

6–87. The stress–strain diagram for a titanium alloy can be approximated by the two straight lines. If a strut made from this material is subjected to bending, determine the moment resisted by the strut if the maximum stress reaches a value of (a) σ_A and (b) σ_B.

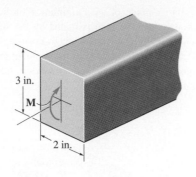

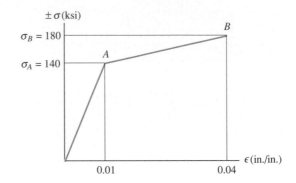

Prob. 6–87

*6–88. A member is made of a polymer having the stress–strain diagram shown. If the curve can be represented by the equation $\sigma = 4.65(10)^3\, \epsilon^{1.35}$ ksi, determine the magnitude of the moment **M** that can be applied without causing the maximum strain in the member to exceed $\epsilon_{max} = 0.005$ in./in.

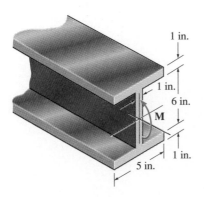

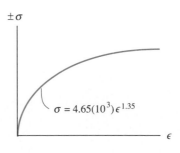

Prob. 6–88

6–89. The beam has a rectangular cross section and is made of an elastic-plastic material having a stress–strain diagram as shown. Determine the magnitude of the moment **M** that must be applied to the beam in order to create a maximum strain in its outer fibers of $\epsilon_{max} = 0.008$.

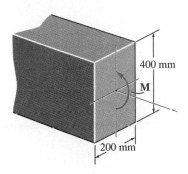

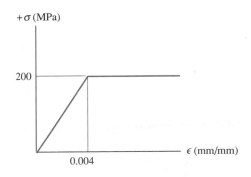

Prob. 6–89

REVIEW PROBLEMS

6–90. A 100-mm-diameter circular rod is bent into an S shape. If it is subjected to the applied moments $M = 125$ N · m at its ends, determine the maximum tensile and compressive stress developed in the rod.

6–91. The beam is subjected to a moment of 15 kip · ft. Determine the resultant force the stress produces on the top flange A and bottom flange B. Also compute the maximum stress developed in the beam.

*6–92. The beam is subjected to a moment of 15 kip · ft. Determine the percent of this moment that is resisted by the web D of the beam.

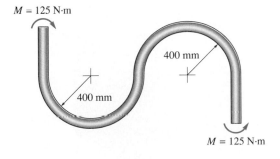

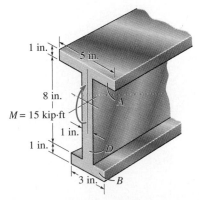

Prob. 6–90   **Probs. 6–91/6–92**

6–93. Determine the moment M that will produce a maximum stress of 12 ksi on the cross section.

6–94. Determine the maximum tensile and compressive bending stress in the beam if it is subjected to a moment of $M = 2$ kip · ft.

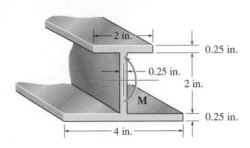

Probs. 6–93/6–94

***6–96.** The steel rod having a diameter of 1 in. is subjected to an internal moment of $M = 300$ lb · ft. Determine the stress created at points A and B. Also, sketch a three-dimensional view of the stress distribution acting over the cross-section.

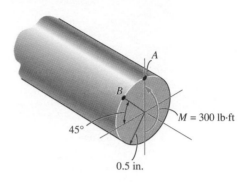

Prob. 6–96

6–95. The beam is made from three boards nailed together as shown. If the moment acting on the cross section is $M = 600$ N · m, determine the maximum bending stress in the beam. Sketch a three-dimensional view of the stress distribution acting over the cross section.

6–97. Determine the residual stress at the top and bottom of a bar having a circular cross section that has been unloaded from a fully plastic moment. The radius of the bar is 3 in., and $\sigma_Y = 36$ ksi.

6–98. Determine the plastic moment $\mathbf{M}_p$ that can be supported by a beam having the cross section shown. $\sigma_Y = 30$ ksi.

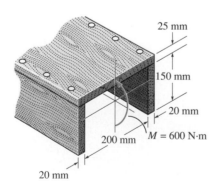

Prob. 6–95

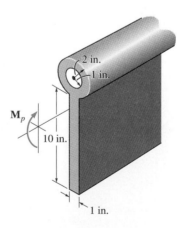

Prob. 6–98

7 Transverse Shear

In this chapter we will develop a method for finding the shear stress in a beam having a prismatic cross section and made from homogeneous material that behaves in a linear-elastic manner. The method of analysis to be developed will be somewhat limited to special cases of cross-sectional geometry. Although this is the case, it has many wide-range applications in engineering design and analysis. The concept of shear flow, along with shear stress, will be discussed for beams and thin-walled members. The chapter ends with a discussion of the shear center.

7.1 Shear in Straight Members

Before we develop a relationship that describes the shear-stress distribution over the cross section of a beam, we will first make some preliminary remarks regarding the way shear stress acts within the beam. Realize that since beams are generally subjected to transverse loadings, these loadings not only cause an internal moment in the beam but *also* an internal shear force. This force **V**, shown in Fig. 7–1*a*, is necessary for translational equilibrium, and it is the result of a *transverse shear-stress* distribution that acts over the beam's cross section, Fig. 7–1*b*. As a result of this distribution, associated *longitudinal shear stresses* will *also* act along longitudinal planes of the beam. For example, a typical element removed from the interior point *A* on the cross section is subjected to both transverse and longitudinal shear stress as shown in Fig. 7–1*b*. In particular, note that the longitudinal shear stress at points *B* and *C*, located on the top and the bottom boundaries of the beam, must be zero since the top and bottom surfaces of the beam are free of any load. Consequently, the transverse shear stress at these points must also be zero.

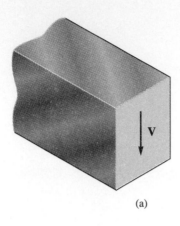

(a)

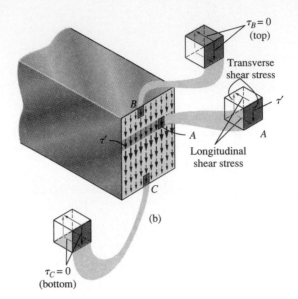

(b)

Fig. 7–1

It is also possible to physically illustrate why shear stress develops on the longitudinal planes of a beam that supports an internal shear loading by considering the beam to be made from three boards, Fig. 7–2a. If the top and bottom surfaces of each board are smooth, and the boards are not bonded together, then application of the load **P** will cause the boards to slide relative to one another, and so the beam will deflect as shown. On the other hand, if the boards are bonded together, then the longitudinal shear stresses between the boards will prevent the relative sliding of the boards, and consequently the beam will act as a single unit, Fig. 7–2b.

As a result of the internal shear-stress distribution, shear strains will be developed and these will tend to distort the cross section in a rather complex manner. To show how the shear strains vary, consider a bar made of a highly deformable material and marked with horizontal and vertical grid lines as shown in Fig. 7–3a. When the shear force is applied, it tends to deform these lines into the pattern shown in Fig. 7–3b. Notice that the squares near the top and bottom of the bar retain their shapes, since there it was shown in Fig. 7–1b that the shear stress and likewise the shear strain are zero (or very small). On the other hand, the shear strain on the square in the center of the bar will cause it to have the greatest deformation. In general then, the nonuniform shear-strain distribution over the cross section will cause the cross section to *warp*, that is, *not* to remain plane.

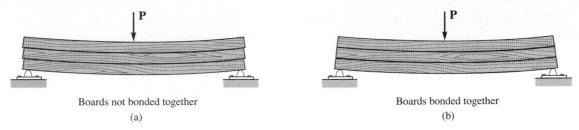

Boards not bonded together
(a)

Boards bonded together
(b)

Fig. 7–2

Recall that in the development of the flexure formula, we assumed that cross sections must *remain plane* and perpendicular to the longitudinal axis of the beam after deformation. Although these assumptions are violated when the beam is subjected to *both* bending and shear, we can generally assume the cross-sectional warping described above is small enough so that it can be neglected. This assumption is particularly true for the most common case of a *slender beam;* that is, one that has a small depth compared with its length.

In the previous chapters we developed the axial load, torsion, and flexure formulas by first determining the strain distribution, based on assumptions regarding the deformation of the cross section. Then using Hooke's law we related the strain to the stress; finally, we used the equilibrium requirements to relate the stress to the loading. Unlike these three cases, however, the shear-strain distribution throughout the depth of a beam *cannot* be easily expressed mathematically; for example, it is not uniform or linear for rectangular cross sections as we have shown. Therefore, the analysis of shear stress will be developed in a manner different from that used to study axial load, torsion, and bending moment.

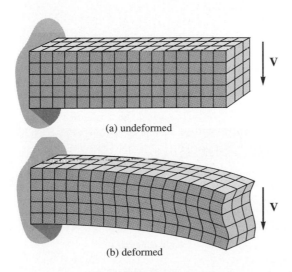

(a) undeformed

(b) deformed

Fig. 7–3

7.2 Differential Relationships Between Load, Shear, and Moment

In this section we will develop two equilibrium relationships that exist between the distributed loading acting on a beam and the internal resultant shear and moment that act on its cross section. These relationships have important applications in mechanical and structural design, as will be discussed in later chapters of the text. In the next section, however, the relationship between the internal shear and moment will be used to derive the shear formula.

Beam Sign Convention. Before we develop the relationships between load, shear, and moment, it is first necessary to establish a sign convention for these quantities. This "beam sign convention" is often used in engineering practice and is shown in Fig. 7–4. Here the *positive directions* are denoted by a *distributed load* that is *directed downward,* an internal *shear force* that causes a *clockwise rotation* of the member on which it acts, and an internal *moment* that causes *compression in the upper fibers* of the member. Loadings that are opposite to these are considered negative.

Equilibrium Relationships Between the Loads. For purposes of generality, consider the beam shown in Fig. 7–5a, which is subjected to an arbitrary distributed load $w = w(x)$ and a series of concentrated forces and couple moments. A free-body diagram for a small segment of the beam, having a length Δx, is shown in Fig. 7–5b. Since this segment has been chosen at a position x along the beam where there is no concentrated force or couple moment, the results to be obtained will *not* apply at these points of concentrated loading.

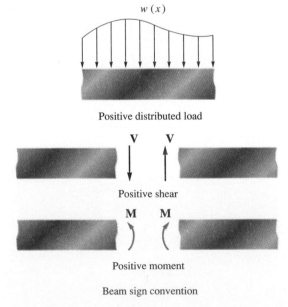

$w(x)$

Positive distributed load

V V

Positive shear

M M

Positive moment

Beam sign convention

Fig. 7–4

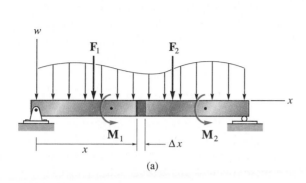

(a)

Fig. 7–5

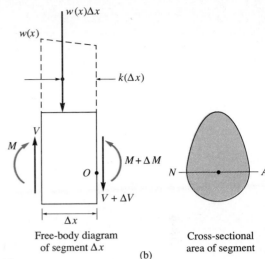

Free-body diagram
of segment Δx

Cross-sectional
area of segment

(b)

Notice that all the loadings shown on the segment act in their positive directions according to the established sign convention. Also, both the internal resultant shear and moment, acting on the right face of the segment, must be increased by a small finite amount in order to keep the segment in equilibrium. The distributed load has been replaced by a resultant force $w(x)\,\Delta x$ that acts at a fractional distance $k(\Delta x)$ from the right end, where $0 < k < 1$ (for example, if $w(x)$ is *uniform,* $k = \frac{1}{2}$). Applying the two equations of equilibrium to the segment, we have

$$+\uparrow \ \Sigma F_y = 0; \qquad\qquad V - w(x)\,\Delta x - (V + \Delta V) = 0$$
$$\Delta V = -w(x)\,\Delta x$$

$$\zeta+ \ \Sigma M_O = 0; \qquad -V\,\Delta x - M + w(x)\,\Delta x[k(\Delta x)] + (M + \Delta M) = 0$$
$$\Delta M = V\,\Delta x - w(x)\,k(\Delta x)^2$$

Dividing by Δx and taking the limit as $\Delta x \to 0$, the above two equations become

$$\boxed{\dfrac{dV}{dx} = -w(x)} \qquad\qquad (7\text{–}1)$$

and

$$\boxed{\dfrac{dM}{dx} = V} \qquad\qquad (7\text{–}2)$$

Equation 7–1 states that *force equilibrium* of a differential beam segment requires that the change of the internal shear be equal to the (negative) intensity of the distributed load at a point, and Eq. 7–2 states that *moment equilibrium* requires that the change in the internal moment be equal to the internal shear at a point.

7.3 The Shear Formula

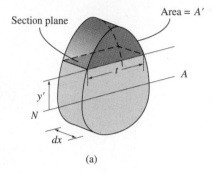

Section plane

Area = A'

t

A

y'

N

dx

(a)

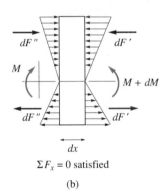

dF'' dF'

M $M + dM$

dF'' dF'

dx

$\Sigma F_x = 0$ satisfied

(b)

Development of a relationship between the shear-stress distribution, acting over the cross section of a beam, and the resultant shear force at the section is based on a study of the *longitudinal shear stress* and the results of Eq. 7–2. To show how this relationship is established, we will consider the *horizontal force equilibrium* of a portion of the element in Fig. 7–5b. First let the element's thickness $\Delta x \rightarrow dx$, Fig. 7–6a, and consider a free-body diagram of the *entire element* that shows *only* the normal-stress distribution acting on the element, Fig. 7–6b. This distribution is caused by the bending moments M and $M + dM$. We have excluded the effects of V, $V + dV$, and $w(x)$ on the free-body diagram since these loadings are vertical and are therefore not involved in a horizontal force summation. The element in Fig. 7–6b will indeed satisfy $\Sigma F_x = 0$ since the stress distribution on each side of the element forms only a couple moment and therefore a zero force resultant.

Now consider the shaded top *segment* of the element that has been sectioned at y' from the neutral axis, Fig. 7–6a. This segment has a width t at the section, and the cross-sectional sides each have an area A'. Because the resultant moments on each side of the element differ by dM, it can be seen in Fig. 7–6c that $\Sigma F_x = 0$ will not be satisfied *unless* a longitudinal shear stress τ acts over the bottom face of the segment. In the following analysis, we will assume this shear stress is *constant* across the width t of the bottom face. It acts on the area $t\,dx$. Applying the equation of horizontal force equilibrium, and using the flexure formula, Eq. 6–6, we have

$$\xrightarrow{+}\ \Sigma F_x = 0; \qquad \int_{A'} \sigma'\, dA - \int_{A'} \sigma\, dA - \tau(t\,dx) = 0$$

$$\int_{A'} \left(\frac{M + dM}{I}\right) y\, dA - \int_{A'} \left(\frac{M}{I}\right) y\, dA - \tau(t\,dx) = 0$$

$$\left(\frac{dM}{I}\right) \int_{A'} y\, dA = \tau(t\,dx) \qquad (7\text{–}3)$$

Fig. 7–6

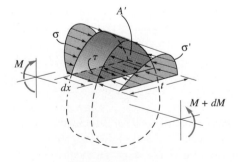

A'

σ

τ

σ'

M

dx

t

$M + dM$

Three dimentional view

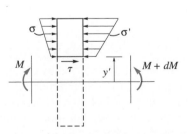

σ σ'

M $M + dM$

τ y'

Profile view

(c)

Solving for τ, we get

$$\tau = \frac{1}{It}\left(\frac{dM}{dx}\right)\int_{A'} y\, dA$$

This equation can be simplified by noting that $V = dM/dx$ (Eq. 7–2). Also, the integral represents the first moment of the area A' about the neutral axis. We will denote it by the symbol Q. Since the location of the centroid of the area A' is determined from $\bar{y}' = \int y\, dA'/A'$, we can also write

$$Q = \int_{A'} y\, dA = \bar{y}'A \qquad (7\text{–}4)$$

The final result is therefore

$$\boxed{\tau = \frac{VQ}{It}} \qquad (7\text{–}5)$$

Here

τ = the shear stress in the member at the point located a distance y' from the neutral axis, Fig. 7–6a. This stress is assumed to be constant and therefore *averaged* across the width t of the member.

V = the internal resultant shear force, determined from the method of sections and the equations of equilibrium.

I = the moment of inertia of the *entire* cross-sectional area computed about the neutral axis.

t = the width of the member's cross-sectional area, measured at the point where τ is to be determined.

Q = $\int_{A'} y\, dA = \bar{y}'A'$, where A' is the top (or bottom) portion of the member's cross-sectional area, defined from the section where t is measured, and $\bar{y}'$ is the distance to the centroid of A', measured from the neutral axis.

The above equation is referred to as the *shear formula*. Although in the derivation we considered only the shear stresses acting on the beam's longitudinal plane, the formula applies as well for finding the transverse shear stress on the beam's cross-sectional area. This, of course, is because the transverse and longitudinal shear stresses are complementary and numerically equal.

Since Eq. 7–5 was derived from the flexure formula, it is necessary that the material behave in a linear-elastic manner and have a modulus of elasticity that is the *same* in tension as it is in compression. The shear stress in composite members, that is, those having cross sections made of different materials, can also be obtained using the shear formula. To do so, however, it is necessary to compute Q and I from the *transformed section* of the member as discussed in Sec. 6.4. The thickness t in the formula, however, remains the actual width t of the cross section at the point where τ is to be calculated.

7.4 Shear Stresses in Beams

In order to develop some insight as to the method of applying the shear formula and also discuss some of its limitations, we will now study the shear-stress distributions in a few common types of beam cross sections. Numerical applications of the shear formula will then be given in the examples that follow.

Rectangular Cross Section. Consider the beam to have a rectangular cross section of width b and height h as shown in Fig. 7–7a. The distribution of the shear stress throughout the cross section can be determined by computing the shear stress at an *arbitrary height y* from the neutral axis, Fig. 7–7b, and then plotting this function. Here the shaded area A' will be used for computing τ.* Hence

$$Q = \bar{y}'A' = \left[y + \frac{1}{2}\left(\frac{h}{2} - y\right)\right]\left(\frac{h}{2} - y\right)b$$

$$= \frac{1}{2}\left(\frac{h^2}{4} - y^2\right)b$$

Applying the shear formula, we have

$$\tau = \frac{VQ}{It} = \frac{V(\frac{1}{2})[(h^2/4) - y^2]b}{(\frac{1}{12}bh^3)b}$$

or

$$\tau = \frac{6V}{bh^3}\left(\frac{h^2}{4} - y^2\right) \tag{7–6}$$

This result indicates that the shear-stress distribution over the cross section is *parabolic*. As shown in Fig. 7–7c, the intensity varies from zero at the top and bottom, $y = \pm h/2$, to a maximum value at the neutral axis, $y = 0$. Specifically, since the area of the cross section is $A = bh$, then at $y = 0$ we have, from Eq. 7–6,

$$\tau_{\max} = 1.5\frac{V}{A} \tag{7–7}$$

This same value for $\tau_{\max}$ can be obtained directly from the shear formula, $\tau = VQ/It$, by realizing that $\tau_{\max}$ occurs where Q is *largest*, since V, I, and t are *constant*. By inspection, Q will be a maximum when the area above (or below) the neutral axis is considered; that is, $A' = bh/2$ and $y' = h/4$. Thus,

---

*The area below y can also be used [$A' = b(h/2 + y)$], but doing so involves a bit more algebraic manipulation.

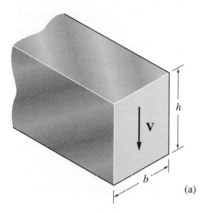

(a)

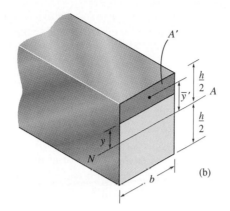

(b)

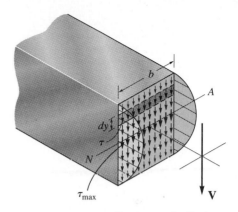

Shear stress distribution

(c)

$$\tau_{max} = \frac{VQ}{It} = \frac{V(h/4)(bh/2)}{[\frac{1}{12}bh^3]b} = 1.5\frac{V}{A}$$

By comparison, τ_{max} is 50% greater than the *average* shear stress computed from Eq. 1–7; that is, $\tau_{avg} = V/A$.

It is important to remember that for every τ acting on the cross-sectional area in Fig. 7–7c, there is a corresponding τ acting in the longitudinal direction along the beam. For example, if the beam is sectioned by a longitudinal plane through its neutral axis, then as noted above, the *maximum shear stress* acts on this plane, Fig. 7–7d. It is this stress that will cause a timber beam to fail as shown in Fig. 7–8. Here horizontal splitting of the wood starts to occur through the neutral axis at the beam's ends, since there the vertical reactions subject the beam to large shear stress and wood has a low resistance to shear along its grains, which are oriented in the longitudinal direction.

Since Eq. 7–6 expresses the shear-stress distribution as a function of position y, it is instructive to show that when integrated over the cross section it yields the resultant shear V. To do this, a differential strip of area $dA = b\,dy$ is chosen, Fig. 7–7c, and since τ acts uniformly over this strip, we have

$$\int_A \tau\,dA = \int_{-h/2}^{h/2} \frac{6V}{bh^3}\left(\frac{h^2}{4} - y^2\right)b\,dy$$

$$= \frac{6V}{h^3}\left[\frac{h^2}{4}y - \frac{1}{3}y^3\right]_{-h/2}^{h/2}$$

$$= \frac{6V}{h^3}\left[\frac{h^2}{4}(h) - \frac{1}{3}\left(\frac{h^3}{8} + \frac{h^3}{8}\right)\right] = V$$

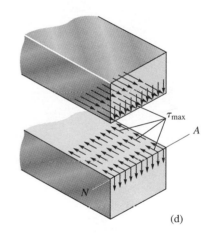

(d)

Fig. 7–7

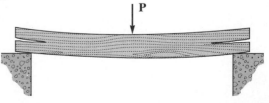

Fig. 7–8

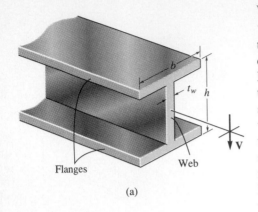

(a)

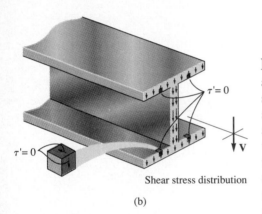

Shear stress distribution

(b)

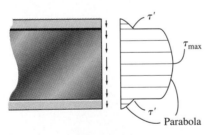

Intensity of shear
stress distribution
(profile view)

(c)

Fig. 7–9

Wide-Flange Beam. A *wide-flange beam* consists of two (wide) "flanges" and a "web" as shown in Fig. 7–9a. Using an analysis similar to that given before we can determine the shear-stress distribution acting over its cross section. The results are depicted graphically in Fig. 7–9b and 7–9c. Like the rectangular cross section, the shear stress varies *parabolically* over the beam's depth, since the cross section can be treated like the rectangular section, first having the width of the top flange, b, then the thickness of the web, t_w, and again the width of the bottom flange, b. In particular, notice that the shear stress will vary *only slightly* throughout the web, and also, a *jump* in shear stress occurs at the flange–web junction since the cross-sectional thickness changes at this point, or in other words, t in the shear formula changes. By comparison, the web will carry significantly more of the shear force than the flanges. This will be illustrated numerically in Example 7–2.

Limitations on the Use of the Shear Formula. One of the major assumptions used in the development of the shear formula is that the shear stress is *uniformly* distributed over the *width t* at the section where the shear stress is computed. In other words, the *average* shear stress is computed across the width. We can test the accuracy of this assumption by comparing it with a more exact mathematical analysis based on the theory of elasticity. In this regard, if the beam's cross section is rectangular, the actual shear-stress distribution across the neutral axis varies as shown in Fig. 7–10. The maximum value, τ'_{max}, occurs at the *edges* of the cross section, and its magnitude depends on the ratio b/h (width/depth). For sections having a b/h = 0.5, τ'_{max} is only about 3% greater than the shear stress calculated from the shear formula, Fig. 7–10a. However, for *flat sections*, say b/h = 2, τ'_{max} is about 40% greater than τ_{max}, Fig. 7–10b. The error becomes even greater as the section becomes flatter, or as the b/h ratio increases. Errors of this magnitude are certainly intolerable if one uses the shear formula to determine the shear stress in the *flange* of a wide-flange beam, as discussed above.

It should also be pointed out that the shear formula will not give accurate results when used to determine the shear stress at the flange–web junction of a wide-flange beam, since this is a point of sudden cross-sectional change and therefore a *stress concentration* occurs here. Furthermore, the inner regions of the flanges are free boundaries, Fig. 7–9b, and as a result the shear stress on these boundaries must be zero. If the shear formula is applied to determine the shear stress at these boundaries, however, one obtains a value of τ' that is *not* equal to zero, Fig. 7–9c. Fortunately, these limitations for applying the shear formula to the flanges of a wide-flange beam are not important in engineering practice. Most often engineers must only calculate the *average maximum shear stress* occurring at the neutral axis, where the b/h (width/depth) ratio is *very small,* and therefore the calculated result is very close to the *actual* maximum shear stress as explained above.

Another important limitation on the use of the shear formula can be pointed out with reference to Fig. 7–11a, which shows a beam having a cross section with an irregular or nonrectangular boundary. If we apply the shear formula to determine the (average) shear stress τ along the line AB, it will be directed as shown in Fig. 7–11b. Consider now an element of material taken from the boundary point B, such that one of its faces is located on the outer surface of the beam, Fig. 7–11c. Here the calculated shear stress τ on the front face of the element is resolved into components, τ' and τ''. By inspection, the component τ' must be equal to zero since its corresponding longitudinal component τ', acting on the stress-free boundary surface, must be zero. To satisfy this boundary condition therefore, the shear stress acting on the element at the boundary must be directed tangent to the boundary. The shear-stress distribution across line AB would then be directed as shown in Fig. 7–11d. Due to the inclination of the shear stresses at the boundaries, the maximum shear stress will occur at points A and B. Specific values for the shear stress must be obtained using the principles of the theory of elasticity. Note that we can, however, apply the shear formula to obtain the shear stress acting across each of the colored lines in Fig. 7–11a. These lines intersect the boundary of the cross section at *right angles,* and so the transverse shear stress at the boundary points will *not* have components normal to the boundary, Fig. 7–11e.

To summarize the above points, the shear formula should *not* be applied to members having cross sections that are *short or flat,* or at points where the cross section suddenly changes. Nor should it be applied across a section that intersects the boundary of the member at an angle other than 90°. Instead, for these cases the shear stress should be determined using more advanced methods based on the theory of elasticity.

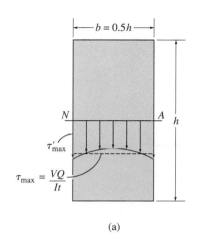

(a)

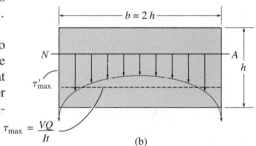

(b)

Fig. 7–10

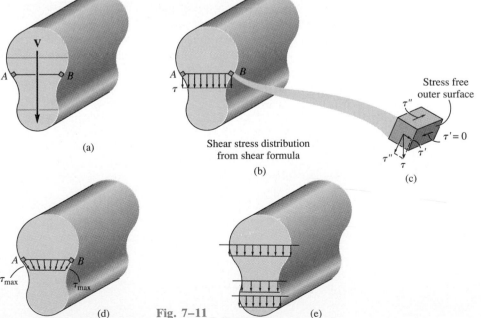

(a)

Shear stress distribution
from shear formula

(b)

Stress free
outer surface

(c)

(d)   Fig. 7–11   (e)

PROCEDURE FOR ANALYSIS

The shear formula can be used to find the shear-stress distribution acting over the cross-sectional area of a straight prismatic member made of homogeneous material that has linear-elastic behavior. It is required that the internal resultant shear force be directed along an axis of symmetry for the cross-sectional area. Also, Saint-Venant's principle requires that the shear formula be applied at points located away from any discontinuities in the cross section and away from points of concentrated loading. In order to apply the equation, the following procedure is suggested.

Internal Shear.   Section the member perpendicular to its axis at the point where the shear stress is to be determined, and use an appropriate free-body diagram and equation of equilibrium to obtain the internal shear **V** at the section.

Section Properties.   Determine the location of the centroid for the cross-sectional area in order to specify the position of the neutral axis. Then compute the moment of inertia I of the *entire area* about the neutral axis. Section the cross-sectional area through the point where the shear stress is to be determined, and measure the width t of the area at this section. The portion of the area lying either above or below this section is A'. Compute Q either by integration, $Q = \int_{A'} y \, dA$, or by using $Q = \bar{y}' A'$. Here $\bar{y}'$ is the distance to the centroid of A', measured from the neutral axis. As noted in the derivation of the shear formula, it may be helpful to realize that A' is the portion of the member's cross-sectional area that is being "held onto the member" by the longitudinal shear stresses, Fig. 7–6c.

Shear Stress.   Using a consistent set of units, substitute the data into the shear formula and compute the shear stress τ.

It is suggested that the proper direction of the transverse shear stress τ be established on a volume element of material located at the point where it is computed. This can be done by realizing that τ acts on the cross section in the same direction as **V.** From this, corresponding shear stresses acting on the other three planes of the element can then be established.

Example 7–1

The beam shown in Fig. 7–12*a* is made of wood and is subjected to a resultant internal vertical shear force of $V = 3$ kip. *(a)* Determine the shear stress in the beam at point *P*, and *(b)* compute the maximum shear stress in the beam.

SOLUTION

Part (a)

Section Properties.   The moment of inertia of the cross-sectional area computed about the neutral axis is

$$I = \frac{1}{12}bh^3 = \frac{1}{12}\,(4 \text{ in.})(5 \text{ in.})^3 = 41.7 \text{ in}^4$$

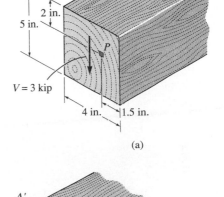

(a)

A horizontal section line is drawn through point *P* and the partial area *A′* is shown shaded in Fig. 7–12*b*. Hence

$$Q = \bar{y}'A' = \left[0.5 \text{ in.} + \frac{1}{2}\,(2 \text{ in.})\right](2 \text{ in.})(4 \text{ in.}) = 12 \text{ in}^3$$

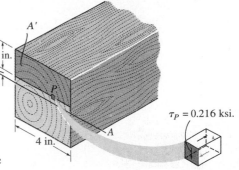

(b)   (c)

Shear Stress.   The shear force at the section is $V = 3$ kip. Applying the shear formula, we have

$$\tau_P = \frac{VQ}{It} = \frac{(3 \text{ kip})(12 \text{ in}^3)}{(41.7 \text{ in}^4)(4 \text{ in.})} = 0.216 \text{ ksi} \qquad \textit{Ans.}$$

Since τ_P contributes to *V*, it acts downward at *P* on the cross section. Consequently, a volume element of the material at this point would have shear stresses acting on it as shown in Fig. 7–12*c*.

Part (b)

Section Properties.   Maximum shear stress occurs at the neutral axis, since *t* is constant throughout the cross section and *Q* is largest for this case. For the shaded area *A′* in Fig. 7–12*d*, we have

$$Q = \bar{y}'A' = \left[\frac{2.5 \text{ in.}}{2}\right](4 \text{ in.})(2.5 \text{ in.}) = 12.5 \text{ in}^3$$

Shear Stress.   Applying the shear formula yields

$$\tau_{max} = \frac{VQ}{It} = \frac{(3 \text{ kip})(12.5 \text{ in}^3)}{(41.7 \text{ in}^4)(4 \text{ in.})} = 0.225 \text{ ksi} \qquad \textit{Ans.}$$

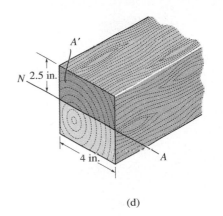

(d)

Fig. 7–12

Example 7–2

A steel wide-flange beam has the dimensions shown in Fig. 7–13a. If it is subjected to a shear of $V = 80$ kN, (a) plot the shear-stress distribution acting over the beam's cross-sectional area, and (b) determine the shear force resisted by the web.

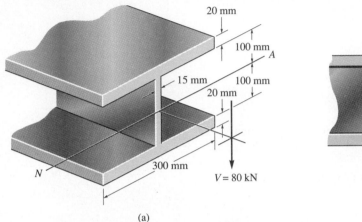

20 mm

100 mm

A

15 mm · 100 mm

20 mm

300 mm

N

$V = 80$ kN

(a)

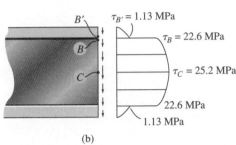

B' $\quad \tau_{B'} = 1.13$ MPa

$\tau_B = 22.6$ MPa

B

C $\quad \tau_C = 25.2$ MPa

22.6 MPa

1.13 MPa

(b)

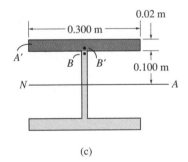

0.02 m

0.300 m

A'

B B' $\quad$ 0.100 m

N $\quad\quad\quad\quad\quad\quad$ A

(c)

Fig. 7–13

SOLUTION

Part (a). The shear-stress distribution will be parabolic and varies in the manner shown in Fig. 7–13b. Due to symmetry, only the shear stresses at points B', B, and C have to be computed. To show how these values are obtained, we must first determine the moment of inertia of the cross-sectional area about the neutral axis. Working in meters, we have

$$I = \left[\frac{1}{12} (0.015)(0.200)^3 \right]$$

$$+ 2\left[\frac{1}{12} (0.300)(0.02)^3 + (0.300)(0.02)(0.110)^2 \right]$$

$$= 155.6(10^{-6}) \ m^4$$

For point B', $t_{B'} = 0.300$ m, and A' is the shaded area shown in Fig. 7–13c. Thus,

$$Q_{B'} = \bar{y}'A' = [0.110](0.300)(0.02) = 0.660(10^{-3}) \ m^3$$

so that

$$\tau_{B'} = \frac{VQ_{B'}}{It_{B'}} = \frac{80 \ kN(0.660(10^{-3}) \ m^3)}{155.6(10^{-6}) \ m^4(0.300 \ m)} = 1.13 \ MPa$$

For point B, $t_B = 0.015$ m and $Q_B = Q_{B'}$, Fig. 7–13c. Hence

$$\tau_B = \frac{VQ_B}{It_B} = \frac{80 \ kN(0.660(10^{-3}) \ m^3)}{155.6(10^{-6}) \ m^4(0.015 \ m)} = 22.6 \ MPa$$

Note from the discussion of "Limitations on the Use of the Shear Formula" that the calculated value for both $\tau_{B'}$ and τ_B will actually be very misleading. Why?

For point C, $t_C = 0.015$ m and A' is the shaded area shown in Fig. 7–13d. Considering this area to be composed of two rectangles, we have

$$Q_C = \Sigma \bar{y}'A' = [0.110](0.300)(0.02) + [0.05](0.015)(0.100)$$
$$= 0.735(10^{-3}) \text{ m}^3$$

Thus,

$$\tau_C = \tau_{max} = \frac{VQ_C}{It_C} = \frac{80 \text{ kN}[0.735(10^{-3}) \text{ m}^3]}{155.6(10^{-6}) \text{ m}^4(0.015 \text{ m})} = 25.2 \text{ MPa}$$

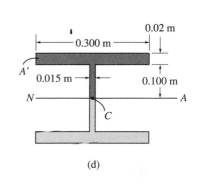

(d)

Part (b).   The shear force in the web will be determined by first computing the shear force in each flange and then subtracting this result from $V = 80$ kN. To obtain the shear force in a flange, we must first determine the shear stress at the *arbitrary* location y, Fig. 7–13e. Using units of meters, we have

$$I = 155.6(10^{-6}) \text{ m}^4$$
$$t = 0.300 \text{ m}$$
$$A' = (0.300)(0.120 - y) \text{ m}^2$$
$$\bar{y}' = y + \frac{1}{2}(0.120 - y) = \frac{1}{2}(0.120 + y) \text{ m}$$
$$Q = \bar{y}'A' = [0.150]((0.120)^2 - y^2) \text{ m}^3$$

so that

$$\tau = \frac{VQ}{It} = \frac{80 \text{ kN}[0.150]((0.120)^2 - y^2) \text{ m}^3}{(155.6(10^{-6}) \text{ m}^4)(0.300 \text{ m})}$$
$$= 257((0.120)^2 - y^2) \text{ MPa}$$

This stress acts on the area strip $dA = 0.300 \ dy$ shown in Fig. 7–13e, and therefore the shear force resisted by the top flange is

$$V_f = \int_{A_f} \tau \ dA = \int_{0.100}^{0.12} 257(10^6)((0.120)^2 - y^2)0.300 \ dy = 3.496 \text{ kN}$$

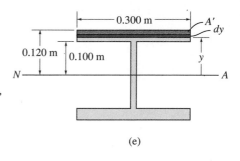

(e)

By symmetry, this force also acts in the bottom flange. Thus the shear force in the web is

$$V_w = V - 2V_f = 80 \text{ kN} - 2(3.496 \text{ kN}) = 73.0 \text{ kN} \qquad Ans.$$

By comparison, the web supports 91% of the total shear (80 kN), whereas the flanges support the remaining 9%.

Example 7–3

The beam shown in Fig. 7–14a is made from two boards. Determine the shear stress in the glue necessary to hold the boards together at point D. The supports at B and C exert only vertical reactions on the beam.

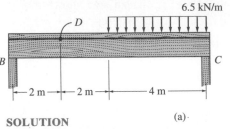
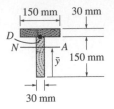

(a)

SOLUTION

Internal Shear. The support reactions on the beam are computed as shown in Fig. 7–14b. From the left segment it is seen that the resultant internal loading at a section through D consists of a shear of $V = 6.5$ kN and a moment of $M = 13$ kN · m.

Section Properties. The centroid and therefore the neutral axis will be determined from the reference axis placed at the bottom of the cross-sectional area, Fig. 7–14a. Working in units of meters, we have

$$\bar{y} = \frac{\Sigma \bar{y}A}{\Sigma A} = \frac{[0.075](0.150)(0.030) + [0.165](0.030)(0.150)}{(0.150)(0.030) + (0.030)(0.150)} = 0.120 \text{ m}$$

The moment of inertia, computed about the neutral axis, Fig. 7–14a, is therefore

$$I = \left[\frac{1}{12} (0.030)(0.150)^3 + (0.150)(0.030)(0.120 - 0.075)^2 \right]$$
$$+ \left[\frac{1}{12} (0.150)(0.030)^3 + (0.030)(0.150)(0.165 - 0.120)^2 \right]$$
$$= 27.0(10^{-6}) \text{ m}^4$$

The top board (flange) is being held onto the bottom board (web) by the glue, which is applied over the thickness $t = 0.03$ m. Consequently A' is defined as the area of the top board, Fig. 7–14a. We have

$$Q = \bar{y}'A' = [0.180 - 0.015 - 0.120](0.03)(0.150) = 0.2025(10^{-3}) \text{ m}$$

Shear Stress. Using the above data and applying the shear formula yields

$$\tau_D = \frac{VQ}{It} = \frac{6.5 \text{ kN}(0.2025(10^{-3}) \text{ m})}{27.0(10^{-6}) \text{ m}^4(0.030 \text{ m})} = 1.62 \text{ MPa} \qquad Ans.$$

The shear stress acting at the top of the bottom board is shown in Fig. 7–14c. Note that it is the glue's resistance to this lateral or *horizontal shear* stress that is necessary to hold the boards from slipping.

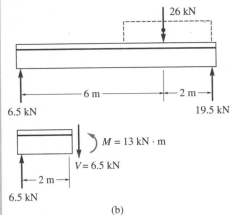

(b)

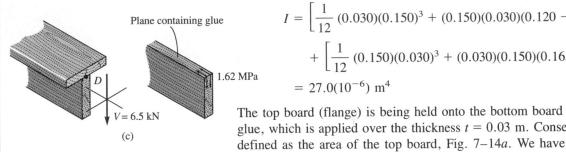

(c)

Fig. 7–14

PROBLEMS

7–1. If the beam is subjected to a shear of $V = 15$ kN, determine the web's shear stress at A and B. Indicate the shear-stress components on a volume element located at these points.

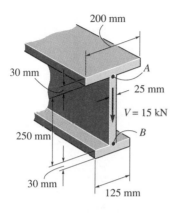

Prob. 7–1

7–3. If the wide-flange beam is subjected to a shear of $V = 25$ kip, determine (a) the maximum shear stress in the beam and (b) the average shear stress in the web.

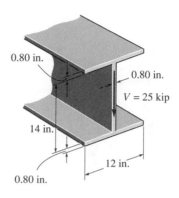

Prob. 7–3

7–2. The beam is made from three boards glued together at the seams A and B. If it is subjected to the loading shown, determine the shear stress developed in the glued joints at section a–a. The supports at C and D exert only vertical reactions on the beam.

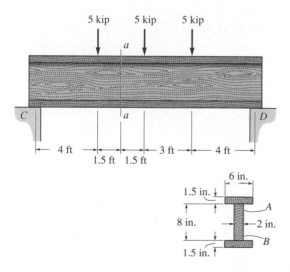

Prob. 7–2

*7–4. Railroad ties must be designed to resist large shear loadings. If the tie is subjected to the 34-kip rail loadings and an assumed uniformly distributed ground reaction, determine the intensity w for equilibrium, and compute the maximum shear stress in the tie at section a–a, which is located just to the left of the rail.

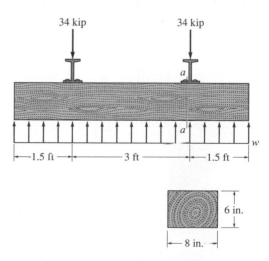

Prob. 7–4

7–5. If the T-beam is subjected to a vertical shear of $V =$ 10 kip, determine the maximum shear stress in the beam. Also, compute the shear-stress jump at the flange–web junction AB. Sketch the variation of the shear-stress intensity over the entire cross section.

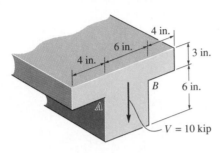

Prob. 7–5

7–6. Determine the maximum shear stress in the strut if it is subjected to a shear force of $V = 15$ kN.

7–7. Determine the maximum shear force V that the strut can support if the allowable shear stress for the material is $\tau_{allow} = 50$ MPa.

*7–8. Determine the intensity of the shear stress distributed over the cross section of the strut if it is subjected to a shear force of $V = 12$ kN.

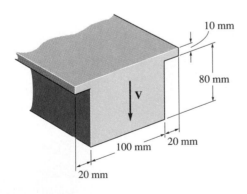

Probs. 7–6/7–7/7–8

7–9. Determine the shear stress at points B and C on the web of the beam located at section a–a.

7–10. Determine the maximum shear stress acting at section a–a in the beam.

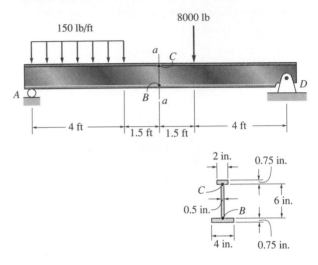

Probs. 7–9/7–10

7–11. Determine the shear stress at point B on the web of the cantilevered strut at section a–a.

*7–12. Determine the maximum shear stress acting at section a–a of the cantilevered strut.

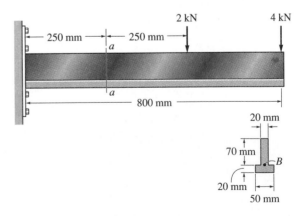

Probs. 7–11/7–12

7–13. Sketch the intensity of the shear-stress distribution acting over the beam's cross-sectional area, and determine the resultant shear force acting on the segment AB. The shear acting at the section is $V = 35$ kip.

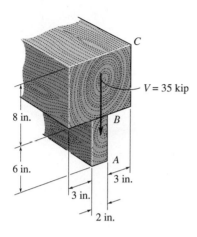

Prob. 7–13

7–15. The strut is subjected to a vertical shear of $V = 130$ kN. Plot the intensity of the shear-stress distribution acting over the cross-sectional area, and compute the resultant shear force developed in the vertical segment AB.

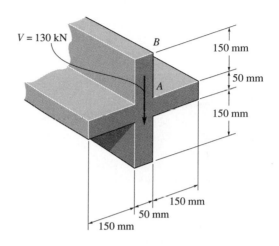

Prob. 7–15

7–14. The T-beam is subjected to a shear force of $V = 150$ kN. Determine the amount of this force that is supported by the web B.

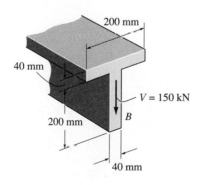

Prob. 7–14

***7–16.** The steel rod has a radius of 1.25 in. If it is subjected to a shear of $V = 5$ kip, determine the maximum shear stress.

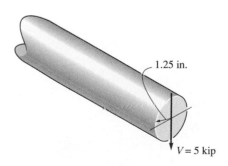

Prob. 7–16

7–17. Show that the maximum shear stress in the shaft, which has a cross section of radius r and area $A = \pi r^2$, is $\tau_{max} = 4V/3A$.

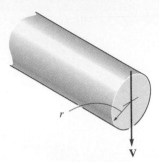

Prob. 7–17

7–18. Determine the largest shear force V that the member can sustain if the allowable shear stress is $\tau_{allow} = 8$ ksi.

7–19. If the applied shear force $V = 18$ kip, determine the maximum shear stress in the member.

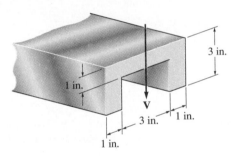

Probs. 7–18/7–19

*7–20. If the pipe is subjected to a shear of $V = 15$ kip, determine the maximum shear stress in the pipe.

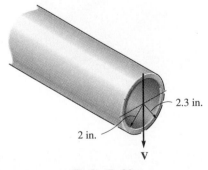

Prob. 7–20

7–21. Develop an expression for the average shear stress acting on the horizontal plane through the strut, located a distance y from the neutral axis.

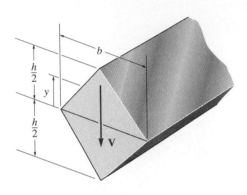

Prob. 7–21

7–22. The beam has a square cross section and is made of wood having an allowable shear stress of $\tau_{allow} = 1.4$ ksi. If it is subjected to a shear of $V = 1.5$ kip, determine the smallest dimension a of its sides.

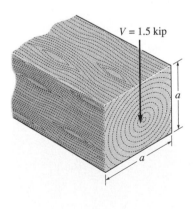

Prob. 7–22

7–23. The supports at A and B exert vertical reactions on the wood beam. If the distributed load $w = 4$ kip/ft, determine the maximum shear stress in the beam at section a–a.

*7–24. The supports at A and B exert vertical reactions on the wood beam. If the allowable shear stress is $\tau_{allow} = 400$ psi, determine the intensity w of the largest distributed load that can be applied to the beam.

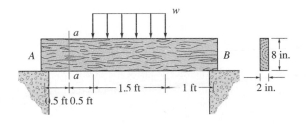

Probs. 7–23/7–24

7–25. The composite beam is made from two types of boards. The web is pine having an $E_p = 2000$ ksi, and the flanges are oak having an $E_o = 1700$ ksi. Use the method of Sec. 6.4 and compute the maximum shear stress in the beam when it is subjected to a vertical shear of $V = 14$ kip.

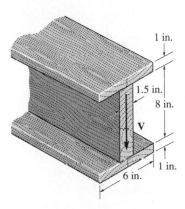

Prob. 7–25

7–26. The composite beam is constructed from wood and reinforced with a steel strap. Use the method of Sec. 6.4 and compute the maximum shear stress in the beam when it is subjected to a vertical shear of $V = 50$ kN. Take $E_{st} = 200$ GPa, $E_w = 15$ GPa.

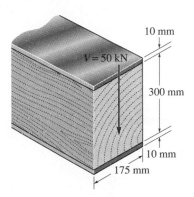

Prob. 7–26

7–27. The beam has a rectangular cross section and is subjected to a load P that is just large enough to develop a fully plastic moment $M_p = PL$ at the fixed support. If the material is clastic-plastic, then at a distance $x < L$ the moment $M = Px$ creates a region of plastic yielding with an associated elastic core having a height $2\,y'$. This situation has been described by Eq. 6–24 and the moment **M** is distributed over the cross section as shown in Fig. 6–33e. Prove that the maximum shear stress developed in the beam is given by $\tau_{max} = (3/2)(P/A')$, where $A' = 2\,y'b$, the cross-sectional area of the elastic core.

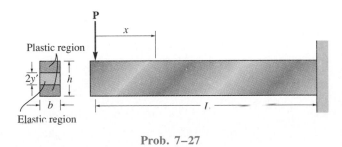

Prob. 7–27

*7–28. The beam in Fig. 6–33f is subjected to a fully plastic moment **M**$_p$. Prove that the longitudinal and transverse shear stresses in the beam are zero. *Hint:* Consider an element of the beam as shown in Fig. 7–6c.

7.5 Shear Flow in Built-up Members

Fig. 7–15

Occasionally in engineering practice members are "built up" from several composite parts in order to achieve a greater resistance to loads. Some examples are shown in Fig. 7–15. If the loads cause the members to bend, fasteners such as nails, bolts, welding material, or glue, may be needed to keep the component parts from sliding relative to one another, Fig. 7–2. In order to design these fasteners it is necessary to know the shear force to be resisted by the fastener along the member's *length*. This loading, when measured as a force per unit length, is referred to as the *shear flow q*.*

The magnitude of the shear flow along any longitudinal section of a beam can be obtained using a development similar to that for finding the shear stress in the beam. To show this, we will consider finding the shear flow along the juncture where the composite part in Fig. 7–16a is connected to the flange of the beam. As shown in Fig. 7–16b, three horizontal forces must act on this part. Two of these forces, F and $F + dF$, are developed by normal stresses caused by the moments M and $M + dM$, respectively. The third force, which for equilibrium equals dF, acts at the juncture and is to be supported by the fastener. Realizing that dF is the result of dM, then, as in the case of the shear formula, Eq. 7–3, we have

$$dF = \frac{dM}{I} \int_{A'} y \, dA'$$

The integral represents Q, that is, the moment of the colored area A' in Fig. 7–16b about the neutral axis for the cross section. Since the segment has a length dx, the shear flow, or force per unit length along the beam, is $q = dF/dx$. Hence dividing both sides by dx and noting that $V = dM/dx$, Eq. 7–2, we can write

$$q = \frac{VQ}{I} \tag{7–8}$$

Here

q = the shear flow, measured as a force per unit length along the beam

V = the internal resultant shear force, determined from the method of sections and the equations of equilibrium

I = the moment of inertia of the *entire* cross-sectional area computed about the neutral axis

$Q = \int_{A'} y \, dA = \bar{y}'A'$, where A' is the cross-sectional area of the segment that is connected to the beam at the juncture where the shear flow is to be calculated, and $\bar{y}'$ is the distance from the neutral axis to the centroid of A'

*The use of the word "flow" in this terminology will become meaningful as it pertains to the discussion in Sec. 7.6.

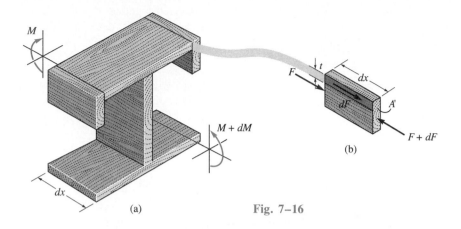

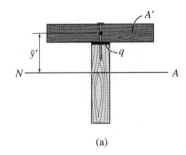

(b)

Fig. 7–16

As previously stated, Eq 7–8 is often used in engineering practice to determine the fastener spacing, glue capacity, or weld size needed to hold the composite parts of a "built-up" member together when the member is subjected to shear caused by bending. Application follows the same "procedure for analysis" as outlined in Sec. 7.4 for the shear formula. In this regard it is very important to identify correctly the proper value for Q when computing the shear flow at a particular junction on the cross section. A proper understanding of this can always be realized if one recalls how the shear-flow equation was derived in reference to Fig. 7–16. A few examples should also serve to illustrate some of these points. Consider the beam cross sections shown in Fig. 7–17. The shaded composite parts are connected to the beam by fasteners such that the necessary shear flow q in Eq. 7–8 is determined by using a value of Q computed from A' and $\bar{y}'$ as indicated in each figure. Notice that this value of q will be resisted by a single fastener in Fig. 7–17a and 7–17b, by two fasteners in Fig. 7–17c, and by three fasteners in Fig. 7–17d. Hence, multiple fasteners will support an equal fraction of the shear flow as indicated in the figure.

The following examples illustrate these concepts numerically.

(a)

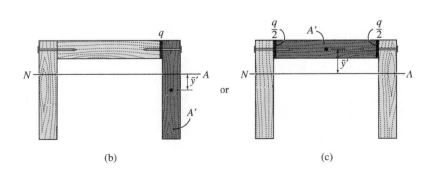

Fig. 7–17

Example 7-4

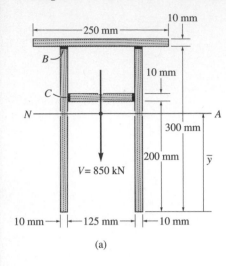

10 mm

250 mm

B

10 mm

C

N ———————————————— A

300 mm

200 mm

$\bar{y}$

V = 850 kN

10 mm →| |← 125 mm →| |← 10 mm

(a)

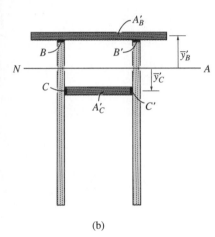

A'_B

B

B'

$\bar{y}'_B$

N ———————————————— A

$\bar{y}'_C$

C

A'_C

C'

(b)

Fig. 7-18

The beam is constructed from four boards glued together as shown in Fig. 7–18a. If it is subjected to a shear of $V = 850$ kN, determine the shear flow at B and C that must be resisted by the glue.

SOLUTION

Section Properties. The neutral axis (centroid) will be located from the bottom of the beam, Fig. 7–18a. Working in units of millimeters, we have

$$\bar{y} = \frac{\Sigma \bar{y}A}{\Sigma A} = \frac{2[150](300)(10) + [205](125)(10) + [305](250)(10)}{2(300)(10) + 125(10) + 250(10)}$$

$$= 196.8 \text{ mm}$$

The moment of inertia computed about the neutral axis is thus

$$I = 2\left[\frac{1}{12}(10)(300)^3 + (10)(300)(196.8 - 150)^2\right]$$

$$+ \left[\frac{1}{12}(125)(10)^3 + (125)(10)(205 - 196.8)^2\right]$$

$$+ \left[\frac{1}{12}(250)(10)^3 + (250)(10)(305 - 196.8)^2\right]$$

$$= 87.52(10^6) \text{ mm}^4 = 87.52(10^{-6}) \text{ m}^4$$

Since the glue at B and B' holds the top board to the beam, Fig. 7–18b, we have

$$Q_B = \bar{y}'_B A'_B = [305 - 196.8](250)(10) = 270.5(10^3) \text{ mm}^3$$

$$= 0.270(10^{-3}) \text{ m}^3$$

Likewise, the glue at C and C' holds the inner board to the beam, Fig. 7–18b, and so

$$Q_C = \bar{y}'_C A'_C = [205 - 196.8](125)(10) = 10.25(10^3) \text{ mm}^3$$

$$= 0.01025(10^{-3}) \text{ m}^3$$

Shear Flow. For B and B' we have

$$q'_B = \frac{VQ_B}{I} = \frac{850 \text{ kN}(0.270(10^{-3}) \text{ m}^3)}{87.52(10^{-6}) \text{ m}^4} = 2.62 \text{ MN/m}$$

And for C and C',

$$q'_C = \frac{VQ_C}{I} = \frac{850 \text{ kN}(0.01025(10^{-3}) \text{ m}^3)}{87.52(10^{-6}) \text{ m}^4} = 0.0995 \text{ MN/m}$$

Since *two seams* are used to secure each board, the glue per meter length of beam at each seam must be strong enough to resist *one-half* of each calculated value of q'. Thus,

$$q_B = 1.32 \text{ MN/m} \quad \text{and} \quad q_C = 0.0498 \text{ MN/m} \quad \textit{Ans.}$$

Example 7–5

A box beam is to be constructed from four boards nailed together as shown in Fig. 7–19a. If each nail can support a shear force of 30 lb, determine the maximum spacing s of nails at B and at C so that the beam will support the vertical force of 80 lb.

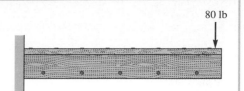

SOLUTION

Internal Shear. If the beam is sectioned at an *arbitrary point* along its length, Fig. 7–19b, the internal shear required for equilibrium is $V = 80$ lb.*

Section Properties. The moment of inertia of the cross-sectional area about the neutral axis can be determined by considering a 7.5-in $\times$ 7.5-in. square minus a 4.5-in. $\times$ 4.5-in. square.

$$I = \frac{1}{12}(7.5)(7.5)^3 - \frac{1}{12}(4.5)(4.5)^3 = 229.5 \text{ in}^4$$

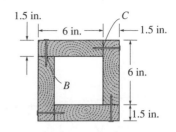

(a)

The shear flow at B is determined using a Q_B defined by the darker shaded area shown in Fig. 7–19c. It is this "symmetric" portion of the beam that is to be "held" onto the rest of the beam by nails on the left side and by the fibers of the board on the right side. Thus,

$$Q_B = \bar{y}'A' = [3](7.5)(1.5) = 33.75 \text{ in}^3$$

Likewise, the shear flow at C can be determined using the "symmetric" shaded area shown in Fig. 7–19d. We have

$$Q_C = \bar{y}'A' = [3](4.5)(1.5) = 20.25 \text{ in}^3$$

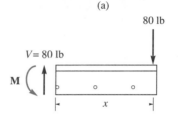

(b)

Shear Flow

$$q_B = \frac{VQ_B}{I} = \frac{80 \text{ lb}(33.75 \text{ in}^3)}{229.5 \text{ in}^4} = 11.76 \text{ lb/in.}$$

$$q_C = \frac{VQ_C}{I} = \frac{80 \text{ lb}(20.25 \text{ in}^3)}{229.5 \text{ in}^4} = 7.059 \text{ lb/in.}$$

These values represent the shear force per unit length of the beam that must be resisted by the nails at B and the fibers at B', Fig. 7–19c, and the nails at C and the fibers at C', Fig. 7–19d, respectively. Since in each case the shear flow is resisted at *two* surfaces and each nail can resist 30 lb, for B the spacing is

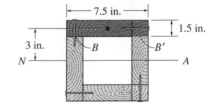

(c)

$$s_B = \frac{30 \text{ lb}}{(11.76/2) \text{ lb/in.}} = 5.10 \text{ in.} \qquad \text{Use } s_B = 5 \text{ in.} \qquad \textit{Ans.}$$

And for C,

$$s_C = \frac{30 \text{ lb}}{(7.059/2) \text{ lb/in.}} = 8.50 \text{ in.} \qquad \text{Use } s_C = 8.5 \text{ in.} \qquad \textit{Ans.}$$

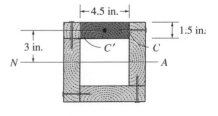

(d)

*The magnitude of the moment **M** depends on the location x of the section. Its value, however, is not needed here.

Fig. 7–19

Example 7–6

Nails having a total shear strength of 40 lb are used in a beam that can be constructed either as in Case I or as in Case II, Fig. 7–20. Determine the largest vertical shear in each case based on a nail spacing of 9 in.

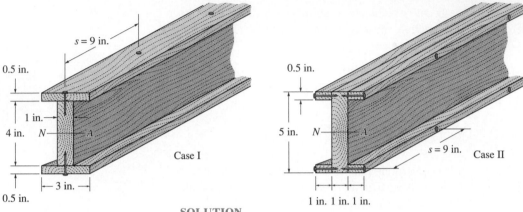

Fig. 7–20

SOLUTION

Since the geometry is the same in both cases, the moment of inertia about the neutral axis is

$$I = \frac{1}{12}\,(3\text{ in.})(5\text{ in.})^3 - 2\left[\frac{1}{12}\,(1\text{ in.})(4\text{ in.})^3\right] = 20.58\text{ in}^4$$

Case I. For this design a single row of nails holds the top or bottom flange onto the web. For one of these flanges,

$$Q = \bar{y}'A' = [2.25\text{ in.}](3\text{ in.}(0.5\text{ in.})) = 3.375\text{ in}^3$$

so that

$$q = \frac{VQ}{I}$$

$$\frac{40\text{ lb}}{9\text{ in.}} = \frac{V(3.375\text{ in}^3)}{20.58\text{ in}^4}$$

$$V = 27.1\text{ lb} \qquad\qquad Ans.$$

Case II. Here a single row of nails holds one of the side boards onto the web. Thus,

$$Q = \bar{y}'A' = [2.25\text{ in.}](1\text{ in.}(0.5\text{ in.})) = 1.125\text{ in}^3$$

$$q = \frac{VQ}{I}$$

$$\frac{40\text{ lb}}{9\text{ in.}} = \frac{V(1.125\text{ in}^3)}{20.58\text{ in}^4}$$

$$V = 81.3\text{ lb} \qquad\qquad Ans.$$

PROBLEMS

7–29. The beam is subjected to a shear of $V = 800$ kN. Determine the average shear stress developed in the nails along the sides A and B if the nails are spaced $s = 100$ mm apart. Each nail has a diameter of 2 mm.

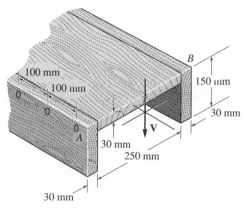

Prob. 7–29

7–30. The beam is constructed from three boards. If it is subjected to a shear of $V = 5$ kip, determine the spacing s of the nails used to hold the top and bottom flanges to the web. Each nail can support a shear force of 500 lb.

7–31. The beam is constructed from three boards. Determine the maximum shear V that it can sustain if the allowable shear stress for the wood is $\tau_{allow} = 400$ psi. What is the required spacing s of the nails if each nail can resist a shear force of 400 lb?

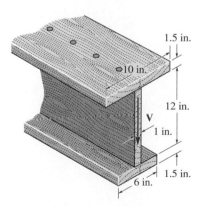

Probs. 7–30/7–31

*7–32. The beam is constructed from two boards fastened together at the top and bottom with two rows of nails spaced every 6 in. If each nail can support a 500-lb shear force, determine the maximum shear force V that can be applied to the beam.

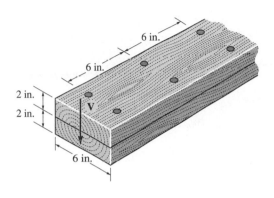

Prob. 7–32

7–33. The beam is fabricated from four boards nailed together as shown. Determine the shear force each nail along the sides C and the top D must resist if the nails are uniformly spaced at $s = 3$ in. Also, compute the maximum shear stress developed in the beam's web AB. The beam is subjected to a shear of $V = 4.5$ kip.

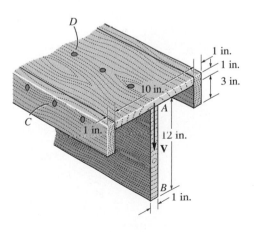

Prob. 7–33

7–34. The beam is made from four boards nailed together as shown. If the nails can each support a shear force of 100 lb, determine their required spacings s' and s if the beam is subjected to a shear of $V = 700$ lb.

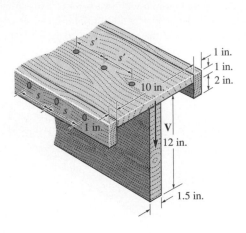

Prob. 7–34

7–35. The beam is fabricated from two equivalent structural tees and two plates; each plate has a width of 6 in. and a thickness of 0.5 in. If a shear of $V = 50$ kip is applied to the cross section, determine the maximum spacing of the bolts. Each bolt can resist a shear force of 15 kip.

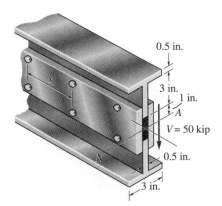

Prob. 7–35

*7–36. A beam is constructed from three boards bolted together as shown. Determine the shear force developed in each bolt if the bolts are spaced $s = 250$ mm apart and the applied shear is $V = 35$ kN.

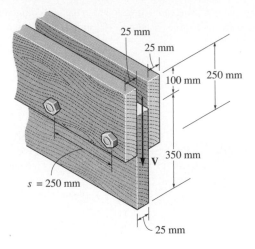

Prob. 7–36

7–37. The beam is constructed from four boards glued together at their seams. Based only on the allowable shear stress for the glue $\tau_{allow} = 150$ lb/in^2, what is the maximum vertical shear V that the beam can support *(a)* if it is used in the position shown, *(b)* if it is rotated 90°?

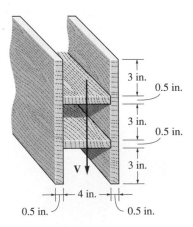

Prob. 7–37

7.6 Shear Flow in Thin-Walled Members

In the previous section we developed the shear-flow equation, $q = VQ/I$, and showed how it can be used to determine the shear flow acting along any longitudinal plane of a member. In this section we will show how to apply this equation to find the shear-flow *distribution* throughout a member's cross-sectional area. Here we will assume that the member has *thin walls*, that is, the wall thickness is small compared with the height or width of the member. As will be shown in the next section, this analysis has important applications in structural and mechanical design.

Before we determine the shear-flow distribution on the cross section, we will first show how the shear flow is related to the shear stress. To do this, consider the differential slice of the wide-flange beam in Fig. 7–21a. A free-body diagram of a segment on the flange is shown in Fig. 7–21b. The force dF is developed along the color-shaded longitudinal section in order to balance the normal force F and $F + dF$ created by the moments M and $M + dM$, respectively. Since the segment has a length dx, then the shear flow along the section is $q = dF/dx$. Because the flange wall is *thin*, the shear stress τ will not vary much across the thickness t of the section; instead we can assume that it is *constant*. Hence, $dF = \tau dA = \tau(t\ dx)$. Since $dF = q\ dx$, we have

$$q = \tau t \qquad (7\text{–}9)$$

This same result can also be determined by comparing the shear-flow equation, $q = VQ/I$, with the shear formula, $\tau = VQ/It$.

Since the shear stress acts on both longitudinal and transverse planes, so does the shear flow. For example, if the corner element at point B in Fig. 7–21b is removed, Fig. 7–21c, the shear flow acts as shown on the colored face of the element—which is also shown on the cross section of the beam in Fig. 7–21g. Notice that in this discussion we have *neglected* the vertical transverse component of shear flow. As shown in Fig. 7–21d, this component, like the shear stress, is approximately zero throughout the thickness of the element since the walls are assumed to be thin and the top and bottom surfaces of the element are free of stress. To summarize, then, only the shear-flow component that acts *parallel* to the walls of the member will be considered.

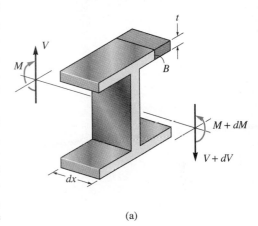

(a)

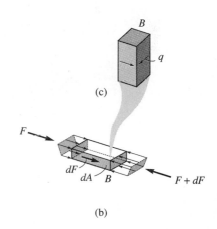

(c)

(b)

Fig. 7–21

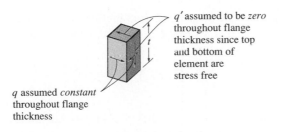

q' assumed to be *zero* throughout flange thickness since top and bottom of element are stress free

q assumed *constant* throughout flange thickness

(d)

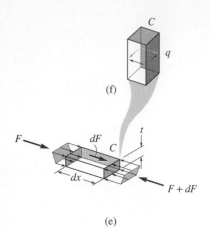

(f)

(e)

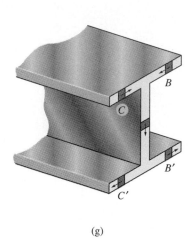

(g)

Fig. 7–21 (e,g)

By a similar analysis, the isolation of the left-hand segment on the top flange, Fig. 7–21e, will establish the correct direction of the shear flow on the corner element C of the segment, Fig. 7–21f and 7–21g. Using this method, show that the shear flow at corresponding points B' and C' on the bottom flange is directed as shown in Fig. 7–21g.

This example illustrates how the *direction* of the shear flow is established at any point on a beam's cross section. Using the shear-flow formula, $q = VQ/I$, we will now show how to determine the distribution of the shear flow throughout the cross section. It is to be expected that this formula will give reasonable results for the shear flow since, as stated in Sec. 7.4, the accuracy of this equation improves for members having thin rectangular cross sections. For any application, however, the shear force **V** must act along an axis of symmetry or principal centroidal axis of inertia for the cross section.

We will begin by finding the distribution of shear flow along the top right flange of the wide-flange beam in Fig. 7–22a. To do this, consider the shear flow q, acting on the colored element, located an arbitrary distance x from the centerline of the cross section, Fig. 7–22b. It is determined using Eq. 7–8 with $Q = \bar{y}'A' = [d/2](b/2 - x)t$. Thus,

$$q = \frac{VQ}{I} = \frac{V[d/2]((b/2) - x)t}{I} = \frac{Vtd}{2I}\left(\frac{b}{2} - x\right) \qquad (7\text{–}10)$$

By inspection, this distribution is *linear,* varying from $q = 0$ at $x = b/2$ to $(q_{max})_f = Vtdb/4I$ at $x = 0$. (The limitation of $x = 0$ is possible here since the member is assumed to have "thin walls" and so the thickness of the web is neglected.) Due to symmetry, a similar analysis yields the same distribution of shear flow for the other flanges. Also, the shear flow acts in the direction established in Fig. 7–21g, and therefore the results are as shown in Fig. 7–22d.

The total force developed in each portion of a flange can be determined by integration. Since the force on the colored element in Fig. 7–22b is $dF = q\,dx$, then

$$F_f = \int q\,dx = \int_0^{b/2} \frac{Vtd}{2I}\left(\frac{b}{2} - x\right) dx = \frac{Vtdb^2}{16I}$$

Also, we can determine this result by computing the area under the triangle in Fig. 7–22d since q is a distribution of force/length. Hence,

$$F_f = \frac{1}{2}\,(q_{max})_f\left(\frac{b}{2}\right) = \frac{Vtdb^2}{16I}$$

All four of these flange forces are shown in Fig. 7–22e, and we can see from the direction of these forces that horizontal force equilibrium of the cross section is maintained.

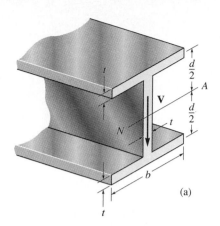

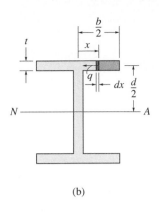

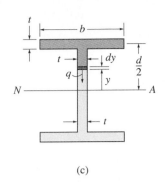

(a)

(b)

(c)

A similar analysis can be performed for the web, Fig. 7–22c. Here we have $Q = \Sigma \bar{y}'A' = [d/2](bt) + [y + (1/2)(d/2 - y)]t(d/2 - y) = btd/2 + (t/2)(d^2/4 - y^2)$, so that

$$q = \frac{VQ}{I} = \frac{Vt}{I}\left[\frac{db}{2} + \frac{1}{2}\left(\frac{d^2}{4} - y^2\right)\right] \qquad (7\text{–}11)$$

For the web, the shear flow, like the shear stress, varies in a *parabolic* manner, from $q = 2(q_{max})_f = Vtdb/2I$ at $y = d/2$ to a maximum of $q = (q_{max})_w = (Vtd/I)(b/2 + d/8)$ at $y = 0$, Fig. 7–22d.

In order to determine the force in the web, F_w, we must integrate Eq. 7–11, that is,

$$F_w = \int q\,dy = \int_{-d/2}^{d/2} \frac{Vt}{I}\left[\frac{db}{2} + \frac{1}{2}\left(\frac{d^2}{4} - y^2\right)\right]dy$$

$$= \frac{Vt}{I}\left[\frac{db}{2}y + \frac{1}{2}\left(\frac{d^2}{4}y - \frac{1}{3}y^3\right)\right]\Bigg|_{-d/2}^{d/2}$$

$$= \frac{Vtd^2}{4I}\left(2b + \frac{1}{3}d\right)$$

Simplification is possible by noting that the moment of inertia for the cross-sectional area is

$$I = 2\left[\frac{1}{12}bt^3 + bt\left(\frac{d}{2}\right)^2\right] + \frac{1}{12}td^3$$

Neglecting the first term, since the thickness of each flange is small, we get

$$I = \frac{td^2}{4}\left(2b + \frac{1}{3}d\right)$$

Substituting into the above equation, we see that $F_w = V$, which is to be expected, Fig. 7–22e.

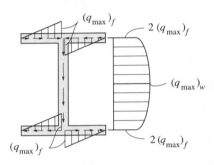

Shear flow distribution

(d)

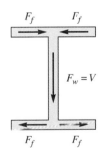

(e)

Fig. 7–22

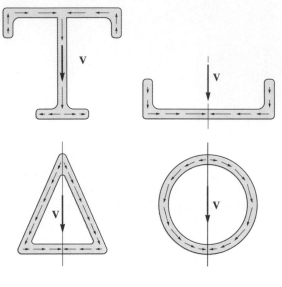

Shear flow q

Fig. 7–23

From the above analysis three important points should be observed. First, the value of q changes over the cross section, since Q will be different for each area segment A' for which it is computed. In particular, q will vary *linearly* along elements (flanges) that are *perpendicular* to the direction of **V,** and *parabolically* along elements (web) that are *inclined or parallel* to **V.** Second, q will *always act parallel to the walls* of the member, since the section on which q is computed is taken perpendicular to the walls. And third, the *directional sense* of q is such that the shear appears to *"flow"* through the cross section, *inward* at the beam's top flange, *"combining"* and then *"flow-ing"* *downward* through the web, since it must contribute to the shear force **V,** and then separating and *"flowing"* *outward* at the bottom flange. If one is able to *"visualize"* this *"flow"* it will provide an easy means for establishing not only the direction of q, but *also* the corresponding direction of τ. Other examples of how q is directed along the segments of thin-walled members are shown in Fig. 7–23. In all cases, symmetry prevails about an axis that is collinear with **V,** and as a result, q *"flows"* in a direction such that it will provide the necessary vertical force components equivalent to **V** and yet also satisfy the horizontal force equilibrium requirements for the cross section.

The following examples numerically illustrate these concepts.

Example 7–7

The thin-walled box beam in Fig. 7–24a is subjected to a shear of 10 kip. Determine the variation of the shear flow throughout the cross section.

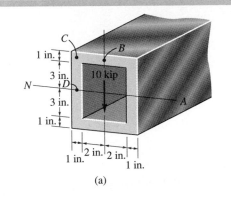

(a)

SOLUTION

By symmetry, the neutral axis passes through the center of the cross section. The moment of inertia is

$$I = \frac{1}{12} \text{ (6 in.)(8 in.)}^3 - \frac{1}{12} \text{ (4 in.)(6 in.)}^3 = 184 \text{ in}^3$$

Only the shear flow at points B, C, and D has to be determined. For point B, the area $A' = 0$, since it can be thought of as being located entirely at point B. Alternatively, A' can represent the entire cross-sectional area, in which case $Q_B = \bar{y}'A' = 0$ since $\bar{y}' = 0$. Because $Q_B = 0$, then

$$q_B = 0$$

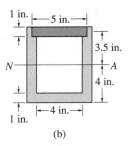

(b)

For point C, the area A' is shown dark shaded in Fig. 7–24b. Here we have used the mean dimensions since point C is on the centerline of each thickness. We have

$$Q_C = \bar{y}'A' = (3.5 \text{ in.})(5 \text{ in.})(1 \text{ in.}) = 17.5 \text{ in}^3$$

Thus,

$$q_C = \frac{VQ_C}{I} = \frac{10 \text{ kip}(17.5 \text{ in}^3)}{184 \text{ in}^3} = 0.951 \text{ kip/in.}$$

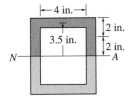

(c)

The shear flow at D is computed using the three dark-shaded rectangles shown in Fig. 7–24c. We have

$$Q_D = \Sigma \bar{y}'A' = 2[2 \text{ in.}](1 \text{ in.})(4 \text{ in.}) + [3.5 \text{ in.}](4 \text{ in.})(1 \text{ in.}) = 30 \text{ in}^3$$

So that

$$q_D = \frac{VQ_D}{I} = \frac{10 \text{ kip}(30 \text{ in}^3)}{184 \text{ in}^4} = 1.63 \text{ kip/in.}$$

Using these results, and the symmetry of the cross section, the shear-flow distribution is plotted in Fig. 7–24d. As expected the distribution is linear along the horizontal walls (perpendicular to $\mathbf{V}$) and parabolic along the vertical walls (parallel to $\mathbf{V}$).

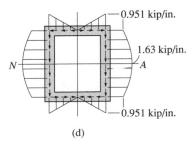

(d)

Fig. 7–24

Example 7–8

The channel shown in Fig. 7–25a is subjected to a shear of 20 kN. Determine the shear flow at points B, C, and D, and plot the shear-flow distribution throughout the cross section. Also, compute the resultant force in each region of the cross section.

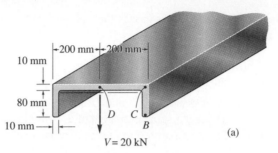

(a)

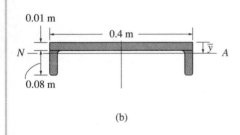

(b)

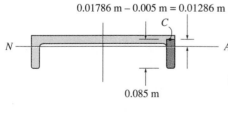

(c)

Fig. 7–25

SOLUTION

First we will locate the neutral axis and determine the moment of inertia of the cross-sectional area. To do this the cross section will be divided into three rectangles, Fig. 7–25b. Working in units of meters, the location of the neutral axis, measured from the top, is

$$\bar{y} = \frac{\Sigma \bar{y}A}{\Sigma A} = \frac{[0.005](0.4)(0.01) + 2[[0.05](0.08)(0.01)]}{0.4(0.01) + 2(0.08)(0.01)} = 0.01786 \text{ m}$$

The moment of inertia about the neutral axis is then

$$I = \left[\frac{1}{12} (0.4)(0.01)^3 + (0.4)(0.01)(0.01786 - 0.005)^2 \right]$$

$$+ 2 \left[\frac{1}{12} (0.01)(0.08)^3 + (0.08)(0.01)(0.05 - 0.01786)^2 \right]$$

$$= 3.20(10^{-6}) \text{ m}^4$$

Since $A' = 0$ for point B, Fig. 7–24a, then $Q_B = 0$, and therefore

$$q_B = 0 \qquad \textit{Ans.}$$

To determine the shear flow at C, which is located on the *centerline* of the wall, we use the shaded area A' shown in Fig. 7–25c. Thus,

$$Q_C = \bar{y}'_C A'_C = [0.085/2 - 0.01286](0.085)(0.01)$$
$$= 25.19(10^{-6}) \text{ m}^3$$

so that

$$q_C = \frac{V Q_C}{I} = \frac{20 \text{ kN}(25.19(10^{-6}) \text{ m}^3)}{3.20(10^{-6}) \text{ m}^4} = 157 \text{ kN/m} \qquad \textit{Ans.}$$

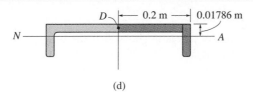

(d)

For point D, the dark-shaded area A' in Fig. 7–25d will be considered as the sum of two rectangles. To compute Q_D, a *negative* value for $\tilde{y}'$ is assigned to the rectangle standing vertical since its centroid is *below* the neutral axis, and a *positive* value for $\tilde{y}'$ is assigned to the other rectangle since $\tilde{y}'$ for its centroid is *above* this axis. Thus,

$$Q'_D = \Sigma\tilde{y}'A' = -[0.09/2 - 0.01786](0.09)(0.01)$$
$$+ [0.01786 - 0.005](0.19)(0.01) = 0$$

Therefore,

$$q_D = \frac{VQ_D}{I} = \frac{20\text{ kN}(0)}{3.20(10^{-6})\text{ m}^4} = 0 \qquad \textit{Ans.}$$

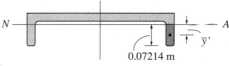

(e)

In accordance with the direction of **V**, the shear flow throughout the cross section is shown in Fig. 7–25e. As discussed previously, the intensity of q varies *linearly* along each *horizontal segment,* then varies *parabolically* along each *vertical segment*. The maximum shear flow is at the neutral axis. It can be determined using the dark-shaded area A' shown in Fig. 7–25f.

$$Q = \tilde{y}'A' = \left(\frac{0.07214}{2}\right)(0.07214)(0.01) = 26.02(10^{-6})\text{ m}^3$$

Hence

$$q_{\max} = \frac{VQ}{I} = \frac{20\text{ kN}(26.02(10^{-6})\text{ m}^3)}{3.20(10^{-6})\text{ m}^4} = 163\text{ kN/m}$$

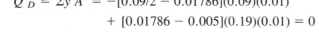

(f)

For equilibrium of the cross section, the force in each vertical segment must be

$$F_v = \frac{V}{2} = 10\text{ kN} \qquad \textit{Ans.}$$

The forces in each horizontal segment can be computed by integration, or in a more direct manner, by finding the area under the triangular distribution for q, Fig. 7–25e. We have

$$F_h = \frac{1}{2}(0.195\text{ m})(157\text{ kN/m}) = 15.4\text{ kN} \qquad \textit{Ans.}$$

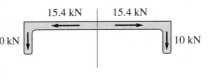

(g)

These results are shown in Fig. 7–25g.

*7.7 Shear Center

In the previous section it was assumed that the internal shear **V** was applied along a principal centroidal axis of inertia that *also* represents an *axis of symmetry* for the cross section. In this section we will consider the effect of applying the shear along a principal centroidal axis that is *not* an axis of symmetry. As before, only thin-walled members will be analyzed, so the dimensions to the centerline of the walls of the members will be used. A typical example of this case is the channel section shown in Fig. 7–26a, which is cantilevered from a fixed support and is subjected to the force **P.** If this force is applied along the once vertical, unsymmetrical axis that passes through the *centroid C* of the cross-sectional area, the channel will not only bend downward, *it will also twist* clockwise as shown. In order to understand why this occurs, it is necessary to study the shear-flow distribution along the channel's flanges and web.

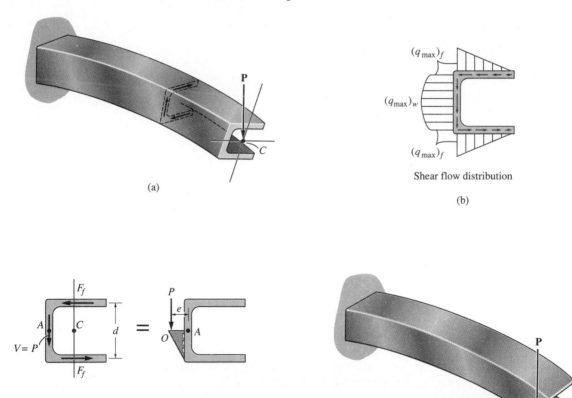

(a)

(b)

Shear flow distribution

(c)   (d)   (e)

Fig. 7–26

Like the case of the wide-flange beam discussed in Sec. 7.6, the shear-flow distribution in each *flange is linear* and in the *web it is parabolic*, Fig. 7–26b. When this distribution is integrated over the flange and web areas, it will give resultant forces of F_f in each flange and a force of $V = P$ in the web, Fig. 7–26c. If the moments of these forces are summed about point A, it can be seen that the couple or torque created by the flange forces is responsible for causing the member to twist. The actual twist is clockwise when viewed from the front of the beam as shown in Fig. 7–26a, since *reactive* internal "equilibrium" forces F_f cause the twisting. In order to *prevent* this twisting it is therefore necessary to apply **P** at a point O located a distance e from the web of the channel, Fig. 7–26d. We require $\Sigma M_A = F_f d = Pe$, or

$$e = \frac{F_f d}{P}$$

Using the method discussed in Sec. 7.6, F_f can be evaluated in terms of $P\ (= V)$ and the dimensions of the flanges and web. Once this is done, then P will cancel upon substitution into the above equation, and it becomes possible to express e simply as a function of the cross-sectional geometry and *not* as a function of P or its location along the length of the beam (see Example 7–9). The point O so located is called the *shear center* or *flexural center*. When **P** is applied at the shear center, the beam will bend *without* twisting as shown in Fig. 7–26e. Design handbooks often list the location of this point for a variety of beams having thin-walled cross sections that are commonly used in practice.

When doing this analysis it should be noted that the shear center will *always lie on an axis of symmetry* of a member's cross-sectional area. For example, if the channel in Fig. 7–26a is rotated 90° and **P** is applied at A, Fig. 7–27a, no twisting will occur since the shear flow in the web and flanges for this case is *symmetrical*, and therefore the force resultants in these elements will create zero moments about A, Fig. 7–27b. (Also see Example 7–8.) Obviously, if a member has a cross section with *two* axes of symmetry, as in the case of a wide-flange beam, the shear center will then coincide with the intersection of these axes (the centroid).

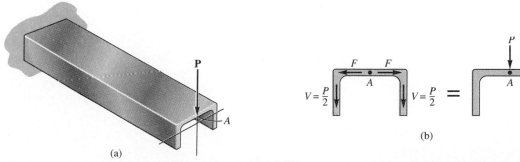

(a)

(b)

Fig. 7–27

PROCEDURE FOR ANALYSIS

The location of the shear center for a thin-walled member for which the internal shear is applied in the *same direction* as a principal centroidal axis for the cross section may be determined by using the following procedure.

Shear-Flow Resultants.   Determine the direction as to how the shear "flows" through the various segments of the cross section, and sketch the force resultants on each segment of the cross section. (For example, see Fig. 7–26*c*.) Since the shear center is determined by taking the moments of these force resultants about a point (*A*), choose this point at a location that eliminates the moments of as many force resultants as possible. The magnitudes of the force resultants that do create a moment about *A* must then be calculated. For any segment this is done by determining the shear flow *q* at an arbitrary point on the segment and then integrating *q* along the segment's length. Recall that **V** will create a *linear* variation of shear flow in segments that are *perpendicular* to **V,** and a *parabolic* variation of shear flow in segments that are *parallel or inclined* to **V.** (See Fig. 7–26*b*.)

Shear Center.   Sum the moments of the shear-flow resultants about point *A* and set this moment equal to the moment of **V** about *A*. By solving this equation one is able to determine the moment-arm distance *e,* which locates the line of action of **V** from *A*. If an *axis of symmetry* for the cross section exists, the shear center lies at the point where this axis intersects the line of action of **V.** If, however, no axes of symmetry exist, rotate the cross section 90° and repeat the process to obtain another line of action for **V.** The shear center then lies at the point of intersection of the two 90° lines of action for **V.**

The following examples illustrate this procedure.

Example 7–9

Determine the location of the shear center for the thin-walled channel section having the dimensions shown in Fig. 7–28a.

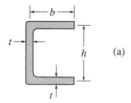

(a)

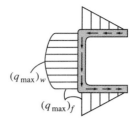

$(q_{max})_w$

$(q_{max})_f$

Shear flow distribution

(b)

SOLUTION

Shear-Flow Resultants. A vertical downward shear $\mathbf{V}$ applied to the section causes the shear to flow through the flanges and web as shown in Fig. 7–28b. This causes force resultants F_f and V in the flanges and web as shown in Fig. 7–28c. We will take moments about point A so that only the force F_f on the lower flange has to be determined.

The cross-sectional area can be divided into three component rectangles—a web and two flanges. Since each component is assumed to be thin, the moment of inertia of the area about the neutral axis is

$$I = \frac{1}{12}th^3 + 2\left[bt\left(\frac{h}{2}\right)^2\right] = \frac{th^2}{2}\left(\frac{h}{6} + b\right)$$

From Fig. 7–25d, q at the arbitrary position x is

$$q = \frac{VQ}{I} = \frac{V(h/2)[b - x]t}{(th^2/2)[(h/6) + b]} = \frac{V(b - x)}{h[(h/6) + b]}$$

Hence, the force F_f is

$$F_f = \int_0^b q\, dx = \frac{V}{h[(h/6) + b]} \int_0^b (b - x)\, dx = \frac{Vb^2}{2h[(h/6) + b]}$$

Try to obtain this same result by first finding $(q_{max})_f$, Fig. 7–28b, then determining the triangular area $\frac{1}{2}b(q_{max})_f = F_f$.

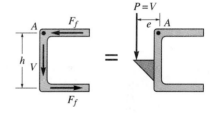

$P = V$

e

A

(c)

Shear Center. Summing moments about point A, Fig. 7–28c, we require

$$Ve = F_f h = \frac{Vb^2 h}{2h[(h/6) + b]}$$

Thus

$$e = \frac{b^2}{[(h/3) + 2b]} \qquad Ans.$$

As stated previously, note that e does not depend on the magnitude of $\mathbf{V}$; rather it depends only on the cross-sectional dimensions of the channel.

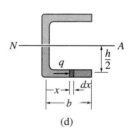

N

A

q

$\frac{h}{2}$

x

dx

b

(d)

Fig. 7–28

Example 7–10

Determine the location of the shear center for the angle having equal legs, Fig. 7–29a. Also, calculate the internal shear force resultant in each leg.

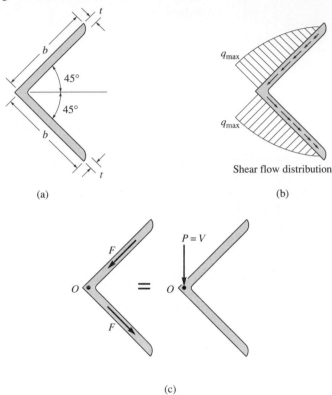

(a)

(b)

Shear flow distribution

(c)

Fig. 7–29

SOLUTION

When a vertical downward shear **V** is applied at the section, the shear flow and shear-flow resultants are directed as shown in Fig. 7–29b and 7–29c, respectively. Note that the force F in each leg must be equal, since for equilibrium the sum of their horizontal components must be equal to zero. Also, since the lines of action of both forces intersect point O, this point *must be the shear center* since the sum of the moments of these forces and **P** about O is zero, Fig. 7–29c.

The magnitude of **F** can be determined by first finding the shear flow at the arbitrary locations s along the top leg, Fig. 7–29d. Here

$$Q = \bar{y}'A' = \frac{1}{\sqrt{2}}\left((b-s)+\frac{s}{2}\right)ts = \frac{1}{\sqrt{2}}\left(b-\frac{s}{2}\right)st$$

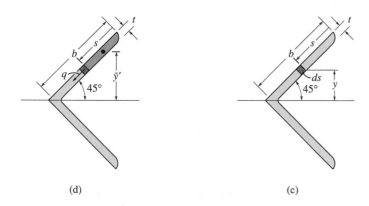

(d)   (c)

The moment of inertia of the angle, computed about the neutral axis, must be determined from "first principles," since the legs are inclined with respect to the neutral axis. For the area element $dA = t\,ds$, Fig. 7–29e, we have

$$I = \int_A y^2\,dA = 2\int_0^b \left[\frac{1}{\sqrt{2}}(b-s)\right]^2 t\,ds = t\left(b^2 s - bs^2 + \frac{1}{3}s^3\right)\Bigg|_0^b = \frac{tb^3}{3}$$

Thus, the shear flow is

$$q = \frac{VQ}{I} = \frac{V}{(tb^3/3)}\left[\frac{1}{\sqrt{2}}\left(b - \frac{s}{2}\right)st\right]$$

$$= \frac{3V}{\sqrt{2}b^3}\,s\left(b - \frac{s}{2}\right)$$

The variation of q is parabolic, which reaches a maximum value when $s = b$ as shown in Fig. 7–29b. The force F is therefore

$$F = \int_0^b q\,ds = \frac{3V}{\sqrt{2}b^3}\int_0^b s\left(b - \frac{s}{2}\right)ds$$

$$= \frac{3V}{\sqrt{2}b^3}\left(b\frac{s^2}{2} - \frac{1}{6}s^3\right)\Bigg|_0^b$$

$$= \frac{1}{\sqrt{2}}V \qquad\qquad\qquad Ans.$$

This result can be verified since the sum of the vertical components of the force F in each leg must equal V and, as stated previously, the sum of the horizontal components equals zero.

PROBLEMS

7–38.   The H-beam is subjected to a shear of $V = 80$ kN. Determine the shear flow at point A.

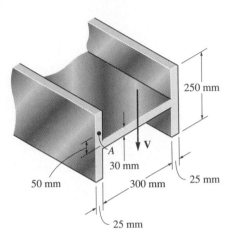

50 mm   30 mm   300 mm   25 mm   25 mm   250 mm

Prob. 7–38

*7–40.   The channel is subjected to a shear of $V = 75$ kN. Determine the shear flow developed at point A and the maximum shear flow in the channel.

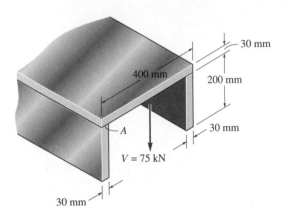

400 mm   30 mm   200 mm   30 mm A $V = 75$ kN   30 mm

Prob. 7–40

7–39.   The beam is made from three thin plates welded together as shown. If it is subjected to a shear of $V = 48$ kN, determine the shear flow at points A and B. Also, calculate the maximum shear stress in the beam.

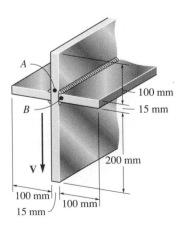

A B 100 mm   15 mm   200 mm V 100 mm   100 mm   15 mm

Prob. 7–39

7–41.   The box girder is subjected to a shear of $V = 15$ kN. Determine (a) the shear flow developed at point B and (b) the maximum shear flow in the girder's web AB.

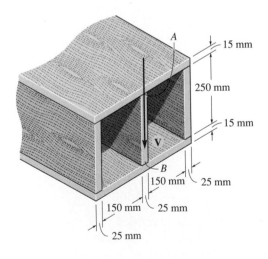

A 15 mm   250 mm   15 mm V B 150 mm   25 mm   150 mm   25 mm   25 mm

Prob. 7–41

7–42. The member is subjected to a shear force of $V = 2$ kN. Determine the shear flow at points A, B, and C. The thickness of each thin-walled segment is 15 mm.

7–43. The member is subjected to a shear force of $V = 2$ kN. Determine the maximum shear flow in the member. All segments of the cross section are 15 mm thick.

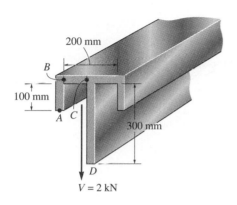

Probs. 7–42/7–43

7–45. The assembly is subjected to a vertical shear of $V = 7$ kip. Determine the shear flow at points A, B, and C and the maximum shear flow in the cross section.

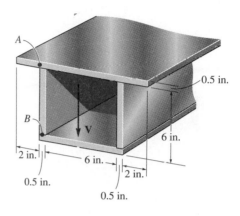

Prob. 7–45

*7–44. An aluminum strut is 10 mm thick and has the cross section shown. If it is subjected to a shear of $V = 150$ N, determine the shear flow at points A and B. Neglect the size of the curve.

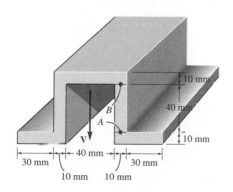

Prob. 7–44

7–46. A steel plate having a thickness of 3 mm is formed into the thin-walled section shown. If it is subjected to a shear force of $V = 50$ N, determine the shear flow at points A and B and the maximum shear flow in the strut.

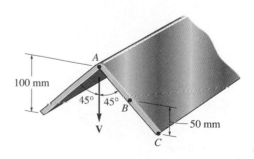

Prob. 7–46

7–47. The pipe is subjected to a shear force of V = 8 kip. Determine the shear flow in the pipe at points A and B.

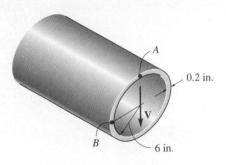

Prob. 7–47

7–49. Determine the shear-stress variation over the cross section of the thin-walled tube as a function of elevation y and show that $\tau_{max} = 2V/A$, where $A = 2\pi rt$. *Hint:* Choose a differential area element $dA = Rt\, d\theta$. Using $dQ = y\, dA$, formulate Q for a circular section from θ to $(\pi - \theta)$ and show that $Q = 2R^2 t \cos \theta$, where $\cos \theta = \sqrt{R^2 - y^2}/R$.

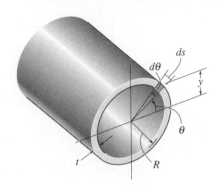

Prob. 7–49

*7–48. The stiffened beam is constructed from plates having a thickness of 0.25 in. If it is subjected to a shear force of V = 8 kip, determine the shear-flow distribution in segments AB and CD of the beam. What is the resultant shear force supported by these segments? Also, sketch how the shear flow passes through the cross section. The vertical dimensions are measured to the centerline of each horizontal segment.

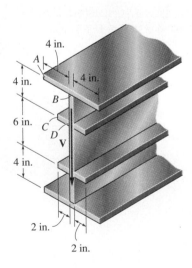

Prob. 7–48

7–50. Determine the location e of the shear center, point O, for the thin-walled member having the cross section shown. The member has a thickness of 0.25 in.

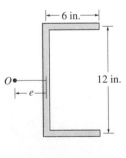

Prob. 7–50

7–51. Determine the location e of the shear center, point O, for the thin-walled member having the cross section shown where $b_2 > b_1$. The member segments have the same thickness t.

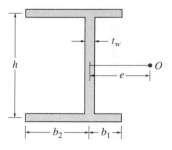

Prob. 7–51

***7–52.** Determine the placement e for the force **P** so that the beam bends downward without twisting. Take $h = 200$ mm.

7–53. A force **P** is applied to the web of the beam as shown. If $e = 250$ mm, determine the height h of the right flange so that the beam will deflect downward without twisting. The member segments have the same thickness t.

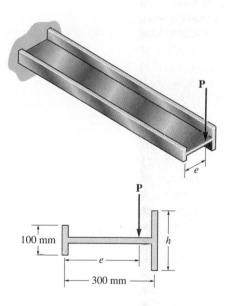

Probs. 7–52/7–53

7–54. Determine the location e of the shear center, point O, for the thin-walled member having the cross section shown. The member segments have the same thickness t.

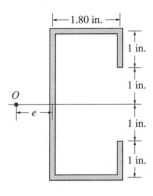

Prob. 7–54

7–55. Determine the location e of the shear center, point O, for the thin-walled member having the cross section shown. The member segments have the same thickness t.

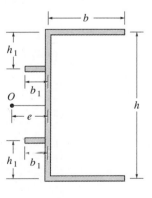

Prob. 7–55

*7–56.   Determine the location e of the shear center, point O, for the thin-walled member having the cross section shown. The member segments have the same thickness t.

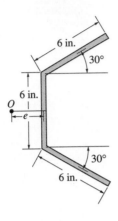

Prob. 7–56

7–58.   Determine the location e of the shear center, point O, for the thin-walled member having the cross section shown.

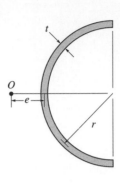

Prob. 7–58

7–57.   Determine the location e of the shear center, point O, for the thin-walled member having the cross section shown. The member segments have the same thickness t.

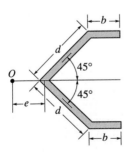

Prob. 7–57

7–59.   Determine the location e of the shear center, point O, for the thin-walled open pipe having the cross section shown.

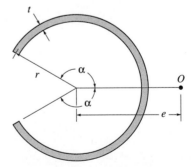

Prob. 7–59

REVIEW PROBLEMS

*7–60. The beam is fabricated from three steel plates, and it is subjected to a shear force of $V = 150$ kN. Determine the shear stress at points A, B, and C where the plates are joined.

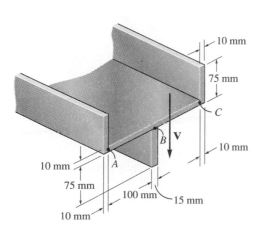

Prob. 7–60

7–61. Develop an expression for the average vertical component of shear stress acting on the horizontal plane through the shaft, located a distance y from the neutral axis.

Prob. 7–61

7–62. Determine the location e of the shear center, point O, for the thin-walled member having a slit along its side.

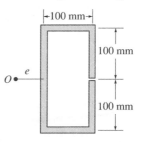

Prob. 7–62

7–63. The box beam is constructed from four boards that are fastened together using nails spaced along the beam every 2 in. If each nail can resist a shear force of 50 lb, determine the greatest shear force V that can be applied to the beam without causing failure of the nails.

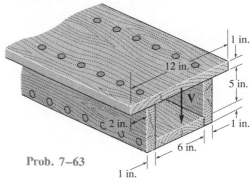

Prob. 7–63

*7–64. The beam supports a vertical shear of $V = 7$ kip. Determine the resultant force this develops in segment AB of the beam.

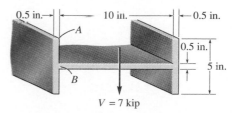

Prob. 7–64

7–65. Determine the location e of the shear center, point O, for the thin-walled member having the cross section shown. The member segments have the same thickness t.

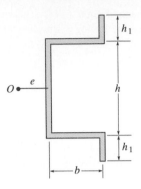

Prob. 7–65

7–66. The member is subjected to a shear force of $V = 10$ kip. Sketch the shear-flow distribution along the vertical plate AB. Indicate numerical values of all peaks.

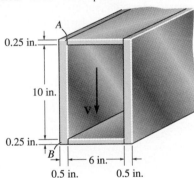

Prob. 7–66

7–67. Determine the location e of the shear center, point O, for the thin-walled member having the cross section shown. The member segments have the same thickness t.

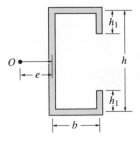

Prob. 7–67

*7–68. A steel plate having a thickness of 0.25 in. is formed into the thin-walled section shown. If it is subjected to a shear force of $V = 250$ lb, determine the shear stress at points A, B, and C. Indicate the results on volume elements located at these points.

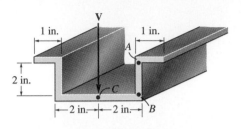

Prob. 7–68

7–69. The member consists of two triangular plastic strips bonded together along AB. If the glue can support an allowable shear stress of $\tau_{allow} = 600$ psi, determine the maximum vertical shear V that can be applied to the member based on the strength of the glue.

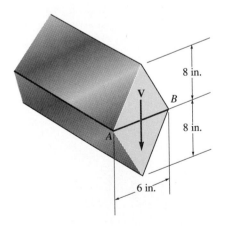

Prob. 7–69

8 Combined Loadings

This chapter serves as a review of the stress analysis that has been developed in the previous chapters regarding axial load, torsion, bending, and shear. We will discuss the solution of problems where several of these internal loads occur simultaneously on a member's cross section. Before doing this, however, the chapter begins with an analysis of stress developed in thin-walled pressure vessels.

8.1 Thin-Walled Pressure Vessels

Cylindrical or spherical vessels are commonly used in industry to serve as boilers or tanks. When under pressure, the material of which they are made is subjected to a loading from all directions. Although this is the case, the vessel can be analyzed in a simple manner provided it has a thin wall. In general, *"thin wall"* refers to a vessel having an inner-radius-to-wall-thickness ratio of 10 or more ($r/t \geq 10$). Specifically, when $r/t = 10$ the results of a thin-wall analysis will predict a stress that is approximately 4% less than the actual maximum stress in the vessel. For larger r/t ratios this error will be even smaller.

When the vessel wall is "thin," the stress distribution throughout its thickness will not vary significantly, and so we will assume that it is *uniform* or *constant*. Using this assumption, we will now analyze the state of stress in thin-walled cylindrical and spherical pressure vessels. In both cases the pressure in the vessel is understood to be the *gauge pressure,* since it measures the pressure *above* atmospheric pressure, which is assumed to exist both inside and outside the vessel's wall.

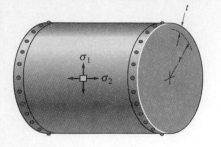

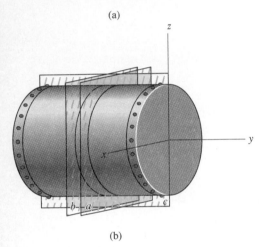

(a)

(b)

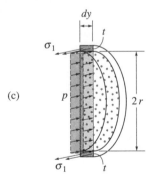

(c)

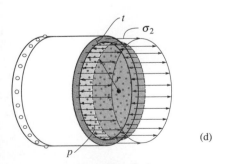

(d)

Fig. 8-1

Cylindrical Vessels. Consider the cylindrical vessel having a wall thickness t and inner radius r as shown in Fig. 8–1a. A gauge pressure p is developed within the vessel by a contained gas or fluid, which is assumed to have negligible weight. Due to the uniformity of this loading, an element of the vessel that is sufficiently removed from the end and oriented as shown in Fig. 8–1a is subjected to normal stresses σ_1 in the *circumferential or hoop direction* and σ_2 in the *longitudinal or axial direction*. Both of these stress components exert tension on the material. We wish to determine the magnitude of each of these components in terms of the vessel's geometry and the internal pressure. To do this requires using the method of sections and applying the equations of force equilibrium.

For the hoop stress, consider the vessel to be sectioned by planes a, b, and c, Fig. 8–1b. A free-body diagram of the back segment along with the contained gas or fluid is shown in Fig. 8–1c. Here only the loadings in the x direction are shown. These loadings are developed by the uniform hoop stress σ_1, acting throughout the vessel's wall, and the pressure acting on the vertical face of the sectioned gas or fluid. For equilibrium in the x direction, we require

$$\Sigma F_x = 0; \qquad 2[\sigma_1(t\, dy)] - p(2r\, dy) = 0$$

$$\boxed{\sigma_1 = \frac{pr}{t}} \qquad (8-1)$$

In order to obtain the longitudinal stress σ_2, we will consider the left portion of section b of the cylinder, Fig. 8–1b. As shown in Fig. 8–1d, σ_2 acts uniformly throughout the wall, and p acts on the section of gas or fluid. Since the mean radius is approximately equal to the vessel's inner radius, equilibrium in the y direction requires

$$\Sigma F_y = 0; \qquad \sigma_2(2\pi rt) - p(\pi r^2) = 0$$

$$\boxed{\sigma_2 = \frac{pr}{2t}} \qquad (8-2)$$

In the above equations,

σ_1, σ_2 = the normal stress in the hoop and longitudinal directions, respectively. Each is assumed to be *constant* throughout the wall of the cylinder, and each subjects the material to tension.

p = the internal gauge pressure developed by the contained gas or fluid.

r = the inner radius of the cylinder.

t = the thickness of the wall ($r/t \geq 10$).

Comparing Eqs. 8–1 and 8–2, it should be noted that the hoop or circumferential stress is twice as large as the longitudinal or axial stress. Consequently, when fabricating cylindrical pressure vessels from rolled-formed plates, the longitudinal joints must be designed to carry twice as much stress as the circumferential joints.

Spherical Vessels.   We can analyze a spherical pressure vessel in a similar manner. For example, consider the vessel to have a wall thickness t and inner radius r and to be subjected to an internal gauge pressure p, Fig. 8–2a. If the vessel is sectioned in half using section a, the resulting free-body diagram is shown in Fig. 8–2b. Like the cylinder, equilibrium in the x direction requires

$$\Sigma F_x = 0; \qquad \sigma_2(2\pi rt) - p(\pi r^2) = 0$$

$$\boxed{\sigma_2 = \frac{pr}{2t}} \qquad (8\text{–}3)$$

By comparison, this is the *same result* as that obtained for the longitudinal stress in the cylindrical pressure vessel. Furthermore, from the analysis, this stress will be the same *regardless* of the orientation of the hemispheric free-body diagram. Consequently, an element of the material is subjected to the state of stress shown in Fig. 8–2a.

The above analysis indicates that an element of material taken from either a cylindrical or a spherical pressure vessel is subjected to *biaxial stress,* i.e., normal stress existing in only two directions, Figs. 8–1a and 8–2c. Actually, the material of the vessel is also subjected to a *radial stress,* σ_3, which acts along a radial line. This stress has a maximum value equal to the pressure p at the interior wall and decreases through the wall to zero at the exterior surface of the vessel, since the gauge pressure there is zero. For thin-walled vessels, however, we will *ignore* the radial-stress component, since our limiting assumption of $r/t = 10$ results in σ_2 and σ_1 being, respectively, 5 and 10 times *higher* than the maximum radial stress, $(\sigma_3)_{\max} = p$. Lastly, realize that the above formulas apply only for vessels subjected to an internal gauge pressure. If the vessel is subjected to an external pressure, it may cause the vessel to become unstable, and collapse may occur by buckling.

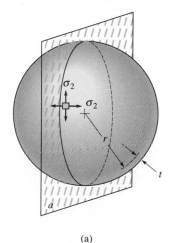

(a)

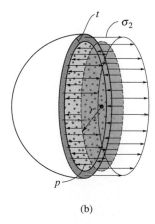

(b)

Fig. 8–2

Example 8–1

A cylindrical pressure vessel has an inner diameter of 4 ft and a thickness of $\frac{1}{2}$ in. Determine the maximum internal pressure it can sustain so that neither its circumferential nor its longitudinal stress component exceeds 20 ksi. Under the same conditions, what is the maximum internal pressure that a similar-size spherical vessel can sustain?

SOLUTION

Cylindrical Pressure Vessel. The maximum stress occurs in the circumferential direction. From Eq. 8–1 we have

$$\sigma_1 = \frac{pr}{t}; \qquad\qquad 20 \text{ kip/in}^2 = \frac{p(24 \text{ in.})}{\frac{1}{2} \text{ in.}}$$

$$p = 417 \text{ psi} \qquad\qquad\qquad \textit{Ans.}$$

Note that when this pressure is reached, from Eq. 8–2, the stress in the longitudinal direction will be $\sigma_2 = \frac{1}{2}(20 \text{ ksi}) = 10 \text{ ksi}$. Furthermore, the *maximum stress* in the *radial direction* occurs on the material at the inner wall of the vessel and is $(\sigma_3)_{max} = p = 417 \text{ psi}$. This value is 48 times smaller than the circumferential stress (20 ksi), and as stated earlier, its effects will be neglected.

Spherical Vessel. Here the maximum stress occurs in any two perpendicular directions on an element of the vessel, Fig. 8–2a. From Eq. 8–3, we have

$$\sigma_2 = \frac{pr}{2t}; \qquad\qquad 20 \text{ kip/in}^2 = \frac{p(24 \text{ in.})}{2(\frac{1}{2} \text{ in.})}$$

$$p = 833 \text{ psi} \qquad\qquad\qquad \textit{Ans.}$$

Although it is more difficult to fabricate, the spherical pressure vessel will carry twice as much internal pressure as a cylindrical vessel.

PROBLEMS

8–1. The spherical gas tank has an inner radius of $r = 1.5$ m. (If it is subjected to an internal pressure of $p = 300$ kPa, determine its required thickness if the maximum normal stress is not to exceed 12 MPa.

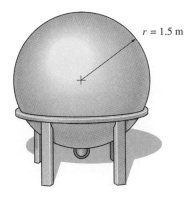

$r = 1.5$ m

Prob. 8–1

8–2. Air pressure in the cylinder is increased by exerting forces of 800 lb on the two pistons, each having a diameter of 4.875 in. If the cylinder has a wall thickness of 0.25 in., determine the state of stress in the wall of the cylinder.

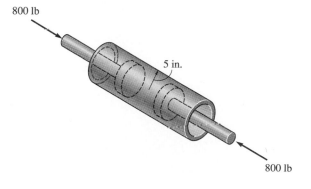

800 lb

5 in.

800 lb

Prob. 8–2

8–3. The cap on the cylindrical tank is bolted to the tank along the flanges. The tank has an inner diameter of 1.5 m and a wall thickness of 18 mm. If the largest normal stress is not to exceed 150 MPa, determine the maximum pressure the tank can sustain. Also, compute the number of bolts required to attach the cap to the tank if each bolt has a diameter of 20 mm. The allowable stress for the bolts is $(\sigma_{\text{allow}})_b = 180$ MPa.

Prob. 8–3

*8–4.** The barrel is filled to the top with water. Determine the distance s that the top hoop should be placed from the bottom hoop so that the tensile force in each hoop is the same. Also, what is the force in each hoop? The barrel has an inner diameter of 4 ft. Neglect its wall thickness. Assume that only the hoops resist the water pressure. *Note:* Water develops pressure in the barrel according to Pascal's law, $p = (62.4z)$ lb/ft^2, where z is the depth from the surface of the water in feet.

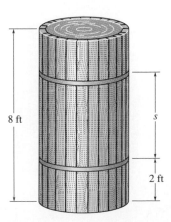

8 ft

s

2 ft

Prob. 8–4

8–5. A boiler is constructed of 8-mm steel plates that are fastened together at their ends using a butt joint consisting of two 8-mm cover plates and rivets having a diameter of 10 mm and spaced 50 mm apart as shown. If the steam pressure in the boiler is 1.35 MPa, determine (*a*) the circumferential stress in the boiler's plate apart from the seam, (*b*) the circumferential stress in the outer cover plate along the rivet line *a–a*, and (*c*) the shear stress in the rivets.

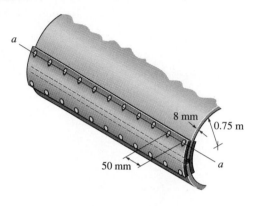

Prob. 8–5

8–6. The staves or vertical members of the wooden tank are held together using semicircular hoops having a thickness of 0.5 in. and a width of 2 in. Determine the normal stress in hoop *AB* if the tank is subjected to an internal gauge pressure of 2 psi and this loading is transmitted directly to the hoops. Also, if 0.25-in.-diameter bolts are used to connect each hoop together, determine the tensile stress in each bolt at *A* and *B*. Assume hoop *AB* supports the pressure loading within a 12-in. length of the tank as shown.

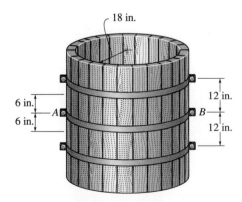

Prob. 8–6

8–7. A wood stave pipe having an inner diameter of 3 ft is bound together using steel hoops having a cross-sectional diameter of 0.5 in. If the allowable stress for the hoops is $\sigma_{allow} = 12$ ksi, determine their maximum spacing *s* along the section of pipe so that the pipe can resist an internal gauge pressure of 4 psi. Assume each hoop supports the pressure loading acting along the length *s* of the pipe.

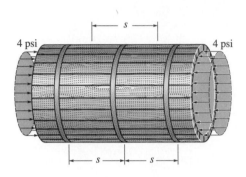

Prob. 8–7

8.2 State of Stress Caused by Combined Loading

In previous chapters we developed methods for determining the stress distributions in a member subjected to either an internal axial force, a shear force, a bending moment, or a torsional moment. Most often, however, the cross section of a member is subjected to several of these types of loadings *simultaneously*, and as a result, the method of superposition, if it applies, can be used to determine the *resultant* stress distribution caused by the loads. For application, the stress distribution due to *each loading* is first determined, and then these distributions are superimposed to determine the resultant stress distribution. As stated in Sec. 4.3, the principle of superposition can be used for this purpose provided a *linear relationship* exists between the *stress* and the *loads*. Also, the geometry of the member should *not* undergo *significant change* when the loads are applied. This is necessary in order to ensure that the stress produced by one load is not related to the stress produced by any other load. The discussion will be confined to meet these two criteria.

PROCEDURE FOR ANALYSIS

The following procedure provides a general means for establishing the normal and shear stress components at a point in a member when the member is subjected to several different types of loadings simultaneously. It will be assumed the material is homogeneous and behaves in a linear-elastic manner. Also, Saint-Venant's principle requires that the point where the stress is to be determined is far removed from any discontinuities in the cross section or points of applied load.

Internal Loadings.   Section the member perpendicular to its axis at the point where the stress is to be determined. Use the necessary free-body diagrams and equations of equilibrium to obtain the resultant internal normal and shear force components and the bending and torsional moment components. The force components should act through the *centroid* of the cross section, and the moment components should be computed about *centroidal axes,* which represent the principal axes of inertia for the cross section.

Stress Components.   Compute the stress component associated with *each* internal loading. For each case, represent the effect either as a distribution of stress acting over the entire cross-sectional area, or show the stress on an element of the material located at a specified point on the cross section.

Normal Force. The internal normal force is developed by a uniform normal-stress distribution determined from $\sigma = P/A$.

Shear Force. The internal shear force in a beam is developed by a shear-stress distribution determined from the shear formula, $\tau = VQ/It$. Special care, however, must be exercised when applying this equation, as noted in Sec. 7.4. In some cases, especially if the member is very short, so that bending can be neglected, direct shear may be applied to the member, and in this case calculation of the *average shear stress* may be justified, $\tau_{\text{avg}} = V/A$.

Bending Moment. For *straight members* the internal bending moment is developed by a normal-stress distribution that varies linearly from zero at the neutral axis to a maximum at the outer boundary of the member. The stress distribution is determined from the flexure formula, $\sigma = -My/I$. If the member is *curved,* the stress distribution is nonlinear and is determined from $\sigma = My/[Ae(R - y)]$.

Torsional Moment. For circular shafts and tubes the internal torsional moment is developed by a shear-stress distribution that varies linearly from the central axis of the shaft to a maximum at the shaft's outer boundary. The shear-stress distribution is determined from the torsion formula, $\tau = T\rho/J$. If the member is a closed thin-walled tube, use $\tau = T/2A_m t$.

Thin-Walled Pressure Vessels. If the vessel is a thin-walled cylinder, the internal pressure p will cause a biaxial state of stress in the material such that the hoop or circumferential stress component is $\sigma_1 = pr/t$ and the longitudinal stress component is $\sigma_2 = pr/2t$. If the vessel is a thin-walled sphere, then the biaxial state of stress is represented by two equivalent components, each having a magnitude of $\sigma_2 = pr/2t$.

Superposition. Once the normal and shear stress components for each loading case have been calculated, use the principle of superposition and determine the resultant normal and shear stress components. Represent the results on an element of material located at the point, or show the results as a distribution of stress acting over the member's cross-sectional area.

Problems in this section, which involve combined loadings, serve as a basic *review* of the application of many of the important stress equations mentioned above. A thorough understanding of how these equations are applied, as indicated in the previous chapters, is necessary if one is to successfully solve the problems at the end of this section. The following examples should be carefully studied before proceeding to solve the problems.

Example 8–2

A force of 150 lb is applied to the edge of the member shown in Fig. 8–3a. Neglect the weight of the member and determine the state of stress at points B and C.

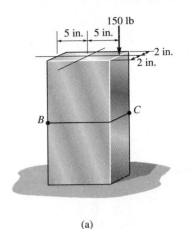

5 in. | 5 in.

150 lb

2 in.
2 in.

(a)

SOLUTION

Internal Loadings. The member is sectioned through B and C. For equilibrium at the section there must be an axial force of 150 lb acting through the centroid and a bending moment of 750 lb · in. about the centroidal or principal axis, Fig. 8–3b.

Stress Components

Normal Force. The uniform normal-stress distribution due to the normal force is shown in Fig. 8–3c. Here

$$\sigma = \frac{P}{A} = \frac{150 \text{ lb}}{(10 \text{ in.})(4 \text{ in.})} = 3.75 \text{ psi}$$

Bending Moment. The normal-stress distribution due to the bending moment is shown in Fig. 8–3d. The maximum stress is

$$\sigma_{max} = \frac{Mc}{I} = \frac{750 \text{ lb} \cdot \text{in.}(5 \text{ in.})}{[\frac{1}{12}(4 \text{ in.})(10 \text{ in.})^3]} = 11.25 \text{ psi}$$

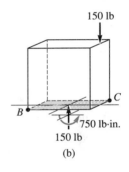

150 lb

C

B

750 lb·in.

150 lb

(b)

Superposition. If the above normal-stress distributions are added algebraically, the resultant stress distribution is shown in Fig. 8–3e. Although it is not needed here, the location of the line of zero stress can be determined by proportional triangles; i.e.,

$$\frac{7.5 \text{ psi}}{x} = \frac{15 \text{ psi}}{(10 \text{ in.} - x)}$$

$$x = 3.33 \text{ in.}$$

Elements of material at B and C are subjected only to normal or *uniaxial stress* as shown in Fig. 8–3f and 8–3g. Hence,

$$\sigma_B = 7.5 \text{ psi} \quad \text{(tension)} \qquad Ans.$$
$$\sigma_C = 15 \text{ psi} \quad \text{(compression)} \qquad Ans.$$

Fig. 8–3

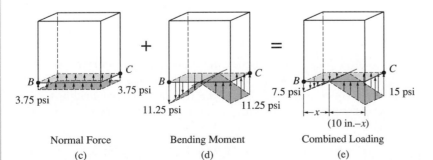

3.75 psi

3.75 psi

Normal Force

(c)

+

11.25 psi

11.25 psi

7.5 psi

Bending Moment

(d)

=

7.5 psi

15 psi

x

(10 in.–x)

Combined Loading

(e)

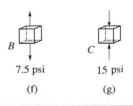

B

7.5 psi

(f)

C

15 psi

(g)

Example 8–3

The member shown in Fig. 8–4a has a rectangular cross section. Determine the state of stress that the loading produces at point C.

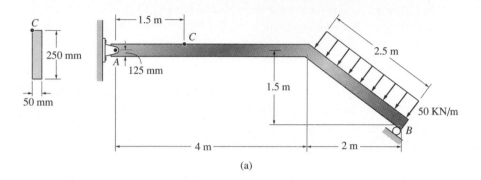

(a)

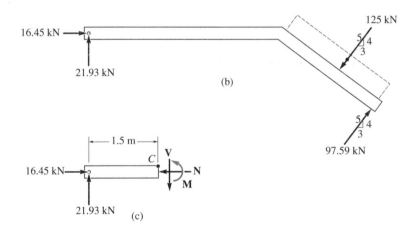

(b)

(c)

Fig. 8–4

SOLUTION

Internal Loadings. The support reactions on the member have been computed and are shown in Fig. 8–4b. If the left segment AC of the member is considered, Fig. 8–4c, the resultant internal loadings at the section consist of an axial force, a shear force, and a bending moment. Solving,

$$N = 16.45 \text{ kN} \qquad V = 21.93 \text{ kN} \qquad M = 32.89 \text{ kN} \cdot \text{m}$$

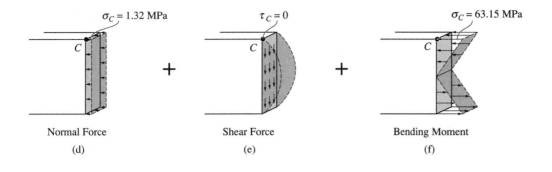

Normal Force
(d)

Shear Force
(e)

Bending Moment
(f)

$\sigma_C = 1.32$ MPa

$\tau_C = 0$

$\sigma_C = 63.15$ MPa

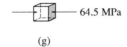

64.5 MPa

(g)

Stress Components

Normal Force.   The force is produced by a uniform normal-stress distribution over the cross section. At point C, Fig. 8–4d, it has a magnitude of

$$\sigma_C = \frac{16.45 \text{ kN}}{(0.050 \text{ m})(0.250 \text{ m})} = 1.32 \text{ MPa}$$

Shear Force.   Here the area $A' = 0$, since point C is located at the top of
the member. Thus $Q = \bar{y}'A' = 0$ and for C, Fig. 8–4e, the shear stress

$$\tau_C = 0$$

Bending Moment.   Point C is located at $y = c = 125$ mm from the neutral
axis, so the normal stress at C, Fig. 8–4f, is

$$\sigma_C = \frac{Mc}{I} = \frac{(32.89 \text{ kN} \cdot \text{m})(0.125 \text{ m})}{[\frac{1}{12}(0.050 \text{ m})(0.250)^3]} - 63.15 \text{ MPa}$$

Superposition.   The shear stress is zero. Adding the normal stresses computed above gives a compressive stress at C having a value of

$$\sigma_C = 1.32 \text{ MPa} + 63.15 \text{ MPa} = 64.5 \text{ MPa} \qquad \textit{Ans.}$$

This result, acting on an element of material, is shown in Fig. 8–4g.

Example 8–4

The solid rod shown in Fig. 8–5a has a radius of 0.75 in. If it is subjected to the loading shown, determine the state of stress at point A.

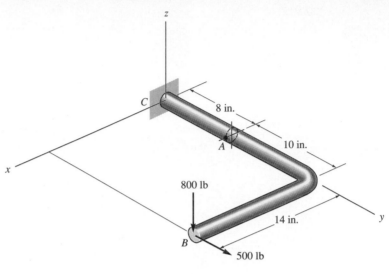

(a)

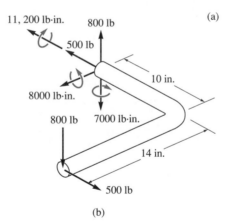

(b)

Fig. 8–5

SOLUTION

Internal Loadings. The rod is sectioned through point A. Using the free-body diagram of segment AB, Fig. 8–5b, the resultant internal loadings can be determined for equilibrium. The normal force (500 lb) and shear force (800 lb) must act through the centroid of the cross section and the bending-moment components (8000 lb · in. and 7000 lb · in.) are applied about centroidal (principal) axes. In order to better "visualize" the stress distributions due to each of these loadings, we will consider the *equal but opposite resultants* acting on segment AC of the rod, Fig. 8–5c.

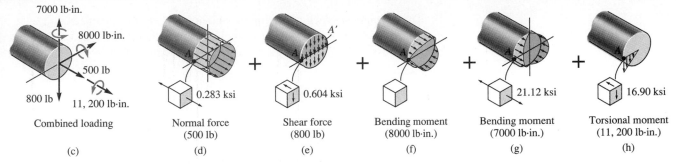

Combined loading	Normal force (500 lb)	Shear force (800 lb)	Bending moment (8000 lb·in.)	Bending moment (7000 lb·in.)	Torsional moment (11, 200 lb·in.)
(c)	(d)	(e)	(f)	(g)	(h)

Stress Components

Normal Force.   The normal-stress distribution is shown in Fig. 8–5*d*. For point *A*, we have

$$\sigma_A = \frac{P}{A} = \frac{500 \text{ lb}}{\pi (0.75 \text{ in.})^2} = 283 \text{ psi} = 0.283 \text{ ksi}$$

Shear Force.   The shear-stress distribution is shown in Fig. 8–5*e*. For point *A*, *Q* is determined from the shaded *semicircular* area. Using the table on the inside front cover, we have

$$Q = \bar{y}'A' = \frac{4(0.75 \text{ in.})}{3\pi}\left[\frac{1}{2}\pi\left(0.75 \text{ in.}\right)^2\right] = 0.2813 \text{ in}^3$$

so that

$$\tau_A = \frac{VQ}{It} = \frac{800 \text{ lb}(0.2813 \text{ in}^3)}{[\frac{1}{4}\pi(0.75 \text{ in.})^4]2(0.75 \text{ in.})} = 604 \text{ psi} = 0.604 \text{ ksi}$$

Bending Moments.   For the 8000-lb · in. component, point *A* lies on the neutral axis, Fig. 8–5*f*, so the normal stress is

$$\sigma_A = 0$$

For the 7000-lb · in. moment, *c* = 0.75 in., so the normal stress at point *A*, Fig. 8–5*g*, is

$$\sigma_A = \frac{Mc}{I} = \frac{7000 \text{ lb} \cdot \text{in.}(0.75 \text{ in.})}{[\frac{1}{4}\pi(0.75 \text{ in.})^4]} = 21{,}126 \text{ psi} = 21.12 \text{ ksi}$$

Torsional Moment.   At point *A*, $\rho_A = c = 0.75$ in., Fig. 8–5*h*. Thus the shear stress is

$$\tau_A = \frac{Tc}{J} = \frac{11{,}200 \text{ lb} \cdot \text{in.}(0.75 \text{ in.})}{[\frac{1}{2}\pi(0.75 \text{ in.})^4]} = 16{,}901 \text{ psi} = 16.90 \text{ ksi}$$

Superposition.   When the above results are superimposed, it is seen that an element of material at *A* is subjected to both normal and shear stress components, Fig. 8–5*i*.

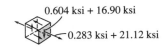

0.604 ksi + 16.90 ksi

0.283 ksi + 21.12 ksi

or

17.5 ksi

21.4 ksi

(i)

■ **Example 8–5** ■

The rectangular block of negligible weight in Fig. 8–6*a* is subjected to a vertical force of 40 kN, which is applied to its corner. Determine the normal-stress distribution acting on a section through *ABCD*.

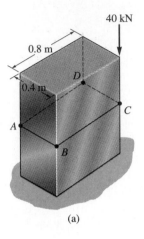

40 kN

0.8 m

D

0.4 m

A

C

B

(a)

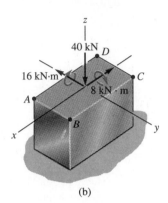

z

40 kN D

16 kN·m

8 kN·m

A

C

x

B

y

(b)

Fig. 8–6

SOLUTION

Internal Loadings.   If we consider the equilibrium of the bottom segment of the block, Fig. 8–6*b*, it is seen that the 40-kN force must act through the centroid of the cross section and *two* bending-moment components must also act about the centroidal or principal axes of inertia for the section.

Stress Components

Normal Force.   The uniform normal-stress distribution is shown in Fig. 8–6*c*. We have

$$\sigma = \frac{P}{A} = \frac{40 \text{ kN}}{(0.8 \text{ m})(0.4 \text{ m})} = 125 \text{ kPa}$$

Bending Moments.   The normal-stress distribution for the 8-kN · m moment is shown in Fig. 8–6*d*. The maximum stress is

$$\sigma_{max} = \frac{M_x c_y}{I_x} = \frac{8 \text{ kN} \cdot \text{m}(0.2 \text{ m})}{[\frac{1}{12}(0.8 \text{ m})(0.4 \text{ m})^3]} = 375 \text{ kPa}$$

Likewise, for the 16-kN · m moment, Fig. 8–6*e*, the maximum normal stress is

$$\sigma_{max} = \frac{M_y c_x}{I_y} = \frac{16 \text{ kN} \cdot \text{m}(0.4 \text{ m})}{[\frac{1}{12}(0.4 \text{ m})(0.8 \text{ m})^3]} = 375 \text{ kPa}$$

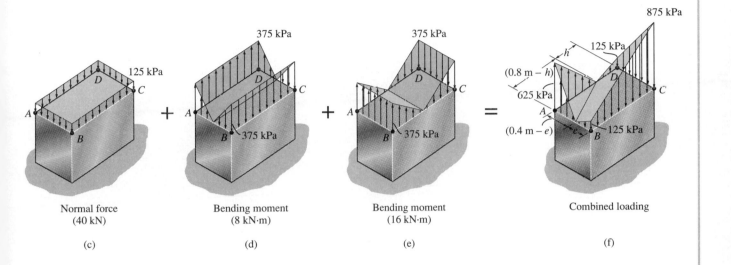

| Normal force (40 kN) | Bending moment (8 kN·m) | Bending moment (16 kN·m) | Combined loading |

(c)   (d)   (e)   (f)

Superposition.   The normal stress at each corner point can be determined by algebraic addition. Assuming that tensile stress is positive, we have

$$\sigma_A = -125 + 375 + 375 = 625 \text{ kPa}$$
$$\sigma_B = -125 - 375 + 375 = -125 \text{ kPa}$$
$$\sigma_C = -125 - 375 - 375 = -875 \text{ kPa}$$
$$\sigma_D = -125 + 375 - 375 = -125 \text{ kPa}$$

Since the stress distributions due to bending moment are linear, the resultant stress distribution is also linear and therefore looks like that shown in Fig. 8–6f. The line of zero stress can be located along each side by proportional triangles. From the figure we require

$$\frac{(0.4 \text{ m} - e)}{625 \text{ kPa}} = \frac{e}{125 \text{ kPa}}$$
$$e = 0.0667 \text{ m}$$

and

$$\frac{(0.8 \text{ m} - h)}{625 \text{ kPa}} = \frac{h}{125 \text{ kPa}}$$
$$h = 0.133 \text{ m}$$

Example 8–6

A rectangular block has a negligible weight and is subjected to a vertical force **P**, Fig. 8–7a. (a) Determine the range of values for the eccentricity e_y of the load along the y axis so that it does not cause any tensile stress in the block. (b) Specify the region on the cross section where **P** may be applied without causing a tensile stress in the block.

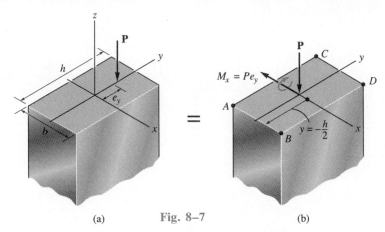

Fig. 8–7

(a)   (b)

SOLUTION

Part (a).   When **P** is moved to the centroid of the cross section, Fig. 8–7b, it is necessary to add a couple moment $M_x = Pe_y$ in order to maintain a statically equivalent loading. The combined normal stress at any coordinate location $\pm y$ on the cross section caused by these two loadings is

$$\sigma = -\frac{P}{A} - \frac{(Pe_y)y}{I_x} = -\frac{P}{A}\left(1 + \frac{Ae_y y}{I_x}\right)$$

Here the negative sign indicates compressive stress. For positive e_y, Fig. 8–7a, the smallest compressive stress will occur along edge AB at $y = -h/2$, Fig. 8–7b. Hence

$$\sigma_{\min} = -\frac{P}{A}\left(1 - \frac{Ae_y h}{2I_x}\right)$$

This stress will remain negative, i.e., compressive, provided the term in parentheses is positive; i.e.,

$$1 > \frac{Ae_y h}{2I_x}$$

Since $A = bh$ and $I_x = \frac{1}{12}bh^3$, then

$$1 > \frac{6e_y}{h}$$

or

$$e_y < \frac{1}{6}h \qquad\qquad \textit{Ans.}$$

In other words, if $-\frac{1}{6}h \le e_y \le \frac{1}{6}h$, the stress in the block along edge AB or CD will be zero or remain *compressive*. This is sometimes referred to as the *"middle-third rule."* It is obviously important to keep this rule in mind when loading columns or arches having a rectangular cross section and made of materials such as stone or concrete, which can support little or no tensile stress.

Part (b). We can extend the above analysis in two directions by assuming that **P** acts in the positive quadrant of the x–y plane, Fig. 8–7c. The equivalent static loading when **P** acts at the centroid is shown in Fig. 8–7d. At any coordinate point x,y on the cross section, the combined normal stress due to both normal and bending loadings is

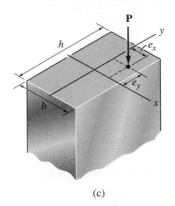

(c)

||

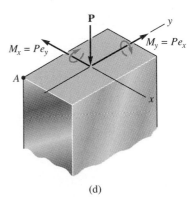

(d)

$$\sigma = -\frac{P}{A} - \frac{Pe_y y}{I_x} - \frac{Pe_x x}{I_y}$$

$$= -\frac{P}{A}\left(1 + \frac{Ae_y y}{I_x} + \frac{Ae_x x}{I_y}\right)$$

By inspection, Fig. 8–7d, the moments both create tensile stress at point A and the normal force creates a compressive stress there. Hence, the smallest compressive stress will occur at point A, for which $x = -b/2$ and $y = -h/2$. Thus,

$$\sigma_A = -\frac{P}{A}\left(1 - \frac{Ae_y h}{2I_x} - \frac{Ae_x b}{2I_y}\right)$$

As before, the normal stress remains negative or compressive at point A, provided the terms in the parentheses remain positive; i.e.,

$$0 < \left(1 - \frac{Ae_y h}{2I_x} - \frac{Ae_x b}{2I_y}\right)$$

Substituting $A = bh$, $I_x = \frac{1}{12}bh^3$, $I_y = \frac{1}{12}hb^3$ yields

$$0 < 1 - \frac{6e_x}{h} - \frac{6e_y}{b} \qquad\qquad \textit{Ans.}$$

Hence, regardless of the magnitude of **P**, if it is applied at any point within the boundary of line GH shown in Fig. 8–7e, the normal stress at point A will remain compressive. In a similar manner, the normal stress at the other corners of the cross section will be compressive if **P** is confined within the boundaries of lines EG, FE, and HF. The shaded parallelogram so defined is referred to as the *core* or *kern* of the section. From the *"middle-third rule"* of part (a), the diagonals of the parallelogram will have lengths of $b/3$ and $h/3$.

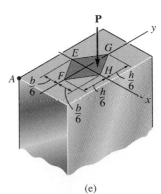

(e)

Example 8–7

(a)

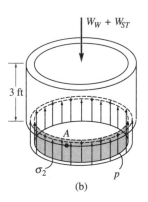

(b)

Fig. 8–8

The tank in Fig. 8–8a has an inner radius of 24 in. and thickness of 0.5 in. It is filled to the top with water having a specific weight of $\gamma_w = 62.4$ lb/ft^3. If it is made of steel having a specific weight of $\gamma_{st} = 490$ lb/ft^3, determine the state of stress at point A. The tank is open at the top.

SOLUTION

Internal Loadings. The free-body diagram of the section of both the tank and the water above point A is shown in Fig. 8–8b. Notice that the weight of the water is supported by the water surface just *below* the section, *not* by the walls of the tank. In the vertical direction, the walls simply hold up the weight of the tank. This weight is

$$W_{st} = \gamma_{st} V_{st} = (490 \text{ lb/ft}^3)\left[\pi\left(\frac{24.5}{12} \text{ ft}\right)^2 - \pi\left(\frac{24}{12} \text{ ft}\right)^2\right](3 \text{ ft}) = 777.7 \text{ lb}$$

The stress in the circumferential direction is developed by the water pressure at level A. To obtain this pressure we must use *Pascal's law*, which states that the pressure at a point located a depth z in the water is $p = \gamma_w z$. Consequently, the pressure on the tank at level A is

$$p = \gamma_w z = (62.4 \text{ lb/ft}^3)(3 \text{ ft}) = 187.2 \text{ lb/ft}^2 = 1.30 \text{ psi}$$

Stress Components

Circumferential Stress. Applying Eq. 8–1, using the inner radius $r = 24$ in., we have

$$\sigma_1 = \frac{pr}{t} = \frac{1.30 \text{ lb/in}^2 (24 \text{ in.})}{(0.5 \text{ in.})} = 62.4 \text{ psi} \qquad Ans.$$

Longitudinal Stress. Since the weight of the tank is supported uniformly by the walls, we have

$$\sigma_2 = \frac{W_{st}}{A_{st}} = \frac{777.7 \text{ lb}}{\pi[(24.5 \text{ in.})^2 - (24 \text{ in.})^2]} = 10.2 \text{ psi} \qquad Ans.$$

Note that Eq. 8–2, $\sigma_2 = Pr/2t$, does *not apply* here, since the tank is open at the top and therefore, as stated previously, the water cannot develop a loading on the walls in the longitudinal direction.

Point A is subjected to the biaxial stress shown in Fig. 8–8c.

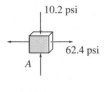

(c)

PROBLEMS

*8–8. The screw of the clamp exerts a compressive force of 500 lb on the wood blocks. Determine the maximum normal stress developed along section a–a. The cross section there is rectangular, 0.75 in. by 0.50 in.

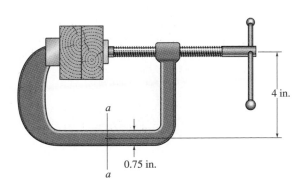

Prob. 8–8

8–9. The clamp is made from members AB and AC, which are pin-connected at A. If the compressive force at C and B is 180 N as shown, determine the maximum compressive stress in the clamp at section a–a. The screw EF is subjected only to a tensile force along its axis.

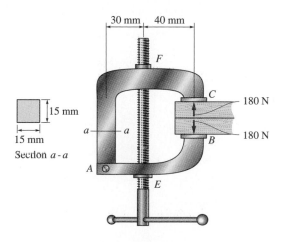

Prob. 8–9

8–10. The joint is subjected to a force of 80 lb as shown. Sketch the normal-stress distribution acting over section a–a if the member has a rectangular cross-sectional area of width 2 in. and thickness 0.5 in.

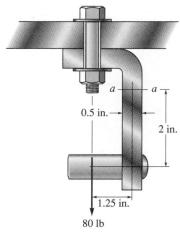

Prob. 8–10

8–11. The offset link supports the loading shown. Determine its required width w if the allowable normal stress is $\sigma_{\text{allow}} = 73$ MPa. The link has a thickness of 40 mm.

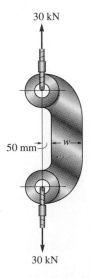

Prob. 8–11

*8–12.   The steel bracket is used to connect the ends of two cables. If the allowable normal stress is $\sigma_{allow} = 24$ ksi, determine the largest tensile force P that can be applied to the cables. The bracket has a thickness of 0.5 in. and a width of 0.75 in.

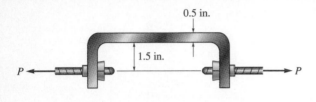

0.5 in.

1.5 in.

P P

Prob. 8–12

8–13.   The gondola and passengers have a weight of 1500 lb and center of gravity at G. The suspender arm AE has a square cross-sectional area of 1.5 in. by 1.5 in., and is pin-connected at its ends A and E. Determine the largest tensile stress developed in regions AB and DC of the arm.

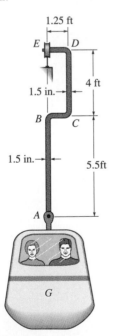

1.25 ft

E D

1.5 in.   4 ft

B C

1.5 in.   5.5ft

A

G

Prob. 8–13

8–14.   A bar having a square cross section of 30 mm by 30 mm is 2 m long and is held upward. If it has a mass of 5 kg/m, determine the largest angle θ at which it can be supported before it is subjected to a tensile stress along its axis near the grip.

8–15.   Solve Prob. 8–14 if the bar has a circular cross section of 30-mm diameter.

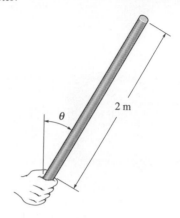

2 m

θ

Probs. 8–14/8–15

*8–16.   The vertical force $\mathbf{P}$ acts on the bottom of the plate having a negligible weight. Determine the shortest distance d to the edge of the plate at which it can be applied so that it produces no compressive stresses on the plate at section a–a. The plate has a thickness of 10 mm and $\mathbf{P}$ acts along the center line of this thickness.

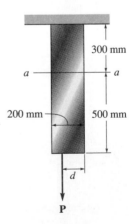

300 mm

a a

200 mm   500 mm

d

$\mathbf{P}$

Prob. 8–16

8–17. The masonry pier is subjected to the 800-kN load. Determine the equation of the line $y = f(x)$ along which the load can be placed without causing a tensile stress in the pier. Neglect the weight of the pier.

8–18. The masonry pier is subjected to the 800-kN load. If $x = 0.25$ m and $y = 0.5$ m, determine the normal stress at each corner A, B, C, D (not shown) and plot the stress distribution over the cross section. Neglect the weight of the pier.

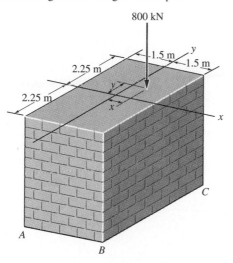

Probs. 8–17/8–18

8–19. The block is subjected to the three axial loads shown. Determine the normal stress developed at points A and B. Neglect the weight of the block.

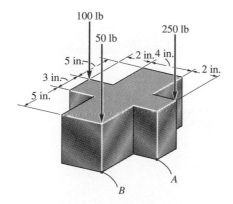

Prob. 8–19

*8–20.** The bar has a diameter of 40 mm. If it is subjected to a force of 800 N as shown, determine the stress components that act at points A and B and show the results on volume elements located at these points.

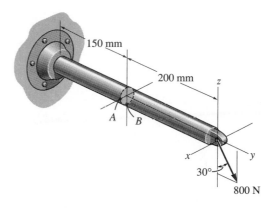

Prob. 8–20

8–21. The cylindrical post is being pulled from the ground using a sling of negligible thickness. If the rope is subjected to a vertical force of 500 N, determine the stress at points A and B. Show the results on a volume element located at each of these points.

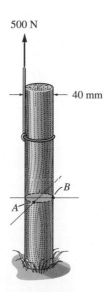

Prob. 8–21

8–22. The beam supports the loading shown. Determine the stresses at points E and F at section $a–a$ and represent the results on a volume element located at each of these points.

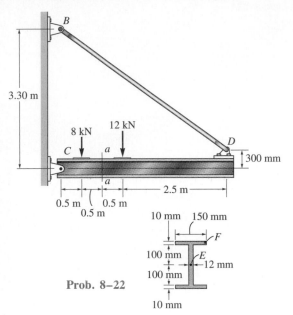

Prob. 8–22

8–23. The cable drum is locked in the position shown while supporting a horizontal cable force of 4 kN. If the drum and wound cable have a weight of 2 kN and center of gravity at G, determine the stress components in the supporting beam at point A.

*8–24. The cable drum is locked in the position shown while supporting a horizontal cable force of 4 kN. If the drum and wound cable have a weight of 2 kN and center of gravity at G, determine the stress components in the supporting beam at point B.

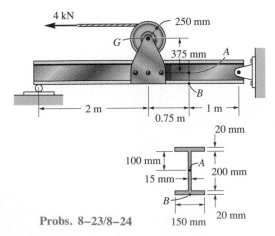

Probs. 8–23/8–24

8–25. The pliers are made from two steel parts pinned together at A. If a smooth bolt is held in the jaws and a gripping force of 10 lb is applied at the handles, determine the stress components developed in the pliers at points B and C. Here the cross section is rectangular, having the dimensions shown in the figure.

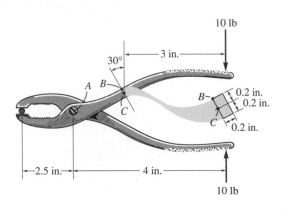

Prob. 8–25

8–26. The pin support is made from a steel rod and has a diameter of 20 mm. Determine the stress components at points A and B and represent the results on a volume element located at each of these points.

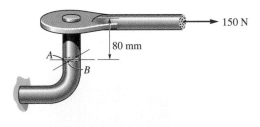

Prob. 8–26

8–27. The wide-flange beam is subjected to the loading shown. Determine the stress components at points A and B and show the results on a volume element located at each of these points. Use the shear formula to compute the shear stress.

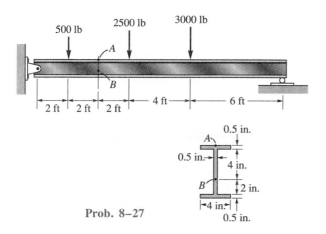

Prob. 8–27

8–30. The crow bar is used to pull out the nail at A. If a force of 8 lb is required, determine the stress components in the bar at points D and E. Show the results on a volume element located at each of these points. The bar has a circular cross section with a diameter of 0.5 in. No slipping occurs at B.

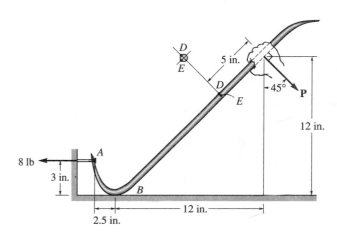

Prob. 8–30

*8–28. Since concrete can support little or no tension, this problem can be avoided by using wires or rods to *prestress* the concrete once it is formed. Consider the simply-supported beam shown, which has a rectangular cross section of 18 in. by 12 in. If concrete weighs 150 lb/ft³, determine the required tension in the rod AB, which runs through the beam so that no tensile stress is developed in the concrete at its center section a–a. Neglect the size of the rod and any deflection of the beam.

8–29. Solve Prob. 8–28 if the rod has a diameter of 0.5 in. Use the transformed-area method discussed in Sec. 6–4. $E_{st} = 29(10^3)$ ksi, $E_c = 3.60(10^3)$ ksi.

8–31. The beveled gear is subjected to the loads shown. Determine the stress components acting on the shaft at point A, and show the results on a volume element located at this point. The shaft has a diameter of 1 in. and is fixed to the wall at C.

*8–32. The beveled gear is subjected to the loads shown. Determine the stress components acting on the shaft at point B, and show the results on a volume element located at this point. The shaft has a diameter of 1 in. and is fixed to the wall at C.

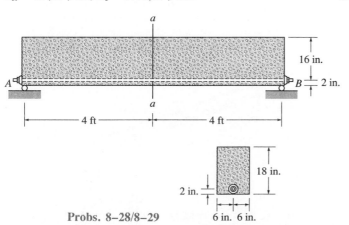

Probs. 8–28/8–29

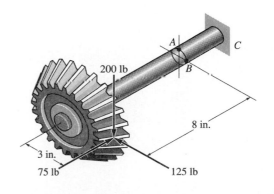

Probs. 8–31/8–32

8–33. The $\frac{3}{4}$-in.-diameter shaft is subjected to the loading shown. Determine the stress components at points A and B. Sketch the results on a volume element located at each of these points. The journal bearing at C can exert only force components $\mathbf{C}_x$ and $\mathbf{C}_z$ on the shaft, and the thrust bearing at D can exert force components $\mathbf{D}_x$, $\mathbf{D}_y$, and $\mathbf{D}_z$ on the shaft.

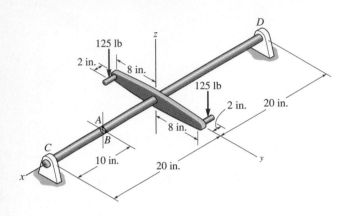

Prob. 8–33

8–34. The circular block of negligible weight rests on a smooth floor. Determine the eccentric distance e_y at which the load can be placed so that the stress at point A is zero.

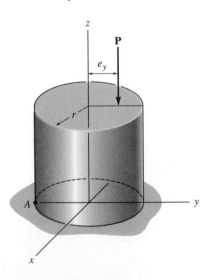

Prob. 8–34

8–35. The hook is used to lift the force of 600 lb. Determine the maximum tensile and compressive stresses at section a–a. The cross section is circular and has a diameter of 1 in. Use the curved-beam formula to compute the bending stress.

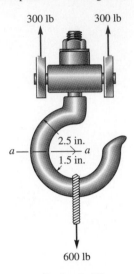

Prob. 8–35

***8–36.** The bent shaft is fixed in the wall at A. If a force $\mathbf{F}$ is applied at B, determine the stress components at points D and E. Show the results on a volume element located at each of these points. Take $F = 12$ lb and $\theta = 0°$.

8–37. The bent shaft is fixed in the wall at A. If a force $\mathbf{F}$ is applied at B, determine the stress components at points D and E. Show the results on a volume element located at each of these points. Take $F = 12$ lb and $\theta = 90°$.

8–38. The bent shaft is fixed in the wall at A. If a force $\mathbf{F}$ is applied at B, determine the stress components at points D and E. Show the results on a volume element located at each of these points. Take $F = 12$ lb and $\theta = 45°$.

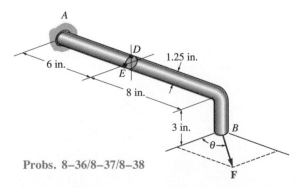

Probs. 8–36/8–37/8–38

REVIEW PROBLEMS

8–39.   The uniform sign has a weight of 1500 lb and is supported by the pipe *AB*, which has an inner radius of 2.75 in. and an outer radius of 3.00 in. If the face of the sign is subjected to a uniform wind pressure of $p = 150$ lb/ft^2, determine the stress components at points *C* and *D*. Show the results on a volume element located at each of these points. Neglect the thickness of the sign, and assume that it is supported along the outside edge of the pipe.

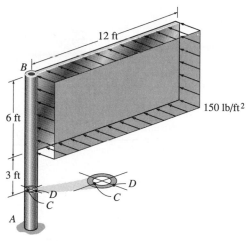

Prob. 8–39

8–41.   The handle of the press is subjected to a force of 20 lb. Due to internal gearing, this causes the block to be subjected to a compressive force of 80 lb. Determine the normal stress acting in the frame at points *A* and *B* caused by these loadings. *Hint:* The curved-beam formula should be used to compute the bending stress.

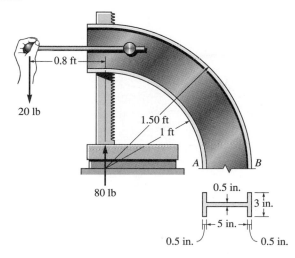

Prob. 8–41

***8–40.**   The column has a circular cross section of radius *c*. Determine the maximum radius *e* at which the load can be applied so that no part of the column experiences a tensile stress. Neglect the weight of the column.

Prob. 8–40

8–42.   The clamp exerts a force of 500 lb on the block. If the cross section at *a–a* is triangular, determine the maximum tensile and compressive stress at the section. Use the curved-beam formula to determine the bending stress.

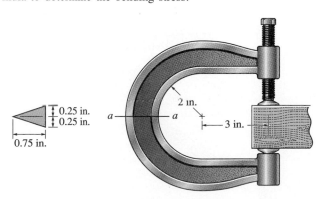

Prob. 8–42

8–43. The coiled spring is subjected to a force P. If we assume the shear stress caused by the shear force at any vertical section of the coil wire to be uniform, show that the maximum shear stress in the coil is $\tau_{max} = P/A + PRr/J$, where J is the polar moment of inertia of the coil wire and A is its cross-sectional area. If the spring has n complete coils and the shear modulus for the material is G, show that the vertical deflection of the spring can be approximated by $\Delta = 4PR^3n/Gr^4$, where Δ is small.

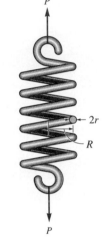

Prob. 8–43

8–45. The C-frame is used in a riveting machine. If the force at the ram on the dolly D is 8 kN, sketch the stress distribution acting over the section $a–a$.

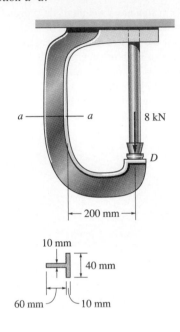

Prob. 8–45

*8–44. The frame supports a centrally applied distributed load of 1.8 kip/ft. Determine the stress components developed at points A and B on member CD and indicate the results on a volume element located at each of these points. The pins at C and D are at the same location as the neutral axis for the cross section.

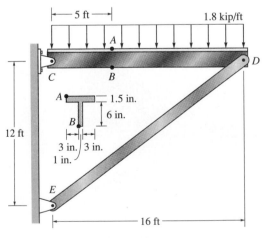

Prob. 8–44

8–46. Plot the distribution of stress acting over the cross section $a–a$ of the offset link.

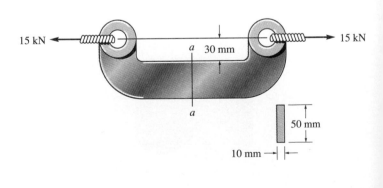

Prob. 8–46

9 Stress Transformation

In this chapter we will show how to transform the stress components that are associated with a particular coordinate system, into components associated with another coordinate system. Once the transformation equations are established, we will then be able to obtain the maximum normal and shear stress components at a point and find the orientation of an element on which they act. Plane-stress transformation will be discussed in the first part of the chapter, since this condition is most common in engineering practice. At the end of the chapter we will discuss a method for finding the absolute maximum shear stress at a point when the material is subjected to both plane and three-dimensional states of stress.

9.1 Plane Stress Transformation

As discussed in Sec. 1.3, in the general case, the state of stress at a point is characterized by *six* independent normal and shear stress components, which act on the faces of an element of material located at the point, Fig. 9–1a. This state of stress, however, is not often encountered in engineering practice. Instead, engineers frequently make approximations or simplifications in order that the stress produced in a structural member or mechanical element can be analyzed in a single plane. When this is the case, the material is said to be subjected to *plane stress,* Fig. 9–1b. For example, if there is no load on the surface of a body, then the normal and shear stress components will be zero on the face of an element that lies on the surface. Consequently, the corresponding stress components on the opposite face will also be zero, and so the material at the point will be subjected to plane stress.

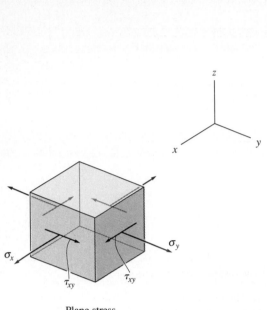

General state of stress

(a)

Fig. 9–1

Plane stress

(b)

Plane stress
(two dimensional view)

(c)

The general state of *plane stress* at a point is therefore represented by a combination of two normal-stress components, σ_x, σ_y, and one shear-stress component, τ_{xy}, which act on four faces of the element as shown in Fig. 9–1b. For convenience, in this text we will view this state of stress in the x–y plane, Fig. 9–1c. It should be realized, however, that if one can determine these three stress components at a point, for an element oriented in the x and y directions, Fig. 9–2a, then the three stress components representing the *same state of stress* at the point on an element oriented in the x' and y' directions will be *different*, Fig. 9–2b. In other words, the state of plane stress at the point is uniquely represented by three components acting on an element that has a *specific orientation* at the point.

In this section, by using numerical examples, we will show how to *transform* the stress components from one orientation of an element to an element having a different orientation. That is, if the state of stress is defined by the components σ_x, σ_y, τ_{xy}, oriented along the x, y axes, Fig. 9–2a, we will show how to obtain the components $\sigma_{x'}$, $\sigma_{y'}$, $\tau_{x'y'}$, oriented along the x', y' axes, Fig. 9–2b, so that they represent the *same* state of stress at the point. This is like knowing two force components, say, $\mathbf{F}_x$ and $\mathbf{F}_y$, directed along the x, y axes, that produce a resultant force $\mathbf{F}_R$, and then trying to find the force components $\mathbf{F}_{x'}$ and $\mathbf{F}_{y'}$, directed along the x', y' axes, so they produce the *same* resultant. The transformation of stress components, however, is more difficult than that of force components, since for stress, the transformation must account for the magnitude and direction of each stress component and

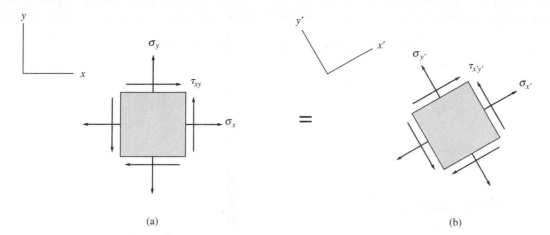

$$=$$

(a)

(b)

Fig. 9–2

the orientation of the area upon which each component acts, whereas for force, the transformation must account only for the force component's magnitude and direction.

Recall from Sec. 1.3 that each face of an element is specified by a coordinate axis that is directed *perpendicular* to the face. For example, the $+x$ face in Fig. 9–2a is the vertical right-hand face, the $-y$ face is the bottom horizontal face, and so on. If we wish to determine the normal and shear stress components acting on the $+x'$ face of the element in Fig. 9–2b, then we must section the x, y element as shown in Fig. 9–2c. The three *known* x, y components of stress, σ_x, σ_y, τ_{xy}, acting on this segment can be related to the two unknown x', y' components, $\sigma_{x'}$, $\tau_{x'y'}$, using the equations of equilibrium. This is done by first *converting* all the stress components to forces by multiplying the stresses by their associated areas. The forces (not the stresses) are then shown on a free-body diagram of the segment and the two unknown x' and y' force components, acting on the sectioned plane ($+x'$ face), are determined from the equations of force equilibrium. Once obtained, these forces are then converted to the required normal and shear stress components, $\sigma_{x'}$ and $\tau_{x'y'}$, by dividing the forces by the sectioned area.

If $\sigma_{y'}$, acting on the $+y'$ face of the element in Fig. 9–2b, is to be determined, then it is necessary to consider a segment of the x, y element as shown in Fig. 9–2d and follow the same procedure just described. Here, however, the shear stress $\tau_{x'y'}$ does not have to be determined if it was previously calculated. Recall that shear stress is complementary, and so equilibrium of the x', y' element requires $\tau_{x'y'}$ to have the same magnitude on each of the four faces of the element, Fig. 9–2b.

The following examples illustrate this procedure numerically.

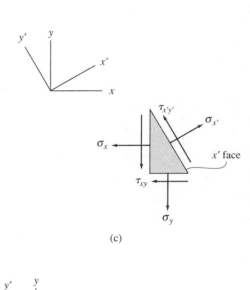

(c)

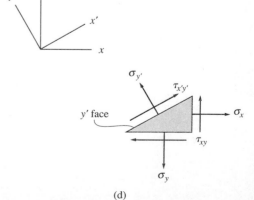

(d)

Example 9–1

An axial force of 600 N acts on the steel bar shown in Fig. 9–3a. Determine the stress components acting on a plane defined by section a–a.

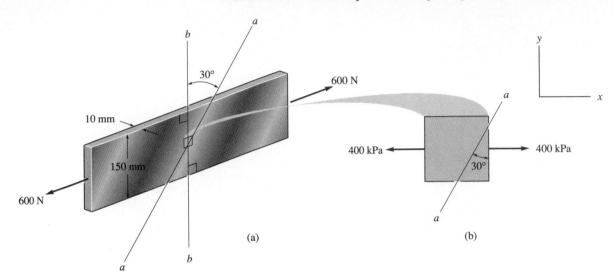

(a)

(b)

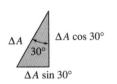

(c)

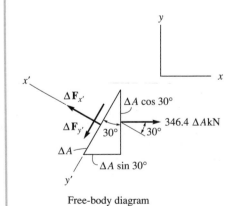

Free-body diagram

(d)

Fig. 9–3

SOLUTION I

Since the force is applied through the centroid of the cross section, the bar is subjected only to a normal stress along section b–b. This stress is

$$\sigma_x = \frac{P}{A} = \frac{600 \text{ N}}{(0.150 \text{ m})(0.01 \text{ m})} = 400 \text{ kPa}$$

The result is shown acting on an element of material in Fig. 9–3b.

Since the stress components along a–a are to be determined, the element will now be sectioned as shown in Fig. 9–3b. If we assume that the inclined face of the segment has an area ΔA, then, as shown in Fig. 9–3c, the horizontal and vertical faces have areas of $\Delta A \sin 30°$ and $\Delta A \cos 30°$. Using these areas, the *free-body diagram* of the segment is shown in Fig. 9–3d. Note that the force on the $+x$ face is

$$\Delta F_x = 400 \text{ kPa } (\Delta A \cos 30°) = 346.4 \ \Delta A \text{ kN}$$

We can obtain a direct solution for the two unknown forces $\Delta F_{x'}$ and $\Delta F_{y'}$ by applying the equations of equilibrium along the x' and y' axes.*

$$+\nwarrow \Sigma F_{x'} = 0; \quad \Delta F_{x'} - (346.4 \ \Delta A) \cos 30° = 0 \quad \Delta F_{x'} = 300 \ \Delta A$$

$$+\nearrow \Sigma F_{y'} = 0; \quad \Delta F_{y'} - (346.4 \ \Delta A) \sin 30° = 0 \quad \Delta F_{y'} = 173 \ \Delta A$$

*If the equations of equilibrium are applied in the x and y directions, they will have to be solved *simultaneously* for $\Delta F_{x'}$ and $\Delta F_{y'}$.

The normal stress on section a–a is therefore

$$\sigma_{x'} = \frac{\Delta F_{x'}}{\Delta A} = \frac{300\ \Delta A}{\Delta A} = 300\ \text{kPa} \qquad \textit{Ans.}$$

And the shear stress on the section is

$$\tau_{x'y'} = \frac{\Delta F_{y'}}{\Delta A} = \frac{173\ \Delta A}{\Delta A} = 173\ \text{kPa} \qquad \textit{Ans.}$$

The results are independent of the area ΔA and are shown distributed uniformly over the entire section a–a in Fig. 9–3e.

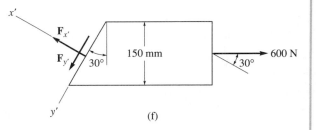

(e)

SOLUTION II

A free-body diagram of the right side of the bar, sectioned along a–a, is shown in plane view, Fig. 9–3f. Applying the equations of force equilibrium in the x' and y' directions, we have

$$+\nwarrow \Sigma F_{x'} = 0; \quad F_{x'} - 600 \cos 30° = 0 \qquad F_{x'} = 519.6\ \text{N}$$
$$+\nearrow \Sigma F_{y'} = 0; \quad F_{y'} - 600 \sin 30° = 0 \qquad F_{y'} = 300\ \text{N}$$

The area of the inclined section a–a is

$$A' = \left(\frac{0.150\ \text{m}}{\cos 30°}\right)(0.01\ \text{m}) = 1.732(10^{-3})\ \text{m}^2$$

(f)

The average normal and shear stresses on the inclined plane are thus

$$\sigma_{x'} = \frac{F_{x'}}{A'} = \frac{519.6\ \text{N}}{1.732(10^{-3})\ \text{m}^2} = 300\ \text{kPa} \qquad \textit{Ans.}$$

$$\tau_{x'y'} = \frac{F_{y'}}{A'} = \frac{300\ \text{N}}{1.732(10^{-3})\ \text{m}^2} = 173\ \text{kPa} \qquad \textit{Ans.}$$

Note that this method of analysis can be used here since the formulas $\sigma = P/A$ and $\tau = V/A$ can be applied in this manner. Stresses generated by beam shear, bending moment, and torsional moment must be determined only on the *cross-sectional area* or on sections that are perpendicular to the member's axis, and as a result, the first method of analysis must be used for these cases.

Example 9–2

The state of plane stress at a point on the surface of the airplane fuselage is represented on the element oriented as shown in Fig. 9–4a. Represent the state of stress at the point on another element that is oriented 30° clockwise from the position shown.

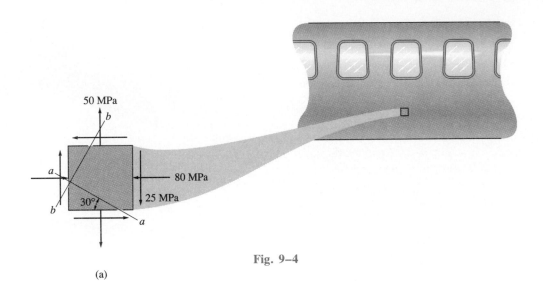

Fig. 9–4

(a)

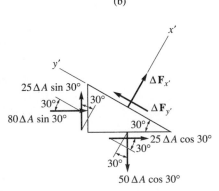

(b)

(c)

SOLUTION

The element is sectioned by the line a–a in Fig. 9–4a, the bottom segment is removed, and assuming the sectioned (inclined) plane has an area ΔA, the horizontal and vertical planes have the areas shown in Fig. 9–4b. The free-body diagram of the segment is shown in Fig. 9–4c. Applying the equations of force equilibrium in the x' and y' directions to avoid a simultaneous solution for the two unknowns $\Delta F_{x'}$ and $\Delta F_{y'}$, we have

$$+\nearrow \Sigma F_{x'} = 0; \quad \Delta F_{x'} - (50 \; \Delta A \cos 30°) \cos 30°$$
$$+(25 \; \Delta A \cos 30°) \sin 30° + (80 \; \Delta A \sin 30°) \sin 30°$$
$$+(25 \; \Delta A \sin 30°) \cos 30° = 0$$
$$\Delta F_{x'} = -4.15 \; \Delta A$$

$$+\searrow \Sigma F_{y'} = 0; \quad -\Delta F_{y'} + (50 \; \Delta A \cos 30°) \sin 30°$$
$$+(25 \; \Delta A \cos 30°) \cos 30° + (80 \; \Delta A \sin 30°) \cos 30°$$
$$-(25 \; \Delta A \sin 30°) \sin 30° = 0$$
$$\Delta F_{y'} = 68.8 \; \Delta A$$

Since $\Delta F_{x'}$ is negative, $\Delta F_{x'}$ acts in the opposite direction of that shown in Fig. 9–4c.

The normal and shear stress components acting on the inclined face along section *a–a* are therefore

$$\sigma_{x'} = \frac{\Delta F_{x'}}{\Delta A} = \frac{4.15 \ \Delta A}{\Delta A} = 4.15 \ \text{MPa} \qquad \textit{Ans.}$$

$$\tau_{x'y'} = \frac{\Delta F_{y'}}{\Delta A} = \frac{68.8 \ \Delta A}{\Delta A} = 68.8 \ \text{MPa} \qquad \textit{Ans.}$$

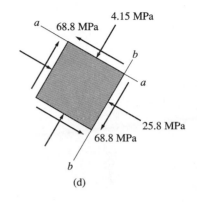

(d)

These results are shown on the *top* of the element in Fig. 9–4*d*, since this surface is the one considered in Fig. 9–4*c*.

We must now repeat the procedure to obtain the stress on the *perpendicular* plane *b–b*. Sectioning the element in Fig. 9–4*a* in the direction of *b–b* results in a segment having sides with areas shown in Fig. 9–4*e*. The associated free-body diagram is shown in Fig. 9–4*f*. Thus,

$$+\searrow \ \Sigma F_{x'} = 0; \quad \Delta F_{x'} - (25 \ \Delta A \cos 30°) \sin 30°$$
$$+(80 \ \Delta A \cos 30°) \cos 30° - (25 \ \Delta A \sin 30°) \cos 30°$$
$$-(50 \ \Delta A \sin 30°) \sin 30° = 0$$
$$\Delta F_{x'} = -25.8 \ \Delta A$$

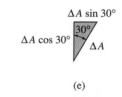

(e)

$$+\nearrow \ \Sigma F_{y'} = 0; \quad -\Delta F_{y'} + (25 \ \Delta A \cos 30°) \cos 30°$$
$$+(80 \ \Delta A \cos 30°) \sin 30° - (25 \ \Delta A \sin 30°) \sin 30°$$
$$+(50 \ \Delta A \sin 30°) \cos 30° = 0$$
$$\Delta F_{y'} = 68.8 \ \Delta A$$

From these results, note that $\Delta F_{x'}$ acts opposite to its direction shown in Fig. 9–4*f*. The stress components are therefore

$$\sigma_{x'} = \frac{\Delta F_{x'}}{\Delta A} = \frac{25.8 \ \Delta A}{\Delta A} = 25.8 \ \text{MPa} \qquad \textit{Ans.}$$

$$\tau_{x'y'} = \frac{\Delta F_{y'}}{\Delta A} = \frac{68.8 \ \Delta A}{\Delta A} = 68.8 \ \text{MPa} \qquad \textit{Ans.}$$

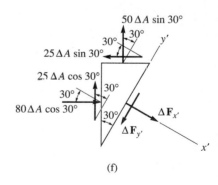

(f)

These stress components are shown acting on the *right side* of the element in Fig. 9–4*d*. Here the direction and magnitude of the shear stress, which have been calculated twice, verifies the requirement for moment equilibrium discussed in Sec. 1.3. From this analysis we may therefore conclude that the state of stress at the point can, for example, be represented by choosing an element oriented as shown in Fig. 9–4*a* or by choosing one oriented as shown in Fig. 9–4*d*.

9.2 General Equations of Plane-Stress Transformation

The method of transforming the normal and shear stress components from the x, y to the x', y' coordinate axes, as discussed in the previous section, will now be developed in a general manner and expressed as a set of stress-transformation equations.

Sign Convention. Before the transformation equations can be developed, we must first establish a sign convention for the stress components. Here we will adopt the same one used in Sec. 1.3. Briefly stated, once the x, y or x', y' axes have been established, a normal or shear stress component is *positive* provided it acts in the *positive* coordinate direction on the *positive* face of the element, or it acts in the *negative* coordinate direction on the *negative* face of the element, Fig. 9–5a. For example, σ_x is positive since it acts to the right ($+x$ direction) on the right-hand vertical face (defined by the $+x$ axis), and it acts to the left ($-x$ direction) on the left-hand vertical face (defined by the $-x$ axis). The shear stress in Fig. 9–5a is shown acting in the positive direction on all four faces of the element. On the right-hand face (defined by the $+x$ axis), τ_{xy} acts upward ($+y$ direction); on the bottom face (defined by the $-y$ axis), τ_{xy} acts to the left ($-x$ direction), and so on.

All the stress components shown in Fig. 9–5a maintain equilibrium of the element as discussed in Sec. 1.3, and because of this, knowing the direction of τ_{xy} on one face of the element defines its direction on the other three faces. Hence, the above sign convention can also be remembered by simply noting that *positive normal stress* acts *outward* and *positive shear stress* acts *upward on the right-hand face* of the element.

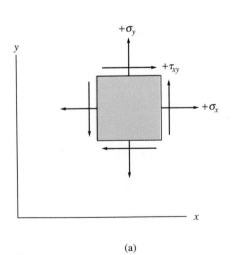

(a)

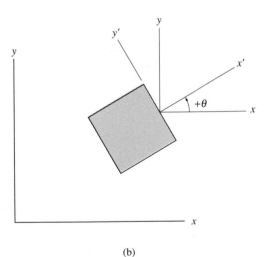

(b)

Positive Sign Convention

Fig. 9–5

Given the state of plane stress shown in Fig. 9–5a, the orientation of the inclined plane on which the normal and shear stress components are to be determined will be defined using the angle θ. To establish this angle properly, it is first necessary to establish a positive x' axis, *directed outward, perpendicular* or normal to the plane, and an associated y' axis, directed along the plane, Fig. 9–5b. Both the unprimed and primed sets of axes form right-handed coordinate systems; that is, the positive z (or z') axis is established by the right-hand rule. Curling the fingers from x (or x') toward y (or y') gives the direction for the positive z (or z') axis that points outward. The *angle* θ is measured from the positive x to the positive x' axis, such that it is *positive* provided it also follows the curl of the right-hand fingers, i.e., counterclockwise as shown in Fig. 9–5b.

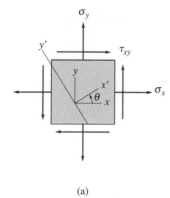

(a)

Fig. 9–6

Normal and Shear Stress Components.

Using the established sign convention, the element in Fig. 9–6a is sectioned along the inclined plane defined by $+x'$ and the segment shown in Fig. 9–6b is isolated. If the sectioned area is ΔA, then the horizontal and vertical faces of the segment have an area of $\Delta A \sin \theta$ and $\Delta A \cos \theta$, respectively.

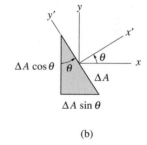

(b)

The resulting *free-body diagram* of the segment is shown in Fig. 9–6c. Applying the equations of force equilibrium to determine the unknown normal and shear stress components $\sigma_{x'}$ and $\tau_{x'y'}$, we obtain

$$+\nearrow \ \Sigma F_{x'} = 0; \quad \sigma_{x'} \, \Delta A - (\tau_{xy} \, \Delta A \sin \theta) \cos \theta$$
$$-(\sigma_y \, \Delta A \sin \theta) \sin \theta - (\tau_{xy} \, \Delta A \cos \theta) \sin \theta$$
$$-(\sigma_x \, \Delta A \cos \theta) \cos \theta = 0$$

$$\sigma_{x'} = \sigma_x \cos^2 \theta + \sigma_y \sin^2 \theta + \tau_{xy}(2 \sin \theta \cos \theta)$$

$$+\nwarrow \ \Sigma F_{y'} = 0; \quad \tau_{x'y'} \, \Delta A + (\tau_{xy} \, \Delta A \sin \theta) \sin \theta$$
$$-(\sigma_y \, \Delta A \sin \theta) \cos \theta - (\tau_{xy} \, \Delta A \cos \theta) \cos \theta$$
$$+(\sigma_x \, \Delta A \cos \theta) \sin \theta = 0$$

$$\tau_{x'y'} = (\sigma_y - \sigma_x) \sin \theta \cos \theta + \tau_{xy}(\cos^2 \theta - \sin^2 \theta)$$

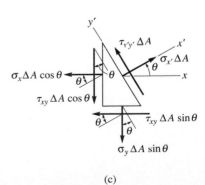

(c)

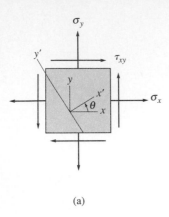

(a)

Fig. 9–6(a)

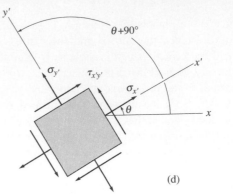

(d)

Fig. 9–6(d)

These two equations may be simplified by using the trigonometric identities $\sin 2\theta = 2 \sin \theta \cos \theta$, $\sin^2 \theta = (1 - \cos 2\theta)/2$, and $\cos^2 \theta = (1 + \cos 2\theta)/2$, in which case,

$$\sigma_{x'} = \frac{\sigma_x + \sigma_y}{2} + \frac{\sigma_x - \sigma_y}{2} \cos 2\theta + \tau_{xy} \sin 2\theta \qquad (9\text{–}1)$$

$$\tau_{x'y'} = -\frac{\sigma_x - \sigma_y}{2} \sin 2\theta + \tau_{xy} \cos 2\theta \qquad (9\text{–}2)$$

To check these equations, note that when $\theta = 0°$, $\sigma_{x'} = \sigma_x$ and $\tau_{x'y'} = \tau_{xy}$ as required, Fig. 9–6a. Also, when $\theta = 90°$, $\sigma_{x'} = \sigma_y$ and $\tau_{x'y'} = -\tau_{xy}$. The negative sign indicates that $\tau_{x'y'}$ acts in the negative y' direction as required, Fig. 9–6a.

If the normal stress acting in the y' direction is needed, it can be obtained by simply substituting ($\theta = \theta + 90°$) for θ into Eq. 9–1, Fig. 9–6d. The result is

$$\sigma_{y'} = \frac{\sigma_x + \sigma_y}{2} - \frac{\sigma_x - \sigma_y}{2} \cos 2\theta - \tau_{xy} \sin 2\theta \qquad (9\text{–}3)$$

If $\sigma_{y'}$ is calculated as a positive quantity, it indicates that it acts in the positive y' direction as shown in Fig. 9–6d.

PROCEDURE FOR ANALYSIS

To apply the stress transformation Eqs. 9–1 and 9–2, it is simply necessary to substitute in the known data for σ_x, σ_y, τ_{xy}, and θ in accordance with the established sign convention, Fig. 9–6a. If $\sigma_{x'}$ and $\tau_{x'y'}$ are calculated as positive quantities, then these stresses act in the positive direction of the x' and y' axes as established in Fig. 9–6c.

For convenience these equations can easily be programmed on a pocket calculator or microcomputer. The following example illustrates their numerical application.

Example 9–3

The state of plane stress at a point is represented by the element shown in Fig. 9–7a. Determine the state of stress at the point on another element oriented 30° clockwise from the position shown.

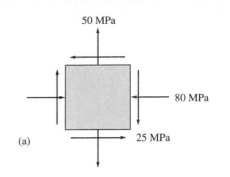

(a)

SOLUTION

This problem was solved in Example 9–2 using basic principles. Here we will apply Eqs. 9–1 and 9–2. From the established sign convention, Fig. 9–5, it is seen that

$$\sigma_x = -80 \text{ MPa} \qquad \sigma_y = 50 \text{ MPa} \qquad \tau_{xy} = -25 \text{ MPa}$$

To obtain the stress components on plane CD, Fig. 9–7b, the positive x' axis is directed outward, perpendicular to CD, and the associated y' axis is directed along CD. The angle θ measured from the x to the x' axis is $\theta = -30°$ (clockwise). Applying Eqs. 9–1 and 9–2 yields

$$\sigma_{x'} = \frac{-80 + 50}{2} + \frac{-80 - 50}{2} \cos 2(-30°) + (-25) \sin 2(-30°)$$

$$= -25.8 \text{ MPa} \qquad\qquad Ans.$$

$$\tau_{x'y'} = -\frac{-80 - 50}{2} \sin 2(-30°) + (-25) \cos 2(-30°)$$

$$= -68.8 \text{ MPa} \qquad\qquad Ans.$$

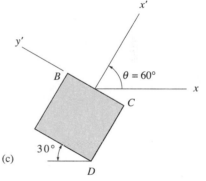

(b)

The negative signs indicate that $\sigma_{x'}$ and $\tau_{x'y'}$ act in the negative x' and y' directions, respectively. The results are shown acting on the element in Fig. 9–7d.

In a similar manner, the stress components acting on face BC, Fig. 9–7b, are obtained using $\theta = 60°$, Fig. 9–7c. Applying Eqs. 9–1 and 9–2,* we get

$$\sigma_{x'} = \frac{-80 + 50}{2} + \frac{-80 - 50}{2} \cos 2(60°) + (-25) \sin 2(60°)$$

$$= -4.15 \text{ MPa} \qquad\qquad Ans.$$

$$\tau_{x'y'} = -\frac{-80 - 50}{2} \sin 2(60°) + (-25) \cos 2(60°)$$

$$= 68.8 \text{ MPa} \qquad\qquad Ans.$$

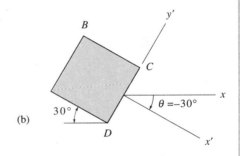

(c)

Here $\tau_{x'y'}$ has been computed twice in order to provide a check. The negative sign for $\sigma_{x'}$ indicates that this stress acts in the negative x' direction shown in Fig. 9–7c. The results are shown on the element in Fig. 9–7d.

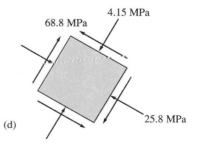

(d)

*Alternatively, we could apply Eq. 9–3 with $\theta = -30°$ rather than Eq. 9–1.

Fig. 9–7

9.3 Principal Stresses and Maximum In-Plane Shear Stress

From Eqs. 9–1 and 9–2, it can be seen that $\sigma_{x'}$ and $\tau_{x'y'}$ depend on the angle of inclination θ of the planes on which these stresses act. In engineering practice it is often important to determine the orientation of the planes that causes the normal stress to be a maximum and a minimum and the orientation of the planes that causes the shear stress to be a maximum. In this section each of these problems will be considered.

In-Plane Principal Stresses. To determine the maximum and minimum normal stress we must differentiate Eq. 9–1 with respect to θ and set the result equal to zero. Thus,

$$\frac{d\sigma_{x'}}{d\theta} = -\frac{\sigma_x - \sigma_y}{2}(2\sin 2\theta) + 2\tau_{xy}\cos 2\theta = 0$$

Solving this equation we obtain the orientation $\theta = \theta_p$ of the planes of maximum and minimum normal stress. At $\theta = \theta_p$,

$$\boxed{\tan 2\theta_p = \frac{\tau_{xy}}{(\sigma_x - \sigma_y)/2}} \tag{9–4}$$

The solution of this equation has two roots, θ_{p_1} and θ_{p_2}. Specifically, the values of $2\theta_{p_1}$ and $2\theta_{p_2}$ are 180° apart, so θ_{p_1} and θ_{p_2} are 90° apart.

The values of θ_{p_1} and θ_{p_2} must be substituted into Eq. 9–1 if we are to obtain the required normal stresses. Rather than doing this, we can instead obtain the necessary sine and cosine of $2\theta_{p_1}$ and $2\theta_{p_2}$ from the triangles shown in Fig. 9–8. The construction of these triangles is based on Eq. 9–4, assuming that τ_{xy} and $(\sigma_x - \sigma_y)$ are both positive or both negative quantities. We have,

for θ_{p_1}, $\qquad \sin 2\theta_{p_1} = \tau_{xy} \Big/ \sqrt{\left(\dfrac{\sigma_x - \sigma_y}{2}\right)^2 + \tau_{xy}{}^2}$

$$\cos 2\theta_{p_1} = \left(\frac{\sigma_x - \sigma_y}{2}\right) \Big/ \sqrt{\left(\frac{\sigma_x - \sigma_y}{2}\right)^2 + \tau_{xy}{}^2}$$

for θ_{p_2}, $\qquad \sin 2\theta_{p_2} = -\tau_{xy} \Big/ \sqrt{\left(\dfrac{\sigma_x - \sigma_y}{2}\right)^2 + \tau_{xy}{}^2}$

$$\cos 2\theta_{p_2} = -\left(\frac{\sigma_x - \sigma_y}{2}\right) \Big/ \sqrt{\left(\frac{\sigma_x - \sigma_y}{2}\right)^2 + \tau_{xy}{}^2}$$

If either of these two sets of trigonometric relations are substituted into Eq. 9–1 and simplified, we obtain

$$\boxed{\sigma_{1,2} = \frac{\sigma_x + \sigma_y}{2} \pm \sqrt{\left(\frac{\sigma_x - \sigma_y}{2}\right)^2 + \tau_{xy}{}^2}} \tag{9–5}$$

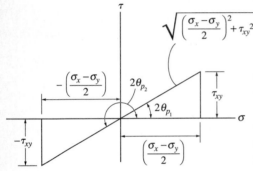

Fig. 9–8

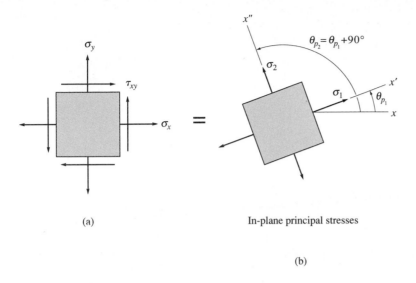

(a)

In-plane principal stresses

(b)

Fig. 9–9

Depending on the sign chosen, this result gives the maximum or minimum in-plane normal stress acting at a point, where $\sigma_1 \geq \sigma_2$. This particular set of values are called in-plane *principal stresses,* and the corresponding planes on which they act are called the *principal planes* of stress, Fig. 9–9*b*. Furthermore, if the above trigonometric relations for θ_{p_1} and θ_{p_2} are substituted into Eq. 9–2, it may be seen that $\tau_{x'y'} = 0$; that is, *no shear stress acts on the principal planes.*

To summarize, then, the state of stress at a point, if initially represented by the stress components acting on an element oriented as shown in Fig. 9–9*a*, can instead be represented by the normal stress components acting on an element oriented as shown in Fig. 9–9*b*. The faces of this element represent principal planes of stress defined by the angles θ_{p_1} and θ_{p_2}. In this position, only the maximum and minimum in-plane normal stresses (principal stresses) σ_1 and σ_2 act on the element; that is, the element in this position is *not* subjected to shear stress.

Maximum In-Plane Shear Stress. The orientation of an element that is subjected to maximum shear stress on its faces can be determined by taking the derivative of Eq. 9–2 with respect to θ and setting the result equal to zero. This gives

$$\tan 2\theta'_s = \frac{-(\sigma_x - \sigma_y)/2}{\tau_{xy}} \qquad (9\text{–}6)$$

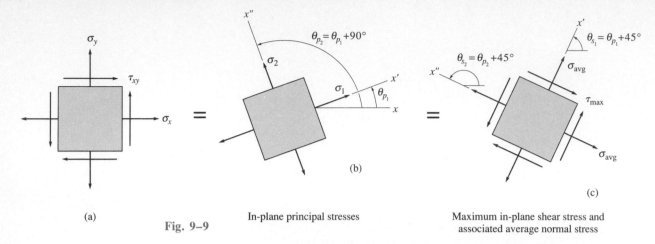

(a)

Fig. 9–9

In-plane principal stresses

Maximum in-plane shear stress and
associated average normal stress

The two roots of this equation, θ_{s_1} and θ_{s_2}, can be determined from the triangles shown in Fig. 9–10. By comparison with Fig. 9–8, each root of $2\theta_s$ is 90° from $2\theta_p$. Thus, the roots θ_s and θ_p are 45° apart, and as a result the planes for maximum shear stress can be determined by orienting an element 45° from the position of an element that defines the planes of principal stress.

Using either one of the roots θ_{s_1} or θ_{s_2}, the maximum shear stress can be determined by taking the trigonometric values of $\sin 2\theta_s$ and $\cos 2\theta_s$ from Fig. 9–10 and substituting them into Eq. 9–2. The result is

$$\tau_{max} = \sqrt{\left(\frac{\sigma_x - \sigma_y}{2}\right)^2 + \tau_{xy}^2} \qquad (9\text{–}7)$$

In Sec. 9.5 we will show that the absolute maximum shear stress on the element may, in some cases, lie in a plane that is perpendicular to the one considered here. For this reason, the value of τ_{max} as calculated by Eq. 9–7 will be referred to as the *maximum in-plane shear stress* because it acts on the element in the x–y plane.

Substituting the values for $\sin 2\theta_s$ and $\cos 2\theta_s$ into Eq. 9–1, we see that there is also a normal stress on the planes of maximum in-plane shear stress. We get

$$\sigma_{avg} = \frac{\sigma_x + \sigma_y}{2} \qquad (9\text{–}8)$$

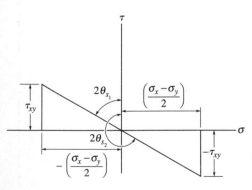

Fig. 9–10

To summarize, the planes of maximum in-plane shear stress are oriented 45° from the planes of principal stress, Fig. 9–9c. The shear stress on these planes is defined by Eq. 9–7; furthermore, on these planes there is also an associated average normal stress σ_{avg}, Eq. 9–8.

Like the stress-transformation equations, it may be convenient to program all the above equations for use on a pocket calculator or a microcomputer. The following examples illustrate their numerical application.

The state of plane stress at a point on a body is shown on the element in Fig. 9–11a. Represent the stress state at the point in terms of the principal stresses.

SOLUTION

From the sign convention established in Fig. 9–5a, we have

$$\sigma_x = -20 \text{ MPa} \qquad \sigma_y = 90 \text{ MPa} \qquad \tau_{xy} = 60 \text{ MPa}$$

The orientation of the principal planes of stress is determined from Eq. 9–4.

$$\tan 2\theta_p = \frac{\tau_{xy}}{(\sigma_x - \sigma_y)/2} = \frac{60}{(-20 - 90)/2}$$

Solving, and referring to this root as θ_{p_2}, for the reason to be stated below, gives

$$2\theta_{p_2} = -47.49° \qquad \theta_{p_2} = -23.7°$$

Since the difference between $2\theta_{p_1}$ and $2\theta_{p_2}$ is 180°, we have

$$2\theta_{p_1} = 180° + 2\theta_{p_2} = 132.51° \qquad \theta_{p_1} = 66.3°$$

Recall that an angle θ is measured positive *counterclockwise* from the x axis to the outward normal (x' axis) on the face of the element, Fig. 9–11b.

The principal stresses are determined from Eq. 9–5. We have

$$\sigma_{1,2} = \frac{\sigma_x + \sigma_y}{2} \pm \sqrt{\left(\frac{\sigma_x - \sigma_y}{2}\right)^2 + \tau_{xy}^2}$$

$$= \left(\frac{-20 + 90}{2}\right) \pm \sqrt{\left(\frac{-20 - 90}{2}\right)^2 + (60)^2}$$

$$= 35.0 \pm 81.4$$

$$\sigma_1 = 116 \text{ MPa} \qquad\qquad\qquad Ans.$$

$$\sigma_2 = -46.4 \text{ MPa} \qquad\qquad\qquad Ans.$$

The principal plane on which each of these normal stresses acts can be determined by applying Eq. 9–1 with, say, $\theta = \theta_{p_2} = -23.7°$. We have

$$\sigma_{x'} = \frac{\sigma_x + \sigma_y}{2} + \frac{\sigma_x - \sigma_y}{2} \cos 2\theta + \tau_{xy} \sin 2\theta$$

$$= \left(\frac{-20 + 90}{2}\right) + \left(\frac{-20 - 90}{2}\right) \cos 2(-23.7°) + 60 \sin 2(-23.7°)$$

$$= -46.4 \text{ MPa}$$

Hence, $\sigma_2 = -46.4$ MPa acts on the plane defined by $\theta_{p_2} = -23.7°$, whereas $\sigma_1 = 116$ MPa acts on the plane defined by $\theta_{p_1} = 66.3°$. The principal stresses are shown on the element in Fig. 9–11c. As expected, no shear stress acts on this element.

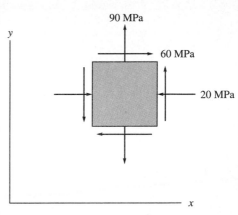

(a)

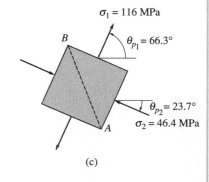

(b)

(c)

Fig. 9–11

Example 9–5

The state of plane stress at a point on a body is represented on the element shown in Fig. 9–12a. Represent the stress state at the point in terms of the maximum in-plane shear stress and average normal stress.

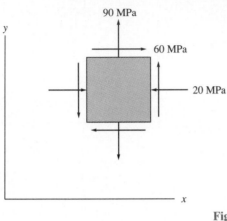

(a)

Fig. 9–12

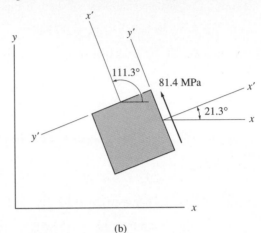

(b)

SOLUTION

The orientation of the planes for maximum in-plane shear stress is determined from Eq. 9–6. Since $\sigma_x = -20$ MPa, $\sigma_y = 90$ MPa, and $\tau_{xy} = 60$ MPa, we have

$$\tan 2\theta_s = \frac{-[(\sigma_x - \sigma_y)/2]}{\tau_{xy}}$$

$$= \frac{-(-20 - 90)/2}{60}$$

$$2\theta_{s_2} = 42.5° \qquad \theta_{s_2} = 21.3°$$
$$2\theta_{s_1} = 180° + 2\theta_{s_2} \qquad \theta_{s_1} = 111.3°$$

These angles can also be obtained in a more direct manner if the directions of the principal planes of stress are known, Example 9–4. Realizing that the planes of maximum shear stress are 45° away from the principal planes of stress, then, from Example 9–4, $\theta_{s_1} = \theta_{p_1} + 45° = 111.3°$ and $\theta_{s_2} = \theta_{p_2} + 45° = 21.3°$, Fig. 9–12b.

The maximum shear stress on these planes is determined from Eq. 9–7; that is,

$$\tau_{\max} = \sqrt{\left(\frac{\sigma_x - \sigma_y}{2}\right)^2 + \tau_{xy}{}^2}$$

$$= \sqrt{\left(\frac{-20 - 90}{2}\right)^2 + (60)^2} = 81.4 \text{ MPa}$$

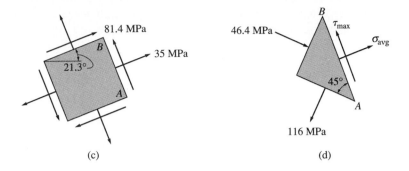

(c)

(d)

The proper direction of τ_{max} on the element can be determined in one of two ways. Applying Eq. 9–2 with, say, $\theta = \theta_{s_2} = 21.3°$, we have

$$\tau_{x'y'} = -\left(\frac{\sigma_x - \sigma_y}{2}\right) \sin 2\theta + \tau_{xy} \cos 2\theta$$

$$= -\left(\frac{-20 - 90}{2}\right) \sin 2(21.3°) + 60 \cos 2(21.3°)$$

$$= 81.4 \text{ MPa}$$

Thus, $\tau_{max} = \tau_{x'y'}$ acts in the *positive y'* direction on this face ($\theta = 21.3°$), Fig. 9–12*b*. The shear stresses on the other three faces are directed as shown in Fig. 9–12*c*.

Assuming that the principal stress and the direction of the principal planes of stress are known, it is also possible to determine the direction of τ_{max} using equilibrium principles. The element in Fig. 9–11*c* is sectioned along one of its diagonals, say *AB* (a plane of maximum shear stress), and one of the segments is then isolated, Fig. 9–12*d*. By observation, to preserve force equilibrium along *AB*, τ_{max} must create a force that acts upward on this face of the element in order to balance the downward force components produced by the principal stresses 46.4 MPa and 116 MPa.

Besides the maximum shear stress, as calculated above, the element is also subjected to an average normal stress determined from Eq. 9–8; that is,

$$\sigma_{avg} = \frac{\sigma_x + \sigma_y}{2}$$

$$= \frac{-20 + 90}{2} = 35 \text{ MPa} \qquad Ans.$$

This is a tensile stress, since it is calculated as a positive quantity. The results are shown in Fig. 9–12*c*.

Example 9–6

Due to the applied loading, the element at point A on the cantilevered beam in Fig. 9–13a is subjected to the state of stress shown. Determine the principal stresses acting at point A.

SOLUTION

According to our sign convention,

$$\sigma_x = 20 \text{ ksi} \qquad \sigma_y = 0 \qquad \tau_{xy} = 3 \text{ ksi}$$

Orientation of the principal planes of stress is determined from Eq. 9–4; that is,

$$\tan 2\theta_p = \frac{\tau_{xy}}{(\sigma_x - \sigma_y)/2}$$

$$= \frac{3}{(20 - 0)/2}$$

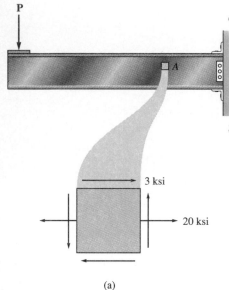

So that

$$2\theta_{p_1} = 16.70° \qquad\qquad \theta_{p_1} = 8.35°$$
$$2\theta_{p_2} = 180° + 16.70° = 196.7° \qquad \theta_{p_2} = 98.4°$$

The principal stresses are found from Eq. 9–5,

$$\sigma_{1,2} = \left(\frac{\sigma_x + \sigma_y}{2}\right) \pm \sqrt{\left(\frac{\sigma_x - \sigma_y}{2}\right)^2 + \tau_{xy}^2}$$

$$= \left(\frac{20 + 0}{2}\right) \pm \sqrt{\left(\frac{20 - 0}{2}\right)^2 + (3)^2}$$

$$= 10 \pm 10.440$$

$$\sigma_1 = 20.4 \text{ ksi} \qquad\qquad\qquad\qquad Ans.$$

$$\sigma_2 = -0.440 \text{ ksi} \qquad\qquad\qquad Ans.$$

(a)

To determine the face over which each of these stresses acts, apply Eq. 9–1, with $\theta = \theta_{p_1} = 8.35°$. We have

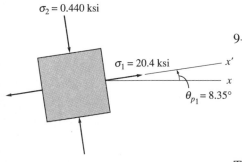

$$\sigma_{x'} = \left(\frac{\sigma_x + \sigma_y}{2}\right) + \left(\frac{\sigma_x - \sigma_y}{2}\right)\cos 2\theta + \tau_{xy}\sin 2\theta$$

$$= \left(\frac{20 + 0}{2}\right) + \left(\frac{20 - 0}{2}\right)\cos 2(8.35°) + 3\sin 2(8.35°)$$

$$= 20.4 \text{ ksi}$$

(b)

Fig. 9–13

Thus, $\sigma_{x'} = \sigma_1 = 20.4$ ksi acts on the principal plane defined by $\theta_{p_1} = 8.35°$. In other words, the notation using θ_{p_1} to represent the angle $8.35°$ is correct. The final results are shown in Fig. 9–13b.

PROBLEMS

9–1. Prove that the sum of the normal stresses $\sigma_x + \sigma_y = \sigma_{x'} + \sigma_{y'}$ is constant.

9–2. The state of stress at a point in a member is shown on the element. Determine the stress components acting on the inclined plane AB. Solve the problem using the method of equilibrium described in Sec. 9.1.

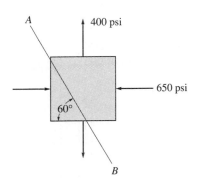

Prob. 9–2

9–3. The state of stress at a point in a member is shown on the element. Determine the stress components acting on the inclined plane AB. Solve the problem using the method of equilibrium described in Sec. 9.1.

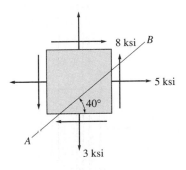

Prob. 9–3

*9–4. The state of stress at a point in a member is shown on the element. Determine the stress components acting on the inclined plane AB. Solve the problem using the method of equilibrium described in Sec. 9.1.

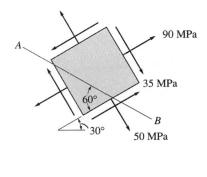

Prob. 9–4

9–5. Solve Prob. 9–2 using the stress-transformation equations developed in Sec. 9.2.

9–6. Solve Prob. 9–3 using the stress-transformation equations developed in Sec. 9.2.

9–7. Solve Prob. 9–4 using the stress-transformation equations developed in Sec. 9.2.

*9–8. Determine the equivalent state of stress on an element if the element is oriented 60° clockwise from the element shown. Use the stress-transformation equations.

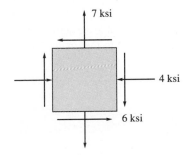

Prob. 9–8

9–9.   Determine the equivalent state of stress on an element if the element is oriented 30° clockwise from the element shown. Use the stress-transformation equations.

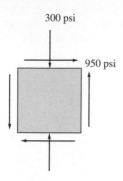

300 psi

950 psi

Prob. 9–9

9–10.   Determine the equivalent state of stress on an element if the element is oriented 50° counterclockwise from its position shown.

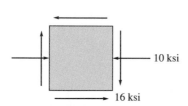

10 ksi

16 ksi

Prob. 9–10

9–11.   The wood beam is subjected to a load of 12 kN. If grains of wood in the beam at point A make an angle of 25° with the horizontal as shown, determine the normal and shear stress that act perpendicular and parallel to the grains due to the loading.

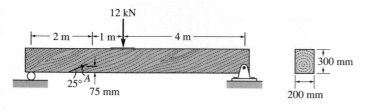

12 kN

2 m   1 m   4 m

25° A
75 mm

300 mm

200 mm

Prob. 9–11

*9–12.   The grains of wood in the board make an angle of 20° with the horizontal as shown. Determine the normal and shear stress that act perpendicular and parallel to the grains if the board is subjected to an axial load of 250 N.

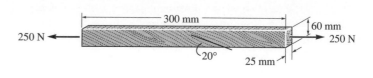

300 mm

60 mm

250 N   250 N

20°

25 mm

Prob. 9–12

9–13.   The state of stress at a point is shown on the element. Determine (a) the principal stresses and (b) the maximum in-plane shear stress and average normal stress at the point. Specify the orientation of the element in each case.

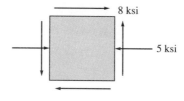

8 ksi

5 ksi

Prob. 9–13

9–14.   The state of stress at a point is shown on the element. Determine (a) the principal stresses and (b) the maximum in-plane shear stress and average normal stresses at the point. Specify the orientation of the element in each case.

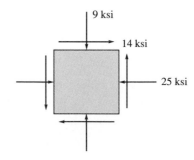

9 ksi

14 ksi

25 ksi

Prob. 9–14

9–15. The state of stress at a point is shown on the element. Determine (*a*) the principal stresses and (*b*) the maximum in-plane shear stress and the average normal stress at the point. Specify the orientation of the element in each case.

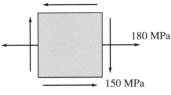

180 MPa

150 MPa

Prob. 9–15

*9–16. The state of stress at a point on the upper surface of the wing is shown on the element. Determine (*a*) the principal stresses and (*b*) the maximum in-plane shear stress and average normal stress at the point. Specify the orientation of the element in each case.

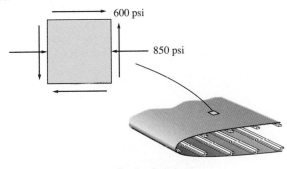

600 psi

850 psi

Prob. 9–16

9–17. The state of stress at a point is shown on the element. Determine (*a*) the principal stresses and (*b*) the maximum in-plane shear stress and average normal stress on the element. Specify the orientation of the element in each case.

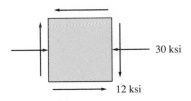

30 ksi

12 ksi

Prob. 9–17

9–18. The square steel plate has a thickness of 10 mm and is subjected to the edge loading shown. Determine the maximum in-plane shear stress and the average normal stress developed in the steel.

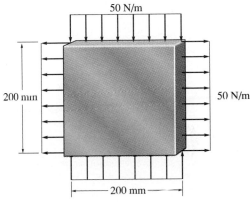
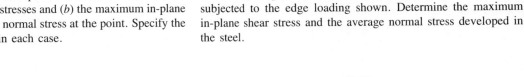

50 N/m

200 mm

50 N/m

200 mm

Prob. 9–18

9–19. The square steel plate has a thickness of 0.5 in. and is subjected to the edge loading shown. Determine the principal stresses developed in the steel.

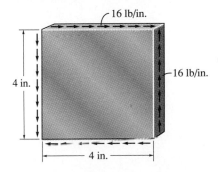

16 lb/in.

16 lb/in.

4 in.

4 in.

Prob. 9–19

*9–20. A point on a thin plate is subjected to the two successive states of stress shown. Determine the resultant state of stress represented on the element orientated as shown on the right.

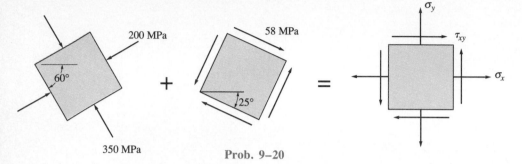

Prob. 9–20

9–21. The internal loadings at a section of the beam consist of an axial force of 500 N, a shear force of 800 N, and two moment components of 30 N · m and 40 N · m. Determine the principal stresses at point A. Also compute the maximum in-plane shear stress at this point.

9–22. The internal loadings at a section of the beam consist of an axial force of 500 N, a shear force of 800 N, and two moment components of 30 N · m and 40 N · m. Determine the principal stresses at point B. Also compute the maximum in-plane shear stress at this point.

9–23. The internal loadings at a section of the beam consist of an axial force of 500 N, a shear force of 800 N, and two moment components of 30 N · m and 40 N · m. Determine the principal stresses at point C. Also compute the maximum in-plane shear stress at this point.

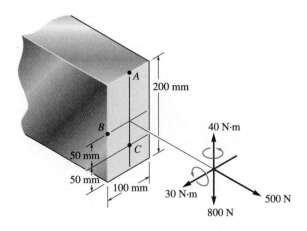

Probs. 9–21/9–22/9–23

*9–24. A steel pipe has an inner diameter of 2.75 in. and an outer diameter of 3 in. If it is fixed at C and subjected to the horizontal force acting on the handle of the pipe wrench at its end, determine the principal stresses in the pipe at point A.

Prob. 9–24

9–25. The clamp bears down on the smooth surfaces at C and D by tightening the bolt. If the tensile force in the bolt is 40 kN, determine the principal stress at points A and B and show the results on elements located at each of these points. The cross-sectional area at A and B is shown in the adjacent figure.

9–27. Determine the principal stresses acting at point A of the supporting frame. Show the results on an element located at the point.

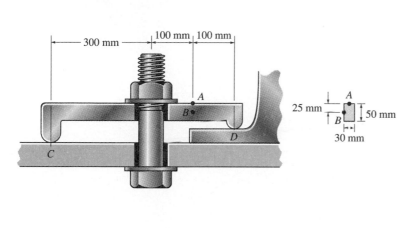

Prob. 9–25

Prob. 9–27

9–26. The cantilevered rectangular bar is subjected to the force of 5 kip. Determine the principal stresses at points A and B.

*9–28. The clamp exerts a force of 150 lb on the boards at G. Determine the axial force in each screw, AB and CD, and then compute the principal stresses at points E and F. Show the results on elements located at these points. The section through EF is rectangular and is 1 in. wide.

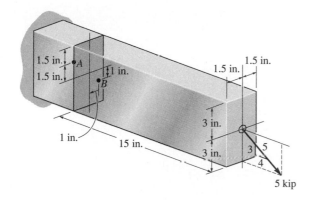

Prob. 9–26

Prob. 9–28

9–29. The bell crank is pinned at *A* and supported by a short link *BC*. If it is subjected to the force of 80 N, determine the principal stresses at (*a*) point *D* and (*b*) point *E*. The crank is constructed from an aluminum plate having a thickness of 20 mm.

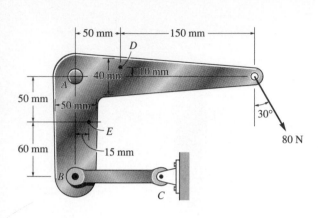

Prob. 9–29

9–31. The cantilevered beam is subjected to the load at its end. Determine the principal stresses in the beam at points *A* and *B*.

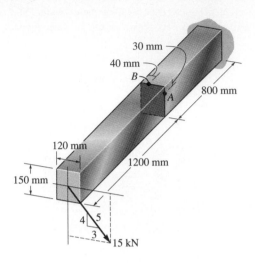

Prob. 9–31

9–30. The box beam is subjected to the 26-kN force that is applied at the center of its width, 75 mm from each side. Determine the principal stresses at points *A* and *B* and show the results on elements located at each of these points. Use the shear formula to compute the shear stress.

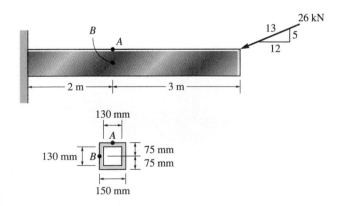

Prob. 9–30

*9–32. The wide-flange beam is subjected to the 50-kN force. Determine the principal stress in the beam at point *A* located on the *web* at the bottom of the upper flange and at point *B* located on the *web* at the top of the bottom flange. Although not very accurate, use the shear formula to compute the shear stress.

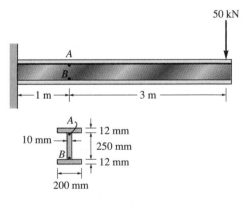

Prob. 9–32

9–33. The bolt is fixed to its support at C. If a force of 18 lb is applied to the wrench to tighten it, determine the principal stresses developed in the bolt shank at points A and B. Represent the results on an element located at each of these points. The shank has a diameter of 0.25 in.

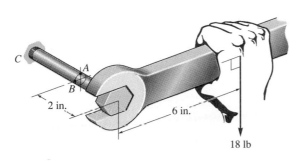

Prob. 9–33

9–34. The solid shaft is subjected to a torque, bending moment, and shear force as shown. Determine the principal stresses acting at points A and B.

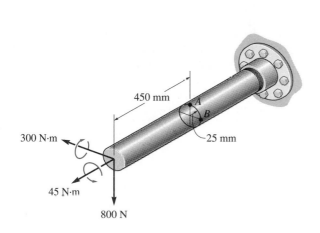

Prob. 9–34

9–35. The internal loadings at a cross section through the 6-in.-diameter drive shaft of a turbine consist of an axial force of 2500 lb, a bending moment of 800 lb · ft, and a torsional moment of 1500 lb · ft. Determine the principal stresses at point A. Also compute the maximum in-plane shear stress at this point.

*9–36. The internal loadings at a cross section through the 6-in.-diameter drive shaft of a turbine consist of an axial force of 2500 lb, a bending moment of 800 lb · ft, and a torsional moment of 1500 lb · ft. Determine the principal stresses at point B. Also compute the maximum in-plane shear stress at this point.

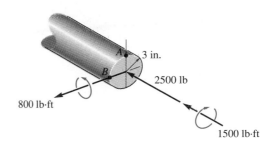

Probs. 9–35/9–36

9–37. The box beam is subjected to the loading shown. Determine the principal stresses in the beam at points A and B.

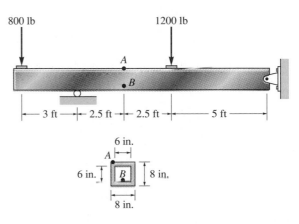

Prob. 9–37

9–38. A bar has a circular cross section with a diameter of 1 in. It is subjected to a torque and a bending moment. At the point of maximum bending stress the principal tensile stresses are 20 ksi and −10 ksi. Determine the torque and the bending moment.

9.4 Mohr's Circle—Plane Stress

In this section we will show that the equations for plane stress transformation have a graphical solution that is often convenient to use and easy to remember. Furthermore, this approach will allow us to "visualize" how the normal and shear stress components vary as the plane on which they act is orientated in different directions.

Equations 9–1 and 9–2 can be rewritten in the form

$$\sigma_{x'} - \left(\frac{\sigma_x + \sigma_y}{2}\right) = \left(\frac{\sigma_x - \sigma_y}{2}\right)\cos 2\theta + \tau_{xy}\sin 2\theta \qquad (9\text{–}9)$$

$$\tau_{x'y'} = -\left(\frac{\sigma_x - \sigma_y}{2}\right)\sin 2\theta + \tau_{xy}\cos 2\theta \qquad (9\text{–}10)$$

The parameter θ can be *eliminated* by squaring each equation and adding the equations together. The result is

$$\left[\sigma_{x'} - \left(\frac{\sigma_x + \sigma_y}{2}\right)\right]^2 + \tau_{x'y'}^2 = \left(\frac{\sigma_x - \sigma_y}{2}\right)^2 + \tau_{xy}^2$$

For a specific problem, σ_x, σ_y, and τ_{xy} are *known constants*. Thus the above equation can be written in a more compact form as

$$(\sigma_{x'} - \sigma_{\text{avg}})^2 + \tau_{x'y'}^2 = R^2 \qquad (9\text{–}11)$$

where

$$\sigma_{\text{avg}} = \frac{\sigma_x + \sigma_y}{2}$$

$$R = \sqrt{\left(\frac{\sigma_x - \sigma_y}{2}\right)^2 + \tau_{xy}^2} \qquad (9\text{–}12)$$

If we establish coordinate axes, σ *positive to the right* and τ *positive downward,* and then plot Eq. 9–11, it will be seen that this equation repre-

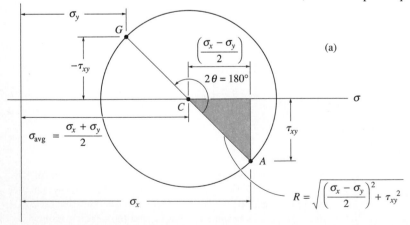

(a)

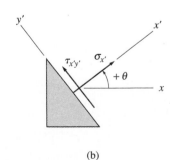

(b)

Fig. 9–14

sents the equation of a *circle* having a radius R and center on the σ axis at point $C(\sigma_{avg}, 0)$, Fig. 9–14a. This circle is called *Mohr's circle,* because it was developed by the German engineer Otto Mohr.

Stress Components on an Arbitrary Plane.

Since Mohr's circle represents the stress-transformation equations graphically, it can be used to determine the normal and shear stress components $\sigma_{x'}$ and $\tau_{x'y'}$ acting on any arbitrary plane, Fig. 9–14b. For example, consider the case when $\theta = 0°$, that is, when the x' axis is coincident with the x axis, Fig. 9–14c. Here the stress-transformation Eqs. 9–1 and 9–2 give the expected values of $\sigma_{x'} = \sigma_x$ and $\tau_{x'y'} = \tau_{xy}$. This "reference" point $A(\sigma_x, \tau_{xy})$ is plotted in Fig. 9–14a. The radius of the circle is therefore CA. Its value can be calculated from the shaded triangle by using the Pythagorean theorem, which gives the expected result of Eq. 9–12. Now, when $\theta = 90°$, Fig. 9–14d, Eqs. 9–1 and 9–2 give $\sigma_{x'} = \sigma_y$ and $\tau_{x'y'} = -\tau_{xy}$. This point is plotted as point $B(\sigma_y, -\tau_{xy})$ on the circle, Fig. 9–14a. By inspection we see that the radial line CG is $2\theta = 180°$ away from the radial line CA.

As a general statement, then, the normal and shear stress components $\sigma_{x'}$, $\tau_{x'y'}$ acting on any specified plane defined by the x' axis and orientated at an angle θ from the x axis, Fig. 9–15a, can be found from the coordinates of point P on the circle, which is orientated at an angle 2θ measured *from* the radial "reference" line CA ($\theta = 0°$) *to* the radial line CP, Fig. 9–15b. It is important to realize that if the τ axis is constructed *positive downward* as shown, then the angle 2θ on the circle is measured in the *same direction* as the angle θ for the orientation of the plane.*

*If instead the τ axis is constructed *positive upwards,* then the angle 2θ on the circle would be measured in the *opposite direction* of the orientation θ of the plane.

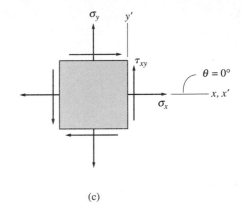

(c)

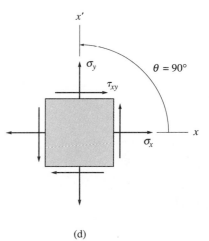

(d)

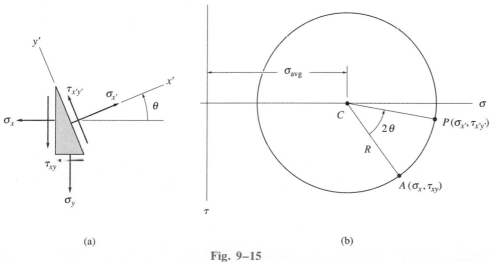

(a) (b)

Fig. 9–15

Principal Stresses. The points B and D, where the circle intersects the σ axis, Fig. 9–16a, give the two values of the principal stress since these points represent the maximum and minimum normal stresses on the element. Note that, as required, the shear stress is zero at these points. From the construction, $\sigma_1 = \sigma_{avg} + R$ and $\sigma_2 = \sigma_{avg} - R$, which is the same as applying Eq. 9–5.

The *counterclockwise* angle $2\theta_{p_1}$ is shown measured from the reference radial line CA ($\theta = 0°$) to the radial line CB. From the shaded triangle, the angle θ_{p_1} is then found from $\tan 2\theta_{p_1} = \tau_{xy} /[(\sigma_x - \sigma_y)/2]$, which is the same as Eq. 9–4. This angle, θ_{p_1}, is then measured in the same direction, i.e., *counterclockwise* from the x (reference) axis to the plane on which σ_1 acts, Fig. 9–16b. Note also on the circle that the angle between the radial lines CB and CD is 180°, and as expected the planes on which σ_1 and σ_2 act are $180°/2 = 90°$ apart.

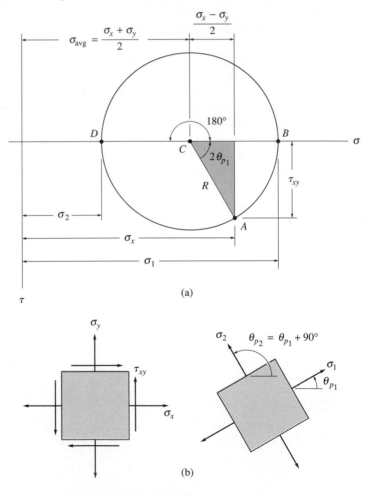

(a)

(b)

Fig. 9–16

Maximum In-Plane Shear Stress. The maximum in-plane shear stress is identified on the circle as either point E or F, Fig. 9–17a. These points have coordinates $E(\sigma_{avg}, R)$ and $F(\sigma_{avg}, -R)$, which represent the same values computed from Eqs. 9–7 and 9–8. Since CE is 90° from CB, Fig. 9–17a, the planes of maximum shear stress are 90°/2 = 45° from the principal planes of stress, as expected. Also, note that the *clockwise* angle, measured from CA to CE, is $2\theta_{s_1}$. The angle θ_{s_1} can be computed from trigonometry using the data on the circle. This angle is then measured *clockwise* from the x (reference) axis to the x' axis, which identifies the plane that has σ_{avg} and $(\tau_{x'y'})_{max}$ acting on it, Fig. 9–17b. Once these components are established, the stress components on the other three faces of the element can then be constructed as shown.

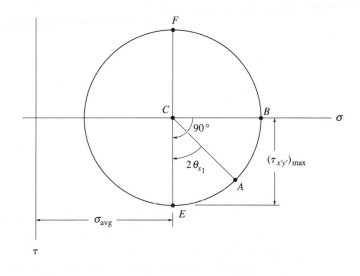

(a)

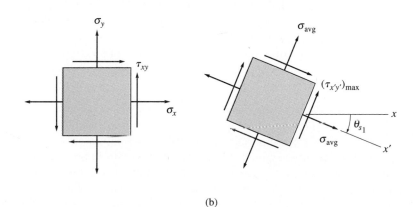

(b)

Fig. 9–17

PROCEDURE FOR ANALYSIS

Mohr's circle represents graphically the state of stress at a point. Once established, it can be used to determine the principal stresses, the maximum in-plane shear stress, or the normal and shear stress components acting on an arbitrary inclined plane. The following steps are required to draw and use the circle.

Construction of the Circle.   Establish a coordinate system such that the abscissa represents the normal stress σ, with *positive to the right,* and the ordinate represents the shear stress τ, with *positive downward,* Fig. 9–18c.

Using the positive sign convention for σ_x, σ_y, τ_{xy}, as shown in Fig. 9–18a, determine the center of the circle C, which is located on the σ axis at a distance $\sigma_{avg} = (\sigma_x + \sigma_y)/2$ from the origin, Fig. 9–18c.

Plot the reference point A having coordinates $A(\sigma_x, \tau_{xy})$. This point represents the normal and shear stress components on the element's right-hand vertical face, and since the x' axis coincides with the x axis, this represents $\theta = 0°$, Fig. 9–18a.

Connect point A with the center C of the circle and determine the hypotenuse CA by trigonometry. This distance represents the radius R of the circle, Fig. 9–18c.

Once R has been determined, sketch the circle.

By constructing the circle in this manner, the stress components $\sigma_{x'}$, $\tau_{x'y'}$ on any plane oriented at an angle θ, Fig. 9–18b, are represented by the coordinates of point P on the circle, measured 2θ from the reference line CA, Fig. 9–18c. Both angles are measured in the *same direction*.

Principal Stresses.   The principal stresses σ_1 and σ_2 are determined from the two points of intersection of the circle with the σ axis, i.e., the points B and D where $\tau = 0$, Fig. 9–16a. The principal stresses act on planes defined by the angles θ_{p_1} and θ_{p_2} which can be determined from the circle using trigonometry. They are represented by angles $2\theta_{p_1}$ and $2\theta_{p_2}$ and are measured *from* the radial reference line CA *to* lines CB and CD. Actually, only one of these angles needs to be calculated, since they are 90° apart. Remember that the direction of rotation $2\theta_p$ on the circle represents the *same* direction of rotation θ_p from the reference axis ($+x$) to the principal plane ($+x'$). The angles $2\theta_{p_1}$ and θ_{p_1} are shown in Fig. 9–16a and 9–16b, respectively.

Maximum In-Plane Shear Stress.   The average normal stress and maximum in-plane shear stress components are determined from the circle as the coordinates of points E and F, Fig. 9–17a. In this case the angles θ_{s_1} and θ_{s_2} give the orientation of the planes that contain these components. The calculations from the circle are performed in the same manner as for finding the principal stresses and the principal planes. The results are shown in Fig. 9–17b for the element oriented on the basis of calculating both θ_{s_1} and the stress components acting on the plane associated with point E on the circle, Fig. 9–17a.

Stresses on Arbitrary Plane. The normal and shear stress components $\sigma_{x'}$ and $\tau_{x'y'}$ acting on a specified plane defined by the angle θ, Fig. 9–18b, can be obtained from the circle using trigonometry to determine the coordinates of point P, Fig. 9–18c. To locate P, the known angle θ for the plane (in this case counterclockwise) is measured on the circle in the *same direction* 2θ (counterclockwise), *from* the radial reference line CA to the radial line CP. Also, if the value of $\sigma_{y'}$ is required, it can be determined by calculating the σ coordinate of point Q in Fig. 9–18c, since Q lies 180° away from P, and thus represents a rotation of 90° of the x' axis on the element.

The following examples illustrate how to apply this procedure to problems that involve each of the three cases just discussed.

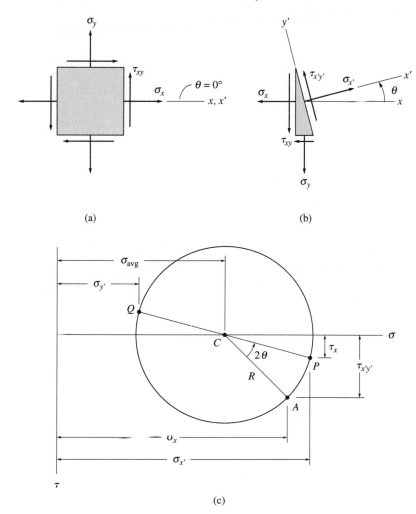

(a)   (b)

(c)

Fig. 9–18

Example 9–7

The state of plane stress at a point is shown on the element in Fig. 9–19a. Determine the principal stresses and the orientation of the element at the point. Also, determine the maximum in-plane shear stresses and the orientation of the element upon which they act.

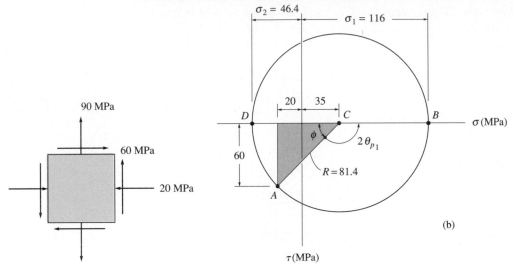

(a)

Fig. 9–19

(b)

SOLUTION

Construction of the Circle.   From the problem data,

$$\sigma_x = -20 \text{ MPa} \qquad \sigma_y = 90 \text{ MPa} \qquad \tau_{xy} = 60 \text{ MPa}$$

The σ, τ axes are established in Fig. 9–19b. Note that the positive τ axis must be directed *downward,* so that rotations of the element correspond to rotations in the same direction around the circle. The center of the circle C is located on the σ axis, at the point

$$\sigma_{\text{avg}} = \frac{-20 + 90}{2} = 35 \text{ MPa}$$

The stress components on the right-hand face of the element are the coordinates of the reference point A $(-20, 60)$, $\theta = 0°$, Fig. 9–19b. Applying the Pythagorean theorem to the shaded triangle to determine the circle's radius CA, we have

$$R = \sqrt{(60)^2 + (55)^2} = 81.4 \text{ MPa}$$

Principal Stresses.   The principal stresses are represented by points B and D in Fig. 9–19b. The σ coordinates of these points are

$$\sigma_1 = 35 + 81.4 = 116 \text{ MPa} \qquad\qquad Ans.$$
$$\sigma_2 = 35 - 81.4 = -46.4 \text{ MPa} \qquad\qquad Ans.$$

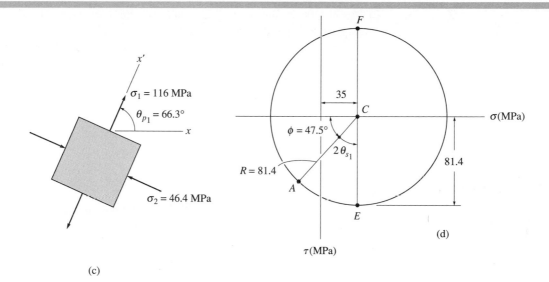

(c)

(d)

The orientation of the element is determined by computing the *counter-clockwise* angle $2\theta_{p_1}$ in Fig. 9–19b, which defines the direction of σ_1 ($\sigma_1 > \sigma_2$) and its associated principal plane. Thus,

$$2\theta_{p_1} = 180° - \phi = 180° - \tan^{-1}\frac{60}{55} = 180° - 47.5° = 132.5°$$

$$\theta_{p_1} = 66.3° \hspace{4cm} \textit{Ans.}$$

The element is oriented such that the x' axis or σ_1 is directed at a *counterclockwise* angle of 66.3° with the horizontal as shown in Fig. 9–19c.

Maximum In-Plane Shear Stress. The maximum in-plane shear stress and the average normal stress are identified by point E or F on the circle in Fig. 9–19d. The coordinates of these points are $E(35, 81.4)$ and $F(35, -81.4)$. Thus,

$$\tau_{max} = 81.4 \text{ MPa} \hspace{3cm} \textit{Ans.}$$

$$\sigma_{avg} = 35 \text{ MPa} \hspace{3cm} \textit{Ans.}$$

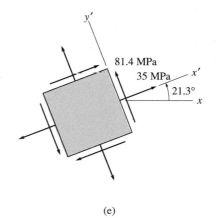

(e)

The *counterclockwise* angle θ_{s_1} can be computed from the circle, identified as $2\theta_{s_1}$. Using the result of $\phi = 47.5°$, computed above, we have

$$2\theta_{s_1} = 90° - 47.5° - 42.5°$$

$$\theta_{s_1} = 21.3° \hspace{3cm} \textit{Ans.}$$

This *counterclockwise* angle defines the direction of σ_{avg} (or the x' axis), Fig. 9–19e. Since point E has *positive* coordinates, then on the associated face of the element, the average normal stress and the maximum in-plane shear stress act in the *positive* x' and y' directions as shown.

Example 9–8

Due to the applied loading, the element at point A on the solid cylinder in Fig. 9–20a is subjected to the state of stress shown. Determine the principal stresses acting at this point.

SOLUTION

Construction of the Circle. From the element's right-hand face,

$$\sigma_x = -12 \text{ ksi} \qquad \sigma_y = 0 \qquad \tau_{xy} = -6 \text{ ksi}$$

The center of the circle is at

$$\sigma_{avg} = \frac{-12 + 0}{2} = -6 \text{ ksi}$$

The initial point $A(-12, -6)$ and the center $C(-6, 0)$ are plotted in Fig. 9–20b. The circle is constructed having a radius of

$$R = \sqrt{(12 - 6)^2 + (6)^2} = 8.49 \text{ ksi}$$

Principal Stresses. The principal stresses are indicated by the coordinates of points B and D. We have, for $\sigma_1 > \sigma_2$,

$$\sigma_1 = 8.49 - 6 = 2.49 \text{ ksi} \qquad\qquad Ans.$$
$$\sigma_2 = -6 - 8.49 = -14.5 \text{ ksi} \qquad\qquad Ans.$$

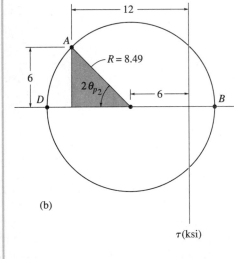

The orientation of the element is determined by computing the *counter-clockwise* angle $2\theta_{p_2}$ in Fig. 9–20b, which defines the direction θ_{p_2} of σ_2 and its associated principal plane. We have

$$2\theta_{p_2} = \tan^{-1}\frac{6}{(12 - 6)} = 45.0°$$
$$\theta_{p_2} = 22.5°$$

The element is oriented such that the x' axis or σ_2 is directed *counterclockwise* 22.5° from the horizontal (x axis) as shown in Fig. 9–20c.

As an exercise, use the circle to show that the maximum in-plane shear stress is shown on the element in Fig. 9–20d.

(b)

Fig. 9–20

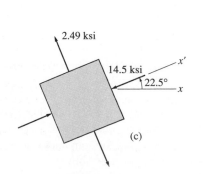

(c)

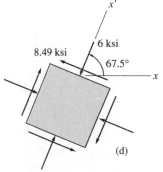

(d)

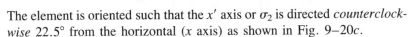

Example 9–9

The state of plane stress at a point is shown on the element in Fig. 9–21a. Represent this state of stress on an element oriented 30° counter-clockwise from the position shown.

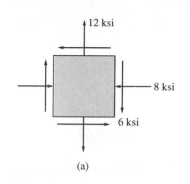

(a)

SOLUTION

Construction of the Circle. From the problem data,

$$\sigma_x = -8 \text{ ksi} \qquad \sigma_y = 12 \text{ ksi} \qquad \tau_{xy} = -6 \text{ ksi}$$

The σ and τ axes are established in Fig. 9–21b. The center of the circle C is on the σ axis at

$$\sigma_{\text{avg}} = \frac{-8 + 12}{2} = 2 \text{ ksi}$$

The initial point for $\theta = 0°$ has coordinates $A(-8, -6)$. Hence from the figure the radius CA is

$$R = \sqrt{(10)^2 + (6)^2} = 11.66$$

Stresses on 30° Element. Since the element is to be rotated 30° *counter-clockwise*, we must construct a radial line CP, $2(30°) = 60°$ *counterclock-wise*, measured from CA ($\theta = 0°$), Fig. 9–21b. The coordinates of point P ($\sigma_{x'}$, $\tau_{x'y'}$) must now be obtained. From the geometry of the circle,

$$\phi = \tan^{-1}\frac{6}{10} = 30.96° \qquad \psi = 60° - 30.96° = 29.04°$$

$$\sigma_{x'} = 2 - 11.66 \cos 29.04° = -8.20 \text{ ksi} \qquad\qquad Ans.$$

$$\tau_{x'y'} = 11.66 \sin 29.04° = 5.66 \text{ ksi} \qquad\qquad Ans.$$

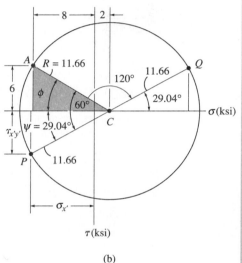

(b)

Using established x', y' axes, these stress components act on face BD of the element shown in Fig. 9–21c.

The stress components acting on an adjacent face are represented by the coordinates of point Q on the circle. This point lies on the radial line CQ, which is 180° from CP (90° from face BD on the element, as expected). The coordinates of point Q are

$$\sigma_{x'} = 2 + 11.66 \cos 29.04° = 12.2 \text{ ksi} \qquad\qquad Ans.$$

$$\tau_{x'y'} = -(11.66 \sin 29.04) = -5.66 \text{ ksi} \qquad \text{(check)}$$

Since CQ is 120° *clockwise* from CA, Fig. 9–21b, the above stress compo-nents act on face DE of the element in Fig. 9–21c since the x' axis for this face is oriented 120°/2 = 60° *clockwise* from the positive x axis.

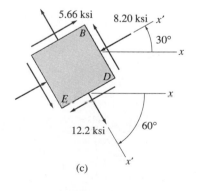

(c)

Fig. 9–21

PROBLEMS

9–39.   Solve Prob. 9–2 using Mohr's circle.

*9–40.   Solve Prob. 9–16 using Mohr's circle.

9–41.   Solve Prob. 9–15 using Mohr's circle.

9–42.   Determine the equivalent state of stress if an element is oriented 60° counterclockwise from the element shown.

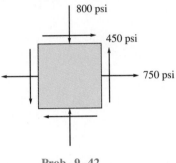

800 psi

450 psi

750 psi

Prob. 9–42

9–43.   Determine the equivalent state of stress if an element is oriented 40° clockwise from the element shown.

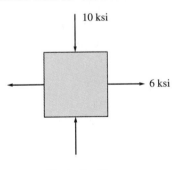

10 ksi

6 ksi

Prob. 9–43

*9–44.   Determine the equivalent state of stress if an element is oriented 25° counterclockwise from the element shown.

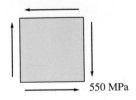

550 MPa

Prob. 9–44

9–45.   Determine (*a*) the principal stress and (*b*) the maximum in-plane shear stress and average normal stress. Specify the orientation of the element in each case.

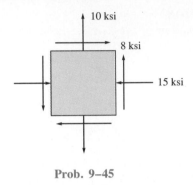

10 ksi

8 ksi

15 ksi

Prob. 9–45

9–46.   Determine (*a*) the principal stress and (*b*) the maximum in-plane shear stress and average normal stress. Specify the orientation of the element in each case.

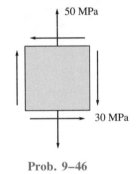

50 MPa

30 MPa

Prob. 9–46

9–47.   Determine (*a*) the principal stress and (*b*) the maximum in-plane shear stress and average normal stress. Specify the orientation of the element in each case.

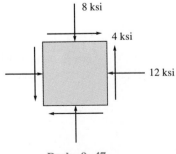

8 ksi

4 ksi

12 ksi

Prob. 9–47

*9–48.   Determine (*a*) the principal stress and (*b*) the maximum in-plane shear stress and average normal stress. Specify the orientation of the element in each case.

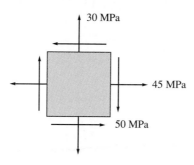

30 MPa

45 MPa

50 MPa

Prob. 9–48

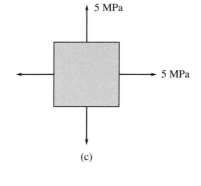

5 MPa

5 MPa

(c)

9–49.   Draw Mohr's circle that describes each of the following states of stress.

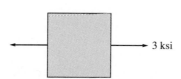

3 ksi

(a)

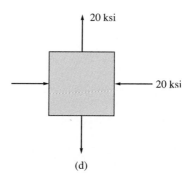

20 ksi

20 ksi

(d)

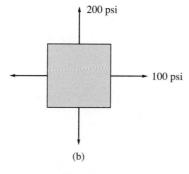

200 psi

100 psi

(b)

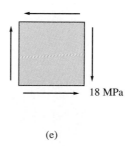

18 MPa

(e)

Prob. 9–49

9–50. A point on a thin plate is subjected to two successive states of stress as shown. Determine the resulting state of stress with reference to an element oriented as shown on the right.

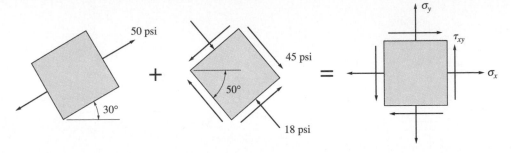

Prob. 9–50

9–51. Mohr's circle for the state of stress in Fig. 9–15a is shown in Fig. 9–15b. Show that finding the coordinates of point $P(\sigma_{x'}, \tau_{x'y'})$ on the circle gives the same value as the stress-transformation Eqs. 9–1 and 9–2.

*9–52. The post has a square cross-sectional area. If it is fixed-supported at its base and a horizontal force is applied at its end as shown, determine (a) the maximum in-plane shear stress developed at A and (b) the principal stresses at A.

9–53. The ladder is supported on the rough surface at A and by a smooth wall at B. If a man weighing 150 lb stands upright at C, determine the principal stress in one of the legs at point D. Each leg is made from a 1-in.-thick board having a rectangular cross section. Assume that the total weight of the man is exerted vertically on the rung at C and is shared equally by each of the ladder's two legs. Neglect the weight of the ladder and the forces developed by the man's arms.

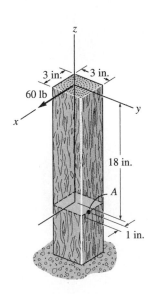

Prob. 9–52

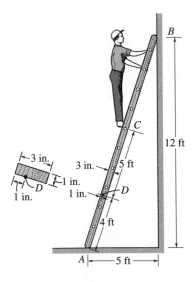

Prob. 9–53

9–54. The crane is used to support the 350-lb load. Determine the principal stresses acting in the boom at points A and B. The cross section is rectangular and has a width of 6 in. and a thickness of 3 in.

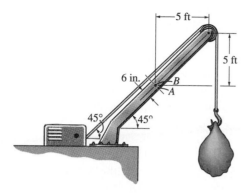

Prob. 9–54

9–55. The thin-walled pipe has an inner diameter of 0.5 in. and a thickness of 0.025 in. If it is subjected to an internal pressure of 50 psi and the axial tension and torsional loadings shown, determine the principal stresses at a point on the surface of the pipe.

Prob. 9–55

*9–56. The pipe has an inner radius of 25 mm and an outer radius of 27 mm. If it is subjected to an internal pressure of 8 MPa and a torsional moment of 500 N · m, determine the principal stresses and the maximum in-plane shear stress at point A, which lies on the pipe's outer surface.

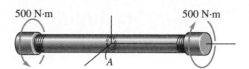

Prob. 9–56

9–57. The spherical pressure vessel has an inner radius of 5 ft and a wall thickness of 0.5 in. Draw Mohr's circle for the state of stress at point A and explain the significance of the result. The vessel is subjected to an internal pressure of 80 psi.

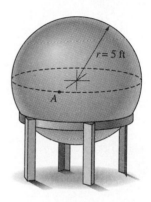

Prob. 9–57

9–58. The cylindrical pressure vessel has an inner radius of 1.25 m and a wall thickness of 15 mm. It is made from steel plates that are welded along the 45° seam. Determine the normal and shear stress components along this seam if the vessel is subjected to an internal pressure of 8 MPa.

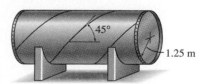

Prob. 9–58

9.5 Absolute Maximum Shear Stress

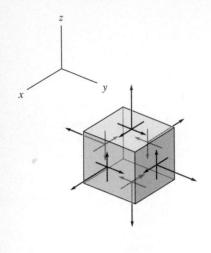

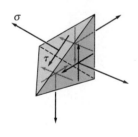

When a point in a body is subjected to a general three-dimensional state of stress, an element of material has a normal-stress and two shear-stress components acting on each of its faces, Fig. 9–22a. Like the case of plane stress, it is possible to develop stress-transformation equations that can be used to determine the normal and shear stress components σ and τ acting on *any* skewed plane of the element, Fig. 9–22b. Furthermore, at the point it is also possible to determine the unique orientation of an element having only principal stresses acting on its faces. As shown in Fig. 9–22c, these principal stresses are assumed to have magnitudes of maximum, intermediate, and minimum intensity, i.e., $\sigma_{max} \geq \sigma_{int} \geq \sigma_{min}$.

A discussion of the transformation of stress in three dimensions is beyond the scope of this text; however, it is discussed in books related to the theory of elasticity. For our purposes, we will assume that the principal orientation of the element and the principal stresses are known, Fig. 9–22c. This is a condition known as *triaxial stress*. If we view the element in two dimensions, that is, in the $y'–z'$, $x'–z'$, and $x'–y'$ planes, Fig. 9–23a, 9–23b, and 9–23c, we can then use Mohr's circle to determine the *maximum in-plane shear stress* for each case. For example, the diameter of Mohr's circle extends between the principal stresses σ_{int} and σ_{min} for the case shown in Fig. 9–23a. From this circle, Fig. 9–23d, the maximum in-plane shear stress is $(\tau_{y'z'})_{max} = (\sigma_{int} - \sigma_{min})/2$, and the associated average normal stress is $(\sigma_{int} + \sigma_{min})/2$. As shown in Fig. 9–23e, the element having these stress components on it must be oriented 45° from the position of the element shown in Fig. 9–23a. Mohr's circles for the elements in Fig. 9–23b and 9–23c have also been constructed in Fig. 9–23d. The corresponding elements having a 45° orientation and subjected to maximum in-plane shear and average normal stress components are shown in Fig. 9–23f and 9–23g, respectively.

(a)

(b)

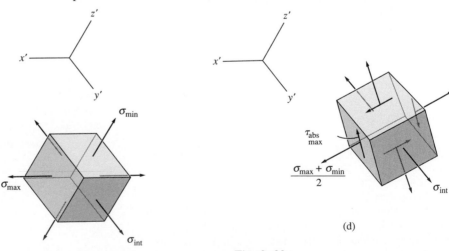

(c)

(d)

Fig. 9–22

Comparing the three circles in Fig. 9–23d, it is seen that the *absolute maximum shear stress*, $\tau_{abs \atop max}$, is defined by the circle having the largest radius, which occurs for the element shown in Fig. 9–23b. In other words, the element in Fig. 9–23f is oriented by a rotation of 45° about the y' axis from the element in Fig. 9–22c. It is shown in Fig. 9–22d. Notice that this condition can also be *determined directly* by simply choosing the maximum and minimum principal stresses from Fig. 9–22c, in which case the absolute maximum shear stress will be

$$\tau_{abs \atop max} = \frac{\sigma_{max} - \sigma_{min}}{2} \qquad (9\text{–}13)$$

And the associated average normal stress will be

$$\sigma_{avg} = \frac{\sigma_{max} + \sigma_{min}}{2} \qquad (9\text{–}14)$$

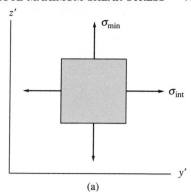

(a)

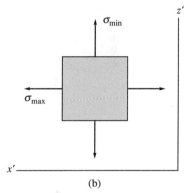

(b)

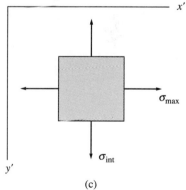

(c)

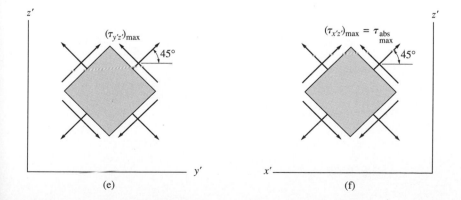

(d)

Fig. 9–23

(e)

(f)

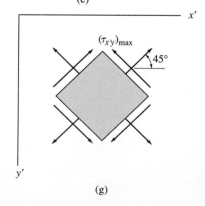

(g)

The analysis considered only the stress components acting on elements located in positions found from rotations about the x', y', or z' axis. If we had used the three-dimensional stress-transformation equations of the theory of elasticity to obtain values of the normal and shear stress components acting on any arbitrary skewed plane at the point, Fig. 9–22b, it could be shown that regardless of the orientation of the plane, specific values of the shear stress τ on the plane will *always be less* than the maximum shear stress computed from Eq. 9–13. Furthermore, the normal stress σ acting on the plane will have a value lying between the maximum and minimum principal stresses, that is, $\sigma_{\max} \geq \sigma \geq \sigma_{\min}$.

Plane Stress. The above results have an important implication for the case of plane stress, particularly when the in-plane principal stresses have the *same sign,* i.e., they are both tensile or both compressive. For example, consider the material to be subjected to plane stress such that the in-plane principal stresses are represented as $\sigma_{\max}$ and σ_{int}, in the x' and y' directions, respectively; while the out-of-plane principal stress in the z' direction is $\sigma_{\min} = 0$, Fig. 9–24a. Mohr's circles that describe this state of stress for element orientations about each coordinate axis are shown in Fig. 9–24b. Here it is seen that although the maximum in-plane shear stress is $(\tau_{x'y'})_{\max} = (\sigma_{\max} - \sigma_{\mathrm{int}})/2$, this value is *not* the absolute maximum shear stress to which the material is subjected. Instead, from Eq. 9–13 or Fig. 9–24b,

$$\tau_{\substack{\mathrm{abs} \\ \max}} = (\tau_{x'z'})_{\max} = \frac{\sigma_{\max} - 0}{2} = \frac{\sigma_{\max}}{2} \qquad (9\text{–}15)$$

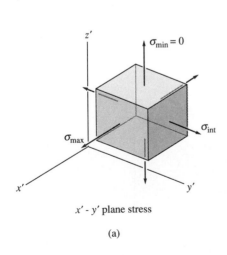

x' - y' plane stress

(a)

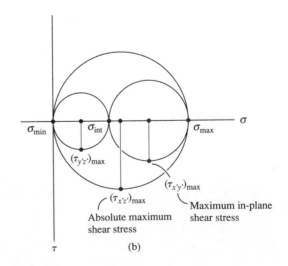

(b)

Fig. 9–24

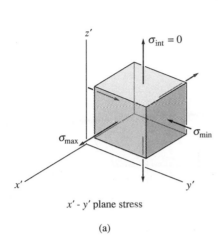

x' - y' plane stress

(a)

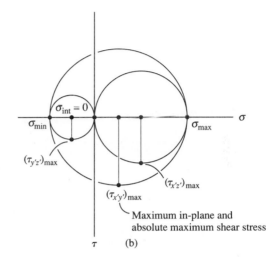

Maximum in-plane and
absolute maximum shear stress

(b)

Fig. 9–25

On the other hand, if one of the in-plane principal stresses had the *opposite sign* of that of the other, then these stresses would be represented as σ_{max} and σ_{min}, and the out-of-plane principal stress $\sigma_{int} = 0$, Fig. 9–25a. Mohr's circles that describe this state of stress for element orientations about each coordinate axis are shown in Fig. 9–25b. Clearly, in this case

$$\tau_{\substack{abs \\ max}} = (\tau_{x'y'})_{max} = \frac{\sigma_{max} - \sigma_{min}}{2} \qquad (9\text{–}16)$$

In summary, then, if the in-plane principal stresses both have the *same sign*, the *absolute maximum shear stress* will occur *out of the plane* and has a value of $\tau_{\substack{abs \\ max}} = \sigma_{max}/2$. However, if the in-plane principal stresses are of *opposite signs*, then the *absolute maximum shear stress equals the maximum in-plane shear stress*; that is, $\tau_{\substack{abs \\ max}} = (\sigma_{max} - \sigma_{min})/2$. Calculation of the absolute maximum shear stress as indicated here is important when designing members made of a ductile material, since the strength of the material depends on its ability to resist shear stress. This situation will be discussed further in Sec. 10.7.

Example 9–10

Due to the applied loading, the element A on the wood frame in Fig. 9–26a is subjected to the state of plane stress shown. Determine the principal stresses and the absolute maximum shear stress at A.

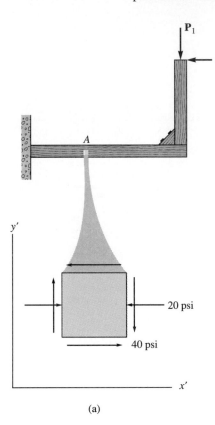

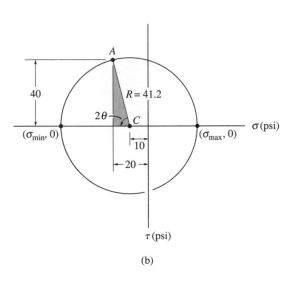

(b)

Fig. 9–26

(a)

SOLUTION

Principal Stresses. The principal planes and principal stresses can be determined from Mohr's circle. The center of the circle is on the σ axis at $\sigma_{\text{avg}} = (-20 + 0)/2 = -10$ psi. Plotting the controlling point $A(-20, -40)$, the circle can be drawn as shown in Fig. 9–26b. The radius is

$$R = \sqrt{(20 - 10)^2 + (40)^2} = 41.2 \text{ psi}$$

The principal stresses are the points where the circle intersects the σ axis; i.e.,

$$\sigma_{\text{max}} = -10 + 41.2 = 31.2 \text{ psi}$$
$$\sigma_{\text{min}} = -10 - 41.2 = -51.2 \text{ psi}$$

From the circle, the *counterclockwise* angle 2θ, measured from CA to the $-\sigma$ axis, is

$$2\theta = \tan^{-1}\left(\frac{40}{(20-10)}\right) = 76.0°$$

Thus,

$$\theta = 38.0°$$

This *counterclockwise* rotation defines the direction of the x' axis or σ_{min} and its associated principal plane, Fig. 9–26c. Since there is no principal stress on the element in the z direction, we have

$$\sigma_{max} = 31.2 \text{ psi} \qquad \sigma_{int} = 0 \qquad \sigma_{min} = -51.2 \text{ psi} \qquad \textit{Ans.}$$

Notice that the notation has been selected so that $\sigma_{max} \geq \sigma_{int} \geq \sigma_{min}$.

Absolute Maximum Shear Stress.   Applying Eqs. 9–13 and 9–14, we have

$$\tau_{\substack{abs \\ max}} = \frac{\sigma_{max} - \sigma_{min}}{2} = \frac{31.2 - (-51.2)}{2} = 41.2 \text{ psi} \qquad \textit{Ans.}$$

$$\sigma_{avg} = \frac{\sigma_{max} + \sigma_{min}}{2} = \frac{31.2 - 51.2}{2} = -10 \text{ psi}$$

These same results can be obtained by drawing Mohr's circle for each orientation of an element about the x', y', and z' axes, Fig. 9–26d. In either case, since σ_{max} and σ_{min} are of *opposite signs,* the state of absolute maximum shear stress results from the 45° rotation of the element in Fig. 9–26c about the z' axis. In other words, the absolute maximum shear stress equals the maximum in-plane shear stress. The properly oriented element is shown in Fig. 9–26e.

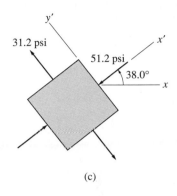

(c)

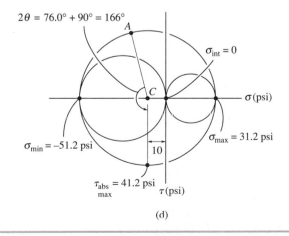

(d)

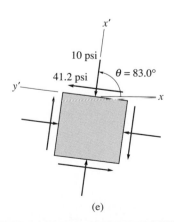

(e)

Example 9–11

Point A on the surface of the cylindrical pressure vessel in Fig. 9–27a is subjected to the state of plane stress. Determine the absolute maximum shear stress at this point.

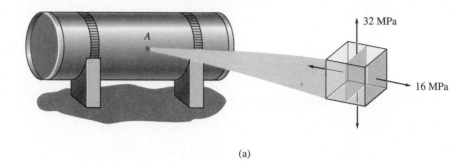

(a)

SOLUTION

The principal stresses are $\sigma_{max} = 32$ MPa, $\sigma_{int} = 16$ MPa, and $\sigma_{min} = 0$. If these stresses are plotted along the σ axis, the three Mohr's circles can be constructed that describe the stress state viewed in each of the three perpendicular planes, Fig. 9–27b. The largest circle has a radius of 16 MPa and describes the state of stress in the plane containing $\sigma_{max} = 32$ MPa and $\sigma_{min} = 0$, shown shaded in Fig. 9–27a. An orientation of an element 45° within this plane yields the state of absolute maximum shear stress and the associated average normal stresses, namely,

$$\tau_{abs \atop max} = 16 \text{ MPa} \qquad\qquad Ans.$$

$$\sigma_{avg} = 16 \text{ MPa}$$

These same results can be obtained from direct application of Eqs. 9–13 and 9–14; that is,

$$\tau_{abs \atop max} = \frac{\sigma_{max} - \sigma_{min}}{2} = \frac{32 - 0}{2} = 16 \text{ MPa} \qquad Ans.$$

$$\sigma_{avg} = \frac{\sigma_{max} + \sigma_{min}}{2} = \frac{32 + 0}{2} = 16 \text{ MPa}$$

By comparison, the maximum in-plane shear stress can be determined by using Mohr's circle drawn between $\sigma_{max} = 32$ MPa and $\sigma_{int} = 16$ MPa, Fig. 9–27b. This gives a value of

$$\tau_{max \atop in\text{-}plane} = 8 \text{ MPa}$$

$$\sigma_{avg} = \frac{32 + 16}{2} = 24 \text{ MPa}$$

Fig. 9–27

PROBLEMS

9–59. Draw the three Mohr's circles that describe each of the following states of stress.

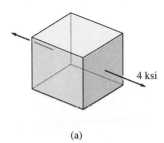

(a)

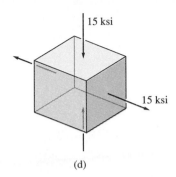

(d)

Prob. 9–59

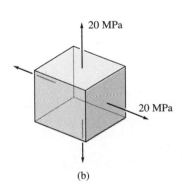

(b)

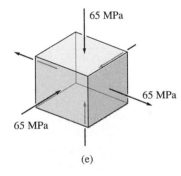

(e)

*9–60. The principal stresses acting at a point in a body are shown. Draw the three Mohr's circles that describe this state of stress and find the maximum in-plane shear stresses and associated average normal stresses for the x–y, y–z, and x–z planes. For each case, show the results on the element oriented in the appropriate direction.

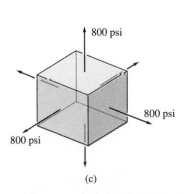

(c)

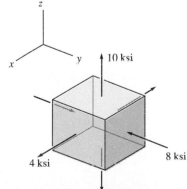

Prob. 9–60

9–61. The stress at a point is shown on the element. Determine the principal stresses and the absolute maximum shear stress.

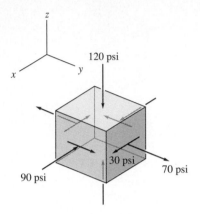

Prob. 9–61

9–63. The state of stress at a point is shown on the element. Determine the principal stresses and the absolute maximum shear stress.

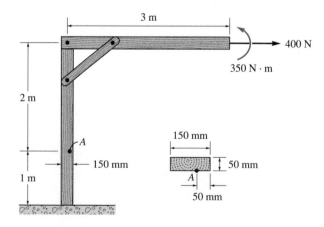

Prob. 9–63

9–62. The stress at a point is shown on the element. Determine the principal stresses and the absolute maximum shear stress.

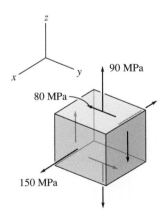

Prob. 9–62

***9–64.** The frame is subjected to a horizontal force and couple moment at its end. Determine the principal stresses and the absolute maximum shear stress at point A. The cross-sectional area at this point is shown.

Prob. 9–64

REVIEW PROBLEMS

9–65. A rod has a circular cross section with a diameter of 2 in. It is subjected to a torque of 12 kip · in. and a bending moment **M**. The greater principal stress at the point of maximum flexural stress is 15 ksi. Determine the magnitude of the bending moment.

Prob. 9–65

9–66. The state of stress at a point is shown on the element. Determine (a) the principal stresses and (b) the maximum in-plane shear stress and average normal stress at the point. Specify the orientation of the element in each case. Use the stress-transformation equations.

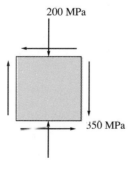

Prob. 9–66

9–67. The principal stresses acting at a point in a body are shown. Draw the three Mohr's circles that describe this state of stress and find the three maximum in-plane shear stresses and associated average normal stresses in the $x–y$, $y–z$, and $x–z$ planes. For each case, show the results on the element oriented in the appropriate direction.

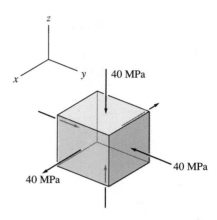

Prob. 9–67

***9–68.** The state of stress at a point is shown on the element. Determine (a) the principal stresses and (b) the maximum in-plane shear stress and average normal stress on the element. Specify the orientation of the element in each case. Use the stress transformation equations.

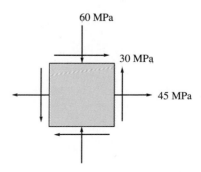

Prob. 9–68

9–69. The drill pipe has an outer diameter of 3 in. and a wall thickness of 0.25 in. If it is subjected to a torque and axial load as shown, determine (a) the principal stresses and (b) the maximum in-plane shear stress at a point on its surface.

Prob. 9–69

9–71. Determine the equivalent state of stress if the element is oriented 30° clockwise from its position shown. Use Mohr's circle.

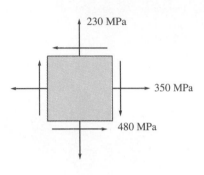

Probs. 9–71

9–70. The beam is subjected to the two forces shown. Determine the principal stresses at point A.

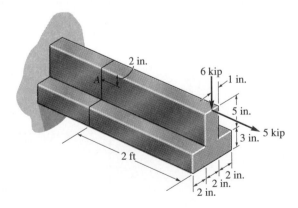

Prob. 9–70

10 Strain Transformation

The transformation of strain at a point is similar to the transformation of stress, and as a result the methods of Chapter 9 will also be used in this chapter. Here we will also discuss various ways for measuring strain and develop some important material-property relationships, including a generalized form of Hooke's law. At the end of the chapter, a few of the theories used to predict the failure of a material will be discussed.

10.1 Plane Strain

As outlined in Sec. 2.2, the general state of strain at a point in a body is represented by a combination of three components of normal strain, ϵ_x, ϵ_y, ϵ_z, and three components of shear strain, γ_{xy}, γ_{xz}, γ_{yz}. These six components tend to deform each face of an element of the material, and like stress, the normal and shear strain *components* at the point will vary according to the orientation of the element. The *strain* components at a point are often determined by using strain gauges, which measure these components in *specified directions*. For both analysis and design, however, engineers must sometimes transform this data in order to obtain the strain components in other directions.

To understand how this is done, we will first confine our attention to a study of *plane strain*. Specifically, we will not consider the effects of the components ϵ_z, γ_{xz}, and γ_{yz}. In general, then, a plane-strained element is subjected to two components of normal strain, ϵ_x, ϵ_y, and one component of shear strain, γ_{xy}. The deformations of an element caused by each of these strains are shown graphically in Fig. 10–1. Note that the normal strains are produced by *changes in length* of the element in the x and y directions, and the

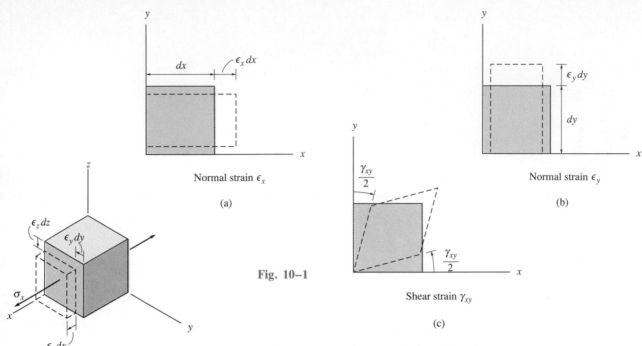

Normal strain ϵ_x

(a)

Fig. 10–1

Normal strain ϵ_y

(b)

Shear strain γ_{xy}

(c)

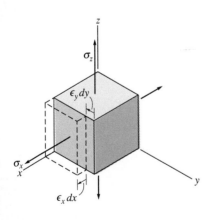

Plane stress, σ_x, does not cause plane strain in the x–y plane since $\epsilon_z \neq 0$

(a)

Plane strain, ϵ_x, ϵ_y, does not cause plane stress in the x–y plane since $\sigma_z \neq 0$

(b)

Fig. 10–2

shear strain is produced by the *relative rotation* of two adjacent sides of the element.

Although plane strain and plane stress both have three components lying in the same plane, realize that plane stress *does not* necessarily cause plane strain or vice versa. The reason for this has to do with the Poisson effect discussed in Sec. 3.6. For example, if the element in Fig. 10–2a is subjected to a uniaxial stress σ_x, not only is a normal strain ϵ_x produced, but there are *also* two associated normal strains, $\epsilon_y = -\nu\epsilon_x$ and $\epsilon_z = -\nu\epsilon_x$. This is obviously *not* a case of plane strain in the x–y plane, since $\epsilon_z \neq 0$. On the other hand, a uniaxial strain of ϵ_x requires application of σ_x, Fig. 10–2b. However, this will also produce contractions of the element in the y and z directions due to the Poisson effect. In order to *require* $\epsilon_z = 0$, and thereby produce plane strain in the x–y plane, it is therefore *necessary* that a normal stress of σ_z *also* be applied to the element, Fig. 10–2b. As a result, plane stress does not exist in the x–y plane. In general, then, unless $\nu = 0$, the Poisson effect will *prevent* the simultaneous occurrence of plane strain and plane stress. It should also be pointed out that since shear stress and shear strain are *not* affected by Poisson's ratio, a condition of $\tau_{xz} = \tau_{yz} = 0$ requires $\gamma_{xz} = \gamma_{yz} = 0$.

As stated above, the measurement of strain at a point is generally determined using strain gauges. These measurements are usually made on the load-free surface of the body, and as a result, only the strain within the surface is determined. Realize, however, that such strains are a consequence of *plane stress*. Although this is the case, the analysis of plane strain, which is to be discussed in this chapter, can also be used to analyze the in-plane strains caused by plane stress.

10.2 General Equations of Plane-Strain Transformation

It is important in plane-strain analysis to establish transformation equations that can be used to determine the x', y' components of normal and shear strain at a point, provided the x, y components of strain are known. Essentially this problem is one of geometry and requires relating the deformations and rotations of differential line segments, which represent the sides of differential elements that are parallel to each set of axes.

Sign Convention. Before the strain-transformation equations can be developed, we must first establish a sign convention for the strains. This convention is the same as that established in Sec. 2.2 and will be restated here for the condition of plane strain. With reference to the differential element shown in Fig. 10–3a, *normal strains* ϵ_x and ϵ_y are *positive* if they cause *elongation* along the x and y axes, respectively, and the *shear strain* γ_{xy} is *positive* if the interior angle *AOB becomes smaller* than 90°. This sign convention also follows the corresponding one used for plane stress, Fig. 9–5a, that is, positive σ_x, σ_y, τ_{xy} will cause the element to *deform* in the positive ϵ_x, ϵ_y, γ_{xy} directions, respectively.

The problem here will be to determine at a point the normal and shear strains $\epsilon_{x'}$, $\epsilon_{y'}$, $\gamma_{x'y'}$, measured relative to the x', y' axes, if we know ϵ_x, ϵ_y, γ_{xy}, measured relative to the x, y axes. If the angle between the x and x' axes is θ, then, like the case of plane stress, θ will be *positive* provided it follows the curl of the right-hand fingers, i.e., counterclockwise, as shown in Fig. 10–3b.

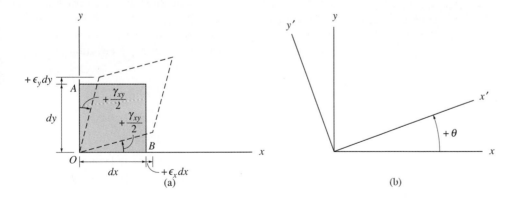

Positive sign convention

Fig. 10–3

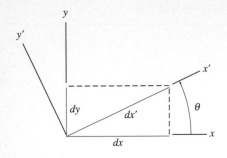

Before deformation

(a)

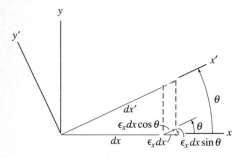

Normal strain ϵ_x

(b)

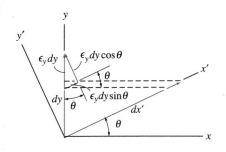

Normal strain ϵ_y

(c)

Fig. 10–4(a, b, c)

Normal and Shear Strains. In order to develop the strain-transformation equation for determining $\epsilon_{x'}$, we must determine the elongation of a line segment dx' that lies along the x' axis and is subjected to strain components ϵ_x, ϵ_y, γ_{xy}. As shown in Fig. 10–4a, the components of the line dx' along the x and y axes are

$$dx = dx' \cos \theta$$
$$dy = dx' \sin \theta \tag{10–1}$$

When the positive normal strain ϵ_x occurs, Fig. 10–4b, the line dx is elongated $\epsilon_x \, dx$, which causes line dx' to elongate $\epsilon_x \, dx \cos \theta$. Likewise, when ϵ_y occurs, Fig. 10–4c, line dy elongates $\epsilon_y \, dy$, which causes line dx' to elongate $\epsilon_y \, dy \sin \theta$. Lastly, assuming that dx remains fixed in position, the shear strain γ_{xy}, which is the change in angle between dx and dy, causes the top of line dy to be displaced $\gamma_{xy} \, dy$ to the right, as shown in Fig. 10–4d. This causes dx' to elongate $\gamma_{xy} \, dy \cos \theta$. If all three of these elongations are added together, the resultant elongation of dx' is then

$$\delta x' = \epsilon_x \, dx \cos \theta + \epsilon_y \, dy \sin \theta + \gamma_{xy} \, dy \cos \theta$$

From Eq. 2–2, the normal strain along the line dx' is $\epsilon_{x'} = \delta x'/dx'$. Using Eqs. 10–1, we therefore have

$$\epsilon_{x'} = \epsilon_x \cos^2 \theta + \epsilon_y \sin^2 \theta + \gamma_{xy} \sin \theta \cos \theta \tag{10–2}$$

The strain-transformation equation for determining $\gamma_{x'y'}$ can be developed by considering the amount of rotation each of the line segments dx' and dy' undergo when subjected to the strain components ϵ_x, ϵ_y, γ_{xy}. First we will consider the rotation of dx', which is defined by the counterclockwise angle α shown in Fig. 10–4e. It can be determined from the displacement $\delta y'$ using $\alpha = \delta y'/dx'$. To obtain $\delta y'$, consider the following three displacement components acting in the y' direction: one from ϵ_x, giving $-\epsilon_x \, dx \sin \theta$, Fig. 10–4b; another from ϵ_y, giving $\epsilon_y \, dy \cos \theta$, Fig. 10–4c; and the last from γ_{xy}, giving $-\gamma_{xy} \, dy \sin \theta$, Fig. 10–4d. Thus, $\delta y'$, as caused by all three strain components, is

$$\delta y' = -\epsilon_x \, dx \sin \theta + \epsilon_y \, dy \cos \theta - \gamma_{xy} \, dy \sin \theta$$

Using Eq. 10–1, with $\alpha = \delta y'/dx'$, we have

$$\alpha = (-\epsilon_x + \epsilon_y) \sin \theta \cos \theta - \gamma_{xy} \sin^2 \theta \tag{10–3}$$

As shown in Fig. 10–4e, the line dy' rotates by an amount β. We can determine this angle by a similar analysis, or by simply substituting $\theta + 90°$ for θ into Eq. 10–3. Using the identities $\sin (\theta + 90°) = \cos \theta$, $\cos(\theta + 90°) = -\sin \theta$, we have

$$\beta = (-\epsilon_x + \epsilon_y) \sin (\theta + 90°) \cos (\theta + 90°) - \gamma_{xy} \sin^2 (\theta + 90°)$$
$$= -(-\epsilon_x + \epsilon_y) \cos \theta \sin \theta - \gamma_{xy} \cos^2 \theta$$

Since α and β represent the rotation of the sides dx' and dy' of a differential element whose sides were originally oriented along the x' and y' axes, and

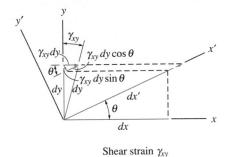

Shear strain γ_{xy}

(d)

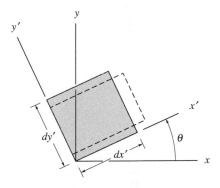

Positive normal strain, $\epsilon_{x'}$

(a)

Fig. 10–4(d, e)

(e)

β is in the opposite direction to α, Fig. 10–4e, the element is then subjected to a shear strain of

$$\gamma_{x'y'} = \alpha - \beta = -2(\epsilon_x - \epsilon_y) \sin \theta \cos \theta + \gamma_{xy}(\cos^2 \theta - \sin^2 \theta) \qquad (10\text{–}4)$$

Using the trigonometric identities $\sin 2\theta = 2 \sin \theta \cos \theta$, $\cos 2\theta = (1 + \cos^2 \theta)/2$, and $\sin^2 \theta + \cos^2 \theta = 1$, we can rewrite Eqs. 10–3 and 10–4 in the final form

$$\epsilon_{x'} = \frac{\epsilon_x + \epsilon_y}{2} + \frac{\epsilon_x - \epsilon_y}{2} \cos 2\theta + \frac{\gamma_{xy}}{2} \sin 2\theta \qquad (10\text{–}5)$$

$$\frac{\gamma_{x'y'}}{2} = -\frac{\epsilon_x - \epsilon_y}{2} \sin 2\theta + \frac{\gamma_{xy}}{2} \cos 2\theta \qquad (10\text{–}6)$$

These strain-transformation equations give the normal strain $\epsilon_{x'}$ in the x' direction and the shear strain $\gamma_{x'y'}$ of an element oriented at an angle θ, as shown in Fig. 10–5. According to the established sign convention, if $\epsilon_{x'}$ is *positive,* the element *elongates* in the positive x' direction, Fig. 10–5a, and if $\gamma_{x'y'}$ is positive, the element deforms as shown in Fig. 10–5b. Note again, as stated earlier, that these deformations occur due to positive normal stress $\sigma_{x'}$ and positive shear stress $\tau_{x'y'}$ acting on the element.

If the normal strain in the y' direction is required, it can be obtained from Eq. 10–5 by simply substituting $(\theta + 90°)$ for θ. The result is

$$\epsilon_{y'} = \frac{\epsilon_x + \epsilon_y}{2} - \frac{\epsilon_x - \epsilon_y}{2} \cos 2\theta - \frac{\gamma_{xy}}{2} \sin 2\theta \qquad (10\text{–}7)$$

The similarity between the above three equations and those for plane-stress transformation, Eqs. 9–1, 9–2, and 9–3, should be noted. By comparison, σ_x, σ_y, $\sigma_{x'}$, $\sigma_{y'}$ correspond to ϵ_x, ϵ_y, $\epsilon_{x'}$, $\epsilon_{y'}$; and τ_{xy}, $\tau_{x'y'}$ correspond to $\gamma_{xy}/2$, $\gamma_{x'y'}/2$.

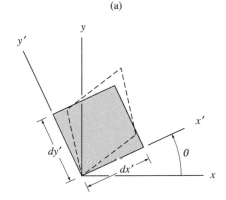

Positive shear strain, $\gamma_{x'y'}$

(b)

Fig. 10–5

Principal Strains. Like stress, the orientation of an element at a point can be determined such that the element's deformation is represented by normal strains, with *no* shear strain. When this occurs the normal strains are referred to as *principal strains,* and if the material is isotropic, the axes along which these strains occur coincide with the axes that define the planes of principal stress.

From Eqs. 9–4 and 9–5, and the correspondence between stress and strain mentioned above, the direction of the axes and the two values of the principal strains ϵ_1 and ϵ_2 are determined from

$$\tan 2\theta_p = \frac{\gamma_{xy}}{\epsilon_x - \epsilon_y} \qquad (10\text{–}8)$$

and

$$\epsilon_{1,2} = \frac{\epsilon_x + \epsilon_y}{2} \pm \sqrt{\left(\frac{\epsilon_x - \epsilon_y}{2}\right)^2 + \left(\frac{\gamma_{xy}}{2}\right)^2} \qquad (10\text{–}9)$$

Maximum In-Plane Shear Strain. Carrying the stress–strain analogy further, the axes along which maximum in-plane shear strain occurs are 45° away from those that define the principal strains. Using Eq. 9–6, they can be found from

$$\tan 2\theta_s = -\frac{(\epsilon_x - \epsilon_y)}{\gamma_{xy}} \qquad (10\text{–}10)$$

By analogy to Eqs. 9–7 and 9–8, the maximum in-plane shear strain and associated average normal strain are determined from the following equations:

$$\frac{(\gamma_{x'y'})_{\max}}{2} = \sqrt{\left(\frac{\epsilon_x - \epsilon_y}{2}\right)^2 + \left(\frac{\gamma_{xy}}{2}\right)^2} \qquad (10\text{–}11)$$

and

$$\epsilon_{\text{avg}} = \frac{\epsilon_x + \epsilon_y}{2} \qquad (10\text{–}12)$$

Application of the above equations is illustrated numerically in the following examples.

Example 10–1

A differential element of material at a point is subjected to a state of plane strain $\epsilon_x = 500(10^{-6})$, $\epsilon_y = -300(10^{-6})$, $\gamma_{xy} = 200(10^{-6})$, which tends to distort the element as shown in Fig. 10–6a. Determine the equivalent strains acting on an element oriented at the point, *clockwise* 30° from the original position.

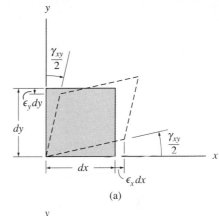

(a)

SOLUTION

The strain-transformation Eqs. 10–5 and 10–6 will be used to solve the problem. Since θ is *positive counterclockwise*, then for this problem $\theta = -30°$. Thus,

$$\epsilon_{x'} = \frac{\epsilon_x + \epsilon_y}{2} + \frac{\epsilon_x - \epsilon_y}{2}\cos 2\theta + \frac{\gamma_{xy}}{2}\sin 2\theta$$

$$= \left[\frac{500 + (-300)}{2}\right](10^{-6}) + \left[\frac{500 - (-300)}{2}\right](10^{-6})\cos(-60°)$$

$$+ \frac{200(10^{-6})}{2}\sin(-60°)$$

$$\epsilon_{x'} = 213(10^{-6}) \qquad\qquad\qquad Ans.$$

$$\frac{\gamma_{x'y'}}{2} = -\frac{\epsilon_x - \epsilon_y}{2}\sin 2\theta + \frac{\gamma_{xy}}{2}\cos 2\theta$$

$$= -\left[\frac{500 - (-300)}{2}\right](10^{-6})\sin(-60°) + \frac{200(10^{-6})}{2}\cos(-60°)$$

$$\gamma_{x'y'} = 793(10^{-6}) \qquad\qquad\qquad Ans.$$

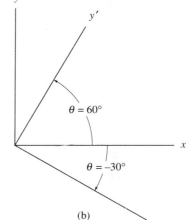

(b)

The strain in the y' direction can be obtained from Eq. 10–7 with $\theta = -30°$. However, we can also obtain $\epsilon_{y'}$ using Eq. 10–5 with $\theta = 60°$ ($\theta = -30° + 90°$), Fig. 10–6b. We have with $\epsilon_{y'}$ replacing $\epsilon_{x'}$,

$$\epsilon_{y'} = \frac{\epsilon_x + \epsilon_y}{2} + \frac{\epsilon_x - \epsilon_y}{2}\cos 2\theta + \frac{\gamma_{xy}}{2}\sin 2\theta$$

$$= \left[\frac{500 + (-300)}{2}\right](10^{-6}) + \left[\frac{500 - (-300)}{2}\right](10^{-6})\cos(120°)$$

$$+ \frac{200(10^{-6})}{2}\sin(120°)$$

$$\epsilon_{y'} = -13.4(10^{-6}) \qquad\qquad\qquad Ans.$$

Also, Eq. 10–7 can be used with $\theta = -30°$ to obtain this same value. These results tend to distort the element as shown in Fig. 10–6c.

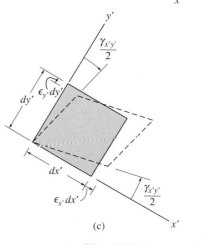

(c)

Fig. 10–6

Example 10-2

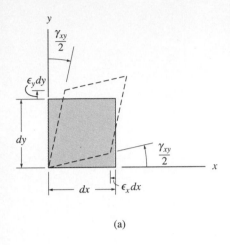

(a)

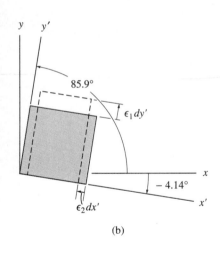

(b)

Fig. 10-7

A differential element of material at a point is subjected to a state of plane strain defined by $\epsilon_x = -350(10^{-6})$, $\epsilon_y = 200(10^{-6})$, $\gamma_{xy} = 80(10^{-6})$, which tends to distort the element as shown in Fig. 10-7a. Determine at the point the orientation of an element and the associated principal strains, and the orientation of an element and the maximum in-plane shear strains.

SOLUTION

Principal Strains. The orientation of the principal planes of strain is determined from Eq. 10-8. We have

$$\tan 2\theta_p = \frac{\gamma_{xy}}{\epsilon_x - \epsilon_y}$$

$$= \frac{80(10^{-6})}{(-350 - 200)(10^{-6})}$$

Thus, $2\theta_p = -8.28°$ and $-8.28° + 180° = 172°$, so that

$$\theta_p = -4.14° \text{ and } 85.9° \qquad \textit{Ans.}$$

Each of these angles is measured *positive counterclockwise*, from the x axis to the outward normals on each face of the element, Fig. 10-7b.

The principal strains are determined from Eq. 10-9. We have

$$\epsilon_{1,2} = \frac{\epsilon_x + \epsilon_y}{2} \pm \sqrt{\left(\frac{\epsilon_x - \epsilon_y}{2}\right)^2 + \left(\frac{\gamma_{xy}}{2}\right)^2}$$

$$= \frac{(-350 + 200)(10^{-6})}{2} \pm \left[\sqrt{\left(\frac{-350 - 200}{2}\right)^2 + \left(\frac{80}{2}\right)^2}\right](10^{-6})$$

$$= -75.0(10^{-6}) \pm 277.9(10^{-6})$$

$$\epsilon_1 = 203(10^{-6}) \qquad \epsilon_2 = -353(10^{-6}) \qquad \textit{Ans.}$$

We can determine which of these two strains deforms the element in the x' direction by applying Eq. 10-5 with $\theta = -4.14°$. Thus,

$$\epsilon_{x'} = \frac{\epsilon_x + \epsilon_y}{2} + \frac{\epsilon_x - \epsilon_y}{2} \cos 2\theta + \frac{\gamma_{xy}}{2} \sin 2\theta$$

$$= \left(\frac{-350 + 200}{2}\right)(10^{-6}) + \left(\frac{-350 - 200}{2}\right)(10^{-6}) \cos 2(-4.14°)$$

$$+ \frac{80(10^{-6})}{2} \sin 2(-4.14°)$$

$$\epsilon_{x'} = -353(10^{-6})$$

Hence $\epsilon_{x'} = \epsilon_2$. When subjected to the principal strains, the element is distorted as shown in Fig. 10-7b.

Maximum In-Plane Shear Strain. The element for maximum in-plane shear strain is oriented 45° from the one shown in Fig. 10-7b. In other

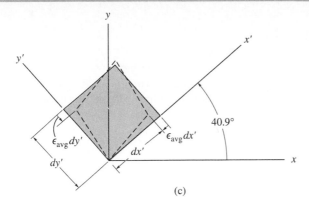

(c)

words, measured from the x axis, $\theta_s = \theta_p + 45° = -4.14° + 45° = 40.9°$, Fig. 10–7c. These same results can also be obtained from Eq. 10–10.

The maximum in-plane shear strains are determined from Eq. 10–11; that is,

$$\frac{(\gamma_{x'y'})_{max}}{2} = \sqrt{\left(\frac{\epsilon_x - \epsilon_y}{2}\right)^2 + \left(\frac{\gamma_{xy}}{2}\right)^2}$$

$$= \left[\sqrt{\left(\frac{-350 - 200}{2}\right)^2 + \left(\frac{80}{2}\right)^2}\right](10^{-6})$$

$$(\gamma_{x'y'})_{max} = 556(10^{-6}) \hspace{3em} \textit{Ans.}$$

Also, there are associated average normal strains imposed on the element that are determined from Eq. 10–12:

$$\epsilon_{avg} = \frac{\epsilon_x + \epsilon_y}{2} = \frac{-350 + 200}{2}(10^{-6})$$

$$\epsilon_{avg} = -75(10^{-6})$$

The proper sign of $(\gamma_{x'y'})_{max}$ can be obtained by applying Eq. 10–6 with $\theta_s = 40.9°$. We have

$$\frac{\gamma_{x'y'}}{2} = -\frac{\epsilon_x - \epsilon_y}{2}\sin 2\theta + \frac{\gamma_{xy}}{2}\sin 2\theta$$

$$= -\left(\frac{-350 - 200}{2}\right)(10^{-6})\sin 2(40.9°) + \frac{80(10^{-6})}{2}\cos 2(40.9°)$$

$$\gamma_{x'y'} = 556(10^{-6})$$

Thus, $(\gamma_{x'y'})_{max}$ tends to distort the element in such a way so that the right angle between dx' and dy' is decreased (positive sign convention). Note that this is in agreement with the distortion caused by positive shear stress acting on the x' face of the element and corresponding shear stresses acting on the element's adjacent faces. The distorted element is shown in Fig. 10–7c.

*10.3 Mohr's Circle—Plane Strain

Since the equations of plane-strain transformation are mathematically similar to the equations of plane-stress transformation, we can also solve problems involving the transformation of strain using Mohr's circle. This approach has the advantage of making it possible to see graphically how the normal and shear strain components at a point vary from one orientation of the element to the next.

Like the case for stress, the parameter θ in Eqs. 10–5 and 10–6 can be eliminated and the result rewritten in the form

$$(\epsilon_x - \epsilon_{avg})^2 + \left(\frac{\gamma_{xy}}{2}\right)^2 = R^2 \qquad (10\text{–}13)$$

where

$$\epsilon_{avg} = \frac{\epsilon_x + \epsilon_y}{2}$$

$$R = \sqrt{\left(\frac{\epsilon_x - \epsilon_y}{2}\right)^2 + \left(\frac{\gamma_{xy}}{2}\right)^2}$$

Equation 10–13 represents the equation of Mohr's circle for strain. It has a center on the ϵ axis at point $C(\epsilon_{avg}, 0)$ and a radius R.

PROCEDURE FOR ANALYSIS

The procedure for drawing Mohr's circle for strain follows the same one established for stress. The following steps are required to draw and use the circle.

Construction of the Circle.   Establish a coordinate system such that the abscissa represents the normal strain ϵ, with *positive to the right,* and the ordinate represents *half* the value of the shear strain, $\gamma/2$, with *positive downward,* Fig. 10–8.

Using the positive sign convention for ϵ_x, ϵ_y, γ_{xy}, as shown in Fig. 10–3, determine the center of the circle C, which is located on the ϵ axis at a distance $\epsilon_{avg} = (\epsilon_x + \epsilon_y)/2$ from the origin, Fig. 10–8.

Plot the reference point A having coordinates $A(\epsilon_x, \gamma_{xy}/2)$. This point represents the case for which the x' axis coincides with the x axis. Hence $\theta = 0°$, Fig. 10–8.

Connect point A with the center C of the circle and determine the hypotenuse CA of the shaded triangle by trigonometry. This distance represents the radius R of the circle, and CA is referred to as the radial reference line, Fig. 10–8.

Once R has been determined, sketch the circle.

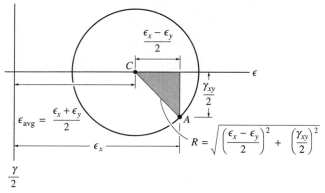

$$\epsilon_{\text{avg}} = \frac{\epsilon_x + \epsilon_y}{2}$$

$$R = \sqrt{\left(\frac{\epsilon_x - \epsilon_y}{2}\right)^2 + \left(\frac{\gamma_{xy}}{2}\right)^2}$$

Fig. 10–8

Principal Strains.   The principal strains ϵ_1 and ϵ_2 are determined from the points of intersection of the circle with the ϵ axis, i.e., where $\gamma/2 = 0$, Fig. 10–9a. These strains act in two directions, defined by the angles θ_{p_1} and θ_{p_2}, which can be determined from the circle using trigonometry. They are represented on the circle by the angles $2\theta_{p_1}$ and $2\theta_{p_2}$ and measured *from* the radial reference line *CA to* lines *CB* and *CD*, Fig. 10–9a. Actually only one of these angles needs to be calculated, since θ_{p_1} and θ_{p_2} are 90° apart. Remember that a *rotation* of 2θ on the *circle* represents a *rotation* of θ in the *same direction,* from the reference axis x to the x' axis.* The angle θ_{p_1} is shown in Fig. 10–9b. Notice that since ϵ_1 and ϵ_2 are indicated as being positive in Fig. 10–9a, the element in Fig. 10–9b will elongate in the x' and y' directions as shown by the dashed outline. This deformation is the result of positive principal stresses σ_1 and σ_2 acting on the element.

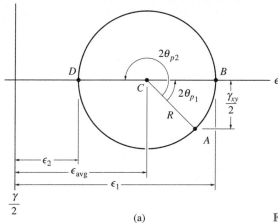

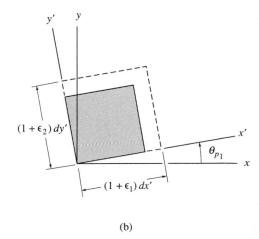

(a)   Fig. 10–9   (b)

*If instead the $\gamma/2$ axis were constructed *positive upward,* then the angle 2θ on the circle would be measured in the *opposite direction* as the orientation θ of the plane.

Maximum In-Plane Shear Strain. The average normal strain and maximum in-plane shear strain components are determined from the circle as the coordinates of points E and F, Fig. 10–10a. In this case the angles θ_{s_1} and θ_{s_2} define the two directions for orienting the planes that experience the strain components. The calculations from the circle are performed in the same manner as for finding the principal strains. The results are shown in Fig. 10–10b for the element oriented on the basis of calculating θ_{s_1} and the strain components associated with point E on the circle. Here $(\gamma_{x'y'})_{\max}$ and ϵ_{avg} are positive quantities, and so the element deforms as shown by the dashed outline.

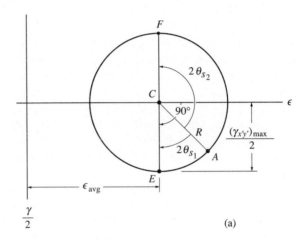

(a)

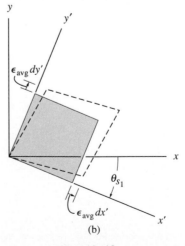

(b)

Fig. 10–10

Strains on Arbitrary Plane. The normal and shear strain components $\epsilon_{x'}$ and $\gamma_{x'y'}$ for a specified plane oriented at an angle θ, Fig. 10–11a, can be obtained from the circle using trigonometry to determine the coordinates of point P, Fig. 10–11b. To locate P, the known angle θ for the x' axis (in this case counterclockwise) is measured in the *same direction* as 2θ (counterclockwise) on the circle. The measurement of 2θ is *from* the radial reference line CA to the radial line CP. Also, if the value of $\epsilon_{y'}$ is required, it can be determined by calculating the ϵ coordinate of point Q in Fig. 10–11b, since CQ lies 180° away from CP and thus represents a rotation of 90° of the x' axis.

The following examples illustrate numerically how to apply this procedure to problems that involve each of the three cases just discussed.

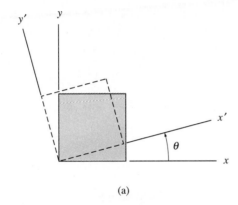

(a)

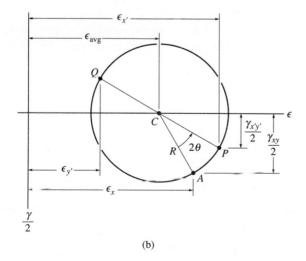

(b)

Fig. 10–11

Example 10–3

The state of plane strain at a point is represented by the components $\epsilon_x = 250(10^{-6})$, $\epsilon_y = -150(10^{-6})$, and $\gamma_{xy} = 120(10^{-6})$. Determine the orientation of an element and the associated principal strains, and the orientation of an element and the associated maximum in-plane shear strains.

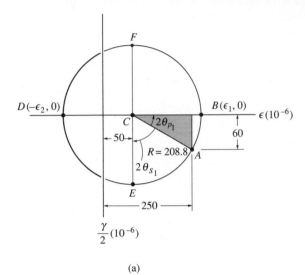

(a)

Fig. 10–12

SOLUTION

Construction of the Circle. The ϵ and $\gamma/2$ axes are established in Fig. 10–12a. Remember that the *positive* $\gamma/2$ axis must be directed *downward* so that *counterclockwise* rotations of the element correspond to *counterclockwise* rotation around the circle, and vice versa. The center of the circle C is located on the ϵ axis at

$$\epsilon_{\text{avg}} = \frac{250 + (-150)}{2}\,(10^{-6}) = 50(10^{-6})$$

Since $\gamma_{xy}/2 = 60(10^{-6})$, the reference point A ($\theta = 0°$) has coordinates $A(250(10^{-6}), 60(10^{-6}))$. From the shaded triangle in Fig. 10–12a, the radius of the circle is CA; that is,

$$R = \sqrt{(250 - 50)^2 + (60)^2} = 208.8$$

Principal Strains. The ϵ coordinates of points B and D represent the principal strains. They are

$$\epsilon_1 = (50 + 208.8)(10^{-6}) = 259(10^{-6}) \qquad \textit{Ans.}$$
$$\epsilon_2 = (50 - 208.8)(10^{-6}) = -159(10^{-6}) \qquad \textit{Ans.}$$

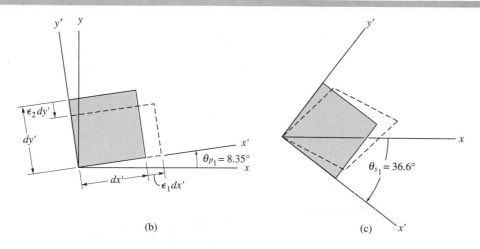

(b)

(c)

The direction of the positive principal strain ϵ_1 is defined by the *counterclockwise* angle $2\theta_{p_1}$, measured from the radial reference line CA to the line CB. We have

$$\tan 2\theta_{p_1} = \frac{60}{(250 - 50)}$$

$$\theta_{p_1} = 8.35° \qquad\qquad Ans.$$

Hence the side dx' of the element is oriented *counterclockwise* 8.35° as shown in Fig. 10–12b. This also defines the direction of ϵ_1. The deformation of the element is also shown in the figure.

Maximum In-Plane Shear Strain.   The maximum in-plane shear strain is represented by points E and F on the circle, Fig. 10–12a. From the coordinates of point E,

$$\frac{(\gamma_{x'y'})_{\text{max}}}{2} = 208.8(10^{-6})$$

$$(\gamma_{x'y'})_{\text{max}} = 418(10^{-6}) \qquad\qquad Ans.$$

$$\epsilon_{\text{avg}} = 50(10^{-6})$$

To orient the element, we can determine the clockwise angle $2\theta_{s_1}$ from the circle.

$$2\theta_{s_1} = 90° - 2(8.35°)$$

$$\theta_{s_1} = 36.6° \qquad\qquad Ans.$$

This angle is shown in Fig. 10–12c. Since the shear strain defined from point E on the circle has a positive value and the average normal strain is also positive, corresponding positive shear and positive average normal stress deform the element into the dashed shape shown in Fig. 10–12c.

Example 10–4

The state of plane strain at a point is represented on an element having components $\epsilon_x = -300(10^{-6})$, $\epsilon_y = -100(10^{-6})$, and $\gamma_{xy} = 100(10^{-6})$. Determine the state of strain on an element oriented 20° clockwise from this reported position.

SOLUTION

Construction of the Circle. The ϵ and $\gamma/2$ axes are established in Fig. 10–13a. The center of the circle is on the ϵ axis at

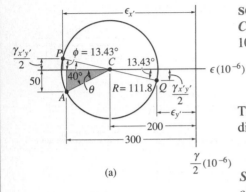

$$\epsilon_{\text{avg}} = \left(\frac{-300 - 100}{2}\right)(10^{-6}) = -200(10^{-6})$$

The reference point A has coordinates $A(-300(10^{-6}), 50(10^{-6}))$. The radius CA determined from the shaded triangle is therefore

$$R = \left[\sqrt{(300-200)^2 + (50)^2}\right](10^{-6}) = 111.8(10^{-6})$$

Strains on Inclined Element. Since the element is to be oriented 20° *clockwise*, we must establish a radial line CP, $2(20°) = 40°$ *clockwise*, measured from CA ($\theta = 0°$), Fig. 10–13a. The coordinates of point P ($\epsilon_{x'}$, $-\gamma_{x'y'}/2$) are obtained from the geometry of the circle. Note that

$$\theta = \tan^{-1}\left(\frac{50}{(300-200)}\right) = 26.57°, \qquad \phi = 40° - 26.57° = 13.43°$$

Thus,

$$\epsilon_{x'} = -(200 + 111.8 \cos 13.43°)(10^{-6})$$
$$= -309(10^{-6}) \qquad\qquad\qquad Ans.$$

$$\frac{\gamma_{x'y'}}{2} = -(111.8 \sin 13.43°)(10^{-6})$$

$$\gamma_{x'y'} = -51.9(10^{-6}) \qquad\qquad\qquad Ans.$$

The normal strain $\epsilon_{y'}$ can be determined from the ϵ coordinate of point Q on the circle, Fig. 10–13a.

$$\epsilon_{y'} = -(200 - 111.8 \cos 13.43°)(10^{-6}) = -91.3(10^{-6}) \quad Ans.$$

As a result of these strains, the element deforms relative to x', y' axes as shown in Fig. 10–13b.

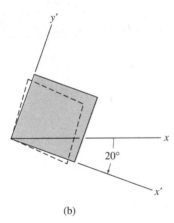

Fig. 10–13

PROBLEMS

10–1. Prove that the sum of the normal strains in perpendicular directions is constant.

10–2. The state of strain at the point on the bracket has components $\epsilon_x = -130(10^{-6})$, $\epsilon_y = 280(10^{-6})$, $\gamma_{xy} = 75(10^{-6})$. Use the strain-transformation equations to determine (a) the in-plane principal strains and (b) the maximum in-plane shear strain and average normal strain. In each case specify the orientation of the element and show how the strains deform the element within the x–y plane.

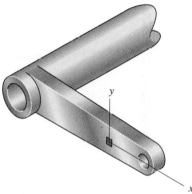

Prob. 10–2

10–3. The state of strain at the point on the bracket has components $\epsilon_x = 350(10^{-6})$, $\epsilon_y = -860(10^{-6})$, $\gamma_{xy} = 250(10^{-6})$. Use the strain-transformation equations to determine the equivalent in-plane strains on an element oriented at an angle of $\theta = 45°$ clockwise from the original position. Sketch the deformed element within the x–y plane due to these strains.

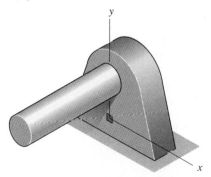

Prob. 10–3

*10–4. The state of strain at the point on the arm has components of $\epsilon_x = 250(10^{-6})$, $\epsilon_y = -450(10^{-6})$, $\gamma_{xy} = -825(10^{-6})$. Use the strain-transformation equations to determine (a) the in-plane principal strains and (b) the maximum in-plane shear strain and average normal strain. In each case specify the orientation of the element and show how the strains deform the element within the x–y plane.

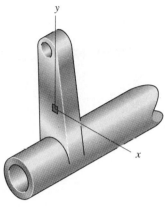

Prob. 10–4

10–5. The state of strain at the point on the gear tooth has components $\epsilon_x = 850(10^{-6})$, $\epsilon_y = 480(10^{-6})$, $\gamma_{xy} = 650(10^{-6})$. Use the strain-transformation equations to determine (a) the in-plane principal strains and (b) the maximum in-plane shear strain and average normal strain. In each case specify the orientation of the element and show how the strains deform the element within the x–y plane.

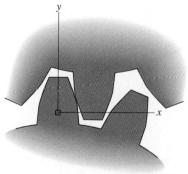

Prob. 10–5

10–6. The state of strain at the point on the bracket has components $\epsilon_x = 400(10^{-6})$, $\epsilon_y = -250(10^{-6})$, $\gamma_{xy} = 310(10^{-6})$. Use the strain-transformation equations to determine the equivalent in-plane strains on an element oriented at an angle of $\theta = 30°$ clockwise from the original position. Sketch the deformed element due to these strains within the x–y plane.

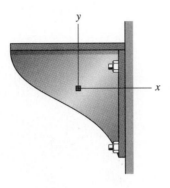

Prob. 10–6

10–7. The state of strain at the point on the bracket has components $\epsilon_x = -200(10^{-6})$, $\epsilon_y = -650(10^{-6})$, $\gamma_{xy} = -175(10^{-6})$. Use the strain-transformation equations to determine the equivalent in-plane strains on an element oriented at an angle of $\theta = 20°$ counterclockwise from the original position. Sketch the deformed element due to these strains within the x–y plane.

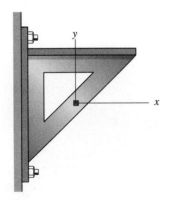

Prob. 10–7

*10–8. Solve Prob. 10–2 using Mohr's circle.

10–9. Solve Prob. 10–3 using Mohr's circle.

10–10. Solve Prob. 10–6 using Mohr's circle.

10–11. Solve Prob. 10–5 using Mohr's circle.

*10–12. Solve Prob. 10–4 using Mohr's circle.

10–13. Solve Prob. 10–7 using Mohr's circle.

*10.4 Absolute Maximum Shear Strain

In Sec. 9.5 it was pointed out that in three dimensions the state of stress at a point can be represented by an element oriented in a specific direction, such that the element is subjected only to *principal stresses* having maximum, intermediate, and minimum values, σ_{max}, σ_{int}, and σ_{min}. These stresses subject the material to associated *principal strains* ϵ_{max}, ϵ_{int}, and ϵ_{min}. Also, if the material is both homogeneous and isotropic, the element will *not* be subjected to shear strains since the shear stress on the principal planes is zero.

Assume that the three principal strains cause elongations along the x', y', z' axes as shown in Fig. 10–14a. If we view the element in two dimensions, that is, in the x'–y', x'–z', and y'–z' planes, Fig. 10–14b, 10–14c, and 10–14d, we can then use Mohr's circle to determine the *maximum in-plane shear strain* for each case. For example, from the view of the element in the x'–y' plane, Fig. 10–14b, the diameter of Mohr's circle extends between ϵ_{max} and ϵ_{int}, Fig. 10–14e. This circle gives the normal and shear strain components on each element oriented about the z' axis. Likewise, Mohr's circles for each element oriented about the y' and x' axes are also shown in Fig. 10–14e.

From these three circles it is seen that the *absolute maximum shear strain* is determined from the circle having the largest radius. It occurs on the element oriented $45°$ about the y' axis from the element shown in its original position, Fig. 10–14a or 10–14c. For this condition,

$$\gamma_{\substack{abs \\ max}} = \epsilon_{max} - \epsilon_{min} \qquad (10\text{–}14)$$

and

$$\epsilon_{avg} = \frac{\epsilon_{max} + \epsilon_{min}}{2}$$

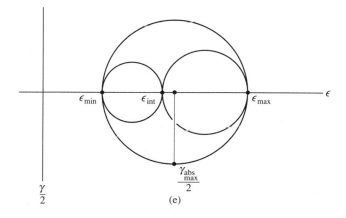

(e)

Fig. 10–14

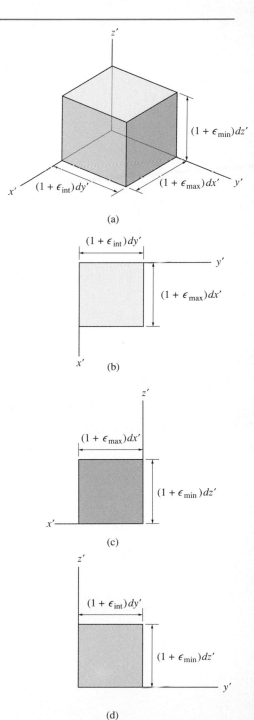

(a)

(b)

(c)

(d)

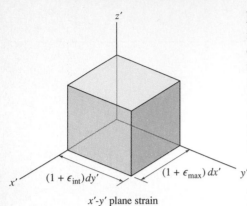

$x'-y'$ plane strain

(a)

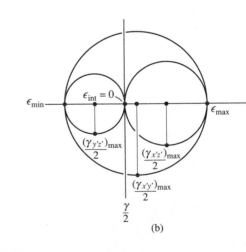

(b)

Fig. 10–15

Plane Strain. Like the case of plane stress, the above analysis has an important implication when the material is subjected to *plane strain*, especially when the *principal strains* have the *same sign*, i.e., both cause elongation or both cause contraction. For example, if the principal in-plane strains are ϵ_{max} and ϵ_{int}, while the out-of-plane principal strain is $\epsilon_{min} = 0$, Fig. 10–15a, then the three Mohr's circles describing the normal and shear strain components for elements oriented about the x', y', and z' axes are shown in Fig. 10–15b. By inspection, the largest circle has a radius $R = (\gamma_{x'z'})_{max}/2$. Hence,

$$\gamma_{abs}_{max} = (\gamma_{x'z'})_{max} = \epsilon_{max}$$

This value represents the *absolute maximum shear strain* for the material. Note that it is *larger* than the maximum in-plane shear strain, which is $(\gamma_{x'y'})_{max} = \epsilon_{max} - \epsilon_{int}$.

On the other hand, if one of the in-plane principal strains is of *opposite sign* to the other in-plane principal strain, then ϵ_{max} causes elongation, ϵ_{min} causes contraction, and the out-of-plane principal strain is $\epsilon_{int} = 0$, Fig. 10–16a. Mohr's circles, which describe the strains on each element's orientation about the x', y', z' axes, are shown in Fig. 10–16b. In this case,

$$\gamma_{abs}_{max} = (\gamma_{x'y'})_{max} = \epsilon_{max} - \epsilon_{min}$$

We may therefore summarize the above two points as follows. If the in-plane principal strains both have the *same sign*, the *absolute maximum shear strain* will occur *out of plane* and has a value of $\gamma_{abs}_{max} = \epsilon_{max}$. However, if the in-plane principal strains are of *opposite signs*, then the absolute maximum shear strain *equals* the maximum in-plane shear strain.

$x'-y'$ plane strain

(a)

(b)

Fig. 10–16

Example 10–5

The state of plane strain at a point is represented by the strain components $\epsilon_x = -400(10^{-6})$, $\epsilon_y = 200(10^{-6})$, $\gamma_{xy} = 150(10^{-6})$. Determine the maximum in-plane shear strain and the absolute maximum shear strain.

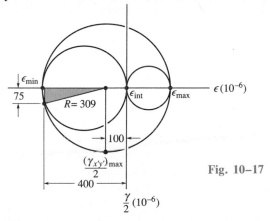

Fig. 10–17

SOLUTION

Maximum In-Plane Strain.   We will solve this problem using Mohr's circle. From the strain components, the center of the circle is on the ϵ axis at

$$\epsilon_{avg} = \frac{-400 + 200}{2}(10^{-6}) = -100(10^{-6})$$

Since $\gamma_{xy}/2 = 75(10^{-6})$, the reference point has coordinates $A(-400(10^{-6}), 75(10^{-6}))$. As shown in Fig. 10–17, the radius of the circle is therefore

$$R = \left[\sqrt{(400 - 100)^2 + (75)^2} \right] (10^{-6}) = 309(10^{-6})$$

Computing the in-plane principal strains, we have

$$\epsilon_{max} = (-100 + 309)(10^{-6}) = 209(10^{-6})$$
$$\epsilon_{min} = (-100 - 309)(10^{-6}) = -409(10^{-6})$$

Also, the maximum in-plane shear strain is

$$(\gamma_{x'y'})_{max} = \epsilon_{max} - \epsilon_{min} = [209 - (-409)](10^{-6}) = 618(10^{-6}) \qquad Ans.$$

Absolute Maximum Shear Strain.   From the above results, we have $\epsilon_{max} = 209(10^{-6})$, $\epsilon_{int} = 0$, $\epsilon_{min} = -409(10^{-6})$. The three Mohr's circles, plotted for element orientations about each of the x', y', z' axes, are also shown in Fig. 10–17. It is seen that since the *principal in-plane strains have opposite signs*, the maximum in-plane shear strain is *also* the absolute maximum shear strain; i.e.,

$$\gamma_{abs \atop max} = 618(10^{-6}) \qquad Ans.$$

10.5 Strain Rosettes

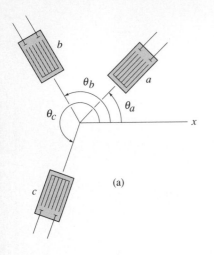

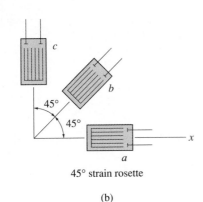

45° strain rosette

(b)

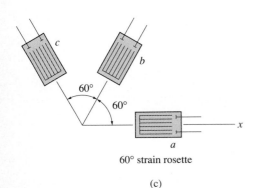

60° strain rosette

(c)

Fig. 10–18

It was mentioned in Sec. 3.1 that the normal strain in a tension-test specimen can be measured using an *electrical-resistance strain gauge,* which consists of a wire grid or piece of metal foil bonded to the specimen. However, for a general loading on a body, the *normal strains* at a point on its free surface are often determined using a cluster of three electrical-resistance strain gauges, arranged in a specified pattern. This pattern is referred to as a *strain rosette,* and once the readings on the three gauges are made, the data can then be used to specify the state of strain at the point. It should be noted, however, that these strains are measured *only* in the plane of the gauges, and since the body is stress-free in a direction perpendicular to the gauges, the gauges may be subjected to *plane stress* but *not* plane strain. In this regard, the normal line to the free surface is a principal axis of strain, and so the principal normal strain along this axis is *not* measured by the strain rosette. What is important here is that the out-of-plane displacement caused by this principal strain will *not* affect the in-plane measurements of the gauges.

In the general case, the axes of the three gauges are arranged at the angles θ_a, θ_b, θ_c as shown in Fig. 10–18a. If the readings ϵ_a, ϵ_b, ϵ_c are taken, we can determine the strain components ϵ_x, ϵ_y, γ_{xy} at the point by applying the strain-transformation equation, Eq. 10–2, for each gauge. We have,

$$\epsilon_a = \epsilon_x \cos^2 \theta_a + \epsilon_y \sin^2 \theta_a + \gamma_{xy} \sin \theta_a \cos \theta_a$$
$$\epsilon_b = \epsilon_x \cos^2 \theta_b + \epsilon_y \sin^2 \theta_b + \gamma_{xy} \sin \theta_b \cos \theta_b \qquad (10\text{–}15)$$
$$\epsilon_c = \epsilon_x \cos^2 \theta_c + \epsilon_y \sin^2 \theta_c + \gamma_{xy} \sin \theta_c \cos \theta_c$$

The values of ϵ_x, ϵ_y, γ_{xy} are determined by solving these three equations simultaneously.

Strain rosettes are often arranged in 45° or 60° patterns. In the case of the 45° or "rectangular" strain rosette shown in Fig. 10–18b, $\theta_a = 0°$, $\theta_b = 45°$, $\theta_c = 90°$, so that Eq. 10–15 gives

$$\epsilon_x = \epsilon_a$$
$$\epsilon_y = \epsilon_c \qquad (10\text{–}16)$$
$$\gamma_{xy} = 2\epsilon_b - (\epsilon_a + \epsilon_c)$$

And for the 60° strain rosette in Fig. 10–18c, $\theta_a = 0°$, $\theta_b = 60°$, $\theta_c = 120°$. Here Eq. 10–15 gives

$$\epsilon_x = \epsilon_a$$
$$\epsilon_y = \frac{1}{3}\left(2\epsilon_b + 2\epsilon_c - \epsilon_a\right) \qquad (10\text{–}17)$$
$$\gamma_{xy} = \frac{2}{\sqrt{3}}\left(\epsilon_b - \epsilon_c\right)$$

Once ϵ_x, ϵ_y, γ_{xy} are determined, the transformation equations of Sec. 10.2 or Mohr's circle can then be used to determine the principal in-plane strains and the maximum in-plane shear strain at the point.

Example 10–6

The state of strain at point A on the bracket in Fig. 10–19a is measured using the strain rosette shown in Fig. 10–19b. Due to the loadings, the readings from the gauges give $\epsilon_a = 60(10^{-6})$, $\epsilon_b = 135(10^{-6})$, and $\epsilon_c = 264(10^{-6})$. Determine the in-plane principal strains at the point and the directions in which they act.

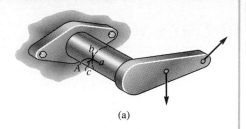

(a)

SOLUTION

We will use Eq. 10–15 for the solution. Establishing an x axis as shown in Fig. 10–19b and measuring the angles from the $+x$ axis to the center-lines of each gauge, we have $\theta_a = 0°$, $\theta_b = 60°$, and $\theta_c = 120°$. Substituting these results, along with the problem data, into Eq. 10–15 gives

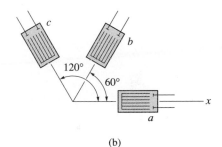

(b)

$$60(10^{-6}) = \epsilon_x \cos^2 0° + \epsilon_y \sin^2 0° + \gamma_{xy} \sin 0° \cos 0°$$
$$= \epsilon_x \qquad (1)$$
$$135(10^{-6}) = \epsilon_x \cos^2 60° + \epsilon_y \sin^2 60° + \gamma_{xy} \sin 60° \cos 60°$$
$$= 0.25\epsilon_x + 0.75\epsilon_y + 0.433\gamma_{xy} \qquad (2)$$
$$264(10^{-6}) = \epsilon_x \cos^2 120° + \epsilon_y \sin^2 120° + \gamma_{xy} \sin 120° \cos 120°$$
$$= 0.25\epsilon_x + 0.75\epsilon_y - 0.433\gamma_{xy} \qquad (3)$$

Using Eq. (1) and solving Eqs. (2) and (3) simultaneously, we get

$$\epsilon_x = 60(10^{-6}) \qquad \epsilon_y = 246(10^{-6}) \qquad \gamma_{xy} = -149(10^{-6})$$

These same results can also be obtained in a more direct manner from Eq. 10–17.

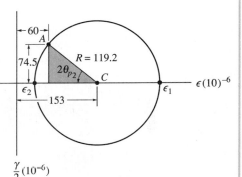

(c)

The in-plane principal strains can be determined using Mohr's circle. The reference point on the circle is at A $(60(10^{-6}), -74.5(10^{-6}))$ and the center of the circle, C, is on the ϵ axis at $\epsilon_{\text{avg}} = 153(10^{-6})$, Fig. 10–19c. From the shaded triangle, the radius is

$$R = \sqrt{(153 - 60)^2 + (74.5)^2} = 119.2$$

The in-plane principal strains are thus

$$\epsilon_1 = 153(10^{-6}) + 119.2(10^{-6}) = 272(10^{-6}) \qquad Ans.$$
$$\epsilon_2 = 153(10^{-6}) - 119.2(10^{-6}) = 33.8(10^{-6}) \qquad Ans.$$

$$2\theta_{p_2} = \tan^{-1}\frac{74.5}{(153 - 60)} = 38.7°$$

$$\theta_{p_2} = 19.3° \qquad Ans.$$

The deformed element is shown in the dashed position in Fig. 10–19d.

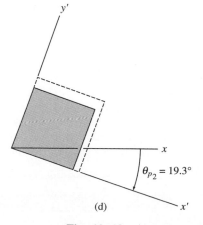

(d)

Fig. 10–19

PROBLEMS

10–14. The strain at point A on the shell has components $\epsilon_x = 250(10^{-6})$, $\epsilon_y = 400(10^{-6})$, $\gamma_{xy} = 275(10^{-6})$, $\epsilon_z = 0$. Determine (a) the principal strains at A, (b) the maximum shear strain in the x–y plane, and (c) the absolute maximum shear strain.

Prob. 10–14

10–15. The strain at point A on the bracket has components $\epsilon_x = 300(10^{-6})$, $\epsilon_y = 550(10^{-6})$, $\gamma_{xy} = -650(10^{-6})$, $\epsilon_z = 0$. Determine (a) the principal strains at A, (b) the maximum shear strain in the x–y plane, and (c) the absolute maximum shear strain.

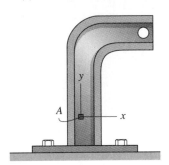

Prob. 10–15

*10–16. The strain at point A on the beam has components $\epsilon_x = 450(10^{-6})$, $\epsilon_y = 825(10^{-6})$, $\gamma_{xy} = 275(10^{-6})$, $\epsilon_z = 0$. Determine (a) the principal strains at A, (b) the maximum shear strain in the x–y plane, and (c) the absolute maximum shear strain.

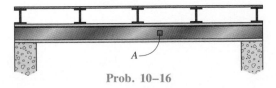

Prob. 10–16

10–17. The strain at point A on the pressure-vessel wall has components $\epsilon_x = 480(10^{-6})$, $\epsilon_y = 720(10^{-6})$, $\gamma_{xy} = 650(10^{-6})$, $\epsilon_z = 0$. Determine (a) the principal strains at A, (b) the maximum shear strain in the x–y plane, and (c) the absolute maximum shear strain.

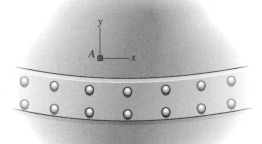

Prob. 10–17

10–18. The steel bar is subjected to the tensile load of 500 lb. If it is 0.5 in. thick determine the absolute maximum shear strain. $E = 29(10^3)$ ksi, $\nu = 0.3$.

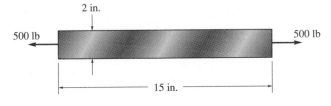

Prob. 10–18

10–19. The strain at point A on the leg of the angle has components $\epsilon_x = -140(10^{-6})$, $\epsilon_y = 180(10^{-6})$, $\gamma_{xy} = -125(10^{-6})$, $\epsilon_z = 0$. Determine (a) the principal strains at A, (b) the maximum shear strain in the x–y plane, and (c) the absolute maximum shear strain.

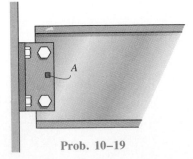

Prob. 10–19

*10–20. The 45° strain rosette is mounted on a machine element. The following readings are obtained for each gauge: $\epsilon_a = 650(10^{-6})$, $\epsilon_b = -300(10^{-6})$, $\epsilon_c = 480(10^{-6})$. Determine (a) the in-plane principal strains and (b) the maximum in-plane shear strain and associated average normal strain. In each case show the deformed element due to these strains.

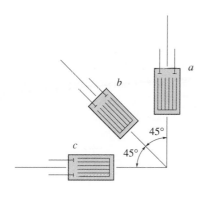

Prob. 10–20

10–22. The 60° strain rosette is mounted on the surface of an aluminum plate. The following readings are obtained for each gauge: $\epsilon_a = 950(10^{-6})$, $\epsilon_b = 380(10^{-6})$, $\epsilon_c = 220(10^{-6})$. Determine the in-plane principal strains and their orientation developed at a point on the plate.

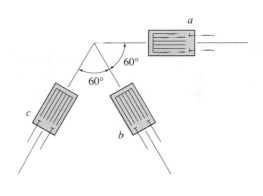

Prob. 10–22

10–21. The 60° strain rosette is mounted on a beam. The following readings are obtained for each gauge: $\epsilon_a = 250(10^{-6})$, $\epsilon_b = -400(10^{-6})$, $\epsilon_c = 280(10^{-6})$. Determine (a) the in-plane principal strains and their orientation, and (b) the maximum in-plane shear strain and average normal strain. In each case show the deformed element due to these strains.

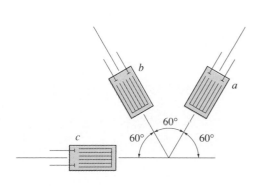

Prob. 10–21

10–23. The 45° strain rosette is mounted on a steel shaft. The following readings are obtained for each gauge: $\epsilon_a = 800(10^{-6})$, $\epsilon_b = 520(10^{-6})$, $\epsilon_c = -450(10^{-6})$. Determine the in-plane principal strains and their orientation developed at a point on the shaft.

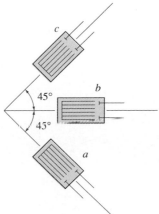

Prob. 10–23

10.6 Material-Property Relationships

Now that the general principles of multiaxial stress and strain have been presented, we will use these principles to develop some important relationships involving the material's properties. To do so we will assume that the material is homogeneous and isotropic and behaves in a linear-elastic manner.

Generalized Hooke's Law. If the material at a point is subjected to a state of triaxial stress, σ_x, σ_y, σ_z, Fig. 10–20a, associated normal strains ϵ_x, ϵ_y, ϵ_z are developed in the material. The stresses can be related to the strains by using the principle of superposition, Poisson's ratio, $\epsilon_{\text{lat}} = -\nu\epsilon_{\text{long}}$, and Hooke's law, as it applies in the uniaxial direction, $\epsilon = \sigma/E$. To show how this is done we will first consider the normal strain of the element in the x direction, caused by separate application of each normal stress. When σ_x is applied, Fig. 10–20b, the element elongates in the x direction and the strain ϵ_x' in this direction is

$$\epsilon_x' = \frac{\sigma_x}{E}$$

Application of σ_y causes the element to contract with a strain ϵ_x'' in the x direction, Fig. 10–20c. Here

$$\epsilon_x'' = -\nu\frac{\sigma_y}{E}$$

Likewise, application of σ_z, Fig. 10–20d, causes a contraction in the x direction such that

$$\epsilon_x''' = -\nu\frac{\sigma_z}{E}$$

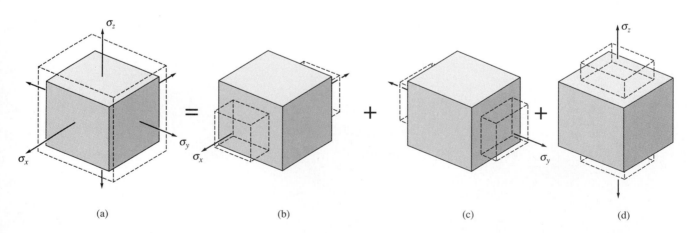

Fig. 10–20

When these three normal strains are superimposed, the normal strain ϵ_x is determined for the state of stress in Fig. 10–20a. Similar equations can be developed for the normal strains in the y and z directions. The final results can be written as

$$
\begin{aligned}
\epsilon_x &= \frac{1}{E}\left[\sigma_x - \nu(\sigma_y + \sigma_z)\right] \\
\epsilon_y &= \frac{1}{E}\left[\sigma_y - \nu(\sigma_x + \sigma_z)\right] \\
\epsilon_z &= \frac{1}{E}\left[\sigma_z - \nu(\sigma_x + \sigma_y)\right]
\end{aligned}
\qquad (10\text{–}18)
$$

These three equations express Hooke's law in general form for a triaxial state of stress. As noted in the derivation, they are valid only if the principle of superposition applies, which requires a *linear-elastic* response of the material and application of strains that do not severely alter the shape of the material—i.e., small deformations are required. When applying these equations, note that tensile stresses are considered positive quantities, and compressive stresses are negative. If a resulting normal strain is *positive,* it indicates that the material *elongates,* whereas a *negative* normal strain indicates the material *contracts.*

Since the material is isotropic, the element in Fig. 10–20a will *remain rectangular* when subjected to the normal stresses; i.e., *no shear strains* will be produced in the material. If we now apply a shear stress τ_{xy} to the element, Fig. 10–21a, experimental observations indicate that the material will deform *only* due to a shear strain γ_{xy}; that is, τ_{xy} will not cause other strains in the material. Likewise, τ_{yz} and τ_{zx} will only cause shear strains γ_{yz} and γ_{zx}, respectively. Hooke's law for shear stress and shear strain can therefore be written as

$$
\gamma_{xy} = \frac{1}{G}\tau_{xy} \qquad \gamma_{yz} = \frac{1}{G}\tau_{yz} \qquad \gamma_{zx} = \frac{1}{G}\tau_{zx} \qquad (10\text{–}19)
$$

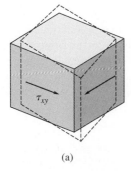

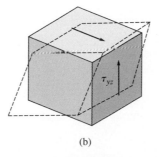

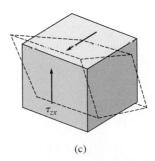

(a)

(b)

(c)

Fig. 10–21

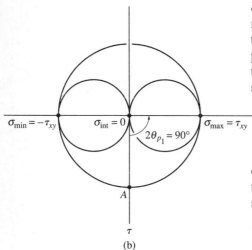

(a)

(b)

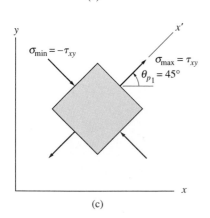

(c)

Fig. 10–22

Relationship Involving E, ν, and G. In Sec. 3.7 we stated that the modulus of elasticity E is related to the shear modulus G by Eq. 3–11, namely,

$$G = \frac{E}{2(1 + \nu)} \qquad (10\text{–}20)$$

One way to derive this relationship is to consider an element of the material to be subjected to pure shear ($\sigma_x = \sigma_y = \sigma_z = 0$), Fig. 10–22a. In accordance with the discussion of Sec. 10.4, Mohr's circles for the stresses acting on each planar view of the element are shown in Fig. 10–22b. Notice that the intermediate principal stress $\sigma_{int} = 0$ since the element is subjected to plane stress. The center of the largest circle is at $\sigma_{avg} = (\sigma_x + \sigma_y)/2 = 0$, and the reference point A has coordinates $A(0, \tau_{xy})$. Hence, the radius of this circle is $R = \tau_{xy}$, and the principal stresses are therefore $\sigma_{max} = \tau_{xy}$ and $\sigma_{min} = -\tau_{xy}$. Since $2\theta_{p_1} = 90°$ counterclockwise, Fig. 10–22b, the element must be oriented $\theta_{p_1} = 45°$ counterclockwise from the x axis in order to define the direction of the plane on which σ_{max} acts, Fig. 10–22c. If the three principal stresses $\sigma_{max} = \tau_{xy}$, $\sigma_{int} = 0$, and $\sigma_{min} = -\tau_{xy}$ are substituted into the first of Eq. 10–18, the principal strain ϵ_{max} can be related to the shear stress τ_{xy}. The result is

$$\epsilon_{max} = \frac{\tau_{xy}}{E}(1 + \nu) \qquad (10\text{–}21)$$

This strain, which deforms the element along the x' axis, Fig. 10–22c, can also be related to the shear strain γ_{xy} caused by τ_{xy}, Fig. 10–22a. To do this, first note that from the first two equations of Hooke's law, Eq. 10–18, since $\sigma_x = \sigma_y = \sigma_z = 0$, then $\epsilon_x = \epsilon_y = 0$. Substituting these results into Eq. 10–5 with $\theta = 45°$, we get

$$\epsilon_1 = \epsilon_{max} = \frac{\gamma_{xy}}{2}$$

By Hooke's law, $\gamma_{xy} = \tau_{xy}/G$, so that $\epsilon_{max} = \tau_{xy}/2G$. Substituting this result into Eq. 10–21 and rearranging terms gives the final result, namely, Eq. 10–20.

Dilatation and Bulk Modulus. When an elastic material is subjected to normal stress, its volume will change. In order to compute this change, consider the volume element shown in Fig. 10–23, which is subjected to the principal stresses σ_x, σ_y, σ_z. The sides of the element are originally dx, dy, dz, Fig. 10–23a; however, after application of the stress they become $(1 + \epsilon_x)\, dx$, $(1 + \epsilon_y)\, dy$, $(1 + \epsilon_z)\, dz$, respectively, Fig. 10–23b. The change in volume of the element is therefore

$$\delta V = (1 + \epsilon_x)(1 + \epsilon_y)(1 + \epsilon_z)\, dx\, dy\, dz - dx\, dy\, dz$$

Neglecting the products of the strains since the strains are very small as discussed in Sec. 2.2, we have

$$\delta V = (\epsilon_x + \epsilon_y + \epsilon_z)\, dx\, dy\, dz$$

The change in volume per unit volume is called the "volumetric strain" or the *dilatation e*. It can be written as

$$e = \frac{\delta V}{V} = \epsilon_x + \epsilon_y + \epsilon_z \qquad (10\text{-}22)$$

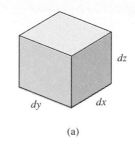

(a)

By comparison, the shear strains will *not* change the volume of the element, rather they will only change its rectangular shape.

If we use the generalized Hooke's law, as defined by Eq. 10–18, we can write the dilatation in terms of the applied stress. After substitution and simplification, we have

$$e = \frac{1 - 2\nu}{E}(\sigma_x + \sigma_y + \sigma_z) \qquad (10\text{-}23)$$

When a volume element of material is subjected to the uniform pressure p of a liquid, the pressure on the body is the same in all directions and is always normal to any surface on which it acts. Shear stresses are *not present*, since the shear resistance of a liquid is zero. This state of "hydrostatic" loading requires the normal stresses to be equal in any and all directions, and therefore an element of the body is subjected to principal stresses $\sigma_x = \sigma_y = \sigma_z = -p$, Fig. 10–24. Substituting into Eq. 10–23 and rearranging terms yields

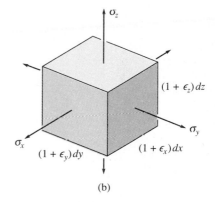

(b)

Fig. 10–23

$$\frac{p}{e} = \frac{E}{3(1 - 2\nu)} \qquad (10\text{-}24)$$

The term on the right consists *only* of the material's properties E and ν. It is equal to the ratio of the uniform normal stress p to the dilatation or "volumetric strain." Since this ratio is *similar* to the ratio of linear-elastic stress to strain, which defines E, i.e., $\sigma/\epsilon = E$, the terms on the right are called the *volume modulus of elasticity* or the *bulk modulus*. It has the same units as stress and will be symbolized by the letter k; that is,

$$k = \frac{E}{3(1 - 2\nu)} \qquad (10\text{-}25)$$

Note that for most metals $\nu \approx \frac{1}{3}$ so $k \approx E$. Also, a *rigid material* requires k to be infinite, since $\delta V = 0$. Hence, from Eq. 10–25, the theoretical *maximum* value for Poisson's ratio is $\nu = 0.5$. It may also be added that during yielding, no actual volume change of the material is observed, and so $\nu = 0.5$ is used when plastic yielding occurs.

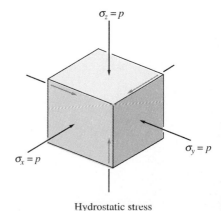

Hydrostatic stress

Fig. 10–24

Example 10–7

The bracket in Example 10–6, Fig. 10–25a, is made of steel for which $E_{st} = 200$ GPa, $\nu_{st} = 0.3$. Determine the principal stresses at point A.

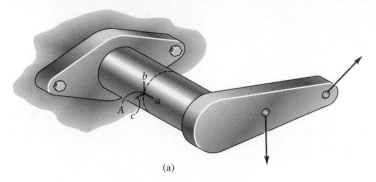

(a)

Fig. 10–25

SOLUTION I

From Example 10–6 the principal strains have been determined as

$$\epsilon_1 = 272(10^{-6})$$
$$\epsilon_2 = 33.8(10^{-6})$$

Since point A is on the *surface* of the bracket for which there is no loading, the stress on the surface is zero, and so point A is subjected to plane stress. Applying Hooke's law with $\sigma_z = 0$, we have

$$\epsilon_x = \frac{\sigma_1}{E} - \frac{\nu}{E}\sigma_2; \qquad 272(10^{-6}) = \frac{\sigma_1}{200(10^9)} - \frac{0.3}{200(10^9)}\sigma_2$$

$$54.4(10^6) = \sigma_1 - 0.3\sigma_2 \qquad (1)$$

$$\epsilon_y = \frac{\sigma_2}{E} - \frac{\nu}{E}\sigma_1; \qquad 33.8(10^{-6}) = \frac{\sigma_2}{200(10^9)} - \frac{0.3}{200(10^9)}\sigma_1$$

$$6.76(10^6) = \sigma_2 - 0.3\sigma_1 \qquad (2)$$

Solving Eqs. (1) and (2) simultaneously yields

$$\sigma_1 = 62.0 \text{ MPa} \qquad\qquad Ans.$$
$$\sigma_2 = 25.4 \text{ MPa} \qquad\qquad Ans.$$

SOLUTION II

It is also possible to solve the problem using the given state of strain,

$$\epsilon_x = 60(10^{-6}) \qquad \epsilon_y = 246(10^{-6}) \qquad \gamma_{xy} = -149(10^{-6})$$

as specified in Example 10–6. Applying Hooke's law in the $x - y$ plane, we have

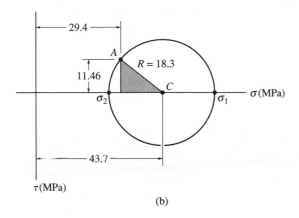

(b)

$$\epsilon_x = \frac{\sigma_x}{E} - \frac{\nu}{E}\sigma_y; \qquad 60(10^{-6}) = \frac{\sigma_x}{200(10^9)} - \frac{0.3\sigma_y}{200(10^9)}$$

$$\epsilon_y = \frac{\sigma_y}{E} - \frac{\nu}{E}\sigma_x; \qquad 246(10^{-6}) = \frac{\sigma_y}{200(10^9)} - \frac{0.3\sigma_x}{200(10^9)}$$

$$\sigma_x = 29.4 \text{ MPa} \qquad \sigma_y = 58.0 \text{ MPa}$$

The shear stress is computed using Hooke's law for shear. First, however, we must calculate G.

$$G = \frac{E}{2(1 + \nu)} = \frac{200 \text{ GPa}}{2(1 + 0.3)} = 76.9 \text{ GPa}$$

Thus,

$$\tau_{xy} = G\gamma_{xy}; \quad \tau_{xy} = 76.9(10^9)[-149(10^{-6})] = -11.46 \text{ MPa}$$

The Mohr's circle for this state of plane stress has a reference point $A(29.4 \text{ MPa}, -11.46 \text{ MPa})$ and center at $\sigma_{\text{avg}} = 43.7 \text{ MPa}$, Fig. 10–25b. The radius is determined from the shaded triangle.

$$R = \sqrt{(43.7 - 29.4)^2 + (11.46)^2} = 18.3 \text{ MPa}$$

Therefore,

$$\sigma_1 = 43.7 + 18.3 = 62.0 \text{ MPa} \qquad \qquad Ans.$$
$$\sigma_2 = 43.7 - 18.3 = 25.4 \text{ MPa} \qquad \qquad Ans.$$

Note that each of these solutions is valid provided the material is both linear-elastic and isotropic, since then the principal planes of stress and strain coincide.

Example 10–8

The copper bar in Fig. 10–26 is subjected to a uniform loading along its edges as shown. If it has a length $a = 300$ mm, width $b = 50$ mm, and thickness $t = 20$ mm before the load is applied, determine its new length, width, and thickness after application of the load. Take $E_{cu} = 120$ GPa, $\nu_{cu} = 0.34$.

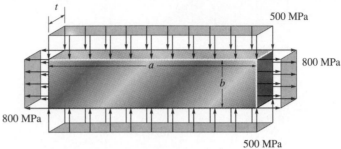

Fig. 10–26

SOLUTION

By inspection, the bar is subjected to a state of plane stress. From the loading we have

$$\sigma_x = 800 \text{ MPa} \qquad \sigma_y = -500 \text{ MPa} \qquad \tau_{xy} = 0 \qquad \sigma_z = 0$$

The associated normal strains are determined from the generalized Hooke's law, Eq. 10–18; that is,

$$\epsilon_x = \frac{\sigma_x}{E} - \frac{\nu}{E}(\sigma_y + \sigma_z)$$

$$= \frac{800 \text{ MPa}}{120(10^3) \text{ MPa}} - \frac{0.34}{120(10^3) \text{ MPa}}(-500 \text{ MPa}) = 0.00808$$

$$\epsilon_y = \frac{\sigma_y}{E} - \frac{\nu}{E}(\sigma_x + \sigma_z)$$

$$= \frac{-500 \text{ MPa}}{120(10^3) \text{ MPa}} - \frac{0.34}{120(10^3) \text{ MPa}}(800 \text{ MPa} + 0) = -0.00643$$

$$\epsilon_z = \frac{\sigma_z}{E} - \frac{\nu}{E}(\sigma_x + \sigma_y)$$

$$= 0 - \frac{0.34}{120(10^3) \text{ MPa}}(800 \text{ MPa} - 500 \text{ MPa}) = -0.000850$$

The new bar length, width, and thickness are therefore

$$a' = 300 \text{ mm} + 0.00808(300 \text{ mm}) = 302.4 \text{ mm} \qquad Ans.$$

$$b' = 50 \text{ mm} - 0.00643(50 \text{ mm}) = 49.68 \text{ mm} \qquad Ans.$$

$$t' = 20 \text{ mm} - 0.000850(20 \text{ mm}) = 19.98 \text{ mm} \qquad Ans.$$

Example 10–9

If the rectangular rubber block shown in Fig. 10–27 is subjected to a uniform pressure of $p = 20$ psi, determine the dilatation and the change in length of each side. Take $E_r = 600$ psi, $\nu_r = 0.45$.

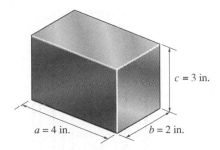

$c = 3$ in.

$a = 4$ in. $b = 2$ in.

Fig. 10–27

SOLUTION

Dilatation. The dilatation can be determined using Eq. 10–23 with $\sigma_x = \sigma_y = \sigma_z = -20$ psi. We have

$$e_r = \frac{1 - 2\nu}{E}(\sigma_x + \sigma_y + \sigma_z)$$

$$= \frac{1 - 2(0.45)}{600 \text{ psi}}[3(-20 \text{ psi})]$$

$$= -0.01 \text{ in}^3/\text{in}^3 \qquad\qquad Ans.$$

Change in Length. The normal strain on each side can be determined from Hooke's law, Eq. 10–18; that is,

$$\epsilon = \frac{1}{E}[\sigma - \nu(\sigma + \sigma)]$$

$$= \frac{1}{600 \text{ psi}}[-20 \text{ psi} - (0.45)(-20 \text{ psi} - 20 \text{ psi})] = -0.00333 \text{ in./in.}$$

Thus the change in length of each side is

$$\Delta a = -0.00333(4 \text{ in.}) = -0.0133 \text{ in.} \qquad Ans.$$
$$\Delta b = -0.00333(2 \text{ in.}) = -0.00667 \text{ in.} \qquad Ans.$$
$$\Delta c = -0.00333(3 \text{ in.}) = -0.0100 \text{ in.} \qquad Ans.$$

The negative signs indicate that each dimension is decreased.

PROBLEMS

*10–24. Use Hooke's law, Eq. 10–18, to develop the strain-transformation equations, Eqs. 10–5 and 10–6, from the stress-transformation equations, Eqs. 9–1 and 9–2.

10–25. Determine the bulk modulus for each of the following materials: (a) rubber, $E_r = 0.4$ ksi, $\nu_r = 0.48$, and (b) glass, $E_g = 8(10^3)$ ksi, $\nu_g = 0.24$.

10–26. A uniform edge load of $w_1 = 60$ N/m and $w_2 = 45$ N/m is applied to the rubber membrane. If it is originally square and has dimensions of $a = 100$ mm and $b = 100$ mm and a thickness of $t = 2$ mm, determine its new dimensions a', b', and t' after the load is applied. $E_r = 4$ MPa, $\nu_r = 0.42$.

10–27. A uniform edge load of $w_1 = 60$ N/m and $w_2 = 0$ is applied to the rubber membrane. If it is originally square and has dimensions of $a = 100$ mm and $b = 100$ mm and a thickness of $t = 2$ mm, determine its new dimensions a', b', and t' after the load is applied. $E_r = 4$ MPa, $\nu_r = 0.42$.

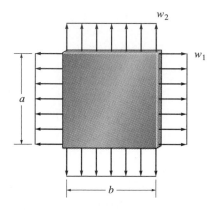

Probs. 10–26/10–27

*10–28. The aluminum bar is subjected to an axial force of 15 kip. If it has the original dimensions shown, determine the *change* in the angle θ after the load is applied. $E_{al} = 10(10^3)$ ksi, $\nu_{al} = 0.33$.

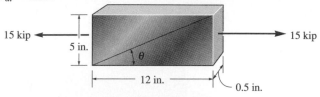

Prob. 10–28

10–29. Determine the principal strains that occur at a point on a steel member where the principal stresses are $\sigma_{max} = 22$ ksi, $\sigma_{int} = 0$, $\sigma_{min} = -14$ ksi. $E_{st} = 29(10^3)$ ksi, $\nu_{st} = 0.3$.

10–30. A bar of copper alloy is loaded in a tension machine and it is determined that $\epsilon_x = 940(10^{-6})$ and $\sigma_x = 14$ ksi, $\sigma_y = 0$, $\sigma_z = 0$. Determine the modulus of elasticity, E_{cu}, and the dilatation, e_{cu}, of the copper. $\nu_{cu} = 0.35$.

10–31. The principal plane stresses and associated strains in a plane at a point are $\sigma_1 = 36$ ksi, $\sigma_2 = 16$ ksi, $\epsilon_1 = 1.02(10^{-3})$, $\epsilon_2 = 0.180(10^{-3})$. Determine the modulus of elasticity and Poisson's ratio.

*10–32. From experiment, the principal strains in a plane at a point on a steel shell are $\epsilon_1 = 350(10^{-6})$ and $\epsilon_2 = -250(10^{-6})$. If $E_{st} = 200$ GPa and $\nu_{st} = 0.28$, determine the principal plane stresses in this plane.

10–33. The principal stresses at a point are shown in the figure. If the material is nylon for which $E_n = 2.5$ GPa and $\nu_n = 0.4$, determine the principal strains.

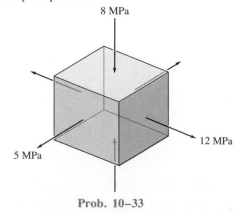

Prob. 10–33

10–34. For the case of plane stress, show that Hooke's law can be written as

$$\sigma_x = \frac{E}{(1 - \nu^2)}(\epsilon_x + \nu\epsilon_y), \qquad \sigma_y = \frac{E}{(1 - \nu^2)}(\epsilon_y + \nu\epsilon_x)$$

10–35. The principal strains in a plane, measured experimentally at a point on the aluminum fuselage of a jet aircraft, are $\epsilon_1 = 780(10^{-6})$ and $\epsilon_2 = 400(10^{-6})$. Determine the associated principal stresses at the point in the same plane. $E_{al} = 10(10^3)$ ksi, $\nu_{al} = 0.33$. *Hint:* See Prob. 10–34.

*10–36. A thin-walled spherical pressure vessel has an inner radius r, thickness t, and is subjected to an internal pressure p. If the material constants are E and ν, determine the strain in the circumferential direction in terms of the stated parameters.

10–37. The spherical pressure vessel has an inner diameter of 2 m and a thickness of 10 mm. A strain gauge having a length of 20 mm is attached to the vessel and it is observed to increase in length by 0.012 mm as the pressure in the vessel is increased. Determine the change in pressure causing this deformation and also compute the maximum in-plane shear stress and the absolute maximum shear stress. The material is steel, for which $E_{st} = 200$ GPa and $\nu_{st} = 0.3$.

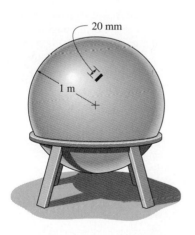

20 mm

1 m

Prob. 10–37

10–38. A strain gauge, placed on the outer surface and at an angle of 45° to the axis of the copper pipe, gives a reading at point A of $\epsilon_a = 250(10^{-6})$. Determine the force P applied to the wrench. The pipe has an outer diameter of 1 in. and an inner diameter of 0.6 in. $E_{cu} = 18(10^3)$ ksi, $\nu_{cu} = 0.33$.

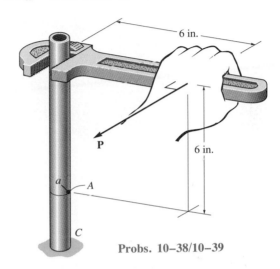

6 in.

P

6 in.

a A

C

Probs. 10–38/10–39

10–39. A strain gauge, placed on the outer surface and at an angle of 45° to the axis of the copper pipe, gives a reading at point A of $\epsilon_a = 250(10^{-6})$. Determine the principal strains in the pipe at point A. The pipe has an outer diameter of 1 in. and an inner diameter of 0.6 in. $E_{cu} = 18(10^3)$ ksi, $\nu_{cu} = 0.33$.

*10–40. The cross section of the rectangular beam is subjected to the bending moment $\mathbf{M}$. Determine an expression for the increase in length of lines AB and CD. The material has a modulus of elasticity E and Poisson's ratio is ν.

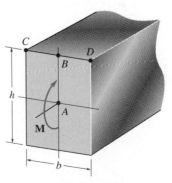

C

D

B

h

A

M

b

Prob. 10–40

10–41.　The aluminum beam has the rectangular cross section shown. If it is subjected to a bending moment of $M = 60$ kip $\cdot$ in., determine the increase in the 2-in. dimension at the top of the beam and the decrease in this dimension at the bottom. $E_{al} = 10(10^3)$ ksi, $\nu_{al} = 0.3$.

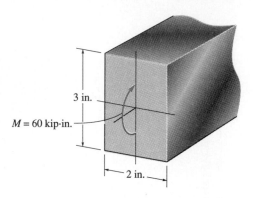

$M = 60$ kip·in.

3 in.

2 in.

Prob. 10–41

10–42.　The thin-walled cylindrical pressure vessel of inner radius r and thickness t is subjected to an internal pressure p. Determine the maximum x,y in-plane shear strain at point A, where the principal stresses are σ_{max} and σ_{int}, and $\sigma_{min} = 0$. Also compute the absolute maximum shear strain at A. The material properties are E and ν.

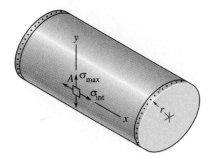

Prob. 10–42

10–43.　Air is pumped into the steel thin-walled pressure vessel at C. If the ends of the vessel are closed using two pistons connected by a rod AB, determine the increase in the diameter of the pressure vessel when the internal gauge pressure is 5 MPa. Also, what is the tensile stress in rod AB if it has a diameter of 100 mm? The inner radius of the vessel is 400 mm, and its thickness is 10 mm. $E_{st} = 200$ GPa and $\nu_{st} = 0.3$.

*10–44.　Determine the increase in the diameter of the pressure vessel in Prob. 10–43 if the pistons are replaced by walls connected to the vessel.

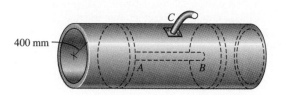

400 mm

Probs. 10–43/10–44

10–45.　The strain gauge is placed on the surface of a thin-walled steel boiler as shown. If it is 0.5 in. long, determine the pressure in the boiler when the gauge elongates $0.2(10^{-3})$ in. The boiler has a thickness of 0.5 in. and inner diameter of 60 in. Also, determine the maximum x, y in-plane shear strain in the material. $E_{st} = 29(10^3)$ ksi, $\nu_{st} = 0.3$.

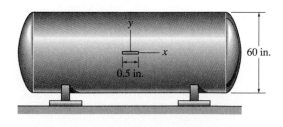

0.5 in.

60 in.

Prob. 10–45

10–46. The steel shaft has a radius of 15 mm. Determine the torque T in the shaft if the two strain gauges, attached to the surface of the shaft, report strains of $\epsilon_{x'} = -80(10^{-6})$ and $\epsilon_{y'} = 80(10^{-6})$. Also, compute the strains acting in the x and y directions. $E_{st} = 200$ GPa, $\nu_{st} = 0.3$.

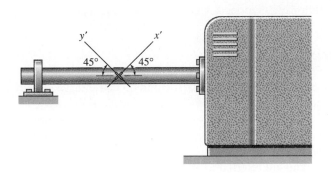

Prob. 10–46

10–47. Two strain gauges are attached to the 6-in.-diameter steel shaft. The gauges are placed at 45° with the axis of the shaft. If the readings are $\epsilon_{x'} = 320(10^{-6})$ and $\epsilon_{y'} = -320(10^{-6})$, determine the torque T developed in the shaft. $E_{st} = 29(10^3)$ ksi, $\nu_{st} = 0.3$.

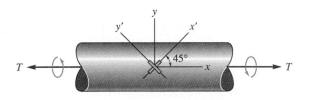

Prob. 10–47

*10–48. The strain in the x direction at point A on the steel beam is measured and found to be $\epsilon_x = -100(10^{-6})$. Determine the applied load P. What is the shear strain γ_{xy} at point A? $E_{st} = 29(10^3)$ ksi, $\nu_{st} = 0.3$.

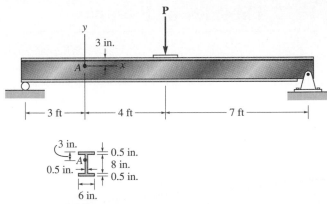

Prob. 10–48

10–49. A thin-walled cylindrical pressure vessel has an inner radius r, thickness t, and length L. If it is subjected to an internal pressure p, show that the increase in its inner radius is $dr = r\epsilon_1 = pr^2(1 - \tfrac{1}{2}\nu)/Et$ and the increase in its length is $\Delta L = pLr(\tfrac{1}{2} - \nu)/Et$. Using these results, show that the change in internal volume becomes $dV = \pi r^2(1 + \epsilon_1)^2(1 + \epsilon_2)L - \pi r^2 L$. Since ϵ_1 and ϵ_2 are small quantities, show further that the change in volume per unit volume, called *volumetric strain,* can be written as $dV/V = Pr(2.5 - 2\nu)/Et$.

10–50. The cylindrical pressure vessel is fabricated using hemispherical end caps in order to reduce the bending stress that would occur if flat ends were used. The bending stresses at the seam where the caps are attached can be eliminated by proper choice of the thickness t_h and t_c of the caps and cylinder, respectively. This requires the radial expansion to be the same for both the hemispheres and cylinder. Show that this ratio is $t_c/t_h = (2 - \nu)/(1 - \nu)$. Assume that the vessel is made of the same material and both the cylinder and hemispheres have the same inner radius. If the cylinder is to have a thickness of 0.5 in., what is the required thickness of the hemispheres? Take $\nu = 0.3$.

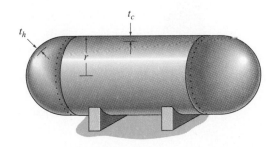

Prob. 10–50

*10.7 Theories of Failure

When an engineer is faced with the problem of design using a specific material, it becomes important to place an upper *limit* on the state of stress that defines the material's failure. If the material is *ductile,* failure is usually specified by the initiation of *yielding,* whereas if the material is *brittle,* it is specified by *fracture.* These modes of failure are readily defined if the member is subjected to a uniaxial state of stress, as in the case of simple tension; however, if the member is subjected to biaxial or triaxial stress, the criterion for failure becomes more difficult to establish.

In this section we will discuss four theories that are often used in engineering practice to predict the failure of a material subjected to a *multiaxial* state of stress. These theories, and others like them, are also used to determine the allowable stresses reported in many design codes. No single theory of failure, however, may apply to a specific material at *all times,* because a material may behave in either a ductile or brittle manner depending on the temperature, rate of loading, chemical environment, or the way the material is shaped or formed. When using a particular theory of failure, it is first necessary to calculate the normal and shear stress components at points where they are the largest in the member. This may be done by using the fundamentals of mechanics of materials and applying stress-concentration factors where applicable, or in complex situations, the largest stress components can be found by using either a mathematical analysis based on the theory of elasticity or by an appropriate experimental technique. In any case, once this state of stress is established, the *principal stresses* at these critical points are then determined, since each of the following theories is based on knowing the principal stress.

Ductile Materials

Maximum-Shear-Stress Theory. The most common cause of *yielding of a ductile material* such as steel is *slipping,* which occurs along the contact planes of randomly ordered crystals that make up the material. This *slipping* is due to *shear stress,* and if we subject the material to a simple tension test we can see how it causes the material to *yield.* As shown in Fig. 10–28, the edges of the planes of slipping are referred to as *Lüder's lines,* and they appear on the surface of a highly polished thin strip of a mild steel subjected to tension. These lines clearly indicate the slip planes in the strip, which occur at approximately 45° with the axis of the strip.

Consider now an element of the material taken from a tension specimen, Fig. 10–29a, which is subjected only to the yield stress σ_Y. The maximum shear stress can be determined by drawing Mohr's circle for the element, Fig. 10–29b. The results indicate that

$$\tau_{\max} = \frac{\sigma_Y}{2} \tag{10–26}$$

Lüder's lines on
mild steel strip

Fig. 10–28

Furthermore, this shear stress acts on planes that are 45° from the planes of principal stress, Fig. 10–29c, and these planes *coincide* with the direction of the Lüder lines shown on the specimen, indicating that indeed failure occurs by shear.

Using this idea, that ductile materials fail by shear, Henri Tresca in 1868 proposed the *maximum-shear-stress theory* or *Tresca yield criterion*. This theory can be used to predict the failure stress of a ductile material subjected to any type of loading. The maximum-shear-stress theory states that yielding of the material begins when the absolute maximum shear stress in the material reaches the shear stress that causes the same material to yield when it is subjected *only* to axial tension. To avoid failure, therefore, the maximum-shear-stress theory requires $\tau_{abs \atop max}$ in the material to be less than or equal to $\sigma_Y/2$, where σ_Y is determined from a simple tension test.

For application we will express the absolute maximum shear stress in terms of the *principal stresses*. The procedure for doing this was discussed in Sec. 9.5 with reference to a condition of *plane stress,* that is, where the out-of-plane principal stress is zero. If the two in-plane principal stresses have the *same sign,* i.e., they are both tensile or both compressive, then failure will occur *out of the plane,* and from Eq. 9–15,

$$\tau_{abs \atop max} = \frac{\sigma_{max}}{2}$$

On the other hand, if the in-plane principal stresses are of *opposite signs,* then failure occurs in the plane, and from Eq. 9–16,

$$\tau_{abs \atop max} = \frac{\sigma_{max} - \sigma_{min}}{2}$$

Using these equations and Eq. 10–26, the maximum-shear-stress theory for *plane stress* can be expressed for any two in-plane principal stresses as σ_1 and σ_2 by the following criteria:

$$\left.\begin{matrix} |\sigma_1| = \sigma_Y \\ |\sigma_2| = \sigma_Y \end{matrix}\right\}\ \sigma_1,\ \sigma_2 \text{ have same signs}$$
$$|\sigma_1 - \sigma_2| = \sigma_Y\ \}\ \sigma_1,\ \sigma_2 \text{ have opposite signs}$$

$$(10\text{--}27)$$

A graph of these equations is given in Fig. 10–30. Clearly, if any point of the material is subjected to plane stress, and its in-plane principal stresses are represented by a coordinate (σ_1, σ_2) plotted *on* or *outside* the shaded hexagonal area shown in this figure, the material will yield at the point and failure is said to occur.

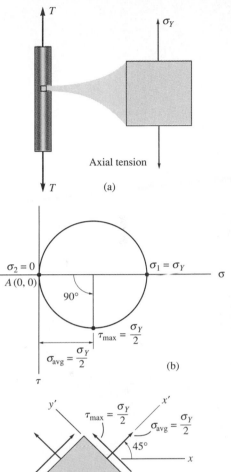

Fig. 10–29

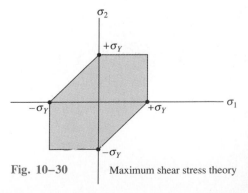

Fig. 10–30   Maximum shear stress theory

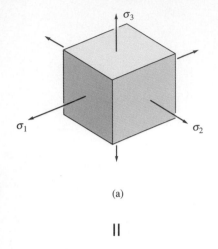

(a)

$$\parallel$$

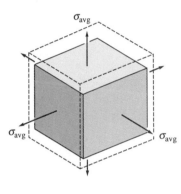

(b)

$$+$$

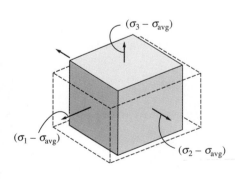

(c)

Fig. 10–31

Maximum-Distortion-Energy Theory. It was stated in Sec. 3.5 that a material, when deformed by an external loading, tends to store energy *internally* throughout its volume. The energy per unit volume of material is called the *strain-energy density,* and if the material is subjected to a uniaxial stress, σ, the strain-energy density, defined by Eq. 3–6, can be written as

$$u = \frac{1}{2}\sigma\epsilon \qquad (10\text{--}28)$$

It is possible to formulate a failure criterion based on the distortions caused by strain energy. Before doing this, however, we need to formulate the strain-energy density in a volume element of material subjected to the three principal stresses σ_1, σ_2, and σ_3, Fig. 10–31a. Here, each principal stress contributes a portion of the total strain-energy density, so that

$$u = \frac{1}{2}\sigma_1\epsilon_1 + \frac{1}{2}\sigma_2\epsilon_2 + \frac{1}{2}\sigma_3\epsilon_3$$

If the material behaves in a linear-elastic manner, then Hooke's law applies. Therefore, substituting Eq. 10–18 into the above equation and simplifying, we get

$$u = \frac{1}{2E}[\sigma_1^2 + \sigma_2^2 + \sigma_3^2 - 2\nu(\sigma_1\sigma_2 + \sigma_1\sigma_3 + \sigma_3\sigma_2)] \quad (10\text{--}29)$$

This strain-energy density can be considered as the sum of two parts, one part representing the energy needed to cause a *volume change* of the element with no change in shape, and the other part representing the energy needed to *distort* the element. Specifically, the energy stored in the element as a result of its volume being changed is caused by application of the average principal stress, $\sigma_{\text{avg}} = (\sigma_1 + \sigma_2 + \sigma_3)/3$, since this stress causes equal principal strains in the material, Fig. 10–31b. The remaining portion of the stress, $(\sigma_1 - \sigma_{\text{avg}})$, $(\sigma_2 - \sigma_{\text{avg}})$, $(\sigma_3 - \sigma_{\text{avg}})$, causes the energy of distortion, Fig. 10–31c.

Experimental evidence has shown that materials do not yield when subjected to a uniform (hydrostatic) stress, such as σ_{avg} discussed above. As a result, in 1904, M. Huber proposed that yielding in a ductile material occurs when the *distortion energy* per unit volume of the material equals or exceeds the distortion energy per unit volume of the same material when it is subjected to yielding in a simple tension test. This theory is called the *maximum-distortion-energy theory,* and since it was later redefined independently by R. von Mises and H. Hencky, it sometimes also bears their names.

To obtain the distortion energy per unit volume, we will substitute the stresses $(\sigma_1 - \sigma_{\text{avg}})$, $(\sigma_2 - \sigma_{\text{avg}})$, and $(\sigma_3 - \sigma_{\text{avg}})$ for σ_1, σ_2, and σ_3, respectively, into Eq. 10–29, realizing that $\sigma_{\text{avg}} = (\sigma_1 + \sigma_2 + \sigma_3)/3$. Expanding and simplifying, we obtain

$$u_d = \frac{1+\nu}{6E}[(\sigma_1 - \sigma_2)^2 + (\sigma_2 - \sigma_3)^2 + (\sigma_3 - \sigma_1)^2]$$

In the case of *plane stress*, $\sigma_3 = 0$, and this equation reduces to

$$u_d = \frac{1 + \nu}{3E}(\sigma_1{}^2 - \sigma_1\sigma_2 + \sigma_2{}^2)$$

For a *uniaxial* tension test, $\sigma_1 = \sigma_Y$, $\sigma_2 = \sigma_3 = 0$, and then the distortion energy becomes

$$(u_d)_Y = \frac{1 + \nu}{3E}\sigma_Y{}^2$$

Since the maximum-distortion-energy theory requires $u_d = (u_d)_Y$, then for the case of plane or biaxial stress, we have

$$\boxed{\sigma_1{}^2 - \sigma_1\sigma_2 + \sigma_2{}^2 = \sigma_Y{}^2} \qquad (10\text{–}30)$$

This equation represents an elliptical curve, Fig. 10–32. Thus, if a point in the material is stressed such that the stress coordinate (σ_1, σ_2) is plotted on or outside the shaded area, the material is said to fail.

A comparison of the above two failure criteria is shown in Fig. 10–33. Note that both theories give the same results when the principal stresses are equal, i.e., from Eqs. 10–27 and 10–30, $\sigma_1 = \sigma_2 = \sigma_Y$, or when one of the principal stresses is zero and the other has a magnitude of σ_Y. On the other hand, if the material is subjected to pure shear, τ, then the theories have the largest discrepancy in predicting failure. The stress coordinates of these points on the curves have been computed by considering the element shown in Fig. 10–34a. From the associated Mohr's circle for this state of stress, Fig. 10–34b, we obtain principal stresses $\sigma_1 = \tau$ and $\sigma_2 = -\tau$. Applying Eqs. 10–27 and 10–30, the maximum-shear-stress theory and maximum-distortion-energy theory yield $\sigma_1 = \sigma_Y/2$ and $\sigma_1 = \sigma_Y/\sqrt{3}$, respectively, Fig. 10–33.

Actual torsion tests, used to develop a condition of pure shear in a ductile specimen, have shown that the maximum-distortion-energy theory gives more accurate results than the maximum-shear-stress theory. In fact, since $(\sigma_Y/\sqrt{3})/(\sigma_Y/2) = 1.15$, the shear stress for yielding, as given by the maximum distortion-energy theory, is 15% more accurate than that given by the maximum-shear-stress theory.

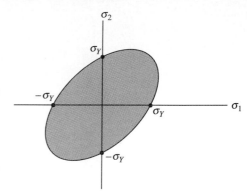

Maximum distortion energy theory

Fig. 10–32

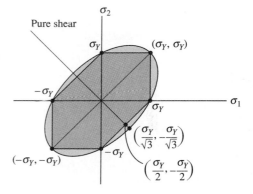

Fig. 10–33

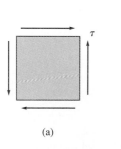

(a)

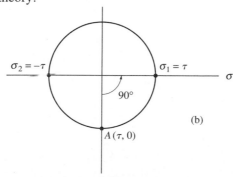

(b)

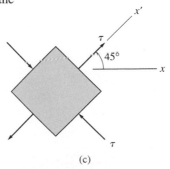

(c)

Fig. 10–34

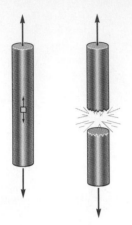

Failure of a brittle material
in tension

(a)

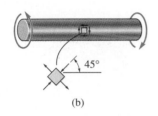

(b)

Failure of a brittle material
in torsion

(c)

Fig. 10–35

Brittle Materials

Maximum-Normal-Stress Theory. It was previously stated that brittle materials, such as gray cast iron, tend to fail suddenly by *fracture* with no apparent yielding. In a *tension test,* the fracture occurs when the normal stress reaches the ultimate stress σ_{ult}, Fig. 10–35a. Also, in a *torsion test,* brittle fracture occurs due to a *maximum tensile stress* since the plane of fracture for an element is at 45° to the shear direction, Fig. 10–35b. The fracture surface is therefore helical as shown.* Experiments have shown further that during torsion the material's strength is somewhat *unaffected* by the presence of the associated principal compressive stress being at right angles to the principal tensile stress. Consequently, the tensile stress needed to fracture a specimen during a torsion test is approximately the same as that needed to fracture a specimen in simple tension. Because of this, the *maximum-normal-stress theory* states that a brittle material will fail when the maximum principal stress σ_1 in the material reaches a limiting value that is equal to the ultimate normal stress the material can sustain when it is subjected to simple tension.

If the material is subjected to *plane stress,* we require that

$$\begin{aligned} |\sigma_1| &= \sigma_{\text{ult}} \\ |\sigma_2| &= \sigma_{\text{ult}} \end{aligned} \qquad (10\text{–}31)$$

These equations are shown graphically in Fig. 10–36. Here it is seen that if the stress coordinate (σ_1, σ_2) at a point in the material falls on or outside the shaded area, the material is said to fracture. This theory is generally credited to W. Rankine, who proposed it in the mid-1800s. Experimentally it has been found to be in close agreement with the behavior of brittle materials that have stress–strain diagrams that are *similar* in both tension and compression.

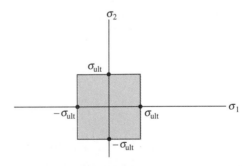

Maximum normal stress theory

Fig. 10–36

* A stick of blackboard chalk fails in this way when its ends are twisted with the fingers.

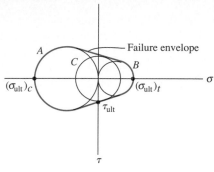

Fig. 10–37

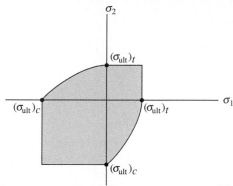

Mohr's failure criteria

Fig. 10–38

Mohr's Failure Criterion. In some brittle materials the tension and compression properties are *different*. When this occurs a criterion based on the use of Mohr's circle may be used to predict failure of the material. This method was developed by Otto Mohr and is sometimes referred to as *Mohr's failure criterion*. To apply it, one first performs *three tests* on the material. A uniaxial tensile test and uniaxial compressive test are used to determine the ultimate tensile and compressive stresses $(\sigma_{ult})_t$ and $(\sigma_{ult})_c$, respectively. Also a torsion test is performed to determine the ultimate shear stress τ_{ult} of the material. Mohr's circle for each of these stress conditions is then plotted as shown in Fig. 10–37. Circle A represents the stress condition $\sigma_1 = \sigma_2 = 0$, $\sigma_3 = -(\sigma_{ult})_c$; circle B represents the stress conditions $\sigma_1 = (\sigma_{ult})_t$, $\sigma_2 = \sigma_3 = 0$; and circle C represents the pure-shear-stress condition caused by τ_{ult}. These three circles are contained in a "failure envelope" indicated by the extrapolated colored curve that is drawn tangent to all three circles. If a plane-stress condition at a point is represented by a circle that is contained within the envelope, the material is said not to fail. If, however, the circle has a point of tangency with the envelope, or if it extends beyond the envelope's boundary, then failure is said to occur.

We may also represent this criterion on a graph of principal stresses σ_1 and σ_2 ($\sigma_3 = 0$). This is shown in Fig. 10–38. Here failure occurs when the absolute value of either one of the principal stresses reaches a value equal to or greater than $(\sigma_{ult})_t$ or $(\sigma_{ult})_c$ or in general, if the state of stress at a point is defined by the stress coordinate (σ_1, σ_2), which is plotted on or outside the shaded area.

Either of the above two criteria can be used in practice to predict the failure of a brittle material. However, it should be realized that their usefulness is quite limited. A tensile fracture occurs very suddenly, and its initiation generally depends on stress concentrations developed at microscopic imperfections of the material such as inclusions or voids, surface indentations, and small cracks. Since each of these irregularities varies from specimen to specimen, it becomes difficult to specify failure on the basis of a single test. On the other hand, cracks and other irregularities tend to close up when the specimen is compressed, and therefore they do not form points of failure as they would when the specimen is subjected to tension.

The steel pipe shown in Fig. 10–39a has an inner diameter of 60 mm and an outer diameter of 80 mm. If it is subjected to a torsional moment of 8 kN · m and a bending moment of 3.5 kN · m, determine if these loadings cause failure as defined by the maximum-distortion-energy theory. The yield stress for the steel found from a tension test is $\sigma_Y = 250$ MPa.

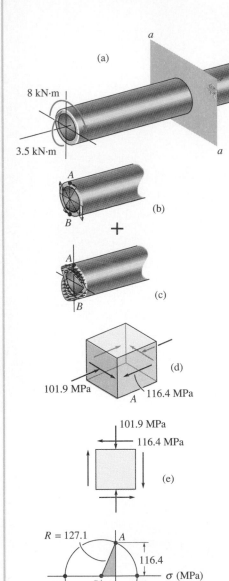

(a)

8 kN·m

3.5 kN·m

a

a

A

B

(b)

+

A

B

(c)

101.9 MPa   116.4 MPa

A

(d)

101.9 MPa

116.4 MPa

(e)

$R = 127.1$

A

116.4

σ_2 C σ_1 σ (MPa)

50.9

(f)

τ (MPa)

Fig. 10–39

SOLUTION

To solve this problem we must investigate a point on the pipe that is subjected to a state of maximum critical stress. Both the torsional and bending moments are uniform throughout the pipe's length. At the arbitrary section a–a, Fig. 10–39a, these loadings produce the stress distributions shown in Fig. 10–39b and 10–39c. By inspection, points A and B are subjected to the same state of critical stress. Here we will investigate the state of stress at A. Thus,

$$\tau_A = \frac{Tc}{J} = \frac{(8000 \text{ N} \cdot \text{m})(0.04 \text{ m})}{(\pi/2)[(0.04 \text{ m})^4 - (0.03 \text{ m})^4]} = 116.4 \text{ MPa}$$

$$\sigma_A = \frac{Mc}{I} = \frac{(3500 \text{ N} \cdot \text{m})(0.04 \text{ m})}{(\pi/4)[(0.04 \text{ m})^4 - (0.03 \text{ m})^4]} = 101.9 \text{ MPa}$$

These results are shown on a three-dimensional view of an element of material at point A, Fig. 10–39d, and also, since the material is subjected to plane stress, it is shown in two dimensions, Fig. 10–39e.

Mohr's circle for this state of plane stress has a center located at

$$\sigma_{avg} = \frac{0 - 101.9}{2} = -50.9 \text{ MPa}$$

The reference point $A(0, -116.4$ MPa$)$ is plotted and the circle is constructed, Fig. 10–39f. Here the radius has been calculated from the shaded triangle to be $R = 127.1$ and so the in-plane principal stresses are

$$\sigma_1 = -50.9 + 127.1 = 76.2 \text{ MPa}$$
$$\sigma_2 = -50.9 - 127.1 = -178.0 \text{ MPa}$$

Using Eq. 10–30, we require

$$(\sigma_1^2 - \sigma_1\sigma_2 + \sigma_2^2) \leq \sigma_Y^2$$
$$[(76.2)^2 - (76.2)(-178.0) + (-178.0)^2] \overset{?}{\leq} (250)^2$$
$$51{,}100 < 62{,}500 \qquad \text{OK}$$

Since the criterion has been met, the material within the pipe will *not* yield ("fail") according to the maximum-distortion-energy theory.

Example 10–11

The solid cast-iron shaft shown in Fig. 10–40a is subjected to a torque of $T = 400$ lb $\cdot$ ft. Determine its smallest radius so that it does not fail according to the maximum-normal-stress theory. A specimen of cast iron, tested in tension, has an ultimate stress of $(\sigma_{ult})_t = 20$ ksi.

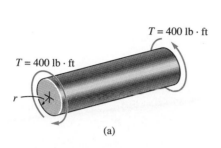

$T = 400$ lb $\cdot$ ft

$T = 400$ lb $\cdot$ ft

r

(a)

Fig. 10–40

(b)

SOLUTION

The maximum or critical stress occurs at a point located on the surface of the shaft. Assuming the shaft to have a radius r, the shear stress is

$$\tau_{max} = \frac{Tc}{J} = \frac{(400 \text{ lb} \cdot \text{ft})(12 \text{ in./ft})r}{(\pi/2)r^4} = \frac{3055.8}{r^3}$$

Mohr's circle for this state of stress (pure shear) is shown in Fig. 10–40b. Since $R = \tau_{max}$, then

$$\sigma_1 = -\sigma_2 = \tau_{max} = \frac{3055.8}{r^3}$$

The maximum-normal-stress theory, Eq. 10–31, requires

$$|\sigma_1| \leq \sigma_{ult}$$

$$\frac{3055.8}{r^3} \leq 20{,}000$$

Thus, the smallest radius of the shaft is determined from

$$\frac{3055.8}{r^3} = 20{,}000$$

$$r = 0.535 \text{ in.} \qquad Ans.$$

■ Example 10–12

The solid shaft shown in Fig. 10–41a has a radius of 0.5 in. and is made of steel having a yield stress of $\sigma_Y = 36$ ksi. Determine if the loadings cause the shaft to fail according to the maximum-shear-stress theory and the maximum-distortion-energy theory.

SOLUTION

The state of stress in the shaft is caused by both the axial force and the torque. Since maximum shear stress caused by the torque occurs in the material at the outer surface, we have

$$\sigma_x = \frac{P}{A} = \frac{15 \text{ kip}}{\pi(0.5 \text{ in.})^2} = 19.10 \text{ ksi}$$

$$\tau_{xy} = \frac{Tc}{J} = \frac{3.25 \text{ kip} \cdot \text{in.}(0.5 \text{ in.})}{\pi(0.5 \text{ in.})^4/2} = 16.55 \text{ ksi}$$

The stress components are shown acting on an element of material at point A in Fig. 10–41b. Rather than using Mohr's circle, the principal stresses can also be obtained using the stress-transformation equations, Eq. 9–5.

$$\sigma_{1,2} = \frac{\sigma_x + \sigma_y}{2} \pm \sqrt{\left(\frac{\sigma_x - \sigma_y}{2}\right)^2 + \tau_{xy}^2}$$

$$= \frac{-19.10 + 0}{2} \pm \sqrt{\left(\frac{-19.10 - 0}{2}\right)^2 + (16.55)^2}$$

$$= -9.55 \pm 19.11$$

$$\sigma_1 = 9.56 \text{ ksi}$$

$$\sigma_2 = -28.66 \text{ ksi}$$

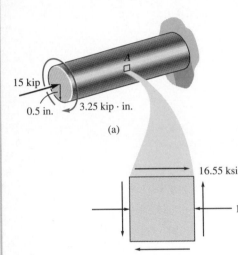

15 kip

0.5 in.   3.25 kip · in.

(a)

16.55 ksi

19.10 ksi

(b)

Fig. 10–41

Maximum-Shear-Stress Theory. Since the principal stresses have *opposite signs,* the absolute maximum shear stress will occur in the plane, and therefore, applying the second of Eq. 10–27, we have

$$|\sigma_1 - \sigma_2| \le \sigma_Y$$

$$|-28.66 - (9.56)| \overset{?}{\le} 36$$

$$38.2 > 36$$

Thus, shear failure of the material will occur according to this theory.

Maximum-Distortion-Energy Theory. Applying Eq. 10–30, we have

$$(\sigma_1^2 - \sigma_1\sigma_2 + \sigma_2^2) \le \sigma_Y$$

$$[(-28.66)^2 - (-28.66)(9.56) + (9.56)^2] \overset{?}{\le} (36)^2$$

$$1187 \le 1296$$

Using this theory, failure will not occur.

PROBLEMS

10–51. A material is subjected to plane stress. Express the maximum-shear-stress theory of failure in terms of σ_x, σ_y, and τ_{xy}. Assume that the principal stresses are of different algebraic signs.

*10–52. A material is subjected to plane stress. Express the distortion-energy theory of failure in terms of σ_x, σ_y, and τ_{xy}.

10–53. The components of plane stress at a critical point on a thin steel shell are shown. Determine if failure (yielding) has occurred on the basis of (a) the maximum-shear-stress theory, (b) the distortion-energy theory. The yield stress for the steel is $\sigma_Y = 650$ MPa.

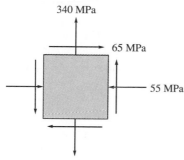

Prob. 10–53

10–54. The state of plane stress at a critical point in a steel machine bracket is shown. If the yield stress for steel is $\sigma_Y = 36$ ksi, determine if yielding occurs using (a) the maximum-shear-stress theory, (b) the maximum-distortion-energy theory.

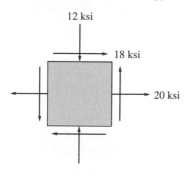

Prob. 10–54

10–55. The yield stress for a certain material is $\sigma_Y = 40$ ksi. If the material is subjected to plane stress and elastic failure occurs when one principal stress is 42 ksi, what is the magnitude of the other principal stress? Use the distortion-energy theory.

*10–56. The plate is made of hard copper, which yields at $\sigma_Y = 105$ ksi. Using the maximum-shear-stress theory, determine the tensile stress σ_x that can be applied to the plate if a tensile stress $\sigma_y = \frac{1}{2}\sigma_x$ is also applied.

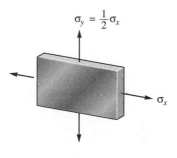

Prob. 10–56

10–57. A bar with a circular cross-sectional area is made of a material having a yield stress of $\sigma_Y = 36$ ksi. If the bar is subjected to a torque of 16 kip · in. and a bending moment of 20 kip · in., determine the required diameter of the bar according to the distortion-energy theory. Use a factor of safety of 2 with respect to yielding.

10–58. Cast iron when tested in tension and compression has an ultimate strength of $(\sigma_{ult})_t = 280$ MPa and $(\sigma_{ult})_c = 420$ MPa, respectively. Also, when subjected to pure torsion it can sustain an ultimate shear stress of $\tau_{ult} = 168$ MPa. Plot the Mohr's circles for each case and establish the failure envelope. If a part made of this material is subjected to the state of plane stress shown, determine if it fails according to Mohr's failure criterion.

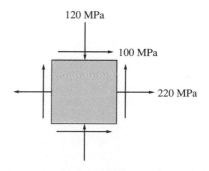

Prob. 10–58

10–59. If a solid shaft having a diameter d is subjected to a torque T and moment M, show that by the maximum-shear-stress theory the maximum allowable shear stress is $\tau_{allow} = (16/\pi d^3)\sqrt{M^2 + T^2}$. Assume the principal stresses to be of opposite algebraic signs.

10–60. If a solid shaft having a diameter d is subjected to a torque T and moment M, show that by the maximum-normal-stress theory the maximum allowable principal stress is $\sigma_{allow} = (16/\pi d^3)(M + \sqrt{M^2 + T^2})$.

10–61. The principal plane stresses acting on a differential element are shown. If the material is machine steel having a yield stress of $\sigma_Y = 700$ MPa, determine the factor of safety with respect to yielding if the maximum-shear-stress theory is considered. $\nu_{st} = 0.3$.

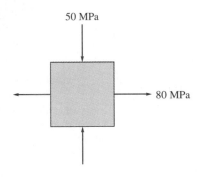

Prob. 10–61

10–62. Derive an expression for an equivalent bending moment M_e that if applied alone to a solid bar with a circular cross section would cause the same energy of distortion as the combination of an applied bending moment M and torque T.

10–63. The yield stress for a zirconium-magnesium alloy is $\sigma_Y = 15.3$ ksi. If a machine part is made of this material and a critical point in the material is subjected to in-plane principal stresses σ_1 and $\sigma_2 = -0.5\sigma_1$, determine the magnitude of σ_1 that will cause yielding according to (a) the maximum-shear-stress theory, (b) the maximum-distortion-energy theory.

10–64. The yield stress for a uranium alloy is $\sigma_Y = 160$ MPa. If a machine part is made of this material and a critical point in the material is subjected to plane stress, such that the principal stresses are σ_1 and $\sigma_2 = 0.25\sigma_1$, determine the magnitude of σ_1 that will cause yielding according to (a) the maximum-shear-stress theory, (b) the maximum-distortion-energy theory.

10–65. The element is subjected to the stresses shown. If $\sigma_Y = 36$ ksi and $\nu = 0.3$, determine the factor of safety for this loading based on (a) the maximum-shear-stress theory, and (b) the maximum-distortion-energy theory.

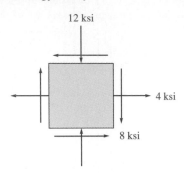

Prob. 10–65

10–66. An aluminum alloy is to be used for a drive shaft such that it transmits 25 hp at 1500 rev/min. Using a factor of safety with respect to yielding of F.S. = 2.5, determine the smallest-diameter shaft that can be selected based on the maximum-distortion-energy theory. $\sigma_Y = 3.5$ ksi.

10–67. An aluminum alloy is to be used for a solid drive shaft such that it transmits 30 hp at 1200 rev/min. Using a factor of safety with respect to yielding of F.S. = 2.5, determine the smallest-diameter shaft that can be selected based on the maximum-shear-stress theory. $\sigma_Y = 10$ ksi.

10–68. The principal stresses acting at a point on a thin-walled cylindrical pressure vessel are $\sigma_1 = pr/t$, $\sigma_2 = pr/2t$, and $\sigma_3 = 0$. If the yield stress is σ_Y and $\nu = 0.3$, determine the maximum value of p based on (a) the maximum-shear-stress theory, and (b) the maximum-distortion-energy theory.

10–69. The state of stress acting at a critical point on a wrench is shown in the figure. Determine the smallest yield stress for a material that might be selected for the part, based on the distortion-energy theory.

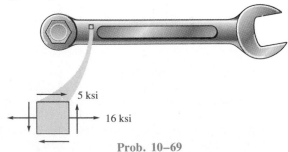

Prob. 10–69

10–70. The state of stress acting at a critical point in a machine element is shown in the figure. Determine the smallest yield stress for a material that might be selected for the part, based on the maximum-shear-stress theory.

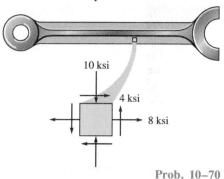

10 ksi

4 ksi

8 ksi

Prob. 10–70

10–71. The short concrete cylinder having a diameter of 50 mm is subjected to a torque of 500 N · m and an axial compressive force of 2 kN. Determine if it fails according to the maximum-normal-stress theory. The ultimate stress of the concrete is σ_{ult} = 28 MPa.

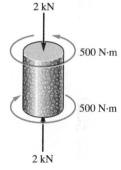

2 kN

500 N·m

500 N·m

2 kN

Prob. 10–71

***10–72.** The internal loadings at a critical section along the steel drive shaft of a ship are calculated to be a torque of 2300 lb · ft, a bending moment of 1500 lb · ft, and an axial thrust of 2500 lb. If the yield points for tension and shear are σ_Y = 100 ksi and τ_Y = 50 ksi, respectively, determine the required diameter of the shaft using the maximum-shear-stress theory.

10–73. The internal loadings at a critical section along the steel drive shaft of a ship are calculated to be a torque of 2300 lb · ft, a bending moment of 1500 lb · ft, and an axial thrust of 2500 lb. If the yield points for tension and shear are σ_Y = 100 ksi and τ_Y = 50 ksi, respectively, determine the required diameter of the shaft using the maximum-distortion-energy theory.

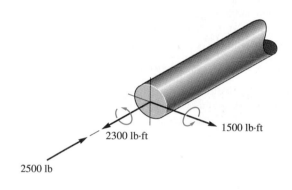

2300 lb·ft

1500 lb·ft

2500 lb

Probs. 10–72/10–73

REVIEW PROBLEMS

10–74. The thin-walled cylindrical pressure vessel of inner radius r and thickness t is subjected to an internal pressure p. If the material constants are E and ν, determine the strains in the circumferential and longitudinal directions. Using these results, compute the increase in both the diameter and the length of a steel pressure vessel filled with air and having an internal gauge pressure of 15 MPa. The vessel is 3 m long, and has an inner radius of 0.5 m and a thickness of 10 mm. E_{st} = 200 GPa, ν_{st} = 0.3.

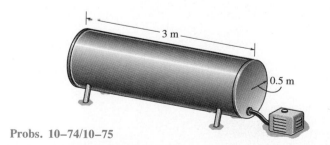

3 m

0.5 m

Probs. 10–74/10–75

10–75. Estimate the increase in volume of the tank in Prob. 10–74. *Suggestion:* Use the results of Prob. 10–49.

*10–76. Determine the bulk modulus for hard rubber if $E_r = 0.68(10^3)$ ksi, $\nu_r = 0.43$, and for gray cast iron if $E_{fe} = 14(10^3)$ ksi, $\nu_{fe} = 0.20$.

10–77. Derive an expression for an equivalent torque T_e that if applied alone to a solid bar with a circular cross section would cause the same energy of distortion as the combination of an applied bending moment M and torque T.

10–78. A differential element is subjected to plane strain that has the following components: $\epsilon_x = 950(10^{-6})$, $\epsilon_y = 420(10^{-6})$, $\gamma_{xy} = -325(10^{-6})$. Use the strain-transformation equations and determine (*a*) the principal strains and (*b*) the maximum in-plane shear strain and the associated average strain. In each case specify the orientation of the element and show how the strains deform the element.

10–79. Solve Prob. 10–78 using Mohr's circle.

*10–80. If a shaft is made of a material for which $\sigma_Y = 50$ ksi, determine the maximum torsional stress required to cause yielding using (*a*) the maximum-shear-stress theory, and (*b*) the maximum-distortion-energy theory.

10–81. The principal stresses at a point are shown in the figure. If the material is aluminum for which $E_{al} = 10(10^3)$ ksi and $\nu_{al} = 0.33$, determine the principal strains.

10–82. A differential element is subjected to plane strain that has the following components: $\epsilon_x = 150(10^{-6})$, $\epsilon_y = 200(10^{-6})$, $\gamma_{xy} = -700(10^{-6})$. Use the strain-transformation equations and determine the equivalent strains on an element oriented at an angle of $\theta = 60°$ counterclockwise from the original position. Sketch the deformed element due to these strains.

10–83. Solve Prob. 10–82 using Mohr's circle.

*10–84. Derive an expression for an equivalent bending moment M_e that if applied alone to a solid bar with a circular cross section would cause the same maximum shear stress as the combination of an applied moment M and torque T. Assume that the principal stresses are of opposite algebraic signs.

10–85. The 60° strain rosette is mounted on the surface of a dome. The following readings are obtained for each gauge: $\epsilon_a = -780(10^{-6})$, $\epsilon_b = 400(10^{-6})$, $\epsilon_c = 500(10^{-6})$. Determine (*a*) the principal strains and (*b*) the maximum in-plane shear strain and associated average strain. In each case, specify the orientation of the element and show how the strain deforms the element.

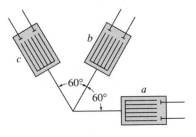

Prob. 10–85

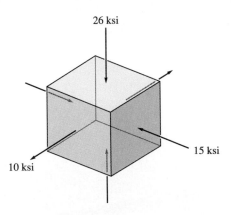

26 ksi

15 ksi

10 ksi

Prob. 10–81

11 Design of Beams and Shafts

A beam is an important structural element in engineering, and in this chapter we will discuss how to design a beam so that it is able to resist bending and shear loads. The chapter begins with a discussion of how to establish the shear and bending moment diagrams for a beam. These diagrams provide a useful means for determining the largest shear and moment in the beam, and they specify where these maximums occur. Methods used for designing prismatic beams and determining the shape of fully stressed beams will then be discussed. Since shafts are an important element in mechanical equipment, at the end of the chapter we will consider their design in terms of resisting both bending and torsional moments.

11.1 Basis for Beam Design

As stated in Sec. 6.1, *beams* are structural members designed to support loadings applied perpendicular to their longitudinal axes. In general, beams are long, straight bars having a constant cross-sectional area. Often they are classified as to how they are supported. For example, a *simply-supported beam* is pinned at one end and roller-supported at the other, Fig. 11–1a, a *cantilevered beam* is fixed at one end and free at the other, Fig. 11–1b, and an *overhanging beam* has one or both of its ends freely extended over the supports, Fig. 11–1c. Certainly beams may be considered among the most important of all structural elements. Examples include members used to support the floor of a building, the deck of a bridge, or the wing of an aircraft. Also, the axle of an automobile, the boom of a crane, even many of the bones of the body act as beams.

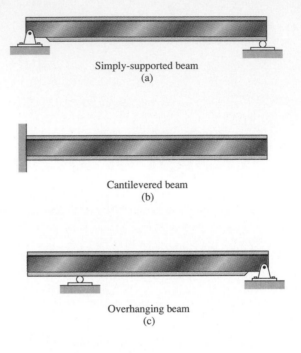

Simply-supported beam
(a)

Cantilevered beam
(b)

Overhanging beam
(c)

Fig. 11–1

Because of the applied loadings, beams develop an internal shear force and bending moment that, in general, vary from point to point along the axis of the beam. Some beams may also be subjected to an internal axial force; however, the effects of this force are often neglected in design, since the axial stress is generally much smaller than the stresses developed by shear and bending. A beam that is chosen to resist both shear and bending stresses is said to be designed on the *basis of strength*. In this chapter we will show how engineers establish design criteria based on the use of the shear and flexure formulas developed in Chapters 6 and 7. Application of these formulas, however, is limited to beams made of a homogeneous material that has linear-elastic behavior. Also, the cross-sectional area must have an axis of symmetry in the plane of the loading. The design approach to be discussed is in a way approximate, as opposed to a more exact theoretical method; however, it does provide an adequate means of obtaining both a safe and economical design.

Although beams are designed mainly for strength, they must also be braced properly along their sides so that they do not buckle or suddenly become unstable. Furthermore, in some cases beams must be designed to resist a limited amount of *deflection*, as when they support ceilings made of brittle materials such as plaster. Methods for computing beam deflections will be discussed in Chapter 12, and limitations placed on beam buckling are often discussed in codes on structural or mechanical design.

11.2 Shear and Moment Diagrams

The design of a beam on the basis of strength first requires finding the *maximum* shear and moment in the beam. One way to do this is to express V and M as functions of the arbitrary position x along the beam's axis. These *shear and moment functions* can then be plotted and represented by graphs called *shear and moment diagrams*. The maximum values of V and M can then be obtained from these graphs. In Chapters 12 and 14 it will be shown that the shear and moment functions can also be used to determine the deflection of a beam and to calculate the amount of internal energy stored in the beam. Furthermore, since the shear and moment diagrams provide detailed information about the *variation* of the shear and moment along the beam's axis, they are often used by engineers to decide where to place reinforcement materials within the beam or how to proportion the size of the beam at various points along its length.

In Sec. 1.2 we used the method of sections to find the shear V and moment M at a *specific point*. However, if we must determine V and M as functions of x along a beam, then it is necessary to locate the imaginary section or cut an *arbitrary distance* x from the end of the beam and compute V and M in terms of x. In this regard, the choice for the origin and the positive direction for any selected x is *arbitrary*. Most often, however, the origin is located at the left end of the beam and the positive direction is to the right.

In general, the internal shear and bending-moment functions obtained as a function of x will be *discontinuous,* or their slope will be discontinuous, at points where a distributed load changes or where concentrated forces or couples are applied. Because of this, shear and bending-moment functions must be determined for *each region* of the beam located *between* any two discontinuities of loading. For example, coordinates x_1, x_2, and x_3 will have to be used to describe the variation of V and M throughout the length of the beam in Fig. 11–2a. These coordinates will be valid *only* within the regions from A to B for x_1, from B to C for x_2, and from C to D for x_3. Although each of these coordinates has the *same* origin, this does not have to be the case. Instead, it may be easier to express V and M as functions of x_1, x_2, and x_3 having origins at A, C, and D as shown in Fig. 11–2b. Here x_1 is positive to the right and x_2 and x_3 are positive to the left.

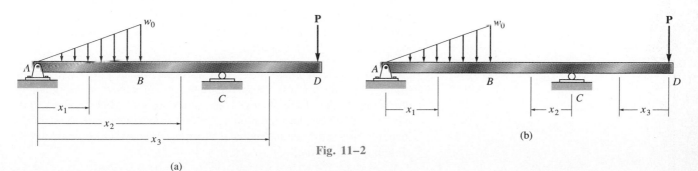

Fig. 11–2

(a)

(b)

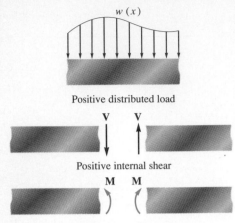

w(*x*)

Positive distributed load

V   **V**

Positive internal shear

M   **M**

Positive internal moment

Beam sign convention

Fig. 11–3

Beam Sign Convention.

Beam Sign Convention. Before presenting a method for determining the shear and moment as functions of x and later plotting these functions (shear and moment diagrams), it is first necessary to establish a *sign convention* so as to define "positive" and "negative" internal shear force and bending moment. [This is analogous to assigning coordinate directions x positive to the right and y positive upward when plotting a function $y = f(x)$.] Although the choice of a sign convention is arbitrary, here we will use the one discussed in Sec. 7.2, which was used to derive two important relationships among distributed load, shear, and moment, namely, $-w(x) = dV/dx$ and $V = dM/dx$. The convention is illustrated in Fig. 11–3. The *positive directions* require the *distributed load* to act *downward* on the beam, the internal *shear force* to cause a *clockwise* rotation of the beam segment on which it acts, and the internal *moment* to cause *compression* in the *top fibers* of the segment.

PROCEDURE FOR ANALYSIS

The following procedure provides a method for determining the shear and moment functions and constructing the shear and moment diagrams for a beam.

Support Reactions. Draw a free-body diagram of the beam and determine all the support reactions. Resolve the forces into components acting perpendicular and parallel to the beam's axis.

Shear and Moment Functions. Select position coordinates x such that each coordinate extends into a region of the beam located *between* concentrated forces, couples, or discontinuities of distributed loading. The *origin* for each coordinate can be established at any suitable point, but usually it is at the beam's left end. Section the beam perpendicular to its axis at each position x and draw the free-body diagram of one of the two segments. Make sure that **V** and **M** are shown acting in their *positive sense,* in accordance with the sign conventions given in Fig. 11–3. Use the equilibrium equation $\Sigma F_y = 0$ to determine V as a function of x. The internal moment M as a function of x is obtained by summing moments about the beam's neutral axis, located on the cross-sectional area at the cut section, $\Sigma M_{NA} = 0$.

Once established, the results for V and M can be *checked* since they are related by Eq. 7–2, $V = dM/dx$. Also, the equation for shear can be checked, since by Eq. 7–1 its derivative must yield the negative of the loading, that is, $-w = dV/dx$.

Shear and Moment Diagrams. Plot the shear function (V versus x) and the moment function (M versus x). If numerical values of the functions describing V and M are *positive,* the values are plotted above the x axis, whereas negative values are plotted below the axis. Generally it is convenient to show the shear and moment diagrams directly below the free-body diagram of the beam.

Example 11–1

Draw the shear and moment diagrams for the beam shown in Fig. 11–4a.

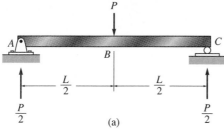

(a)

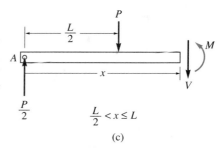

$$0 \leq x < \frac{L}{2}$$

(b)

SOLUTION

Support Reactions.   The support reactions have been computed, Fig. 11–4a.

Shear and Moment Functions.   The beam is sectioned at an arbitrary distance x from the support A, extending within region AB, and the free-body diagram of the left segment is shown in Fig. 11–4b. The unknowns **V** and **M** are indicated acting in the *positive sense* on the right-hand face of the segment according to the established sign convention. Applying the equilibrium equations yields

$$+\uparrow \ \Sigma F_y = 0; \qquad\qquad V = \frac{P}{2} \qquad\qquad (1)$$

$$\downarrow\!\!+ \ \Sigma M_{NA} = 0; \qquad\qquad M = \frac{P}{2}x \qquad\qquad (2)$$

A free-body diagram for a left segment of the beam extending a distance x within region BC is shown in Fig. 11–4c. As always, **V** and **M** are shown acting in the positive sense. Hence,

$$+\uparrow \ \Sigma F_y = 0; \qquad\qquad \frac{P}{2} - P - V = 0$$

$$V = -\frac{P}{2} \qquad\qquad (3)$$

$$\downarrow\!\!+ \ \Sigma M_{NA} = 0; \qquad\qquad M + P\!\left(x - \frac{L}{2}\right) - \frac{P}{2}x = 0$$

$$M = \frac{P}{2}(L - x) \qquad\qquad (4)$$

The shear diagram represents a plot of Eqs. 1 and 3, and the moment diagram represents a plot of Eqs. 2 and 4, Fig. 11–4d. Note that the equations can be checked, since $V = dM/dx$ in each case. Also, $-w = dV/dx = 0$, since there is no distributed load on the beam between A and B and B and C.

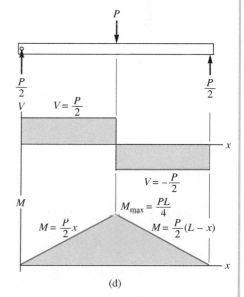

(c)

(d)

Fig. 11–4

Example 11–2

Draw the shear and moment diagrams for the beam shown in Fig. 11–5a.

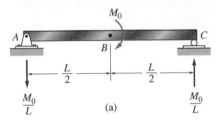

(a)

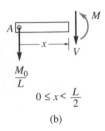

$$0 \le x < \frac{L}{2}$$

(b)

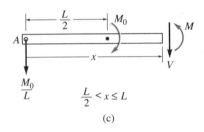

$$\frac{L}{2} < x \le L$$

(c)

SOLUTION

Support Reactions. The support reactions have been computed in Fig. 11–5a.

Shear and Moment Functions. This problem is similar to the previous example, where two x coordinates must be used to express the shear and moment in the beam throughout its length. For the segment within region AB, Fig. 11–5b, we have

$$+\uparrow \ \Sigma F_y = 0; \qquad V = -\frac{M_0}{L}$$

$$\downarrow+ \ \Sigma M_{NA} = 0; \qquad M = -\frac{M_0}{L}x$$

And for the segment within region BC, Fig. 11–5c,

$$+\uparrow \ \Sigma F_y = 0; \qquad V = -\frac{M_0}{L}$$

$$\downarrow+ \ \Sigma M_{NA} = 0; \qquad M = M_0 - \frac{M_0}{L}x$$

$$M = M_0\left(1 - \frac{x}{L}\right)$$

Shear and Moment Diagrams. When the above functions are plotted, the shear and moment diagrams shown in Fig. 11–5d are obtained. In this case, notice that the shear is constant over the entire length of the beam; i.e., it is not affected by the couple moment $\mathbf{M}_0$ acting at the center of the beam. Just as a force creates a jump in the shear diagram, Example 11–1, a couple moment creates a jump in the moment diagram.

Fig. 11–5

(d)

Example 11–3

Draw the shear and moment diagrams for the beam shown in Fig. 11–6a.

SOLUTION

Support Reactions. The support reactions have been computed in Fig. 11–6a.

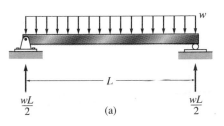

Shear and Moment Functions. A free-body diagram of the left segment of the beam is shown in Fig. 11–6b. The distributed loading on this segment is represented by its resultant force only *after* the segment is isolated as a free-body diagram. Since the segment has a length x, the *magnitude* of the *resultant force* is wx. This force acts through the centroid of the area comprising the distributed loading, a distance of $x/2$ from the right end. Applying the two equations of equilibrium yields

$$+\uparrow \ \Sigma F_y = 0; \qquad \frac{wL}{2} - wx - V = 0$$

$$V = w\left(\frac{L}{2} - x\right) \qquad (1)$$

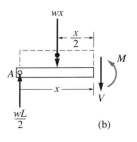

$$\stackrel{\curvearrowright}{+} \ \Sigma M_{NA} = 0; \qquad -\left(\frac{wL}{2}\right)x + (wx)\left(\frac{x}{2}\right) + M = 0$$

$$M = \frac{w}{2}(Lx - x^2) \qquad (2)$$

These results for V and M can be checked by noting that $dV/dx = -w$. This is indeed correct, since positive w acts downward. Also, notice that $dM/dx = V$, as expected.

Shear and Moment Diagrams. The shear and moment diagrams shown in Fig. 11–6c are obtained by plotting Eqs. 1 and 2. The point of *zero shear* can be found from Eq. 1:

$$V = w\left(\frac{L}{2} - x\right) = 0$$

$$x = \frac{L}{2}$$

From the moment diagram, this value of x happens to represent the point on the beam where the *maximum moment* occurs, since by Eq. 7–2, the slope $V = dM/dx = 0$. From Eq. 2, we have

$$M_{\max} = \frac{w}{2}\left[L\left(\frac{L}{2}\right) - \left(\frac{L}{2}\right)^2\right]$$

$$= \frac{wL^2}{8}$$

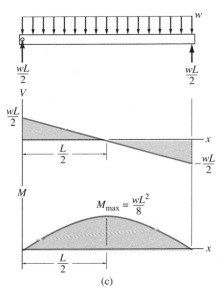

Fig. 11–6

Example 11–4

Draw the shear and moment diagrams for the beam shown in Fig. 11–7a.

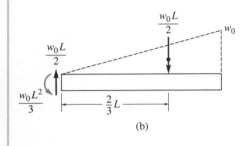

(b)

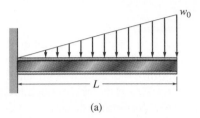

(a)

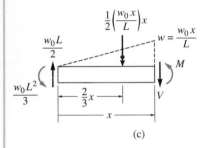

(c)

(d)

Fig. 11–7

SOLUTION

Support Reactions.   The distributed load is replaced by its resultant force and the reactions have been computed as shown in Fig. 11–7b.

Shear and Moment Functions.   A free-body diagram of a beam segment of length x is shown in Fig. 11–7c. Note that the intensity of the triangular load at the section is found by proportion, that is, $w/x = w_0/L$ or $w = w_0 x/L$. With the load intensity known, the resultant of the distributed loading is determined from the area under the diagram, Fig. 11–7c. Thus,

$$+\uparrow \ \Sigma F_y = 0; \qquad \frac{w_0 L}{2} - \frac{1}{2}\left(\frac{w_0 x}{L}\right)x - V = 0$$

$$V = \frac{w_0}{2L}(L^2 - x^2) \qquad (1)$$

$$\zeta+ \ \Sigma M_{NA} = 0; \quad \frac{w_0 L^2}{3} - \frac{w_0 L}{2}(x) + \frac{1}{2}\left(\frac{w_0 x}{L}\right)x\left(\frac{1}{3}x\right) + M = 0$$

$$M = \frac{w_0}{6L}(-2L^3 + 3L^2 x - x^3) \qquad (2)$$

These results can be checked by applying Eqs. 7–1 and 7–2, that is,

$$w = -\frac{dV}{dx} = -\frac{w_0}{2L}(0 - 2x) = \frac{w_0 x}{L} \qquad \text{OK}$$

$$V = \frac{dM}{dx} = \frac{w_0}{6L}(-0 + 3L^2 - 3x^2) = \frac{w_0}{2L}(L^2 - x^2) \qquad \text{OK}$$

Shear and Moment Diagrams.   The graphs of Eqs. 1 and 2 are shown in Fig. 11–7d.

Example 11–5

Draw the shear and moment diagrams for the beam shown in Fig. 11–8a.

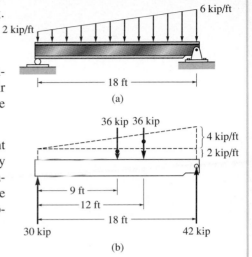

(a)

SOLUTION

Support Reactions. The distributed load is divided into component triangular and rectangular loadings and these loadings are then replaced by their resultant forces. The reactions have been computed as shown on the beam's free-body diagram, Fig. 11–8b.

Shear and Moment Functions. A free-body diagram of the left segment is shown in Fig. 11–8c. As above, the trapezoidal loading is replaced by rectangular and triangular distributions. Note that the intensity of the triangular load at the section is found by proportion. The resultant force and the location of each distributed loading are also shown. Applying the equilibrium equations, we have

$$+\uparrow \ \Sigma F_y = 0; \qquad 30 - 2x - \frac{1}{2}(4)\left(\frac{x}{18}\right)x - V = 0$$

$$V = 30 - 2x - \frac{x^2}{9} \qquad (1)$$

$$\zeta+ \ \Sigma M_{NA} = 0; \quad -30(x) + 2x\left(\frac{x}{2}\right) + \frac{1}{2}(4)\left(\frac{x}{18}\right)x\left(\frac{x}{3}\right) + M = 0$$

$$M = 30x - x^2 - \frac{x^3}{27} \qquad (2)$$

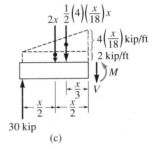

(b)

(c)

Equation 2 may be checked by noting that $dM/dx = V$, that is, Eq. 1. Also, $w = -dV/dx = 2 + \frac{2}{9}x$. This equation checks, since when $x = 0$, $w = 2$ kip/ft, and when $x = 18$ ft, $w = 6$ kip/ft, Fig. 11–8a.

Shear and Moment Diagrams. Equations 1 and 2 are plotted in Fig. 11–8d. Since the point of maximum moment occurs when $dM/dx = V = 0$, then, from Eq. 1,

$$0 = 30 - 2x - \frac{x^2}{9}$$

Choosing the positive root,

$$x = 9.73 \text{ ft}$$

Thus, from Eq. 2,

$$M_{\max} = 30(9.73) - (9.73)^2 - \frac{(9.73)^3}{27}$$

$$= 163 \text{ kip} \cdot \text{ft}$$

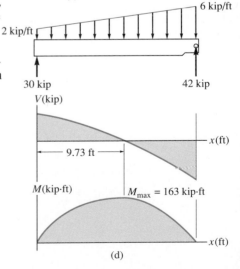

(d)

Fig. 11–8

Example 11–6

Draw the shear and moment diagrams for the beam shown in Fig. 11–9a.

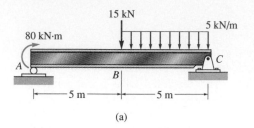

(a)

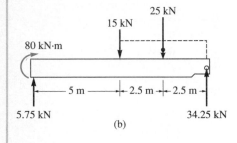

(b)

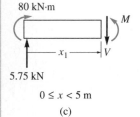

$0 \leq x < 5 \text{ m}$

(c)

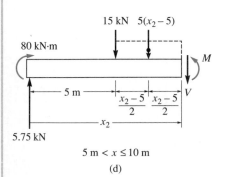

$5 \text{ m} < x \leq 10 \text{ m}$

(d)

Fig. 11–9

SOLUTION I

Support Reactions. The reactions at the supports have been computed and are shown on the free-body diagram of the beam, Fig. 11–9b.

Shear and Moment Functions. Since there is a discontinuity of distributed load and also a concentrated load at the beam's center, two regions of x must be considered in order to describe the shear and moment functions for the entire beam. Both coordinates will have their origin at A, and x_1 will extend from A towards B, whereas x_2 will extend from B towards C.

$0 \leq x_1 < 5$ m, Fig. 11–9c:

$$+\uparrow \ \Sigma F_y = 0; \qquad\qquad 5.75 - V = 0$$
$$V = 5.75 \qquad\qquad (1)$$

$$\downarrow^+ \ \Sigma M_{NA} = 0; \qquad -80 - 5.75x_1 + M = 0$$
$$M = 5.75x_1 + 80 \qquad\qquad (2)$$

5 m $< x_2 \leq 10$ m, Fig. 11–9d:

$$+\uparrow \ \Sigma F_y = 0; \qquad 5.75 - 15 - 5(x_2 - 5) - V = 0$$
$$V = 15.75 - 5x_2 \qquad\qquad (3)$$

$$\downarrow^+ \ \Sigma M_{NA} = 0; \quad -80 - 5.75x_2 + 15(x_2 - 5) + 5(x_2 - 5)\left(\frac{x_2 - 5}{2}\right) + M = 0$$
$$M = -2.5x_2^2 + 15.75x_2 + 92.5 \qquad\qquad (4)$$

These results can be checked by applying $w = -dV/dx$ and $V = dM/dx$. Also, when $x_1 = 0$, Eqs. 1 and 2 give $V = 5.75$ kN and $M = 80$ kN · m; when $x_2 = 10$ m, Eqs. 3 and 4 give $V = -34.25$ kN and $M = 0$. These values check with the support reactions shown on the free-body diagram, Fig. 11–9b.

Shear and Moment Diagrams. Equations 1 through 4 are plotted in Fig. 11–9e. Note the discontinuities that occur at the points of concentrated force and moment.

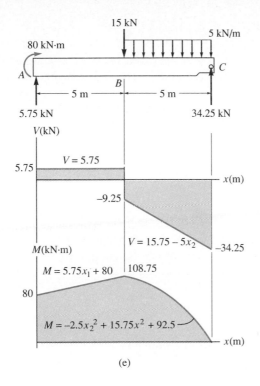

15 kN

80 kN·m

5 kN/m

A

B

C

5 m · 5 m

5.75 kN

34.25 kN

V(kN)

$V = 5.75$

5.75

x(m)

−9.25

$V = 15.75 − 5x_2$

−34.25

M(kN·m)

$M = 5.75x_1 + 80$

108.75

80

$M = −2.5x_2^2 + 15.75x^2 + 92.5$

x(m)

(e)

SOLUTION II

By inspection, it is simpler to choose x_2 with origin at C and extend it positive to the left from C toward B, Fig. 11–9f. Here **V** and **M** act on the left face of the segment and according to our sign convention, Fig. 11–3, must be directed as shown. We have

$0 \le x_2 < 5$ m

$+\uparrow \ \Sigma F_y = 0;$

$$V − 5x_2 + 34.25 = 0$$

$$V = 5x_2 − 34.25 \qquad (5)$$

$\zeta+ \ \Sigma M_{NA} = 0;$

$$-M − 5x_2\left(\frac{x_2}{2}\right) + 34.25x_2 = 0$$

$$M = 34.25x_2 − 2.5x_2^2 \qquad (6)$$

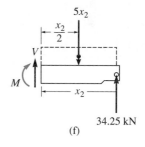

$5x_2$

$\dfrac{x_2}{2}$

V

M

x_2

34.25 kN

(f)

These two equations define the straight-line and parabolic portions of the shear and moment diagrams for region BC in Fig. 11–9f. They are *different functions* than those given by Eqs. 3 and 4, since they describe the curves from a different origin and positive direction. Notice, however, that they do indeed give the same values obtained previously, e.g., at $x_2 = 0$, $V = −34.25$ kN, $M = 0$, and at $x_2 = 5$ m, $V = −9.25$ kN and $M = 108.75$ kN. Can you explain, in this case, with x directed positive to the *left*, why $dM/dx = −V$? (Refer to the derivation of Eq. 7–2).

509

11.3 Graphical Method for Constructing Shear and Moment Diagrams

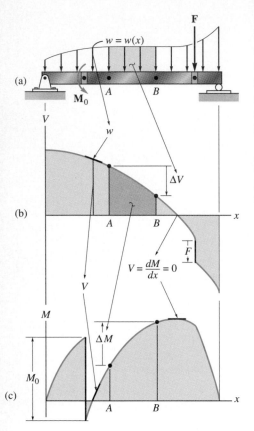

Fig. 11–10

In cases where a beam is subjected to *several* concentrated forces, couples, and distributed loads, determining V and M as functions of x and then plotting these equations can become quite tedious. In this section a simpler method for constructing the shear and moment diagrams is discussed—a method based on the two differential relations that exist among distributed load, shear, and moment; namely, $dV/dx = -w$ and $dM/dx = V$. As stated previously, these equations were developed in Sec. 7.2 in accordance with our established sign convention for $w(x)$, V, and M, with x extending *positive* to the *right*, Fig. 11–3.

Regions of Distributed Load. Since the shear and moment diagrams graphically represent V versus x and M versus x, the two differential equations relating w and V and V and M can be interpreted as

$$\frac{dV}{dx} = -w(x)$$

$$\begin{array}{c} \text{slope of} \quad = \quad -\text{distributed} \\ \text{shear diagram} \quad \text{load intensity} \end{array} \tag{11–1}$$

$$\frac{dM}{dx} = V$$

$$\begin{array}{c} \text{slope of} \quad = \quad \text{shear} \\ \text{moment diagram} \end{array} \tag{11–2}$$

These two equations provide a convenient means for quickly plotting the shear and moment diagrams for a beam. The shear diagram can be constructed by realizing that at each point along the beam the *slope* of the *shear diagram* equals the (negative) intensity of the *distributed loading* at the point (Eq. 11–1). In other words, if w is positive, i.e., acting downward, then the slope of the shear diagram will be negative, Fig. 11–10a and 11–10b. In a similar manner, the moment diagram can be constructed using data from the shear diagram, since the *slope* of the *moment diagram* at each point along the beam is equal to the *shear* at the point (Eq. 11–2), Fig. 11–10b and 11–10c. In particular, if the shear is equal to zero, then $dM/dx = 0$, and therefore a point of zero shear corresponds to a point of maximum (or possibly minimum) moment, Fig. 11–10b and 11–10c.

Equations 11–1 and 11–2 may also be rewritten in the form $dV = -w(x)\,dx$ and $dM = V\,dx$. Noting that $w(x)\,dx$ and $V\,dx$ represent differential areas under the distributed loading and shear diagram, respectively, we can integrate these areas between two points A and B on the beam, Fig. 11–10a, and write

$$\Delta V = -\int w(x)\, dx$$

change in $=$ $-$area under
shear distributed loading

(11–3)

$$\Delta M = \int V(x)\, dx$$

change in $=$ area under
moment shear diagram

(11–4)

Equation 11–3 states that the (negative) *area* under the distributed-loading curve between points A and B, Fig. 11–10a, is equal to the *change in shear* between these two points. Or stated another way, if the area under the loading curve is positive, i.e., due to a positive w, then the change in shear will be negative, Fig. 11–10b. Similarly, from Eq. 11–4, the area under the shear diagram within the region from A to B is equal to the change in moment between A and B, Fig. 11–10c.

Regions of Concentrated Force and Moment. From the derivations given in Sec. 7.2, it should be apparent that Eqs. 11–1 and 11–3 cannot be used at points where an external force $\mathbf{F}$ acts, since these equations do not account for the sudden change in shear that occurs at these points. Similarly, Eq. 11–2 and Eq. 11–4 cannot be used at points where an external couple moment $\mathbf{M}_O$ is applied because of a discontinuity of moment. In order to account for these two cases, we must consider the free-body diagrams of differential segments of the beam that are located at a concentrated force and couple moment, Fig. 11–10a. These diagrams are shown in Fig. 11–11a and Fig. 11–11b, respectively.

From Fig. 11–11a it is seen that force equilibrium requires the change in shear to be

$+\uparrow \ \Sigma F_y = 0;$ $\qquad V - F - (V + \Delta V) = 0$

$$\Delta V = -F \qquad (11\text{–}5)$$

Thus, when $\mathbf{F}$ acts *downward* on the beam, as in Fig. 11–10a, ΔV is *negative* so the shear "jumps" *downward,* Fig. 11–10b. Likewise, if $\mathbf{F}$ acts *upward,* the jump (ΔV) is *upward.*

From Fig. 11–11b, moment equilibrium requires the change in moment to be

$\zeta^+ \ \Sigma M_O = 0;$ $\qquad M + \Delta M + M_0 - V\,\Delta x - M = 0$

Letting $\Delta x \rightarrow 0$, we get

$$\Delta M = -M_0 \qquad (11\text{–}6)$$

In this case, if $\mathbf{M}_0$ is applied *counterclockwise,* as in Fig. 11–10a, ΔM is *negative* so the moment diagram jumps *downward,* Fig. 11–10c. Likewise, when $\mathbf{M}_0$ acts *clockwise,* the jump (ΔM) must be *upward.*

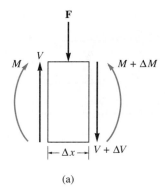

(a)

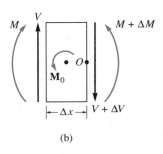

(b)

Fig. 11–11

Table 11–1 illustrates application of Eqs. 11–1, 11–2, 11–5, and 11–6 to some common loading cases. None of these results should be memorized; rather, each should be *carefully studied* so that you become fully aware of how the shear and moment diagrams can be constructed on the basis of knowing the *variation of the slope* from the load and shear diagrams, respectively. It would be well worth the time and effort to self-test your understanding of these concepts by covering over the shear and moment diagram columns in the table and then trying to reconstruct these diagrams on the basis of knowing the loading.

Loading	Shear Diagram $\frac{dV}{dx} = -w$	Moment Diagram $\frac{dV}{dx} = V$

Table 11–1

PROCEDURE FOR ANALYSIS

The following procedure provides a method for constructing the shear and moment diagrams for a beam based on the relations among distributed load, shear, and moment.

Support Reactions. Draw the free-body diagram of the beam and determine the support reactions. Resolve the forces acting on the beam into components that are perpendicular and parallel to the beam's axis.

Shear Diagram. Establish the V and x axes and plot the known values of the shear at the two *ends* of the beam.

Since $dV/dx = -w$, the *slope* of the *shear diagram* at any point is equal to the (negative) intensity of the *distributed loading* at the point; for example, if w acts downward, the slope of the shear diagram will be negative. See Table 11–1.

If a numerical value of the shear is to be determined at a point, one can find this value either by using the method of sections and the equation of force equilibrium, or by using $\Delta V = -\int w(x)\, dx$, which states that the *change in the shear* between any two points is equal to the (negative) *area under the load diagram* between the two points; for example, if w acts downward, the area under the curve gives a negative change in shear.

Since $w(x)$ must be *integrated* to obtain ΔV, then if $w(x)$ is a curve of degree n, $V(x)$ will be a curve of degree $n + 1$; for example, if $w(x)$ is uniform, $V(x)$ will be linear.

Moment Diagram. Establish the M and x axes and plot the known values of the moment at the *ends* of the beam.

Since $dM/dx = V$, the *slope* of the moment diagram at any point is equal to the *shear* at the point. See Table 11–1. In particular, note that at the point where the shear is zero, $dM/dx = 0$, and therefore this may be a point of maximum or minimum moment.

If a numerical value of the moment is to be determined at the point, one can find this value either by using the method of sections and the equation of moment equilibrium, or by using $\Delta M = \int V(x)\, dx$, which states that the *change in moment* between any two points is equal to the *area under the shear diagram* between the two points.

Since $V(x)$ must be *integrated* to obtain ΔM, then if $V(x)$ is a curve of degree n, $M(x)$ will be a curve of degree $n + 1$; for example, if $V(x)$ is linear, $M(x)$ will be parabolic.

The following examples illustrate this procedure. After working through them it is recommended that Examples 11–1 through 11–6 also be solved using this method. Also, it would be instructive to verify the solutions given in Figs. 11–26, 11–27, and 11–28.

Example 11–7

Draw the shear and moment diagrams for the beam in Fig. 11–12a.

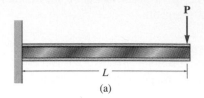

(a)

SOLUTION

Support Reactions.   The reactions are calculated and shown on a free-body diagram, Fig. 11–12b.

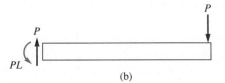

(b)

Shear Diagram.   The shear at each end of the beam is plotted first, i.e., at $x = 0$, $V = +P$ and at $x = L$, $V = +P$, Fig. 11–12c. Since $w = 0$ for $0 < x < L$, Fig. 11–12b, the slope of the shear diagram will be zero ($dV/dx = -w = 0$), and therefore a horizontal straight line connects the end points.

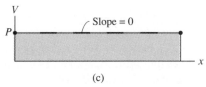

(c)

Moment Diagram.   The moment at each end of the beam is plotted first, i.e., at $x = 0$, $M = -PL$ and at $x = L$, $M = 0$, Fig. 11–12d. The shear diagram, Fig. 11–12c, indicates the slope of the moment diagram will be *constant-positive* for $0 < x < L$, such that $dM/dx = V = +P$. Hence, the end points are connected by a straight positive-sloped line as shown in the figure.

(d)

Fig. 11–12

Example 11–8

Draw the shear and moment diagrams for the beam shown in Fig. 11–13a.

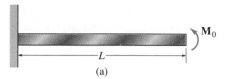

(a)

SOLUTION

Support Reactions.   The reaction at the fixed support has been calculated and is shown on the free-body diagram, Fig. 11–13b.

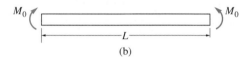

(b)

Shear Diagram.   The shear at each end point, $x = 0$ and $x = L$, is plotted first, Fig. 11–13c. Since no load exists on the beam for $0 < x < L$, Fig. 11–13b, the shear diagram will have zero slope, $dV/dx = 0$. Therefore, a horizontal line connects the end points, which indicates that the shear is zero throughout the beam.

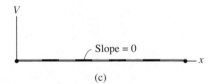

(c)

Moment Diagram.   The moment M_0 at the beam's end points $x = 0$ and $x = L$ is plotted first, Fig. 11–13d. The shear diagram, Fig. 11–13c, indicates that the slope of the moment diagram will be zero for $0 < x < L$ since $dM/dx = V = 0$. Therefore, a horizontal line connects the end points as shown.

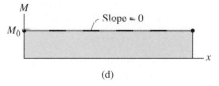

(d)

Fig. 11–13

Example 11–9

Draw the shear and moment diagrams for the beam shown in Fig. 11–14a.

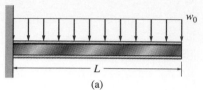

(a)

SOLUTION

Support Reactions. The reactions at the fixed support have been calculated and are shown on the free-body diagram, Fig. 11–14b.

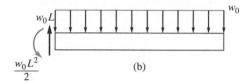

(b)

Shear Diagram. The shear at each end point, $x = 0$ and $x = L$, is plotted first, Fig. 11–14c. The distributed loading on the beam is constant-positive, and since $dV/dx = -w_0$, the slope of the shear diagram will be constant-negative. Thus, a straight negative-sloped line connects the end points.

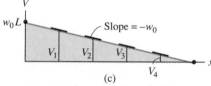

(c)

Moment Diagram. The moment at each end point, $x = 0$ and $x = L$, is plotted first, Fig. 11–14d. Successive values of shear on the shear diagram indicate that the slope of the moment diagram will always be positive, yet it decreases linearly, from $dM/dx = w_0L$ at $x = 0$ to $dM/dx = 0$ at $x = L$. Since the shear diagram is *linear,* the moment diagram will be *parabolic,* having a linear decreasing slope as shown in the figure.

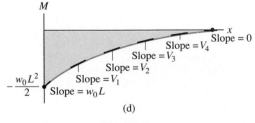

(d)

Fig. 11–14

Draw the shear and moment diagrams for the beam shown in Fig. 11–15a.

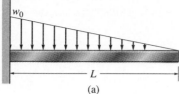

(a)

SOLUTION

Support Reactions. The reactions at the fixed support have been calculated and are shown on the free-body diagram, Fig. 11–15b.

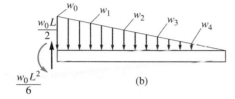

(b)

Shear Diagram. The shear at each end point, $x = 0$ and $x = L$, is plotted first, Fig. 11–15c. The distributed loading on the beam is positive yet linearly decreasing. Therefore the slope of the shear diagram will be *negatively decreasing* from $dV/dx = -w_0$ at $x = 0$ to $dV/dx = 0$ at $x = L$. Since the loading has a *linear distribution,* the shear diagram is a *parabola* having a negatively decreasing slope.

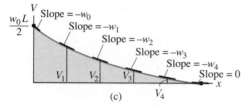

(c)

Moment Diagram. The moment at each end point, $x = 0$ and $x = L$, is plotted first, Fig. 11–15d. From the shear diagram, the slope of the moment diagram will vary parabolically; i.e., it will always be positive yet decreasing, from $dM/dx = +w_0L/2$ at $x = 0$ to $dM/dx = 0$ at $x = L$. The curve connecting the plotted end points that has this characteristic is a *cubic* function of x, as shown in the figure.

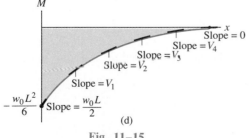

(d)

Fig. 11–15

517

Example 11–11

Draw the shear and moment diagrams for the beam in Fig. 11–16a.

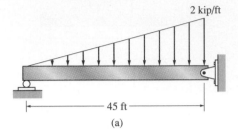

(a)

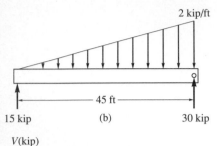

2 kip/ft

45 ft

(b)

15 kip   30 kip

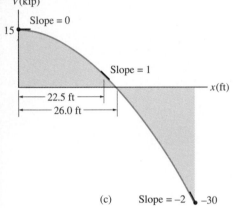

V(kip)

Slope = 0

15

Slope = 1

22.5 ft

26.0 ft

x(ft)

(c)   Slope = –2   –30

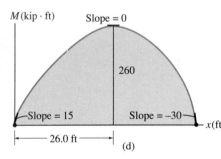

M(kip · ft)   Slope = 0

260

Slope = 15   Slope = –30

26.0 ft x(ft)

(d)

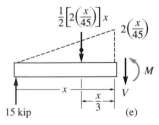

$\frac{1}{2}\left[2\left(\frac{x}{45}\right)\right]x$ $2\left(\frac{x}{45}\right)$

M

x

$\frac{x}{3}$ V

15 kip   (e)

Fig. 11–16

SOLUTION

Support Reactions.   The reactions have been calculated and are shown on the free-body diagram of the beam, Fig. 11–16b.

Shear Diagram.   The end points $x = 0$, $V = +15$ and $x = 45$, $V = -30$ are plotted first. As shown from the beam's distributed loading, the slope of the shear diagram will vary from $dV/dx = 0$ at $x = 0$ to $dV/dx = -2$ at $x = 45$. In general, for $0 \le x \le 45$, the slope of the shear diagram will be *increasingly negative* since the distributed load is increasingly positive ($dV/dx = -w$), Fig. 11–16c. As a result, the shear diagram is a parabola having the slope shown.

    The point of zero shear can be found by using the method of sections for a beam segment of length x, Fig. 11–16e. We require that $V = 0$, so

$$+\uparrow \ \Sigma F_y = 0; \quad 15 - \frac{1}{2}\left[2\left(\frac{x}{45}\right)\right]x = 0; \quad x = 26.0 \text{ ft}$$

Moment Diagram.   The end points $x = 0$, $M = 0$ and $x = 45$, $M = 0$ are plotted first, Fig. 11–16d. From the shear diagram, the slope of the moment diagram will be $dM/dx = 15$ at $x = 0$ and $dM/dx = -30$ at $x = 45$. In general, for $0 \le x < 26.0$ the slope will be *decreasingly positive,* since the shear is decreasingly positive. Likewise, for $26.0 < x \le 45$ the slope will be *increasingly negative,* Fig. 11–16d. Here the moment diagram is a cubic function of x. Why?

    Notice that the maximum moment is at $x = 26.0$, since $dM/dx = V = 0$ at this point. From the free-body diagram in Fig. 11–16e we have

$$\downarrow\!\!+ \ \Sigma M_{NA} = 0; \quad -15(26.0) + \frac{1}{2}\left[2\left(\frac{26.0}{45}\right)\right](26.0)\left(\frac{26.0}{3}\right) + M = 0$$

$$M = 260 \text{ kip} \cdot \text{ft}$$

Example 11–12

Draw the shear and moment diagrams for the overhanging beam shown in Fig. 11–17a.

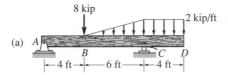

(a)

SOLUTION

Support Reactions.   The free-body diagram with the calculated support reactions is shown in Fig. 11–17b.

Shear Diagram.   As usual we start by plotting the end shears $V_A = +4.40$ kip and $V_D = 0$, Fig. 11–17c. Since there is no load from A to B, the shear diagram will have zero slope. Furthermore, since w is increasingly positive from B to C, the slope of the shear diagram will be *increasingly negative*. And, lastly, from C to D, w is constant, so the slope of the shear diagram will be constant yet negative, Fig. 11–17c. Try to "follow the load" on the free-body diagram and establish the peak values on the shear diagram. Use the appropriate areas under the load diagram (w curve) to find the change in shear. For example, $\Delta V_{BC} = -(1/2)(6 \text{ ft})(2 \text{ kip/ft}) = -6 \text{ kip}$, so that $V_C = -3.60 \text{ kip} - 6 \text{ kip} = -9.60$ kip just to the left of point C.

Moment Diagram.   The end moments $M_A = 0$ and $M_D = 0$ are plotted first. Study the diagram and note how the slopes and therefore the various curves are established from the shear diagram using $dM/dx = V$. Verify the numerical values for the peaks using the method of sections and statics or by computing the appropriate areas under the shear diagram to find the change in moment. Note that the point of zero moment can be determined by establishing M as a function of x, where, for convenience, x extends *from* point B into region BC, Fig. 11–17e. Hence,

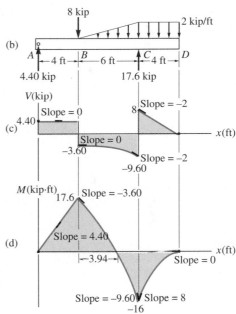

$$\downarrow^+ \ \Sigma M_{NA} = 0; \quad -4.40(4 + x) + 8(x) + \frac{1}{2}\left(\frac{2}{6}\right)x(x)\left(\frac{x}{3}\right) + M = 0$$

$$M = -\frac{1}{18}x^3 - 3.60x + 17.6 = 0$$

$$x = 3.94 \text{ ft}$$

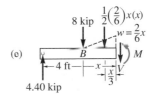

(e)

Reviewing these diagrams, we see that for region AB the load is zero, shear is constant, and moment is linear; for region BC the load is linear, shear is parabolic, and moment is cubic; and for region CD the load is constant, the shear is linear, and the moment is parabolic.

Fig. 11–17

Example 11–13

Draw the shear and moment diagrams for the beam shown in Fig. 11–18a.

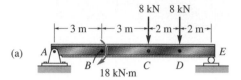

(a)

Fig. 11–18

SOLUTION

Support Reactions. The reactions are calculated and indicated on the free-body diagram, Fig. 11–18b.

Shear Diagram. The values of the shear at the end points A and E are plotted first. At $x = 0$, $V_A = +3$ kN, and at $x = 10$, $V_E = -13$ kN, Fig. 11–18c. At an intermediate point between A and C, $w(x) = 0$, so the slope of the shear diagram will be zero, $dV/dx = -w(x) = 0$. Hence the shear retains its value of $+3$ kN within this region up to point C. At C the shear is *discontinuous,* since there is a *concentrated force* of 8 kN there. The value of the shear just to the right of C (-5 kN) can be found by sectioning the beam at this point. This yields the free-body diagram shown in equilibrium in Fig. 11–18e. This point ($V = -5$ kN) is plotted on the shear diagram. As before, $w(x) = 0$ from C to D and from D to E, Fig. 11–18b, so the slope of the shear diagram will be zero in these regions. The diagram "jumps" again at D, as shown, then closes to the value of -13 kN at E. Notice that no "jump" or discontinuity in shear occurs at B, the point where the 18-kN · m couple moment is applied, Fig. 11–18b. The reason for this can be shown by considering the equilibrium of the free-body diagram in Fig. 11–18f.

It should be noted that based on Eq. 11–5, the shear diagram can also be constructed by "following the load" on the free-body diagram, Fig. 11–18b. In this regard, beginning at A the 3-kN force acts upward, so $V_A = +3$ kN. No distributed load acts between A and C, so the shear remains constant ($dV/dx = 0$). At C the 8-kN force is down, so the shear jumps down 8 kN, from $+3$ kN to -5 kN. Again the shear is constant from C to D (no distributed load), then at D it jumps down another 8 kN to -13 kN. Finally, with no distributed load between D and E, it ends at -13 kN at point E.

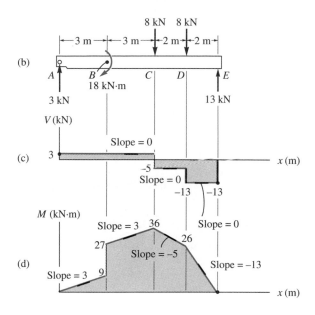

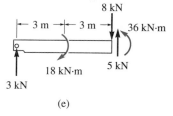

Moment Diagram. The moment at each end of the beam is zero. These two points are plotted first, Fig. 11–18*d*. From the shear diagram, the slope of the moment diagram from *A* to *C* is constant yet positively increasing at +3. The value of the moment just to the *left* of *B* can be determined by using the method of sections and statics or by computing the area under the shear diagram between *A* and *B*, that is, $\Delta M_{AB} = M_B - M_A = (3 \text{ kN})(3 \text{ m}) = 9 \text{ kN} \cdot \text{m}$. Since $M_A = 0$, then $M_B = 0 + 9 \text{ kN} \cdot \text{m} = 9 \text{ kN} \cdot \text{m}$. A jump occurs at point *B* due to the concentrated couple moment of 18 kN · m. Using Eq. 11–6, this jump is 18 kN · m *upward,* since the couple moment is *clockwise*. Also, the method of sections, Fig. 11–18*f*, gives the same value of $M_{B+} = +27 \text{ kN} \cdot \text{m}$ just to the right of *B*. From this point, the slope of $dM/dx = +3$ is maintained until the diagram reaches a peak of 36 kN · m. Again, this value can be obtained by using the method of sections or by finding the area under the shear diagram from *B* to *C*, that is, $\Delta M_{BC} = (3 \text{ kN})(3 \text{ m}) = 9 \text{ kN} \cdot \text{m}$, so that $M_C = 27 \text{ kN} \cdot \text{m} + 9 \text{ kN} \cdot \text{m} = 36 \text{ kN} \cdot \text{m}$. Continuing in this manner, verify the value of 26 kN · m at *D* and closure to zero at *E*.

PROBLEMS

11–1. Draw the shear and moment diagrams for the beam.

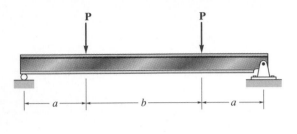

Prob. 11–1

11–2. Draw the shear and moment diagrams for the beam.

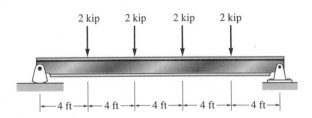

Prob. 11–2

11–3. Draw the shear and moment diagrams for the shaft. The bearings at A and B exert only vertical reactions on the shaft.

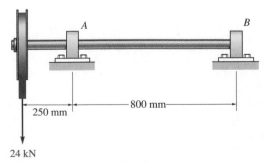

Prob. 11–3

*11–4. Draw the shear and moment diagrams for the shaft, and compute the maximum bending stress in the shaft if it has a diameter of 1.25 in. The bearings at A and D exert only vertical reactions on the shaft. The loading is applied to the pulleys at B, C, and E.

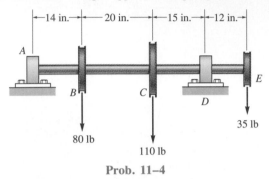

Prob. 11–4

11–5. Draw the shear and moment diagrams for the shaft if it is subjected to the vertical loadings of the belt, gear, and flywheel. The bearings at A and B exert only vertical reactions on the shaft.

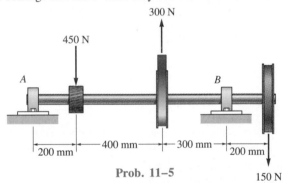

Prob. 11–5

11–6. Draw the shear and moment diagrams for the shaft, and compute the maximum bending stress in the shaft if it has a diameter of 20 mm. The bearings at A and B exert only vertical reactions on the shaft. Also, express the shear and moment in the shaft as a function of x within the region 125 mm $< x <$ 725 mm.

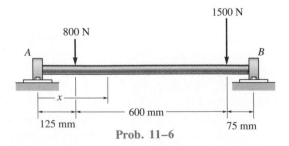

Prob. 11–6

11–7. Draw the shear and moment diagrams for the beam.

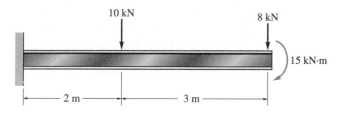

Prob. 11–7

*11–8. Draw the shear and moment diagrams for the pipe. The end screw is subjected to a horizontal force of 5 kN. *Hint:* The reactions at the pin C must be replaced by equivalent loadings at point B on the axis of the pipe.

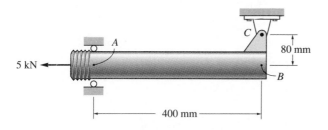

Prob. 11–8

11–9. The walking beam of an oil-field pumping unit is subjected to the draw force of 800 lb. Determine the force P in the pitman arm and the reactions at the pin C. Then draw the shear and moment diagrams for portion AB of the beam. *Hint:* The reactions at C must be replaced by equivalent loadings at point C' on the axis of the beam.

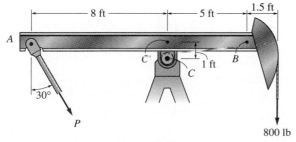

Prob. 11–9

11–10. Draw the shear and moment diagrams for the beam. *Hint:* The 20-kip load must be replaced by equivalent loadings at point C on the axis of the beam.

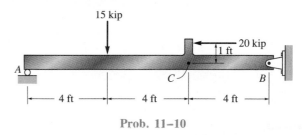

Prob. 11–10

11–11. The shaft is subjected to the loadings caused by belts passing over the two pulleys. If the bearings at A and B exert only vertical reactions on the shaft, draw the shear and moment diagrams and determine the maximum bending stress in the shaft. The shaft has a diameter of 1 in.

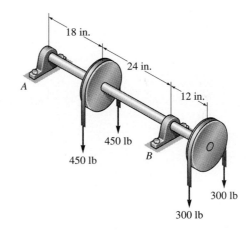

Prob. 11–11

*11–12. Draw the shear and moment diagrams for the beam. There is a short vertical link at B and a pin at C.

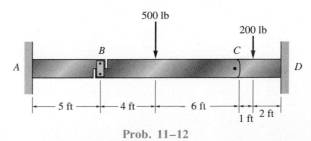

Prob. 11–12

11–13. Draw the shear and moment diagrams for the beam. Also, determine the shear and moment in the beam as functions of x, where 3 ft $< x <$ 15 ft.

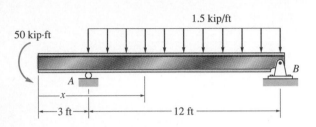

Prob. 11–13

11–14. Draw the shear and moment diagrams for the beam.

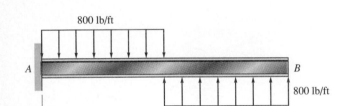

Prob. 11–14

11–15. The 150-lb man sits in the center of the boat, which has a uniform width and a weight per linear foot of 3 lb/ft. Determine the maximum bending moment exerted on the boat. Assume that the water exerts a uniform distributed load upward on the bottom of the boat.

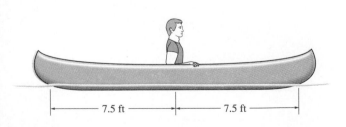

Prob. 11–15

*11–16. Draw the shear and moment diagrams for the beam. The two segments are joined together at B.

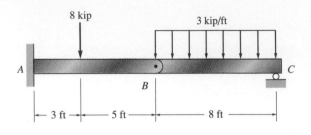

Prob. 11–16

11–17. Draw the shear and moment diagrams for the beam.

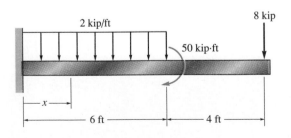

Prob. 11–17

11–18. Draw the shear and moment diagrams for the beam and determine the shear and moment in the beam as functions of x, where $0 \leq x <$ 6 ft.

Prob. 11–18

11–19. Draw the shear and moment diagrams for the beam and determine the shear and moment in the beam as functions of x, where 4 ft $< x <$ 10 ft.

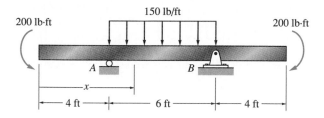

Prob. 11–19

*11–20. Draw the shear and bending-moment diagrams for the compound beam. The three segments are connected by pins at B and E.

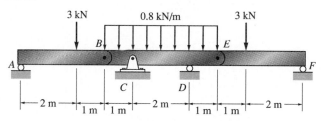

Prob. 11–20

11–21. The sheave on a traveling block supports a load of 15 kip. Draw the moment diagram for the supporting pin AB if the force is (a) assumed concentrated as a single force acting at the *center E* of the pin, and (b) uniformly distributed over the sheave contact area along CD.

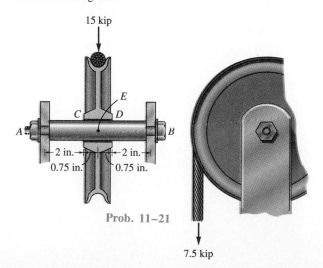

Prob. 11–21

11–22. The T-beam is subjected to the loading shown. Draw the shear and moment diagrams and then determine the maximum tensile and compressive stress in the beam caused by bending.

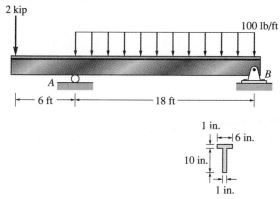

Prob. 11–22

11–23. Draw the shear and moment diagrams for the beam and determine the shear and moment in the beam as functions of x.

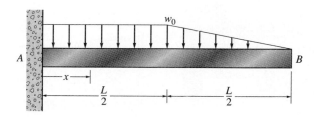

Prob. 11–23

*11–24. Draw the shear and moment diagrams for the beam.

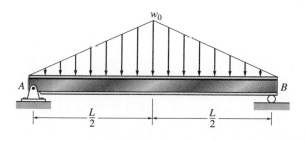

Prob. 11–24

11–25. Draw the shear and moment diagrams for the beam.

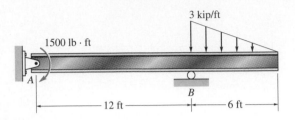

Prob. 11–25

11–26. The smooth pin is supported by two leaves A and B and subjected to a compressive load of 0.4 kN/m caused by bar C. Determine the intensity of the distributed load w_0 of the leaves on the pin and draw the shear and moment diagrams for the pin.

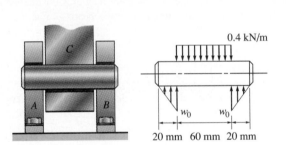

Prob. 11–26

11–27. Draw the shear and moment diagrams for the beam.

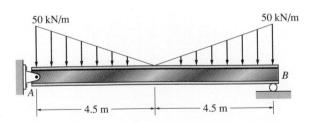

Prob. 11–27

*11–28. Draw the shear and moment diagrams for the beam.

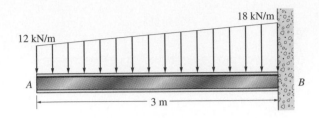

Prob. 11–28

11–29. Draw the shear and moment diagrams for the beam.

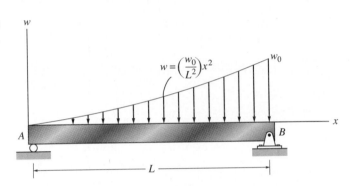

Prob. 11–29

11–30. Draw the shear and moment diagrams for the beam.

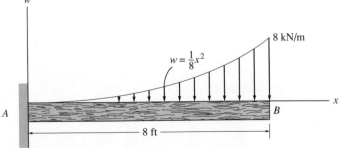

Prob. 11–30

11–31. The connecting rod is subjected to the distributed loading shown. If only vertical reactions occur at its ends A and B, determine the maximum bending stress developed in the rod. The cross-sectional area of the uniform portion of the rod is shown in the figure.

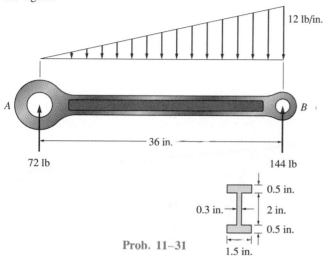

Prob. 11–31

*11–32. Draw the shear and moment diagrams for the beam.

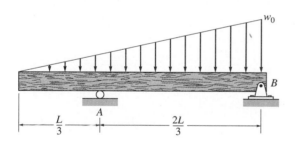

Prob. 11–32

11–33. Draw the shear and moment diagrams for the beam.

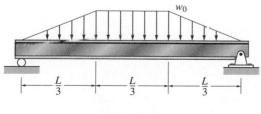

Prob. 11–33

11–34. Draw the shear and moment diagrams for the beam.

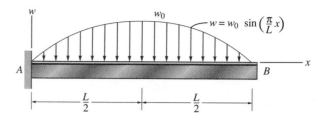

Prob. 11–34

11–35. Determine the placement distance a of the roller support so that the largest absolute value of the moment is a minimum. Draw the shear and moment diagrams for this condition.

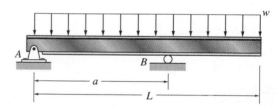

Prob. 11–35

*11–36. Determine the placement distance a of the roller support so that the largest absolute value of the moment is a minimum. Draw the shear and moment diagrams for this condition.

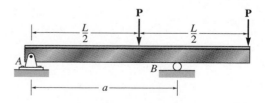

Prob. 11–36

11.4 Stress Variations Throughout a Prismatic Beam

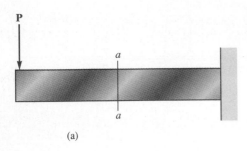

(a)

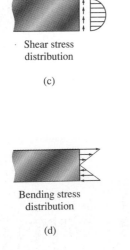

(b)

In Sec. 11.1 we stated that the stress analysis of a beam requires application of the shear and flexure formulas, since beams resist both internal shear and moment loadings. Specific application of these formulas has been treated in Chapters 6 and 7. Here we will discuss the general results obtained when these equations are applied to various points in a cantilevered beam that has a rectangular cross section and supports a load **P** at its end, Fig. 11–19a.

In general, at an arbitrary section a–a along the beam's axis, Fig. 11–19b, the internal shear V and moment M are developed by a *parabolic* shear-stress distribution, Fig. 11–19c, and a *linear* normal-stress distribution, Fig. 11–19d. If these results are applied to specific elements located at points 1 through 5 along the section, Fig. 11–19b, the stresses acting on these elements will be as shown in Fig. 11–19e. In particular, elements 1 and 5 are subjected only to the maximum normal stress, whereas element 3, which is on the neutral axis, is subjected only to the maximum shear stress. The intermediate elements 2 and 4 resist *both* normal and shear stress.

In each case the state of stress can be transformed into *principal stresses,* using either the stress-transformation equations or Mohr's circle. The results are shown in Fig. 11–19f. Notice that each successive element, 1 through 5, undergoes a slight counterclockwise orientation. Specifically, relative to element 1, considered to be at the 0° position, element 3 is oriented at 45° and element 5 is oriented at 90°. Also, the *maximum tensile stress* acting on the

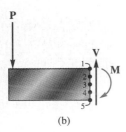

Shear stress
distribution

(c)

Bending stress
distribution

(d)

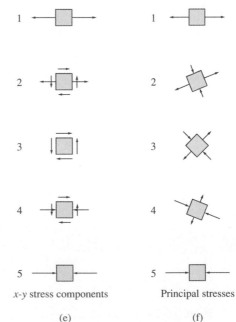

x-y stress components

Principal stresses

Fig. 11–19

(e)

(f)

vertical faces of element 1 becomes smaller on the corresponding faces of each of the successive elements, until it is zero on the horizontal faces of element 5. In a similar manner, the *maximum compressive stress* on the vertical faces of element 5 reduces to zero on the horizontal faces of element 1.

If this analysis is extended to many vertical sections along the beam other than *a–a*, a profile of the results can be represented by curves called *stress trajectories*. Each of these curves indicates the *direction* of a principal stress having a constant magnitude. Some of them are shown for the cantilevered beam in Fig. 11–20. Here the solid lines represent the direction of the tensile principal stresses and the dashed lines represent the direction of the compressive principal stresses. As expected, the lines intersect the neutral axis at 45° angles, and the solid and dashed lines always intersect at 90°. Why? Knowing the direction of these lines can help engineers decide where to reinforce a beam so that it does not crack or become unstable.

Localized Stresses.

The above stress analysis neglects the effects caused by external distributed loadings and concentrated forces applied to the beam. As shown in Fig. 11–21, these loadings will create additional stresses in the beam directly under the load. Notably, a compressive stress σ_y will be developed, in addition to the bending stress σ_x and shear stress τ_{xy} discussed previously. Using advanced methods of analysis, as treated in the theory of elasticity, it can be shown, however, that the stress σ_y diminishes rapidly throughout the beam's depth, and for *most* beam span-to-depth ratios used in engineering practice, the maximum value of σ_y generally represents only a small percentage of the bending stress σ_x, that is, $\sigma_x \gg \sigma_y$. Furthermore, the direct application of concentrated loads is generally avoided in beam design. Instead, *bearing pads* or plates are used to spread these loads more evenly along the surface of the beam.

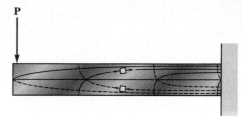

Stress trajectories for
cantilevered beam

Fig. 11–20

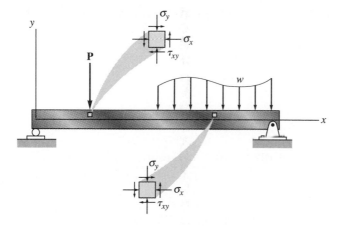

Fig. 11–21

Example 11–14

The wide-flange beam shown in Fig. 11–22a is subjected to the distributed loading of $w = 120$ kN/m. Determine the principal stresses in the beam at points 1 through 5. Neglect the fillets and stress concentrations at points 2 and 4, which lie at the top and bottom of the web at the web–flange junction.

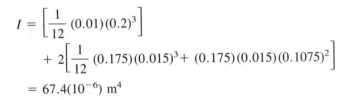

$w = 120$ kN/m

B

0.3 m

2 m

(a)

15 mm

200 mm N — A
          10 mm

15 mm   175 mm

36 kN

0.15 m

$M = 30.6$ kN·m

0.3 m $V = 84$ kN

120 kN   (b)

Fig. 11–22

SOLUTION

Equilibrium Equation. The support reaction on the beam at B is determined, and equilibrium of the sectioned beam shown in Fig. 11–22b yields

$$V = 84 \text{ kN} \qquad M = 30.6 \text{ kN} \cdot \text{m}$$

Section Properties. The beam's moment of inertia about the neutral axis is

$$I = \left[\frac{1}{12}(0.01)(0.2)^3\right]$$
$$+ 2\left[\frac{1}{12}(0.175)(0.015)^3 + (0.175)(0.015)(0.1075)^2\right]$$
$$= 67.4(10^{-6}) \text{ m}^4$$

Stresses. At points 1 and 5,

$$\sigma_1 = \sigma_5 = \frac{Mc}{I} = \frac{30.6(10^3) \text{ N} \cdot \text{m} (0.115 \text{ m})}{67.4(10^{-6}) \text{ m}^4} = 52.2 \text{ MPa} \qquad \textit{Ans.}$$

$$\tau_1 = \tau_5 = 0 \qquad \textit{Ans.}$$

At points 2 and 4,

$$\sigma_2 = \sigma_4 = \frac{My}{I} = \frac{30.6(10^3) \text{ kN} \cdot \text{m} (0.100 \text{ m})}{67.4(10^{-6}) \text{ m}^4} = 45.4 \text{ MPa} \qquad \textit{Ans.}$$

$$\tau_2 = \tau_4 = \frac{VQ}{It} = \frac{84(10^3) \text{ N} [(0.1075 \text{ m})(0.175 \text{ m})(0.015 \text{ m})]}{67.4(10^{-6}) \text{ m}^4 (0.010 \text{ m})}$$
$$= 35.2 \text{ MPa} \qquad \textit{Ans.}$$

At point 3,

$$\sigma_3 = 0 \qquad \textit{Ans.}$$

$$\tau_3 = \frac{VQ}{It}$$
$$= \frac{84(10^3) \text{ N} [(0.1075 \text{ m})(0.175 \text{ m})(0.015 \text{ m}) + (0.050 \text{ m})(0.100 \text{ m})(0.010 \text{ m})]}{67.4(10^{-6}) \text{ m}^4 (0.010 \text{ m})}$$
$$= 41.4 \text{ MPa} \qquad \textit{Ans.}$$

These results are shown in Fig. 11–22c.

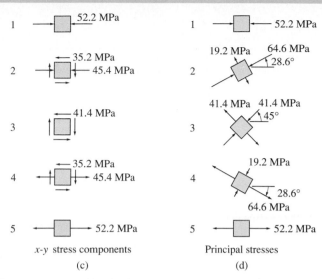

	x-y stress components	Principal stresses
	(c)	(d)

x-y stress components   Principal stresses

(c)   (d)

Using Mohr's circle, the principal stresses at 2 (and 4) can be determined. As shown in Fig. 11–22e, for element 2 the center of the circle is at $(-45.4 + 0)/2 = -22.7$, and the radius is calculated to be $R = 41.9$. Thus $\sigma_1 = (41.9 - 22.7) = 19.2$ MPa, $\sigma_2 = -(22.7 + 41.9) = -64.6$ MPa. The element is rotated counterclockwise $2\theta_{p_2} = 57.2°$, so that $\theta_{p_2} = 28.6°$. In a similar manner, the principal stresses for element 3, $\sigma_1 = 41.4$ MPa and $\sigma_2 = -41.4$ MPa, are calculated from Mohr's circle, Fig. 11–22f. Here $2\theta_{p_2} = 90°$, so that $\theta_{p_2} = 45°$ counterclockwise.

The results for each element are shown in Fig. 11–22d. By comparison, it can be seen that the elements at points 2 and 4 located at the web-flange junction are subjected to the largest principal stress (64.6 MPa). This occurs because at the section we have considered, the shear stress (35.2 MPa) is *significant* compared with the normal stress (45.4 MPa), Fig. 11–22c. In practice, however, beams are relatively long and are designed to resist the maximum internal moment. At this "critical" section, where the internal moment is a maximum, the shear stress at the web–flange junction *will often be rather small* compared with the bending stress.* Consequently, the maximum principal stress at these points will *not* be greater than the maximum bending stress, which occurs at the beam's outer fibers. For this reason, design codes generally require the beam's maximum bending stress to be compared with the allowable bending stress as stated in the codes. In other words, the allowable bending stress is thought to have enough of a factor of safety to compensate for cases when the maximum principal stress at points between the beam's web and flange exceeds the maximum bending stress at points on the beam's top or bottom surface.

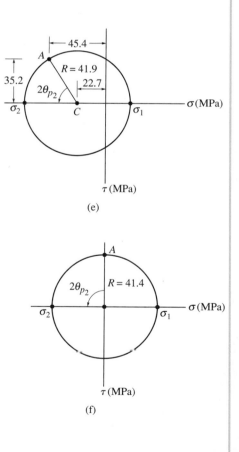

*For the beam in Fig. 11–22a, the shear and moment diagrams reveal that maximum bending occurs at the *center* of the beam, and the internal shear force (and shear stress) is actually zero at this point. See Fig. 11–6.

11.5 Prismatic Beam Design

The design of a prismatic beam requires selecting its cross-sectional area to have a practical size and shape so that the material is not overstressed or the beam does not deflect excessively or buckle when subjected to a given loading. Here we will discuss the design of a beam based only on its *strength*. This requires that the actual bending and shear stresses in the beam do not exceed allowable bending and shear stresses for the material as defined by structural or mechanical codes.

If the suspended span of the beam is relatively long, so that the internal moments become large, the engineer will first consider a design based upon bending. This requires a determination of the beam's *section modulus*, which is the ratio of the unknowns I and c, that is, $S = I/c$. Using the flexure formula, $\sigma = Mc/I$, we have

$$S_{req'd} = \frac{M}{\sigma_{allow}}$$

(11–7)

Here M is determined from the beam's moment diagram, and the allowable bending stress, σ_{allow}, is specified in a design code. In most cases the beam's unknown weight will be small and can be neglected in comparison with the loads the beam must carry. However, if the additional moment caused by the weight is to be included in the design, a selection for S is made so that it slightly *exceeds* $S_{req'd}$.

Once $S_{req'd}$ is known, if the beam has a simple cross-sectional shape, such as a square, a circle, or a rectangle of known width-to-height proportions, its *dimensions* can be determined directly from $S_{req'd}$, since by definition $S_{req'd} = I/c$. However, if the cross section is made from several elements, such as a wide-flange section, then an infinite number of web and flange dimensions can be determined that satisfy the value of $S_{req'd}$. In practice, however, engineers choose a particular beam meeting the requirement that $S > S_{req'd}$ from a handbook that lists the standard shapes available from manufacturers. Often several beams that have the same section modulus can be selected from these tables. If deflections are not restricted, usually the beam having the smallest cross-sectional area is chosen, since it is made of less material and is therefore both lighter and more economical than the others.

The above discussion assumes that the material's allowable bending stress is the *same* for both tension and compression. If this is the case, then a beam having a cross section that is *symmetric* with respect to the neutral axis should be chosen. However, if the allowable tensile and compressive bending stresses are *not* the same, then the choice of an unsymmetric cross section may be more efficient. Under these circumstances the beam must be designed to resist *both* the largest positive and the largest negative moment in the span.

Once the beam has been selected, the shear formula $\tau_{allow} \geq VQ/It$ can then be used to check that the allowable shear stress is not exceeded. Often this requirement will not present a problem. However, if the beam is "short" and supports large concentrated loads, the shear-stress limitation may dictate the size of the beam. This limitation is particularly important in the design of wood beams, because wood tends to split along its grain due to shear (see Fig. 7–8).

Fabricated Beams. Since beams are often made of steel and wood, we will now discuss some of the tabulated properties of beams made of these materials.

Steel Sections. Most manufactured steel beams are produced by rolling a hot ingot of steel until the desired shape is formed. These so-called *rolled shapes* have properties that are tabulated in the American Institute of Steel Construction (AISC) manual. A representative listing for wide-flange beams taken from this manual is given in Appendix *B*. As noted in this appendix, the wide-flange shapes are designated by their depth and weight per unit length; for example, $W18 \times 46$ indicates a wide-flange cross section (W) having a depth of 18 in. and a weight of 46 lb/ft, Fig. 11–23. For any given section, the weight per length, dimensions, cross-sectional area, moment of inertia, and section modulus are reported. Also included is the radius of gyration r, which is a geometric property related to the section's buckling strength. This will be discussed in Chapter 13. Although not included in Appendix *B*, the AISC manual also lists data on other members such as channels, angles, and circular and square tubes.

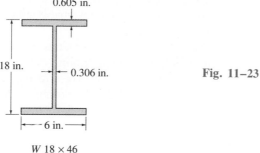

Fig. 11–23

$W\ 18 \times 46$

Wood Sections. Most beams made of wood have rectangular cross sections because such beams are easy to manufacture and handle. Manuals, such as that of the American Forest Products Association, list the dimensions of lumber often used in the design of wood beams. Often both the nominal and net dimensions are reported. Lumber is identified by its *nominal* dimensions, such as 2×4 (2 in. by 4 in.); however, its actual or "dressed" dimensions are smaller, being 1.5 in. by 3.5 in. The reduction in the dimensions occurs due to the requirement of obtaining smooth surfaces from lumber that is rough-sawn. Obviously, the *actual dimensions* must be used whenever stress calculations are performed on wood beams.

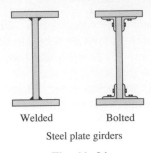

Welded   Bolted

Steel plate girders

Fig. 11–24

Wood box–beam

(a)

Glulam beam

(b)

Fig. 11–25

Built-up Sections. A *built-up section* is constructed from two or more parts joined together to form a single unit. As indicated by Eq. 11–7, the capacity of the beam to resist a moment will vary directly with its section modulus S, and since $S = I/c$, then S is *increased* if I is *increased*. In order to increase I, *most of the material* should be placed as far *away* from the neutral axis as practical. This, of course, is what makes a deep wide-flange beam so efficient in resisting a moment. For very large loads, however, an available rolled-steel section may not have a section modulus great enough to support a given moment. Rather than using several available beams to support the load, engineers will usually "build up" a beam made from plates and angles. A deep I-shaped section having this form is called a *plate girder*. For example, the steel plate girder in Fig. 11–24 has two flange plates that are either welded or, using angles, bolted to the web plate.

Wood beams are also "built up," usually in the form of a box-beam section, Fig. 11–25a. They may be made having plywood webs and larger boards for the flanges. For very large spans, *glulam beams* are used. These members are made from several boards glue-laminated together to form a single unit, Fig. 11–25b.

Just as in the case of rolled sections or beams made from a single piece, the design of built-up sections requires that the bending and shear stresses be checked. In addition the shear stress in the fasteners, such as weld, glue, nails, etc., must be checked to be certain the beam acts as a single unit. The principles for doing this were outlined in Sec. 7.6.

PROCEDURE FOR ANALYSIS

Based on the previous discussion, the following procedure provides a rational method for the design of a beam on the basis of strength.

Shear and Moment Diagrams. Determine the maximum shear and moment in the beam. Often this is done by constructing the beam's shear and moment diagrams. For built-up beams these diagrams are also useful for identifying *regions* where the shear and moment are excessively large and may require additional structural reinforcement or fasteners to hold the beam together.

Bending Stress.   If the beam is relatively long, it is designed on the basis of determining its section modulus using the flexure formula, $S_{req'd} = M_{max}/\sigma_{allow}$, where M_{max} is the maximum moment in the beam and σ_{allow} is the allowable bending stress for the material. Once $S_{req'd}$ is determined, the cross-sectional dimensions for simple shapes can then be computed, since $S_{req'd} = I/c$. If rolled-steel sections are to be used, several possible values of S may be selected from the tables in Appendix B. Of these, choose the one having the smallest cross-sectional area, since this beam has the least weight and is therefore the most economical. Make sure that the selected section modulus, S, is *slightly greater* than $S_{req'd}$, so that the additional moment created by the beam's weight is considered.

Shear Stress.   Using the shear formula, check to see that the allowable shear stress is not exceeded; that is, use $\tau_{allow} \geq V_{max}\ Q/It$. Here V_{max} is the maximum internal shear as determined from the shear diagram. In particular, if the beam has a solid *rectangular* cross section, this equation becomes $\tau_{allow} \geq 1.5(V_{max}/A)$, Eq. 7–7, and if the cross section is a *wide flange,* it is generally appropriate to assume that the shear stress is *constant* over the cross-sectional area of the beam's web so that $\tau_{allow} \geq V_{max}/A_{web}$, where A_{web} is determined from the product of the beam's depth and the web's thickness. (See Sec. 7.4.)

When a beam that has been designed on the basis of bending stress fails to meet the criterion for shear, it must be redesigned to resist the shear stress. Normally beams that are short and carry large loads, especially those made of wood, are first designed to resist shear and then later checked against the allowable-bending-stress requirements.

Adequacy of Fasteners.   Built-up beams made from several elements must be fastened together properly so the beam acts as a single unit when it is loaded. In this regard, the adequacy of a fastener depends on the shear stress it can resist. Specifically, the required spacing of nails or bolts of a particular size is determined from the allowable shear flow, $q_{allow} = VQ/I$, calculated at points on the cross section where the fasteners are located. See Sec. 7.5.

The following examples illustrate application of these principles. Although this procedure for analysis is generally followed by most codes in structural and mechanical design, it should be mentioned that often further analysis must be performed to be certain that a selected beam is adequate. A complete analysis would *also* include checks to see that the beam is properly braced to prevent unstable sidesway, that built-up elements, such as the web and flanges, are not too thin so that they may buckle, that the beam does not severely deflect under loading, and that the localized stresses at points of stress concentrations and concentrated loads are reduced.

Example 11–15

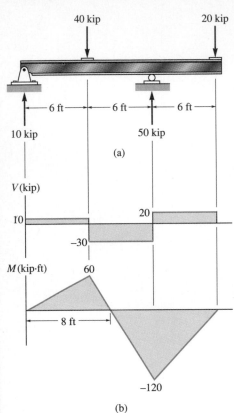

40 kip   20 kip

├── 6 ft ──┤── 6 ft ──┤── 6 ft ──┤

10 kip   50 kip

(a)

V (kip)

10   20

      x (ft)

−30

M (kip·ft)   60

├── 8 ft ──┤   x (ft)

−120

(b)

Fig. 11–26

A beam is to be made of steel that has an allowable bending stress of $\sigma_{allow} = 24$ ksi and an allowable shear stress of $\tau_{allow} = 14.5$ ksi. Select an appropriate W shape that will carry the loading shown in Fig. 11–26a.

SOLUTION

Shear and Moment Diagrams. The support reactions have been calculated in Fig. 11–26a, and the shear and moment diagrams are shown in Fig. 11–26b. From these diagrams, $V_{max} = 30$ kip and $M_{max} = 120$ kip · ft.

Bending Stress. The required section modulus for the beam is determined from the flexure formula,

$$S_{req'd} = \frac{M_{max}}{\sigma_{allow}} = \frac{120 \text{ kip} \cdot \text{ft}(12 \text{ in./ft})}{24 \text{ kip/in}^2} = 60 \text{ in}^3$$

Using the table in Appendix B, the following beams are adequate:

$W18 \times 40$	$S = 68.4$ in³
$W16 \times 45$	$S = 72.7$ in³
$W14 \times 43$	$S = 62.7$ in³
$W12 \times 50$	$S = 64.7$ in³
$W10 \times 54$	$S = 60.0$ in³
$W8 \times 67$	$S = 60.4$ in³

The beam having the least weight per foot is chosen, i.e.,

$$W18 \times 40$$

The *actual* maximum moment M_{max}, which includes the weight of the beam, can be computed and the adequacy of the selected beam can be checked. In comparison with the applied loads, however, the beam's weight, $(0.040 \text{ kip/ft})(18 \text{ ft}) = 0.720$ kip, will only *slightly increase* $S_{req'd}$. In spite of this,

$$S_{req'd} = 60 \text{ in}^3 < 68.4 \text{ in}^3 \qquad \text{OK}$$

Shear Stress. Since the beam is a *wide-flange section,* the *average shear stress* within the web will be considered. Here the web is assumed to extend from the very top to the very bottom of the beam. From Appendix B, for a $W18 \times 40$, $d = 17.90$ in., $t_w = 0.315$ in. Thus,

$$\tau_{avg} = \frac{V_{max}}{A_w} = \frac{30 \text{ kip}}{(17.90 \text{ in.})(0.315 \text{ in.})} = 5.32 \text{ ksi} < 14.5 \text{ ksi} \qquad \text{OK}$$

Use a $W18 \times 40$. *Ans.*

Example 11–16

The laminated wooden beam shown in Fig. 11–27a supports a uniform distributed loading of 12 kN/m. If the beam is to have a height-to-width ratio of 1.5, determine its smallest width. The allowable bending stress is $\sigma_{allow} = 9$ MPa and the allowable shear stress is $\tau_{allow} = 0.6$ MPa. Neglect the weight of the beam.

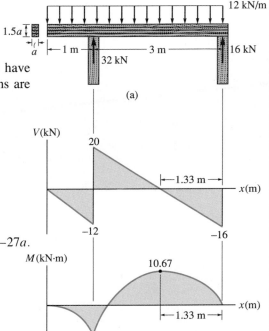

(a)

(b)

Fig. 11–27

SOLUTION

Shear and Moment Diagrams.   The support reactions at A and B have been calculated in Fig. 11–27a, and the shear and moment diagrams are shown in Fig. 11–27b. Here $V_{max} = 20$ kN, $M_{max} = 10.7$ kN · m.

Bending Stress.   Applying the flexure formula yields

$$S_{req'd} = \frac{M_{max}}{\sigma_{allow}} = \frac{10.67 \text{ kN} \cdot \text{m}}{9(10^3) \text{ kN/m}^2} = 0.00119 \text{ m}^3$$

Assuming that the width is a, then the height is $h = 1.5a$, Fig. 11–27a. Thus,

$$S_{req'd} = \frac{I}{c} = \frac{\frac{1}{12}(a)(1.5a)^3}{(0.75a)} = 0.00119 \text{ m}^3$$

$$a^3 = 0.003160 \text{ m}^3$$

$$a = 0.147 \text{ m}$$

Shear Stress.   Applying the shear formula for rectangular sections, we have

$$\tau_{max} = \frac{3}{2}\frac{V_{max}}{A} = \frac{3}{2}\frac{20 \text{ kN}}{(0.147 \text{ m})(1.5)(0.147 \text{ m})} = 0.926 \text{ MPa} > 0.6 \text{ MPa}$$

Since the shear criterion fails, the beam must be redesigned on the basis of shear.

$$\tau_{allow} = \frac{3}{2}\frac{V_{max}}{A}$$

$$600 \text{ kN/m}^2 = \frac{3}{2}\frac{20 \text{ kN}}{(a)(1.5a)}$$

$$a = 0.183 \text{ m} = 183 \text{ mm} \qquad \qquad Ans.$$

This larger section will also adequately resist the normal stress.

Example 11–17

The wooden T-beam shown in Fig. 11–28a is made from two 200 mm × 30 mm boards. If the allowable bending stress is $\sigma_{allow} = 12$ MPa and the allowable shear stress is $\tau_{allow} = 0.8$ MPa, determine if the beam can safely support the loading shown. Also specify the maximum spacing of nails needed to hold the two boards together if each nail can safely resist 1.50 kN in shear.

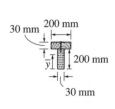

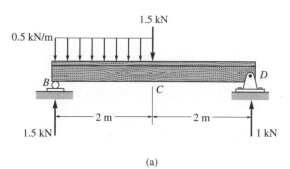

(a)

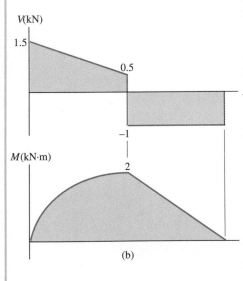

(b)

Fig. 11–28

SOLUTION

Shear and Moment Diagrams. The reactions on the beam are shown in Fig. 11–28a, and the shear and moment diagrams are drawn in Fig. 11–28b. Here $V_{max} = 1.5$ kN, $M_{max} = 2$ kN · m.

Bending Stress. The neutral axis (centroid) will be located from the bottom of the beam. Working in units of meters, we have

$$\bar{y} = \frac{\Sigma \bar{y}A}{\Sigma A} = \frac{(0.1)(0.03)(0.2) + 0.215(0.03)(0.2)}{0.03(0.2) + 0.03(0.2)} = 0.1575 \text{ m}$$

Thus

$$I = \left[\frac{1}{12}(0.03)(0.2)^3 + (0.03)(0.2)(0.1575 - 0.1)^2\right]$$

$$+ \left[\frac{1}{12}(0.2)(0.03)^3 + (0.03)(0.2)(0.215 - 0.1575)^2\right]$$

$$= 60.125(10^{-6}) \text{ m}^4$$

Since $c = 0.1575$ m (not 0.230 m − 0.1575 m = 0.0725 m), we require

$$\sigma_{allow} \geq \frac{M_{max}c}{I}$$

$$12(10^3) \text{ kPa} \geq \frac{2 \text{ kN} \cdot \text{m}(0.1575 \text{ m})}{60.125(10^{-6}) \text{ m}^4} = 5.24(10^3) \text{ kPa}   \text{OK}$$

Shear Stress.   Maximum shear stress in the beam depends upon the magnitude of Q and t. It occurs at the neutral axis, since Q is a maximum there and the neutral axis is in the web, where the thickness $t = 0.03$ m is smallest for the cross section. For simplicity we will use the rectangular area below the neutral axis to calculate Q, rather than a two-part composite area above this axis, Fig. 11–28c. We have

$$Q = \bar{y}'A' = \left(\frac{0.1575}{2}\right)[(0.1575)(0.03)] = 0.372(10^{-3}) \text{ m}^3$$

So that

$$\tau_{\text{allow}} \geq \frac{V_{\text{max}}Q}{It}$$

$$800 \text{ kPa} \geq \frac{1.5 \text{ kN}[0.372(10^{-3})] \text{ m}^3}{60.125(10^{-6}) \text{ m}^4(0.03 \text{ m})} = 309 \text{ kPa} \qquad \text{OK}$$

Nail Spacing.   From the shear diagram it is seen that the shear varies over the entire span. Since the nail spacing depends on the magnitude of shear in the beam, for simplicity (and to be conservative), we will design the spacing on the basis of $V = 1.5$ kN for region BC and $V = 1$ kN for region CD. Since the nails join the flange to the web, Fig. 11–28d, we have

$$Q = \bar{y}'A' = (0.0725 - 0.015)[(0.2)(0.03)] = 0.345(10^{-3}) \text{ m}^3$$

The shear flow for each region is therefore

$$q_{BC} = \frac{V_{BC}Q}{I} = \frac{1.5 \text{ kN}[0.345(10^{-3})] \text{ m}^3}{60.125(10^{-6}) \text{ m}^4} = 8.61 \text{ kN/m}$$

$$q_{CD} = \frac{V_{CD}Q}{I} = \frac{1 \text{ kN}[0.345(10^{-3})] \text{ m}^3}{60.125(10^{-6}) \text{ m}^4} = 5.74 \text{ kN/m}$$

One naii can resist 1.50 kN in shear, so the spacing becomes

$$s_{BC} = \frac{1.50 \text{ kN}}{8.61 \text{ kN/m}} = 0.174 \text{ m}$$

$$s_{CD} = \frac{1.50 \text{ kN}}{5.74 \text{ kN/m}} - 0.261 \text{ m}$$

For ease of measuring, use

$$s_{BC} = 150 \text{ mm} \qquad \textit{Ans.}$$

$$s_{CD} = 250 \text{ mm} \qquad \textit{Ans.}$$

PROBLEMS

Neglect the weight of the beam in the following problems.

11–37. The simply-supported beam is made of timber that has an allowable bending stress of $\sigma_{allow} = 6.5$ MPa and an allowable shear stress of $\tau_{allow} = 500$ kPa. Determine its dimensions if it is to be rectangular and have a height-to-width ratio of 1.25.

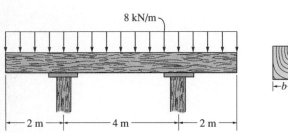

Prob. 11–37

11–38. The beam is made of cypress having an allowable bending stress of $\sigma_{allow} = 850$ psi and an allowable shear stress of $\tau_{allow} = 80$ psi. Determine the width b of the beam if the height $h = 1.5b$.

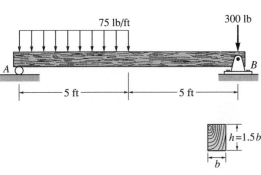

Prob. 11–38

11–39. The beam is made of Douglas fir having an allowable bending stress of $\sigma_{allow} = 1.1$ ksi and an allowable shear stress of $\tau_{allow} = 0.70$ ksi. Determine the width b of the beam if the height $h = 2b$.

Prob. 11–39

*11–40. The joists of a floor in a warehouse are to be selected using square timber beams made of oak. If each beam is to be designed to carry 90 lb/ft over a simply-supported span of 25 ft, determine the dimension a of its square cross section. The allowable bending stress is $\sigma_{allow} = 4.5$ ksi and the allowable shear stress is $\tau_{allow} = 125$ psi.

Prob. 11–40

11–41. The simply-supported beam is made of timber that has an allowable bending stress of $\sigma_{allow} = 960$ psi and an allowable shear stress of $\tau_{allow} = 75$ psi. Determine its dimensions if it is to be rectangular and have a height-to-width ratio of 1.25.

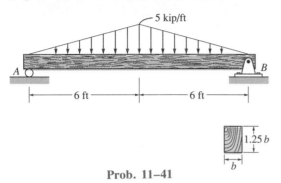

Prob. 11–41

11–42. The timber beam has a rectangular cross section. If the width of the beam is 6 in., determine its height h so that it simultaneously reaches its allowable bending stress of $\sigma_{allow} = 1.50$ ksi and an allowable shear stress of $\tau_{allow} = 50$ psi. Also, what is the maximum load P that the beam can then support?

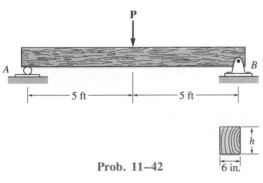

Prob. 11–42

11–43. Select the lightest-weight steel wide-flange beam from Appendix B that will safely support the loading shown. The allowable bending stress is $\sigma_{allow} = 24$ ksi and the allowable shear stress is $\tau_{allow} = 14$ ksi.

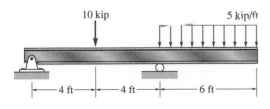

Prob. 11–43

*11–44. Select the lightest-weight steel wide-flange beam from Appendix B that will safely support the loading shown. The allowable bending stress is $\sigma_{allow} = 24$ ksi and the allowable shear stress is $\tau_{allow} = 14$ ksi.

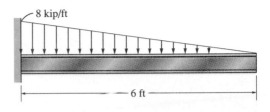

Prob. 11–44

11–45. Select the lightest-weight steel wide-flange beam from Appendix B that will safely support the loading shown. The allowable bending stress is $\sigma_{allow} = 24$ ksi and the allowable shear stress is $\tau_{allow} = 14$ ksi.

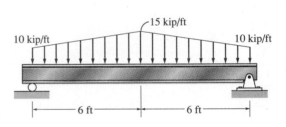

Prob. 11–45

11–46. Select the lightest-weight steel wide-flange beam from Appendix B that will safely support the loading shown. The allowable bending stress is $\sigma_{allow} = 22$ ksi and the allowable shear stress is $\tau_{allow} = 12$ ksi.

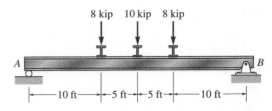

Prob. 11–46

11–47. The simply-supported beam is composed of two W12 × 22 sections built up as shown. Determine the maximum uniform loading w the beam will support if the allowable bending stress is $\sigma_{allow} = 22$ ksi and the allowable shear stress is $\tau_{allow} = 14$ ksi.

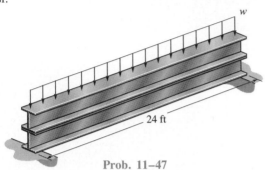

Prob. 11–47

*11–48. The beam is to be used to support the machine, which exerts the forces of 6 kip and 8 kip as shown. If the maximum bending stress is not to exceed $\sigma_{allow} = 22$ ksi, determine the required width b of the flanges.

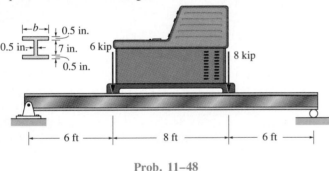

Prob. 11–48

11–49. The steel cantilevered beam is made from two 150-mm by 15-mm plates welded together as shown. Determine the maximum loads P that can be safely supported on the beam if the allowable bending stress is $\sigma_{allow} = 170$ MPa and the allowable shear stress is $\tau_{allow} = 95$ MPa.

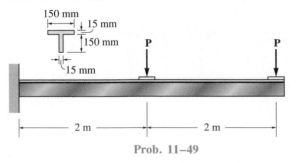

Prob. 11–49

11–50. The steel beam has an allowable bending stress of $\sigma_{allow} = 140$ MPa and an allowable shear stress of $\tau_{allow} = 90$ MPa. Determine the maximum load that it can safely support.

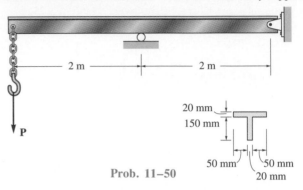

Prob. 11–50

11–51. The simply-supported joist is used in the construction of a floor for a building. In order to keep the floor low with respect to the sill beams C and D, the ends of the joists are notched as shown. If the allowable shear stress for the wood is $\tau_{allow} = 350$ psi and the allowable bending stress is $\sigma_{allow} = 1500$ psi, determine the height h that will cause the beam to reach both allowable stresses at the same time. Also, what load P causes this to happen? Neglect the stress concentration at the notch.

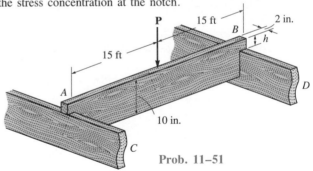

Prob. 11–51

*11–52. The wooden box beam has an allowable bending stress of $\sigma_{allow} = 10$ MPa and an allowable shear stress of $\tau_{allow} = 775$ kPa. Determine the maximum intensity w of the distributed loading that it can safely support. Also, determine the maximum safe nail spacing for each third of the length of the beam. Each nail can resist a shear force of 200 N.

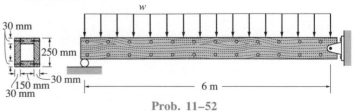

Prob. 11–52

11–53. The beam is constructed from three boards as shown. If each nail can support a shear force of 300 N, determine the maximum spacing of the nails, s and s', within regions AB and BC.

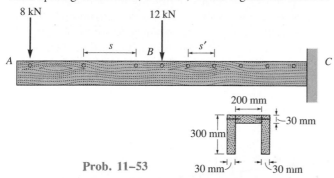

Prob. 11–53

11–54. The beam is constructed from three boards as shown. If each nail can support a shear force of 50 lb, determine the maximum spacing of the nails, s, s', and s'', for regions AB, BC, and CD, respectively.

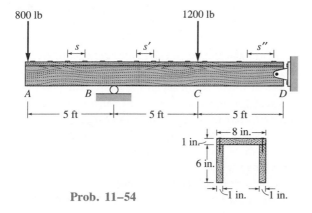

Prob. 11–54

11–55. The compound beam is made from two sections, which are pinned together at B. Use Appendix B and select the lightest wide-flange beam that would be safe for each section if the allowable bending stress is $\sigma_{allow} = 24$ ksi and the allowable shear stress is $\tau_{allow} = 14$ ksi. The beam supports a pipe loading of 4 kip and 6 kip as shown.

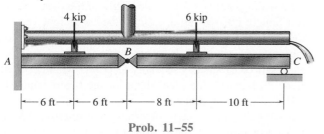

Prob. 11–55

*11–56. The joist AB used in housing construction is to be made from 8-in. by 1.5-in. Southern-pine boards. If the design loading on each board is placed as shown, determine the largest room width L that the boards can span. The allowable bending stress for the wood is $\sigma_{allow} = 2$ ksi and the allowable shear stress is $\tau_{allow} = 180$ psi. Assume that the beam is simply-supported from the walls at A and B.

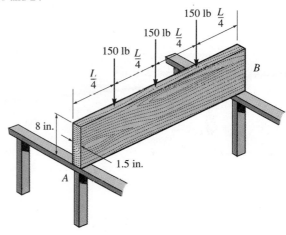

Prob. 11–56

11–57. The overhang beam is constructed using two 2-in. by 4-in. pieces of wood braced as shown. If the allowable bending stress is $\sigma_{allow} = 600$ psi, determine the largest load P that can be applied. Also, determine the associated maximum spacing of nails, s, along the beam section AC if each nail can resist a shear force of 800 lb. Assume the beam is pin-connected at A, B, and D. Neglect the axial force developed in the beam along DA.

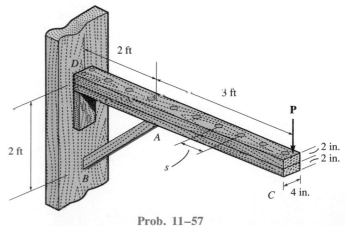

Prob. 11–57

*11.6 Fully Stressed Beams

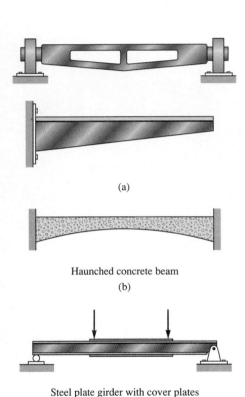

Haunched concrete beam

(b)

Steel plate girder with cover plates

(c)

Fig. 11–29

In Sec. 11.5 we developed a method for determining the dimensions of the cross section of a *prismatic beam* so that it resists the maximum moment, M_{max}, within its span. Since the moment in the beam generally *varies* over the beam's length, the choice of a prismatic beam is usually inefficient since it is never fully stressed at points along the beam where $M < M_{max}$. In order to refine the design so as to reduce a beam's weight, engineers sometimes choose a beam having a *variable* cross-sectional area, such that at each cross section along the beam the bending stress reaches its maximum allowable value. Beams having a variable cross-sectional area are called *nonprismatic beams*. They are often used in machines since they can be readily formed by casting. Examples are shown in Fig. 11–29a. In structures such beams may be "haunched" at their ends as shown in Fig. 11–29b. Also, beams may be "built up" or fabricated in a shop using plates. An example is a girder made from a rolled-shaped prismatic beam and having cover plates welded to it in the region where the moment is a maximum, Fig. 11–29c.

The stress analysis of a nonprismatic beam is generally very difficult to perform and is beyond the scope of this text. Most often these shapes are analyzed by using experimental methods or the theory of elasticity. The results obtained from such an analysis, however, do indicate that the assumptions used in the derivation of the flexure formula are approximately correct for predicting the bending stresses in nonprismatic sections, provided the taper or slope of the upper or lower boundary of the beam is not too severe. On the other hand, the shear formula cannot be used for nonprismatic beam design, since the results obtained from it are very misleading.

Although caution is advised when applying the flexure formula to nonprismatic beam design, we will show here, in principle, how this formula can be used as an approximate means for obtaining the beam's general shape. In this regard, the *size* of the cross section of a nonprismatic beam that supports a given loading can be determined using the flexure formula written as

$$S = \frac{M}{\sigma_{allow}}$$

If we express the internal moment M in terms of its position x along the beam, then since σ_{allow} is a known constant, the section modulus S or the beam's dimensions become a function of x. For example, if the cross-sectional area is to be rectangular and the width is held *constant*, the height of the cross section can be determined as a function of x using $S = I/c$. A beam designed in this manner is called a *fully stressed beam*. Although *only* bending stresses have been considered in approximating its final shape, attention must also be given to ensure that the beam will resist shear, especially at points where concentrated loads are applied. As a result, the ideal shape of the beam cannot be determined from the flexure formula alone.

The following examples illustrate how to determine the approximate dimensions of fully stressed beams using the flexure formula.

Example 11–18

Determine the shape of a fully stressed, simply-supported beam that supports a concentrated force at its center, Fig. 11–30a. The beam has a rectangular cross section of constant width b, and the allowable stress is σ_{allow}.

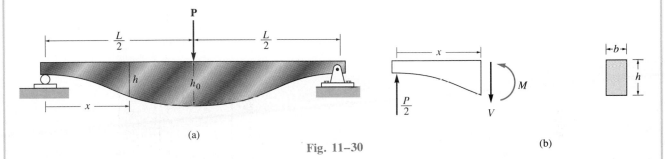

(a)

Fig. 11–30

(b)

SOLUTION

The internal moment in the beam, Fig. 11–30b, expressed as a function of position, $0 \leq x < L/2$, is

$$M = \frac{P}{2}x$$

Hence the beam's section modulus is

$$S = \frac{M}{\sigma_{\text{allow}}} = \frac{P}{2\sigma_{\text{allow}}}x$$

Since $S = I/c$, then for a cross-sectional area h by b we have

$$\frac{I}{c} = \frac{\frac{1}{12}bh^3}{h/2} = \frac{P}{2\sigma_{\text{allow}}}x$$

$$h^2 = \frac{3P}{\sigma_{\text{allow}}b}x$$

If $h = h_0$ at $x = L/2$, then

$$h_0{}^2 = \frac{3PL}{2\sigma_{\text{allow}}b}$$

so that

$$h^2 = \left(\frac{2h_0{}^2}{L}\right)x \qquad\qquad Ans.$$

By inspection, the depth h must therefore vary in a *parabolic* manner with the distance x. In practice this *shape* is the basis for the design of leaf springs used to support the rear-end axles of most heavy trucks.

Example 11–19

The cantilevered beam shown in Fig. 11–31a is formed into a trapezoidal shape having a depth h_0 at A and a depth $3h_0$ at B. If it supports a load **P** at its end, determine the absolute maximum normal stress in the beam. The beam has a rectangular cross section of constant width b.

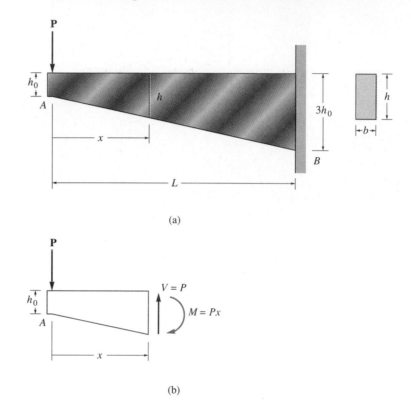

(a)

(b)

Fig. 11–31

SOLUTION

At any cross section, the maximum normal stress occurs at the top and bottom surface of the beam. However, since $\sigma_{max} = M/S$ and the section modulus S increases as x increases, the absolute maximum normal stress does *not* necessarily occur at the wall B, where the moment is maximum. Using the flexure formula we can express the maximum normal stress at an arbitrary section in terms of its position x, Fig. 11–31b. Here the internal moment has a magnitude of $M = Px$. Since the slope of the bottom of the beam is $2h_0/L$, Fig. 11–31a, the depth of the beam at position x is

$$h = \frac{2h_0}{L}x + h_0 = \frac{h_0}{L}(2x + L)$$

Applying the flexure formula, we have

$$\sigma = \frac{Mc}{I} = \frac{Px(h/2)}{(\frac{1}{12} bh^3)} = \frac{6PL^2x}{bh_0^2(2x + L)^2} \tag{1}$$

To determine the position x where the absolute maximum normal stress occurs, we must take the derivative of σ with respect to x and set it equal to zero. This gives

$$\frac{d\sigma}{dx} = \left(\frac{6PL^2}{bh_0^2}\right) \frac{1(2x + L)^2 - x(2)(2x + L)(2)}{(2x + L)^4} = 0$$

Thus,

$$4x^2 + 4xL + L^2 - 8x^2 - 4xL = 0$$

$$L^2 - 4x^2 = 0$$

$$x = \frac{1}{2} L$$

Substituting into Eq. 1 and simplifying, the absolute maximum normal stress is therefore

$$\sigma_{\substack{abs \\ max}} = \frac{3}{4} \frac{PL}{bh_0^2} \qquad \qquad Ans.$$

Note that at the wall, B, the maximum normal stress is

$$(\sigma_{max})_B = \frac{Mc}{I} = \frac{PL(1.5h_0)}{[\frac{1}{12} b(3h_0)^3]} = \frac{2}{3} \frac{PL}{bh_0^2}$$

which is 11.1% smaller than $\sigma_{\substack{abs \\ max}}$. Realize, however, that this difference decreases as the beam's taper becomes smaller. The stresses become equivalent only when the beam becomes prismatic and has a constant depth h_0.

It should be recalled that the flexure formula was derived on the basis of assuming the beam to be *prismatic*. Since this is not the case here, some error is to be expected in this analysis and that of Example 11–18. A more exact mathematical analysis, using the theory of elasticity, reveals that application of the flexure formula as in the above example gives only small errors in the normal stress if the angle of beam taper is small. For example, if this angle is 15°, the stress calculated above is about 5% greater than that calculated by the more exact analysis. It may also be worth noting that the calculation of $(\sigma_{max})_B$ was done only for illustrative purposes, since the actual stress distribution at the support (wall) is highly irregular. See Saint-Venant's Principle, Sec. 4.1.

11.7 Shaft Design

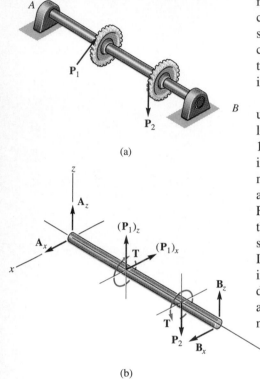

(a)

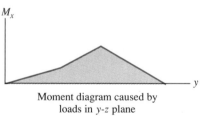
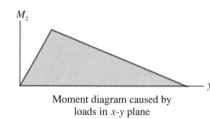

(b)

Shafts that have circular cross sections are frequently used in many types of mechanical equipment and machinery. As a result, they are often subjected to cyclic or fatigue stress, which is caused by the combined bending and torsional loads they must transmit or resist. In addition to these loadings, stress concentrations may exist on a shaft due to keys, couplings, and sudden transitions in its cross-sectional area (Sec. 5.8). In order to design a shaft properly, it is therefore necessary to take all of these effects into account.

In this section we will discuss some of the important aspects of the design of uniform shafts required to transmit power. These shafts are often subjected to loads applied to attached pulleys and gears, such as the one shown in Fig. 11–32a. Since the loads can be applied to the shaft at various angles, the internal bending and torsional moments at any cross section can be determined by first replacing the loads by their statically equivalent counterparts and then resolving these loads into components in two perpendicular planes, Fig. 11–32b. The bending-moment diagrams for the loads *in each plane* can then be drawn, and the resultant internal moment at any section along the shaft is then determined by vector addition, $M = \sqrt{M_x^2 + M_z^2}$, Fig. 11–32c. In addition to the moment, segments of the shaft are also subjected to different internal torques, Fig. 11–32b. Here we see that for equilibrium the torque developed at one gear must balance the torque developed at the other gear. To account for the general variation of torque along the shaft, a *torque diagram* may also be drawn, Fig. 11–32d.

Moment diagram caused by
loads in y-z plane

Moment diagram caused by
loads in x-y plane

(c)

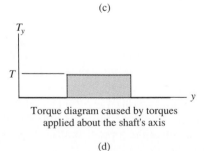

Torque diagram caused by torques
applied about the shaft's axis

(d)

Fig. 11–32(a–d)

Once the moment and torque diagrams have been established, it is then possible to investigate certain critical sections along the shaft where the *combination* of a resultant moment **M** and a torque **T** creates the worst stress situation. In this regard, the moment of inertia of the shaft is the *same* about *any* diametrical axis, and since this axis represents a *principal axis of inertia* for the cross section, we can apply the flexure formula using the *resultant moment* to obtain the maximum bending stress. As shown in Fig. 11–32e, this stress will occur on two elements, C and D, each located on the outer boundary of the shaft. If a torque **T** is also resisted at this section, then a maximum shear stress is also developed on these elements, Fig. 11–32f. Furthermore, the external forces may also create shear stress in the shaft determined from $\tau = VQ/It$; however, this stress will generally contribute a much smaller stress distribution on the cross section compared with that developed by bending and torsion. In some cases it must be investigated, but for simplicity we will neglect its effect in the following analysis. In general, then, the critical element C (or D) on the shaft is subjected to *plane stress* as shown in Fig. 11–32g, where

$$\sigma = \frac{Mc}{I} \qquad \text{and} \qquad \tau = \frac{Tc}{J}$$

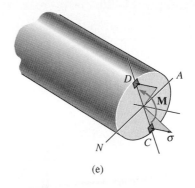

(e)

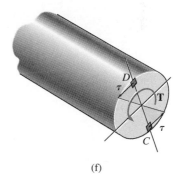

(f)

Knowing the allowable normal and shear stress for the material, the size of the shaft is then based on the use of these equations and selection of an appropriate theory of failure. For example, if the material is known to be ductile, then the maximum-shear-stress theory may be appropriate. As stated in Sec. 10.7, this theory requires the allowable shear stress, which is determined from the results of a simple tension test, to be equal to the maximum shear stress in the element. Using the stress-transformation equation, Eq. 9–7, for the stress state in Fig. 11–32g, we have

$$\tau_{\text{allow}} = \sqrt{\left(\frac{\sigma}{2}\right)^2 + \tau^2}$$

$$= \sqrt{\left(\frac{Mc}{2I}\right)^2 + \left(\frac{Tc}{J}\right)^2}$$

Since $I = \pi c^4/4$ and $J = \pi c^4/2$, this equation becomes

$$\tau_{\text{allow}} = \frac{2}{\pi c^3} \sqrt{M^2 + T^2}$$

Solving for the radius of the shaft, we get

$$c = \left(\frac{2}{\pi \tau_{\text{allow}}} \sqrt{M^2 + T^2}\right)^{1/3} \tag{11–8}$$

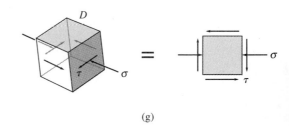

(g)

Fig. 11–32(e–g)

Application of any other theory of failure will, of course, lead to a different formulation for c. However, in all cases it may be necessary to apply this formulation at various ''critical sections'' along the shaft in order to determine the particular combination of M and T that gives the largest value for c.

The following example illustrates the procedure numerically.

■ Example 11–20

The shaft in Fig. 11–33a is supported by smooth journal bearings at A and B. Due to the transmission of power to and from the shaft, the belts on the pulleys are subjected to the tensions shown. Determine the smallest diameter of the shaft using the maximum-shear-stress theory, with $\tau_{allow} = 50$ MPa.

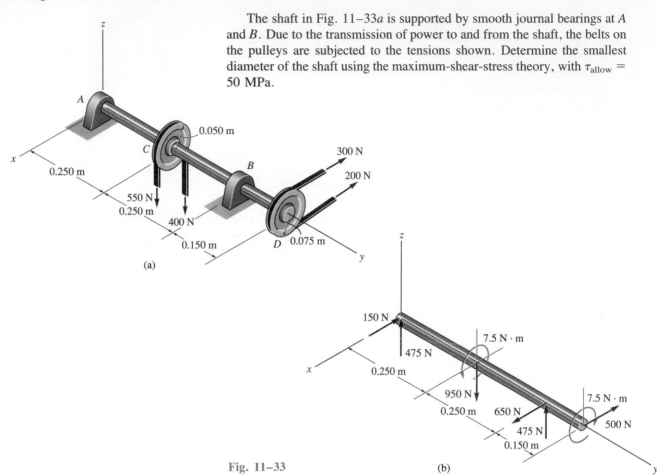

Fig. 11–33

(a)

(b)

SOLUTION

The support reactions have been calculated and are shown on the free-body diagram of the shaft, Fig. 11–33b. Bending-moment diagrams for M_x and M_z are shown in Fig. 11–33c and 11–33d, respectively. The torque diagram is shown in Fig. 11–33e. By inspection, critical points for bending moment occur either at C or B. Also, just to the right of C and at B the torsional moment is 7.5 N · m. At C, the resultant moment is

$$M_C = \sqrt{(118.75)^2 + (37.5)^2} = 124.5 \text{ N} \cdot \text{m}$$

whereas at B it is smaller, namely

$$M_B = 75 \text{ N} \cdot \text{m}$$

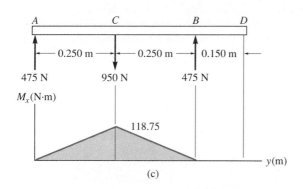

(c)

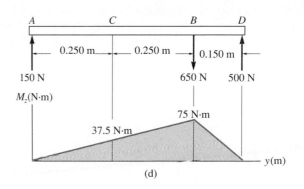

(d)

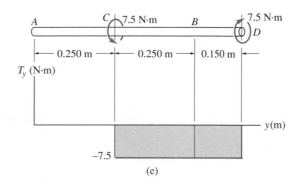

(e)

Since the design is based on the maximum-shear-stress theory, Eq. 11–8 applies. The radical $\sqrt{M^2 + T^2}$ will therefore be largest at a section just to the right of C. We have

$$c = \left(\frac{2}{\pi \tau_{\text{allow}}} \sqrt{M^2 + T^2} \right)^{1/3}$$

$$= \left(\frac{2}{\pi (50)(10^6) \ \text{N/m}^2} \sqrt{(124.5 \ \text{N} \cdot \text{m})^2 + (7.5 \ \text{N} \cdot \text{m})^2} \right)^{1/3}$$

$$= 0.0117 \ \text{m}$$

Thus, the smallest allowable diameter is

$$d = 2(0.0117 \ \text{m}) = 23.3 \ \text{mm} \qquad \textit{Ans.}$$

PROBLEMS

11–58.   Determine the variation in the width w as a function of x for the cantilevered beam that supports a concentrated force $\mathbf{P}$ at its end so that it has a maximum bending stress σ_{allow} throughout its length. The beam has a constant thickness t.

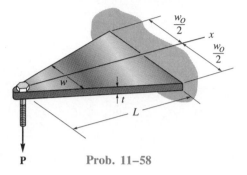

Prob. 11–58

11–59.   Determine the variation in the depth d of a cantilevered beam that supports a concentrated force $\mathbf{P}$ at its end so that it has a constant maximum bending stress σ_{allow} throughout its length. The beam has a constant width b_0.

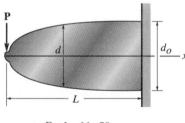

Prob. 11–59

*11–60.   The beam is made from a plate that has a constant thickness b_0. If it is simply-supported and carries a uniform load w, determine the variation of its depth as a function of x so that it maintains a constant maximum bending stress σ_{allow} throughout its length.

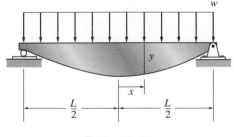

Prob. 11–60

11–61.   The cantilevered beam has a circular cross section. If it supports a force $\mathbf{P}$ at its end, determine its radius y as a function of x so that it is subjected to a constant maximum bending stress σ_{allow} throughout its length.

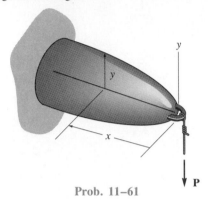

Prob. 11–61

11–62.   The tapered beam supports a concentrated force $\mathbf{P}$ at its center. If it is made from a plate and has a constant width b_0, determine the absolute maximum bending stress in the beam.

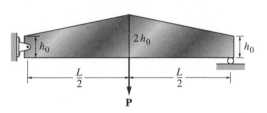

Prob. 11–62

11–63.   The tapered beam supports a uniform distributed load w. If it is made from a plate and has a constant width b_0, determine the absolute maximum bending stress in the beam.

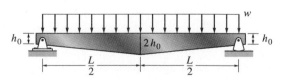

Prob. 11–63

*11–64.   The beam is made into the shape of a frustum and has a diameter of 6 in. at A and a diameter of 12 in. at B. If it supports a force of 800 lb at A, determine the absolute maximum bending stress in the beam and specify its location x.

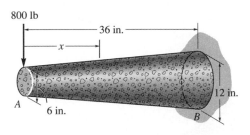

Prob. 11–64

11–65.   The beam is made in the shape of a frustum that has a diameter of 1 ft at A and a diameter of 2 ft at B. If it supports a couple moment of 8 kip · ft at its end, determine the absolute maximum bending stress in the beam and specify its location x.

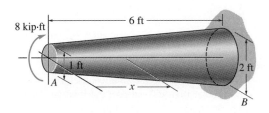

Prob. 11–65

11–66.   The 10-mm-wide bracket is used to support a force of 70 N at its end A. Determine the absolute maximum bending stress in the bracket and specify its location x.

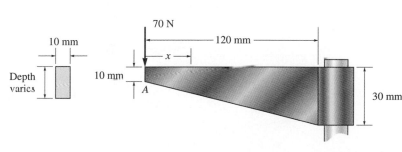

Prob. 11–66

11–67.   The two pulleys attached to the shaft are loaded as shown. If the bearings at A and B exert only vertical forces on the shaft, determine the required diameter of the shaft to the nearest $\frac{1}{8}$ in. using the maximum-shear-stress theory. $\tau_{allow} = 12$ ksi.

*11–68.   Solve Prob. 11–67 using the maximum-distortion-energy theory. $\sigma_{allow} = 67$ ksi.

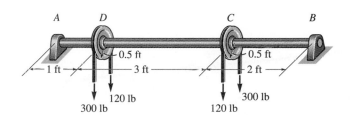

Probs. 11–67/11–68

11–69.   The bearings at A and D exert only y and z components of force on the shaft. If $\tau_{allow} = 60$ MPa, determine to the nearest millimeter the smallest-diameter shaft that will support the loading. Use the maximum-shear-stress theory of failure.

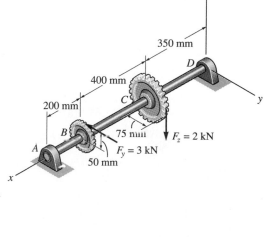

Prob. 11–69

11–70. The shaft is supported by bearings at A and B that exert force components only in the x and z directions on the shaft. If the allowable normal stress for the shaft is $\sigma_{allow} = 15$ ksi, determine to the nearest $\frac{1}{8}$ in. the smallest diameter of the shaft that will support the loading. Use the maximum-distortion-energy theory of failure.

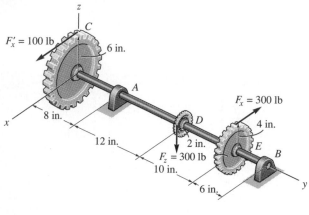

Prob. 11–70

11–71. The tubular shaft has an inner diameter of 15 mm. Determine to the nearest millimeter its outer diameter if it is subjected to the gear loading shown. The bearings at A and B exert force components only in the y and z directions on the shaft. Use an allowable shear stress of $\tau_{allow} = 70$ MPa, and base the design on the maximum-shear-stress theory of failure.

***11–72.** The motor at A is turning the shaft at a constant rate such that the torque $\mathbf{T}$ it develops creates horizontal reactions on the gears at C and D of 250 lb and 350 lb, respectively. The shaft is supported by a bearing in the motor at A and a bearing at B, which exert force components only in the x and z directions on the shaft. If the allowable shear stress for the shaft is $\tau_{allow} = 10$ ksi, determine to the nearest $\frac{1}{8}$ in. the smallest diameter of the shaft that will support the loading. Use the maximum-shear-stress theory of failure.

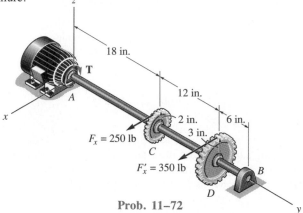

Prob. 11–72

11–73. The gears connected to the shaft are subjected to the loadings shown. If the bearings at A and B exert only y and z components of force on the shaft, determine the equilibrium torque T at gear C and then determine the smallest diameter of the shaft to the nearest millimeter that will support the loading. Use the maximum-shear-stress theory of failure with $\tau_{allow} = 60$ MPa.

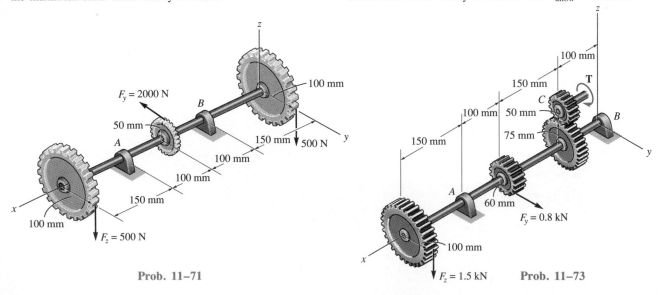

Prob. 11–71

Prob. 11–73

REVIEW PROBLEMS

11–74. The cam shaft has two lobes on it, each of which opens and closes valves at the same instant. The torque for turning it is negligible, and as shown it is subjected to the 110-lb loadings. Draw the moment diagram for the shaft and compute the maximum bending stress in the shaft. The bearings at A and B exert only vertical forces on the shaft. Neglect their size and the size of the cams. The shaft has a diameter of 0.45 in.

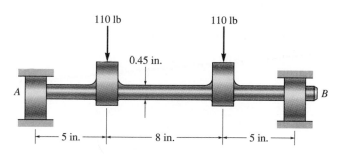

Prob. 11–74

11–75. The overhanging beam has been fabricated with a projected arm BD on it. Draw the shear and moment diagrams for the beam if it supports a load of 800 lb. *Hint:* The loading in the supporting strut DE must be replaced by equivalent loads at point B on the axis of the beam.

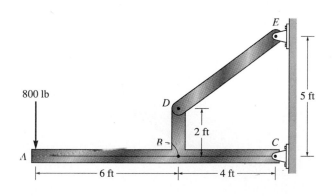

Prob. 11–75

*11–76. The beam is bolted or pinned at A and rests on a bearing pad at B that exerts a uniform distributed loading on the beam over its 2-ft length. Draw the shear and moment diagrams for the beam if it supports a uniform loading of 2 kip/ft.

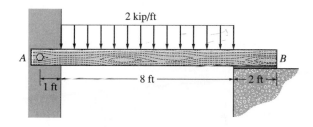

Prob. 11–76

11–77. The beam is made from a plate having a constant thickness t and a width that varies as shown. If it supports a concentrated force **P** at its end, determine the absolute maximum bending stress in the beam and specify its location x.

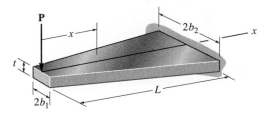

Prob. 11–77

11–78. Draw the shear and moment diagrams for the beam and determine the shear and moment as functions of x.

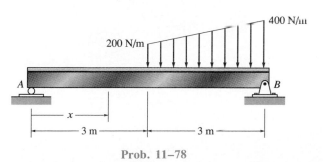

Prob. 11–78

11–79. The bearings at A and B exert only x and z components of force on the steel shaft. Determine its diameter to the nearest millimeter so that it can resist the loadings of the gears without exceeding an allowable shear stress of $\tau_{allow} = 80$ MPa. Use the maximum-shear-stress theory of failure.

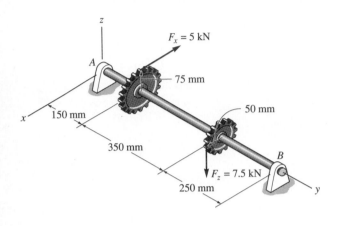

Prob. 11–79

11–81. Select the lightest-weight steel wide-flange beam from Appendix B that will safely support the machine loading shown. The allowable bending stress is $\sigma_{allow} = 24$ ksi and the allowable shear stress is $\tau_{allow} = 14$ ksi.

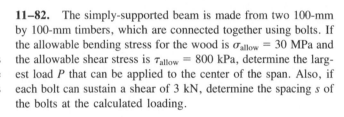

Prob. 11–81

*11–80. The beam has a width w and a depth that varies as shown. If it supports a concentrated force P at its end, determine the absolute maximum bending stress in the beam and specify its location x.

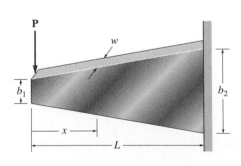

Prob. 11–80

11–82. The simply-supported beam is made from two 100-mm by 100-mm timbers, which are connected together using bolts. If the allowable bending stress for the wood is $\sigma_{allow} = 30$ MPa and the allowable shear stress is $\tau_{allow} = 800$ kPa, determine the largest load P that can be applied to the center of the span. Also, if each bolt can sustain a shear of 3 kN, determine the spacing s of the bolts at the calculated loading.

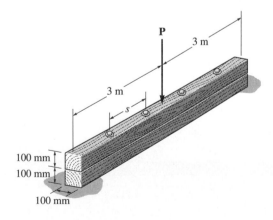

Prob. 11–82

12 Deflections of Beams and Shafts

Often limits must be placed on the amount of deflection a beam or shaft may undergo when it is subjected to a load, and so in this chapter we will discuss various methods for determining the deflection and slope at specific points on beams and shafts. The analytical methods include the integration method, the use of discontinuity functions, and the method of superposition. Also, a semigraphical technique, called the moment-area method, will be presented. At the end of the chapter, we will use these methods to solve for the support reactions on a beam or shaft that is statically indeterminate.

12.1 The Elastic Curve

Before the slope or the displacement at a point on a beam (or shaft) is determined, it is often helpful to sketch the deflected shape of the beam when it is loaded, in order to "visualize" any computed results and thereby partially check these results. The deflection diagram of the longitudinal axis that passes through the centroid of each cross-sectional area of the beam is called the *elastic curve*. For most beams the elastic curve can be sketched without much difficulty. When doing so, however, it is necessary to know the type of restrictions to slope or displacement that often occur at a support. In general, supports that resist a *force*, such as a pin, restrict *displacement*, and those that resist a *moment*, such as a fixed wall, restrict *rotation* or *slope*. With this in mind, two typical examples of the elastic curves for loaded beams (or shafts), sketched to a greatly exaggerated scale, are shown in Fig. 12–1.

If the elastic curve for a beam seems difficult to establish, it is suggested that the moment diagram for the member be drawn first. Using the beam sign

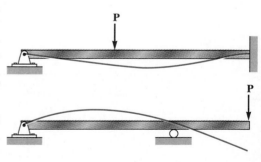

Fig. 12–1

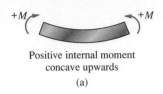

Positive internal moment
concave upwards

(a)

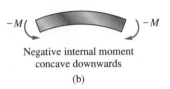

Negative internal moment
concave downwards

(b)

Fig. 12–2

convention established in Chapter 7, a positive internal moment tends to bend the beam concave upward, Fig. 12–2a. Likewise, a negative moment tends to bend the beam concave downward, Fig. 12–2b. Therefore, if the moment diagram is *known*, it will be easy to construct the elastic curve. For example, consider the beam in Fig. 12–3a with its associated moment diagram shown in Fig. 12–3b. Due to the roller and pin supports, the displacement at B and D must be zero. Within the region of negative moment, AC, Fig. 12–3b, the elastic curve must be concave downward, and within the region of positive moment, CD, the elastic curve must be concave upward. Hence, there must be an *inflection point* at point C, where the curve changes from concave up to concave down, since this is a point of zero moment. Using these facts, the beam's elastic curve is sketched to a greatly exaggerated scale in Fig. 12–3c. It should also be noted that the displacements Δ_A and Δ_E are especially critical. At point E the *slope* of the elastic curve is *zero*, and there the beam's *deflection* may be a *maximum*. Whether Δ_E is actually greater than Δ_A depends on the relative magnitudes of $\mathbf{P}_1$ and $\mathbf{P}_2$ and the location of the roller at B.

Following these same principles, note how the elastic curve in Fig. 12–4 was constructed. Here the beam is cantilevered from a fixed support at A and therefore the elastic curve must have both zero displacement and zero slope at this point. Also, the largest displacement will occur either at D, where the slope is zero, or at C.

Moment–Curvature Relationship.

We will now develop an important relation between the internal moment in the beam and the radius of curvature ρ (rho) of the elastic curve at a point. The resulting equation will be used throughout the chapter as a basis for establishing each of the methods for finding the slope and deflection of the elastic curve for a beam (or shaft).

The following analysis, here and in the next section, will require the use of three coordinates. As shown in Fig. 12–5a, the x axis extends positive to

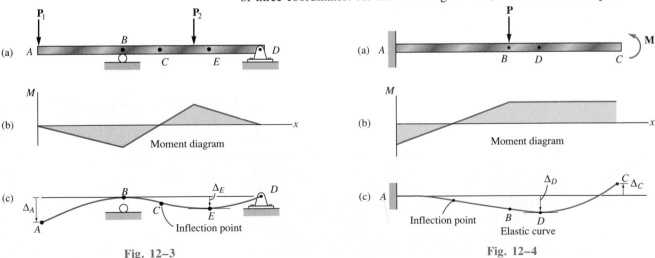

Fig. 12–3

Fig. 12–4

the right, along the initially straight longitudinal axis of the beam. It is used to locate the differential element, having an undeformed width dx. The v axis extends *positive upward* from the x axis. It measures the *displacement* of the centroid on the cross-sectional area of the element. With these two coordinates, we will later define the equation of the elastic curve, v, as a function of x. Lastly, a "localized" y coordinate is used to specify the position of a fiber of the beam element. It is measured *positive upward* from the neutral axis, as shown in Fig. 12–5b. Recall that this same sign convention for x and y was used in the derivation of the flexure formula.

In order to derive the relationship between the internal moment and ρ, we will limit the analysis to the most common case of an initially straight beam that is elastically deformed by loads applied perpendicular to the beam's x axis and lying in the x–v plane of symmetry for the beam's cross-sectional area. Due to the loading, the deformation of the beam is caused by both the internal shear force and bending moment. If the beam has a length that is much greater than its depth, the greatest deformation will be caused by bending, and therefore we will direct our attention to its effects.*

In Sec. 6.1 we described what happens when the internal moment M deforms the element of the beam in Fig. 12–5b. Recall that the cross sections of the element remain plane, such that the angle between them becomes $d\theta$. The arc that represents a portion of the elastic curve intersects the neutral axis for each cross section, and since it does *not* stretch, it maintains a length dx. The *radius of curvature* for this arc is defined as the distance ρ, which is measured from the *center of curvature* O' to dx. Any arc on the element other than dx is subjected to a normal strain. For example, it was shown in Sec. 6.1 that the strain in arc ds, located at a position y from the neutral axis, can be determined from Eq. 6–1, namely,

$$\frac{1}{\rho} = -\frac{\epsilon}{y} \qquad (12\text{–}1)$$

If the material is homogeneous and behaves in a linear-elastic manner, then $\epsilon = \sigma/E$. Also, since the flexure formula applies, $\sigma = -My/I$. Combining these equations and substituting into Eq. 12–1, we have

$$\boxed{\frac{1}{\rho} = \frac{M}{EI}} \qquad (12\text{–}2)$$

where

$\rho =$ the radius of curvature at a specific point on the elastic curve
$\quad$ ($1/\rho$ is referred to as the *curvature*)
$M =$ the internal moment in the beam at the point where ρ is to be determined
$E =$ the material's modulus of elasticity
$I =$ the beam's moment of inertia computed about the neutral axis

*Methods used to determine deflections caused by shear will be discussed in Chapter 14.

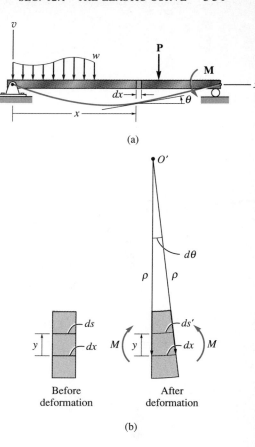

(a)

Before deformation

After deformation

(b)

Fig. 12–5

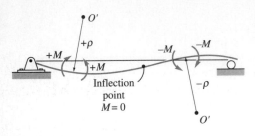

Fig. 12–6

The product EI in this equation is referred to as the *flexural rigidity,* and it is always a positive quantity. The sign for ρ therefore depends on the direction of the moment. As shown in Fig. 12–6, when M is *positive,* ρ extends *above* the beam, i.e., in the positive v direction; when M is *negative,* ρ extends *below* the beam, or in the negative v direction.

Using the flexure formula, $\sigma = -My/I$, we can also express the curvature in terms of the stress in the beam, namely,

$$\frac{1}{\rho} = -\frac{\sigma}{Ey} \qquad (12\text{–}3)$$

Both Eqs. 12–2 and 12–3 are valid for either small or large radii of curvature. However, the value of ρ is almost always calculated as a *very large quantity.* For example, consider a steel beam made from a $W14 \times 53$ (Appendix B), where $E_{st} = 29(10^3)$ ksi and $\sigma_Y = 36$ ksi. When the material at the outer fibers, $y = \pm 7$ in., is about to *yield,* then, from Eq. 12–3, $\rho = \pm 5639$ in. Values of ρ calculated at other points along the beam's elastic curve may be even *larger,* since σ cannot exceed σ_Y at the outer fibers.

12.2 Slope and Displacement by Integration

The elastic curve for a beam can be expressed mathematically as $v = f(x)$. To obtain this equation, we must first represent the curvature $(1/\rho)$ in terms of v and x. In most calculus books it is shown that this relationship is

$$\frac{1}{\rho} = \frac{d^2v/dx^2}{[1 + (dv/dx)^2]^{3/2}}$$

Substituting into Eq. 12–2, we get

$$\frac{d^2v/dx^2}{[1 + (dv/dx)^2]^{3/2}} = \frac{M}{EI} \qquad (12\text{–}4)$$

This equation represents a nonlinear second-order differential equation. Its solution, which is called the *elastica,* gives the exact shape of the elastic curve, assuming, of course, that beam deflections occur only due to bending. Through the use of higher mathematics, elastica solutions have been obtained only for simple cases of beam geometry and loading.

In order to facilitate the solution of a greater number of deflection problems, Eq. 12–4 can be modified. Most engineering design codes specify *limitations* on deflections for tolerance or esthetic purposes, and as a result the elastic deflections for the majority of beams and shafts form a shallow curve. Consequently, the slope of the elastic curve which is determined from dv/dx will be *very small,* and its square will be negligible compared with unity.* Therefore the curvature, as defined above, can be approximated by $1/\rho = d^2v/dx^2$. Using this simplification, Eq. 12–4 can now be written as

$$\frac{d^2v}{dx^2} = \frac{M}{EI} \qquad (12\text{–}5)$$

*See Example 10–1.

It is also possible to write this equation in two alternative forms. If we differentiate each side with respect to x and substitute $V = dM/dx$ (Eq. 7–2), we get

$$\frac{d}{dx}\left(EI\,\frac{d^2v}{dx^2}\right) = V(x) \tag{12–6}$$

Differentiating again, using $-w = dV/dx$ (Eq. 7–1), yields

$$\frac{d^2}{dx^2}\left(EI\,\frac{d^2v}{dx^2}\right) = -w(x) \tag{12–7}$$

For most problems the flexural rigidity will be constant along the length of the beam. Assuming this to be the case, the above results may be reordered into the following set of equations:

$$EI\,\frac{d^4v}{dx^4} = -w(x) \tag{12–8}$$

$$EI\,\frac{d^3v}{dx^3} = V(x) \tag{12–9}$$

$$EI\,\frac{d^2v}{dx^2} = M(x) \tag{12–10}$$

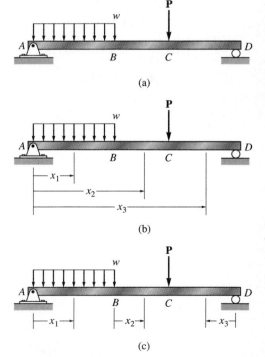

(a)

(b)

(c)

Fig. 12–7

Solution of any of these equations requires successive integrations to obtain the deflection v of the elastic curve. For each integration it is necessary to introduce a "constant of integration" and then solve for all the constants to obtain a unique solution for a particular problem. For example, if the distributed load is expressed as a function of x and Eq. 12–8 is used, then four constants of integration must be evaluated; however, if the internal moment M is determined and Eq. 12–10 is used, only two constants of integration must be found. The choice of which equation to start with depends on the problem. Generally, however, it is easier to determine the internal moment M as a function of x, integrate twice, and evaluate only two integration constants.

Recall from Sec. 11.2 that if the loading on a beam is discontinuous, that is, consists of a series of several distributed and concentrated loads, then several functions must be written for the internal moment, each valid within the region between the discontinuities. Also, for convenience in writing each moment expression, *the origin for each x coordinate can be selected arbitrarily*. For example, consider the beam shown in Fig. 12–7a. The internal moment in regions AB, BC, and CD can be written in terms of the x_1, x_2, and x_3 coordinates selected, as shown in either Fig. 12–7b or 12–7c, or in fact in any manner that will yield $M = f(x)$ in as simple a form as possible. Once these functions are integrated through the use of Eq. 12–10 and the constants of integration determined, the functions will give the slope and deflection (elastic curve) for each region of the beam for which they are valid.

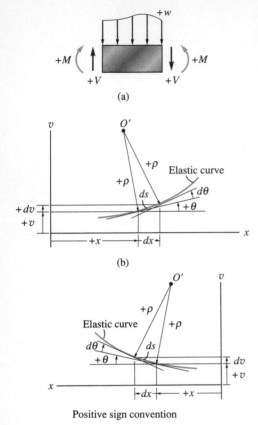

(a)

(b)

(c)

Positive sign convention

Fig. 12–8

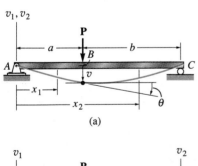

(a)

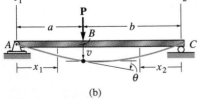

(b)

Fig. 12–9

Sign Convention and Coordinates. When applying Eqs. 12-8 through 12-10, it is important to use the proper signs for M, V, or w as established by the sign convention that was used in the derivation of these equations. For review, these terms are shown in their *positive directions* in Fig. 12–8a. Furthermore, recall that *positive deflection, v, is upward*, and as a result, the *positive slope angle θ will be measured counterclockwise* from the x axis when x is *positive to the right*. The reason for this is shown in Fig. 12–8b. Here positive increases dx and dv in x and v create an increased θ that is counterclockwise. On the other hand, if *positive x is directed to the left*, then θ will be *positive clockwise*, Fig. 12–8c.

It should be pointed out that by assuming dv/dx to be very small, the original horizontal length of the beam's axis and the arc of its elastic curve will be about the same. In other words, ds in Fig. 12–8b and 12–8c is approximately equal to dx, since $ds = \sqrt{(dx)^2 + (dv)^2} = \sqrt{1 + (dv/dx)^2}\ dx \approx dx$. As a result, points on the elastic curve will only be *displaced vertically*, and not horizontally. Also, since the *slope angle θ will be very small*, its value in radians can be determined *directly* from $\theta \approx \tan\theta = dv/dx$.

Boundary and Continuity Conditions. The constants of integration are determined by evaluating the functions for shear, moment, slope, or displacement at a particular point on the beam where the value of the function is known. These values are called *boundary conditions*. Several possible boundary conditions that are often used to solve beam (or shaft) deflection problems are listed in Table 12–1. For example, if the beam is supported by a roller or pin (1, 2, 3, 4), then it is required that the displacement be *zero* at these points. Furthermore, if these supports are located at the *ends of the beam* (1, 2), the internal moment in the beam must also be zero. At the fixed support (5) the slope and displacement are both zero, whereas the free-ended beam (6) has both zero moment and zero shear. Lastly, if two segments of a beam are connected by an "internal" pin or hinge (7), the moment must be zero at this connection.

If a single x coordinate cannot be used to express the equation for the beam's slope or the elastic curve, then *continuity conditions* must be used to evaluate some of the integration constants. For example, consider the beam in Fig. 12–9a. Here the x coordinates are both chosen with origins at A. Each is valid only within the regions $0 \leq x_1 \leq a$ and $a \leq x_2 \leq (b + a)$. Once the functions for the slope and deflection are obtained, they must give the *same values* for the slope and deflection at point B so the elastic curve is physically *continuous*. Expressed mathematically, this requires that $\theta_1(a) = \theta_2(a)$ and $v_1(a) = v_2(a)$. These equations can then be used to evaluate two constants of integration. The elastic curve can also be expressed in terms of the coordinates $0 \leq x_1 \leq a$ and $0 \leq x_2 \leq b$, shown in Fig. 12–9b. Here continuity of slope and deflection at B requires $\theta_1(a) = -\theta_2(b)$ and $v_1(a) = v_2(b)$. In this particular case, a *negative* sign is necessary to match the slopes at B since x_1 extends positive to the right, whereas x_2 extends positive to the left. Consequently, θ_1 is positive counterclockwise, and θ_2 is positive clockwise. See Fig. 12–8b and 12–8c.

PROCEDURE FOR ANALYSIS

The following procedure provides a method for determining the slope and deflection of a beam (or shaft) using the method of integration. It should be realized that this method is suitable only for *elastic deflections* such that the beam's slope is very small. Furthermore, the method considers *only deflections due to bending*. Additional deflection due to shear generally represents only a few percent of the bending deflection, and so it is usually neglected in engineering practice.

Elastic Curve.   Draw an exaggerated view of the beam's elastic curve. Recall that points of zero slope and zero displacement occur at a fixed support, and zero displacement occurs at all pin and roller supports.

Establish the x and v coordinate axes. The x axis must be parallel to the undeflected beam and can have an origin at any point along the beam, with a positive direction either to the right or to the left. If several discontinuous loads are present, establish x coordinates that are valid for each region of the beam between the discontinuities. Choose these coordinates so that they will simplify subsequent algebraic work. In all cases, the associated positive v axis should be directed upward.

Load or Moment Function.   For each region in which there is an x coordinate, express the loading w or the internal moment M as a function of x. In particular, *always* assume that M acts in the positive direction when applying the equation of moment equilibrium to determine $M = f(x)$.

Slope and Elastic Curve.   Provided EI is constant, apply either the load equation $EI \, d^4v/dx^4 = -w(x)$, which requires four integrations to get $v = v(x)$, or the moment equation $EI \, d^2v/dx^2 = M(x)$, which requires only two integrations. For each integration it is important to include a constant of integration. The constants are evaluated using the boundary conditions for the supports (Table 12-1) and the continuity conditions that apply to slope and displacement at points where two functions meet. Once the constants are evaluated and substituted back into the slope and deflection equations, the slope and displacement at *specific points* on the elastic curve can then be determined. The numerical values obtained can be checked graphically by comparing them with the sketch of the elastic curve. Realize that *positive* values for *slope* are *counterclockwise* if the x axis extends *positive* to the *right,* and *clockwise* if the x axis extends *positive* to the *left.* In either of these cases, *positive displacement* is *upward,* Fig. 12–8.

The following examples illustrate application of this procedure.

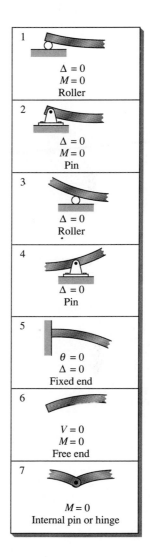

Table 12–1

Example 12–1

The cantilevered beam shown in Fig. 12–10a is subjected to a vertical load **P** at its end. Determine the equation of the elastic curve. *EI* is constant.

SOLUTION I

Elastic Curve.   The load tends to deflect the beam as shown in Fig. 12–10a. By inspection, the internal moment can be represented throughout the beam using a single *x* coordinate.

Moment Function.   From the free-body diagram, with **M** acting in the *positive direction*, Fig. 12–10b, we have

$$M = -Px$$

Slope and Elastic Curve.   Applying Eq. 12–10 and integrating twice yields

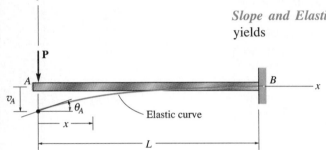

(a)

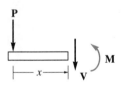

(b)

Fig. 12–10

$$EI \frac{d^2v}{dx^2} = -Px \qquad (1)$$

$$EI \frac{dv}{dx} = -\frac{Px^2}{2} + C_1 \qquad (2)$$

$$EI\, v = -\frac{Px^3}{6} + C_1 x + C_2 \qquad (3)$$

Using the boundary conditions $dv/dx = 0$ at $x = L$ and $v = 0$ at $x = L$, Eqs. 2 and 3 become

$$0 = -\frac{PL^2}{2} + C_1$$

$$0 = -\frac{PL^3}{6} + C_1 L + C_2$$

Thus, $C_1 = PL^2/2$ and $C_2 = -PL^3/3$. Substituting these results into Eqs. 2 and 3 with $\theta = dv/dx$, we get

$$\theta = \frac{P}{2EI}\,(L^2 - x^2)$$

$$v = \frac{P}{6EI}\,(-x^3 + 3L^2x - 2L^3) \qquad \text{Ans.}$$

Maximum slope and displacement occur at $A(x = 0)$, for which

$$\theta_A = \frac{PL^2}{2EI} \qquad (4)$$

$$v_A = -\frac{PL^3}{3EI} \qquad (5)$$

The *positive* result for θ_A indicates *counterclockwise* rotation and the *negative* result for v_A indicates that v_A is *downward*. This agrees with the results sketched in Fig. 12–10a.

In order to obtain some idea as to the actual *magnitude* of the slope and displacement at the end A, consider the beam in Fig. 12–10a to have a length of 15 ft, support a load of $P = 6$ kip, and be made of steel having $E_{st} = 29(10^3)$ ksi. Using the methods of Sec. 11–5, if this beam was designed without a factor of safety by assuming the allowable normal stress is equal to the yield stress $\sigma_{allow} = 36$ ksi, then a $W12 \times 26$ would be found to be adequate ($I = 204$ in^4). From Eqs. 4 and 5 we get

$$\theta_A = \frac{6 \text{ kip}(15 \text{ ft})^2(12 \text{ in./ft})^2}{2[29(10^3) \text{ kip/in}^2](204 \text{ in}^4)} = 0.0164 \text{ rad}$$

$$v_A = -\frac{6 \text{ kip}(15 \text{ ft})^3(12 \text{ in./ft})^3}{3[29(10^3) \text{ kip/in}^2](204 \text{ in}^4)} = -1.97 \text{ in.}$$

Since $\theta_A^2 = 0.000269$ rad$^2 \ll 1$, this justifies the use of Eq. 12–10, rather than applying the more exact Eq. 12–4, for computing the deflection of beams. Also, since this numerical application is for a *cantilevered beam,* we have obtained *larger values* for θ and v than would have been obtained if the beam was supported using pins, rollers, or other fixed supports.

SOLUTION **II**

This problem can also be solved using Eq. 12–8, $EI \, d^4v/dx^4 = -w(x)$. Here $w(x) = 0$ for $0 \leq x \leq L$, Fig. 12–10a, so that upon integrating once we get the form of Eq. 12–9, i.e.,

$$EI \frac{d^4v}{dx^4} = 0$$

$$EI \frac{d^3v}{dx^3} = C_1' = V$$

The shear constant C_1' can be evaluated at $x = 0$, since $V_A = -P$ (negative according to the beam sign convention, Fig. 12–8a.) Thus, $C_1' = -P$. Integrating again yields the form of Eq. 12–10, i.e.,

$$EI \frac{d^3v}{dx^3} - P$$

$$EI \frac{d^2v}{dx^2} = -Px + C_2' = M$$

Here $M = 0$ at $x = 0$, so $C_2' = 0$, and as a result one obtains Eq. 1 and the solution proceeds as before.

Example 12–2

The simply-supported beam shown in Fig. 12–11a supports the triangular distributed loading. Determine its maximum deflection. *EI* is constant.

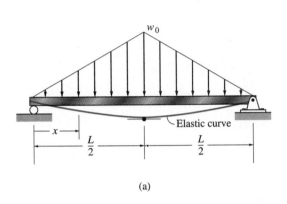

(a)

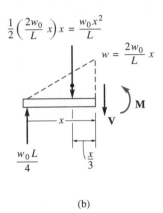

(b)

Fig. 12–11

SOLUTION I

Elastic Curve. Due to symmetry, only one *x* coordinate is needed for the solution. The beam deflects as shown in Fig. 12–11a. Notice that maximum deflection occurs at the center since the slope is zero at this point.

Moment Function. The distributed load acts downward, and therefore it is positive according to our sign convention. A free-body diagram of the segment on the left is shown in Fig. 12–11b. The equation for the distributed loading is

$$w = \frac{2w_0}{L}x$$

Hence,

$$\zeta + \Sigma M_{NA} = 0; \qquad M + \frac{w_0 x^2}{L}\left(\frac{x}{3}\right) - \frac{w_0 L}{4}(x) = 0$$

$$M = \frac{-w_0 x^3}{3L} + \frac{w_0 L}{4}x$$

Slope and Elastic Curve. Using Eq. 12–10 and integrating twice, we have

$$EI \frac{d^2v}{dx^2} = M = -\frac{w_0}{3L}x^3 + \frac{w_0 L}{4}x \qquad (1)$$

$$EI \frac{dv}{dx} = -\frac{w_0}{12L}x^4 + \frac{w_0 L}{8}x^2 + C_1$$

$$EIv = -\frac{w_0}{60L}x^5 + \frac{w_0 L}{24}x^3 + C_1 x + C_2$$

The constants of integration are obtained by applying the boundary condition $v = 0$ at $x = 0$ and the symmetry condition that $dv/dx = 0$ at $x = L/2$. This leads to

$$C_1 = -\frac{5w_0 L^3}{192} \qquad C_2 = 0$$

Hence for $0 \le x \le L/2$,

$$EI \frac{dv}{dx} = -\frac{w_0}{12L}x^4 + \frac{w_0 L}{8}x^2 - \frac{5w_0 L^3}{192}$$

$$EIv = -\frac{w_0}{60L}x^5 + \frac{w_0 L}{24}x^3 - \frac{5w_0 L^3}{192}x$$

Computing the maximum deflection at $x = L/2$, we have

$$v_{max} = -\frac{w_0 L^4}{120EI} \qquad\qquad \text{Ans.}$$

SOLUTION **II**

Starting with the distributed loading, Eq. 1, and applying Eq. 12–8, we have

$$EI \frac{d^4v}{dx^4} = -\frac{2w_0}{L}x$$

$$EI \frac{d^3v}{dx^3} = V = -\frac{w_0}{L}x^2 + C_1'$$

Since $V = +w_0 L/4$ at $x = 0$, then $C_1' = w_0 L/4$. Integrating again yields

$$EI \frac{d^3v}{dx^3} = V = -\frac{w_0}{L}x^2 + \frac{w_0 L}{4}$$

$$EI \frac{d^2v}{dx^2} = M = -\frac{w_0}{3L}x^3 + \frac{w_0 L}{4}x + C_2'$$

Here $M = 0$ at $x = 0$, so $C_2' = 0$. This yields Eq. (1). The solution now proceeds as before.

Example 12–3

The simply-supported beam shown in Fig. 12–12a is subjected to the concentrated force **P**. Determine the maximum deflection of the beam. *EI* is constant.

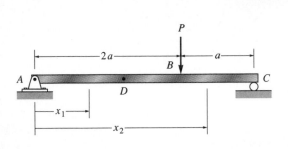

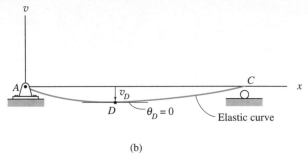

(b)

(a)

SOLUTION

Elastic Curve. The beam deflects as shown in Fig. 12–12b. Two coordinates must be used, since the moment becomes discontinuous at P. Here we will take x_1 and x_2 having the *same origin* at A, so that $0 \le x_1 < 2a$ and $2a < x_2 \le 3a$.

Moment Function. From the free-body diagrams shown in Fig. 12–12c,

$$M_1 = \frac{P}{3}x_1$$

$$M_2 = \frac{P}{3}x_2 - P(x_2 - 2a) = \frac{2P}{3}(3a - x_2)$$

Slope and Elastic Curve. Applying Eq. 12–10 for M_1 and integrating twice yields

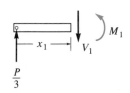

$$EI\frac{d^2v_1}{dx_1^2} = \frac{P}{3}x_1$$

$$EI\frac{dv_1}{dx_1} = \frac{P}{6}x_1^2 + C_1 \tag{1}$$

$$EIv_1 = \frac{P}{18}x_1^3 + C_1x_1 + C_2 \tag{2}$$

Likewise for M_2,

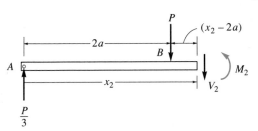

(c)

Fig. 12–12

$$EI\frac{d^2v_2}{dx_2^2} = \frac{2P}{3}(3a - x_2)$$

$$EI\frac{dv_2}{dx_2} = \frac{2P}{3}\left(3ax_2 - \frac{x_2^2}{2}\right) + C_3 \tag{3}$$

$$EIv_2 = \frac{2P}{3}\left(\frac{3}{2}ax_2^2 - \frac{x_2^3}{6}\right) + C_3x_2 + C_4 \tag{4}$$

The four constants are evaluated using *two* boundary conditions, namely, $x_1 = 0$, $v_1 = 0$ (Eq. 2) and $x_2 = 3a$, $v_2 = 0$ (Eq. 4). Also, *two* continuity conditions must be applied at B, that is, $dv_1/dx_1 = dv_2/dx_2$ at $x_1 = x_2 = 2a$ (Eqs. 1 and 3) and $v_1 = v_2$ at $x_1 = x_2 = 2a$ (Eqs. 2 and 4). Substitution as specified results in the following four equations:

$$v_1 = 0 \text{ at } x_1 = 0; \qquad 0 = 0 + 0 + C_2$$

$$v_2 = 0 \text{ at } x_2 = 3a; \qquad 0 = \frac{2P}{3}\left(\frac{3}{2}a(3a)^2 - \frac{(3a)^3}{6}\right) + C_3(3a) + C_4$$

$$\frac{dv_1(2a)}{dx} = \frac{dv_2(2a)}{dx}; \quad \frac{P}{6}(2a)^2 + C_1 = \frac{2P}{3}\left(3a(2a) - \frac{(3a)^2}{2}\right) + C_3$$

$$v_1(2a) = v_2(2a); \qquad \frac{P}{18}(2a)^3 + C_1(2a) + C_2 = \frac{2P}{3}\left(\frac{3}{2}a(2a)^2 - \frac{(2a)^3}{6}\right) + C_3(2a) + C_4$$

Solving these equations we get

$$C_1 = -\frac{4}{9}Pa^2 \qquad C_2 = 0$$

$$C_3 = -\frac{22}{9}Pa^2 \qquad C_4 = \frac{4}{3}Pa^3$$

Thus Eqs. 1–4 become

$$\frac{dv_1}{dx_1} = \frac{P}{6EI}x_1{}^2 - \frac{4}{9}\frac{Pa^2}{EI} \tag{5}$$

$$v_1 = \frac{P}{18EI}x_1{}^3 - \frac{4}{9}\frac{Pa^2}{EI}x_1 \tag{6}$$

$$\frac{dv_2}{dx_2} = \frac{2Pa}{EI}x_2 - \frac{P}{3EI}x_2{}^2 - \frac{22}{9}\frac{Pa^2}{EI} \tag{7}$$

$$v_2 = \frac{Pa}{EI}x_2{}^2 - \frac{P}{9EI}x_2{}^3 - \frac{22}{9}\frac{Pa^2}{EI}x_2 + \frac{4}{3}\frac{Pa^3}{EI} \tag{8}$$

By inspection of the elastic curve, Fig. 12–12*b*, the maximum deflection occurs at D, somewhere within region AB. Here the slope must be zero. From Eq. 5,

$$\frac{1}{6}x_1{}^2 - \frac{4}{9}a^2 = 0$$

$$x_1 = 1.633a$$

Substituting into Eq. 6,

$$v_{\max} = -0.484\frac{Pa^3}{EI} \qquad\qquad Ans.$$

The negative sign indicates that the deflection is downward.

Example 12–4

The beam in Fig. 12–13a is subjected to a load **P** at its end. Determine the displacement at *C*. *EI* is constant.

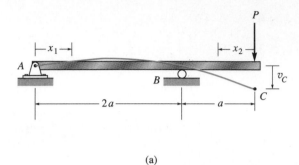

(a)

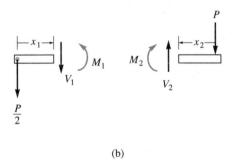

(b)

Fig. 12–13

SOLUTION

Elastic Curve.   The beam deflects into the shape shown in Fig. 12–13a. Due to the loading, two *x* coordinates must be considered. x_2 is directed to the left from *C*, since the internal moment is easy to formulate.

Moment Functions.   Using the free-body diagrams shown in Fig. 12–13b, we have

$$M_1 = -\frac{P}{2}x_1 \qquad M_2 = -Px_2$$

Slope and Elastic Curve.   Applying Eq. 12–10,

for x_1,

$$EI \frac{d^2v_1}{dx_1{}^2} = -\frac{P}{2}x_1$$

$$EI \frac{dv_1}{dx_1} = -\frac{P}{4}x_1{}^2 + C_1 \tag{1}$$

$$EIv_1 = -\frac{P}{12}x_1{}^3 + C_1 x + C_2 \tag{2}$$

For x_2,

$$EI \frac{d^2v_2}{dx_2{}^2} = -Px_2$$

$$EI \frac{dv_2}{dx_2} = -\frac{P}{2}x_2{}^2 + C_3 \tag{3}$$

$$EIv_2 = -\frac{P}{6}x_2{}^3 + C_3 x_2 + C_4 \tag{4}$$

The *four* constants of integration are determined using *three* boundary conditions, namely, $v_1 = 0$ at $x_1 = 0$, $v_1 = 0$ at $x_1 = 2a$, and $v_2 = 0$ at $x_2 = a$ and *one* continuity equation. Here the continuity of slope at the roller requires $dv_1/dx_1 = -dv_2/dx_2$ at $x_1 = 2a$ and $x_2 = a$. Why is there a negative sign in this equation? Note that continuity of displacement at B has been indirectly considered in the boundary conditions, since $v_1 = v_2 = 0$ at $x_1 = 2a$ and $x_2 = a$. Applying these four conditions yields

$v_1 = 0$ at $x_1 = 0$; $0 = 0 + 0 + C_2$

$v_1 = 0$ at $x_1 = 2a$; $0 = -\dfrac{P}{12}(2a)^3 + C_1(2a) + C_2$

$v_2 = 0$ at $x_2 = a$; $0 = -\dfrac{P}{6}a^3 + C_3 a + C_4$

$\dfrac{dv_1(2a)}{dx_1} = -\dfrac{dv_2(a)}{dx_2}$; $-\dfrac{P}{4}(2a)^2 + C_1 = -\left(-\dfrac{P}{2}(a)^2 + C_3\right)$

Solving, we obtain

$$C_1 = \frac{Pa^2}{3} \qquad C_2 = 0 \qquad C_3 = \frac{7}{6}Pa^2 \qquad C_4 = -Pa^3$$

Substituting C_3 and C_4 into Eq. 4 gives

$$v_2 = -\frac{P}{6EI}x_2{}^3 + \frac{7Pa^2}{6EI}x_2 - \frac{Pa^3}{EI}$$

The displacement at C is determined by setting $x_2 = 0$. We get

$$v_C = -\frac{Pa^3}{EI} \qquad\qquad\qquad \textit{Ans.}$$

PROBLEMS

12–1.   A steel strap having a thickness of 0.125 in. and a width of 2 in. is bent into a circular arc of radius $\rho = 600$ in. Determine the maximum bending stress in the strap. $E_{st} = 29(10^3)$ ksi.

12–2.   The steel blade of the band saw wraps around the pulley having a radius of 12 in. Determine the maximum normal stress in the blade. The blade is made of steel having a width of 0.75 in. and a thickness of 0.0625 in. $E_{st} = 29(10^3)$ ksi.

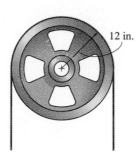

Prob. 12–2

12–3.   Determine the equation of the elastic curve for the beam using the x coordinate that is valid for $0 < x < L/2$. Specify the slope at A and the maximum deflection. EI is constant.

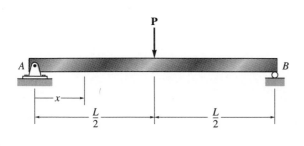

Prob. 12–3

*12–4.   The shaft is supported at A by a journal bearing that exerts only vertical reactions on the shaft, and at B by a thrust bearing that exerts horizontal and vertical reactions on the shaft. Draw the bending-moment diagram for the shaft and then, from this diagram, sketch the deflection or elastic curve for the shaft's centerline. Determine the equations of the elastic curve using the coordinates x_1 and x_2.

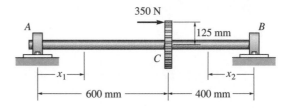

Prob. 12–4

12–5.   The steel beam has a depth of 10 in. and is subjected to a constant moment $\mathbf{M_0}$, which causes the stress at the outer fibers to become $\sigma_Y = 36$ ksi. Determine the radius of curvature of the beam and the deflection at its end B. $E_{st} = 29(10^3)$ ksi. *Comment:* Notice how *small* the maximum elastic deflection actually is.

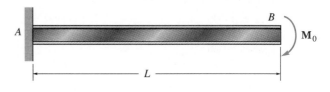

Prob. 12–5

12–6.   Determine the equations of the elastic curve for the beam using the x_1 and x_2 coordinates. Specify the slope at A and the maximum deflection. EI is constant.

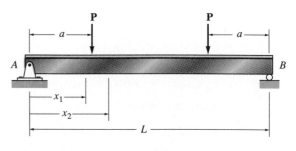

Prob. 12–6

12–7. The shaft is supported at A by a journal bearing that exerts only vertical reactions on the shaft, and at B by a thrust bearing that exerts horizontal and vertical reactions on the shaft. Draw the bending-moment diagram for the shaft and then, from this diagram, sketch the deflection or elastic curve for the shaft's centerline. Determine the equations of the elastic curve using the coordinates x_1 and x_2.

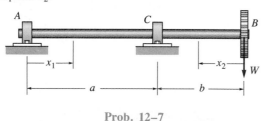

Prob. 12–7

12–10. Determine the elastic curve for the cantilevered beam, which is subjected to the couple moment M_0. Also compute the maximum slope and maximum deflection of the beam. EI is constant.

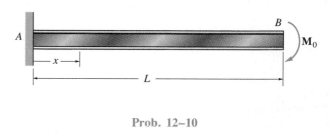

Prob. 12–10

*12–8. The shaft is supported at A by a journal bearing that exerts only vertical reactions on the shaft, and at B by a thrust bearing that exerts both horizontal and vertical reactions on the shaft. Draw the bending-moment diagram for the shaft and then, from this diagram, sketch the deflection or elastic curve for the shaft's centerline. Determine the equations of the elastic curve using the coordinates x_1 and x_2. EI is constant.

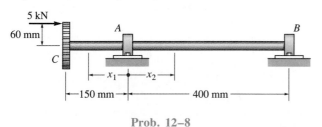

Prob. 12–8

12–11. Determine the elastic curve for the simply-supported beam, which is subjected to a couple moment M_0. Also, compute the slope at each end A and B and the maximum deflection of the beam. EI is constant.

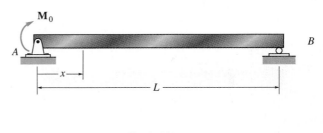

Prob. 12–11

12–9. The shaft is supported by a journal bearing at A, which exerts only vertical reactions on the shaft, and by a thrust bearing at B, which exerts both horizontal and vertical reactions on the shaft. Draw the bending-moment diagram for the shaft and then, from this diagram, sketch the deflection or elastic curve for the shaft's centerline. Determine the equations of the elastic curve using the coordinates x_1 and x_2. EI is constant.

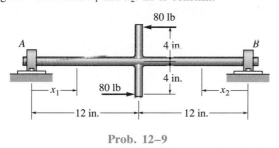

Prob. 12–9

*12–12. Determine the elastic curve for the simply-supported beam, which is subjected to the couple moments M_0. Also, compute the maximum slope and the maximum deflection of the beam. EI is constant.

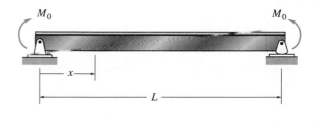

Prob. 12–12

12–13. Determine the equations of the elastic curve for the beam using the x_1 and x_2 coordinates. Specify the slope at A and the deflection at C. EI is constant.

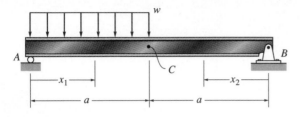

Prob. 12–13

12–14. Determine the equations of the elastic curve for the beam using the x_1 and x_2 coordinates. Specify the slope at A and the maximum deflection. EI is constant.

12–15. Determine the maximum deflection between the supports A and B. EI is constant.

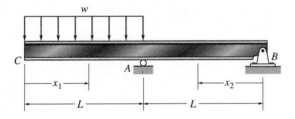

Probs. 12–14/12–15

*12–16. Determine the equations of the elastic curve using the coordinate x and specify the deflection and slope at point C. EI is constant.

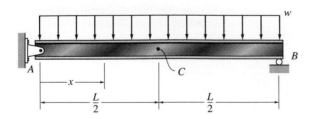

Prob. 12–16

12–17. The floor beam of the airplane is subjected to the loading shown. Assuming that the fuselage exerts only vertical reactions on the ends of the beam, determine the maximum deflection of the beam. EI is constant.

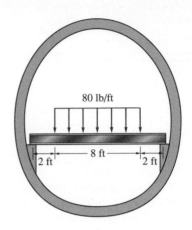

Prob. 12–17

12–18. Determine the equations of the elastic curve for the beam using the x coordinate. Specify the slope and deflection at A. EI is constant.

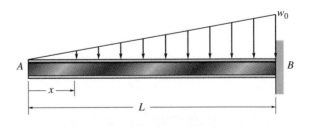

Prob. 12–18

12–19. Wooden posts used for a retaining wall have a diameter of 3 in. If the soil pressure along a post varies uniformly from zero at the top A to a maximum of 300 lb/ft at the bottom B, determine the slope and deflection at the top of the post. $E_w = 1.6(10^3)$ ksi.

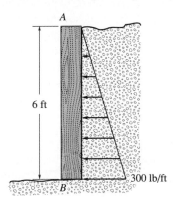

Prob. 12–19

*12–20. At what distance a should the bearing supports at A and B be placed so that the deflection at the center of the shaft is equal to the deflection at its ends? EI is constant.

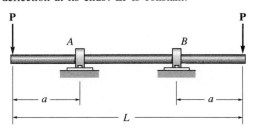

Prob. 12–20

12–21. The two wooden meter sticks are separated at their centers by a smooth rigid cylinder having a diameter of 50 mm. Determine the force $\mathbf{F}$ that must be applied at each end in order to just make their ends touch. Each stick has a width of 20 mm and a thickness of 5 mm. $E_w = 11$ GPa.

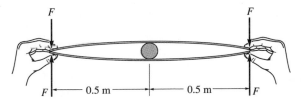

Prob. 12–21

12–22. The fence board weaves between the three smooth fixed posts, each of which has a diameter of 3 in. If the posts remain along the same line, determine the maximum bending stress in the board. The board has a width of 6 in. and a thickness of 0.5 in. $E_w = 1.60(10^3)$ ksi. Assume the deflection of each end of the board relative to its center is 3 in.

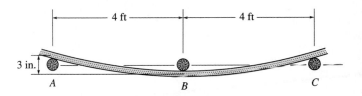

Prob. 12–22

12–23. The tapered beam has a rectangular cross section. Determine the deflection of its end in terms of the load P, length L, modulus of elasticity E, and the moment of inertia I_0 of its end A.

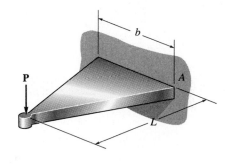

Prob. 12–23

*12–24. The tapered beam has a rectangular cross section. Determine the deflection of its center in terms of the load P, length L, modulus of elasticity E, and the moment of inertia I_c of its center.

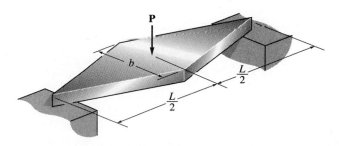

Prob. 12–24

*12.3 Discontinuity Functions

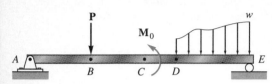

Fig. 12–14

The method of integration, to find the equation of the elastic curve for a beam or shaft, is convenient if the internal moment can be expressed as a *continuous function* throughout the beam's *entire length*. If several loadings act on the beam, however, the method becomes more tedious to apply, because separate moment functions must be written for each region between these loadings. Furthermore, integration of these moment functions requires the evaluation of integration constants using boundary conditions and/or continuity conditions. For example, the beam shown in Fig. 12–14 requires *four* moment functions to be written. They describe the moment in regions *AB, BC, CD,* and *DE.* Applying the moment-curvature relationship, $EI\, d^2v/dx^2 = M$, and integrating each moment equation twice, requires the evaluation of eight constants of integration. These are evaluated using *two* boundary conditions that require zero displacement at points *A* and *E,* and *six* continuity conditions of both slope and displacement at points *B, C,* and *D.*

In this section we will discuss a method for finding the equation of the elastic curve for a multiply-loaded beam using a *single expression* to define the beam's internal moment. When this expression is substituted into the moment-curvature relationship and integrated twice, the *two* constants of integration will be determined *only* from the *boundary conditions.* Since the continuity equations are not involved, the analysis will be greatly simplified.

In order to express the internal moment in the beam as a single function, we will use a special mathematical operator known as a *discontinuity function,* which for purposes of beam deflection may be written as

$$\langle x - a \rangle^n = \begin{cases} 0 & \text{for } x < a \\ (x - a)^n & \text{for } x \geq a \end{cases} \qquad (12\text{–}11)$$
$$n \geq 0$$

Here x represents the coordinate position of a point along the beam and a is the location on the beam where a "discontinuity" occurs, such as the point of application of an external force or couple moment or the start of a distributed loading. Note that the discontinuity function $\langle x - a \rangle^n$ is written with angle brackets to distinguish it from the ordinary function $(x - a)^n$, written with parentheses. As stated in Eq. 12–11, only when $x \geq a$ is $\langle x - a \rangle^n = (x - a)^n$, otherwise it is zero. In particular, integration of the discontinuity function follows the same rules as for ordinary functions, i.e.,

$$\int \langle x - a \rangle^n \, dx = \frac{\langle x - a \rangle^{n+1}}{n + 1} + C \qquad (12\text{–}12)$$

Discontinuity functions can be developed for various types of loadings applied to a beam. We will now consider four that are often encountered in engineering practice.

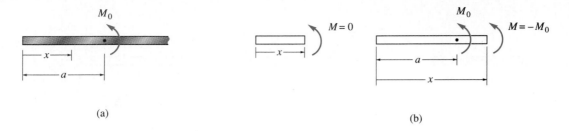

Fig. 12–15

Couple Moment M_0.

When an *external* couple moment $\mathbf{M_0}$ is applied to a beam at $x = a$, Fig. 12–15a, the *internal moment M* in the beam can be represented by the discontinuity function

$$M = -M_0\langle x - a \rangle^0 \qquad (12\text{–}13)$$

This equation states that when $x < a$, $M = 0$, and when $x \geq a$, $M = -M_0(x - a)^0 = -M_0$. The validity of these values can be checked by using the method of sections, shown in Fig. 12–15b. Note that application of the equation assumes that the external *couple moment* $\mathbf{M_0}$ is *positive counterclockwise*.

Concentrated Force P.

If a concentrated force $\mathbf{P}$ is applied to the beam at $x = a$, Fig. 12–16a, then the discontinuity function for the internal moment in the beam is

$$M = -P\langle x - a \rangle \qquad (12\text{–}14)$$

Here the force $\mathbf{P}$ is *positive* when it acts *downward*. Equation 12–14 states that when $x < a, M = 0$, and when $x \geq a, M = -P(x - a)$. Indeed this is the case, as shown by the method of sections, Fig. 12–16b.

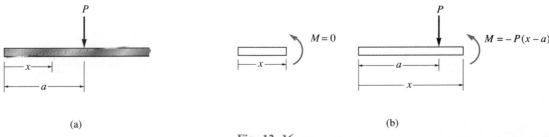

Fig. 12–16

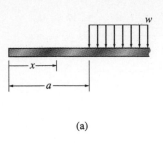

(a)

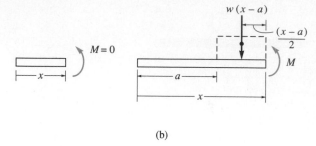

(b)

Fig. 12–17

Uniform Distributed Loading, w. If a uniform distributed loading acts on the beam, starting at $x = a$ and extending all the way to the right end of the beam, Fig. 12–17a, then the discontinuity function that describes the internal moment is

$$M = -\frac{1}{2}w\langle x - a\rangle^2 \qquad (12\text{–}15)$$

This equation states that when $x < a$, $M = 0$, and when $x \ge a$, $M = -\frac{1}{2}w(x - a)^2$. These values may be checked using the method of sections as shown in Fig. 12–17b. Here w_0 *is positive* when it acts *downward* on the beam.

Linear Distributed Loading. If the load on the beam is triangular and has a slope m, Fig. 12–18a, then the internal moment in the beam is defined by the discontinuity function

$$M = -\frac{1}{6}m\langle x - a\rangle^3 \qquad (12\text{–}16)$$

Here the values $M = 0$ for $x < a$ and $M = -\frac{1}{6}m(x - a)^3$ for $x \ge a$ can be verified by the method of sections, Fig. 12–18b. As above, the *distributed loading* is *positive* when it acts *downward* on the beam.

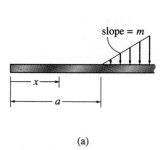

(a)

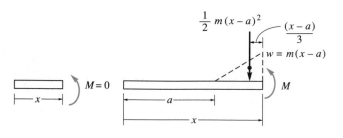

Fig. 12–18

(b)

This analysis can be extended to distributed loadings that have other forms. Also, it is possible to use superposition with the uniform and triangular loadings to create the discontinuity function for a trapezoidal loading.

Application of Eqs. 12–13 through 12–16 provides a rather direct means for writing the internal moment in a beam as a function of x. When doing so, close attention must be paid to the signs of the *external loadings*. As stated above, *concentrated forces* and *distributed loads* are *positive downward,* and *couple moments* are *positive counterclockwise*. If this sign convention is followed, then the *internal moment* **M** is in accordance with the beam sign convention established in Sec. 12.2; i.e., positive moment bends the beam concave upward.

As an example of how to apply the above functions, consider the beam loaded as shown in Fig. 12–19*a*. Here the reactive force **R**$_1$ created by the pin, Fig. 12–19*b*, is negative since it acts upward, and **M**$_0$ is negative since it acts clockwise. Applying Eqs. 12–13 through 12–15, the internal moment at *any point* x in the beam, Fig. 12–19*a*, is therefore

$$M = R_1\langle x - 0\rangle - P\langle x - a\rangle + M_0\langle x - b\rangle^0 - \frac{1}{2}w\langle x - c\rangle^2 \qquad (12\text{–}17)$$

The validity of this expression may be checked by using the method of sections, say, within the region $b < x < c$, Fig. 12–19*b*. Moment equilibrium requires that

$$M = R_1 x - P(x - a) + M_0$$

This result agrees with that of Eq. 12–17, since by Eq. 12–11 only the last term in Eq. 12–17 is zero when $x < c$.

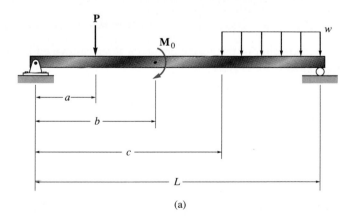

(a)

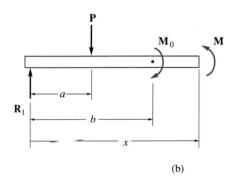

(b)

Fig. 12–19

PROCEDURE FOR ANALYSIS

The following procedure provides a method for using discontinuity functions to determine a beam's elastic curve. This method is particularly advantageous for solving problems involving beams or shafts subjected to *several loadings,* since the two constants of integration can be evaluated by using *only* the boundary conditions.

Elastic Curve.   Sketch the beam's elastic curve and identify the boundary conditions at the supports. Recall that zero displacement occurs at all pin and roller supports, and zero slope and zero displacement occur at fixed supports. Establish the x axis so that it extends to the right and has its origin at the beam's left end.

Moment Function.   Calculate the support reactions and then use the discontinuity functions to express the internal moment M as a function of x. When doing so, make sure to follow the sign convention for each loading as it applies for Eqs. 12–13 through 12–16. Also, note that the distributed loadings must extend all the way to the beam's right end for Eqs. 12–15 and 12–16 to be valid. If this does not occur, use the method of superposition, which is illustrated in Example 12–5.

Slope and Elastic Curve.   Substitute M into the moment-curvature relation $EI \, d^2v/dx^2 = M$, and integrate twice to obtain the equations for the beam's slope and deflection. Evaluate the two constants of integration using the boundary conditions and substitute the constants into the slope and deflection equations to obtain the final results. When these functions are evaluated at any point on the beam, a *positive slope* is *counterclockwise,* and a *positive displacement* is *upward.*

The following examples illustrate application of this procedure.

Example 12–5

Determine the equation of the elastic curve for the cantilevered beam shown in Fig. 12–20a. *EI* is constant.

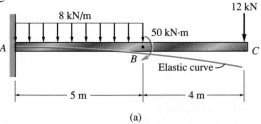

Fig. 12–20(a)

(a)

SOLUTION

Elastic Curve.   The loads cause the beam to deflect as shown in Fig. 12–20a. The boundary conditions require zero slope and displacement at A.

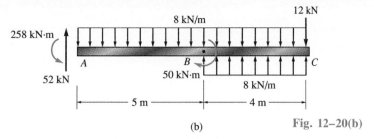

(b)   **Fig. 12–20(b)**

Moment Function.   The support reactions at A have been calculated by statics and are shown on the free-body diagram in Fig. 12–20b. Since the distributed loading in Fig. 12–20a does not extend to C as required, we can use the superposition of loadings shown in Fig. 12–20b to represent the same effect. By our sign convention, the 50-kN · m couple moment, the 52-kN force at A, and the portion of distributed loading from B to C on the bottom of the beam are all negative. With reference to Fig. 12–20b, applying Eqs. 12–13 through 12–15, the beam's internal moment is therefore

$$M = -258\langle x - 0\rangle^0 - (-52)\langle x - 0\rangle - \frac{1}{2}(8)\langle x - 0\rangle^2$$

$$-\frac{1}{2}(-8)\langle x - 5\rangle^2 - (-50)\langle x - 5\rangle^0$$

$$= -258 + 52x - 4x^2 + 4\langle x - 5\rangle^2 + 50\langle x - 5\rangle^0$$

The moment of the 12-kN load at C is *not included* here, since x cannot be greater than 9 m.

Slope and Elastic Curve.   Applying Eq. 12–10 and integrating twice, using Eq. 12–12, we have

$$EI\,\frac{d^2v}{dx^2} = -258 + 52x - 4x^2 + 4\langle x - 5\rangle^2 + 50\langle x-5\rangle^0$$

$$EI\,\frac{dv}{dx} = -258x + 26x^2 - \frac{4}{3}x^3 + \frac{4}{3}\langle x - 5\rangle^3 + 50\langle x - 5\rangle + C_1$$

$$EIv = -129x^2 + \frac{26}{3}x^3 - \frac{1}{3}x^4 + \frac{1}{3}\langle x - 5\rangle^4 + 25\langle x - 5\rangle^2 + C_1x + C_2$$

Since $dv/dx = 0$ at $x = 0$, $C_1 = 0$; and $v = 0$ at $x = 0$, so $C_2 = 0$. Thus,

$$v = \frac{1}{EI}\left(-129x^2 + \frac{26}{3}x^3 - \frac{1}{3}x^4 + \frac{1}{3}\langle x - 5\rangle^4 + 25\langle x - 5\rangle^2\right)\ Ans.$$

Note: If we *did not* use discontinuity functions, two x coordinates would be needed for regions AB and BC, and the problem would require evaluating two more constants of integration involving the continuity of slope and displacement at point B.

Example 12–6

Determine the maximum deflection of the beam shown in Fig. 12–21a. EI is constant.

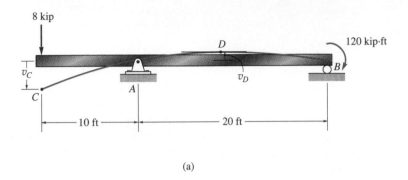

(a)

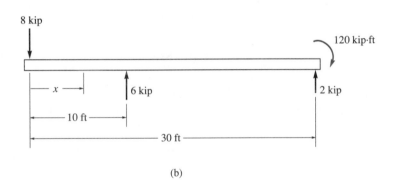

(b)

Fig. 12–21

SOLUTION

Elastic Curve.   The beam deflects as shown in Fig. 12–21a. The boundary conditions require zero displacement at A and B.

Moment Function.   The reactions have been calculated and are shown on the free-body diagram in Fig. 12–21b. Using Eqs. 12–13 and 12–14, we have

$$M = -8\langle x - 0\rangle - (-6)\langle x - 10\rangle$$
$$= -8x + 6\langle x - 10\rangle$$

The couple moment and force at B are not included here, since they are located at the right end of the beam, and x cannot be greater than 30 ft.

Slope and Elastic Curve.   Integrating Eq. 12–10 twice yields

$$EI \frac{d^2v}{dx^2} = -8x + 6\langle x - 10 \rangle$$

$$EI \frac{dv}{dx} = -4x^2 + 3\langle x - 10 \rangle^2 + C_1$$

$$EIv = -\frac{4}{3}x^3 + \langle x - 10 \rangle^3 + C_1 x + C_2 \qquad (1)$$

From Eq. 1, the boundary condition $v = 0$ at $x = 10$ ft and $v = 0$ at $x = 30$ ft gives

$$0 = -1333 + (10 - 10)^3 + C_1(10) + C_2$$
$$0 = -36,000 + (30 - 10)^3 + C_1(30) + C_2$$

Solving these equations simultaneously for C_1 and C_2, we get $C_1 = 1333$ and $C_2 = -12,000$. Thus,

$$EI \frac{dv}{dx} = -4x^2 + 3\langle x - 10 \rangle^2 + 1333 \qquad (2)$$

$$EIv = -\frac{4}{3}x^3 + \langle x - 10 \rangle^3 + 1333x - 12,000 \qquad (3)$$

From Fig. 12–21a, maximum displacement may occur at either C or D, where the slope $dv/dx = 0$. To obtain the displacement of C, set $x = 0$ in Eq. 3. We get

$$v_C = -\frac{12,000 \text{ kip} \cdot \text{ft}^3}{EI}$$

The *negative* sign indicates that the displacement is *downward* as shown in Fig. 12–21a. To locate point D, use Eq. 2 with $x > 10$. This gives

$$0 = -4x_D^2 + 3(x_D - 10)^2 + 1333$$
$$x_D^2 + 60x_D - 1633 = 0$$

Solving for the positive root,

$$x_D = 20.3 \text{ ft}$$

Hence, from Eq. 3,

$$EIv_D - -\frac{4}{3}(20.3)^3 + (20.3 - 10)^3 + 1333(20.3) - 12,000$$

$$v_D = \frac{29,000 \text{ kip} \cdot \text{ft}^3}{EI} \qquad \qquad \textit{Ans.}$$

Comparing this value with v_C, we see that $v_{max} = v_D$.

PROBLEMS

12–25. The beam is subjected to the load shown. Using discontinuity functions, determine the equation of the elastic curve. EI is constant.

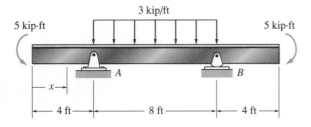

Prob. 12–25

12–26. The beam is subjected to the load shown. Using discontinuity functions, determine the equations of the slope and elastic curve. EI is constant.

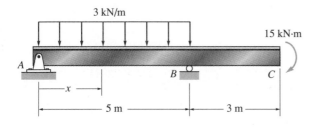

Prob. 12–26

12–27. The beam is subjected to the load shown. Using discontinuity functions, determine the equation of the elastic curve. EI is constant.

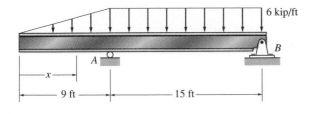

Prob. 12–27

*12–28. The beam is subjected to the load shown. Using discontinuity functions, determine the equation of the elastic curve. EI is constant.

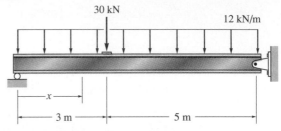

Prob. 12–28

12–29. The wooden beam is subjected to the load shown. Using discontinuity functions, determine the equation of the elastic curve. If $E_w = 12$ GPa, determine the deflection and the slope at end B.

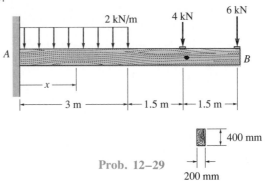

Prob. 12–29

12–30. The shaft supports the two pulley loads shown. Using discontinuity functions, determine the equation of the elastic curve. The bearings at A and B exert only vertical reactions on the shaft. EI is constant.

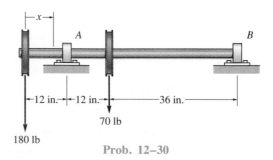

Prob. 12–30

12–31. The beam is subjected to the load shown. Using discontinuity functions, determine the equation of the elastic curve. EI is constant.

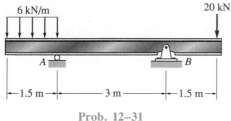

Prob. 12–31

*12–32. The shaft supports the two pulley loads shown. Using discontinuity functions, determine the equation of the elastic curve. The bearings at A and B exert only vertical reactions on the shaft. EI is constant.

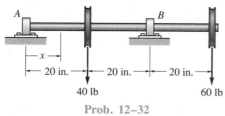

Prob. 12–32

12–33. The beam is subjected to the load shown. Using discontinuity functions, determine the equation of the elastic curve. EI is constant.

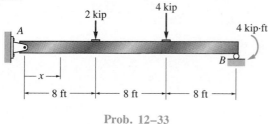

Prob. 12–33

12–34. The wooden beam is subjected to the load shown. Using discontinuity functions, determine the equation of the elastic curve. Specify the deflection at the end C. $E_w = 1.6(10^3)$ ksi.

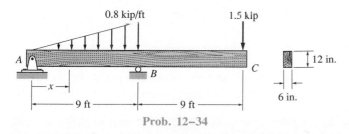

Prob. 12–34

12–35. Determine the slope at B and the deflection at C for the $W10 \times 45$. Solve the problem using discontinuity functions. $E_{st} = 29(10^3)$ ksi.

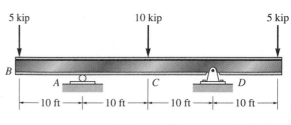

Prob. 12–35

*12–36. Using discontinuity functions, determine the deflection at each of the pulleys C, D, and E. The shaft is made of steel and has a diameter of 30 mm. The bearings at A and B exert only vertical reactions on the shaft. $E_{st} = 200$ GPa.

12–37. Using discontinuity functions, determine the slope of the shaft at the bearings at A and B. The shaft is made of steel and has a diameter of 30 mm. The bearings at A and B exert only vertical reactions on the shaft. $E_{st} = 200$ GPa.

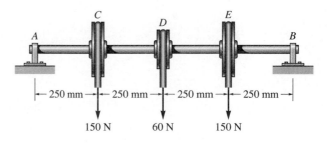

Probs. 12–36/12–37

12–38. The shaft is made of steel and has a diameter of 15 mm. Using discontinuity functions, determine its maximum deflection. The bearings at A and B exert only vertical reactions on the shaft. $E_{st} = 200$ GPa.

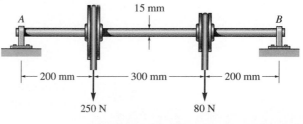

Prob. 12–38

*12.4 Slope and Displacement by the Moment-Area Method

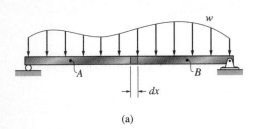

(a)

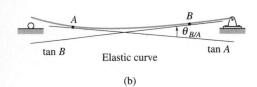

Elastic curve

(b)

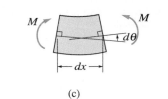

(c)

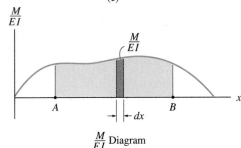

$\frac{M}{EI}$ Diagram

(d)

Fig. 12–22

The moment-area method provides a semigraphical technique for finding the slope and displacement at *specific points* on the elastic curve of a beam or shaft. Application of the method requires computing areas associated with the beam's moment diagram; so if this diagram consists of simple shapes, the method is very convenient to use. Normally this is the case when the beam is loaded with concentrated forces or couple moments or regions of the beam have different moments of inertia.

To develop the moment-area method we will make the same assumptions as we used for the method of integration: The beam is initially straight, it is elastically deformed by the loads, such that the slope and deflection of the elastic curve are very small, and the deformations are caused by bending. Before we can apply the moment-area method we must first develop two theorems. The first theorem provides a means of obtaining the slope of the elastic curve, and the second theorem provides the means of obtaining the displacement of a point.

Theorem 1. To develop the first moment-area theorem, consider the simply-supported beam shown in Fig. 12–22a with its associated elastic curve, Fig. 12–22b. A differential segment dx of the beam is isolated in Fig. 12–22c. It is seen that the beam's internal moment M deforms the element such that the *tangents* to the elastic curve at each side of the element intersect at an angle $d\theta$. This angle can be determined from Eq. 12–10, written as

$$EI \frac{d^2v}{dx^2} = EI \frac{d}{dx}\left(\frac{dv}{dx}\right) = M$$

Since the *slope* is *small*, $\theta = dv/dx$, and therefore

$$d\theta = \frac{M}{EI}\, dx \tag{12–18}$$

If the moment diagram for the beam is constructed and divided by both the beam's moment of inertia I and modulus of elasticity E, Fig. 12–22d, Eq. 12–18 indicates that $d\theta$ is equal to the *area* under the "M/EI diagram" for the beam segment dx. Integrating from a selected point A on the elastic curve to another point B, Fig. 12–22b, we have

$$\theta_{B/A} = \int_A^B \frac{M}{EI}\, dx \tag{12–19}$$

This equation forms the basis for the first moment-area theorem.

Theorem 1: *The angle between the tangents at any two points on the elastic curve equals the area under the M/EI diagram between these two points.*

The notation $\theta_{B/A}$ is referred to as the angle of the tangent at B measured *with respect to* the tangent at A. From the proof it should be evident that this angle is measured *counterclockwise,* from tangent A to tangent B, if the area under the M/EI diagram is *positive,* Fig. 12–22d. Conversely, if the area is *negative,* or lies below the x axis, the angle $\theta_{B/A}$ is measured clockwise from tangent A to tangent B. Furthermore, from the dimensions of Eq. 12–19, $\theta_{B/A}$ will be *measured* in *radians.*

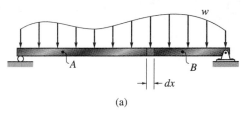

(a)

Theorem 2. The second moment-area theorem is based on the relative deviation of tangents to the elastic curve. Shown in Fig. 12–23b is a greatly exaggerated view of the vertical deviation dt of the tangents on each side of the differential element dx. This deviation is caused by the curvature of the element and has been measured along a vertical line passing through point A located on the elastic curve. Since the slope of the elastic curve and its deflection are assumed to be very small, it is satisfactory to approximate the length of each tangent line by x and the arc ds' by dt. Using the circular-arc formula $s = \theta r$, where r is the length x and s is dt, we can write $dt = x\ d\theta$. Substituting Eq. 12–18 into this equation and integrating from A to B, the vertical deviation of the tangent at A *with respect to* the tangent at B can then be determined; i.e.,

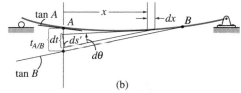

(b)

$$t_{A/B} = \int_A^B x\ \frac{M}{EI}\ dx \qquad (12\text{–}20)$$

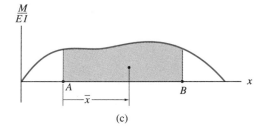

(c)

Since the centroid of an area is found from $\bar{x} \int dA = \int x\ dA$, and $\int (M/EI)dx$ represents the area under the M/EI diagram, we can also write

$$t_{A/B} = \bar{x} \int_A^B \frac{M}{EI}\ dx \qquad (12\text{–}21)$$

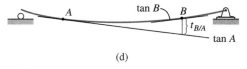

(d)

Here $\bar{x}$ is the distance from the vertical axis through A to the *centroid* of the area under the M/EI diagram between A and B, Fig. 12–23c.

The second moment-area theorem can now be stated as follows in reference to Fig. 12–23b.

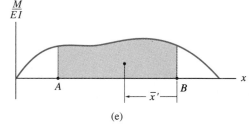

(e)

Theorem 2: *The vertical deviation of the tangent at a point (A) on the elastic curve with respect to the tangent extended from another point (B) equals the moment of the area under the M/EI diagram between the two points (A and B). This moment is computed about a vertical axis passing through the point (A) where the vertical deviation ($t_{A/B}$) is to be determined.*

Fig. 12–23

The distance $t_{A/B}$ used in the theorem can also be interpreted as the vertical displacement from the point located on the extended tangent drawn from B to the point A on the elastic curve. Note that $t_{A/B}$ is *not* equal to $t_{B/A}$, which is shown in Fig. 12–23d. Specifically, the moment of the area under the M/EI diagram between A and B is computed about a vertical axis through point A to determine $t_{A/B}$, Fig. 12–23c, and it is computed about a vertical axis through point B to determine $t_{B/A}$, Fig. 12–23e.

If the moment of a *positive M/EI* area from *A* to *B* is computed for $t_{A/B}$, as in Fig. 12–23c, it indicates that point *A* is *above* the tangent extended from point *B*, Fig. 12–23b. Similarly, *negative M/EI* areas indicate that point *A* is *below* the tangent extended from point *B*.

PROCEDURE FOR ANALYSIS

The following procedure provides a method that may be used to determine the slope and displacement at a specific point on a beam (or shaft) using the two moment-area theorems.

M/EI Diagram.   Determine the support reactions and draw the beam's *M/EI* diagram. If the beam is loaded with concentrated forces, the *M/EI* diagram will consist of a series of straight line segments, and the areas and their moments required for the moment-area theorems will be relatively easy to compute. If the loading consists of a series of distributed loads, the *M/EI* diagram will consist of parabolic or perhaps higher-order curves, and it is suggested that the table on the inside front cover be used to locate the area and centroid under each curve.

Elastic Curve.   Draw an exaggerated view of the beam's elastic curve. Recall that points of zero slope and zero displacement always occur at a fixed support, and zero displacement occurs at all pin and roller supports. If it becomes difficult to draw the general shape of the elastic curve, use the moment (or *M/EI*) diagram. Realize that when the beam is subjected to a *positive moment*, the beam bends *concave up*, whereas *negative moment* bends the beam *concave down*, Fig. 12–2. Furthermore, an inflection point or change in curvature occurs where the moment in the beam (or *M/EI*) is zero.

The unknown displacement and slope to be determined should be indicated on the curve. Since the moment-area theorems apply *only between two tangents*, attention should be given as to which tangents should be constructed on the curve so that the angles or deviations between them will lead to the solution of the problem. In this regard, *the tangents at the supports should be considered*, since the beam usually has zero displacement and/or zero slope at the supports.

Moment-Area Theorems.   Apply Theorem 1 to determine the *angle* between any two tangents on the elastic curve and Theorem 2 to determine the *tangential deviation*. The algebraic sign of the answer can be checked from the angle or deviation indicated on the elastic curve. Specifically, a *positive* $\theta_{B/A}$ represents a *counterclockwise* rotation of the tangent at *B* with respect to the tangent at *A*, and a *positive* $t_{A/B}$ indicates that point *A* on the elastic curve lies *above* the extended tangent from point *B*.

The following examples illustrate application of this procedure.

Example 12–7

Determine the slope of the beam shown in Fig. 12–24a at points B and C. EI is constant.

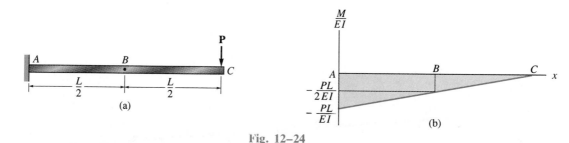

Fig. 12–24

SOLUTION

M/EI Diagram.   This diagram is shown in Fig. 12–24b.

Elastic Curve.   The force **P** causes the beam to deflect as shown in Fig. 12–24c. (The elastic curve is concave downward, since M/EI is negative.) The tangents at B and C are indicated since we are required to find θ_B and θ_C. Also, the tangent at the support (A) is shown. This tangent has a *known* zero slope. By the construction, the angle between tan A and tan B, that is, $\theta_{B/A}$, is equivalent to θ_B, or

$$\theta_B = \theta_{B/A}$$

Also

$$\theta_C = \theta_{C/A}$$

Moment-Area Theorem.   Applying Theorem 1, $\theta_{B/A}$ is equal to the area under the M/EI diagram between points A and $B;$ that is,

$$\theta_B = \theta_{B/A} = \left(-\frac{PL}{2EI}\right)\left(\frac{L}{2}\right) + \frac{1}{2}\left(-\frac{PL}{2EI}\right)\left(\frac{L}{2}\right)$$

$$= -\frac{3PL^2}{8EI} \qquad\qquad Ans.$$

The *negative sign* indicates that the angle measured from the tangent at A to the tangent at B is *clockwise*. Also, the beam slopes downward at A.

In a similar manner, the area under the M/EI diagram between points A and C equals $\theta_{C/A}$. We have

$$\theta_C = \theta_{C/A} = \frac{1}{2}\left(-\frac{PL}{EI}\right)L$$

$$= -\frac{PL^2}{2EI} \qquad\qquad Ans.$$

Example 12-8

Determine the displacement of points B and C of the beam shown in Fig. 12–25a. EI is constant.

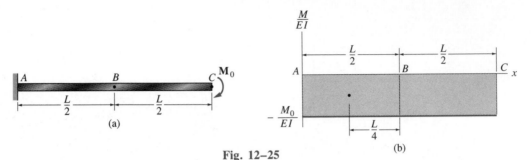

Fig. 12–25

SOLUTION

M/EI Diagram. See Fig. 12–25b.

Elastic Curve. The couple moment at C causes the beam to deflect as shown in Fig. 12–25c. The tangents at B and C are indicated since we are required to find Δ_B and Δ_C. Also, the tangent at the support (A) is shown since it is horizontal. The required displacements can now be related directly to the deviations between the tangents at B and A and C and A. Specifically, Δ_B is equal to the deviation of tan A from tan B; that is,

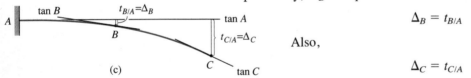

$$\Delta_B = t_{B/A}$$

Also,

$$\Delta_C = t_{C/A}$$

Moment-Area Theorem. Applying Theorem 2, $t_{B/A}$ is equal to the moment of the shaded area under the M/EI diagram between A and B computed about a vertical axis passing through point B (the point on the elastic curve), since this is the point where the tangential deviation is to be determined. Hence, from Fig. 12–25b,

$$\Delta_B = t_{B/A} = \left(\frac{L}{4}\right)\left[\left(-\frac{M_0}{EI}\right)\left(\frac{L}{2}\right)\right] = -\frac{M_0 L^2}{8EI} \qquad Ans.$$

Likewise, for $t_{C/A}$ we must compute the moment of the area under the *entire M/EI* diagram from A to C about a vertical axis passing through point C (the point on the elastic curve). We have

$$\Delta_C = t_{C/A} = \left(\frac{L}{2}\right)\left[\left(-\frac{M_0}{EI}\right)(L)\right] = -\frac{M_0 L^2}{2EI} \qquad Ans.$$

Since both answers are *negative*, they indicate that points B and C lie *below* the tangent at A. This checks with Fig. 12–25c.

Example 12–9

Determine the slope at point C of the beam in Fig. 12–26a. EI is constant.

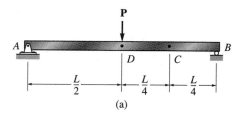

(a)

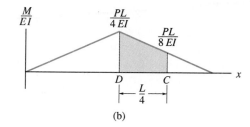

(b)

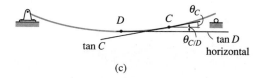

(c)

Fig. 12–26

SOLUTION

M/EI Diagram.   See Fig. 12–26b.

Elastic Curve.   Since the loading is applied symmetrically to the beam, the elastic curve is symmetric and the tangent at D is horizontal as shown in Fig. 12–26c. Tangents are drawn at C, since we must find the slope θ_C, and at D. By the construction, the angle $\theta_{C/D}$, between tan D and tan C, is equal to θ_C, that is,

$$\theta_C = \theta_{C/D}$$

Moment-Area Theorem.   Using Theorem 1, $\theta_{C/D}$ is equal to the shaded area under the M/EI diagram between points D and C. We have

$$\theta_C = \theta_{C/D} = \left(\frac{PL}{8EI}\right)\left(\frac{L}{4}\right) + \frac{1}{2}\left(\frac{PL}{4EI} - \frac{PL}{8EI}\right)\left(\frac{L}{4}\right) = \frac{3PL^2}{64EI} \qquad Ans.$$

What does the positive result indicate?

Example 12–10

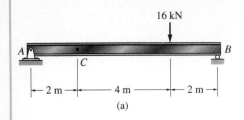

16 kN

A B

C

2 m   4 m   2 m

(a)

Determine the slope at point C for the steel beam in Fig. 12–27a. Take $E_{st} = 200$ GPa, $I = 17(10^6)$ mm^4.

SOLUTION

M/EI Diagram.   See Fig. 12–27b.

Elastic Curve.   The elastic curve is shown in Fig. 12–27c. The tangent at C is shown since we are required to find θ_C. To do this, tangents at the *supports, A and B,* are also constructed as shown. Angle $\theta_{C/A}$ is the angle between the tangents at A and C. The slope at A, θ_A, in Fig. 12–27c can be found using $|\theta_A| = |t_{B/A}|/L_{AB}$. This equation is valid since $t_{B/A}$ is actually very small, so that θ_A in radians can be approximated by the length of a circular arc defined by a radius of $L_{AB} = 8$ m and a sweep of θ_A. (Recall that $s = \theta r$.) From the geometry of Fig. 12–27c, we have

$$|\theta_C| = |\theta_A| - |\theta_{C/A}| = \left|\frac{t_{B/A}}{8}\right| - |\theta_{C/A}| \qquad (1)$$

$\dfrac{M}{EI}$

$\dfrac{8}{EI}$ $\dfrac{24}{EI}$

A B x

C

2 m   4 m   2 m

(b)

Moment-Area Theorems.   Using Theorem 1, $\theta_{C/A}$ is equivalent to the area under the M/EI diagram between points A and C; that is,

$$\theta_{C/A} = \frac{1}{2}(2)\left(\frac{8}{EI}\right) = \frac{8}{EI}$$

Applying Theorem 2, $t_{B/A}$ is equivalent to the moment of the area under the M/EI diagram between B and A about a vertical axis passing through point B (the point on the elastic curve), since this is the point where the tangential deviation is to be determined. We have

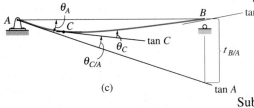

θ_A

A B tan B

C

θ_C tan C

$\theta_{C/A}$ $t_{B/A}$

(c)   tan A

$$t_{B/A} = \left(2 + \frac{1}{3}(6)\right)\left[\frac{1}{2}(6)\left(\frac{24}{EI}\right)\right] + \left(\frac{2}{3}(2)\right)\left[\frac{1}{2}(2)\left(\frac{24}{EI}\right)\right]$$

$$= \frac{320}{EI}$$

Substituting these results into Eq. 1, we get

$$\theta_C = \frac{320}{8EI} - \frac{8}{EI} = \frac{32}{EI} \; \downarrow$$

We have computed this result in units of kN and m, so converting EI into these units, we have

Fig. 12–27

$$\theta_C = \frac{32 \text{ kN} \cdot \text{m}^2}{200(10^6) \text{ kN/m}^2 \; 17(10^{-6}) \text{ m}^4} = 0.00941 \text{ rad} \; \downarrow \qquad \textit{Ans.}$$

Example 12–11

Determine the displacement at C for the beam shown in Fig. 12–28a. EI is constant.

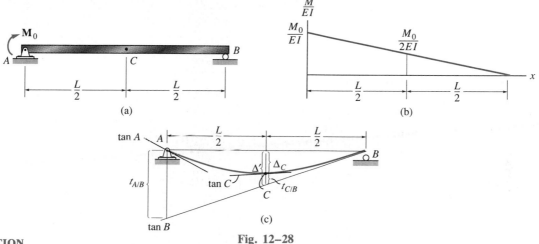

(a)

(b)

(c)

Fig. 12–28

SOLUTION

M/EI Diagram. See Fig. 12–28b.

Elastic Curve. The tangent at C is drawn on the elastic curve since we are required to find Δ_C, Fig. 12–28c. Note that C is *not* the location of the maximum deflection of the beam, because the loading and hence the elastic curve are *not symmetric*. Also indicated in Fig. 12–28c are the tangents at the supports A and B. It is seen that $\Delta_C = \Delta' - t_{C/B}$. If $t_{A/B}$ is determined, then Δ' can be found from proportional triangles, that is, $\Delta'/(L/2) = t_{A/B}/L$ or $\Delta' = t_{A/B}/2$. Hence,

$$\Delta_C = \frac{t_{A/B}}{2} - t_{C/B} \qquad (1)$$

Moment-Area Theorem. Applying Theorem 2 to determine $t_{A/B}$ and $t_{C/B}$, we have

$$t_{A/B} = \left(\frac{1}{3}(L)\right)\left[\frac{1}{2}(L)\left(\frac{M_0}{EI}\right)\right] = \frac{M_0 L^2}{6EI}$$

$$t_{C/B} = \left(\frac{1}{3}\left(\frac{L}{2}\right)\right)\left[\frac{1}{2}\left(\frac{L}{2}\right)\left(\frac{M_0}{2EI}\right)\right] = \frac{M_0 L^2}{48EI}$$

Substituting these results into Eq. 1 gives

$$\Delta_C = \frac{1}{2}\left(\frac{M_0 L^2}{6EI}\right) - \left(\frac{M_0 L^2}{48EI}\right)$$

$$= \frac{M_0 L^2}{16EI} \downarrow \qquad\qquad\qquad \textit{Ans.}$$

Example 12–12

Determine the displacement at point C for the steel overhanging beam shown in Fig. 12–29a. Take $E_{st} = 29(10^3)$ ksi, $I = 125$ in^4.

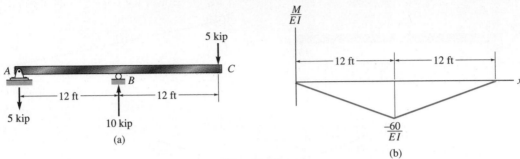

(a)

(b)

Fig. 12–29

SOLUTION

M/EI Diagram. See Fig. 12–29b.

Elastic Curve. The loading causes the beam to deflect as shown in Fig. 12–29c. We are required to find Δ_C. By constructing tangents at C and at the supports A and B, it is seen that $\Delta_C = |t_{C/A}| - \Delta'$. However, Δ' can be related to $t_{B/A}$ by proportional triangles; that is, $\Delta'/24 = |t_{B/A}|/12$ or $\Delta' = 2|t_{B/A}|$. Hence

$$\Delta_C = |t_{C/A}| - 2|t_{B/A}| \tag{1}$$

Moment-Area Theorem. Applying Theorem 2 to determine $t_{C/A}$ and $t_{B/A}$, we have

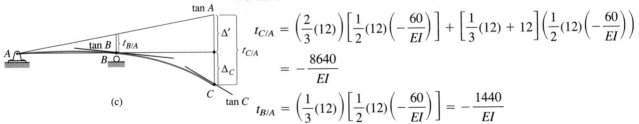

$$t_{C/A} = \left(\frac{2}{3}(12)\right)\left[\frac{1}{2}(12)\left(-\frac{60}{EI}\right)\right] + \left[\frac{1}{3}(12) + 12\right]\left(\frac{1}{2}(12)\left(-\frac{60}{EI}\right)\right)$$

$$= -\frac{8640}{EI}$$

$$t_{B/A} = \left(\frac{1}{3}(12)\right)\left[\frac{1}{2}(12)\left(-\frac{60}{EI}\right)\right] = -\frac{1440}{EI}$$

Why are these terms negative? Substituting the results into Eq. 1 yields

$$\Delta_C = \frac{8640}{EI} - 2\left(\frac{1440}{EI}\right) = \frac{5760}{EI} \;\downarrow$$

Realizing that the computations were made in units of kip and ft, we have

$$\Delta_C = \frac{5760 \text{ kip} \cdot \text{ft}^3 \ (1728 \text{ in}^3/\text{ft}^3)}{[29(10^3) \text{ kip/in}^2](125 \text{ in}^4)} = 2.75 \text{ in.} \;\downarrow \qquad \textit{Ans.}$$

PROBLEMS

12–39. Determine the slope and deflection at *B*. *EI* is constant.

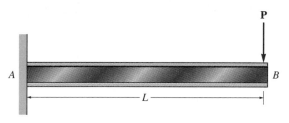

Prob. 12–39

*12–40. Determine the slope and deflection at *C*. *EI* is constant.

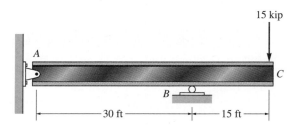

Prob. 12–40

12–41. If the bearings at *A* and *B* exert only vertical reactions on the shaft, determine the slope at *B* and the deflection at *C*. *EI* is constant.

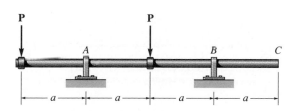

Prob. 12–41

12–42. If the bearings at *A* and *B* exert only vertical reactions on the shaft, determine the slope at *B* and the deflection at *C*. *EI* is constant.

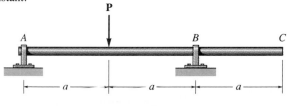

Prob. 12–42

12–43. The composite simply-supported steel shaft is subjected to a force of 10 kN at its center. Determine its maximum deflection. $E_{st} = 200$ GPa.

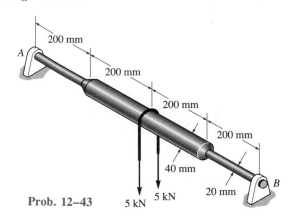

Prob. 12–43

*12–44. Determine the value of *a* so that the slope at *A* is equal to zero. *EI* is constant.

12–45. Determine the value of *a* so that the deflection at *C* is equal to zero. *EI* is constant.

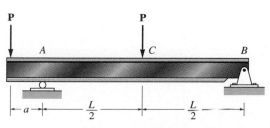

Probs. 12–44/12–45

12–46. Determine the deflection and slope at C. EI is constant.

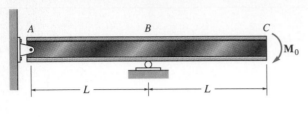

Prob. 12–46

12–47. Determine the maximum deflection of the beam and the slope at A. EI is constant.

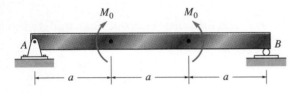

Prob. 12–47

*12–48. Determine the slope at B and the deflection at C. EI is constant.

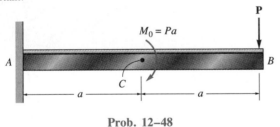

Prob. 12–48

12–49. The beam is subjected to the loading shown. Determine the slope at C and deflection at B. EI is constant.

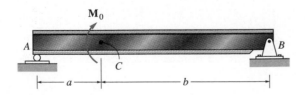

Prob. 12–49

12–50. If the bearings at A and B exert only vertical reactions on the shaft, determine the slope at A and the maximum deflection.

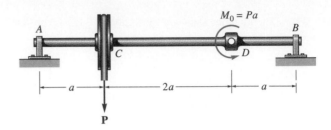

Prob. 12–50

12–51. Determine the slope at B and deflection at C. EI is constant.

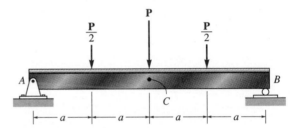

Prob. 12–51

*12–52. The two steel bars have a thickness of 1 in. and a width of 4 in. They are designed to act as a spring for the machine, which exerts a force of 4 kip on them at A and B. If the supports exert only vertical forces on the bars, determine the maximum deflection of the bottom bar. $E_{st} = 29(10^3)$ ksi.

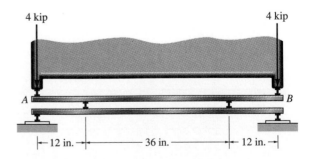

Prob. 12–52

12–53. The flat spring is made of steel and has a rectangular cross section as shown. Determine the maximum elastic load P that can be applied. What is the deflection at B when P reaches its maximum value? Assume that the spring is fixed-supported at A. $E_{st} = 29(10^3)$ ksi, $\sigma_Y = 36$ ksi.

Prob. 12–53

12–54. The bar is supported by a roller constraint at B, which allows vertical displacement but resists axial load and moment. If the bar is subjected to the loading shown, determine the slope at A and the deflection at C. EI is constant.

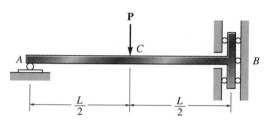

Prob. 12–54

12–55. The two force components act on the tire of the automobile as shown. The tire is fixed to the axle, which is supported by bearings at A and B. Draw the moment diagram of the axle and determine its maximum deflection. Assume that the bearings resist only vertical loads. The thrust on the axle is resisted at C. The axle has a diameter of 1.25 in. and is made of steel for which $E_{st} = 29(10^3)$ ksi. Neglect the effect of axial load on deflection.

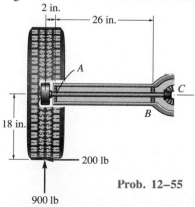

Prob. 12–55

*12–56. Determine the maximum deflection of the shaft. EI is constant.

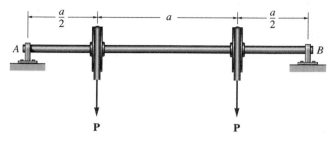

Prob. 12–56

12–57. Determine the slope at C and deflection at B. EI is constant.

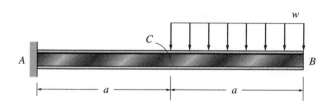

Prob. 12–57

12–58. The shaft is used to support a rotor that exerts a uniform load of 5 kN/m within the region CD of the shaft. Determine the slope of the shaft at the bearings A and B. The bearings exert only vertical reactions on the shaft. $E_{st} = 200$ GPa.

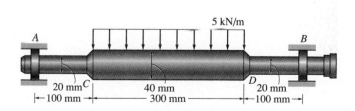

Prob. 12–58

12–59. The beam is subjected to the loading shown. Determine the slope at B and deflection at C. EI is constant.

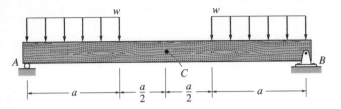

Prob. 12–59

12–61. Determine the maximum deflection of the beam. EI is constant.

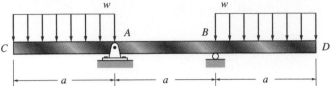

Prob. 12–61

*12–60. Determine the slope at A and the deflection at C. EI is constant.

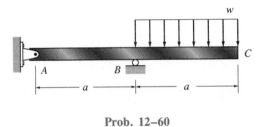

Prob. 12–60

12–62. The two bars are pin-connected at D. Determine the slope at A and the deflection at D. EI is constant.

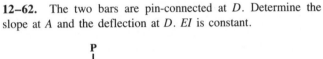

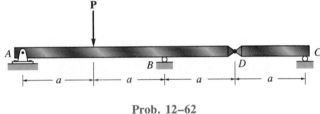

Prob. 12–62

12.5 Method of Superposition

The differential equation $EId^4v/dx^4 = -w(x)$ satisfies the two necessary requirements for applying the principle of superposition as stated in Sec. 4.3. To be specific, the load $w(x)$ is linearly related to the deflection $v(x)$, and the load is assumed not to change significantly the original geometry of the beam or shaft, since the slope is considered very small for this equation to apply. As a result, the deflections for a series of separate loadings acting on a beam may be superimposed. For example, if v_1 is the deflection for one load and v_2 is the deflection for another load, the total deflection for both loads acting together is the algebraic sum $v_1 + v_2$. Using tabulated results for various beam loadings, such as the ones listed on the inside back cover of this book, or those found in various engineering handbooks, it is therefore possible to find the slope and displacement at a point on a beam once the actual loading has been represented by its various component parts.

The following examples illustrate how to use the method of superposition to solve deflection problems, where the deflection is caused not only by beam deformations, but also by rigid-body displacements, which can occur when the beam is supported by springs or portions of a segmented beam are supported by hinges.

Example 12–13

Determine the displacement at point C and the slope at the support A of the beam shown in Fig. 12–30a. EI is constant.

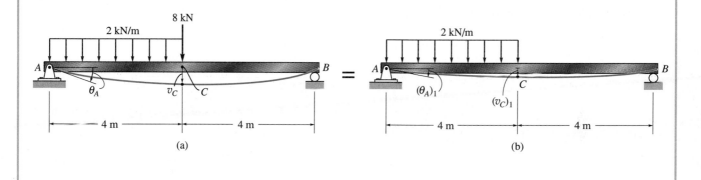

(a)   (b)

$+$

SOLUTION

The loading can be separated into two component parts as shown in Fig. 12–30b and 12–30c. The displacement at C and slope at A are computed using the table on the inside back cover for each part.

For the distributed loading, Fig. 12–30b,

$$(\theta_A)_1 = \frac{9wL^3}{384EI} = \frac{9(2 \text{ kN/m})(8 \text{ m})^3}{384EI} = \frac{24}{EI} \downarrow$$

$$(v_C)_1 = \frac{5wL^4}{768EI} = \frac{5(2 \text{ kN/m})(8 \text{ m})^4}{768EI} = \frac{53.33}{EI} \downarrow$$

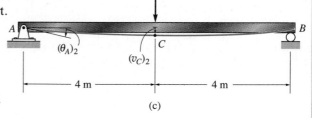

(c)

For the 8-kN concentrated force, Fig. 12–30c,

$$(\theta_A)_2 = \frac{PL^2}{16EI} = \frac{8 \text{ kN}(8 \text{ m})^2}{16EI} = \frac{32}{EI} \downarrow$$

$$(v_C)_2 = \frac{PL^3}{48EI} = \frac{8 \text{ kN}(8 \text{ m})^3}{48EI} = \frac{85.33}{EI} \downarrow$$

The total displacement at C and the slope at A, Fig. 12–30a, are the algebraic sums of these components. Hence

$(+\downarrow)$
$$\theta_A = (\theta_A)_1 + (\theta_A)_2 = \frac{56}{EI} \downarrow \qquad Ans.$$

$(+\downarrow)$
$$v_C = (v_C)_1 + (v_C)_2 = \frac{139}{EI} \downarrow \qquad Ans.$$

Fig. 12–30

Example 12–14

Determine the displacement at the end C of the overhanging beam shown in Fig. 12–31a. EI is constant.

SOLUTION

Since the table on the back cover *does not* include beams with overhangs, the beam will be separated into a simply-supported and a cantilevered portion. First we will calculate the slope at B, as caused by the distributed load acting on the simply-supported span, Fig. 12–31b.

$$(\theta_B)_1 = \frac{wL^3}{24EI} = \frac{5 \text{ kN/m}(4 \text{ m})^3}{24EI} = \frac{13.33}{EI} \nwarrow$$

Since this angle is *small*, $\theta_B \approx \tan \theta_B$, and the vertical displacement at point C is

$$(v_C)_1 = (2 \text{ m})\left(\frac{13.33}{EI}\right) = \frac{26.67}{EI} \uparrow$$

Next, the 10-kN load on the overhang causes a statically equivalent force of 10 kN and couple moment of 20 kN · m at the support B of the simply-supported span, Fig. 12–31c. The 10-kN force does not cause a displacement or slope at B; however, the 20-kN · m couple moment does. The slope at B due to this moment is

$$(\theta_B)_2 = \frac{M_0 L}{3EI} = \frac{20 \text{ kN} \cdot \text{m}(4 \text{ m})}{3EI} = \frac{26.67}{EI} \downarrow$$

So that the extended point C is displaced

$$(v_C)_2 = (2 \text{ m})\left(\frac{26.7}{EI}\right) = \frac{53.33}{EI} \downarrow$$

Finally, the cantilevered portion BC is displaced by the 10-kN force, Fig. 12–31d. We have

$$(v_C)_3 = \frac{PL^3}{3EI} = \frac{10 \text{ kN}(2 \text{ m})^3}{3EI} = \frac{26.67}{EI} \downarrow$$

Summing these results algebraically, we obtain the final displacement of point C.

$$(+\downarrow) \qquad v_C = -\frac{26.7}{EI} + \frac{53.3}{EI} + \frac{26.7}{EI} = \frac{53.3}{EI} \downarrow \qquad \textit{Ans.}$$

(a)
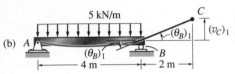

||

(b)
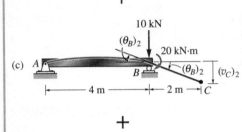

+

(c)

+

(d)

Fig. 12–31

Example 12–15

Determine the displacement at the end C of the cantilever beam shown in Fig. 12–32. EI is constant.

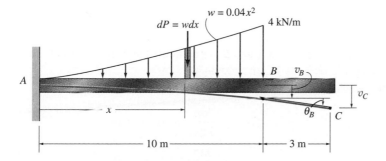

Fig. 12–32

SOLUTION

Distributed loadings that are parabolic are not included in the table on the back cover. In order to solve this problem we can consider the load as an infinite series of concentrated forces dP, and then integrate this result over the region where the loading acts. Specifically, the differential force $dP = wdx = 0.04x^2\,dx$ acts at a distance x from A, Fig. 12–32. Using the table for a single concentrated force dP, the slope and displacement at point B are therefore

$$\theta_B = \int \frac{dP x^2}{2EI} = \int_0^{10} \frac{0.04x^2(x^2)}{2EI}\,dx = \frac{400}{EI}$$

$$v_B = \int \frac{dP x^2}{6EI}(3L - x) = \int_0^{10} \frac{(0.04x^2)x^2(3(10) - x)}{6EI}\,dx = \frac{2888.9}{EI}$$

The unloaded region BC of the beam remains straight, as shown in Fig. 12–32. Since θ_B is small, the displacement at C becomes

$(+\downarrow)$
$$v_C = v_B + \theta_B\,(3\text{ m})$$

$$= \frac{2888.9}{EI} + \frac{400}{EI}(3) = \frac{4089}{EI}\ \downarrow \qquad\qquad \textit{Ans.}$$

Example 12–16

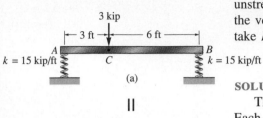

3 kip

3 ft | 6 ft

A   B
k = 15 kip/ft   C   k = 15 kip/ft

(a)

∥

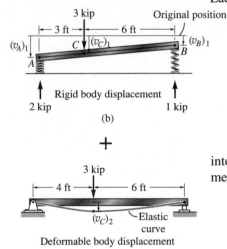

3 kip   Original position

3 ft | 6 ft

$(v_A)_1$ C $(v_C)_1$ $(v_B)_1$
A   B

Rigid body displacement

2 kip   1 kip

(b)

+

3 kip

4 ft | 6 ft

$(v_C)_2$ Elastic curve

Deformable body displacement

(c)

Fig. 12–33

The steel bar shown in Fig. 12–33a is supported by two springs at its ends A and B. Each spring has a stiffness of $k = 15$ kip/ft and is originally unstretched. If the bar is loaded with a force of 3 kip at point C, determine the vertical displacement of the force. Neglect the weight of the bar and take $E_{st} = 29(10^3)$ ksi, $I = 12$ in⁴.

SOLUTION

The end reactions at A and B are computed and shown in Fig. 12–33b. Each spring deflects by an amount

$$(v_A)_1 = \frac{2 \text{ kip}}{15 \text{ kip/ft}} = 0.1333 \text{ ft}$$

$$(v_B)_1 = \frac{1 \text{ kip}}{15 \text{ kip/ft}} = 0.0667 \text{ ft}$$

If the bar is considered to be rigid, these displacements cause it to move into the position shown in Fig. 12–33b. For this case, the vertical displacement at C is

$$(v_C)_1 = (v_B)_1 + \frac{6 \text{ ft}}{9 \text{ ft}} [(v_A)_1 - (v_B)_1]$$

$$= 0.0667 \text{ ft} + \frac{2}{3} [0.1333 \text{ ft} - 0.0667 \text{ ft}] = 0.1111 \text{ ft} \downarrow$$

We can compute the displacement at C caused by the *deformation* of the bar, Fig. 12–33c, by using the table on the inside back cover. We have

$$(v_C)_2 = \frac{Pab}{6EIL} (L^2 - b^2 - a^2)$$

$$= \frac{3 \text{ kip}(6 \text{ ft})(3 \text{ ft})[(9 \text{ ft})^2 - (6 \text{ ft})^2 - (3 \text{ ft})^2]}{6(9 \text{ ft}) [29(10^3)] \text{ kip/in}^2(144 \text{ in}^2/1 \text{ ft}^2) \, 12 \text{ in}^4(1 \text{ ft}^4/20{,}736 \text{ in}^4)}$$

$$= 0.0149 \text{ ft} \downarrow$$

Adding the two displacement components, we get

$$(+\downarrow) \qquad v_C = 0.1111 \text{ ft} + 0.0149 \text{ ft} = 0.126 \text{ ft} = 1.51 \text{ in.} \downarrow \qquad \textit{Ans.}$$

PROBLEMS

12–63. The $W8 \times 48$ steel cantilevered beam is subjected to the loading shown. Using the method of superposition, determine the deflection at its end A. $E_{st} = 29(10^3)$ ksi.

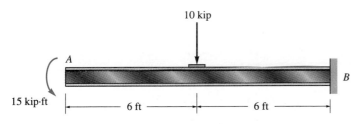

Prob. 12–63

*12–64. The $W12 \times 45$ simply-supported beam is made of steel and is subjected to the loading shown. Using the method of super-position, determine the deflection at its center C. $E_{st} = 29(10^3)$ ksi.

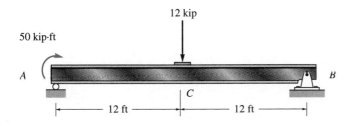

Prob. 12–64

12–65. The wide-flange beam acts as a cantilever. Due to an error it is installed at an angle θ with the vertical. Determine the ratio of its deflection in the x direction to its deflection in the y direction at A when a load $\mathbf{P}$ is applied at this point. The moments of inertia are I_x and I_y. For the solution, resolve $\mathbf{P}$ into components and use the method of superposition. *Note:* The result indicates that large lateral deflections (x direction) can occur in narrow beams, $I_y \ll I_x$, when they are improperly installed in this manner. To show this numerically, compute the deflections in the x and y directions for a $W10 \times 15$, with $P = 1.5$ kip, $\theta = 10°$, and $L = 12$ ft. $E_{st} = 29(10^3)$ ksi.

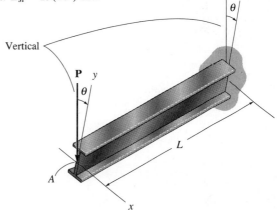

Prob. 12–65

12–66. The $W10 \times 30$ steel cantilevered beam is subjected to unsymmetrical bending caused by the applied moment. Determine the deflection of the centroid at its end A due to the loading. $E_{st} = 29(10^3)$ ksi. *Hint:* Resolve the moment into components and use superposition.

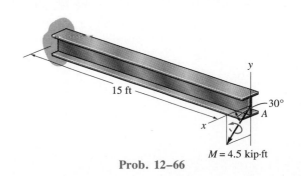

Prob. 12–66

12–67. The $W8 \times 24$ simply-supported beam is subjected to the loading shown. Using the method of superposition, determine the deflection at its center C. The beam is made of steel for which $E_{st} = 29(10^3)$ ksi.

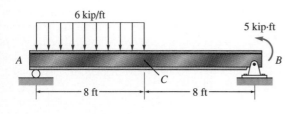

Prob. 12–67

12–69. The shaft for an electric motor and generator supports the weights of the rotor, the armature, and the commutator. Using the method of superposition, determine the deflection at C of the shaft due to these loadings. The bearings exert vertical forces on the shaft at A and B. EI is constant.

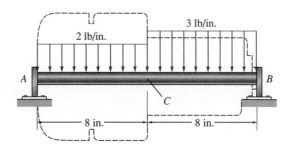

Probs. 12–69

*12–68. Using the method of superposition, determine the magnitude of $\mathbf{M}_0$ in terms of the distributed load w and dimension a so that the deflection at the center of the beam is zero. EI is constant.

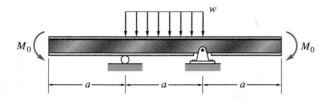

Prob. 12–68

12–70. The shaft for an electric motor and generator supports the weights of the rotor, the armature, and the commutator. Using the method of superposition, determine the slope of the shaft at the bearings at A and B. The bearings exert vertical forces on the shaft at A and B. EI is constant.

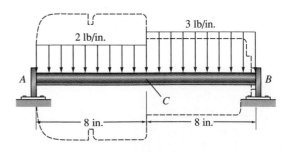

12–70

12–71. Using the method of superposition, determine the vertical deflection at the end A of the bracket. Assume that the bracket is fixed-supported at its base B. EI is constant.

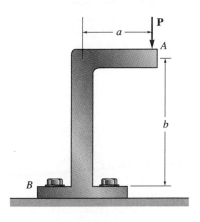

Prob. 12–71

*12–72. Using the method of superposition, determine the vertical deflection and slope at the end A of the bracket. Assume that the bracket is fixed-supported at its base, and neglect the axial deformation of segment AB. EI is constant.

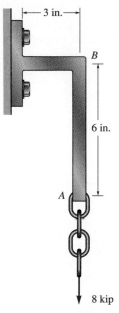

Prob. 12–72

12–73. The pipe assembly consists of three equal-sized pipes with flexibility stiffness EI and torsional stiffness GJ. Using the method of superposition, determine the vertical deflection at point A.

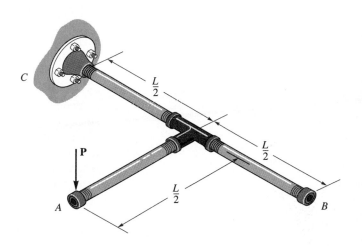

Prob. 12–73

12–74. The framework consists of two cantilevered beams CD and BA and a simply-supported beam CB. If each beam is made of steel and has a moment of inertia about its principal axis of $I_x = 118 \text{ in}^4$, determine the deflection at the center G of beam CB. $E_{st} = 29(10^3)$ ksi. Use the method of superposition.

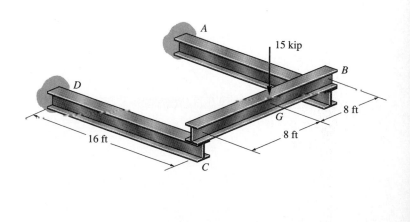

Prob. 12–74

12.6 Statically Indeterminate Beams and Shafts

The analysis of statically indeterminate axially loaded bars and torsionally loaded shafts has been discussed in Sec. 4.4 and Sec. 5.5, respectively. In this section we will illustrate a general method for determining the reactions on statically indeterminate beams and shafts. Recall that a member of any type is classified as *statically indeterminate* if the number of unknown reactions *exceeds* the available number of equilibrium equations.

The additional support reactions on the beam or shaft that are *not needed* to keep it in stable equilibrium are called *redundants*. The number of these redundants is referred to as the *degree of indeterminacy*. For example, consider the beam shown in Fig. 12–34a. If the free-body diagram is drawn, Fig. 12–34b, there will be four unknown support reactions, and since three equilibrium equations are available for solution, the beam is classified as being indeterminate to the first degree. Either A_y, B_y, or M_A can be classified as the redundant, for if any one of these reactions is removed, the beam remains stable and in equilibrium. (A_x cannot be classified as the redundant, for if it were removed, $\Sigma F_x = 0$ would not be satisfied.) In a similar manner, the *continuous beam* in Fig. 12–35a is indeterminate to the second degree, since there are five unknown reactions and only three available equilibrium equations, Fig. 12–35b. Here the two redundant support reactions can be chosen among A_y, B_y, C_y, and D_y.

To determine the reactions on a beam (or shaft) that is statically indeterminate, it is first necessary to specify the redundant reactions. We can determine these redundants from conditions of geometry known as *compatibility conditions*. Once determined, the redundants are then applied to the beam, and the remaining reactions are determined from the equations of equilibrium.

In the following sections we will illustrate the procedure for solution using the method of integration, Sec. 12.7; the moment-area method, Sec. 12.8; and the method of superposition, Sec. 12.9.

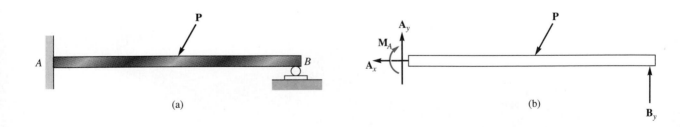

(a)

(b)

Fig. 12–34

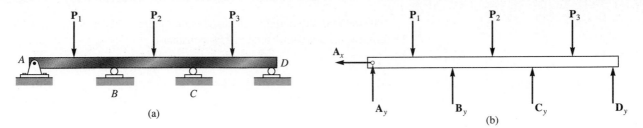

Fig. 12–35

12.7 Statically Indeterminate Beams and Shafts—Method of Integration

The method of integration, discussed in Sec. 12.2, requires two integrations of the differential equation $d^2v/dx^2 = M/EI$ once the internal moment M in the beam is expressed as a function of position x. If the beam is statically indeterminate, however, M can also be expressed in terms of the *unknown* redundants. After integrating this equation twice, there will be two constants of integration and the redundants to be determined. Although this is the case, these unknowns can always be found from the boundary and/or continuity conditions for the problem. For example, the beam in Fig. 12–36 has one redundant. Once it is chosen, the other reactions can be written in terms of the redundant using the equations of equilibrium. Expressing the internal moment M in terms of the redundant, and integrating the moment–displacement relationship, we can then determine the two constants of integration and the redundant from the *three* boundary conditions $v = 0$ at $x = 0$, $dv/dx = 0$ at $x = 0$, and $v = 0$ at $x = L$.

The following example problems illustrate specific applications of this method using the procedure for analysis outlined in Sec. 12.2.

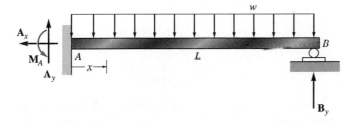

Fig. 12–36

Example 12–17

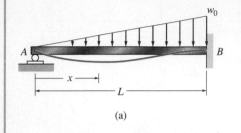

(a)

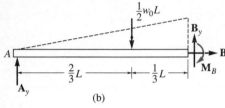

(b)

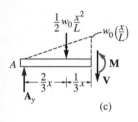

(c)

Fig. 12–37

The beam is subjected to the distributed loading shown in Fig. 12–37a. Determine the reactions at A. EI is constant.

SOLUTION

Elastic Curve. The beam deflects as shown in Fig. 12–37a. Only one coordinate x is needed. For convenience we will take it directed to the right, since the internal moment is easy to formulate.

Moment Function. The beam is indeterminate to the first degree as indicated from the free-body diagram, Fig. 12–37b. We can express the internal moment M in terms of the redundant force at A using the segment shown in Fig. 12–37c. Here

$$M = A_y x - \frac{1}{6} w_0 \frac{x^3}{L}$$

Slope and Elastic Curve. Applying Eq. 12–10, we have

$$EI \frac{d^2v}{dx^2} = A_y x - \frac{1}{6} w_0 \frac{x^3}{L}$$

$$EI \frac{dv}{dx} = \frac{1}{2} A_y x^2 - \frac{1}{24} w_0 \frac{x^4}{L} + C_1$$

$$EIv = \frac{1}{6} A_y x^3 - \frac{1}{120} w_0 \frac{x^5}{L} + C_1 x + C_2$$

The three unknowns A_y, C_1, and C_2 are determined from the boundary conditions $x = 0$, $v = 0$; $x = L$, $dv/dx = 0$; and $x = L$, $v = 0$. Applying these conditions yields

$$x = 0, \ v = 0; \qquad 0 = 0 - 0 + 0 + C_2$$

$$x = L, \ \frac{dv}{dx} = 0; \qquad 0 = \frac{1}{2} A_y L^2 - \frac{1}{24} w_0 L^3 + C_1$$

$$x = L, \ v = 0; \qquad 0 = \frac{1}{6} A_y L^3 - \frac{1}{120} w_0 L^4 + C_1 L + C_2$$

Solving,

$$A_y = \frac{1}{10} w_0 L \qquad\qquad\qquad Ans.$$

$$C_1 = -\frac{1}{120} w_0 L^3 \qquad C_2 = 0$$

Using the result for A_y, the reactions at B can be determined from the equations of equilibrium, Fig. 12–37b. Show that $B_x = 0$, $B_y = 0.4 w_0 L$, and $M_B = w_0 L^2/15$.

Example 12–18

The beam in Fig. 12–38a is fixed-supported at both ends and is subjected to the uniform loading shown. Determine the reactions at the supports. Neglect the effect of axial load.

SOLUTION

Elastic Curve. The beam deflects as shown in Fig. 12–38a. As in the previous problem, only one x coordinate is necessary for the solution since the loading is continuous across the span.

Moment Function. From the free-body diagram, Fig. 12–38b, the respective shear and moment reactions at A and B must be equal, since there is symmetry of both loading and geometry. Because of this, the equation of equilibrium, $\Sigma F_y = 0$, requires

$$V_A = V_B = \frac{wL}{2} \qquad\qquad Ans.$$

The beam is indeterminate to the first degree, where M' is redundant. Using the beam segment shown in Fig. 12–38c, the internal moment M can be expressed in terms of M' as follows:

$$M = \frac{wL}{2}x - \frac{w}{2}x^2 - M'$$

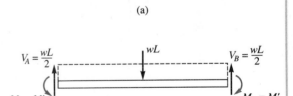

(a)

Slope and Elastic Curve. Applying Eq. 12–10, we have

$$EI\,\frac{d^2v}{dx} = \frac{wL}{2}x - \frac{w}{2}x^2 - M'$$

$$EI\,\frac{dv}{dx} = \frac{wL}{4}x^2 - \frac{w}{6}x^3 - M'x + C_1$$

$$EIv = \frac{wL}{12}x^3 - \frac{w}{24}x^4 - \frac{M'}{2}x^2 + C_1x + C_2$$

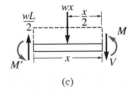

(b)

The three unknowns, M', C_1, and C_2, can be determined from the *three* boundary conditions $v = 0$ at $x = 0$, which yields $C_2 = 0$; $dv/dx = 0$ at $x = 0$, which yields $C_1 = 0$; and $v = 0$ at $x = L$, which yields

$$M' = \frac{wL^2}{12} \qquad\qquad Ans.$$

(c)

Using these results, notice that the remaining boundary condition $dv/dx = 0$ at $x = L$ is automatically satisfied.

Fig. 12–38

It should be realized that this method of solution is generally suitable when only one x coordinate is needed to describe the elastic curve. If several x coordinates are needed, equations of continuity must be written, thus complicating the solution process.

PROBLEMS

12–75. Determine the reactions at the supports A and B, then draw the shear and moment diagrams. EI is constant.

12–77. The loading on a floor beam used in the airplane is shown. Determine the reactions at the supports A and B and then draw the moment diagram for the beam. The beam is made of aluminum and has a moment of inertia of $I = 320 \text{ in}^4$.

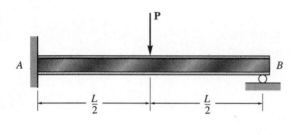

Prob. 12–75

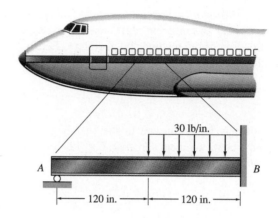

Prob. 12–77

*12–76. Determine the reactions at the supports A and B, then draw the shear and moment diagrams. EI is constant.

12–78. Determine the reactions at the supports A and B. EI is constant.

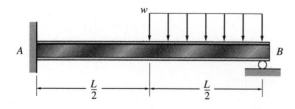

Prob. 12–76

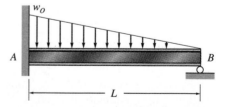

Prob. 12–78

12–79. Determine the reactions on the beam, then draw the shear and moment diagrams. *EI* is constant.

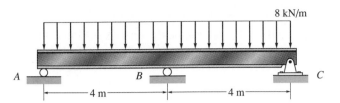

8 kN/m

A   B   C

4 m   4 m

Prob. 12–79

12–82. Determine the moment reactions at the supports *A* and *B* and then draw the shear and moment diagrams. *EI* is constant.

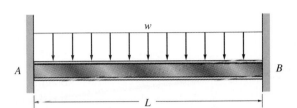

w

A   B

L

Prob. 12–82

*12–80. Determine the reactions at the supports, then draw the shear and moment diagrams. *EI* is constant.

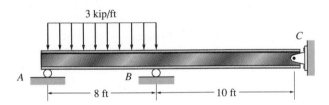

3 kip/ft

C

A   B

8 ft   10 ft

Prob. 12–80

12–83. Determine the moment reactions at the supports *A* and *B*. *EI* is constant.

12–81. The beam has a constant E_1I_1 and is supported by the fixed wall at *B* and the rod *AC*. If the rod has a cross-sectional area A_2 and the material has a modulus of elasticity E_2, determine the force in the rod.

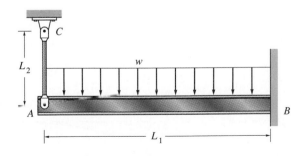

C

L_2

w

A   B

L_1

Prob. 12–81

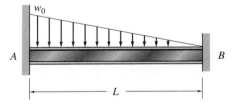

w_0

A   B

L

Prob. 12–83

*12.8 Statically Indeterminate Beams and Shafts— Moment-Area Method

If the moment-area method is used to determine the unknown redundants of a statically indeterminate beam or shaft, then the M/EI diagram must be drawn such that the redundants are represented as *unknowns* on this diagram. Once the M/EI diagram is established, the two moment-area theorems can then be applied to obtain the proper relationships between the tangents on the elastic curve in order to meet the conditions of displacement and/or slope at the supports of the beam or shaft. In all cases the number of these compatibility conditions will be equivalent to the number of redundants, and so a solution for the redundants can be obtained.

Moment Diagrams Constructed by the Method of Superposition. Since application of the moment-area theorems requires calculation of both the area under the M/EI diagram and the centroidal location of this area, it is often convenient to use *separate M/EI diagrams* for *each* of the known loads and redundants rather than using the resultant diagram to compute these geometric quantities. This is especially true if the resultant moment diagram has a complicated shape. The method for drawing the moment diagram in parts is based on the principle of superposition.

Most loadings on beams or shafts will be a combination of the four loadings shown in Fig. 12–39. Construction of the associated moment diagrams, also shown in this figure, has been discussed in the examples of Chapter 11. Based on these results, we will now show how to use the method of superposition to represent the resultant moment diagram for the cantilevered beam shown in Fig. 12–40a, by a series of separate moment diagrams. To do this, we will first replace the loads by a system of statically equivalent loads. For example, the three cantilevered beams shown in Fig. 12–40a are statically equivalent to the resultant beam, since the load at each point on the resultant beam is equal to the superposition or addition of the loadings on the three

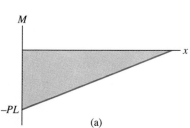

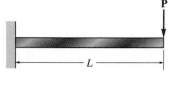

$-PL$

(a)

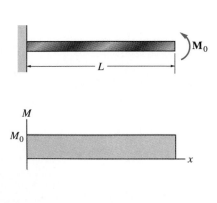

(b)

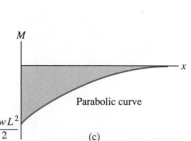

Parabolic curve

$\dfrac{-wL^2}{2}$

(c)

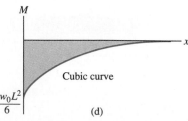

Cubic curve

$\dfrac{-w_0 L^2}{6}$

(d)

Fig. 12–39

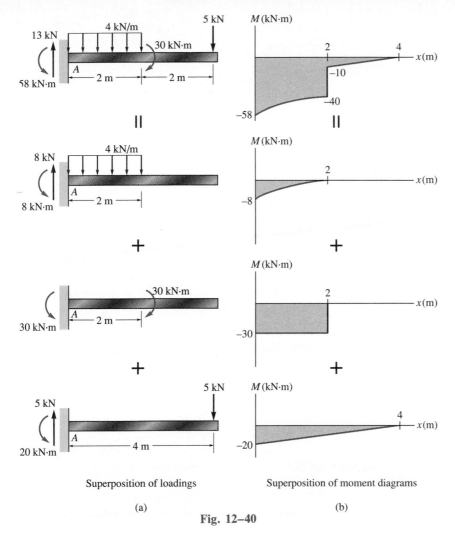

Superposition of loadings

(a)

Superposition of moment diagrams

(b)

Fig. 12–40

separate beams. Indeed, the shear reaction at end A is 13 kN when the reactions on the separate beams are added together. In the same manner, the internal moment at any point on the resultant beam is equal to the sum of the internal moments at any point on the separate beams. Thus, if the moment diagrams for each separate beam are drawn, Fig. 12–40b, the superposition of these diagrams will yield the moment diagram for the resultant beam, shown at the top. For example, from each of the separate moment diagrams, the moment at end A is $M_A = -8 \text{ kN} \cdot \text{m} - 30 \text{ kN} \cdot \text{m} - 20 \text{ kN} \cdot \text{m} = -58 \text{ kN} \cdot \text{m}$, as verified by the top moment diagram. This example demonstrates that it is sometimes easier to construct a series of separate statically equivalent moment diagrams for the beam, *rather* than construct its more complicated resultant moment diagram. Obviously, the area and location of the centroid for each part are easier to establish, compared with those for the resultant diagram.

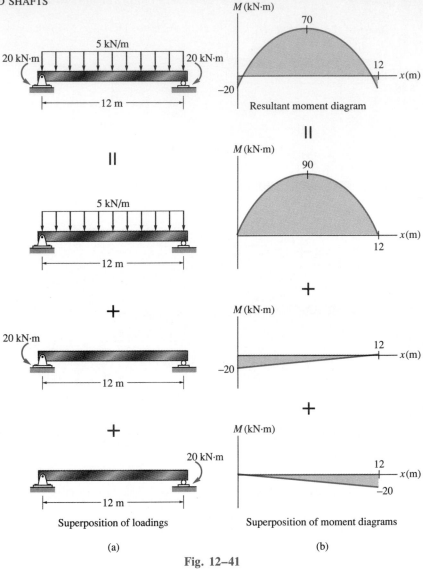

Superposition of loadings

Superposition of moment diagrams

(a)

(b)

Fig. 12–41

In a similar manner, we can also represent the resultant moment diagram for a beam by using a superposition of moment diagrams for a series of simply-supported beams. For example, the beam loading shown at the top of Fig. 12–41a is equivalent to the sum of the beam loadings shown below it. Consequently, the sum of the moment diagrams for each of these three loadings can be used rather than the resultant moment diagram shown at the top of Fig. 12–41b. For complete understanding, these results should be verified.

The examples that follow should also clarify some of these points and illustrate how to use the moment-area theorems to obtain the redundant reactions on statically indeterminate beams and shafts. The solutions follow the procedure for analysis outlined in Sec. 12.4.

Example 12–19

The beam is subjected to the concentrated loading shown in Fig. 12–42a. Determine the reactions at the supports. EI is constant.

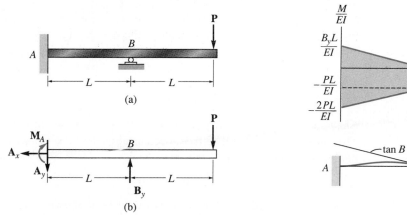

Fig. 12–42

SOLUTION

M/EI Diagram. The free-body diagram is shown in Fig. 12–42b. Using the method of superposition, the separate *M/EI* diagrams for the redundant reaction B_y and the load P are shown in Fig. 12–42c.

Elastic Curve. The elastic curve for the beam is shown in Fig. 12–42d. The tangents at the supports A and B have been constructed. The equation of compatibility requires

$$\Delta_B = t_{B/A} = 0$$

Moment-Area Theorem. Applying Theorem 2, we have

$$t_{B/A} = \left(\frac{2}{3}L\right)\left[\frac{1}{2}\left(\frac{B_y L}{EI}\right)L\right] + \left(\frac{L}{2}\right)\left[\frac{-PL}{EI}(L)\right]$$

$$+ \left(\frac{2}{3}L\right)\left[\frac{1}{2}\left(\frac{-PL}{EI}\right)(L)\right] = 0$$

$$B_y = 2.5P \qquad\qquad Ans.$$

Equations of Equilibrium. Using this result, the reactions at A on the free-body diagram, Fig. 12–42b, are determined as follows:

$$\xrightarrow{+} \Sigma F_x = 0; \qquad\qquad A_x = 0 \qquad\qquad Ans.$$

$$+\uparrow \ \Sigma F_y = 0; \qquad\qquad -A_y + 2.5P - P = 0$$

$$A_y = 1.5P \qquad\qquad Ans.$$

$$\downarrow+ \ \Sigma M_A = 0; \qquad -M_A + 2.5P(L) - P(2L) = 0$$

$$M_A = 0.5PL \qquad\qquad Ans.$$

Example 12–20

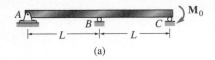

(a)

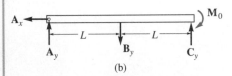

(b)

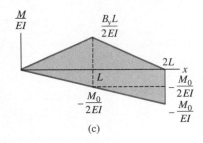

(c)

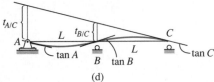

(d)

The beam is subjected to the couple moment at its end C as shown in Fig. 12–43a. Determine the reaction at B. EI is constant.

SOLUTION

M/EI Diagram. The free-body diagram is shown in Fig. 12–43b. By inspection, the beam is indeterminate to the first degree. In order to obtain a direct solution, we will choose $\mathbf{B}_y$ as the redundant. Using superposition, the M/EI diagrams for $\mathbf{B}_y$ and $\mathbf{M}_0$, each applied to a simply-supported beam, are shown in Fig. 12–43c. (Note that for such a beam A_x, A_y, and C_y do not contribute an M/EI diagram.)

Elastic Curve. The elastic curve for the beam is shown in Fig. 12–43d. The tangents at A, B, and C have been established. The equation of compatibility for the redundant requires

$$\Delta_B = 0 \qquad (1)$$

To satisfy this condition, the tangential deviations shown must be proportional; i.e.,

$$t_{B/C} = \frac{1}{2} t_{A/C} \qquad (2)$$

From Fig. 12–43c, we have

$$t_{B/C} = \left(\frac{1}{3}L\right)\left[\frac{1}{2}\left(\frac{B_y L}{2EI}\right)(L)\right] + \left(\frac{2}{3}L\right)\left[\frac{1}{2}\left(\frac{-M_0}{2EI}\right)(L)\right]$$
$$+ \left(\frac{L}{2}\right)\left[\left(\frac{-M_0}{2EI}\right)(L)\right]$$

$$t_{A/C} = (L)\left[\frac{1}{2}\left(\frac{B_y L}{2EI}\right)(2L)\right] + \left(\frac{2}{3}(2L)\right)\left[\frac{1}{2}\left(\frac{-M_0}{EI}\right)(2L)\right]$$

Substituting into Eq. 2 and simplifying yields

$$B_y = \frac{3M_0}{2L} \qquad \qquad Ans.$$

Equations of Equilibrium. The reactions at A and C can now be determined from the equations of equilibrium, Fig. 12–43b. Show that $A_x = 0$, $C_y = 5M_0/4L$, and $A_y = M_0/4L$.

Note from Fig. 12–43e that this problem can also be worked by expressing Eq. 1 in terms of the tangential deviations,

$$t_{B/A} = \frac{1}{2} t_{C/A}$$

Furthermore, since $\theta = -t_{A/B}/L = t_{C/B}/L$, we could also use

$$t_{A/B} = -t_{C/B}$$

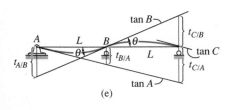

(e)

Fig. 12–43

PROBLEMS

*12–84. Determine the moment reactions at the supports A and B. EI is constant.

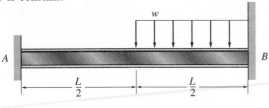

Prob. 12–84

12–85. Determine the moment reactions at the supports A and B, then draw the shear and moment diagrams. EI is constant.

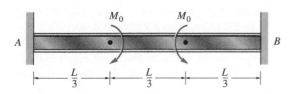

Prob. 12–85

12–86. Determine the reactions at the supports A, B, and C and then draw the shear and moment diagrams. EI is constant.

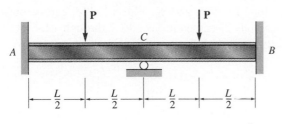

Prob. 12–86

12–87. Determine the vertical reactions at the bearing supports A, B and C of the shaft, then draw the shear and moment diagrams. EI is constant.

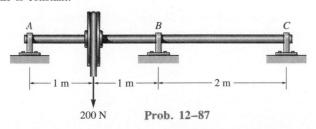

200 N   Prob. 12–87

*12–88. Determine the reactions on the beam, then draw the shear and moment diagrams. EI is constant.

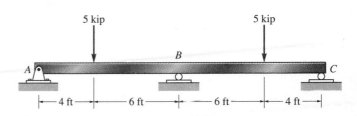

Prob. 12–87

12–89. Determine the reactions at the supports A and B, then draw the shear and moment diagrams. EI is constant.

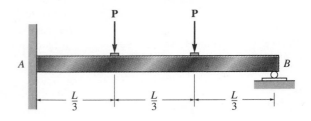

Prob. 12–89

12–90. Determine the value of a for which the maximum positive moment has the same magnitude as the maximum negative moment. EI is constant.

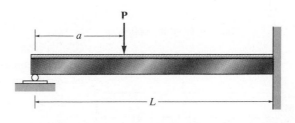

Prob. 12–90

12.9 Statically Indeterminate Beams and Shafts— Method of Superposition

The method of superposition has been used previously to solve for the redundant loadings on axially loaded bars and torsionally loaded shafts. In order to apply this method to the solution of statically indeterminate beams (or shafts), it is first necessary to identify the redundant support reactions as explained in Sec. 12.6. By *removing* them from the beam we obtain the so-called *primary beam,* which is statically determinate and stable, and is subjected *only* to the external load. If we add to this beam a succession of similarly supported beams, each loaded with a *separate* redundant, then by the principle of superposition, we obtain the actual loaded beam. Finally, in order to solve for the redundants, we must write the *conditions of compatibility* that exist at the supports where each of the redundants act. Since the redundant forces are determined directly in this manner, this method of analysis is sometimes called the *force method.* Once the redundants are obtained, the other reactions on the beam are then determined from the three equations of equilibrium.

To clarify these concepts, consider the beam shown in Fig. 12–44a. If we choose the reaction $\mathbf{B}_y$ at the roller as the redundant, then the primary beam is shown in Fig. 12–44b, and the beam with the redundant $\mathbf{B}_y$ acting on it is shown in Fig. 12–44c. The displacement at the roller is to be zero, and since the displacement of point B on the primary beam is v_B, and $\mathbf{B}_y$ causes point B to be displaced upward v'_B, we can write the compatibility equation at B as

$$(+\uparrow) \qquad\qquad 0 = -v_B + v'_B$$

The displacements v_B, Fig. 12–44b, and v'_B, Fig. 12–44c, can be obtained using any one of the methods discussed in Sec. 12.2 through 12.4. Here we will obtain them directly from the table on the inside back cover. We have

$$v_B = \frac{5PL^3}{48EI} \qquad \text{and} \qquad v'_B = \frac{B_y L^2}{3EI}$$

Substituting into the compatibility equation, we get

$$0 = -\frac{5PL^3}{48EI} + \frac{B_y L^3}{3EI}$$

$$B_y = \frac{5}{16}P$$

Now that B_y is known, the reactions at the wall are determined from the three equations of equilibrium applied to the entire beam, Fig. 12–44d. The results are

$$A_x = 0$$

$$A_y = \frac{11}{16}P$$

$$M_A = \frac{3}{16}PL$$

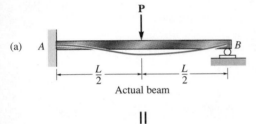

(a)   Actual beam

$\|$

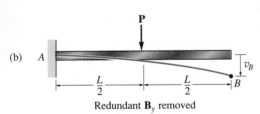

(b)   Redundant $\mathbf{B}_y$ removed

$+$

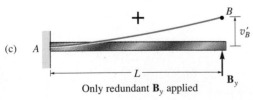

(c)   Only redundant $\mathbf{B}_y$ applied

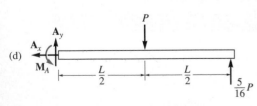

(d)

Fig. 12–44

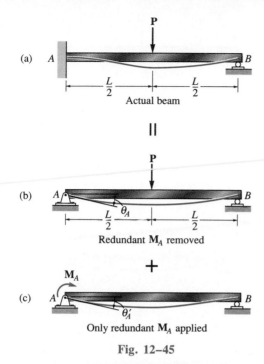

Fig. 12–45

As stated in Sec. 12.6, choice of the redundant is *arbitrary,* provided the primary beam remains stable. For example, the moment at A for the beam in Fig. 12–45a can also be chosen as the redundant. In this case the capacity of the beam to resist M_A is removed, and so the primary beam is then pin-supported at A, Fig. 12–45b. Also, the redundant at A acts alone on this beam, Fig. 12–45c. Referring to the slope at A caused by the load $\mathbf{P}$ as θ_A, Fig. 12–45b, and the slope at A caused by the redundant $\mathbf{M}_A$ as θ'_A, Fig. 12–45c, the compatibility equation for slope at A requires

$$(\curvearrowright +) \qquad 0 = \theta_A + \theta'_A$$

Again using the table on the inside back cover, we have

$$\theta_A = \frac{PL^2}{16EI} \qquad \text{and} \qquad \theta'_A = \frac{M_A L}{3EI}$$

Thus

$$0 = \frac{PL^2}{16EI} + \frac{M_A L}{3EI}$$

$$M_A = -\frac{3}{16}PL$$

This is the same result computed previously. Here the negative sign for M_A simply means that $\mathbf{M}_A$ acts in the opposite sense of direction of that shown in Fig. 12–45c.

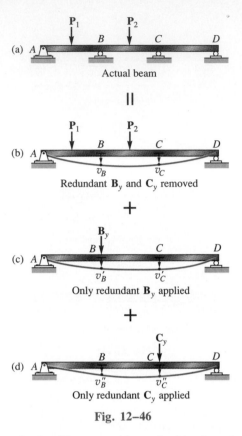

Fig. 12–46

Another example that illustrates this method is given in Fig. 12–46a. In this case the beam is indeterminate to the second degree and therefore *two* compatibility equations will be necessary for the solution. We will choose the forces at the roller supports B and C as redundants. The primary (statically determinate) beam deflects as shown in Fig. 12–46b when the redundants are removed. Each redundant force deflects this beam as shown in Fig. 12–46c and 12–46d, respectively. By superposition, the compatibility equations for the displacements at B and C are

$$(+\downarrow) \qquad\qquad 0 = v_B + v_B' + v_B''$$
$$(+\downarrow) \qquad\qquad 0 = v_C + v_C' + v_C'' \qquad\qquad (12\text{–}22)$$

Here the displacement components v_B' and v_C' will be expressed in terms of the unknown B_y, and the components v_B'' and v_C'' will be expressed in terms of the unknown C_y. When these displacements have been computed and substituted into Eq. 12–22, these equations may then be solved simultaneously for the two unknowns B_y and C_y.

PROCEDURE FOR ANALYSIS

The following procedure provides a means for applying the method of superposition (or the force method) to determine the reactions on statically indeterminate beams or shafts.

Principle of Superposition.   Specify the unknown redundant forces or moments that must be removed from the beam in order to make it statically determinate and stable. Then, using the principle of superposition, draw the statically indeterminate beam and show it equal to a sequence of corresponding *statically determinate beams*. The first of these beams, the primary beam, supports the same external loads as the statically indeterminate beam, and each of the other beams "added" to the primary beam shows the beam loaded with a separate redundant force or moment. Also, sketch the deflection curve on each beam and indicate symbolically the displacement or slope at the point of each redundant force or moment. (See Figs. 12–44 through 12–46.)

Compatibility Equations.   Write a compatibility equation for the displacement or slope at each point where there is a redundant force or moment. Determine all the displacements or slopes using an appropriate method as explained in Sec. 12.2 through 12.4. Substitute the results into the compatibility equations and solve for the unknown redundants. If a numerical value for a redundant is *positive*, it has the *same sense of direction* as originally assumed. Similarly, a *negative* numerical value indicates the redundant acts *opposite* to its assumed *sense of direction*.

Equilibrium Equations.   Once the redundant forces and/or moments have been determined, the remaining unknown reactions can be found from the equations of equilibrium applied to the loadings shown on the beam's free-body diagram.

The following examples illustrate application of this procedure. For brevity, all displacements and slopes have been computed using the table on the inside back cover.

Example 12–21

Determine the reactions at the roller support B of the beam shown in Fig. 12–47a, then draw the shear and moment diagrams. EI is constant.

SOLUTION

Principle of Superposition. By inspection, the beam is statically indeterminate to the first degree. The roller support at B will be chosen as the redundant so that $\mathbf{B}_y$ will be determined *directly*. Figures 12–47b and 12–47c show application of the principle of superposition. Here we have assumed that $\mathbf{B}_y$ acts upward on the beam.

Compatibility Equation. Taking positive displacement as downward, the compatibility equation at B is

$$(+\downarrow) \qquad\qquad 0 = v_B - v_B' \qquad\qquad (1)$$

These displacements can be obtained directly from the table on the inside back cover.

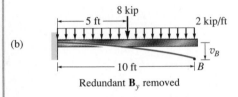

(a) Actual beam

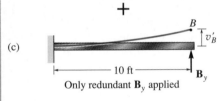

(b) Redundant $\mathbf{B}_y$ removed

+

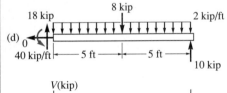

(c) Only redundant $\mathbf{B}_y$ applied

$$v_B = \frac{wL^4}{8EI} + \frac{5PL^3}{48EI} = \frac{2(10)^4}{8EI} + \frac{5(8)(10)^3}{48EI} = \frac{3{,}333}{EI} \downarrow$$

$$v_B' = \frac{PL^3}{3EI} = \frac{B_y(10^3)}{3EI} = \frac{333.3B_y}{EI} \uparrow$$

Substituting into Eq. 1 and solving yields

$$0 = \frac{3{,}333}{EI} + \frac{333.3B_y}{EI}$$

$$B_y = 10 \text{ kip} \qquad\qquad\qquad \textit{Ans.}$$

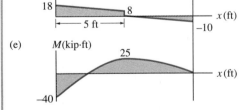

(d)

(e)

Equilibrium Equations. Using this result and applying the three equations of equilibrium, we obtain the results shown on the beam's free-body diagram in Fig. 12–47d. The shear and moment diagrams are shown in Fig. 12–47e.

Fig. 12–47

Example 12–22

Determine the reactions on the beam shown in Fig. 12–48a. Due to the loading and poor construction, the roller support at B settles 12 mm. Take $E = 200$ GPa and $I = 80\ (10^6)$ mm^4.

SOLUTION

Principle of Superposition.   By inspection, the beam is indeterminate to the first degree. The roller support at B will be chosen as the redundant. The principle of superposition is shown in Fig. 12–48b and 12–48c. Here $\mathbf{B}_y$ is assumed to act upward on the beam.

Compatibility Equation.   With reference to point B, using units of meters, we require

$$(+\downarrow) \qquad\qquad 0.012 = v_B - v_B' \qquad\qquad (1)$$

Using the table on the inside back cover, the displacements are

$$v_B = \frac{5wL^4}{768EI} = \frac{5(24)(8)^4}{768EI} = \frac{640}{EI} \downarrow$$

$$v_B' = \frac{PL^3}{48EI} = \frac{B_y(8)^3}{48EI} = \frac{10.67B_y}{EI} \uparrow$$

Thus Eq. 1 becomes

$$0.012EI = 640 - 10.67B_y$$

Expressing E and I in units of kN/m^2 and m^4, respectively, we have

$$0.012(200)(10^6)[80(10^{-6})] = 640 - 10.67B_y$$
$$B_y = 42.0 \text{ kN} \uparrow \qquad\qquad Ans.$$

Equilibrium Equations.   Applying this result to the beam, Fig. 12–48c, we can compute the reactions at A and C using the equations of equilibrium. We obtain

$$\zeta+ \ \Sigma M_A = 0; \qquad -96(2) + 42.0(4) + C_y(8) = 0$$
$$C_y = -3.00 \text{ kN} = 3.00 \text{ kN} \downarrow \qquad\qquad Ans.$$
$$+\uparrow \ \Sigma F_y = 0; \qquad A_y - 96 + 42.0 - 3.00 = 0$$
$$A_y = 57.0 \text{ kN} \uparrow \qquad\qquad Ans.$$

(a)

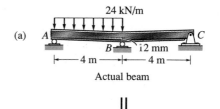

24 kN/m

Actual beam

||

(b)

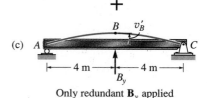

24 kN/m

Redundant $\mathbf{B}_y$ removed

+

(c)

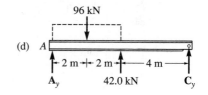

Only redundant $\mathbf{B}_y$ applied

(d)

96 kN

42.0 kN

Fig. 12–48

Example 12–23

The beam in Fig. 12–49a is fixed-supported to the wall at A and pin-connected to a $\frac{1}{2}$-in.-diameter rod BC. If $E = 29(10^3)$ ksi for both members, determine the force developed in the rod due to the loading. The moment of inertia of the beam about its neutral axis is $I = 475$ in^4.

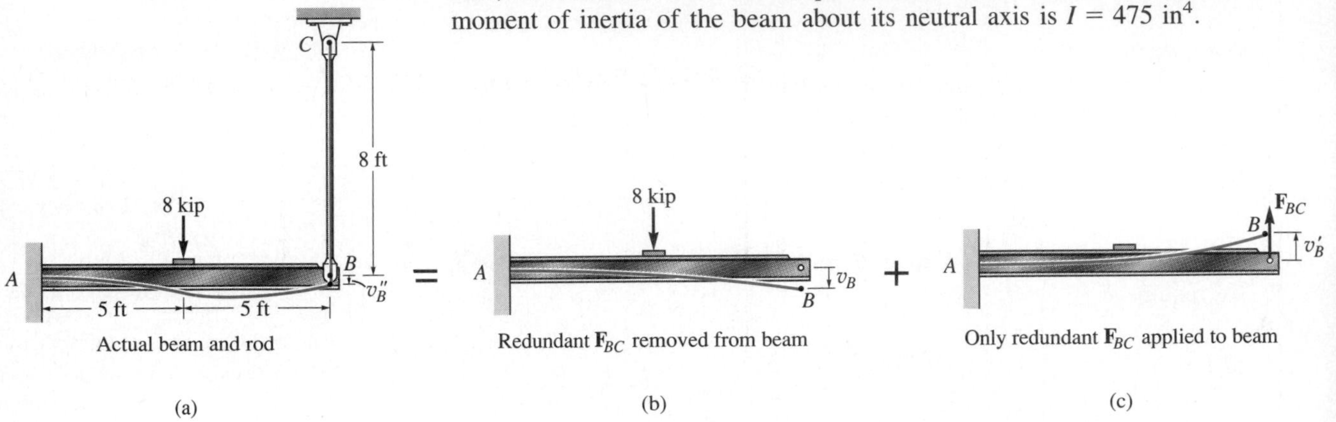

Actual beam and rod	Redundant $\mathbf{F}_{BC}$ removed from beam	Only redundant $\mathbf{F}_{BC}$ applied to beam
(a)	(b)	(c)

Fig. 12–49

SOLUTION I

Principle of Superposition.   By inspection, this problem is indeterminate to the first degree. Here B will undergo an unknown displacement v_B'', since the rod will stretch. The rod will be treated as the redundant and hence the force of the rod is removed from the beam at B, Fig. 12–49b, and then reapplied, Fig. 12–38c.

Compatibility Equation.   At point B we require

$$(+\downarrow) \qquad\qquad v_B'' = v_B - v_B' \qquad\qquad (1)$$

The displacements v_B and v_B' are determined from the table on the back cover. v_B'' is computed from Eq. 4-2. Working in kilopounds and inches, we have

$$v_B'' = \frac{PL}{AE} = \frac{F_{BC}(8\ \text{ft})(12\ \text{in./ft})}{(\pi/4)(\frac{1}{2}\ \text{in.})^2[29(10^3)\ \text{kip/in}^2]} = 0.01686F_{BC}\ \downarrow$$

$$v_B = \frac{5PL^3}{48EI} = \frac{5(8\ \text{kip})(10\ \text{ft})^3(12\ \text{in./ft})^3}{48[29(10^3)\text{kip/in}^2](475\ \text{in}^4)} = 0.1045\ \text{in.}\ \downarrow$$

$$v_B' = \frac{PL^3}{3EI} = \frac{F_{BC}(10\ \text{ft})^3(12\ \text{in./ft})^3}{3[29(10^3)\ \text{kip/in}^2](475\ \text{in}^4)} = 0.04181F_{BC}\ \uparrow$$

Thus, Eq. 1 becomes

$$(+\downarrow) \qquad\qquad 0.01686F_{BC} = 0.1045 - 0.04181F_{BC}$$

$$F_{BC} = 1.78\ \text{kip} \qquad\qquad\qquad Ans.$$

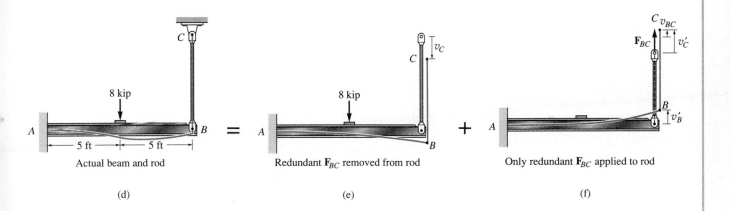

Actual beam and rod

(d)

Redundant $\mathbf{F}_{BC}$ removed from rod

(e)

Only redundant $\mathbf{F}_{BC}$ applied to rod

(f)

SOLUTION II

Principle of Superposition. We can also solve this problem by removing the pin support at C and keeping the rod attached to the beam. In this case the 8-kip load will cause points B and C to be displaced downward the *same amount* v_C, Fig. 12–49e, since no force exists in rod BC. When the redundant force $\mathbf{F}_{BC}$ is applied at point C, it causes the end C of the rod to be displaced upward v_C' and the end B of the beam to be displaced upward v_B', Fig. 12–49f. The difference in these two displacements, v_{BC}, represents the stretch of the rod due to $\mathbf{F}_{BC}$, so that $v_C' = v_{BC} + v_B'$. Hence, from Fig. 12–49d, 12–49e, and 12–49f, the compatibility of displacement at point C is

$$(+\downarrow) \qquad\qquad 0 = v_C - (v_{BC} + v_B') \qquad\qquad (2)$$

From Solution I, we have

$$v_{BC} = v_B'' = 0.01686 F_{BC} \uparrow$$
$$v_C - v_B = 0.1045 \text{ in.} \downarrow$$
$$v_B' = 0.04181 F_{BC} \uparrow$$

Therefore Eq. 2 becomes

$$(+\downarrow) \qquad 0 = 0.1045 - (0.01686 F_{BC} + 0.04181 F_{BC})$$
$$F_{BC} = 1.78 \text{ kip} \qquad\qquad\qquad \textit{Ans.}$$

Example 12–24

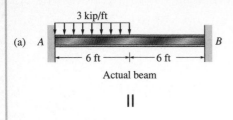

(a)

6 ft ·|· 6 ft

Actual beam

||

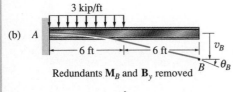

(b)

6 ft ·|· 6 ft

Redundants M_B and B_y removed

+

(c)

12 ft

Only redundant B_y applied

+

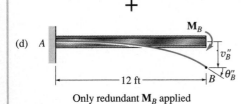

(d)

12 ft

Only redundant M_B applied

Fig. 12–50

Determine the moment at B for the beam shown in Fig. 12–50a. EI is constant. Neglect the effects of axial load.

SOLUTION

Principle of Superposition. If the axial load on the beam is neglected, there will be a vertical force and moment at A and B. Since there are only two available equations of equilibrium ($\Sigma M = 0$, $\Sigma F_y = 0$), the problem is indeterminate to the second degree. We will assume that $\mathbf{B}_y$ and $\mathbf{M}_B$ are redundant, so that by the principle of superposition, the beam is represented as a cantilever, loaded *separately* by the distributed load and reactions $\mathbf{B}_y$ and $\mathbf{M}_B$, Fig. 12–50b, 12–50c, and 12–50d.

Compatibility Equations. Referring to the displacement and slope at B in Fig. 12–50b, 12–50c, and 12–50d, we require

$$(\curvearrowleft +) \qquad\qquad 0 = \theta_B + \theta'_B + \theta''_B \qquad\qquad (1)$$
$$(+\downarrow) \qquad\qquad 0 = v_B + v'_B + v''_B \qquad\qquad (2)$$

Using the table on the inside back cover to compute the slopes and displacements, we have

$$\theta_B = \frac{wL^3}{48EI} = \frac{3(12)^3}{48EI} = \frac{108}{EI} \;\searrow$$

$$v_B = \frac{7wL^4}{384EI} = \frac{7(3)(12)^4}{384EI} = \frac{1134}{EI} \;\downarrow$$

$$\theta'_B = \frac{PL^2}{2EI} = \frac{B_y(12)^2}{2EI} = \frac{72B_y}{EI} \;\searrow$$

$$v'_B = \frac{PL^3}{3EI} = \frac{B_y(12)^3}{3EI} = \frac{576B_y}{EI} \;\downarrow$$

$$\theta''_B = \frac{ML}{EI} = \frac{M_B(12)}{EI} = \frac{12M_B}{EI} \;\searrow$$

$$v''_B = \frac{ML^2}{2EI} = \frac{M_B(12)^2}{2EI} = \frac{72M_B}{EI} \;\downarrow$$

Substituting these values into Eqs. 1 and 2 and canceling out the common factor EI, we get

$$(\curvearrowleft +) \qquad\qquad 0 = 108 + 72B_y + 12M_B$$
$$(+\downarrow) \qquad\qquad 0 = 1134 + 576B_y + 72M_B$$

Solving these equations simultaneously gives

$$B_y = -3.375 \text{ kip}$$
$$M_B = 11.25 \text{ kip} \cdot \text{ft}$$

Ans.

PROBLEMS

12–91. The beam has a moment of inertia I and is supported at its ends by a pin and roller and at its center by a rod having a cross-sectional area A and length L. Determine the tension in the rod when a uniform load w is placed on the beam. The modulus of elasticity is E.

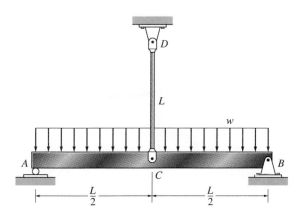

Prob. 12–91

*12–92. The beam AB has a moment of inertia $I = 475$ in^4 and rests on the smooth supports at its ends. A 0.75-in.-diameter rod CD is welded to the center of the beam and to the fixed support at D. If the temperature of the rod is decreased by 150°F, determine the force developed in the rod. The beam and rod are both made of steel, for which $E_{st} = 29(10^3)$ ksi and $\alpha_{st} = 6.5(10^{-6})$/°F.

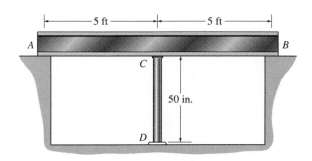

Prob. 12–92

12–93. Determine the reactions at the supports A and B. EI is constant.

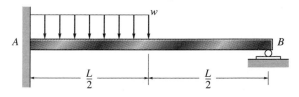

Prob. 12–93

12–94. Determine the deflection at the end B of the clamped steel strip. The spring has a stiffness of $k = 2$ N/mm. The strip is 5 mm wide and 10 mm thick. Also, draw the shear and moment diagrams for the strip. $E_{st} = 200$ GPa..

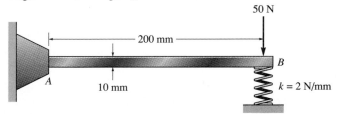

Prob. 12–94

12–95. The assembly consists of a steel and an aluminum bar, each of which is 1 in. thick, fixed at its ends A and B, and pin-connected to the *rigid* short link CD. If a horizontal force of 80 lb is applied to the link as shown, determine the moments created at A and B. $E_{st} = 29(10^3)$ ksi, $E_{al} = 10(10^3)$ ksi.

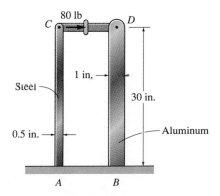

Prob. 12–95

*12–96.   Determine the reaction at support C. EI is constant for both beams.

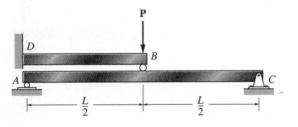

Prob. 12–96

12–98.   The 1-in.-diameter steel shaft is supported by unyielding bearings at A and C. The bearing at B rests on a simply-supported steel wide-flange beam having a moment of inertia of $I = 500$ in^4. If the belt loads on the pulley are 400 lb each, determine the vertical reactions at A, B, and C. $E_{st} = 29(10^3)$ ksi.

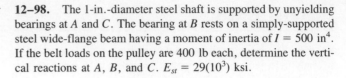

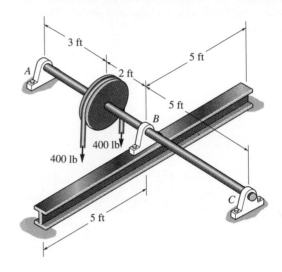

Prob. 12–98

12–97.   Determine the reactions at A and B. The support at A can exert only a moment on the beam. EI is constant.

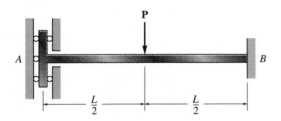

Prob. 12–97

12–99.   Determine the reactions on the beam. EI is constant.

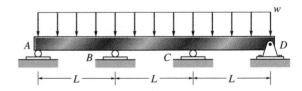

Prob. 12–99

REVIEW PROBLEMS

*12–100.   Determine the equations of the elastic curve for the beam using the method of integration and the x_1 and x_2 coordinates. Specify the slope at A and the maximum deflection. EI is constant.

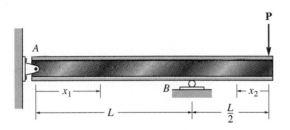

Prob. 12–100

12–101.   The simply-supported shaft has a moment of inertia of $2I$ for region BC and a moment of inertia I for regions AB and CD. Determine the maximum deflection of the beam due to the load **P**. Use the moment-area method.

12–102.   Solve Prob. 12–101 using the method of integration.

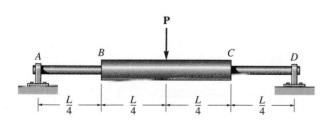

Probs. 12–101/12–102

12–103.   The beam is subjected to the load shown. Using discontinuity functions, determine the equation of the elastic curve. EI is constant.

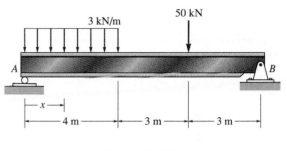

Prob. 12–103

*12–104.   Determine the equation of the elastic curve for the beam using the x coordinate. Specify the slope at A and the maximum deflection. EI is constant.

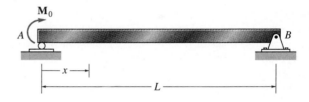

Prob. 12–104

12–105.   The beam is subjected to the load **P** as shown. Determine the magnitude of force **F** that must be applied at the end of the overhang C so that the deflection at C is zero. Use the moment-area method. EI is constant.

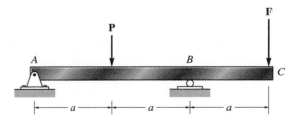

Prob. 12–105

12–106. The shaft supports the two pulley loads shown. Using discontinuity functions, determine the equation of the elastic curve. The bearings at A and B exert only vertical reactions on the shaft. What is the maximum deflection? EI is constant.

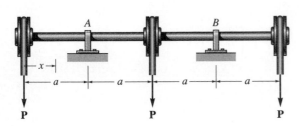

Prob. 12–106

12–107. The steel shaft is subjected to the loadings developed in the belts passing over the two pulleys. If the bearings at A and B exert vertical reactions on the shaft, determine the slope at A. The shaft has a diameter of 0.75 in. Use the moment-area method. $E_{st} = 29(10^3)$ ksi.

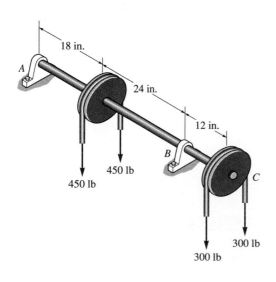

Prob. 12–107

*12–108.** Determine the deflection at C and the slope of the beam at A, B, and C. Use the moment-area method. EI is constant.

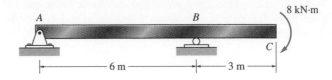

Prob. 12–108

12–109. The beam is subjected to the linearly varying distributed load. Determine the maximum deflection of the beam. Use the method of integration. EI is constant.

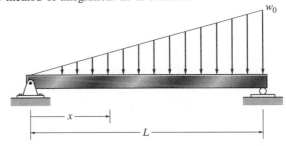

Prob. 12–109

12–110. The 25-mm-diameter steel shaft is supported at A and B by bearings. If the tension in the belt on the pulley at C is 0.75 kN, determine the largest belt tension T on the pulley at D so that the slope of the shaft at A or B does not exceed 0.02 rad. The bearings exert only vertical reactions on the shaft. Use the moment-area method. $E_{st} = 200$ GPa.

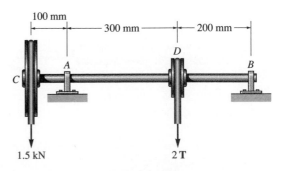

Prob. 12–110

13 Buckling of Columns

In this chapter we will discuss the behavior of columns and indicate some of the methods used for their design. The chapter begins with a general discussion of buckling, followed by a determination of the axial load needed to buckle a so-called ideal column. Afterwards a more realistic analysis is considered, which accounts for any bending of the column. Also, inelastic buckling of a column is presented as a special topic. At the end of the chapter we will discuss some of the methods used to design both concentric and eccentric loaded columns made from common engineering materials.

13.1 Critical Load

Whenever a member is designed, it is necessary that it satisfy specific strength, deflection, and stability requirements. In the preceding chapters we have discussed some of the methods used to determine a member's strength and deflection, while assuming that the member was always in stable equilibrium. Some members, however, may be subjected to compressive loadings, and if these members are long and slender the loading may be large enough to cause the member to deflect laterally or sidesway. To be specific, long slender members subjected to an axial compressive force are called *columns*, and the lateral deflection that occurs is called *buckling*. Quite often the buckling of a column can lead to a sudden and dramatic failure of a structure or mechanism, and as a result, special attention must be given to the design of columns so that they can safely support their intended loadings without buckling.

$P > P_{cr}$

P_{cr}

$P > P_{cr}$

(a)

(b)

Fig. 13–1

The maximum axial load that a column can support when it is on the *verge* of buckling is called the *critical load, P_{cr}* Fig. 13–1*a*. Any additional loading will cause the column to buckle and therefore deflect laterally as shown in Fig. 13–1*b*. In order to better understand the nature of this instability, consider a two-bar mechanism consisting of weightless bars that are rigid and pin-connected at their ends, Fig. 13–2*a*. When the bars are in the vertical position, the spring, having a stiffness k, is unstretched, and a *small* vertical force **P** is applied at the top of one of the bars. We can upset this equilibrium position by displacing the pin at A by a small amount Δ, Fig. 13–2*b*. As shown on the free-body diagram of the pin, Fig. 13–2*c*, the spring will produce a restoring force $F = k\Delta$, while the applied load **P** develops two horizontal components, $P_x = P \tan \theta$, which tend to push the pin (and the bars) further out of equilibrium. Since θ is small, $\Delta = \theta(L/2)$ and $\tan \theta \approx \theta$. Thus the restoring spring force becomes $F = k\theta L/2$, and the disturbing force is $2P_x = 2P\theta$.

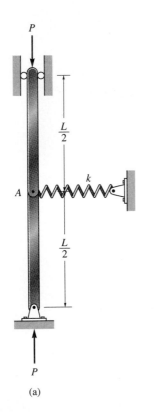

(a)

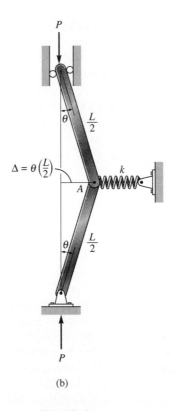

(b)

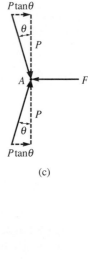

(c)

Fig. 13–2

If the restoring force is greater than the disturbing force, that is, $k\theta L/2 > 2P\theta$, then, noticing that θ cancels out, we can solve for P, which gives

$$P < \frac{kL}{4} \qquad \text{stable equilibrium}$$

This is a condition for *stable equilibrium,* since the force developed by the spring would be adequate in restoring the bars back to their vertical position. On the other hand, if $kL\theta/2 < 2P\theta$, or

$$P > \frac{kL}{4} \qquad \text{unstable equilibrium}$$

then the mechanism would be in *unstable equilibrium.* In other words, if this load P is applied, and a slight displacement occurs at A, the mechanism will tend to move out of equilibrium and not be restored to its original position.

The intermediate value of P, defined by requiring $kL\theta/2 = 2P\theta$, is the *critical load.* Here

$$P_{cr} = \frac{kL}{4} \qquad \text{neutral equilibrium}$$

This loading represents a case of the mechanism being in *neutral equilibrium.* Since P_{cr} is *independent* of the (small) displacement θ of the bars, any slight disturbance given to the mechanism will not cause it to move further out of equilibrium, nor will it be restored to its original position. Instead, the bars will *remain* in the deflected position.

These three different states of equilibrium are represented graphically in Fig. 13–3. The transition point where the load is equal to the critical value $P = P_{cr}$ is called the *bifurcation point.* At this point the mechanism will be in equilibrium for any *small value* of θ, measured either to the right or to the left of the vertical. Physically, P_{cr} represents the load for which the mechanism is on the verge of buckling. It is quite valid to determine this value by assuming *small displacements* as done here; however, it should be understood that P_{cr} may *not* be the largest value of P that the mechanism can support. Indeed, if a larger load is placed on the bars, then the mechanism may have to undergo a further deflection before the spring is compressed or elongated enough to hold the mechanism in equilibrium.

Like the two-bar mechanism just discussed, the critical buckling loads on columns supported in various ways can be obtained, and the method used to do this will be explained in the next section. Although in engineering design the critical load may be considered to be the largest load the column can support, realize that, like the two-bar mechanism in the deflected or buckled position, a column may actually support an even greater load than P_{cr}. Unfortunately, however, this loading may require the column to undergo a *large* deflection, which is generally not tolerated in engineering structures or machines. For example, it may take only a few newtons of force to buckle a meterstick, but the additional load it may support can be applied only after the stick undergoes a relatively large lateral deflection.

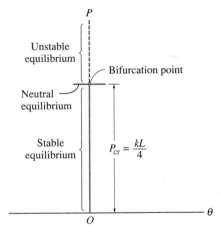

Fig. 13–3

13.2 Ideal Column with Pin Supports

In this section we will determine the critical buckling load for a column that is pin-supported as shown in Fig. 13–4a. The column to be considered is an *ideal column,* meaning one that is perfectly straight before loading, is made of homogeneous material, and upon which the load is applied through the centroid of the column's cross section. It is further assumed that the material behaves in a linear-elastic manner and that the column buckles or bends in a single plane. In reality, the conditions of column straightness and load application are never accomplished; however, the analysis to be performed on an "ideal column" is similar to that used to analyze initially crooked columns or those having an eccentric load application. These more realistic cases will be discussed later in this chapter.

Since an ideal column is straight, theoretically the axial load P could be increased until failure occurs by either fracture or yielding of the material. However, when the critical load P_{cr} is reached, the column is on the verge of becoming unstable, so that a small lateral force F, Fig. 13–4b, will cause the column to remain in the deflected position when F is removed, Fig. 13–4c. Any slight reduction in the axial load P from P_{cr} will allow the column to straighten out, and any slight increase in P, beyond P_{cr}, will cause further increases in lateral deflection.

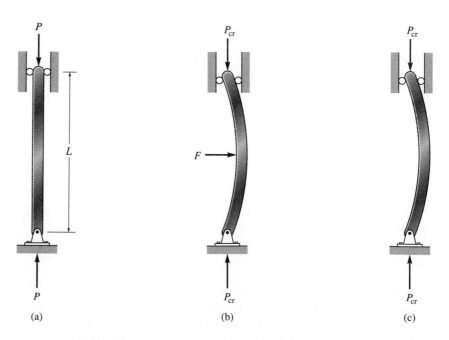

(a)

(b)

(c)

Fig. 13–4

Whether or not a column will remain stable or become unstable when subjected to an axial load will depend on its ability to restore itself, which is based on its resistance to bending. Hence, in order to determine the critical load and the buckled shape of the column, we will apply Eq. 12–10, which relates the internal moment in the column to its deflected shape, i.e.,

$$EI \frac{d^2v}{dx^2} = M \qquad (13\text{–}1)$$

Recall that this equation assumes that the slope of the elastic curve is small* and that deflections occur only by bending. When the column is in its deflected position, Fig. 13–5a, the internal bending moment can be determined by using the method of sections. The free-body diagram shown in Fig. 13–5b will be considered. Here both the deflection v and the internal moment M are shown in the *positive direction* according to the sign convention used to establish Eq. 13–1. Summing moments, the internal moment is $M = -Pv$. Thus Eq. 13–1 becomes

$$EI \frac{d^2v}{dx^2} = -Pv$$

$$\frac{d^2v}{dx^2} + \left(\frac{P}{EI}\right)v = 0 \qquad (13\text{–}2)$$

This is a homogeneous, second-order, linear differential equation with constant coefficients. It can be shown by using the methods of differential equations, or by direct substitution into Eq. 13–2, that the general solution is

$$v = C_1 \sin\left(\sqrt{\frac{P}{EI}}\, x\right) + C_2 \cos\left(\sqrt{\frac{P}{EI}}\, x\right) \qquad (13\text{–}3)$$

The two constants of integration are determined from the boundary conditions at the ends of the column. Since $v = 0$ at $x = 0$, then $C_2 = 0$. And since $v = 0$ at $x = L$, then

$$C_1 \sin\left(\sqrt{\frac{P}{EI}}\, L\right) = 0$$

This equation is satisfied if $C_1 = 0$; however, then $v = 0$, which is a *trivial solution* that requires the column to always remain straight, even though the load causes the column to become unstable. The other possibility is for

$$\sin\left(\sqrt{\frac{P}{EI}}\, L\right) = 0$$

which is satisfied if

$$\sqrt{\frac{P}{EI}}\, L = n\pi$$

*If large deflections are to be considered, the more accurate differential equation, Eq. 12–4, $EI(d^2v/dx^2)/[1 + (dv/dx)^2]^{3/2} = M$ must be used.

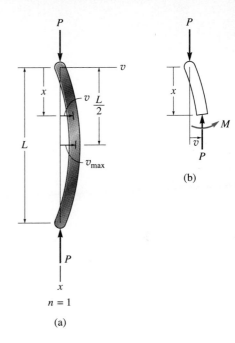

$n = 1$

(a)

(b)

Fig. 13–5

or

$$P = \frac{n^2 \pi^2 EI}{L^2} \qquad n = 1, 2, 3, \ldots \qquad (13\text{--}4)$$

The *smallest value* of P is obtained when $n = 1$, so the *critical load* for the column is therefore

$$P_{cr} = \frac{\pi^2 EI}{L^2}$$

This load is sometimes referred to as the *Euler load,* after the Swiss mathematician Leonhard Euler, who originally solved this problem in 1757. The corresponding buckled shape is defined by the equation

$$v = C_1 \sin \frac{\pi x}{L}$$

Here the constant C_1 represents the maximum deflection, v_{max}, which occurs at the midpoint of the column, Fig. 13–5a. Specific values for C_1 cannot be obtained, since the exact deflected form for the column is unknown once it has buckled. It has been assumed, however, that this deflection is small.

Realize that n in Eq. 13–4 represents the number of waves in the deflected shape of the column. For example, if $n = 2$, then from Eqs. 13–3 and 13–4, *two waves* will appear in the buckled shape, Fig. 13–5c, and the column will support a critical load that is $4P_{cr}$ just prior to buckling. Since this value is

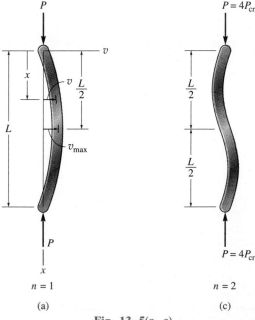

Fig. 13–5(a, c)

four times the critical load, and the deflected shape is unstable, this form of buckling, practically speaking, will not exist, unless the column has a lateral brace at its center.

Like the two-bar mechanism discussed in Sec. 13–1, we can represent the load-deflection characteristics of the ideal column by the graph shown in Fig. 13–6. The bifurcation point represents the state of *neutral equilibrium,* at which point the *critical load* acts on the column. Here the column is on the verge of impending buckling.

It should be noted that the critical load is independent of the strength of the material; rather it depends only on the column's dimensions (*I* and *L*) and the material's stiffness or modulus of elasticity *E*. For this reason, as far as elastic buckling is concerned, columns made, for example, of high-strength steel offer no advantage over those made of lower-strength steel, since the modulus of elasticity for both is approximately the same. Also note that the load-carrying capacity of a column will increase as the moment of inertia of the cross section increases. Thus, efficient columns are designed so that most of the column's cross-sectional area is located as far away as possible from the principal centroidal axes for the section. This is why hollow sections such as tubes are more economical than solid sections. Furthermore, wide-flange sections, and columns that are "built up" from channels, angles, plates, etc., are better than sections that are solid and rectangular.

It is also important to realize that a column will buckle about the principal axis of the cross section having the *least moment of inertia* (the weakest axis). For example, a column having a rectangular cross section, like a meter stick, as shown in Fig. 13–7, will buckle about the *a–a* axis, not the *b–b* axis. As a result, engineers usually try to achieve a balance, keeping the moments of inertia the same in all directions. Geometrically, then, circular tubes would make excellent columns. Also, square tubes or those shapes having $I_x \approx I_y$ are often selected for columns.

Summarizing the above discussion, the buckling equation for a pin-supported column can be rewritten and the terms defined as follows:

$$P_{cr} = \frac{\pi^2 EI}{L^2} \tag{13–5}$$

where

P_{cr} = critical or maximum axial load on the column just before it begins to buckle. This load must *not* cause the stress in the column to exceed the proportional limit

E = modulus of elasticity for the material

I = *least* moment of inertia for the column's cross-sectional area

L = unsupported length of the column, whose ends are pinned

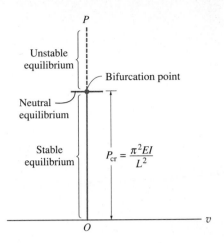

Fig. 13–6

Fig. 13–7

For purposes of design, Eq. 13–5 can also be written in a more useful form by expressing $I = Ar^2$, where A is the cross-sectional area and r is the *radius of gyration* of the cross-sectional area. Thus,

$$P_{cr} = \frac{\pi^2 E (Ar^2)}{L^2}$$

$$\left(\frac{P}{A}\right)_{cr} = \frac{\pi^2 E}{(L/r)^2}$$

or

$$\sigma_{cr} = \frac{\pi^2 E}{(L/r)^2} \qquad (13\text{–}6)$$

Here

σ_{cr} = critical stress, which is an *average stress* in the column just before the column buckles. This stress is an *elastic stress* and therefore $\sigma_{cr} \leq \sigma_Y$
E = modulus of elasticity for the material
L = unsupported length of the column, whose ends are pinned
r = *smallest* radius of gyration of the column, determined from $r = \sqrt{I/A}$, where I is the *least* moment of inertia of the column's cross-sectional area A

The geometric ratio L/r in Eq. 13–6 is known as the *slenderness ratio*. It is a measure of the column's flexibility, and as will be discussed later, it serves as a means of classifying columns as long, intermediate, or short.

It is possible to graph Eq. 13–6 using axes that represent the critical stress versus the slenderness ratio. Examples of this graph for columns made of a typical structural steel and aluminum alloy are shown in Fig. 13–8. Note that the curves are hyperbolic and are valid only for critical stresses below the material's yield point (proportional limit), since the material must behave elastically. For the steel the yield stress is $(\sigma_Y)_{st} = 36$ ksi $[E_{st} = 29(10^3)$ ksi] and for the aluminum it is $(\sigma_Y)_{al} = 27$ ksi $[E_{al} = 10(10^3)$ ksi]. Substituting $\sigma_{cr} = \sigma_Y$ into Eq. 13–6, the *smallest* acceptable slenderness ratios for the steel and aluminum columns are therefore $(L/r)_{st} = 89$ and $(L/r)_{al} = 60.5$. Thus, for a steel column, if $(L/r)_{st} \geq 89$, Euler's formula can be used to determine the buckling load since the stress in the column remains elastic. On the other hand, if $(L/r)_{st} < 89$, the column's stress will exceed the yield point before buckling can occur, and therefore the Euler formula is not valid in this case.

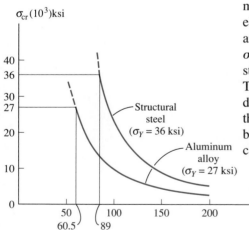

Fig. 13–8

Example 13–1

A 24-ft-long steel tube having the cross section shown in Fig. 13–9 is to be used as a pin-ended column. If $E_{st} = 29(10^3)$ ksi, determine the maximum allowable load the column can support so that it does not buckle. Take $\sigma_Y = 36$ ksi.

SOLUTION

Using Eq. 13–5 to obtain the critical load,

$$P_{cr} = \frac{\pi^2 EI}{L^2}$$

$$= \frac{\pi^2[29(10^3) \text{ kip/in}^2](\frac{1}{4}\pi(3)^4 - \frac{1}{4}\pi(2.75)^4) \text{ in}^4}{[24 \text{ ft}(12 \text{ in./ft})]^2}$$

$$= 64.5 \text{ kip} \qquad\qquad\qquad \textit{Ans.}$$

This force creates an average compressive stress in the column of

$$\sigma_{cr} = \frac{P_{cr}}{A} = \frac{64.5 \text{ kip}}{[\pi(3)^2 - \pi(2.75)^2] \text{ in}^2} = 14.3 \text{ ksi}$$

Since $\sigma_{cr} < \sigma_Y = 36$ ksi, application of Euler's equation is appropriate.

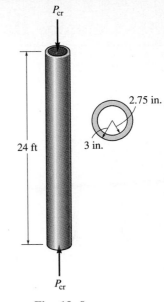

Fig. 13–9

Example 13–2

The steel $W\ 8 \times 31$ shown in Fig. 13–10 is to be used as a pin-connected column. Determine the largest load it can support before it either begins to buckle or the steel yields. $E_{st} = 29(10^3)$ ksi, and $\sigma_Y = 36$ ksi.

SOLUTION

From the table in Appendix B, the column's cross-sectional area and moments of inertia are $A = 9.13$ in^2, $I_x = 110$ in^4, and $I_y = 37.1$ in^4. By inspection, buckling will occur about the y–y axis. Why? Applying Eq. 13–5, we have

$$P_{cr} = \frac{\pi^2 EI}{L^2} = \frac{\pi^2[29(10^3) \text{ kip/in}^2](37.1 \text{ in}^4)}{[12 \text{ ft}(12 \text{ in./ft})]^2} = 512 \text{ kip}$$

When fully loaded, the average compressive stress in the column is

$$\sigma_{cr} = \frac{P_{cr}}{A} = \frac{512 \text{ kip}}{9.13 \text{ in}^2} - 56.1 \text{ ksi}$$

Since this stress exceeds the yield stress (36 ksi), the load P is determined from simple compression:

$$36 \text{ ksi} = \frac{P}{9.13 \text{ in}^2}; \qquad\qquad P = 329 \text{ kip} \qquad\qquad \textit{Ans.}$$

In actual practice, a factor of safety would be placed on this loading.

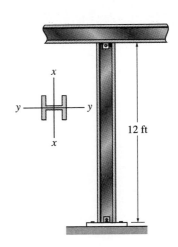

Fig. 13–10

13.3 Columns Having Various Types of Supports

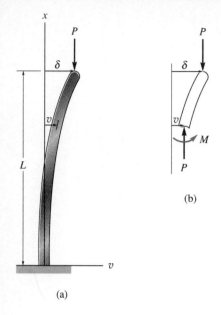

x

P

δ

v

L

v

(a)

Fig. 13–11

P

δ

v

M

P

(b)

In Sec. 13.2 we derived the Euler load for a column that is pin-connected or free to rotate at its ends. Oftentimes, however, columns may be supported in some other way. For example, consider the case of a column fixed at its base and free at the top, Fig. 13–11a. Determination of the buckling load on this column follows the same procedure as that used for the pinned column. From the free-body diagram in Fig. 13–11b, the internal moment at the arbitrary section is $M = P(\delta - v)$. Consequently, the differential equation for the deflection curve is

$$EI \frac{d^2v}{dx^2} = P(\delta - v)$$

$$\frac{d^2v}{dx^2} + \frac{P}{EI}v = \frac{P}{EI}\delta \qquad (13\text{–}7)$$

Unlike Eq. 13–2, this equation is nonhomogeneous because of the non-zero term on the right side. The solution consists of both a complementary and particular solution, namely,

$$v = C_1 \sin\left(\sqrt{\frac{P}{EI}}\, x\right) + C_2 \cos\left(\sqrt{\frac{P}{EI}}\, x\right) + \delta$$

The constants are determined from the boundary conditions. At $x = 0$, $v = 0$, so that $C_2 = -\delta$. Also,

$$\frac{dv}{dx} = C_1 \sqrt{\frac{P}{EI}} \cos\left(\sqrt{\frac{P}{EI}}\, x\right) - C_2 \sqrt{\frac{P}{EI}} \sin\left(\sqrt{\frac{P}{EI}}\, x\right)$$

At $x = 0$, $dv/dx = 0$, so that $C_1 = 0$. The deflection curve is therefore

$$v = \delta\left[1 - \cos\left(\sqrt{\frac{P}{EI}}\, x\right)\right] \qquad (13\text{–}8)$$

Since the deflection at the top of the column is δ, that is, at $x = L$, $v = \delta$, we require

$$\delta \cos\left(\sqrt{\frac{P}{EI}}\, L\right) = 0$$

The trivial solution $\delta = 0$ indicates that no buckling occurs, regardless of the load P. Instead,

$$\cos\left(\sqrt{\frac{P}{EI}}\, L\right) = 0 \qquad \text{or} \qquad \sqrt{\frac{P}{EI}}\, L = \frac{n\pi}{2}$$

The smallest critical load occurs when $n = 1$, so that

$$P_{cr} = \frac{\pi^2 EI}{4L^2} \qquad (13-9)$$

By comparison with Eq. 13–5 it is seen that a column fixed-supported at its base will carry only one-fourth the critical load that can be applied to a pin-supported column.

Other types of supported columns are analyzed in much the same way and will not be covered in detail here. Instead, we will tabulate the results for the most common types of column support and show how to apply these results by writing Euler's formula in a general form.

Effective Length. As stated previously, the Euler formula, Eq. 13–5, was developed for the case of a column having ends that are pinned or free to rotate. In other words, L in the equation represents the unsupported distance between the points of zero moment. If the column is supported in other ways, then Euler's formula can be used to determine the critical load provided "L" represents the distance between the zero-moment points. This distance is called the column's *effective length*, L_e. Obviously, for a pin-ended column $L_e = L$, Fig. 13–12a. For the fixed and free-ended column analyzed above, the deflection curve was found to be one-half that of a column that is pin-connected and has a length of $2L$, Fig. 13–12b. Thus the effective length between the points of zero moment is $L_e = 2L$. Examples for two other columns with different end supports are also shown in Fig. 13–12. The column fixed at its ends, Fig. 13–12c, has inflection points or points of zero moment $L/4$ from each support. The effective length is therefore represented by the middle half of its length, that is, $L_e = 0.5L$. Lastly, the pin- and fixed-ended column, Fig. 13–12d, has an inflection point at approximately $0.7L$ from its pinned end, so that $L_e = 0.7L$.

Rather than specifying the column's effective length, many design codes provide column formulas that employ a dimensionless coefficient K called the *effective-length factor*. K is defined from

$$L_e = KL \qquad (13-10)$$

Specific values of K are also given in Fig. 13–12. Based on this generality, we can therefore write Euler's formula as

$$P_{cr} = \frac{\pi^2 EI}{(KL)^2} \qquad (13-11)$$

or

$$\sigma_{cr} = \frac{\pi^2 E}{(KL/r)^2} \qquad (13-12)$$

Here (KL/r) is the column's *effective-slenderness ratio*. For example, note that for the column fixed at its base and free at its end, $K = 2$, and therefore Eq. 13–11 gives the same result as Eq. 13–9.

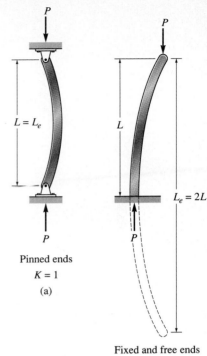

Pinned ends
$K = 1$
(a)

Fixed and free ends
$K = 2$
(b)

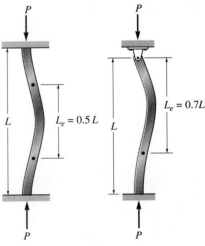

Fixed ends
$K = 0.5$
(c)

Pinned and fixed ends
$K = 0.7$
(d)

Fig. 13–12

Example 13–3

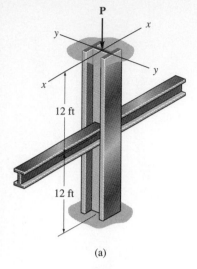

P

x

y

y

x

12 ft

12 ft

(a)

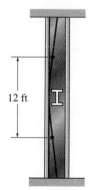

12 ft

I

x–x axis buckling

(b)

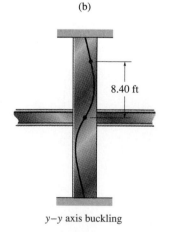

8.40 ft

y–y axis buckling

(c)

A $W\,6 \times 15$ steel column is 24 ft long and is fixed at its ends as shown in Fig. 13–13a. Its load-carrying capacity is increased by bracing it about the *y–y* (weak) axis using struts that are assumed to be pin-connected to its midheight. Determine the load it can support so that the column does not buckle nor the material exceed the yield stress. Take $E_{st} = 29(10^3)$ ksi and $\sigma_Y = 60$ ksi.

SOLUTION

The buckling behavior of the column will be *different* about the *x* and *y* axes due to the *y–y* axis bracing. The buckled shape for each of these cases is shown in Fig. 13–13b and 13–13c. From Fig. 13–13b, the effective length for the *x–x* axis is $(KL)_x = 0.5(24\text{ ft}) = 12\text{ ft} = 144$ in., and from Fig. 13–13c, for the *y–y* axis, $(KL)_y = 0.7(24\text{ ft}/2) = 8.40\text{ ft} = 100.8$ in. The moments of inertia for a $W\,6 \times 15$ are determined from the table in Appendix B. We have $I_x = 29.1$ in^4, $I_y = 9.32$ in^4.

Applying Eq. 13–12, we have

$$(P_{cr})_x = \frac{\pi^2 E I_x}{(KL)_x^2} = \frac{\pi^2 [29(10^3)\text{ ksi}]29.1\text{ in}^4}{(144\text{ in.})^2} = 401.7\text{ kip} \qquad (1)$$

$$(P_{cr})_y = \frac{\pi^2 E I_y}{(KL)_y^2} = \frac{\pi^2 [29(10^3)\text{ ksi}]9.32\text{ in}^4}{(100.8\text{ in.})^2} = 262.5\text{ kip} \qquad (2)$$

By comparison, buckling will occur about the *y–y* axis, Fig. 13–13c.

The area of the cross section is 4.43 in^2, so the average compressive stress in the column will be

$$\sigma_{cr} = \frac{P_{cr}}{A} = \frac{262.5\text{ kip}}{4.43\text{ in}^2} = 59.3\text{ ksi}$$

Since this stress is less than the yield stress, buckling will occur before the material yields. Thus,

$$P_{cr} = 263\text{ kip} \qquad \qquad \textit{Ans.}$$

Note: From Eq. 13–11 it can be seen that buckling will always occur about the column axis having the *largest* slenderness ratio, since a large slenderness ratio will give a small critical load. Thus, using the data for the radius of gyration from the table in Appendix B, we have

$$\left(\frac{KL}{r}\right)_x = \frac{144}{2.56} = 56.2$$

$$\left(\frac{KL}{r}\right)_y = \frac{100.8}{1.46} = 69.0$$

Hence, *y–y* axis buckling will occur, the same conclusion as reached by comparing Eqs. 1 and 2.

Fig. 13–13

Example 13–4

The aluminum column is fixed at its bottom and is braced at its top by cables so as to prevent movement at the top in the x–z plane, Fig. 13–14a. If it is assumed to be fixed at its base, determine the largest allowable load P that can be applied. Use a factor of safety for buckling of F.S. = 3.0. Take $E_{al} = 70$ GPa, $\sigma_Y = 215$ MPa, $A = 7.5(10^{-3})$ m^2, $I_x = 61.3(10^{-6})$ m^4, $I_y = 23.2(10^{-6})$ m^4.

SOLUTION

Buckling about the x and y axes is shown in Fig. 13–14b and 13–14c, respectively. Using Fig. 13–12, for x–x axis buckling, $K = 2$, so $(KL)_x = 2(5$ m$) = 10$ m. Also, for y–y axis buckling, $K = 0.7$, so $(KL)_y = 0.7(5$ m$) = 3.5$ m.

Applying Eq. 13–11, the critical loads for each case are

$$(P_{cr})_x = \frac{\pi^2 E I_x}{(KL)_x^2} = \frac{\pi^2 [70(10^9) \text{ N/m}^2](61.3(10^{-6}) \text{ m}^4)}{(10 \text{ m})^2}$$

$$= 424 \text{ kN}$$

$$(P_{cr})_y = \frac{\pi^2 E I_y}{(KL)_y^2} = \frac{\pi^2 [70(10^9) \text{ N/m}^2](23.2(10^{-6}) \text{ m}^4)}{(3.5 \text{ m})^2}$$

$$= 1.31 \text{ MN}$$

By comparison, as P is increased the column will buckle about the x–x axis. The allowable load is therefore

$$P_{allow} = \frac{P_{cr}}{\text{F.S.}} = \frac{424 \text{ kN}}{3} = 141 \text{ kN} \qquad \textit{Ans.}$$

Since

$$\sigma_{cr} = \frac{P_{cr}}{A} = \frac{424 \text{ kN}}{7.5(10^{-3}) \text{ m}^2} = 56.5 \text{ MPa} < 215 \text{ MPa}$$

Euler's equation can be applied.

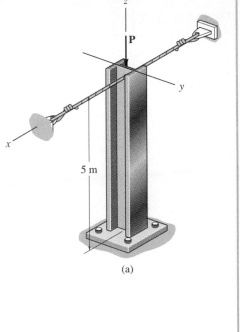

(a)

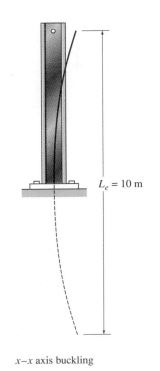

$L_e = 10$ m

x–x axis buckling

(b)

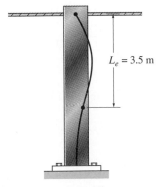

$L_e = 3.5$ m

y–y axis buckling

(c)

Fig. 13–14

PROBLEMS

13–1. The aircraft link is made from a steel rod. Determine the smallest diameter of the rod, to the nearest $\frac{1}{16}$ in., that will support the load of 4 kip without buckling. The ends are pin-connected. $E_{st} = 29(10^3)$ ksi, $\sigma_Y = 36$ ksi.

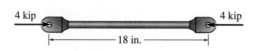

4 kip   4 kip

|←————— 18 in. —————→|

Prob. 13–1

13–2. A steel column has a length of 5 m and is fixed at both ends. If the cross-sectional area has the dimensions shown, determine the critical load. $E_{st} = 200$ GPa, $\sigma_Y = 360$ MPa.

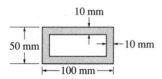

10 mm

50 mm   ←10 mm

|←—100 mm—→|

Prob. 13–2

13–3. A steel column has a length of 15 ft and is pinned at both ends. If the cross-sectional area has the dimensions shown, determine the critical load. $E_{st} = 29(10^3)$ ksi, $\sigma_Y = 36$ ksi.

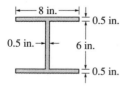

|←— 8 in. —→|
0.5 in.
0.5 in. →|←   6 in.
0.5 in.

Prob. 13–3

*13–4. The steel pipe is fixed-supported at its ends. If it is 4 m long and has an outer diameter of 50 mm and a thickness of 10 mm, determine the maximum axial load P that it can carry without buckling. Use a factor of safety with respect to buckling of F.S. = 1.5. $E_{st} = 200$ GPa, $\sigma_Y = 360$ MPa.

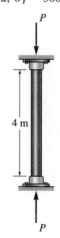

P

4 m

P

Prob. 13–4

13–5. The handle is used to operate a machine. If the machine is jammed, however, determine the maximum force P that can be applied to the handle so that the steel control rod AB does not buckle. The rod is made of steel and has a diameter of 1.25 in. It is pin-connected at its ends. $E_{st} = 29(10^3)$ ksi, $\sigma_Y = 36$ ksi.

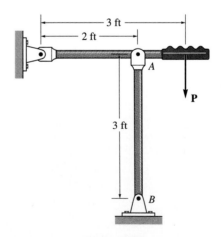

|←————— 3 ft —————→|
|←— 2 ft —→|
A
P

3 ft

B

Prob. 13–5

13–6. The $W\,12 \times 87$ column has a length of 12 ft. If its bottom end is fixed-supported while its top is free, and it is subjected to an axial load of 380 kip, determine the factor of safety with respect to buckling. $E_{st} = 29(10^3)$ ksi, $\sigma_Y = 36$ ksi.

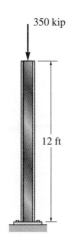

350 kip

12 ft

Prob. 13–6

13–7. The 12-ft pipe column has an outer diameter of 3 in. and a thickness of 0.25 in. Determine the critical load if (a) the ends are assumed to be pin-connected, (b) the bottom is fixed and the top is pinned. $E_{st} = 29(10^3)$ ksi, $\sigma_Y = 36$ ksi.

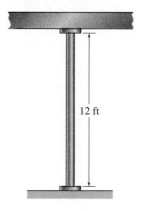

12 ft

Prob. 13–7

*13–8. An acrobat weighing 150 lb sits in the chair at the top of the 18-ft-long wooden pole, which is assumed fixed at the ground. If his center of gravity is positioned directly over the pole's longitudinal axis, determine the smallest diameter of the pole so that it will not buckle. $E_w = 1.6(10^3)$ ksi, $\sigma_Y = 5$ ksi.

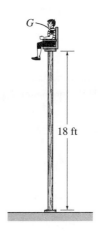

G

18 ft

Prob. 13–8

13–9. The steel pipe has an outer diameter of 2 in. and a thickness of 0.5 in. If it is held in place by a guywire, determine the largest horizontal force P that can be applied without causing the pipe to buckle. Assume that the ends of the pipe are pin-connected. $E_{st} = 29(10^3)$ ksi, $\sigma_Y = 36$ ksi.

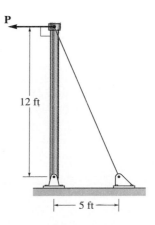

P

12 ft

5 ft

Prob. 13–9

13–10. The members of the truss are assumed to be pin-connected. If member *BD* is a steel rod of radius $r = 2$ in., determine the maximum load *P* that can be supported by the truss without causing the member to buckle. $E_{st} = 29(10^3)$ ksi, $\sigma_Y = 36$ ksi.

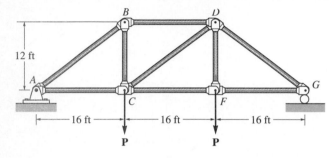

Prob. 13–10

13–11. The truss is made from steel bars, each of which has a circular cross section with a diameter of 1.5 in. Determine the maximum force *P* that can be applied without causing any of the members to buckle. The members are pin-supported at their ends. $E_{st} = 29(10^3)$ ksi, $\sigma_Y = 36$ ksi.

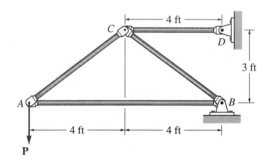

Prob. 13–11

*13–12. The column is supported at *B* by a support that does not permit rotation but allows vertical deflection. Determine the critical load P_{cr}. Assume elastic buckling.

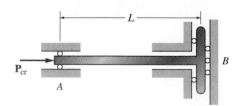

Prob. 13–12

13–13. The linkage is made using two steel rods, each having a circular cross section. Determine the diameter of each rod to the nearest $\frac{1}{8}$ in. that will support the 6-kip load. Assume that the rods are pin-connected at their ends. Use a factor of safety with respect to buckling of F.S. = 1.8. $E_{st} = 29(10^3)$, $\sigma_Y = 36$ ksi.

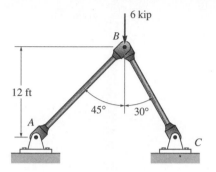

Prob. 13–13

13–14. The steel bar *AB* of the frame is pin-connected at its ends. Determine the factor of safety with respect to buckling about the *y–y* axis due to the applied loading. $E_{st} = 200$ GPa, $\sigma_Y = 360$ MPa.

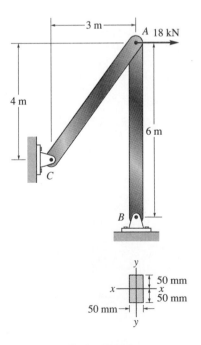

Prob. 13–14

13–15. The steel bar *AB* has a square cross section. If it is pin-connected at its ends, determine the maximum allowable load *P* that can be applied to the frame. Use a factor of safety with respect to buckling of F.S. = 2. E_{st} = 29(10³) ksi, σ_Y = 36 ksi.

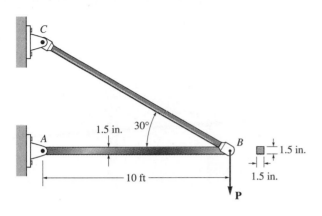

Prob. 13–15

13–17. The steel bar *AB* has a rectangular cross section. If it is pin-connected at its ends, determine the maximum allowable intensity *w* of the distributed load that can be applied to *BC* without causing bar *AB* to buckle. Use a factor of safety with respect to buckling of F.S. = 1.5. E_{st} = 200 GPa, σ_Y = 360 MPa.

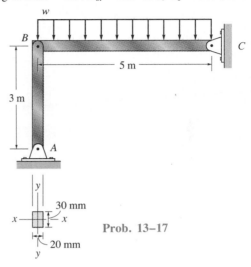

Prob. 13–17

*13–16. The steel bar *AB* of the frame is pin-connected at its ends. Determine the factor of safety with respect to buckling about the *y–y* axis due to the applied loading. E_{st} = 29(10³) ksi, σ_Y = 36 ksi.

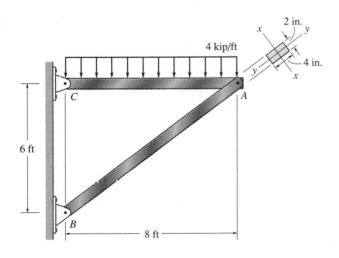

Prob. 13–16

13–18. Determine the maximum allowable intensity *w* of the distributed load that can be applied to member *BC* without causing member *AB* to buckle. Assume that *AB* is made of steel and is pinned at its ends for *x–x* axis buckling and fixed at its ends for *y–y* axis buckling. Use a factor of safety with respect to buckling of F.S. = 3. E_{st} = 200 GPa, σ_Y = 360 MPa.

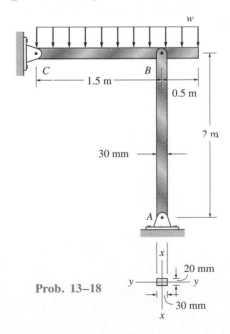

Prob. 13–18

13–19. The two steel channels are to be laced together to form a 30-ft-long bridge column assumed to be pin-connected at its ends. Each channel has a cross-sectional area of $A = 3.10$ in^2 and moments of inertia $I_{xx} = 55.4$ in^4, $I_{yy} = 0.382$ in^4. The centroid C of its area is located in the figure. Determine the proper distance d between the centroids of the channels so that buckling occurs about the x–x and y'–y' axes due to the same load. What is the value of this critical load? Neglect the effect of the lacing. $E_{st} = 29(10^3)$ ksi, $\sigma_Y = 50$ ksi.

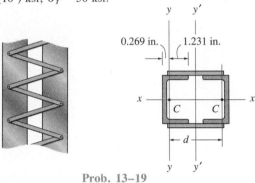

Prob. 13–19

*13–20. A column is constructed using four angles that are laced together as shown in Prob. 13–19. The length of the column is to be 25 ft and the ends are assumed to be pin-connected. Each angle, such as B, has an area of $A = 2.75$ in^2 and moments of inertia of $I_{xx} = I_{yy} = 2.22$ in^4. Determine the distance d between the centroids C of the angles so that the column can support an axial load of $P = 350$ kip without buckling. Neglect the effect of the lacing. $E_{st} = 29(10^3)$ ksi, $\sigma_Y = 36$ ksi.

13–21. A column is constructed using four angles that are laced together as shown in Prob. 10–19. The length of the column is to be 40 ft and the ends are assumed to be fixed-connected. Each angle, such as B, has an area of $A = 2.75$ in^2 and moments of inertia of $I_{xx} = I_{yy} = 2.22$ in^4. Determine the distance d between the centroids C of the angles so that the column can support an axial load of $P = 350$ kip without buckling. Neglect the effect of the lacing. $E_{st} = 29(10^3)$ ksi, $\sigma_Y = 36$ ksi.

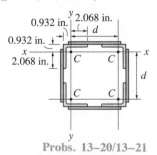

Probs. 13–20/13–21

13–22. The distributed loading is supported by two pin-connected columns, each having a circular cross section. If AB is made of aluminum and CD of steel, determine the required diameter of each column so that both will be on the verge of buckling at the same time. $E_{st} = 200$ GPa, $E_{al} = 70$ GPa, $(\sigma_Y)_{st} = 250$ MPa, $(\sigma_Y)_{al} = 100$ MPa.

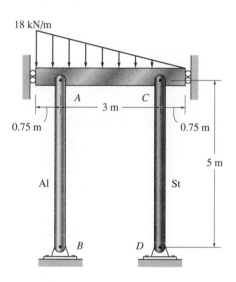

Prob. 13–22

13–23. The steel angle has a cross-sectional area of $A = 2.48$ in^2 and a radius of gyration about the x axis of $r_x = 1.26$ in. and about the y axis of $r_y = 0.879$ in. The smallest radius of gyration occurs about the z axis and is $r_z = 0.644$ in. If the angle is to be used as a pin-connected 10-ft-long column, determine the largest axial load that can be applied through its centroid C without causing it to buckle. $E_{st} = 29(10^3)$ ksi, $\sigma_Y = 36$ ksi.

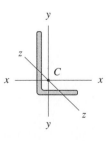

Prob. 13–23

*13–24. The steel angle has a cross-sectional area of $A = 2.48$ in^2 and a radius of gyration about the x axis of $r_x = 1.26$ in. and about the y axis of $r_y = 0.879$ in. The smallest radius of gyration occurs about the z axis and is $r_z = 0.644$ in. If the angle is to be used as a fixed-connected 10-ft-long column, determine the largest axial load that can be applied through its centroid C without causing it to buckle. $E_{st} = 29(10^3)$ ksi, $\sigma_Y = 36$ ksi.

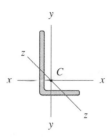

Prob. 13–24

13–25. The wood column has a length of 16 ft. It is fixed at its base A and pinned at its top C. Also, the two members are pin connected at B. If it has the cross section shown, determine the maximum axial force P that can be applied without causing it to buckle about the $y–y$ axis. The column is braced such that it will not buckle about the $x–x$ axis. $E_w = 1.6(10^3)$ ksi, $\sigma_Y = 8$ ksi.

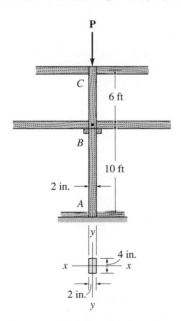

Prob. 13–25

13–26. The rod of radius r is pin-supported at each end as shown when the temperature is T_1. Determine the increased temperature T_2 that will cause the rod to buckle. The coefficient of thermal expansion is α, and the modulus of elasticity is E.

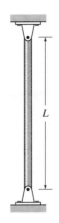

Prob. 13–26

13–27. The steel pipe is constrained between the walls at A and B. If there is a gap of 3 mm at B when $T_1 = 4°C$, determine the temperature required to cause the pipe to become unstable and begin to buckle. The pipe has an outer diameter of 40 mm and a wall thickness of 10 mm. Assume that the collars at A and B provide fixed connections for the pipe. Neglect their size. $\alpha_{st} = 12(10^{-6})/°C$, $E_{st} = 200$ GPa, $\sigma_Y = 250$ MPa.

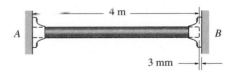

Prob. 13–27

*13.4 The Secant Formula

The Euler formula was derived with the assumptions that the load P is always applied through the centroid of the column's cross-sectional area and that the column is perfectly straight. This is actually quite unrealistic, since manufactured columns are *never* perfectly straight, nor is the application of the load known with great accuracy. In reality, then, columns never *suddenly buckle;* instead they begin to bend, although ever so slightly, immediately upon application of the load. As a result, the actual criterion for load application will be limited either to a specified deflection of the column or by not allowing the maximum stress in the column to exceed an allowable stress.

To study this effect, we will apply the load P to the column at a short *eccentric distance e* from the centroid of the cross section, Fig. 13–15a. This loading on the column is statically equivalent to the axial load P and bending moment $M' = Pe$ shown in Fig. 13–15b. As shown, in both cases, the ends A and B are supported so that they are free to rotate (pin-supported). As before, we will only consider small slopes and deflections and linear-elastic material behavior. Furthermore, the x–v plane is a plane of symmetry for the cross-sectional area.

From the free-body diagram of the arbitrary section, Fig. 13–15c, the internal moment in the column is

$$M = -P(e + v) \qquad (13-13)$$

The differential equation for the deflection curve is therefore

$$EI \frac{d^2v}{dx^2} = -P(e + v)$$

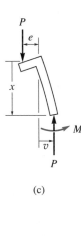

(c)

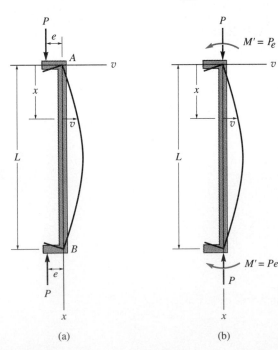

(a)

(b)

Fig. 13–15

or

$$\frac{d^2v}{dx^2} + \frac{P}{EI}v = -\frac{P}{EI}e$$

This equation is similar to Eq. 13–7 and has a general solution consisting of the complementary and particular solutions, namely,

$$v = C_1 \sin\sqrt{\frac{P}{EI}}\,x + C_2 \cos\sqrt{\frac{P}{EI}}\,x - e \qquad (13–14)$$

To evaluate the constants we must apply the boundary conditions. At $x = 0$, $v = 0$, so $C_2 = e$. And at $x = L$, $v = 0$, which gives

$$C_1 = \frac{e[1 - \cos(\sqrt{P/EI}\,L)]}{\sin(\sqrt{P/EI}\,L)}$$

Since $1 - \cos(\sqrt{P/EI}\,L) = 2\sin^2(\sqrt{P/EI}\,L/2)$ and $\sin(\sqrt{P/EI}\,L) = 2\sin(\sqrt{P/EI}\,L/2)\cos(\sqrt{P/EI}\,L/2)$, we have

$$C_1 = e\tan\left(\sqrt{\frac{P}{EI}}\,\frac{L}{2}\right)$$

Hence, the deflection curve, Eq. 13–14, can be written as

$$v = e\left(\tan\left(\sqrt{\frac{P}{EI}}\,\frac{L}{2}\right)\sin\left(\sqrt{\frac{P}{EI}}\,x\right) + \cos\left(\sqrt{\frac{P}{EI}}\,x\right) - 1\right) \quad (13–15)$$

Maximum Deflection. Due to symmetry of loading, both the maximum deflection and maximum stress occur at the column's midpoint. Therefore, when $x = L/2$, $v = v_{max}$, so

$$v_{max} = e\left[\sec\left(\sqrt{\frac{P}{EI}}\,\frac{L}{2}\right) - 1\right] \qquad (13–16)$$

Notice that if e approaches zero, then v_{max} approaches zero. However, if the terms in the brackets approach infinity as e approaches zero, then v_{max} will have a nonzero value. Mathematically, this would represent the behavior of an axially loaded column at failure when subjected to the critical load P_{cr}. Therefore, to find P_{cr} we require

$$\sec\left(\sqrt{\frac{P_{cr}}{EI}}\,\frac{L}{2}\right) = \infty$$

$$\sqrt{\frac{P_{cr}}{EI}}\,\frac{L}{2} = \frac{\pi}{2}$$

$$P_{cr} = \frac{\pi^2 EI}{L^2} \qquad (13–17)$$

which is the same result found from the Euler formula, Eq. 13–5.

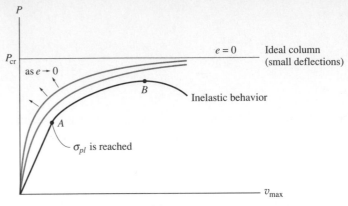

Fig. 13–16

If Eq. 13–16 is plotted as load P versus deflection v_{max} for various values of eccentricity e, the family of colored curves shown in Fig. 13–16 results. Here the critical load becomes an asymptote to the curves, and, of course represents the unrealistic case of an ideal column ($e = 0$). As stated earlier, e is *never zero,* due to imperfections in initial column straightness and load application. What is seen to happen is that as $e \to 0$, the curves tend to approach the ideal case. Furthermore, these curves are appropriate only for *small deflections,* since the curvature was approximated by d^2v/dx^2 when Eq. 13–16 was developed. Had a more exact analysis been performed, all these curves would tend to turn upward, intersecting and then rising above the line $P = P_{cr}$. This, of course, indicates that a larger load P is needed to create larger column deflections. We have not considered this analysis here, however, since most often engineering design restricts the deflection of columns to small values.

It should also be noted that the colored curves in Fig. 13–16 apply only for linear-elastic material behavior. Such is the case if the column is long and slender. However, if a short or intermediate-length stocky column is considered, then the applied load, as it is increased, may eventually cause the material to yield, and the column will begin to behave in an *inelastic manner.* This occurs at point A for the black curve in Fig. 13–16. As the load is further increased, the curve never reaches the critical load and instead the load reaches a maximum value at B. Afterwards, a sudden decrease in load-carrying capacity occurs as the column continues to deflect by larger amounts.

Lastly, the colored curves in Fig. 13–16 also illustrate that a *nonlinear* relationship occurs between the load P and the deflection v. As a result, the principle of superposition *cannot be used* to compute the total deflection of a column caused by applying *successive loads* to the column. Instead, the loads must first be added and then the corresponding deflection due to their resultant can be computed. Physically, the reason that successive loads and deflections cannot be superimposed is that the column's internal moment *depends* on both the load P and the *deflection v,* that is, $M = -P(e + v)$, Eq. 13–13. In other words, any *deflection* caused by a component load increases the moment. This behavior is unlike beam bending, where the actual deflection caused by the load *does not* increase the internal moment.

The Secant Formula. The maximum stress in the column can be determined by realizing that it is caused by both the axial load and the moment, Fig. 13–17a. Maximum moment occurs at the column's midpoint and, using Eq. 13–13, it has a magnitude of

$$M = |P(e + v_{max})|$$

$$M = Pe \sec\left(\sqrt{\frac{P}{EI}} \frac{L}{2}\right) \qquad (13\text{–}18)$$

As shown in Fig. 13–17b, the maximum stress in the column is compressive, and it has a value of

$$\sigma_{max} = \frac{P}{A} + \frac{Mc}{I}$$

$$\sigma_{max} = \frac{P}{A} + \frac{Pec}{I} \sec\left(\sqrt{\frac{P}{EI}} \frac{L}{2}\right)$$

Since the radius of gyration is defined as $r^2 = I/A$, the above equation can be written in a form called the *secant formula*:

$$\sigma_{max} = \frac{P}{A}\left[1 + \frac{ec}{r^2} \sec\left(\frac{L}{2r}\sqrt{\frac{P}{EA}}\right)\right] \qquad (13\text{–}19)$$

Here

σ_{max} = maximum *elastic stress* in the column, which occurs at the inner concave side at the column's midpoint. This stress is compressive.

P = vertical load applied to the column. $P < P_{cr}$ unless $e = 0$, then $P = P_{cr}$ (Eq. 13–5)

e = eccentricity of the load P, measured from the neutral axis of the column's cross-sectional area to the line of action of P

c = distance from the neutral axis to the outer fiber of the column where the maximum compressive stress σ_{max} occurs

A = cross-sectional area of the column

L = unsupported length of the column *in the plane of bending*. For supports other than pins, the effective length L_e should be used. See Fig. 13–12

E = modulus of elasticity for the material

r = radius of gyration, $r = \sqrt{I/A}$, where I is computed about the neutral or bending axis

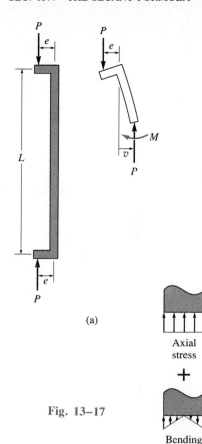

(a)

Fig. 13–17

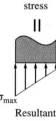

Axial
stress

+

Bending
stress

=

σ_{max}

Resultant
stress

(b)

Like Eq. 13–16, Eq. 13–19 indicates that there is a nonlinear relationship between the load and the stress. Hence, the principle of superposition does not apply, and therefore the loads have to be added *before* the stress is determined. Furthermore, due to this nonlinear relationship, any factor of safety used for design purposes applies to the load, not to the stress.

For a given value of σ_{max}, graphs of Eq. 13–19 can be plotted as KL/r versus P/A for various values of the *eccentricity ratio* ec/r^2. A specific set of graphs for a structural-grade steel having a yield point of $\sigma_{max} = \sigma_Y = 36$ ksi and a modulus of elasticity of $E_{st} = 29(10^3)$ ksi is shown in Fig. 13–18. Here the abscissa and ordinate represent the slenderness ratio KL/r and the average stress P/A, respectively. Note that when $e \rightarrow 0$, or when $ec/r^2 \rightarrow 0$, Eq. 13–19 gives $\sigma_{max} = P/A$, where P is the critical load on the column, defined by Euler's formula. This results in Eq. 13–6, which has been plotted in Fig. 13–8 and repeated in Fig. 13–18. Since both Eqs. 13–6 and 13–19 are valid only for elastic loadings, the stresses shown in Fig. 13–18 cannot exceed $\sigma_Y = 36$ ksi, represented here by the horizontal line.

The curves in Fig. 13–18 indicate that differences in the eccentricity ratio have a marked effect on the load-carrying capacity of columns that have small slenderness ratios. On the other hand, columns that have large slenderness ratios tend to fail at or near the Euler critical load regardless of the eccentricity ratio. When using Eq. 13–19 for design purposes, it is therefore important to have a somewhat accurate value for the eccentricity ratio for shorter-length columns.

Design.   Once the eccentricity ratio has been determined, the column data can be substituted into Eq. 13–19. If a value of $\sigma_{max} = \sigma_Y$ is chosen, then the corresponding load P_Y can be determined from a trial-and-error procedure, since the equation is transcendental and cannot be solved explicitly for P_Y. As a design aid, computer software, or graphs such as those in Fig. 13–18, can also be used to determine P_Y directly.

Realize that P_Y is the load that will cause the column to develop a maximum compressive stress of σ_Y at its inner concave fibers. Due to the eccentric application of P_Y, this load will *always be smaller* than the critical load P_{cr}, which is determined from the Euler formula and assumes unrealistically that the column is axially loaded. Once P_Y is obtained, an appropriate factor of safety can then be applied in order to specify the column's safe load.

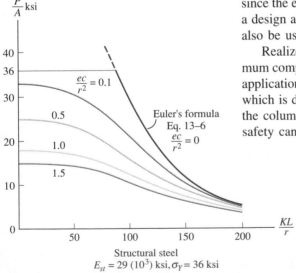

Fig. 13–18

Example 13–5

The $W\ 8 \times 40$ steel column shown in Fig. 13–19a is fixed at its base and braced at the top so that it is fixed from displacement, yet free to rotate about the y–y axis. Also, it can sway to the side in the y–z plane. Determine the maximum eccentric load the column can support before it either begins to buckle or the steel yields. Take $\sigma_Y = 36$ ksi and $E_{st} = 29(10^3)$ ksi.

SOLUTION

From the support conditions it is seen that about the y–y axis the column behaves as if it were pinned at its top and fixed at the bottom and subjected to an axial load P, Fig. 13–19b. About the x–x axis the column is free at the top and fixed at the bottom, and it is subjected to both an axial load P and moment $M = P(9\ \text{in.})$, Fig. 13–19c.

y–y Axis Buckling. From Fig. 13–12d the effective length factor is $K_y = 0.7$, so $(KL)_y = 0.7(12)$ ft $= 8.40$ ft $= 100.8$ in. Using the table in Appendix B to determine I_y for the $W\ 8 \times 40$ section and applying Eq. 13–11, we have

$$(P_{cr})_y = \frac{\pi^2 E I_y}{(KL)_y^2} = \frac{\pi^2 [29(10^3)\ \text{ksi}](49.1\ \text{in}^4)}{(100.8\ \text{in.})^2} = 1383\ \text{kip}$$

x–x Axis Yielding. From Fig. 13–12b, $K_x = 2$, so $(KL)_x = 2(12)$ ft $= 24$ ft $= 288$ in. Again using the table in Appendix B to determine A, c, and r_x, and applying the secant formula, we have

$$\sigma_Y = \frac{P_x}{A}\left[1 + \frac{ec}{r_x^2}\sec\left(\frac{(KL)_x}{2r_x}\sqrt{\frac{P_x}{EA}}\right)\right]$$

$$36 = \frac{P_x}{11.7}\left\{1 + \frac{9(8.25/2)}{(3.53)^2}\sec\left[\frac{288}{2(3.53)}\sqrt{\frac{P_x}{29(10^3)(11.7)}}\right]\right\}$$

or

$$421.2 = P_x[1 + 2.979\sec(0.0700\sqrt{P_x})]$$

Solving for P_x by trial and error, we get

$$P_x = 88.4\ \text{kip} \qquad\qquad Ans.$$

Since this value is less than $(P_{cr})_y = 1383$ kip, failure will occur about the x–x axis.

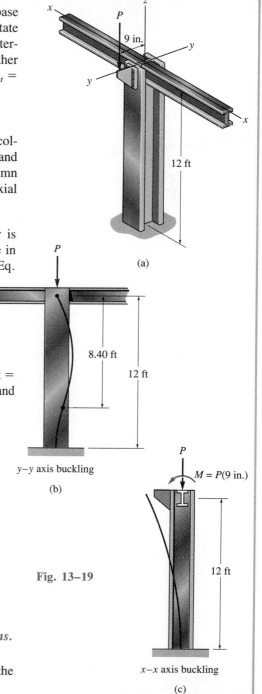

(a)

P

y–y axis buckling

(b)

$M = P(9\ \text{in.})$

x–x axis buckling

(c)

Fig. 13–19

Example 13–6

The steel column shown in Fig. 13–18 is assumed to be pin-connected at its top and base. Determine the allowable eccentric load P that can be applied. Also, what is the maximum deflection of the column due to this loading? Assume buckling does not occur about the y axis. Take $E_{st} = 29(10^3)$ ksi, $\sigma_Y = 36$ ksi.

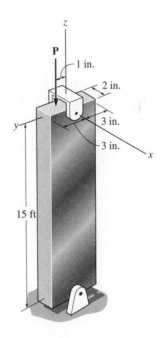

Fig. 13–20

SOLUTION

Computing the necessary geometrical properties, we have

$$I_x = \frac{1}{12}\,(2\text{ in.})(6\text{ in.})^3 = 36\text{ in}^4$$

$$A = (2\text{ in.})(6\text{ in.}) = 12\text{ in}^2$$

$$r_x = \sqrt{\frac{36\text{ in}^4}{12\text{ in}^2}} = 1.732\text{ in.}$$

$$e = 1\text{ in.}$$

$$L_e = KL = 1(15\text{ ft})(12\text{ in./ft}) = 180\text{ in.}$$

$$\frac{L_e}{r_x} = \frac{180\text{ in.}}{1.732\text{ in.}} = 104$$

Since the curves in Fig. 13–18 have been established for $E_{st} = 29(10^3)$ ksi, and $\sigma_Y = 36$ ksi, we can use them to determine the value of P/A and thus avoid a trial-and-error solution of the secant formula. Here $L_e/r_x = 104$. Using the curve defined by the eccentricity ratio $ec/r^2 = 1(3)/(1.732)^2 = 1$, we get

$$\frac{P}{A} \approx 12 \text{ ksi}$$

$$P = (12 \text{ ksi})(12 \text{ in}^2) = 144 \text{ kip} \qquad \textbf{\textit{Ans.}}$$

We can check this value by showing that it satisfies the secant formula, Eq. 13–19:

$$\sigma_{max} = \frac{P}{A}\left[1 + \frac{ec}{r^2} \sec\left(\frac{L}{2r}\sqrt{\frac{P}{EA}}\right)\right]$$

$$36 \overset{?}{=} \frac{144}{12}\left[1 + (1) \sec\left(\frac{180}{2(1.732)}\sqrt{\frac{144}{29(10^3)(12)}}\right)\right]$$

$$36 \overset{?}{=} 12[1 + \sec(1.0570 \text{ rad})]$$

$$36 \overset{?}{=} 12(1 + \sec 60.56°)$$

$$36 \approx 36.4$$

The maximum deflection occurs at the column's center, where $\sigma_{max} = 36$ ksi. Applying Eq. 13–16, we have

$$v_{max} = e\left[\sec\left(\sqrt{\frac{P}{EI}}\frac{L}{2}\right) - 1\right]$$
$$= 1\left[\sec\left(\sqrt{\frac{144}{29(10^3)(36)}}\frac{180}{2}\right) - 1\right]$$
$$= 1[\sec 1.057 \text{ rad} - 1]$$
$$= 1[\sec 60.56° - 1]$$
$$= 1.03 \text{ in.} \qquad \textbf{\textit{Ans.}}$$

*13.5 Inelastic Buckling

In engineering practice, columns are generally classified according to the type of stresses developed within the column at the time of failure. *Long slender columns* will become unstable when the compressive stress remains elastic. The failure that occurs is referred to as *elastic instability*. *Intermediate columns* fail due to *inelastic instability*, meaning that the compressive stress at failure is greater than the material's yield stress. And *short columns*, sometimes called *posts*, do not become unstable; rather the material simply yields or fractures.

Application of the Euler equation requires that the stress in the column remain *below* the material's yield point (actually the proportional limit) when the column buckles, and so this equation applies only to long columns. In practice, however, most columns are selected to have intermediate lengths. The behavior of these columns can be studied by modifying the Euler equation so that it applies for inelastic buckling. To show how this can be done, consider the material to have a stress–strain diagram as shown in Fig. 13–21a. Here the yield stress is σ_Y, and the modulus of elasticity, or slope of the line AB, is E. A plot of Euler's hyperbola, Fig. 13–8, is shown in Fig. 13–21b. This equation is valid for a column having a slenderness ratio as small as $(KL/r)_Y$, since at this point the axial stress in the column becomes $\sigma_{cr} = \sigma_Y$. If the column has a slenderness ratio that is *less* than $(KL/r)_Y$, then the critical stress in the column must be greater than σ_Y. For example, suppose a column has a slenderness ratio of $(KL/r)_1 < (KL/r)_Y$, with corresponding critical stress $\sigma_D > \sigma_Y$ needed to cause instability. When the column is *about to buckle*, the change in strain that occurs in the column is within a *small range* $\Delta\epsilon$, so that the modulus of elasticity or stiffness for the material can be taken as the *tangent modulus* E_t, defined as the slope of the σ–ϵ diagram at point D, Fig. 13–21a. In other words, at the time of failure, the column

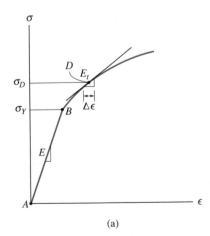

(a)

Fig. 13–21(a)

behaves as if it was made from a material that has a *lower stiffness* than when it behaves elastically, $E_t < E$. In general, therefore, as the slenderness ratio decreases, the *critical stress* for a column continues to rise; and from the $\sigma-\epsilon$ diagram, the *tangent modulus* for the material *decreases*. Using this idea, we can modify Euler's equation to include these cases of inelastic buckling by substituting the material's tangent modulus E_t for E, so that

$$\sigma_{cr} = \frac{\pi^2 E_t}{(KL/r)^2} \qquad (13\text{--}20)$$

This is the so-called *tangent modulus* or *Engesser equation,* proposed by F. Engesser in 1889. A plot of this equation for intermediate and short-length columns of a material defined by the $\sigma-\epsilon$ diagram in Fig. 13–21a is shown in Fig. 13–21b.

No *actual column* can be considered to be either perfectly straight or loaded along its centroidal axis, as assumed here, and therefore it is indeed very difficult to develop an expression that will provide a full analysis of this phenomenon. It should also be pointed out that other methods of describing the inelastic buckling of columns have been considered. One of these methods was developed by the aeronautical engineer F. R. Shanley and is called the *Shanley theory* of inelastic buckling. Although it provides a better description of the phenomenon than the tangent modulus theory, as explained here, experimental testing of a large number of columns, each of which approximates the ideal column, has shown that Eq. 13–20 is *reasonably accurate* in predicting the column's critical stress. Furthermore, the tangent modulus approach to modeling inelastic column behavior is relatively easy to apply.

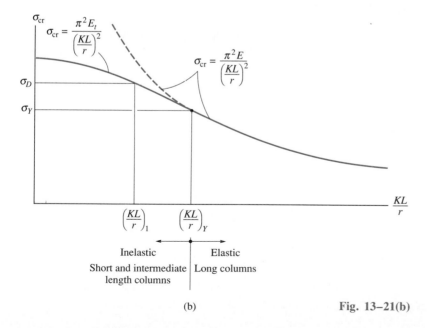

(b)   Fig. 13–21(b)

Example 13–7

A solid rod has a diameter of 30 mm and is 600 mm long. It is made of a material that can be modeled by the stress–strain diagram shown in Fig. 13–22. If it is used as a pin-supported column, determine the critical load.

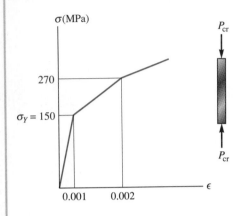

Fig. 13–22

SOLUTION

The radius of gyration is

$$r = \sqrt{\frac{I}{A}} = \sqrt{\frac{(\pi/4)(15)^4}{\pi(15)^2}} = 7.5 \text{ mm}$$

and therefore the slenderness ratio is

$$\frac{KL}{r} = \frac{1(600)}{7.5} = 80$$

Applying Eq. 13–20 yields

$$\sigma_{cr} = \frac{\pi^2 E_t}{(KL/r)^2} = \frac{\pi^2 E_t}{(80)^2} = 1.542(10^{-3})E_t \qquad (1)$$

First we will assume that the critical stress is elastic. From Fig. 13–22,

$$E = \frac{150 \text{ MPa}}{0.001} = 150 \text{ GPa}$$

Thus, Eq. 1 becomes

$$\sigma_{cr} = 1.542(10^{-3})[150(10^3)] \text{ MPa} = 231.3 \text{ MPa}$$

Since $\sigma_{cr} > \sigma_Y = 150$, inelastic buckling occurs.

From the second line segment of the σ–ϵ diagram, Fig. 13–22, we have

$$E_t = \frac{\Delta\sigma}{\Delta\epsilon} = \frac{270 - 150}{0.002 - 0.001} = 120 \text{ GPa}$$

Applying Eq. 1 yields

$$\sigma_{cr} = 1.542(10^{-3})[120(10^3)] \text{ MPa} = 185 \text{ MPa}$$

Since this value falls within the limits of 150 MPa and 270 MPa, it is indeed the critical stress.

The critical load on the rod is therefore

$$P_{cr} = \sigma_{cr}A = 185 \text{ MPa}[\pi(0.015 \text{ m})^2] = 131 \text{ kN} \qquad \textit{Ans.}$$

PROBLEMS

*13–28. Determine the load P required to cause the steel $W\,8 \times 15$ column to fail either by buckling or by yielding. Take $E_{st} = 29(10^3)$ ksi, $\sigma_Y = 36$ ksi.

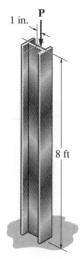

Prob. 13–28

13–29. The wood column has a square cross section with dimensions 100 mm by 100 mm. It is fixed at its base and free at its top. Determine the load P that can be applied to the edge of the column without causing the column to fail either by buckling or by yielding. $E_w = 12$ GPa, $\sigma_Y = 55$ MPa.

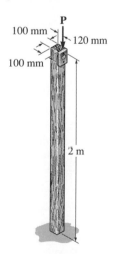

Prob. 13–29

13–30. The $W\,10 \times 30$ column is pinned at its top and bottom. If it is subjected to the eccentric load of 85 kip, determine the factor of safety with respect to the initiation of yielding. $E_{st} = 29(10^3)$ ksi, $\sigma_Y = 36$ ksi.

13–31. The $W\,10 \times 30$ column is fixed at its bottom and free at its top. If it is subjected to the eccentric load of 85 kip, determine if the column fails by yielding. The column is braced so that it does not buckle about the y-y axis. $E_{st} = 29(10^3)$ ksi, $\sigma_Y = 36$ ksi.

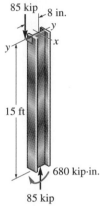

Probs. 13–30/13–31

*13–32. A $W\,12 \times 26$ is to be used as a pin-connected column with $L = 20$ ft. Determine the maximum eccentric load P that can be applied so the column does not buckle or yield. Compare this value with an axial critical load P' applied through the centroid of the column. $E_{st} = 29(10^3)$ ksi, $\sigma_Y = 36$ ksi.

Prob. 13–32

13–33. A $W\,12 \times 26$ is to be used as a fixed-connected column with $L = 23$ ft. Determine the maximum eccentric load P that can be applied so the column does not buckle or yield. Compare this value with an axial critical load P' applied through the centroid of the column. $E_{st} = 29(10^3)$ ksi, $\sigma_Y = 36$ ksi.

Prob. 13–33

13–34. The wood column is fixed at its base and can be assumed pin-connected at its top. Determine the maximum eccentric load P that can be applied at its top without causing the column to buckle or yield. $E_w = 1.8(10^3)$ ksi, $\sigma_Y = 8$ ksi.

13–35. The wood column is fixed at its base and can be assumed fixed-connected at its top. Determine the maximum eccentric load P that can be applied at its top without causing the column to buckle or yield. $E_w = 1.8(10^3)$ ksi, $\sigma_Y = 8$ ksi.

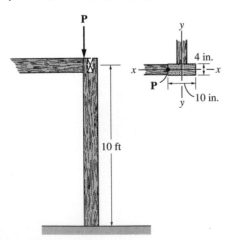

Probs. 13–34/13–35

*13–36.** The $W\,14 \times 53$ column is fixed at its base and free at its top. If $P = 75$ kip, determine the sidesway deflection at its top and the maximum stress in the column. $E_{st} = 29(10^3)$ ksi, $\sigma_Y = 36$ ksi.

13–37. The $W\,14 \times 53$ column is fixed at its base and free at its top. Determine the maximum eccentric load P that it can support without causing it to buckle or yield. $E_{st} = 29(10^3)$ ksi, $\sigma_Y = 50$ ksi.

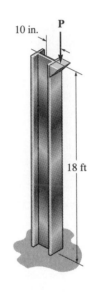

Probs. 13–36/13–37

13–38. The tube is made of copper and has an outer diameter of 30 mm and a wall thickness of 5 mm. Using a factor of safety with respect to yielding of F.S. = 2.5, determine the allowable eccentric load P. The tube is pin-supported at its ends. $E_{cu} = 120$ GPa, $\sigma_Y = 750$ MPa.

13–39. The tube is made of copper and has an outer diameter of 30 mm and a wall thickness of 5 mm. Using a factor of safety with respect to yielding of F.S. = 2.5, determine the allowable eccentric load P that it can support without failure. The tube is fixed-supported at its ends. $E_{cu} = 120$ GPa, $\sigma_Y = 750$ MPa.

P

13 mm

1.5 m

P

Probs. 13–38/13–39

13–42.   The aluminum rod is fixed at its base and free at its top. If the eccentric load $P = 200$ kN is applied, determine the greatest allowable length L of the rod so that it does not buckle or yield. $E_{al} = 72$ GPa, $\sigma_Y = 410$ MPa.

13–43.   The aluminum rod is fixed at its base and free at its top. If the length of the rod is $L = 2$ m, determine the greatest allowable load P that can be applied so that the rod does not buckle or yield. Also, determine the largest sidesway deflection of the rod due to the loading. $E_{al} = 72$ GPa, $\sigma_Y = 410$ MPa.

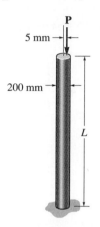

P

5 mm

200 mm

L

Probs. 13–42/13–43

*13–40.   The $W\,10 \times 45$ is used as an 18-ft-long column that is assumed to be pinned at its top and fixed at its bottom. If the 12-kip load is applied at an eccentric distance of 8 in., determine the maximum stress in the column. $E_{st} = 29(10^3)$ ksi, $\sigma_Y = 36$ ksi.

13–41.   The $W\,10 \times 45$ is used as a column that is assumed to be fixed at its top and bottom. If the 12-kip load is applied at an eccentric distance of 8 in., determine the maximum stress in the column. $E_{st} = 29(10^3)$ ksi, $\sigma_Y = 36$ ksi.

*13–44.   The square structural tubing has outer dimensions of 8 in. by 8 in. Its cross-sectional area is 14.40 in^2 and its moments of inertia are $I_x = I_y = 131$ in^4. If a load of 120 kip is applied at its top as shown, determine the factor of safety of the tube with respect to yielding. The column can be assumed fixed at its base and free at its top. $E_{st} = 29(10^3)$ ksi, $\sigma_Y = 36$ ksi.

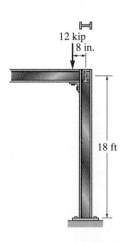

12 kip

8 in.

18 ft

Probs. 13–40/13–41

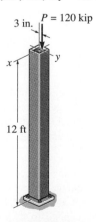

3 in. $P = 120$ kip

x   y

12 ft

Prob. 13–44

13–45. The aluminum column has the cross section shown. If it is fixed at the bottom and free at the top, determine the maximum force P that can be applied at A without causing it to buckle or yield. Use a factor of safety of F.S. = 3 with respect to buckling and yielding. $E_{al} = 70$ GPa, $\sigma_Y = 95$ MPa.

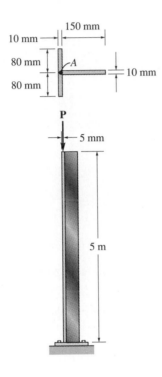

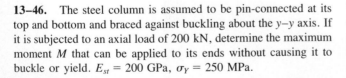

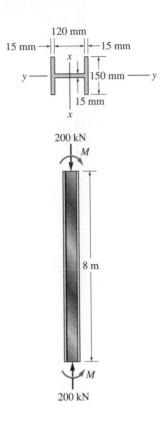

Prob. 13–45

Probs. 13–46/13–47

13–46. The steel column is assumed to be pin-connected at its top and bottom and braced against buckling about the y–y axis. If it is subjected to an axial load of 200 kN, determine the maximum moment M that can be applied to its ends without causing it to buckle or yield. $E_{st} = 200$ GPa, $\sigma_Y = 250$ MPa.

13–47. The steel column is assumed to be fixed-connected at its top and bottom and braced against buckling about the y–y axis. If it is subjected to an axial load of 200 kN, determine the maximum moment M that can be applied to its ends without causing it to yield. $E_{st} = 200$ GPa, $\sigma_Y = 250$ MPa.

*13–48. The column supports the two eccentric loadings. If it is assumed to be pinned at its top, fixed at the bottom, and braced against buckling about the y–y axis, determine the maximum deflection of the column and the maximum stress in the column. $E_{st} = 200$ GPa, $\sigma_Y = 360$ MPa.

13–49. The column supports the two eccentric loadings. If it is assumed to be fixed at its top and bottom, and braced against buckling about the y–y axis, determine the maximum deflection of the column and the maximum stress in the column. $E_{st} = 200$ GPa, $\sigma_Y = 360$ MPa.

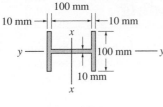

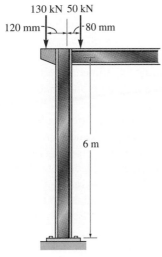

Probs. 13–48/13–49

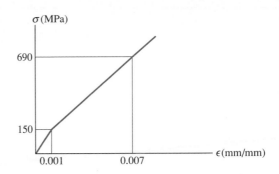

Probs. 13–50/13–51/13–52

13–53. A column of intermediate length buckles when the compressive stress is 40 ksi. If the slenderness ratio is 60, determine the tangent modulus. The material's stress–strain curve is shown in the figure.

13–54. Construct the buckling curve, P/A versus L/r, for a column that has a bilinear stress–strain curve in compression as shown.

13–50. The stress–strain diagram for a material can be approximated by the two line segments shown. If a bar having a diameter of 80 mm and a length of 1.5 m is made of this material, determine the critical load provided the ends are pinned. Assume that the load acts through the axis of the bar. Use Engesser's equation.

13–51. The stress–strain diagram for a material can be approximated by the two line segments shown. If a bar having a diameter of 80 mm and a length of 1.5 m is made of this material, determine the critical load provided the ends are fixed. Assume that the load acts through the axis of the bar. Use Engesser's equation.

*13–52. The stress–strain diagram for a material can be approximated by the two line segments shown. If a bar having a diameter of 80 mm and length of 1.5 m is made of this material, determine the critical load provided one end is pinned and the other is fixed. Assume that the load acts through the axis of the bar. Use Engesser's equation.

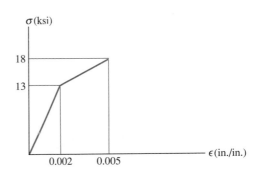

Prob. 13–54

*13.6 Design of Columns for Concentric Loading

The theory presented thus far applies to columns that are perfectly straight, made of homogeneous material, and originally stress-free. Practically speaking, though, as stated previously, columns are not perfectly straight, and most have residual stresses in them, primarily due to nonuniform cooling during manufacture. Also, the supports for columns are less than exact, and the points of application and directions of loads are not known with absolute certainty. In order to compensate for these effects, which actually vary from one column to the next, many design codes specify the use of column formulas that are empirical. By performing experimental tests on a large number of *axially loaded columns,* the results may be plotted and a design formula developed by curve-fitting the mean of the data.

An example of such tests for wide-flange steel columns is shown in Fig. 13–23. Notice the similarity between these results and those of the family of curves determined from the secant formula, Fig. 13–18. The reason for this similarity has to do with the influence of an "accidental" eccentricity ratio on the column's strength. As stated in Sec. 13.4, this ratio has more of an effect on the strength of short and intermediate-length columns than on those that are long. Tests have indicated that ec/r^2 can range from 0.1 to 0.6 for most axially loaded columns.

In order to account for the behavior of different-length columns, design codes usually specify several formulas that will best fit the data within the short, intermediate, and long column range. Hence, each formula will apply only for a specific *range* of slenderness ratios, and so it is important that the engineer carefully observe the *KL/r* limits for which a particular formula is valid. Examples of design formulas for steel, aluminum, and wood columns that are currently in use will now be discussed. The purpose is to give some idea as to how columns are designed in practice. These formulas should not, however, be used for the design of actual columns, unless the code from which they are referenced is consulted.

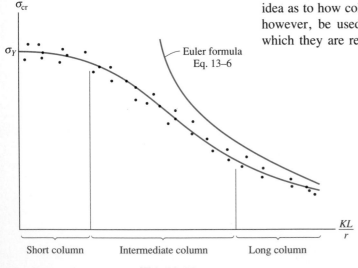

Fig. 13–23

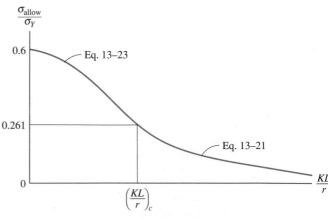

Fig. 13–24

Steel Columns. Columns made of structural steel are currently designed on the basis of formulas proposed by the Structural Stability Research Council (SSRC). Factors of safety have been applied to these formulas and adopted as specifications for building construction by the American Institute of Steel Construction (AISC). Basically these specifications provide two formulas for column design, each of which gives the maximum allowable stress in the column for a specific range of slenderness ratios. For long columns the Euler formula is proposed, i.e., $\sigma_{max} = \pi^2 E/(KL/r)^2$.

Application of this formula requires that a factor of safety F.S. $= \frac{23}{12} \approx$ 1.92 be applied. Thus, for design,

$$\sigma_{allow} = \frac{12\pi^2 E}{23(KL/r)^2} \qquad \left(\frac{KL}{r}\right)_c \le \frac{KL}{r} \le 200 \qquad (13\text{-}21)$$

As stated, this equation is applicable for a slenderness ratio bounded by 200 and $(KL/r)_c$. A specific value of $(KL/r)_c$ is obtained by requiring the Euler formula to be used only for elastic material behavior. Through experiments it has been determined that compressive residual stresses can exist in rolled-formed steel sections that may be as much as one-half the yield stress. Consequently, if the stress in the Euler formula is greater than $\frac{1}{2}\sigma_Y$, the equation will not apply. Therefore the value of $(KL/r)_c$ can be determined as follows:

$$\frac{1}{2}\sigma_Y = \frac{\pi^2 E}{(KL/r)^2}$$

$$\left(\frac{KL}{r}\right)_c = \sqrt{\frac{2\pi^2 E}{\sigma_Y}} \qquad (13\text{-}22)$$

Columns having slenderness ratios less than $(KL/r)_c$ are designed on the basis of an empirical formula that is parabolic and has the form

$$\sigma_{max} = \left[1 - \frac{(KL/r)^2}{(KL/r)_c^2}\right]\sigma_Y$$

Since there is more uncertainty in the use of this formula for longer columns, it is divided by a factor of safety defined as follows:

$$\text{F.S.} = \frac{5}{3} + \frac{3}{8}\frac{(KL/r)}{(KL/r)_c} - \frac{(KL/r)^3}{8(KL/r)_c^3}$$

Here it is seen that F.S. $= \frac{5}{3}$ at $KL/r = 0$ and increases to F.S. $= \frac{23}{12}$ at $(KL/r)_c$. Hence, for design purposes,

$$\sigma_{allow} = \frac{\left[1 - \dfrac{(KL/r)^2}{2(KL/r)_c^2}\right]\sigma_Y}{\{(5/3) + [(3/8)(KL/r)/(KL/r)_c] - [(KL/r)^3/8(KL/r)_c^3]\}} \qquad \begin{array}{c}\left(\dfrac{KL}{r}\right) < \left(\dfrac{KL}{r}\right)_c \\ \\ (13\text{-}23)\end{array}$$

Equations 13–21 and 13–23 are plotted in Fig. 13–24. When applying any of these equations, either FPS or SI units can be used for the calculations.

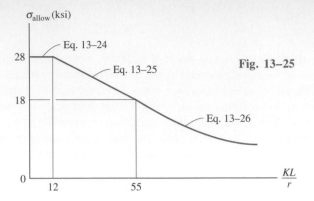

Fig. 13–25

Aluminum Columns.

Column design for structural aluminum is specified by the Aluminum Association using three equations, each applicable for a specific range of slenderness ratios. Since several types of aluminum alloy exist, there is a unique set of formulas for each type. For a common alloy (2014-T6) used in building construction, the formulas are

$$\sigma_{\text{allow}} = 28 \text{ ksi} \qquad 0 \le \frac{KL}{r} \le 12 \tag{13–24}$$

$$\sigma_{\text{allow}} = \left[30.7 - 0.23\left(\frac{KL}{r}\right) \right] \text{ksi} \qquad 12 < \frac{KL}{r} < 55 \tag{13–25}$$

$$\sigma_{\text{allow}} = \frac{54{,}000 \text{ ksi}}{(KL/r)^2} \qquad 55 \le \frac{KL}{r} \tag{13–26}$$

These equations are plotted in Fig. 13–25. As shown, the first two represent straight lines and are used to model the effects of columns in the short and intermediate range. The third formula has the same form as the Euler formula and is used for long columns.

Timber Columns.

Columns used in timber construction are designed on the basis of formulas published by the National Forest Products Association (NFPA) and the American Institute of Timber Construction (AITC). For example, the NFPA formulas for the allowable stress in short, intermediate, and long columns having a rectangular cross section of dimensions b and d, where d is the *smallest* dimension of the cross section, are

$$\sigma_{\text{allow}} = 1.20 \text{ ksi} \qquad 0 \le \frac{KL}{d} \le 11 \tag{13–27}$$

$$\sigma_{\text{allow}} = 1.20\left[1 - \frac{1}{3}\left(\frac{KL/d}{26.0}\right)^2 \right]\text{ksi} \qquad 11 < \frac{KL}{d} \le 26 \tag{13–28}$$

$$\sigma_{\text{allow}} = \frac{540 \text{ ksi}}{(KL/d)^2} \qquad 26 < \frac{KL}{d} \le 50 \tag{13–29}$$

Here the wood has a modulus of elasticity of $E_w = 1.8(10^3)$ ksi and an allowable compressive stress of 1.2 ksi parallel to the grain. In particular, Eq. 13–29 is simply Euler's equation having a factor of safety of 3. These three equations are plotted in Fig. 13–26.

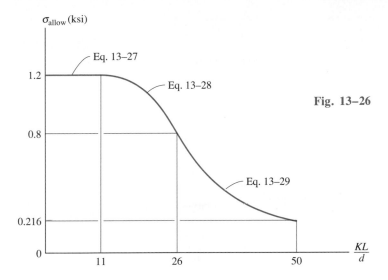

Fig. 13–26

PROCEDURE FOR ANALYSIS

Column Analysis. When using any of the previous formulas to *analyze* a column, that is, to find its allowable load, it is first necessary to calculate the slenderness ratio in order to determine which column formula applies. Then, once the average allowable stress has been computed, the allowable load in the column is determined from $P = \sigma_{\text{allow}}A$.

Column Design. If the previous formulas are used to *design* a column, that is, to determine the column's cross-sectional area for a given loading and effective length, then a trial-and-check procedure generally must be followed if the column has a composite shape, such as a wide-flange section. This is necessary since the allowable stress depends upon the slenderness ratio as indicated by the formulas. Furthermore the slenderness ratio depends upon the column's cross-sectional area ($KL/r = KL/\sqrt{I/A}$), which has to be determined. One possible way to solve this problem would be to *assume* the column's cross-sectional area, A', and calculate the corresponding stress $\sigma' = P/A'$. Also, use an appropriate design formula to determine the allowable stress σ_{allow}. From this, compute the *required* column area $A_{\text{req'd}} = P/\sigma_{\text{allow}}$. If $A' > A_{\text{req'd}}$, the design is safe. Note, however, that when making the comparison, it is economical to require A' to be close to but greater than $A_{\text{req'd}}$, usually within 2–3%. A redesign is necessary if $A' < A_{\text{req'd}}$. In each case, whenever a trial-and-check procedure is repeated, the choice of an area is determined by the previously computed required area. In engineering practice this method for design is usually shortened through the use of computer software or published tables and graphs.

The following examples illustrate some of the methods for column design and analysis.

Example 13–8

A steel $W\ 10 \times 100$ member is used as a pin-supported column, Fig. 13–27. Using the AISC column design formulas, determine the largest load that it can safely support. $E_{st} = 29(10^3)$ ksi, $\sigma_Y = 36$ ksi.

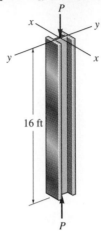

Fig. 13–27

SOLUTION

The following data for a $W\ 10 \times 100$ is taken from the table in Appendix B.

$$A = 29.4 \text{ in}^2 \qquad r_x = 4.60 \text{ in.} \qquad r_y = 2.65 \text{ in.}$$

Since $K = 1$ for both x and y axis buckling, the slenderness ratio is largest if r_y is used. Thus,

$$\frac{KL}{r} = \frac{1(16 \text{ ft})(12 \text{ in./ft})}{(2.65 \text{ in.})} = 72.45$$

From Eq. 13–22, we have

$$\left(\frac{KL}{r}\right)_c = \sqrt{\frac{2\pi^2 E}{\sigma_Y}} = \sqrt{\frac{2\pi^2[29(10^3) \text{ ksi}]}{36 \text{ ksi}}} = 126.1$$

Here $0 < KL/r < (KL/r)_c$, so Eq. 13–23 applies.

$$\sigma_{\text{allow}} = \frac{[1 - (72.45)^2/2(126.1)^2]36 \text{ ksi}}{\{(5/3) + [(3/8)(72.45/126.1)] - [(72.45)^3/8(126.1)^3]\}}$$

$$= 16.17 \text{ ksi}$$

The allowable load P on the column is therefore

$$\sigma_{\text{allow}} = \frac{P}{A}; \qquad 16.17 \text{ kip/in}^2 = \frac{P}{29.4 \text{ in}^2}$$

$$P = 476 \text{ kip} \qquad\qquad \textit{Ans.}$$

Example 13–9

The steel rod in Fig. 13–28 is to be used to support an axial load of 18 kip. If $E_{st} = 29(10^3)$ ksi and $\sigma_Y = 50$ ksi, determine the smallest diameter of the rod as allowed by the AISC specification. The rod is fixed at both ends.

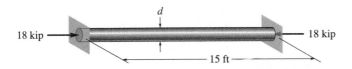

Fig. 13–28

SOLUTION

For a circular cross section the radius of gyration becomes

$$r = \sqrt{\frac{I}{A}} = \sqrt{\frac{(1/4)\pi(d/2)^4}{(1/4)\pi d^2}} = \frac{d}{4}$$

Applying Eq. 13–22, we have

$$\left(\frac{KL}{r}\right)_c = \sqrt{\frac{2\pi^2[29(10^3)\text{ ksi}]}{50 \text{ ksi}}} = 107.0$$

Since the rod's radius of gyration is unknown, KL/r is unknown, and therefore a choice must be made as to whether Eq. 13–21 or Eq. 13–23 applies. We will consider Eq. 13–21. For a fixed-end column $K = 0.5$, so

$$\sigma_{\text{allow}} = \frac{12\pi^2 E}{23(KL/r)^2}$$

$$\frac{18 \text{ kip}}{(1/4)\pi d^2} = \frac{12\pi^2[29(10^3) \text{ kip/in}^2]}{23[0.5(15 \text{ ft})(12 \text{ in./ft})/(d/4)]^2}$$

$$\frac{22.92}{d^2} = 1.152 d^2$$

$$d = 2.11 \text{ in.}$$

Use

$$d = 2.25 \text{ in.} = 2\tfrac{1}{4} \text{ in.} \qquad\qquad \textit{Ans.}$$

For this design, we must check the slenderness-ratio limits; i.e.,

$$\frac{KL}{r} = \frac{0.5(15)(12)}{(2.25/4)} = 160$$

Since $107.0 < 160 < 200$, use of Eq. 13–21 is appropriate.

Example 13–10

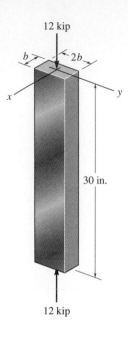

12 kip

b $2b$

x y

30 in.

12 kip

Fig. 13–29

A bar having a length of 30 in. is used to support an axial compressive load of 12 kip, Fig. 13–29. It is pin-supported at its ends and made from an aluminum alloy 2014-T6. Determine the dimensions of its cross-sectional area if its width is to be twice its thickness.

SOLUTION

Since $KL = 30$ in. is the same for both x–x and y–y axis buckling, the largest slenderness ratio is determined using the smallest radius of gyration, i.e., using I_{min}:

$$\frac{KL}{r_y} = \frac{KL}{\sqrt{I_y/A}} = \frac{1(30)}{\sqrt{(1/12)\,2b(b)^3/[2b\,(b)]}} = \frac{103.9}{b}$$

Here we must apply Eq. 13–24, 13–25, or 13–26. Since we do not as yet know the slenderness ratio, we will begin by using Eq. 13–24.

$$\frac{P}{A} = 28 \text{ ksi}$$

$$\frac{12 \text{ kip}}{2b(b)} = 28 \text{ kip/in}^2$$

$$b = 0.463 \text{ in.}$$

Checking the slenderness ratio, we have

$$\frac{KL}{r} = \frac{103.9}{0.463} = 224.5 > 12$$

Try Eq. 13–26, which is valid for $KL/r \geq 55$:

$$\frac{P}{A} = \frac{54{,}000 \text{ ksi}}{(KL/r)^2}$$

$$\frac{12}{2b(b)} = \frac{54{,}000}{(103.9/b)^2}$$

$$b = 1.05 \text{ in.} \qquad\qquad Ans.$$

Here

$$\frac{KL}{r} = \frac{103.9}{1.05} = 99.3 > 55 \qquad \text{OK}$$

Note: It would be satisfactory to choose the cross section with dimensions 1 in. by 2 in.

Example 13–11

A board having cross-sectional dimensions of 5.5 in. by 1.5 in. is used to support an axial load of 5 kip, Fig. 13–30. If the board is assumed to be pin-supported at its top and bottom, determine its *greatest* allowable length L as specified by the NFPA.

SOLUTION

By inspection, the board will buckle about the y axis. In the NFPA equations, $d = 1.5$ in. Assuming that Eq. 13–29 applies, we have

$$\frac{P}{A} = \frac{540}{(KL/d)^2}$$

$$\frac{5 \text{ kip}}{(5.5 \text{ in.})(1.5 \text{ in.})} = \frac{540}{(1\ L/1.5 \text{ in.})^2}$$

$$L = 44.8 \text{ in.} \qquad \textit{Ans.}$$

Here

$$\frac{KL}{d} = \frac{1(44.8 \text{ in.})}{1.5 \text{ in.}} = 29.8$$

Since $26 < KL/d \le 50$, the solution is valid.

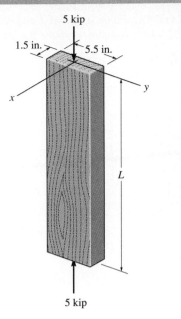

Fig. 13–30

PROBLEMS

13–55. Determine the largest length of a $W\ 10 \times 12$ section if it is pin-supported and is subjected to an axial load of 28 kip. Use the AISC equations. $E_{st} = 29(10^3)$ ksi, $\sigma_Y = 36$ ksi.

*13–56. Determine the largest length of a $W\ 10 \times 12$ section if it is fixed-supported and is subjected to an axial load of 28 kip. Use the AISC equations. $E_{st} = 29(10^3)$ ksi, $\sigma_Y = 36$ ksi.

13–57. Determine the largest length of a $W\ 8 \times 31$ section if it is pin-supported and is subjected to an axial load of 130 kip. Use the AISC equations. $E_{st} = 29(10^3)$ ksi, $\sigma_Y = 36$ ksi.

13–58. Determine the largest length of a $W\ 8 \times 31$ section if it is pin-supported and is subjected to an axial load of 80 kip. Use the AISC equations. $E_{st} = 29(10^3)$ ksi, $\sigma_Y = 36$ ksi.

13–59. Using the AISC equations, select from Appendix B the lightest-weight column that is 12 ft long and supports an axial load of 20 kip. The ends are pinned. $E_{st} = 29(10^3)$ ksi, $\sigma_Y = 36$ ksi.

*13–60. Using the AISC equations, select from Appendix B the lightest-weight column that is 14 ft long and supports an axial load of 40 kip. The ends are pinned. $E_{st} = 29(10^3)$ ksi, $\sigma_Y = 50$ ksi.

13–61. Using the AISC equations, select from Appendix B the lightest-weight column that is 12 ft long and supports an axial load of 40 kip. The ends are fixed. $E_{st} = 29(10^3)$ ksi, $\sigma_Y = 36$ ksi.

13–62. Using the AISC equations, select from Appendix B the lightest-weight column that is 14 ft long and supports an axial load of 40 kip. The ends are fixed. $E_{st} = 29(10^3)$ ksi, $\sigma_Y = 50$ ksi.

13–63. The 1-in.-diameter rod is used to support an axial load of 5 kip. Determine its greatest allowable length L if it is made of 2014-T6 aluminum. Assume that the ends are pin-connected.

*13–64. The 1-in.-diameter rod is used to support an axial load of 5 kip. Determine its greatest allowable length L if it is made of 2014-T6 aluminum. Assume that the ends are fixed-connected.

13–67. The bar is made of aluminum alloy 2014-T6. Determine its thickness b if its width is $1.5b$. Assume that it is pin-connected at its ends.

*13–68. The bar is made of aluminum alloy 2014-T6. Determine its thickness b if its width is $1.5b$. Assume that it is fixed-connected at its ends.

Probs. 13–63/13–64

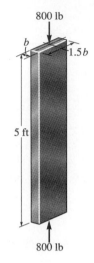

Probs. 13–67/13–68

13–65. The tube is 0.5 in. thick, is made from aluminum alloy 2014-T6, and is assumed to be pin-connected at its ends. Determine the largest axial load that it can support.

13–66. The tube is 0.5 in. thick, is made of aluminum alloy 2014-T6, and is assumed to be fixed-connected at its ends. Determine the largest axial load that it can support.

13–69. The member has a symmetric cross section. Assuming that it is pin-connected at its ends, determine the largest force it can support. It is made of the aluminum alloy 2014-T6.

Probs. 13–65/13–66

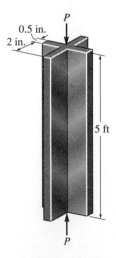

Prob. 13–69

13–70. The timber column has a square cross section and is assumed to be pin-connected at its top and bottom. If it supports an axial load of 50 kip, determine its side dimensions a to the nearest $\frac{1}{2}$ in. Use the NFPA formulas.

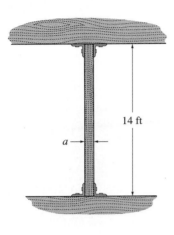

14 ft

a

Prob. 13–70

13–71. The timber column is used to support an axial load of $P = 30$ kip. If it is fixed at the bottom and free at the top, determine the minimum width of the column based on the NFPA formulas.

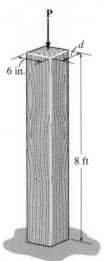

P

d

6 in.

8 ft

Prob. 13–71

*13–72. The timber column has a length of 18 ft and is pin-connected at its ends. Use the NFPA formulas to determine the largest axial force P that it can support.

13–73. The timber column has a length of 18 ft and is fixed-connected at its ends. Use the NFPA formulas to determine the largest axial force P that it can support.

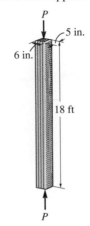

P

5 in.

6 in.

18 ft

P

Probs. 13–72/13–73

13–74. The column is made of wood. It is fixed at its bottom and free at its top. Use the NFPA formulas to determine its greatest allowable length if it supports an axial load of $P = 6$ kip.

13–75. The column is made of wood. It is fixed at its bottom and free at its top. Use the NFPA formulas to determine the largest allowable axial load P that it can support if it has a length $L = 6$ ft.

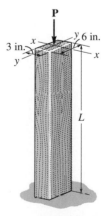

P

x y 6 in.

3 in. x

y

L

Probs. 13–74/13–75

*13.7 Design of Columns for Eccentric Loading

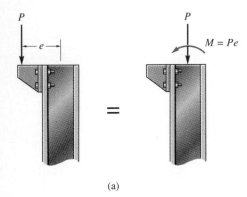

(a)

σ_{max}

(b)

Fig. 13–31

Occasionally a column may be required to support a load acting either at its edge or on an angle bracket attached to its side, such as is shown in Fig. 13–31a. The bending moment $M = Pe$, which is caused by the eccentric loading, must be accounted for when the column is designed. There are several acceptable ways in which this is done in engineering practice. We will discuss two of the most common methods.

Use of Available Column Formulas. The stress distribution acting over the cross-sectional area of the column shown in Fig. 13–31a is determined from both the axial force P and the bending moment $M = Pe$. In particular, the maximum compressive stress is

$$\sigma_{max} = \frac{P}{A} + \frac{Mc}{I} \qquad (13\text{–}30)$$

A typical stress profile is shown in Fig. 13–31b. If we conservatively *assume* that the entire cross section is subjected to the uniform stress σ_{max} as determined from Eq. 13–30, then we can compare σ_{max} with σ_{allow}, which is determined using the formulas given in Sec. 13–6. Calculation of σ_{allow} is usually done using the *largest* slenderness ratio for the column, regardless of the axis about which the column experiences bending. This requirement is normally specified in design codes and will in most cases lead to a conservative design. If

$$\sigma_{max} \le \sigma_{allow}$$

then the column can carry the specified loading. If this inequality does not hold, then the column's area A must be increased, and a new σ_{max} and σ_{allow} must be calculated. This method of design is rather simple to apply and works well for columns that are short or of intermediate length.

Interaction Formula. When *designing* an eccentrically loaded column it is desirable to see how the bending and axial loads *interact,* so that a balance between these two effects can be achieved. To do this, we will consider the separate contributions made to the total column area by the axial force and moment. If the allowable stress for the axial load is $(\sigma_a)_{allow}$, then the required area for the column needed to support the load P is

$$A_a = \frac{P}{(\sigma_a)_{allow}}$$

Similarly, if the allowable bending stress is $(\sigma_b)_{allow}$, then since $I = Ar^2$, the required area of the column needed to support the eccentric moment is determined from the flexure formula, i.e.,

$$A_b = \frac{Mc}{(\sigma_b)_{allow}r^2}$$

The total area A for the column needed to resist *both* the axial load and moment requires that

$$A_a + A_b = \frac{P}{(\sigma_a)_{\text{allow}}} + \frac{Mc}{(\sigma_b)_{\text{allow}}r^2} \leq A$$

or

$$\frac{P/A}{(\sigma_a)_{\text{allow}}} + \frac{Mc/Ar^2}{(\sigma_b)_{\text{allow}}} \leq 1$$

$$\frac{\sigma_a}{(\sigma_a)_{\text{allow}}} + \frac{\sigma_b}{(\sigma_b)_{\text{allow}}} \leq 1 \qquad (13\text{–}31)$$

Here

σ_a = axial stress caused by the force P and computed from $\sigma_a = P/A$, where A is the cross-sectional area of the column

σ_b = bending stress caused by an eccentric load or applied moment M; σ_b is computed from $\sigma_b = Mc/I$, where I is the moment of inertia of the cross-sectional area computed about the bending or neutral axis

$(\sigma_a)_{\text{allow}}$ = allowable axial stress as defined by formulas given in Sec. 13.6 or by other design code specifications. For this purpose, always use the *largest* slenderness ratio for the column, regardless of the axis about which the column experiences bending

$(\sigma_b)_{\text{allow}}$ = allowable bending stress as defined by code specifications

In particular, if the column is subjected only to an axial load, then the bending-stress ratio in Eq. 13–31 would be equal to zero and the design will be based only on the allowable axial stress. Likewise, when no axial load is present, the axial-stress ratio is zero and the stress requirement will be based on the allowable bending stress. Hence, each stress ratio indicates the contribution of axial load or bending moment. Since Eq. 13–31 shows how these loadings interact, this equation is sometimes referred to as the *interaction formula*. This design approach requires a trial-and-check procedure, where it is required that the designer *pick* an available column and then check to see if the inequality is satisfied. If it is not, a larger section is then picked and the process repeated. An economical choice is made when the left side is close to but less than 1.

The interaction method is often specified in codes for the design of members made of steel, aluminum, or timber. In particular, the American Institute of Steel Construction specifies the use of this equation only when the axial-stress ratio $\sigma_a/(\sigma_a)_{\text{allow}} \leq 0.15$. For other values of this ratio, a modified form of Eq. 13–31 is used.

The following examples illustrate the above methods for design and analysis of eccentrically loaded columns.

Example 13–12

The column in Fig. 13–32 is made of aluminum alloy 2014-T6 and is used to support an eccentric load **P**. Determine the magnitude of **P** that can be supported if the column is fixed at its base and free at its top. Use Eq. 13–30.

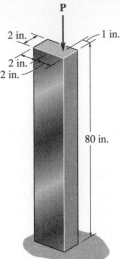

Fig. 13–32

SOLUTION

From Fig. 13–12b, $K = 2$. The largest slenderness ratio for the column is therefore

$$\frac{KL}{r} = \frac{2(80 \text{ in.})}{\sqrt{[(1/12)(4 \text{ in.})(2 \text{ in.})^3]/[(2 \text{ in.})4 \text{ in.}]}} = 277.1$$

By inspection, Eq. 13–26 must be used (277.1 > 55). Thus,

$$\sigma_{\text{allow}} = \frac{54,000}{(277.1)^2} = 0.703 \text{ ksi}$$

The actual maximum compressive stress in the column is determined from the combination of axial load and bending. We have

$$\sigma_{\text{max}} = \frac{P}{A} + \frac{(Pe)c}{I}$$

$$= \frac{P}{2 \text{ in.}(4 \text{ in.})} + \frac{P(1 \text{ in.})(2 \text{ in.})}{(1/12)(2 \text{ in.})(4 \text{ in.})^3}$$

$$= 0.3125P$$

Assuming that this stress is *uniform* over the cross section, instead of just at the outer boundary, we require

$$0.703 = 0.3125P$$

$$P = 2.25 \text{ kip} \qquad\qquad Ans.$$

Example 13–13

The steel $W\ 6 \times 20$ column in Fig. 13–33 is pin-connected at its ends and is subjected to the eccentric load **P**. Determine the maximum allowable value of P using the interaction method if the allowable bending stress is $(\sigma_b)_{\text{allow}} = 22$ ksi, and $E_{st} = 29(10^3)$ ksi, $\sigma_Y = 36$ ksi.

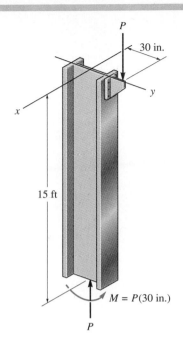

SOLUTION

Here $K = 1$. The necessary geometric properties for the $W\ 6 \times 20$ are taken from the table in Appendix B.

$$A = 5.87 \text{ in.} \qquad I_x = 41.4 \text{ in}^4 \qquad r_y = 1.50 \text{ in.} \qquad d = 6.20 \text{ in.}$$

We have considered r_y because this will lead to the *largest* value of the slenderness ratio. Also, I_x is needed since bending occurs about the x axis ($c = 6.20$ in./2 $= 3.10$ in.). To compute the allowable compressive stress, we have

$$\frac{KL}{r} = \frac{1(15 \text{ ft})(12 \text{ in./ft})}{1.50 \text{ in.}} = 120$$

Since

$$\left(\frac{KL}{r}\right)_c = \sqrt{\frac{2\pi^2[29(10^3) \text{ ksi}]}{36 \text{ ksi}}} = 126.1$$

Fig. 13–33

then $KL/r < (KL/r)_c$ and so Eq. 13–23 must be used.

$$
\begin{aligned}
\sigma_{\text{allow}} &= \frac{[1 - (KL/r)^2/2(KL/r)_c^2]\sigma_Y}{\{(5/3) + [(3/8)(KL/r)/(KL/r)_c] - [(KL/r)^3/8(KL/r)_c^3]\}} \\[2mm]
&= \frac{[1 - (120)^2/2(126.1)^2]36}{\{(5/3) + [(3/8)(120)/(126.1)] - [(120)^3/8(126.1)^3]\}} \\[2mm]
&= 10.28 \text{ ksi}
\end{aligned}
$$

Applying the interaction Eq. 13–31 yields

$$\frac{\sigma_a}{(\sigma_a)_{\text{allow}}} + \frac{\sigma_b}{(\sigma_b)_{\text{allow}}} \leq 1$$

$$\frac{P/5.87 \text{ in}^2}{10.28 \text{ ksi}} + \frac{P(30 \text{ in.})(3.10 \text{ in.})/(41.4 \text{ in}^4)}{22 \text{ ksi}} = 1$$

$$P = 8.43 \text{ kip} \qquad\qquad\qquad \textit{Ans.}$$

Checking the application of the interaction method for a steel section, we require

$$\frac{\sigma_a}{(\sigma_a)_{\text{allow}}} = \frac{(8.43 \text{ kip})/(5.87 \text{ in.})}{10.28 \text{ kip/in}^2} = 0.140 < 0.15 \qquad \text{OK}$$

The timber column in Fig. 13–34 is made from two boards nailed together so that the cross section has the dimensions shown. If the column is fixed at its base and free at its top, use Eq. 13–30 to determine the eccentric load **P** that can be supported.

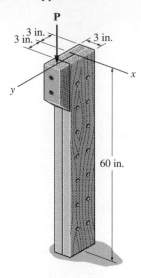

Fig. 13–34

SOLUTION

From Fig. 13–12b, $K = 2$. Here we must compute KL/d to determine which equation from Eqs. 13–27 through 13–29 should be used. Since σ_{allow} is determined using the largest slenderness ratio, we choose $d = 3$ in. This is done to make this ratio as large as possible, and thereby yields the lowest possible allowable axial stress. This assumption is made even though bending due to P is about the x axis. We have

$$\frac{KL}{d} = \frac{2(60 \text{ in.})}{3 \text{ in.}} = 40$$

The allowable axial stress is determined using Eq. 13–29 since $26 < KL/d < 50$. Thus,

$$\sigma_{\text{allow}} = \frac{540 \text{ ksi}}{(40)^2} = 0.3375 \text{ ksi}$$

Applying Eq. 13–30 with $\sigma_{\text{allow}} = \sigma_{\text{max}}$, we have

$$\sigma_{\text{allow}} = \frac{P}{A} + \frac{Mc}{I}$$

$$0.3375 \text{ ksi} = \frac{P}{3 \text{ in.}(6 \text{ in.})} + \frac{P(4 \text{ in.})(3 \text{ in.})}{(1/12)(3 \text{ in.})(6 \text{ in.})^3}$$

$$P = 1.22 \text{ kip} \qquad \qquad \textit{Ans.}$$

PROBLEMS

*13–76. A 16-ft-long column is made of aluminum alloy 2014-T6. If it is fixed at its top and bottom, and a compressive load **P** is applied at point *A*, determine the maximum allowable magnitude of **P** using the equations of Sec. 13.6 and Eq. 13–30.

13–77. A 16-ft-long column is made of aluminum alloy 2014-T6. If it is fixed at its top and bottom, and a compressive load **P** is applied at point *A*, determine the maximum allowable magnitude of **P** using the equations of Sec. 13.6 and the interaction formula with $(\sigma_b)_{allow} = 20$ ksi.

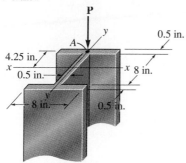

Probs. 13–76/13–77

13–78. The 5-ft-long bar is made of aluminum alloy 2014-T6. If it is fixed at its bottom and free at the top, determine the maximum allowable eccentric load *P* that can be applied using the formulas in Sec. 13.6 and Eq. 13–30.

13–79. The 5-ft-long bar is made of aluminum alloy 2014-T6. If it is fixed at its bottom and free at the top, determine the maximum allowable eccentric load *P* that can be applied using the equations of Sec. 13.6 and the interaction formula with $(\sigma_b)_{allow} = 18$ ksi.

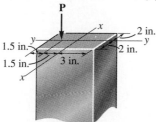

Probs. 13–78/13–79

*13–80. The $W\ 8 \times 15$ column is fixed at its top and bottom. If it supports end moments of $M = 5$ kip · ft, determine the axial force P that can be applied. Bending is about the x–x axis and use the AISC equations of Sec. 13.6 and Eq. 13–30. $E_{st} = 29(10^3)$ ksi, $\sigma_Y = 36$ ksi.

13–81. The $W\ 8 \times 15$ column is fixed at its top and bottom. If it supports end moments of $M = 23$ kip · ft, determine the axial force P that can be applied. Bending is about the x–x axis and use the interaction formula with $(\sigma_b)_{allow} = 24$ ksi. $E_{st} = 29(10^3)$ ksi, $\sigma_Y = 36$ ksi.

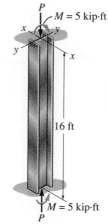

Probs. 13–80/13–81

13–82. The $W\ 10 \times 45$ is used as a column and supports an axial load of 60 kip in addition to an eccentric load P. Determine the maximum allowable value of P based on the AISC equations of Sec. 13.6 and Eq. 13–30. Assume that the column is fixed at its base and at its top it is free to sway in the x–z plane while it is pinned in the y–z plane. $E_{st} = 29(10^3)$ ksi, $\sigma_Y = 50$ ksi.

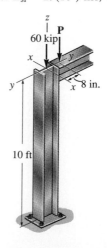

Prob. 13–82

13–83. The $W\ 10 \times 19$ column is assumed to be pinned at its top and bottom. Determine the largest eccentric load P that can be applied using Eq. 13–30 and the AISC equations of Sec. 13.6. $E_{st} = 29(10^3)$ ksi, $\sigma_Y = 36$ ksi.

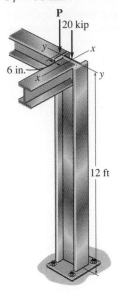

Prob. 13–83

*13–84. The $W\ 12 \times 50$ column is fixed at its bottom and free at its top. Determine the greatest eccentric load P that can be applied using Eq. 13–30 and the AISC equations of Sec. 13.6. $E_{st} = 29(10^3)$ ksi, $\sigma_Y = 36$ ksi.

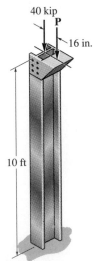

Prob. 13–84

13–85. Check if the wood column is adequate for supporting the eccentric load of 800 lb applied at its top. It is fixed at its base and free at its top. Use the NFPA equations of Sec. 13.6 and Eq. 13–30.

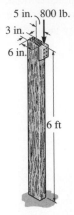

Prob. 13–85

13–86. The wood column has a thickness of 4 in. and a width of 6 in. Using the NFPA equations of Sec. 13.6, and Eq. 13–30, determine the maximum allowable eccentric load P that can be applied. Assume that the column is pinned at both its top and bottom.

13–87. The wood column has a thickness of 4 in. and a width of 6 in. Using the NFPA equations of Sec. 13.6 and Eq. 13–30, determine the maximum allowable eccentric load P that can be applied. Assume that the column is pinned at the top and fixed at the bottom.

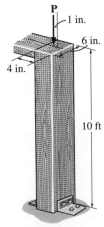

Probs. 13–86/13–87

REVIEW PROBLEMS

*13–88. The $W \ 8 \times 67$ is used as a column that can be assumed fixed at its base and pinned at its top. Determine the largest axial force P that can be applied without causing it to buckle. Use the Euler equation. $E_{st} = 29(10^3)$ ksi, $\sigma_Y = 50$ ksi.

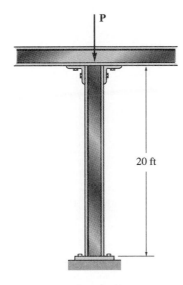

Prob. 13–88

13–89. Construct the buckling curve, P/A versus L/r, for a column that has a bilinear stress–strain curve in compression as shown.

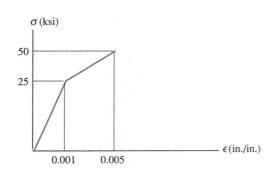

Prob. 13–89

13–90. The $W \ 10 \times 45$ is used as a column that has a length of 15 ft. If its ends are assumed pin-supported and it is subjected to an axial load of 100 kip, determine the factor of safety with respect to buckling. Use the Euler equation. $E_{st} = 29(10^3)$ ksi, $\sigma_Y = 36$ ksi.

Prob. 13–90

13–91. The $W \ 10 \times 12$ column is used to support a load of 4 kip. If the column is fixed at the base and free at the top, determine the deflection at the top of the column due to the loading. $E_{st} = 29(10^3)$ ksi, $\sigma_Y = 36$ ksi.

*13–92. The $W \ 10 \times 12$ column is used to support a load of 4 kip. If the column is fixed at its base and free at its top, determine the maximum stress in the column due to this loading. $E_{st} = 29(10^3)$ ksi, $\sigma_Y = 36$ ksi.

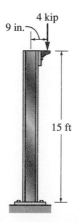

Probs. 13–91/13–92

13–93. Using the AISC equations, check if a $W\,6 \times 4$ column that is 10 ft long can support an axial load of 40 kip. The ends are fixed. $E_{st} = 29(10^3)$ ksi, $\sigma_Y = 36$ ksi.

13–94. A 5-ft-long rod is used in a machine to transmit an axial compressive load of 3 kip. Determine its diameter if it is pin-connected at its ends and is made of a 2014-T6 aluminum alloy.

13–95. Use the AISC equations and check if a $W\,6 \times 15$ column can support the axial load of 16 kip. The column is fixed at its base and free at its top. $E_{st} = 29(10^3)$ ksi, $\sigma_Y = 36$ ksi.

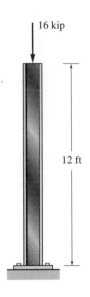

16 kip

12 ft

Prob. 13–95

*13–96. The wood column is 4 m long and is required to support the load of 25 kN. If the cross section is square, determine the dimension a of each of its sides using a factor of safety against buckling of F.S. = 2.5. The column is assumed to be pinned at its top and bottom. Use the Euler equation. $E_w = 11$ GPa, $\sigma_Y = 10$ MPa.

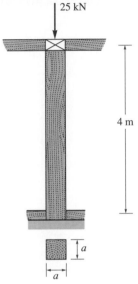

25 kN

4 m

a

a

Prob. 13–96

13–97. Determine if the column can support the eccentric compressive load of 1.5 kip. Assume that the ends are pin-connected. Use the NFPA equations in Sec. 13.6 and Eq. 13-30.

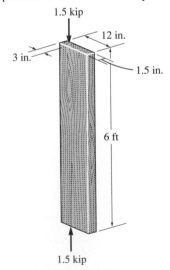

1.5 kip

12 in.

3 in.

1.5 in.

6 ft

1.5 kip

Prob. 13–97

14 Energy Methods

In this chapter we will show how to apply energy methods to solve problems involving deflection. The chapter begins with a discussion of work and strain energy, followed by a development of the principle of conservation of energy. Using this principle, the stress and deflection of a member are determined when the member is subjected to impact. The method of virtual work and Castigliano's theorem are then developed, and these methods are used to determine the displacement and slope at points on various structural and mechanical elements.

14.1 External Work and Strain Energy

Before developing any of the energy methods that will be used throughout this chapter, we will first define the work caused by an external force and couple moment and show how to express the work in terms of a body's strain energy. The formulations to be presented here and in the next section will provide the basis for applying the work and energy methods that follow throughout the chapter.

Work of a Force. In mechanics a force does *work* when it undergoes a displacement dx that is in the *same direction* as the force. The work done is a scalar, defined as $dU_e = F \, dx$. If the total displacement is x, the work becomes

$$U_e = \int_0^x F \, dx \qquad (14\text{–}1)$$

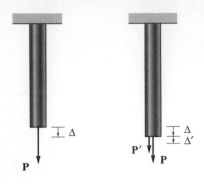

(a)   (b)

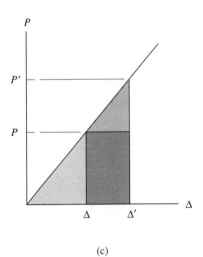

(c)

Fig. 14–1

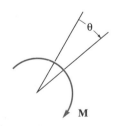

Fig. 14–2

To show how to apply this equation, we will calculate the work done by an axial force applied to the end of the bar shown in Fig. 14–1a. As the magnitude of **F** is *gradually* increased from zero to some limiting value $F = P$, the final displacement of the end of the bar becomes Δ. If the material behaves in a linear-elastic manner, then the force will be directly proportional to the displacement; that is, $F = (P/\Delta)x$. Substituting into Eq. 14–1 and integrating from 0 to Δ, we get

$$U_e = \frac{1}{2}P\Delta \qquad (14\text{–}2)$$

Therefore, as the force is gradually applied to the bar, its magnitude builds from zero to some value P, and consequently, the work done is equal to the *average force magnitude*, $P/2$, times the total displacement Δ. We can represent this graphically as the color-shaded area of the triangle in Fig. 14–1c.

Suppose, however, that **P** is already applied to the bar and that *another force* **P'** is now applied, so that the end of the bar is displaced *further* by an amount Δ', Fig. 14–1b. The work done by **P** (not **P'**) when the bar undergoes this further displacement Δ' is then

$$U_e' = P\Delta' \qquad (14\text{–}3)$$

Here the work represents the dark color-shaded *rectangular area* in Fig. 14–1c. In this case **P** does not change its magnitude, since the bar's displacement Δ' is caused only by **P'**. Therefore, work is simply the force magnitude P times the displacement Δ'.

In summary, then, when a force **P** is applied to the bar, followed by application of the force **P'**, the total work done by both forces is represented by the area of the entire triangle in Fig. 14–1c. The light-colored triangular area represents the work of **P** that is caused by its displacement Δ. The light-shaded triangular area represents the work of **P'**, since this force is displaced Δ'; and lastly, the dark color-shaded rectangular area represents the additional work done by **P** when **P** is displaced Δ', as caused by **P'**.

Work of a Couple Moment. A couple moment **M** does work when it undergoes a rotational displacement $d\theta$ along its line of action. The work done is defined as $dU_e = M\, d\theta$, Fig. 14–2. If the total angle of rotational displacement is θ rad, the work becomes

$$U_e = \int_0^\theta M\, d\theta \qquad (14\text{–}4)$$

As in the case of force, if the couple moment is applied to a *body* having linear-elastic material behavior, such that its magnitude is increased gradually from zero at $\theta = 0$ to M at θ, then the work is

$$U_e = \frac{1}{2}M\theta \qquad (14\text{-}5)$$

However, if the couple moment is already applied to the body and other loadings further rotate the body by an amount θ', then the work is

$$U'_e = M\theta'$$

Strain Energy. When loads are applied to a body, they will deform the material. Provided no energy is lost in the form of heat, the external work done by the loads will be converted into internal work called *strain energy*. This energy, which is *always positive,* is stored in the body and is caused by the action of either normal or shear stress. We will now develop the strain energy for each of these cases separately.

Normal Stress. If the volume element shown in Fig. 14–3 is subjected to the normal stress σ_z, then the force created on the top and bottom faces is $dF_z = \sigma_z \, dA = \sigma_z \, dx \, dy$. If this force is applied gradually to the element, like the force **P** discussed previously, its magnitude is increased from zero to dF_z, while the element undergoes a displacement $d\Delta_z = \epsilon_z \, dz$. The work done by dF_z is therefore

$$dU_i = \frac{1}{2}\, dF_z \, d\Delta_z = \frac{1}{2}[\sigma_z \, dx \, dy]\epsilon \, dz$$

Since the volume of the element is $dV = dx \, dy \, dz$, we have

$$dU_i = \frac{1}{2}\, \sigma_z\epsilon_z \, dV \qquad (14\text{-}6)$$

Notice that U_i is *always positive,* even if σ_z is compressive, since σ_z and ϵ_z will always be in the same direction.

In general then, if the body is subjected only to a uniaxial *normal stress σ,* acting in a specified direction, the strain energy in the body is then

$$U_i = \int_V \frac{\sigma\epsilon}{2} \, dV \qquad (14\text{-}7)$$

Also, if the material behaves in a linear-elastic manner, Hooke's law applies, $\sigma = E\epsilon$, and therefore we can express the strain energy in terms of the normal stress as

$$\boxed{U_i = \int_V \frac{\sigma^2}{2E} \, dV} \qquad (14\text{-}8)$$

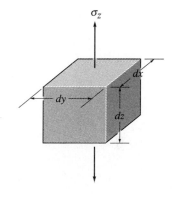

Fig. 14–3

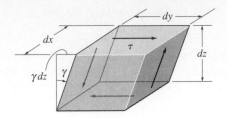

Fig. 14–4

Shear Stress. A strain-energy expression similar to that for normal stress can also be established for the material when it is subjected to shear stress. Consider the volume element shown in Fig. 14–4. Here the shear stress causes the element to deform such that only the shear force $dF = \tau\,(dx\;dy)$, acting on the top face of the element, is displaced $\gamma\,dz$ relative to the bottom face. The *vertical faces* only rotate, and therefore the shear forces on these faces do no work. Hence, the strain energy stored in the element is

$$dU_i = \frac{1}{2}[\tau(dx\;dy)]\gamma\,dz$$

or

$$dU_i = \frac{1}{2}\,\tau\gamma\,dV \qquad (14\text{–}9)$$

where $dV = dx\;dy\;dz$ is the volume of the element.

Integrating over the body's entire volume to obtain the strain energy stored in the body, we have

$$U_i = \int_V \frac{\tau\gamma}{2}\,dV \qquad (14\text{–}10)$$

Like the case of normal stress, shear strain energy is always positive since τ and γ are always in the same direction. If the material is linear-elastic, then, applying Hooke's law, $\gamma = \tau/G$, we can express the strain energy in terms of the shear stress as

$$U_i = \int_V \frac{\tau^2}{2G}\,dV \qquad (14\text{–}11)$$

In the next section we will use Eqs. 14–8 and 14–11 to obtain formal expressions for the strain energy stored in members subjected to several types of loads. Once this is done we will then be able to develop the energy methods necessary to determine the displacement and slope at points on a body.

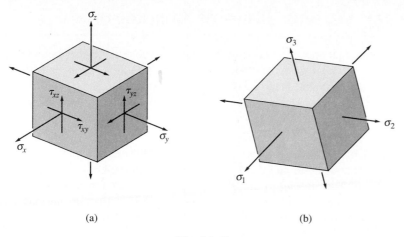

Fig. 14–5

Multiaxial Stress. The previous development may be expanded to determine the strain energy in a body when it is subjected to a general state of stress, Fig. 14–5a. The strain energies associated with each of the normal and shear stress components can be obtained from Eqs. 14–6 and 14–9. Since energy is a scalar, the total strain energy in the body is therefore

$$U_i = \int_V \left[\frac{1}{2}\sigma_x\epsilon_x + \frac{1}{2}\sigma_y\epsilon_y + \frac{1}{2}\sigma_z\epsilon_z \right.$$
$$\left. + \frac{1}{2}\tau_{xy}\gamma_{xy} + \frac{1}{2}\tau_{yz}\gamma_{yz} + \frac{1}{2}\tau_{xz}\gamma_{xz} \right] dV \qquad (14\text{--}12)$$

The strains can be eliminated by using the generalized form of Hooke's law given by Eqs. 10–18 and 10–19. After substituting and combining terms, we have

$$U_i = \int_V \left[\frac{1}{2E}(\sigma_x{}^2 + \sigma_y{}^2 + \sigma_z{}^2) - \frac{\nu}{E}(\sigma_x\sigma_y + \sigma_y\sigma_z + \sigma_z\sigma_x) \right.$$
$$\left. + \frac{1}{2G}(\tau_{xy}{}^2 + \tau_{yz}{}^2 + \tau_{zx}{}^2) \right] dV \qquad (14\text{--}13)$$

If only the principal stresses σ_1, σ_2, σ_3 act on the element, Fig. 14–5b, this equation reduces to a simpler form, namely,

$$U_i = \int_V \left[\frac{1}{2E}(\sigma_1{}^2 + \sigma_2{}^2 + \sigma_3{}^2) - \frac{\nu}{E}(\sigma_1\sigma_2 + \sigma_2\sigma_3 + \sigma_3\sigma_1) \right] dV \quad (14\text{--}14)$$

Recall that we used this equation in Sec. 10.7 as a basis for developing the maximum-distortion-energy theory.

14.2 Elastic Strain Energy for Various Types of Loading

Using the equations for elastic strain energy developed in the previous section, we will now formulate the strain energy stored in a member when it is subjected to an axial load, bending moment, transverse shear, and torsional moment. Examples will be given to show how to calculate the strain energy in members subjected to each of these loadings.

Axial Load. Consider a bar of variable yet slightly tapered cross section,* which is subjected to an axial load coincident with the bar's centroidal axis, Fig. 14–6. The *internal axial force* at a section located a distance x from one end is N. If the cross-sectional area at this section is A, then the normal stress on the section is $\sigma = N/A$. Applying Eq. 14–8, we have

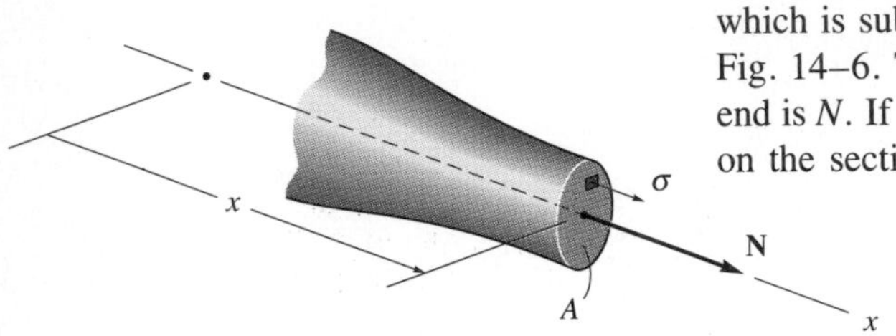

$$U_i = \int_V \frac{\sigma_x^2}{2E}\, dV = \int_V \frac{N^2}{2EA^2}\, dV$$

Fig. 14–6

If we choose an element or differential slice having a volume $dV = A\, dx$, the general formula for the strain energy in the bar is therefore

$$U_i = \int_0^L \frac{N^2}{2EA}\, dx \qquad (14\text{--}15)$$

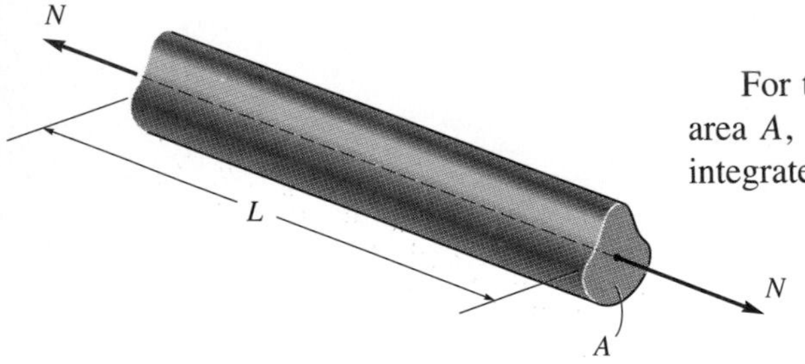

For the more common case of a prismatic bar of constant cross-sectional area A, length L, and constant axial load N, Fig. 14–7, Eq. 14–15, when integrated, gives

$$U_i = \frac{N^2 L}{2AE} \qquad (14\text{--}16)$$

Fig. 14–7

From this equation it can be seen that the bar's elastic strain energy will *increase* if the length of the bar is increased, or if the modulus of elasticity or cross-sectional area is decreased. For example, an aluminum rod [$E_{al} = 10(10^3)$ ksi] will store approximately three times as much energy as a steel rod [$E_{st} = 29(10^3)$ ksi] having the same size and subjected to the same load. On the other hand, doubling the cross-sectional area of a given rod will decrease its ability to store energy by one-half. The following example illustrates this point numerically.

---

*A slight taper to the bar will not greatly affect the results, as discussed in Sec. 4.2.

One of the two high-strength steel bolts *A* and *B* shown in Fig. 14–8 is to be chosen to support a sudden tensile loading. For the choice it is necessary to determine the greatest amount of elastic strain energy that each bolt can absorb. Bolt *A* has a diameter of 0.875 in. for 2 in. of its length and a root (or smallest) diameter of 0.731 in. within the 0.25-in. threaded region. Bolt *B* has "upset" threads, such that the diameter throughout its 2.25-in. length can be taken as 0.731 in. In both cases, neglect the extra material that makes up the threads. Take $E_{st} = 29(10^3)$ ksi, $\sigma_Y = 44$ ksi.

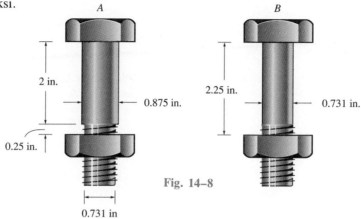

Fig. 14–8

SOLUTION

Bolt A. If the bolt is subjected to its maximum tension, the maximum stress of $\sigma_Y = 44$ ksi will occur within the 0.25-in. region. This tension force is

$$P_{max} = \sigma_Y A = 44 \text{ ksi}\left[\pi\left(\frac{0.731 \text{ in.}}{2}\right)^2\right] = 18.47 \text{ kip}$$

Applying Eq. 14–16 to each region of the bolt, we have

$$U_i = \sum \frac{N^2 L}{2AE}$$

$$= \frac{(18.47 \text{ kip})^2(2 \text{ in.})}{2[\pi(0.875 \text{ in.}/2)^2][29(10^3) \text{ ksi}]} + \frac{(18.47 \text{ kip})^2(0.25 \text{ in.})}{2[\pi(0.731 \text{ in.}/2)^2]29(10^3) \text{ ksi}}$$

$$= 0.0231 \text{ in.} \cdot \text{kip} \qquad\qquad\qquad\qquad\qquad\qquad\qquad \textit{Ans.}$$

Bolt B. Here the bolt is assumed to have a uniform diameter of 0.731 in. throughout its 2.25-in. length. Also, from the calculation above, it can support a maximum tension force of $P_{max} = 18.47$ kip. Thus,

$$U_i = \frac{N^2 L}{2AE} = \frac{(18.47 \text{ kip})^2(2.25 \text{ in.})}{2[\pi(0.731 \text{ in.}/2)^2]29(10^3) \text{ ksi}} = 0.0315 \text{ in.} \cdot \text{kip} \qquad \textit{Ans.}$$

By comparison, bolt *B* can absorb 36% more elastic energy than bolt *A* even though it has a smaller cross section along its shank.

Bending Moment. Since a bending moment applied to a straight prismatic member develops *normal stress* in the member, we can use Eq. 14–8 to determine the strain energy stored in the member due to bending. For example, consider the axisymmetric beam shown in Fig. 14–9. Here the internal moment is M, and the normal stress acting on the arbitrary element a distance y from the neutral axis is $\sigma = My/I$. If the volume of the element is $dV = dA\ dx$, where dA is the area of its exposed face and dx is its length, the elastic strain energy in the beam is

$$U_i = \int_V \frac{\sigma^2}{2E}\ dV = \int_V \frac{1}{2E}\left(\frac{My}{I}\right)^2 dA\ dx$$

The integral over the volume can be expressed as the product of an integral over the beam's cross-sectional area A and an integral over its length L. Thus,

$$U_i = \int_0^L \frac{M^2}{2EI^2}\ dx \int_A y^2\ dA$$

Realizing that the area integral represents the moment of inertia of the beam about the neutral axis, the final result can be written as

$$U_i = \int_0^L \frac{M^2\ dx}{2EI} \tag{14-17}$$

To evaluate the strain energy, therefore, we must first express the internal moment as a function of its position x along the beam, and then perform the integration over the beam's entire length.* The following examples illustrate this procedure.

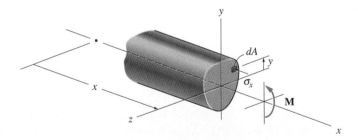

Fig. 14–9

*Recall that the flexure formula, as used here, can also be used with justifiable accuracy to determine the stress in slightly tapered beams. (See Sec. 6.2.) So in the general sense I in Eq. 14–17 may also have to be expressed as a function of x.

■ Example 14–2 ■

Determine the elastic strain energy due to bending of the cantilevered beam if the beam is subjected to the uniform distributed load w, Fig. 14–10a. EI is constant.

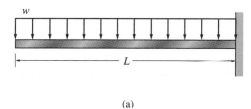

(a)

Fig. 14–10

SOLUTION

The internal moment in the beam is determined by establishing the x coordinate with origin at the left side. The left-hand segment of the beam is shown in Fig. 14–10b. We have

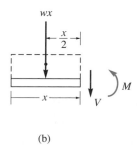

(b)

$$\zeta^+ \ \Sigma M_{NA} = 0; \qquad M + wx\left(\frac{x}{2}\right) = 0$$

$$M = -w\left(\frac{x^2}{2}\right)$$

Applying Eq. 14–17 yields

$$U_i = \int_0^L \frac{M^2 dx}{2EI} = \int_0^L \frac{[-w(x^2/2)]^2 \, dx}{2EI} = \frac{w^2}{8EI} \int_0^L x^4 \, dx$$

or

$$U_i = \frac{w^2 L^5}{40EI} \qquad\qquad Ans.$$

We can also obtain the strain energy using an x coordinate having its origin at the right side of the beam and extending positive to the left, Fig. 14–10c. In this case,

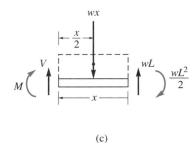

(c)

$$\zeta^+ \ \Sigma M_{NA} = 0; \quad -M - wx\left(\frac{x}{2}\right) + wL(x) - \frac{wL^2}{2} = 0$$

$$M = -\frac{wL^2}{2} + wLx - w\left(\frac{x^2}{2}\right)$$

Applying Eq. 14–17, we obtain the same result as before.

Example 14–3

Determine the bending strain energy in region AB of the beam shown in Fig. 14–11*a*. *EI* is constant.

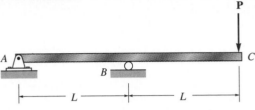

(a)

(b)

Fig. 14–11

SOLUTION

A free-body diagram of the beam is shown in Fig. 14–11*b*. To obtain the answer we can express the internal moment in terms of any one of the indicated three "x" coordinates and then apply Eq. 14–17. Each of these solutions will now be considered.

$0 \leq x_1 \leq L$.   From the free-body diagram of the section in Fig. 14–11*c*, we have

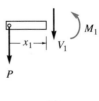

(c)

$$\zeta + \ \Sigma M_{NA} = 0; \qquad \qquad M_1 + P x_1 = 0$$

$$M_1 = -P x_1$$

$$U_i = \int \frac{M^2 \, dx}{2EI} = \int_0^L \frac{(-P x_1)^2 \, dx_1}{2EI}$$

$$= \frac{P^2 L^3}{6EI} \qquad \qquad \qquad \qquad Ans.$$

$0 \leq x_2 \leq L.$ Using the free-body diagram of the section in Fig. 14–11*d* gives

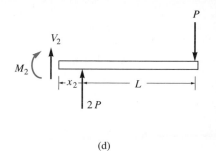

(d)

$\overset{\downarrow}{\LARGE+}\;\Sigma M_{NA} = 0; \qquad -M_2 + 2P(x_2) - P(x_2 + L) = 0$

$$M_2 = P(x_2 - L)$$

$$U_i = \int \frac{M^2 \, dx}{2EI} = \int_0^L \frac{[P(x_2 - L)]^2 \, dx_2}{2EI}$$

$$= \frac{P^2 L^3}{6EI} \qquad\qquad\qquad \textit{Ans.}$$

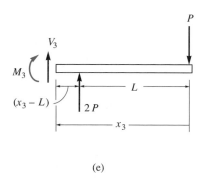

(e)

$L < x_3 < 2L.$ From the free-body diagram in Fig. 14–11*e*, we have

$\overset{\downarrow}{\LARGE+}\;\Sigma M_{NA} = 0; \qquad -M_3 + 2P(x_3 - L) - P(x_3) = 0$

$$M_3 = P(x_3 - 2L)$$

$$U_i = \int \frac{M^2 \, dx}{2EI} = \int_L^{2L} \frac{[P(x_3 - 2L)]^2 \, dx_3}{2EI}$$

$$= \frac{P^2 L^3}{6EI} \qquad\qquad\qquad \textit{Ans.}$$

This and the previous example indicate that the strain energy for the beam can be computed using *any* suitable *x* coordinate. It is only necessary to integrate over the range of the coordinate where the internal energy is to be determined. Here the choice of x_1 provides the simplest solution.

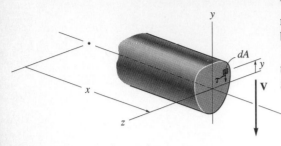

Fig. 14–12

Transverse Shear. The strain energy due to shear stress in a beam element can be determined by applying Eq. 14–11. Here we will consider the beam to be prismatic and to have an axis of symmetry as shown in Fig. 14–12. If the internal shear at the section x is V, then the shear stress acting on the volume element of material, having a length dx and area dA, is $\tau = VQ/It$. Substituting into Eq. 14–11, the strain energy for shear becomes

$$U_i = \int_V \frac{\tau^2}{2G}\, dV = \int_V \frac{1}{2G}\left(\frac{VQ}{It}\right)^2 dA\ dx$$

or

$$U_i = \int_0^L \frac{V^2}{2GI^2}\left(\int_A \frac{Q^2}{t^2}\, dA\right) dx$$

The integral in parentheses is evaluated over the beam's cross-sectional area. To simplify this expression we will define the *form factor* for shear as

$$f_s = \frac{A}{I^2}\int_A \frac{Q^2}{t^2}\, dA \tag{14–18}$$

Substituting into the above equation, we get

$$\boxed{U_i = \int_0^L \frac{f_s V^2\ dx}{2GA}} \tag{14–19}$$

The form factor defined by Eq. 14–18 is a dimensionless number that is unique for each specific cross-sectional area. For example, if the beam has a rectangular cross section of width b and height h, Fig. 14–13, then

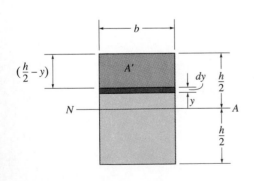

Fig. 14–13

$$t = b$$
$$A = bh$$
$$I = \frac{1}{12}bh^3$$
$$Q = \bar{y}'A' = \left(y + \frac{(h/2) - y}{2}\right)b\left(\frac{h}{2} - y\right) = \frac{b}{2}\left(\frac{h^2}{4} - y^2\right)$$

Substituting these terms into Eq. 14–18, we get

$$f_s = \frac{bh}{(\frac{1}{12}bh^3)^2}\int_{-h/2}^{h/2}\frac{b^2}{4b^2}\left(\frac{h^2}{4} - y^2\right)^2 b\ dy = \frac{6}{5} \tag{14–20}$$

The form factor for other sections can be determined in a similar manner. Once obtained, this number is substituted into Eq. 14–19 and the strain energy for transverse shear can then be evaluated.

Example 14–4

Determine the strain energy in the cantilevered beam due to shear if the beam is subjected to a uniform distributed load w, Fig. 14–14a. EI is constant. The beam has a square cross section of side dimension a.

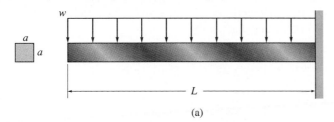

(a)

Fig. 14–14

(b)

SOLUTION

From the free-body diagram of an arbitrary section, Fig. 14–14b, we have

$$+\uparrow \ \Sigma F_y = 0; \qquad\qquad -V - wx = 0$$

$$V = -wx$$

Since the cross section is square, the form factor $f_s = \frac{6}{5}$ (Eq. 14–20) and therefore Eq. 14–19 becomes

$$(U_i)_s = \int_0^L \frac{\frac{6}{5}(-wx)^2 \, dx}{2GA} = \frac{3w^2}{5GA} \int_0^L x^2 \, dx$$

or

$$(U_i)_s = \frac{w^2 L^3}{5GA} \qquad\qquad Ans.$$

Using the results of Example 14–2, with $A = a^2$, $I = \frac{1}{12}a^4$, the ratio of shear to bending strain energy is

$$\frac{(U_i)_s}{(U_i)_b} = \frac{w^2 L^3/5Ga^2}{w^2 L^5/40E(\frac{1}{12}a^4)} = \frac{2}{3}\left(\frac{a}{L}\right)^2 \frac{E}{G}$$

Since $G = E/2(1 + \nu)$ and $\nu \leq \frac{1}{2}$ (Sec. 10.6), then as an *upper* bound, $E = 3G$, so that

$$\frac{(U_i)_s}{(U_i)_b} = 2\left(\frac{a}{L}\right)^2$$

It can be seen that this ratio will increase as L decreases. However, even for very short beams, where, say, $L = 5a$, the contribution due to shear strain energy is only 8% of the bending strain energy. For this reason, the shear strain energy stored in beams is usually neglected in engineering analysis.

Torsional Moment. To determine the internal strain energy in a circular shaft or tube due to an applied torsional moment, we must apply Eq. 14–11. Consider the tapered shaft in Fig. 14–15. A section of the shaft taken a distance x from one end is subjected to an internal torque T. The stress distribution that causes this torque varies linearly from the center of the shaft. On the arbitrary element of length dx and area dA, the stress is $\tau = T\rho/J$. The strain energy stored in the shaft is thus

$$U_i = \int_V \frac{\tau^2}{2G}\, dV = \int_V \frac{1}{2G}\left(\frac{T\rho}{J}\right)^2 dA\, dx$$

$$= \int_0^L \frac{T^2}{2GJ^2}\, dx \int_A \rho^2\, dA$$

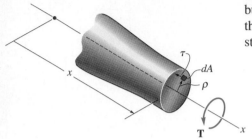

Fig. 14–15

Since the area integral represents the polar moment of inertia J for the shaft at the section, the final result can be written as

$$U_i = \int_0^L \frac{T^2}{2GJ}\, dx \tag{14–21}$$

The most common case occurs when the shaft (or tube) has a constant cross-sectional area and the applied torque is constant, Fig. 14–16. Integration of Eq. 14–21 then gives

$$U_i = \frac{T^2L}{2GJ} \tag{14–22}$$

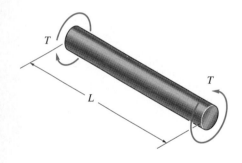

Fig. 14–16

From this equation we may conclude that like an axially loaded member, Eq. 14–16, the energy-absorbing capacity of a torsionally loaded shaft is *decreased* by increasing the diameter of the shaft, since this increases J.

If the cross section of the shaft is some shape other than circular or tubular, Eq. 14–22 must be modified. For example, if it is rectangular, having dimensions $h > b$, then using a mathematical analysis based on the theory of elasticity, it can be shown that the strain energy in the shaft is determined from

$$U_i = \frac{T^2L}{2Cb^3hG} \tag{14–23}$$

where

$$C = \frac{hb^3}{16}\left[\frac{16}{3} - 3.336\,\frac{b}{h}\left(1 - \frac{b^4}{12h^4}\right)\right] \tag{14–24}$$

The following example illustrates how to determine the strain energy in a shaft due to a torsional loading.

Example 14–5

The tubular shaft in Fig. 14–17a is fixed at the wall and subjected to two torques as shown. Determine the strain energy stored in the shaft due to this loading. $G = 75$ GPa.

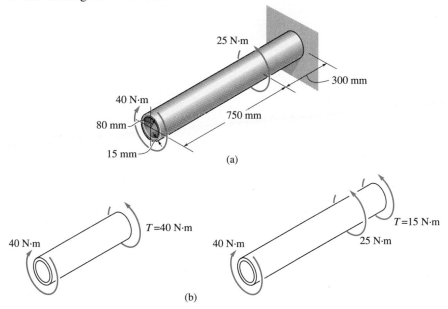

Fig. 14–17

SOLUTION

Using the method of sections, the internal torque is first determined within the two regions of the shaft where it is constant, Fig. 14–17b. Although these torques (40 N · m and 15 N · m) are in opposite directions, this will be of no consequence in determining the strain energy, since the torque is squared in Eq. 14–22. In other words, the strain energy is always positive. The polar moment of inertia for the shaft is

$$J = \frac{\pi}{2}\,[(0.08\text{ m})^4 - (0.065\text{ m})^4] = 36.3(10^{-6})\text{ m}^4$$

Applying Eq. 14–22, we have

$$U_i = \sum \frac{T^2 L}{2GJ}$$

$$= \frac{(40\text{ N} \cdot \text{m})^2(0.750\text{ m})}{2[75(10^9)\text{ N/m}^2]36.3(10^{-6})\text{ m}^4} + \frac{(15\text{ N} \cdot \text{m})^2(0.300\text{ m})}{2[75(10^9)\text{ N/m}^2]36.3(10^{-6})\text{ m}^4}$$

$$= 233\ \mu\text{J} \qquad\qquad\qquad Ans.$$

PROBLEMS

14–1. A 3220-lb automobile has a velocity of 100 ft/s. Determine the required number of cubic inches of steel if the energy from the car's motion could be converted into internal strain energy of the steel caused by a uniaxial stress of 30 ksi. *Hint:* The kinetic energy is determined from $\frac{1}{2}mv^2$, where m is the automobile's mass and v is its velocity. Take $g = 32.2$ ft/s^2, $E_{st} = 29(10^3)$ ksi, $\sigma_Y = 36$ ksi.

14–2. The strain-energy density must be the same whether the state of stress is represented by σ_x, σ_y, and τ_{xy}, or by the principal stresses σ_1 and σ_2. This being the case, equate the strain-energy expressions for these two cases and show that $G = E/[2(1 + \nu)]$.

14–3. For what ratio of σ_x to σ_y is the strain-energy density a minimum?

*14–4. Determine the torsional strain energy in the steel shaft. The shaft has a radius of 30 mm. $G_{st} = 75$ GPa.

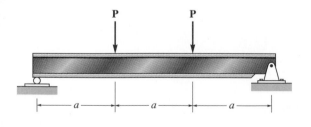

Prob. 14–4

14–5. Determine the bending strain energy in the beam due to the loading shown. *EI* is constant.

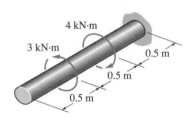

Prob. 14–5

14–6. Determine the bending strain energy in the beam due to the loading shown. *EI* is constant.

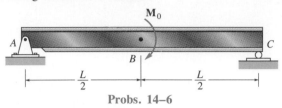

Probs. 14–6

14–7. Determine the bending strain energy in the cantilevered beam due to a uniform load w. Solve the problem two ways. (a) Apply Eq. 14–17. (b) The load $w\,dx$ acting on a segment dx of the beam is displaced a distance y, where $y = w(-x^4 + 4L^3x - 3L^4)/(24EI)$, the equation of the elastic curve. Hence the internal strain energy in the differential segment dx of the beam is equal to the external work, i.e., $dU_i = \frac{1}{2}(w\,dx)(-y)$. Integrate this equation to obtain the total strain energy in the beam. *EI* is constant.

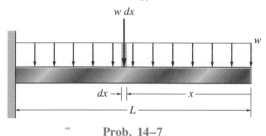

Prob. 14–7

*14–8. Determine the bending strain energy in the simply-supported beam due to a uniform load w. Solve the problem two ways. (a) Apply Eq. 14–17. (b) The load $w\,dx$ acting on the segment dx of the beam is displaced a distance y, where $y = w(-x^4 + 2Lx^3 - L^3x)/(24EI)$, the equation of the elastic curve. Hence the internal strain energy in the differential segment dx of the beam is equal to the external work, i.e., $dU_i = \frac{1}{2}(w\,dx)(-y)$. Integrate this equation to obtain the total strain energy in the beam. *EI* is constant.

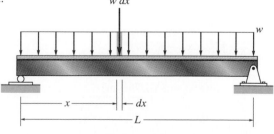

Prob. 14–8

14–9. Determine the total axial and bending strain energy in the steel beam. $A = 2300$ mm^2, $I = 9.5(10^6)$ mm^4, $E_{st} = 200$ GPa.

*14–12. Determine the bending strain energy in the steel beam due to the loading shown. Obtain the answer using the coordinates (a) x_1 and x_4, and (b) x_2 and x_3. $E_{st} = 29(10^3)$ ksi. $I = 53.8$ in^4.

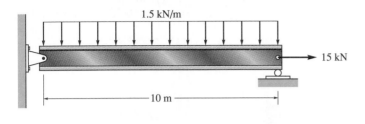

1.5 kN/m

15 kN

10 m

Prob. 14–9

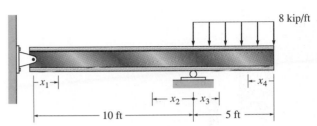

8 kip/ft

x_1 x_4

x_2 x_3

10 ft   5 ft

Prob. 14–12

14–10. Determine the shear strain energy in the beam. The beam has a rectangular cross section of area A. The shear modulus is G.

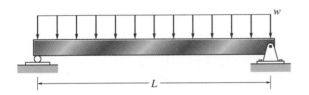

w

L

Prob. 14–10

14–13. The concrete column contains six 1-in.-diameter steel reinforcing rods. If the column supports a load of 300 kip, determine the strain energy in the column. $E_{st} = 29(10^3)$ ksi, $E_c = 3.6(10^3)$ ksi.

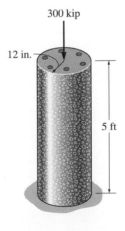

300 kip

12 in.

5 ft

Prob. 14–13

14–11. The simply-supported beam is subjected to the loading shown. Determine the bending strain energy in the beam.

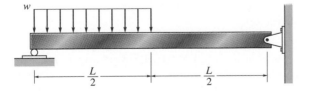

w

$\frac{L}{2}$ $\frac{L}{2}$

Prob. 14–11

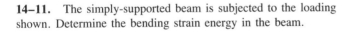

14–14. Determine the bending strain energy in the beam and the axial strain energy in each of the two posts. All members are made of aluminum and have a square cross section 50 mm by 50 mm. $E_{al} = 70$ GPa. Assume the posts only support an axial load.

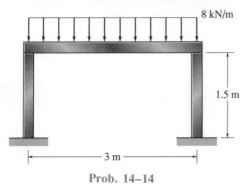

8 kN/m

1.5 m

3 m

Prob. 14–14

14–15. A load of 5 kN is applied to the center of the steel beam, for which $I = 4.5(10^6)$ mm^4 and $E_{st} = 200$ GPa. If the beam is supported on two springs, each having a stiffness of $k = 8$ MN/m, determine the strain energy in each of the springs and the bending strain energy in the beam.

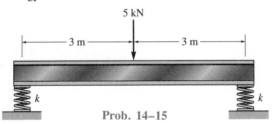

5 kN

3 m   3 m

k k

Prob. 14–15

*14–16. Determine the total strain energy in the steel assembly. Consider the axial strain energy in the two 0.5-in.-diameter rods and the bending strain energy in the beam, which has a moment of inertia of $I = 43.4$ in^4 about its neutral axis. $E_{st} = 29(10^3)$ ksi.

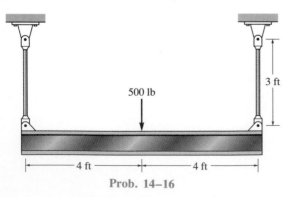

3 ft

500 lb

4 ft   4 ft

Prob. 14–16

14–17. The bolt has a diameter of 0.25 in., and the link AB has a rectangular cross section that is 0.5 in. wide by 0.2 in. thick. Determine the strain energy in the link AB due to bending, and in the bolt due to axial force. The bolt is tightened so that it has a tension of 350 lb. $E_{st} = 29(10^3)$ ksi. Neglect the hole in link AB.

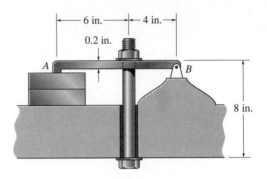

6 in.   4 in.

0.2 in.

A B

8 in.

Prob. 14–17

14–18. Determine the strain energy in the *horizontal* curved bar due to torsion. There is a *vertical* force **P** acting at its end. JG is constant.

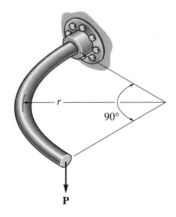

r

90°

P

Prob. 14–18

14–19. Consider the thin-walled tube of Fig. 5–30. Use the formula for shear stress, $\tau = T/2tA_m$, Eq. 5–18, and the general equation of shear strain energy, Eq. 14–11, to show that the twist of the tube is given by Eq. 5–20. *Hint:* Equate the work done by the torque T to the strain energy in the tube, determined from integrating the strain energy for a differential element, Fig. 14–4, over the volume of material.

14.3 Conservation of Energy

All energy methods used in mechanics are based on a balance of energy, often referred to as the conservation of energy. In this chapter, only mechanical energy will be considered in the energy balance; that is, the energy developed by heat, chemical reactions, and electromagnetic effects will be neglected. As a result, if a loading is applied *slowly* to a body, so that kinetic energy can also be neglected, then physically the external loads tend to deform the body so that the loads do *external work* U_e as they are displaced. This external work caused by the loads is transformed into *internal work* or strain energy U_i, which is stored in the body. Furthermore, when the loads are removed, the strain energy restores the body back to its original undeformed position, provided the material's elastic limit is not exceeded. The conservation of energy for the body can therefore be stated mathematically as

$$\boxed{U_e = U_i} \qquad (14\text{--}25)$$

We will now show three examples of how this equation can be applied to determine the displacement of a point on a deformable member or structure. As the first example, consider the truss in Fig. 14–18 subjected to the known load **P**. Provided **P** is applied gradually, the external work done by **P** is determined from Eq. 14–2, that is, $U_e = \frac{1}{2}P\Delta$, where Δ is the vertical displacement of the truss at the joint where **P** is applied. Assuming that **P** develops an axial force **N** in a particular member, the strain energy stored in this member is determined from Eq. 14–16, that is, $U_i = N^2L/2AE$. Summing the strain energies for all the members of the truss, we can write Eq. 14–25 as

$$\frac{1}{2}P\Delta = \sum \frac{N^2L}{2AE} \qquad (14\text{--}26)$$

Once the internal forces in all the members of the truss are determined and the terms on the right computed, it is then possible to determine the unknown displacement Δ.

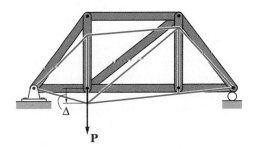

Fig. 14–18

As a second example, consider finding the vertical displacement Δ under the known load **P** acting on the beam in Fig. 14–19. Again, the external work is $U_e = \frac{1}{2}P\Delta$. However, in this case the strain energy would be the result of internal shear and moment loadings caused by **P**. In particular, the contribution of strain energy due to shear is generally *neglected* in most beam deflection problems unless the beam is short and supports a very large load. (See Example 14–4.) Consequently, the beam's strain energy is determined only by the internal bending moment M, and therefore, using Eq. 14–17, Eq. 14–25 can be written symbolically as

$$\frac{1}{2}\,P\Delta = \int_0^L \frac{M^2}{2EI}\,dx \qquad (14\text{–}27)$$

Fig. 14–19

Once M is expressed as a function of position and the integral is evaluated, Δ can then be determined.

As the last example, we will consider a beam loaded by a couple moment $\mathbf{M}_0$ as shown in Fig. 14–20. This moment causes the rotational displacement θ at the point of application of the couple moment. Since the couple moment only does work when it *rotates,* using Eq. 14–5, the external work is $U_e = \frac{1}{2}M_0\theta$. Therefore Eq. 14–25 becomes

$$\frac{1}{2}\,M_0\theta = \int_0^L \frac{M^2}{2EI}\,dx \qquad (14\text{–}28)$$

Fig. 14–20

Here the strain energy is determined as a result of the internal bending moment M caused by application of the couple moment $\mathbf{M}_0$. Once M has been expressed as a function of x and the strain energy evaluated, then θ can be calculated.

In each of the above examples, it should be noted that application of Eq. 14–25 is *quite limited,* because only a *single* external force or couple moment must act on the member or structure. Also, the displacement can *only* be calculated at the point and in the direction of the external force or couple moment. If more than one external force or couple moment were applied, then the external work of each loading would involve its associated unknown displacement. As a result, *all* these unknown displacements could not be determined, since only the single Eq. 14–25 is available for the solution. Although application of the conservation of energy as described here has these restrictions, it does serve as an introduction to more general energy methods, which we will consider throughout the rest of this chapter. Specifically, it will be shown in later sections of this chapter that by modifying the method for applying the conservation-of-energy principle, we will be able to perform a completely general deflection analysis of a member or structure.

Example 14–6

The three-bar truss in Fig. 14–21a is subjected to a horizontal force of 5 kip. If the cross-sectional area of each member is 0.20 in², determine the horizontal displacement at point B. $E = 29(10^3)$ ksi.

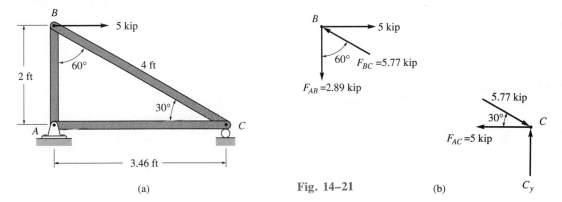

Fig. 14–21

(a)

(b)

SOLUTION

We can apply the method of work and energy to solve this problem because only a *single* external force acts on the truss and the required displacement is in the *same direction* as the force. Furthermore, the reactive forces on the truss do no work since they are not displaced.

Using the method of joints, the force in each member is determined and the results are shown on the free-body diagrams of the pins at B and C, Fig. 14–21b.

Applying Eq. 14–26, we have

$$\frac{1}{2} P\Delta = \sum \frac{N^2 L}{2AE}$$

$$\frac{1}{2} (5 \text{ kip})(\Delta_B)_h = \frac{(2.89 \text{ kip})^2(2 \text{ ft})}{2AE} + \frac{(-5.77 \text{ kip})^2(4 \text{ ft})}{2AE} + \frac{(-5 \text{ kip})^2(3.46 \text{ ft})}{2AE}$$

$$(\Delta_B)_h = \frac{47.28 \text{ kip} \cdot \text{ft}}{AE}$$

Notice that since N is squared, it does not matter if a particular member is in tension or compression. Substituting in the numerical data for A and E and solving, we get

$$(\Delta_B)_h = \frac{47.28 \text{ kip} \cdot \text{ft}(12 \text{ in./ft})}{(0.2 \text{ in}^2)[29(10^3) \text{ kip/in}^2]}$$

$$= 0.0978 \text{ in.} \quad \rightarrow \qquad \qquad \text{Ans.}$$

The cantilevered beam in Fig. 14–22a has a rectangular cross section and is subjected to a load **P** at its end. Determine the displacement of the load. EI is constant.

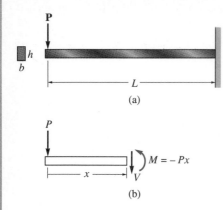

Fig. 14–22

SOLUTION

The internal shear and moment in the beam as a function of x are determined using the method of sections, Fig. 14–22b.

When applying Eq. 14–25 we will consider the strain energy due to shear and bending. Using Eqs. 14–19 and 14–17, we have

$$\frac{1}{2} P\Delta = \int_0^L \frac{f_s V^2 \, dx}{2GA} + \int_0^L \frac{M^2 \, dx}{2EI}$$

$$= \int_0^L \frac{(\frac{6}{5})(-P)^2 \, dx}{2GA} + \int_0^L \frac{(-Px)^2 \, dx}{2EI} = \frac{3P^2 L}{5GA} + \frac{P^2 L^3}{6EI} \qquad (1)$$

The first term on the right side of this equation represents the strain energy due to shear, while the second is the strain energy due to bending. As stated in Example 14–4, for most beams the shear strain energy is much smaller than the bending strain energy. To show when this is the case for the beam in Figure 14–22, we require

$$\frac{3}{5} \frac{P^2 L}{GA} \overset{?}{<} \frac{P^2 L^3}{6EI}$$

$$\frac{3}{5} \frac{P^2 L}{G(bh)} \overset{?}{<} \frac{P^2 L^3}{6E[\frac{1}{12}(bh^3)]}$$

$$\frac{3}{5G} \overset{?}{<} \frac{2L^2}{Eh^2}$$

Since $E \le 3G$ (see Example 14–4), then

$$0.9 \overset{?}{<} \left(\frac{L}{h}\right)^2$$

Hence if h is small and L relatively long (compared with h), the beam becomes slender and the shear strain energy can be neglected. In other words, the *shear strain energy* becomes important *only* for *short, deep beams*. For example, beams for which $L = 5h$ have more than 28 times more bending strain energy than shear strain energy, so neglecting the shear strain energy represents an error of about 3.5%. With this in mind, Eq. 1 can be simplified to

$$\frac{1}{2} P\Delta = \frac{P^2 L^3}{6EI}$$

So that

$$\Delta = \frac{PL^3}{3EI} \qquad \qquad Ans.$$

PROBLEMS

*14–20. Determine the vertical displacement of joint C. AE is constant.

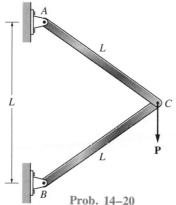

Prob. 14–20

14–21. Determine the horizontal displacement of joint D. AE is constant.

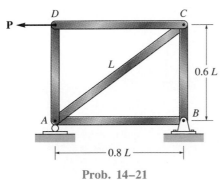

Prob. 14–21

14–22. Determine the vertical displacement of joint A. Each bar is made of steel and has a cross-sectional area of 600 mm². E_{st} = 200 GPa.

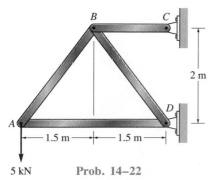

5 kN   Prob. 14–22

14–23. Determine the displacement of point B on the aluminum beam. E_{al} = 10.6(10³) ksi.

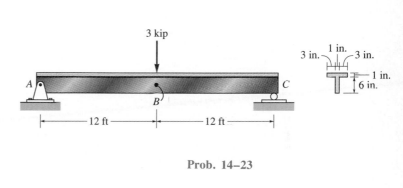

Prob. 14–23

*14–24. Determine the displacement of point B on the steel beam. I = 80(10⁶) mm⁴, E_{st} = 200 GPa.

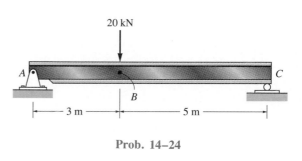

Prob. 14–24

14–25. Use the method of work and energy and determine the slope of the beam at point B in Prob. 14–6. EI is constant.

14–26. Determine the slope at point A of the beam. EI is constant.

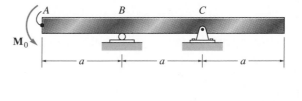

Prob. 14–26

14–27.   Determine the slope at point C of the steel beam. $I = 9.50(10^6)$ mm^4, $E_{st} = 200$ GPa.

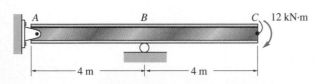

Prob. 14–27

*14–28.   The cantilevered beam has a rectangular cross-sectional area A, a moment of inertia I, and a modulus of elasticity E. If a load P acts at point B as shown, determine the displacement at B in the direction of P, accounting for bending, axial force, and shear.

Prob. 14–28

14–29.   The steel bars are pin-connected at C. If they each have a diameter of 2 in., determine the displacement at E. $E_{st} = 29(10^3)$ ksi.

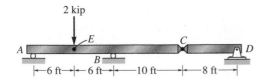

Prob. 14–29

14–30.   The steel bars are pin-connected at B and C. If they each have a diameter of 30 mm, determine the slope at E. $E_{st} = 200$ GPa.

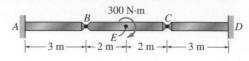

Prob. 14–30

14–31.   The rod has a circular cross section with a polar moment of inertia J and moment of inertia I. If a vertical force P is applied at A, determine the vertical displacement at this point. Consider the strain energy due to bending and torsion. The material constants are E and G.

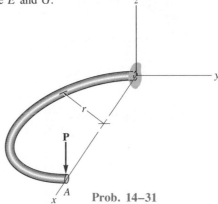

Prob. 14–31

*14–32.   The rod has a circular cross section with a polar moment of inertia J and moment of inertia I. If a vertical force P is applied at A, determine the vertical displacement at this point. Consider the strain energy due to bending and torsion. The material constants are E and G.

Prob. 14–32

14.4   Impact Loading

Throughout this text we have considered all loadings to be applied to a body in a gradual manner, such that when they reach a maximum value they remain constant or static. Some loadings, however, are dynamic; that is, they vary with time. A typical example would be caused by the collision of objects. This is called an impact loading. Specifically, *impact* occurs when one object strikes another, such that large forces are developed between the objects during a very short period of time.

If we assume no energy is lost during impact, we can study the mechanics of impact using the conservation of energy. To show how this is done, we will first analyze the motion of a simple block-and-spring system as shown in Fig. 14–23. When the block is released from rest, it falls a distance h, striking the spring and compressing it a distance Δ_{max} before momentarily coming to rest. If we neglect the mass of the spring and assume that the spring responds *elastically,* then the conservation of energy requires that the energy of the falling block be transformed into stored (strain) energy in the spring; or in other words, the work done by the block's weight, falling $h + \Delta_{max}$, is equal to the work needed to displace the end of the spring by an amount Δ_{max}. Since the force in a spring is related to Δ_{max} by the equation $F = k\Delta_{max}$, where k is the spring stiffness, then applying the conservation of energy, we have

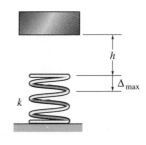

Fig. 14–23

$$U_e = U_i$$

$$W(h + \Delta_{max}) = \frac{1}{2}(k\Delta_{max})\,\Delta_{max}$$

$$W(h + \Delta_{max}) = \frac{1}{2}k\Delta_{max}^2 \qquad (14\text{–}29)$$

$$\Delta_{max}^2 - \frac{2W}{k}\Delta_{max} - 2\left(\frac{W}{k}\right)h = 0$$

This quadratic equation may be solved for Δ_{max}. The maximum root is

$$\Delta_{max} = \frac{W}{k} + \sqrt{\left(\frac{W}{k}\right)^2 + 2\left(\frac{W}{k}\right)h}$$

If the weight W is applied statically (or gradually) to the spring, the end displacement of the spring is $\Delta_{st} = W/k$. Using this simplification, the above equation becomes

$$\Delta_{max} = \Delta_{st} + \sqrt{(\Delta_{st})^2 + 2\Delta_{st}h}$$

or

$$\Delta_{max} = \Delta_{st}\left[1 + \sqrt{1 + 2\left(\frac{h}{\Delta_{st}}\right)}\right] \qquad (14\text{–}30)$$

Once Δ_{max} is computed, the maximum force applied to the spring can be determined from

$$F_{max} = k\Delta_{max} \qquad (14\text{--}31)$$

It should be realized, however, that this force and associated displacement occur only at an *instant*. Provided the block does not rebound off the spring, it will continue to vibrate until the motion dampens out and the block assumes the static position, Δ_{st}. Note also that if the block is held just above the spring, $h = 0$, and *dropped*, then, from Eq. 14–29, the maximum displacement of the block is

$$\Delta_{max} = 2\Delta_{st}$$

In other words, when the block is dropped from the top of the spring (dynamically applied load), the displacement is *twice* what it would be if it were set on the spring (statically applied load).

Using a similar analysis, it is also possible to determine the maximum displacement of the end of the spring if the block is sliding on a smooth horizontal surface with a known velocity **v** just before it collides with the spring, Fig. 14–24. Here the block's kinetic energy,* $\frac{1}{2}(W/g)v^2$, is transformed into stored energy in the spring. Hence,

$$U_e = U_i$$

$$\frac{1}{2}\left(\frac{W}{g}\right)v^2 = \frac{1}{2}k\,\Delta_{max}^2$$

$$\Delta_{max} = \sqrt{\frac{Wv^2}{gk}} \qquad (14\text{--}32)$$

Since the static displacement at the top of the spring caused by the weight W resting on it is $\Delta_{st} = W/k$, then

$$\Delta_{max} = \sqrt{\frac{\Delta_{st}v^2}{g}} \qquad (14\text{--}33)$$

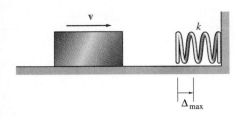

Fig. 14–24

The results of this simplified analysis can be used to determine both the approximate deflection and the stress developed in a deformable member when it is subjected to impact. To do this we must make the necessary assumptions regarding the collision, so that the behavior of the colliding bodies is similar to the response of the block-and-spring models discussed above. Hence we will consider the moving body to be *rigid* like the block and the stationary body to be deformable like the spring. It is assumed that the material behaves in a linear-elastic manner. Furthermore, during collision no energy is lost due to heat, sound, or localized plastic deformations. When collision occurs, the bodies remain in contact until the elastic body reaches its maximum deformation, and during the motion the inertia or mass of the elastic body is neglected. Realize that each of these assumptions will lead to a *conservative* estimate of both the stress and deflection of the elastic body. That is, their values will be larger than those that actually occur.

---

*Recall from physics that kinetic energy is "energy of motion." For the translation of a body it is determined from $\frac{1}{2}mv^2$, where m is the body's mass, $m = W/g$.

A few examples of when the above theory can be applied are shown in Fig. 14–25. Here a known weight (block) is dropped onto a post or beam, causing it to deform a maximum amount Δ_{max}. The energy of the falling block is transformed momentarily into axial strain energy in the post and bending strain energy in the beam.* Although vibrations are established in each member after impact, they will tend to dissipate as time passes. In order to determine the deformation Δ_{max}, we could use the same approach as the block–spring system, and that is to write the conservation-of-energy equation for the block and post or block and beam, and then solve for Δ_{max}. However, we can also solve these problems in a more direct manner by modeling the post and beam by an *equivalent spring*. For example, if a force **P** displaces the top of the post $\Delta = PL/AE$, then a spring having a stiffness $k = AE/L$ would be displaced the same amount by **P**, that is, $\Delta = P/k$. In a similar manner, from the inside back cover, a force **P** applied to the center of a simply-supported beam displaces the center $\Delta = PL^3/48EI$, and therefore an equivalent spring would have a stiffness of $k = 48EI/L^3$. It is not necessary, however, to actually compute the equivalent spring stiffness to apply Eqs. 14–30 or 14–32. All that is needed to determine the dynamic displacement, Δ_{max}, is to compute the *static displacement*, Δ_{st}, due to the weight W of the block resting on the member.

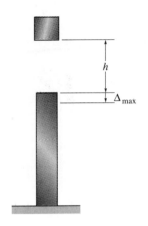

Once Δ_{max} is determined, the maximum dynamic force can then be calculated from $P_{max} = k\Delta_{max}$. If we consider P_{max} to be an *equivalent static load* then the maximum stress in the member can be determined using statics and the theory of mechanics of materials. Recall that this stress acts only for an *instant*. In reality, vibrational waves pass through the material, and the stress in the post or the beam, for example, does not remain constant.

The ratio of the equivalent static load P_{max} to the load W is called the *impact factor*, n. Since $P_{max} = k\Delta_{max}$ and $W = k\Delta_{st}$, then from Eq. 14–30, we can express it as

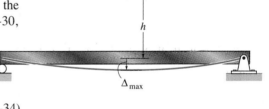

$$ n = 1 + \sqrt{1 + 2\left(\frac{h}{\Delta_{st}}\right)} \qquad (14\text{–}34) $$

This factor represents the magnification of a statically applied load so that it can be treated dynamically. Using Eq. 14–34, n can be computed for any member that has a linear relationship between load and deflection. For a complicated system of connected members, however, impact factors are determined by experience or by experimental testing. Once n is determined, the dynamic stress and deflections are easily computed from the static stress σ_{st} and static deflection Δ_{st} caused by the load W, that is, $\sigma_{max} = n\sigma_{st}$ and $\Delta_{max} = n\Delta_{st}$.

Fig. 14–25

---

*Strain energy due to shear is neglected for reasons discussed in Example 14–4.

Example 14–8

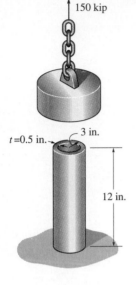

150 kip

$t = 0.5$ in.

3 in.

12 in.

Fig. 14–26

The aluminum pipe shown in Fig. 14–26 is used to support a load of 150 kip. Determine the displacement at the top of the pipe and the maximum stress in the pipe if the load is (a) applied gradually, and (b) applied suddenly by releasing it from the top of the pipe. Take $E_{al} = 10(10^3)$ ksi and assume that the aluminum behaves elastically.

SOLUTION

Part (a).   When the load is applied gradually, the work done by the weight is transformed into elastic strain energy in the pipe. Applying the conservation of energy, we have

$$U_e = U_i$$

$$\frac{1}{2}\,W\Delta_{st} = \frac{W^2 L}{2AE}$$

$$\Delta_{st} = \frac{WL}{AE} = \frac{150 \text{ kip}(12 \text{ in.})}{\pi[(3 \text{ in.})^2 - (2.5 \text{ in.})^2]10(10^3) \text{ kip/in}^2}$$

$$= 0.0208 \text{ in.} \qquad \textit{Ans.}$$

This same result also follows from using $\Delta = PL/AE$, Eq. 4–2.
The static stress in the pipe is

$$\sigma_{st} = \frac{W}{A} = \frac{150 \text{ kip}}{\pi[(3 \text{ in.})^2 - (2.5 \text{ in.})^2]} = 17.4 \text{ ksi} \qquad \textit{Ans.}$$

Part (b).   Here Eq. 14–30 can be applied, with $h = 0$. Hence,

$$\Delta_{max} = 2\Delta_{st} = 2(0.0208 \text{ in.})$$
$$= 0.0416 \text{ in.} \qquad \textit{Ans.}$$

Since elastic behavior occurs,

$$\sigma_{max} = 2\sigma_{st} = 2(17.4 \text{ ksi})$$
$$= 34.8 \text{ ksi} \qquad \textit{Ans.}$$

Hence, the displacement of the weight and the stress in the pipe are both twice as great as when the load is applied statically. In other words, the impact factor is $n = 2$.

Example 14–9

The steel beam shown in Fig. 14–27a is a $W10 \times 39$. Determine the maximum bending stress in the beam and the beam's maximum deflection if the weight $W = 1.50$ kip is dropped from a height $h = 2$ in. onto the beam. $E_{st} = 29(10^3)$ ksi.

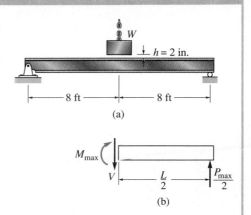

Fig. 14–27

SOLUTION I

We will apply Eq. 14–30. First, however, we must calculate Δ_{st}. Using the table on the inside back cover, and the data in Appendix B for the properties of a $W10 \times 39$, we have

$$\Delta_{st} = \frac{WL^3}{48EI} = \frac{1.50 \text{ kip}(16 \text{ ft})^3(12 \text{ in./ft})^3}{48[29(10^3) \text{ ksi}](209 \text{ in}^4)} = 0.0365 \text{ in.}$$

$$\Delta_{max} = \Delta_{st}\left[1 + \sqrt{1 + 2\left(\frac{h}{\Delta_{st}}\right)}\right]$$

$$= 0.0365 \text{ in.} \left[1 + \sqrt{1 + 2\left(\frac{2 \text{ in.}}{0.0365 \text{ in.}}\right)}\right] = 0.420 \text{ in.} \qquad Ans.$$

This deflection is caused by an equivalent static load P_{max}, computed from $P_{max} = (48EI/L^3)\,\Delta_{max}$.

The internal moment caused by this load is maximum at the center of the beam, such that by the method of sections, Fig. 14–27b, $M_{max} = P_{max}L/4$. Applying the flexure formula, we have

$$\sigma_{max} = \frac{M_{max}c}{I} = \frac{P_{max}Lc}{4I} = \frac{12E\Delta_{max}c}{L^2}$$

$$= \frac{12[29(10^3) \text{ kip/in}^2](0.4202 \text{ in.})(9.92 \text{ in./2})}{(16 \text{ ft})^2(12 \text{ in./ft})^2} = 19.7 \text{ ksi} \qquad Ans.$$

SOLUTION II

It is also possible to obtain the dynamic or maximum deflection Δ_{max} from first principles. The external work of the falling weight W is $U_e = W(h + \Delta_{max})$. Since the beam deflects Δ_{max}, and $P_{max} = 48EI\Delta_{max}/L^3$ then

$$U_e = U_i$$

$$W(h + \Delta_{max}) = \frac{1}{2}\left(\frac{48EI\Delta_{max}}{L^3}\right)\Delta_{max}$$

$$(1.50 \text{ kip})(2 \text{ in.} + \Delta_{max}) = \frac{1}{2}\left[\frac{48[29(10^3) \text{ kip/in}^2]209 \text{ in}^4}{(16 \text{ ft})^3(12 \text{ in./ft})^3}\right]\Delta_{max}^2$$

$$20.55\Delta_{max}^2 - 1.50\Delta_{max} - 3.00 = 0$$

Solving and choosing the positive root yields

$$\Delta_{max} = 0.420 \text{ in.} \qquad Ans.$$

A railroad car that is assumed to be rigid and has a mass of 80 Mg is moving forward at a speed of $v = 0.2$ m/s when it strikes a steel 200-mm by 200-mm post at A, Fig. 14–28a. If the post is fixed to the ground at C, determine the maximum horizontal displacement of its top B due to the impact. Take $E_{st} = 200$ GPa.

SOLUTION

Here the kinetic energy of the railroad car is transformed into internal bending strain energy only for region AC of the post. Assuming that point A is displaced $(\Delta_A)_{max}$, then the force P_{max} that causes this displacement can be determined from the table on the inside back cover. We have

$$P_{max} = \frac{3EI(\Delta_A)_{max}}{L^3_{AC}} \qquad (1)$$

$$U_e = U_i; \qquad \frac{1}{2} mv^2 = \frac{1}{2} P_{max}(\Delta_A)_{max}$$

$$\frac{1}{2} mv^2 = \frac{1}{2} \frac{3EI}{L^3_{AC}} (\Delta_A)^2_{max}; \quad (\Delta_A)_{max} = \sqrt{\frac{mv^2 L^3_{AC}}{3EI}}$$

Substituting in the numerical data yields

$$(\Delta_A)_{max} = \sqrt{\frac{80(10^3) \text{ kg}(0.2 \text{ m/s})^2(1.5 \text{ m})^3}{3[200(10^9) \text{ N/m}^2][\frac{1}{12}(0.20 \text{ m})^4]}} = 0.0116 \text{ m} = 11.6 \text{ mm}$$

Using Eq. 1, the force P_{max} is therefore

$$P_{max} = \frac{3[200(10^9) \text{ N/m}^2][\frac{1}{12}(0.20 \text{ m})^4](0.0116 \text{ m})}{(1.5 \text{ m})^3} = 275.4 \text{ kN}$$

With reference to Fig. 14–28b, segment AB of the post remains straight. To determine the maximum displacement at B, we must first determine the slope at A. Using the appropriate formula from the table on the inside back cover to determine θ_A, we have

$$\theta_A = \frac{P_{max}L^2_{AC}}{2EI} = \frac{275.4(10^3) \text{ N } (1.5 \text{ m})^2}{2[200(10^9) \text{ N/m}^2][\frac{1}{12} (0.20 \text{ m})^4]} = 0.011619 \text{ rad}$$

The maximum displacement at B is thus

$$(\Delta_B)_{max} = (\Delta_A)_{max} + \theta_A L_{AB} = 11.6 \text{ mm} + (0.011619 \text{ rad}) \, 1(10^3) \text{ mm} = 23.2 \text{ mm}$$

The maximum bending stress in the post occurs at C and has a value of

$$\sigma_{max} = \frac{Mc}{I} = \frac{P_{max}L_{AC}c}{(\frac{1}{12}a^4)} = \frac{275.4(10^3) \text{ N } (1.5 \text{ m})(0.1 \text{ m})}{[\frac{1}{12} (0.2 \text{ m})^4]} = 310 \text{ MPa}$$

For the solution to be valid, this stress should be *below* or equal to the yield stress for the steel. If it is not, the post will permanently deform.

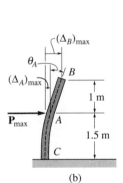

(a)

(b)

Fig. 14–28

$v = 0.2$ m/s

200 mm

B 1 m

A 1.5 m

C

$(\Delta_B)_{max}$

θ_A

$(\Delta_A)_{max}$

B

1 m

$\mathbf{P}_{max}$

A

1.5 m

C

PROBLEMS

14–33. A bar is 4 m long and has a diameter of 30 mm. If it is to be used to absorb energy in tension from an impact loading, determine the total amount of elastic energy that it can absorb if (*a*) it is made of steel for which E_{st} = 200 GPa, σ_Y = 800 MPa, and (*b*) it is made from an aluminum alloy for which E_{al} = 70 GPa, σ_Y = 405 MPa.

14–34. The collar has a weight of 50 lb and falls down the titanium bar. If the bar has a diameter of 0.5 in., determine the maximum stress developed in the bar if the weight is (*a*) dropped from a height of *h* = 1 ft, (*b*) released from a height *h* ≈ 0, and (*c*) placed slowly on the flange at *A*. E_{ti} = 16(10³) ksi, σ_Y = 60 ksi.

14–35. The collar has a weight of 50 lb and falls down the titanium bar. If the bar has a diameter of 0.5 in., determine the largest height *h* at which the weight can be released and not permanently damage the bar after striking the flange at *A*. E_{ti} = 16(10³) ksi, σ_Y = 60 ksi.

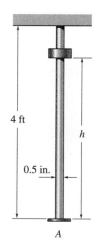

Probs. 14–34/14–35

*14–36. The composite aluminum bar is made from two segments having diameters of 5 mm and 10 mm. Determine the maximum axial stress developed in the bar if the 5-kg collar is dropped from a height of *h* = 100 mm. E_{al} = 70 GPa, σ_Y = 410 MPa.

14–37. The composite aluminum bar is made from two segments having diameters of 5 mm and 10 mm. Determine the maximum height *h* from which the 5-kg collar should be dropped so that it produces a maximum axial stress in the bar of σ_{max} = 300 MPa. E_{al} = 70 GPa, σ_Y = 410 MPa.

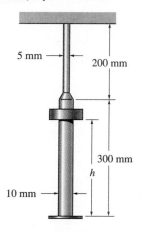

Probs. 14–36/14–37

14–38. The mass of 50 Mg is held just over the top of the steel post having a length of 2 m and a cross-sectional area of 0.01 m². If the mass is released, determine the maximum stress developed in the bar and its maximum deflection. E_{st} = 200 GPa, σ_Y = 600 MPa.

Prob. 14–38

14–39. A stepped magnesium post having the dimensions shown is struck by a rigid body having a weight of 800 lb and traveling at 2 ft/s. Determine the maximum stress in the post. Neglect the mass of the post. $E_{mg} = 6.4(10^3)$ ksi, $\sigma_Y = 40$ ksi.

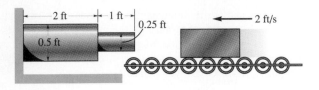

Prob. 14–39

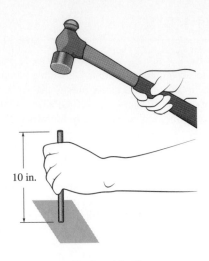

Probs. 14–41

*14–40. The drop hammer of a pile driver has a mass of 150 kg. It is dropped from rest, 0.75 m from the top of a wooden pile having a diameter of 300 mm and a length of 6 m. Determine the maximum axial stress developed in the pile if the pile absorbs 70% of the impacting energy. Assume the material responds elastically. $E_w = 11.5$ GPa.

14–42. A steel cable having a diameter of 0.4 in. wraps over the drum and is used to lower an elevator having a weight of 800 lb. The elevator is 150 ft below the drum and is descending at the rate of 2 ft/s when the drum suddenly stops. Determine the maximum stress developed in the cable when this occurs. $E_{st} = 29(10^3)$ ksi, $\sigma_Y = 50$ ksi.

14–43. Solve Prob. 14–42 if the elevator is ascending at the rate of 3 ft/s.

Prob. 14–40

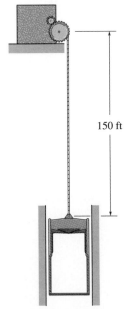

14–41. The steel chisel has a diameter of 0.5 in. and a length of 10 in. It is struck by a hammer that weighs 3 lb, and at the instant of impact it is moving at 12 ft/s. Determine the maximum compressive stress in the chisel, assuming that 80% of the impacting energy goes into the chisel. $E_{st} = 29(10^3)$ ksi, $\sigma_Y = 100$ ksi.

Probs. 14–42/14–43

*14–44. A 20-lb weight is dropped from a height of 4 ft onto the end of a cantilevered steel beam. If the beam is a $W\ 12 \times 50$, determine the maximum stress developed in the beam. $E_{st} = 29(10^3)$ ksi, $\sigma_Y = 36$ ksi.

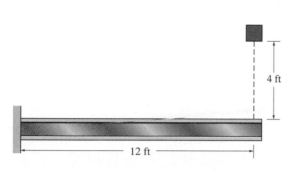

Prob. 14–44

14–46. The diver weighs 150 lb and, while holding himself rigid, strikes the end of a wooden diving board with a downward velocity of 2 ft/s. Determine the maximum stress developed in the board. The board has a thickness of 1.5 in. and width of 1.5 ft. $E_w = 1.8(10^3)$ ksi, $\sigma_Y = 8$ ksi.

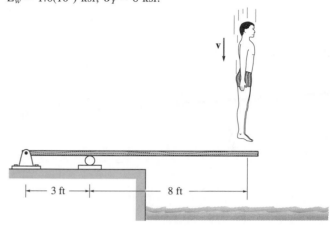

Prob. 14–46

14–45. The wide-flange beam has a length of $2L$, a depth $2c$, and a constant EI. Determine the maximum height h at which a weight W can be dropped on its end without exceeding a maximum elastic stress σ_{max} in the beam.

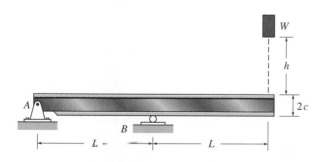

Prob. 14–45

14–47. The steel beam AB acts to stop the oncoming railroad car, which has a mass of 10 Mg and is coasting toward it at $v = 0.5$ m/s. Determine the maximum stress developed in the beam if it is struck at its center by the car. The beam is simply supported and only horizontal forces occur at A and B. Assume that the railroad car and the supporting framework for the beam remains rigid. Also, compute the maximum deflection of the beam. $E_{st} = 200$ GPa, $\sigma_Y = 250$ MPa.

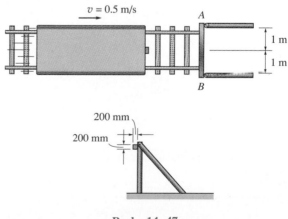

Prob. 14–47

*14–48. The 200-lb box has a downward velocity of 4 ft/s when it is 3 ft from the top of the wooden beam. Determine the maximum stress in the beam due to the impact and compute the maximum deflection of its end C. $E_w = 1.9(10^3)$ ksi, $\sigma_Y = 6$ ksi.

14–49. The 100-lb box has a downward velocity of 4 ft/s when it is 3 ft from the top of the wood beam. Determine the maximum stress in the beam due to the impact and compute the maximum deflection of point B. $E_w = 1.9(10^3)$ ksi, $\sigma_Y = 8$ ksi.

*14–52. The W 10 × 12 is cantilevered from the wall at B. The spring mounted on the beam has a stiffness of $k = 1000$ lb/in. If a weight of 8 lb is dropped onto the spring from a height of 3 ft, determine the maximum bending stress developed in the beam. The beam is made of steel for which $E_{st} = 29(10^3)$ ksi, $\sigma_Y = 36$ ksi.

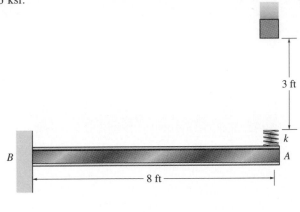

Prob. 14–52

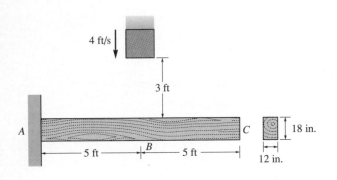

Probs. 14–48/14–49

14–53. The simply supported W 10 × 15 steel beam lies in the horizontal plane and acts as a shock absorber for the 500-lb block, which is traveling toward it at 5 ft/s. Determine the maximum deflection of the beam and the maximum stress in the beam due to the impact. The spring has a stiffness of $k = 1000$ lb/in. $E_{st} = 29(10^3)$ ksi, $\sigma_Y = 36$ ksi.

14–50. The weight of 175 lb is dropped from a height of 4 ft from the top of the steel beam. Determine the maximum deflection and maximum stress in the beam if the supporting springs at A and B each have a stiffness of $k = 500$ lb/in. The beam is 3 in. thick and 4 in. wide. $E_{st} = 29(10^3)$ ksi, $\sigma_Y = 36$ ksi.

14–51. The weight of 175 lb is dropped from a height of 4 ft from the top of the steel beam. Determine the load factor n if the supporting springs at A and B each have a stiffness of $k = 300$ lb/in. The beam is 3 in. thick and 4 in. wide. $E_{st} = 29(10^3)$ ksi, $\sigma_Y = 36$ ksi.

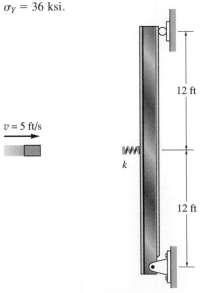

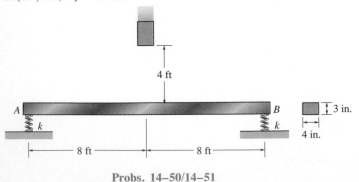

Probs. 14–50/14–51

Prob. 14–53

*14.5 Principle of Virtual Work

The principle of virtual work was developed by John Bernoulli in 1717, and like other energy methods of analysis, it is based on the conservation of energy. Although the principle of virtual work has many applications in mechanics, in this text we will use it to obtain the displacement and slope at various points on a deformable body. Before doing this, however, we will need to make some preliminary remarks that apply to the development of this method.

Whenever a body is fixed from moving, it is necessary that the loadings satisfy the equilibrium conditions and the displacements satisfy the compatibility conditions. Specifically, *equilibrium conditions* require the external loads to be uniquely related to the internal loads, and the *compatibility conditions* require the external displacements to be uniquely related to the internal deformations. For example, if we consider a deformable body of any shape or size and apply a series of external loads P to it, these loadings will cause internal loadings u within the body. Here the external and internal loads are related by the equations of equilibrium. Furthermore, since the body is deformable, the external loads will be displaced Δ, and the internal loadings will undergo displacements δ. In general, the material does *not* have to behave elastically, and so the displacements may *not* be related to the loads. However, if the external displacements are known, the corresponding internal displacements are uniquely defined since the body is continuous. For this case, the conservation of energy states that

$$U_e = U_i; \qquad\qquad \Sigma\, P\, \Delta = \Sigma\, u\, \delta \qquad\qquad (14\text{--}35)$$

Based on this concept, we will now develop the principle of virtual work so that it can be used to determine the displacement and slope at *any point* on a body. To do this, we will consider the body to be of arbitrary shape as shown in Fig. 14–29b, and to be subjected to the "real loads" P_1, P_2, and P_3. It is to be understood that these loads cause no movement of the supports; however, in general they can strain the material *beyond* the elastic limit. Suppose that it is necessary to determine the displacement Δ of point A on the body caused by these loads. To do this we will consider applying the conservation-of-energy principle, Eq. 14–35. In this case, however, there is no force acting at A, and so the unknown displacement Δ will *not* be included as an external "work term" in the equation.

In order to get around this limitation, we will place an *imaginary* or "virtual" force $\mathbf{P}'$ on the body at point A, such that P' acts in the *same direction* as Δ. Furthermore, this load is applied to the body *before* the real loads are applied, Fig. 14–29a. For convenience, which will be made clear later, we will choose $\mathbf{P}'$ to have a "unit" magnitude; that is, $P' = 1$. It is to be emphasized that the term "virtual" is used to describe the load because it is *imaginary* and does not actually exist as part of the real loading. This external virtual load, however, does create an internal virtual load $\mathbf{u}$ in a

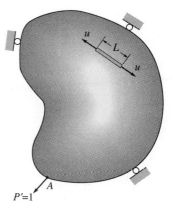

Application of virtual unit load
(a)

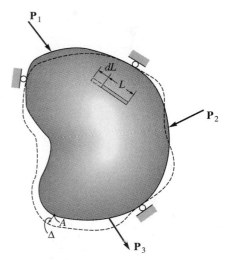

Application of real loads
(b)

Fig. 14–29

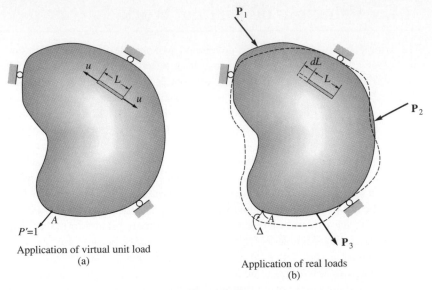

Application of virtual unit load
(a)

Application of real loads
(b)

Fig. 14–29

representative element or fiber of the body, as shown in Fig. 14–29a. As expected, P' and u can be related by the equations of equilibrium. Also, because of these loadings, the body and the element will each undergo a virtual displacement, although we will *not* be concerned with their magnitudes. Once the virtual load is applied and *then* the body is subjected to the *real loads* $\mathbf{P}_1$, $\mathbf{P}_2$, and $\mathbf{P}_3$, point A will be displaced a real amount Δ, which causes the element to be displaced dL, Fig. 14–29b. As a result, the external virtual force $\mathbf{P}'$ and internal virtual load $\mathbf{u}$ "ride along" by Δ and dL, respectively; consequently these loads perform *external virtual work* of 1Δ on the body and *internal virtual work* of $u\,dL$ on the element. Considering *only* the conservation of *virtual* energy, the external virtual work is then equal to the internal virtual work done on all the elements of the body. Therefore, we can write the virtual-work equation as

$$
1 \cdot \Delta = \Sigma u \cdot dL \tag{14–36}
$$

virtual loadings

real displacements

Here

$P' = 1 =$ external virtual unit load acting in the direction of Δ

$u =$ internal virtual load acting on the element

$\Delta =$ external displacement caused by the real loads

$dL =$ internal displacement of the element in the direction of $\mathbf{u}$, caused by the real loads

By choosing $P' = 1$, it can be seen that the solution for Δ follows directly, since $\Delta = \Sigma u\,dL$.

In a similar manner, if the rotational displacement or slope of the tangent at a point on a body is to be determined, a virtual *couple moment* **M′**, having a "unit" magnitude, is applied at the point. As a consequence, this couple moment causes a virtual load u_θ in one of the elements of the body. Assuming that the real loads deform the element an amount dL, the rotation θ can be found from the virtual-work equation

$$1 \cdot \theta = \Sigma u_\theta \, dL \qquad\qquad (14\text{--}37)$$

with "virtual loadings" labeling the right-hand term and "real displacements" labeling the terms below.

Here

$M' = 1 =$ external virtual unit couple moment acting in the direction of θ

$\quad u_\theta =$ internal virtual load acting on an element

$\quad\;\; \theta =$ external rotational displacement in radians caused by the real loads

$\quad dL =$ internal displacement of the element in the direction of u_θ, caused by the real loads

This method for applying the principle of virtual work is often referred to as the *method of virtual forces,* since a *virtual force* is applied, resulting in a calculation of an external *real displacement*. The equation of virtual work in this case represents a statement of *compatibility requirements* for the body. Although it is not important here, realize that we can also apply the principle of virtual work as a *method of virtual displacements*. In this case, *virtual displacements* are imposed on the body when the body is subjected to *real loadings*. This method can be used to determine the external reactive force on the body or an unknown internal loading in the body. When it is used in this manner, the equation of virtual work is a statement of the *equilibrium requirements* for the body.*

Internal Virtual Work   The terms on the right-hand side of Eqs. 14–36 and 14–37 represent the internal virtual work developed in the body. The real internal displacements dL in these terms can be produced in several different ways. For example, these displacements may result from geometric fabrication errors, from temperature, or more commonly from stress. In particular, no restriction has been placed on the magnitude of the external loading, so the stress may be large enough to cause yielding or even strain hardening of the material.

---

*See *Engineering Mechanics: Statics,* 5th edition, R.C. Hibbeler, Macmillan Publishing Co., 1989.

Deformation caused by	Strain energy	Internal virtual work
Axial load N	$\int_0^L \dfrac{N^2}{2EA}\,dx$	$\int_0^L \dfrac{nN}{EA}\,dx$
Shear V	$\int_0^L \dfrac{f_s V^2}{2GA}\,dx$	$\int_0^L \dfrac{f_s vV}{GA}\,dx$
Bending moment M	$\int_0^L \dfrac{M^2}{2EI}\,dx$	$\int_0^L \dfrac{mM}{EI}\,dx$
Torsional moment T	$\int_0^L \dfrac{T^2}{2GJ}\,dx$	$\int_0^L \dfrac{tT}{GJ}\,dx$

Table 14–1

If we assume that the material behavior is linear-elastic and the stress does not exceed the proportional limit, we can formulate the expressions for internal virtual work caused by stress using the equations of elastic strain energy developed in Sec. 14.2 and listed in the second column of Table 14–1. Recall that each of these expressions assumes that the stress resultant N, V, M, or T was applied gradually from zero to its full value. As a result, the work done by the stress resultant is shown in these expressions as *one-half* the product of the stress resultant and its displacement. In the case of the virtual-force method, however, the ''full'' virtual loading is applied *before* the real loads cause displacements, and therefore the work of the internal virtual loading is simply the product of the internal virtual load and its real displacement. Referring to these internal virtual loadings (u) by the corresponding lowercase symbols n, v, m, and t, the virtual work due to axial load, shear, bending moment, and torsional moment is listed in the right-hand column of Table 14–1. Using these results, the virtual-work equation for a body subjected to a general loading can therefore be written as

$$1 \cdot \Delta = \int \frac{nN}{AE}\,dx + \int \frac{mM}{EI}\,dx + \int \frac{f_s vV}{GA}\,dx + \int \frac{tT}{GJ}\,dx \quad (14\text{–}38)$$

In the following sections we will apply the above equation to problems involving the deflections of trusses, beams, and mechanical elements. We will also include a discussion of how to handle the effects of fabrication errors and differential temperature. For application it is important that a consistent set of units be used for all the terms. For example, if the real loads are expressed in kilonewtons and the body's dimensions are in meters, a 1-kN virtual force or 1-kN · m virtual couple should be applied to the body. By doing so a calculated displacement Δ will be in meters, and a calculated slope will be in radians.

*14.6 Method of Virtual Forces Applied to Trusses

In this section we will apply the method of virtual forces to determine the displacement of a truss joint. To illustrate the principles, the vertical displacement of joint A of the truss shown in Fig. 14–30b will be determined. This displacement is caused by the "real loads" P_1 and P_2, and since these loads cause only axial force in the members, it is only necessary to consider the internal virtual work due to axial load. To obtain this virtual work, we will assume that each member has a constant cross-sectional area A, and the virtual load n and real load N are constant throughout the member's length. As a result, the internal virtual work for a member is

$$\int_0^L \frac{nN}{AE}\, dx = \frac{nNL}{AE}$$

And the virtual-work equation for the entire truss is therefore

$$1 \cdot \Delta = \sum \frac{nNL}{AE} \qquad (14\text{–}39)$$

Here

1 = external virtual unit load acting on the truss joint in the stated direction of Δ

Δ = joint displacement caused by the real loads on the truss

n = internal virtual force in a truss member caused by the external virtual unit load

N = internal force in a truss member caused by the real loads

L = length of a member

A = cross-sectional area of a member

E = modulus of elasticity of a member

The formulation of this equation follows naturally from the development in Sec. 14.5. Here the external virtual unit load creates internal virtual "n" forces in each of the truss members, Fig. 14–30a. When the real loads are applied to the truss, they cause the truss joint to be displaced Δ in the same direction as the virtual unit load, Fig. 14–30b, and each member undergoes a displacement NL/AE, in the same direction as its respective n force. Consequently, the external virtual work $1 \cdot \Delta$ equals the internal virtual work or the internal (virtual) strain energy stored in all the truss members, i.e., Eq. 14–39.

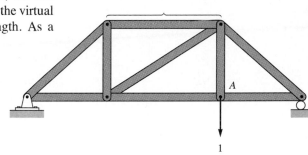

Application of virtual unit load

(a)

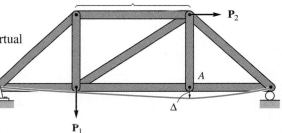

Application of real loads

(b)

Fig. 14–30

Temperature Change. Truss members can change their length due to a change in temperature. If α is the coefficient of thermal expansion for a member and ΔT is the change in temperature, the change in length of a member is $\Delta L = \alpha \, \Delta T L$ (Eq. 4–4). Hence, we can determine the displacement of a selected truss joint due to this temperature change from Eq. 14–36, written as

$$\boxed{1 \cdot \Delta = \Sigma n\alpha \, \Delta TL} \qquad (14\text{–}40)$$

Here

1 = external virtual unit load acting on the truss joint in the stated direction of Δ

n = internal virtual force in a truss member caused by the external virtual unit load

Δ = external joint displacement caused by the temperature change

α = coefficient of thermal expansion of member

ΔT = change in temperature of member

L = length of member

Fabrication Errors. Occasionally errors in fabricating the lengths of the members of a truss may occur. If this happens, the displacement in a particular direction of a truss joint from its expected position can be determined from direct application of Eq. 14–36 written as

$$\boxed{1 \cdot \Delta = \Sigma n \, \Delta L} \qquad (14\text{–}41)$$

Here

1 = external virtual unit load acting on the truss joint in the stated direction of Δ

n = internal virtual force in a truss member caused by the external virtual unit load

Δ = external joint displacement caused by the fabrication errors

ΔL = difference in length of the member from its intended length caused by a fabrication error

A combination of the right-hand sides of Eqs. 14–39 through 14–41 will be necessary if external loads act on the truss and some of the members undergo a temperature change or have been fabricated with the wrong dimensions.

PROCEDURE FOR ANALYSIS

The following procedure provides a method that may be used to determine the displacement of any joint on a truss using the method of virtual force.

Virtual Forces **n.**   Place the virtual unit load on the truss at the joint where the desired displacement is to be determined. The load should be directed along the line of action of the displacement. For example, if a vertical displacement is to be determined, the unit load should then act vertically. With the unit load so placed and all the real loads *removed* from the truss, calculate the internal n force in each truss member. Assume that tensile forces are positive and compressive forces are negative.

Real Forces **N.**   Determine the N forces in each member. These forces are caused only by the real loads acting on the truss. Again, assume that tensile forces are positive and compressive forces are negative. This is necessary since positive or negative internal virtual work depends on the directional sense of both the virtual load, defined by $\pm n$, and displacement, caused by $\pm N$.

Virtual-Work Equation.   Apply the equation of virtual work, Eq. 14–39, to determine the desired displacement Δ. It is important to retain the algebraic sign for each of the corresponding n and N forces when substituting these terms into the equation. If the resultant sum $\Sigma nNL/AE$ is positive, the displacement Δ is in the same direction as the virtual unit load. If a negative value results, Δ is opposite to the virtual unit load.

In a similar manner, when applying Eq. 14–40, realize that if any of the members undergo an *increase* in temperature, ΔT will be *positive;* whereas a *decrease* in temperature will result in a *negative* value for ΔT. Likewise, for Eq. 14–41, when a fabrication error *increases* the length of a member, ΔL is *positive,* whereas a *decrease* in length is *negative.*

When applying this method, attention should be paid to the units of each numerical quantity. Notice, however, that the virtual unit load can be assigned any arbitrary unit, pounds, kips, newtons, etc., since the n forces will have these *same* units, and as a result, the units for both the virtual unit load and the n forces will cancel from both sides of the equation.

The following examples illustrate application of this method.

Example 14–11

Determine the vertical displacement of joint C of the steel truss shown in Fig. 14–31a. The cross-sectional area of each member is $A = 400$ mm^2 and $E_{st} = 200$ GPa.

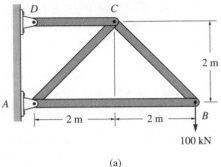

(a)

SOLUTION

Virtual Forces **n.**   *Only* a vertical 1-kN virtual load is placed at joint C, and the force in each member is calculated using the method of joints. The results of this analysis are shown in Fig. 14–31b. Using our sign convention, positive numbers indicate tensile forces and negative numbers indicate compressive forces.

Real Force **N.**   The applied load of 100 kN causes forces in the members that can be calculated using the method of joints. The results of this analysis are shown in Fig. 14–31c.

Virtual-Work Equation.   Arranging the data in tabular form, we have

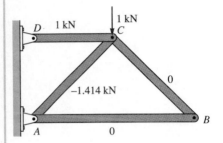

Virtual forces

(b)

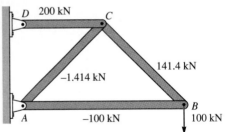

Real forces

(c)

Fig. 14–31

Member	n	N	L	nNL
AB	0	−100	4	0
BC	0	141.4	2.828	0
AC	−1.414	−141.4	2.828	565.7
CD	1	200	2	400
				Σ 965.7 kN$^2 \cdot$ m

Thus

$$1 \cdot \Delta_{C_v} = \sum \frac{nNL}{AE} = \frac{965.7}{AE}$$

Substituting the numerical values for A and E, we have

$$1 \text{ kN} \cdot \Delta_{C_v} = \frac{965.7 \text{ kN}^2 \cdot \text{m}}{[400(10^{-6}) \text{ m}^2] 200(10^6) \text{ kN/m}^2}$$

$$\Delta_{C_v} = 0.01207 \text{ m} = 12.1 \text{ mm} \qquad \textit{Ans.}$$

Example 14–12

The cross-sectional area of each member of the steel truss shown in Fig. 14–32a is $A = 0.5$ in^2, and the modulus of elasticity for the steel members is $E_{st} = 29(10^3)$ ksi. (a) Determine the horizontal displacement of joint C if a force of 12 kip is applied to the truss at B. (b) If no external loads act on the truss, what is the horizontal displacement of joint C if member AC is fabricated 0.25 in. too short?

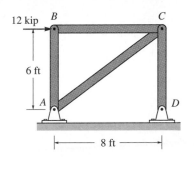

(a)

SOLUTION
Part (a)

Virtual Forces **n.**   Since the *horizontal displacement* of joint C is to be determined, a horizontal virtual force of 1 kip is applied at C. The **n** force in each member is determined by the method of joints and shown on the truss in Fig. 14–32b. As usual, a positive number represents a tensile force and a negative number represents a compressive force.

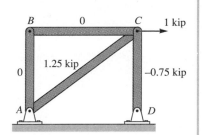

Virtual forces

(b)

Real Forces **N.**   The force in each member as caused by the externally applied 12-kip force is shown in Fig. 14–32c.

Virtual-Work Equation.   Since AE is constant, ΣnNL is computed as follows:

Member	n	N	L	nNL
AB	0	0	6	0
AC	1.25	15	10	187.5
CB	0	−12	8	0
CD	−0.75	−9	6	40.5
				Σ 228.0 kip$^2 \cdot$ ft

$$1 \text{ kip} \cdot \Delta_{C_h} = \sum \frac{nNL}{AE} = \frac{228.0 \text{ kip}^2 \cdot \text{ft}}{AE}$$

$$1 \text{ kip} \cdot \Delta_{C_h} = \frac{228.0 \text{ kip} \cdot \text{ft}(12 \text{ in./ft})}{(0.5 \text{ in}^2)\, 29(10^3) \text{ kip/in}^2}$$

$$\Delta_{C_h} = 0.189 \text{ in.} \qquad\qquad Ans.$$

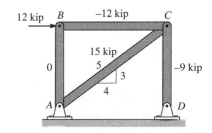

Real forces

(c)

Fig. 14–32

Part (b).   Here we must apply Eq. 14–41. Since the horizontal displacement of C is to be determined, we can use the results of Fig. 14–32b. Realizing that member AC is *shortened* by $\Delta L = -0.25$ in., we have

$$1 \cdot \Delta = \Sigma u \Delta L; \quad 1 \text{ kip} \cdot \Delta_{C_h} = (1.25 \text{ kip})(-0.25 \text{ in.})$$
$$\Delta_{C_h} = -0.312 \text{ in.} = 0.312 \text{ in.} \leftarrow \qquad Ans.$$

The negative sign indicates that joint C is displaced to the left, opposite to the 1-kip horizontal load.

Example 14–13

Determine the horizontal displacement of joint B of the truss shown in Fig. 14–33a. Due to radiant heating, member AB is subjected to an *increase* in temperature of $\Delta T = +60°C$. The members are made of steel, for which $\alpha_{st} = 12(10^{-6})/°C$ and $E_{st} = 200$ GPa. The cross-sectional area of each member is 250 mm².

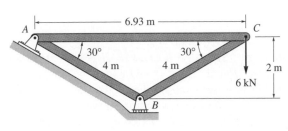

(a)

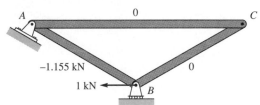

Virtual forces

(b)

SOLUTION

Virtual Forces n. A horizontal 1-kN virtual load is applied to the truss at joint B, and the forces in each member are computed, Fig. 14–33b.

Real Forces N. Since the **n** forces in members AC and BC are *zero*, the N forces in these members do *not* have to be computed. Why? For completeness, though, the entire "real" force analysis is shown in Fig. 14–33c.

Virtual-Work Equation. Both loads and temperature affect the deformation; therefore, Eqs. 14–39 and 14–40 are combined, which gives

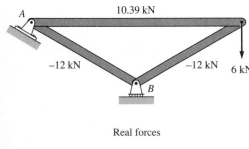

Real forces

(c)

Fig. 14–33

$$1 \text{ kN} \cdot \Delta_{B_h} = \sum \frac{nNL}{AE} + \Sigma u\alpha \, \Delta TL$$

$$= 0 + 0 + \frac{(-1.155 \text{ kN})(-12 \text{ kN})(4 \text{ m})}{[250(10^{-6}) \text{ m}^2][200(10^6) \text{ kN/m}^2]} +$$

$$0 + 0 + (-1.155 \text{ kN})[12(10^{-6})/°C](60°C)(4 \text{ m})$$

$$\Delta_{B_h} = -0.00222 \text{ m}$$

$$= 2.22 \text{ mm} \rightarrow \qquad \qquad Ans.$$

The negative sign indicates that roller B moves to the right, opposite to the direction of the virtual load, Fig. 14–33b.

PROBLEMS

14–54. Determine the horizontal displacement of point B on the two-member frame. Each steel member has a cross-sectional area of 2 in². $E_{st} = 29(10^3)$ ksi.

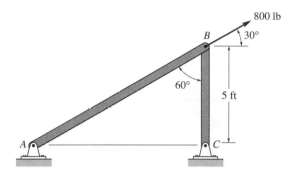

Prob. 14–54

14–55. Determine the vertical displacement of point B of the truss. Each steel member has a cross-sectional area of 300 mm². $E_{st} = 200$ GPa.

*14–56. Determine the horizontal displacement of point B of the truss. Each steel member has a cross-sectional area of 300 mm². $E_{st} = 200$ GPa.

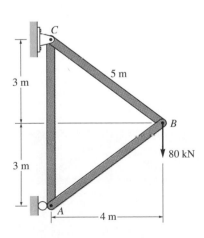

Probs. 14–55/14–56

14–57. Determine the horizontal displacement of point C on the truss. Each steel member has a cross-sectional area of 3 in². $E_{st} = 29(10^3)$ ksi.

14–58. Determine the horizontal displacement of point B on the truss. Each steel member has a cross-sectional area of 3 in². $E_{st} = 29(10^3)$ ksi.

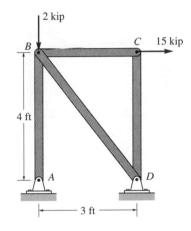

Probs. 14–57/14–58

14–59. Determine the vertical displacement of point B on the truss. For each member $A = 400$ mm², $E = 200$ GPa.

*14–60. Determine the vertical displacement of point E on the truss. For each member $A = 400$ mm², $E = 200$ GPa.

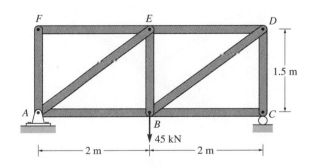

Probs. 14–59/14–60

14–61. Determine the horizontal displacement of point B of the truss. Each steel member has a cross-sectional area of 400 mm². E_{st} = 200 GPa.

14–62. Determine the vertical displacement of point C of the truss. Each steel member has a cross-sectional area of 400 mm². E_{st} = 200 GPa.

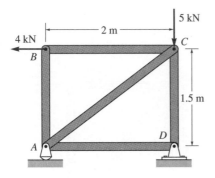

Probs. 14–61/14–62

14–63. Determine the vertical displacement of point C on the truss. For each member A = 300 mm², E = 200 GPa.

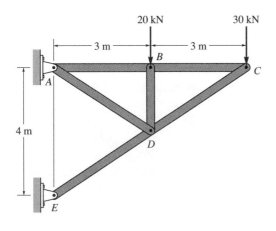

Prob. 14–63

*14–64. Determine the vertical displacement of point D on the truss. For each member A = 300 mm², E_{st} = 200 GPa.

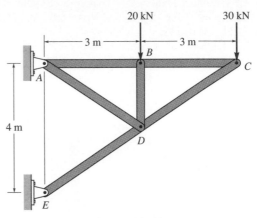

Prob. 14–64

14–65. Determine the vertical displacement of point C on the truss. For each steel member A = 4.5 in², E_{st} = 29(10³) ksi.

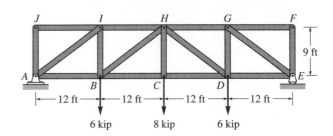

Prob. 14–65

14–66. Determine the vertical displacement of point H on the truss. For each steel member A = 4.5 in², E_{st} = 29(10³) ksi.

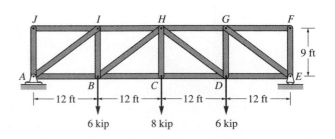

Prob. 14–66

*14.7 Method of Virtual Forces Applied to Beams

In this section we will apply the method of virtual forces to determine the displacement and slope at a point on a beam. To illustrate the principles, the displacement Δ of point A on the beam shown in Fig. 14–34b will be determined. This displacement is caused by the "real distributed load" w, and since this load causes both a shear and moment within the beam, we must actually consider the internal virtual work due to both of these loadings. In Example 14–7, however, it was shown that beam deflections due to shear are negligible compared with those caused by bending, particularly if the beam is long and slender. Since this type of beam is most often used in practice, we will consider only the virtual strain energy due to bending. Applying Eq. 14–36, the virtual-work equation for the beam is therefore

$$1 \cdot \Delta = \int_0^L \frac{mM}{EI}\, dx \qquad (14\text{--}42)$$

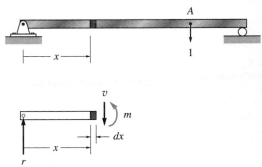

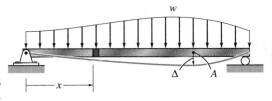

Virtual loads

(a)

Here

1 = external virtual unit load acting on the beam in the direction of Δ

Δ = displacement caused by the real loads acting on the beam

m = internal virtual moment in the beam, expressed as a function of x and caused by the external virtual unit load

M = internal moment in the beam, expressed as a function of x and caused by the real loads

E = modulus of elasticity of the material

I = moment of inertia of the cross-sectional area, computed about the neutral axis

In a similar manner, if the slope θ of the tangent at a point on the beam's elastic curve is to be determined, a virtual unit couple moment must be applied at the point, and the corresponding internal virtual moment m_θ has to be determined. If we apply Eq. 14–37 for this case and neglect the effect of shear deformations, we have

$$1 \cdot \theta = \int_0^L \frac{m_\theta M}{EI}\, dx \qquad (14\text{--}43)$$

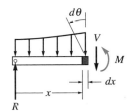

Real loads

(b)

Notice that the formulation of the above equations follows naturally from the development in Sec. 14.5. For example, the external virtual unit load creates an internal virtual moment m in the beam at position x, Fig. 14–34a. When the real load w is applied, it causes the element dx at x to deform or rotate by an angle $d\theta$, Fig. 14–34b. Provided the material responds elastically, then $d\theta$ equals $(M/EI)dx$. Consequently, the external virtual work $1 \cdot \Delta$ equals the internal virtual work for the entire beam, $\int m(M/EI)dx$, Eq. 14–42.

Unlike beams, as discussed here, some members may also be subjected to significant virtual strain energy caused by axial load, shear, and torsional moment. When this is the case, we must include in the above equations the energy terms for these loadings as formulated in Eq. 14–38.

Fig. 14–34

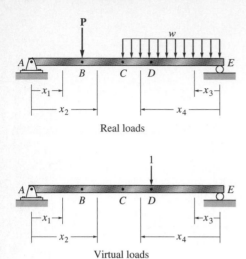

Real loads

Virtual loads

Fig. 14–35

When applying Eqs. 14–42 and 14–43, it is important to realize that the integrals on the right side represent the amount of virtual bending strain energy that is *stored* in the beam. If concentrated forces or couple moments act on the beam or the distributed load is discontinuous, a single integration *cannot* be performed across the beam's entire length. Instead, separate x coordinates must be chosen within regions that have no discontinuity of loading. Also, it is not necessary that each x have the same origin; however, the x selected for determining the real moment M in a particular region must be the *same x* as that selected for determining the virtual moment m or m_θ within the same region. For example, consider the beam shown in Fig. 14–35. In order to determine the displacement at D, we can use x_1 to determine the strain energy in region AB, x_2 for region BC, x_3 for region DE, and x_4 for region DC. In any case, each x coordinate should be selected so that both M and m (or m_θ) can easily be formulated.

PROCEDURE FOR ANALYSIS

The following procedure provides a method that may be used to determine the displacement and slope at a point on the elastic curve of a beam using the method of virtual work.

Virtual Moments **m** *or* **m**$_\theta$. Place a *virtual unit load* on the beam at the point and directed along the line of action of the desired displacement. If the slope is to be determined, place a virtual *unit couple moment* at the point. Establish appropriate x coordinates that are valid within regions of the beam where there is no discontinuity of real or virtual load. With the virtual load so placed, and all the real loads *removed* from the beam, calculate the internal moment m or m_θ as a function of each x coordinate. To be consistent, assume that m or m_θ acts in the positive direction according to the established beam sign convention, Fig. 7–4.

Real Moments. Using the *same x* coordinates as those established for m or m_θ, determine the internal moments M caused by the real loads. Since positive m or m_θ was assumed to act in the conventional "positive direction," it is important that positive M acts in this same direction. This is necessary since positive or negative internal virtual work depends on the directional sense of both the virtual load, defined by $\pm m$ or $\pm m_\theta$, and displacement, caused by $\pm M$.

Virtual-Work Equation. Apply the equation of virtual work, Eq. 14–42 or 14–43, to determine the desired displacement Δ or slope θ. It is important to retain the algebraic sign of each integral calculated within its specified region. If the algebraic sum of all the integrals for the entire beam is positive, Δ or θ is in the same direction as the virtual unit load or virtual unit couple moment, respectively. If a negative value results, Δ or θ is opposite to the virtual unit load or couple moment.

The following examples illustrate this method of solution.

Example 14–14

Determine the displacement of point B on the beam shown in Fig. 14–36a. EI is constant.

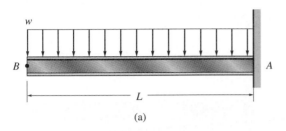

(a)

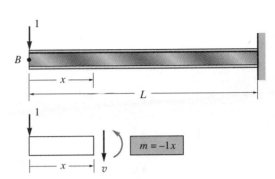

Fig. 14–36

$m = -1x$

Virtual loads
(b)

SOLUTION

Virtual Moment **m.**   The vertical displacement of point B is obtained by placing a virtual unit load at B, Fig. 14–36b. By inspection, there are no discontinuities of loading on the beam for *both* the real and virtual loads. Thus, a *single x* coordinate can be used to determine the virtual strain energy. This coordinate will be selected with its origin at B, since then the reactions at A do not have to be determined in order to find the internal moments m and M. Using the method of sections, the internal moment m is computed as shown in Fig. 14–36b.

Real Moments **M.**   Using the *same x* coordinate, the internal moment M is computed as shown in Fig. 14–36c.

Virtual-Work Equation.   The vertical displacement at B is thus

$$1 \cdot \Delta_B = \int \frac{mM}{EI}\, dx = \int_0^L \frac{(-1x)(-wx^2/2)\, dx}{EI}$$

$$\Delta_B = \frac{wL^4}{8EI}$$

Ans.

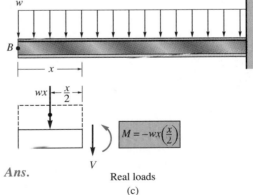

$M = -wx\left(\dfrac{x}{2}\right)$

Real loads
(c)

This value agrees with that given in the table on the inside back cover. *Note:* Here we have neglected the virtual strain energy due to shear, since its effect is small compared with that of bending. See Example 14–4.

Example 14–15

Determine the slope at point B of the beam shown in Fig. 14–37a. EI is constant.

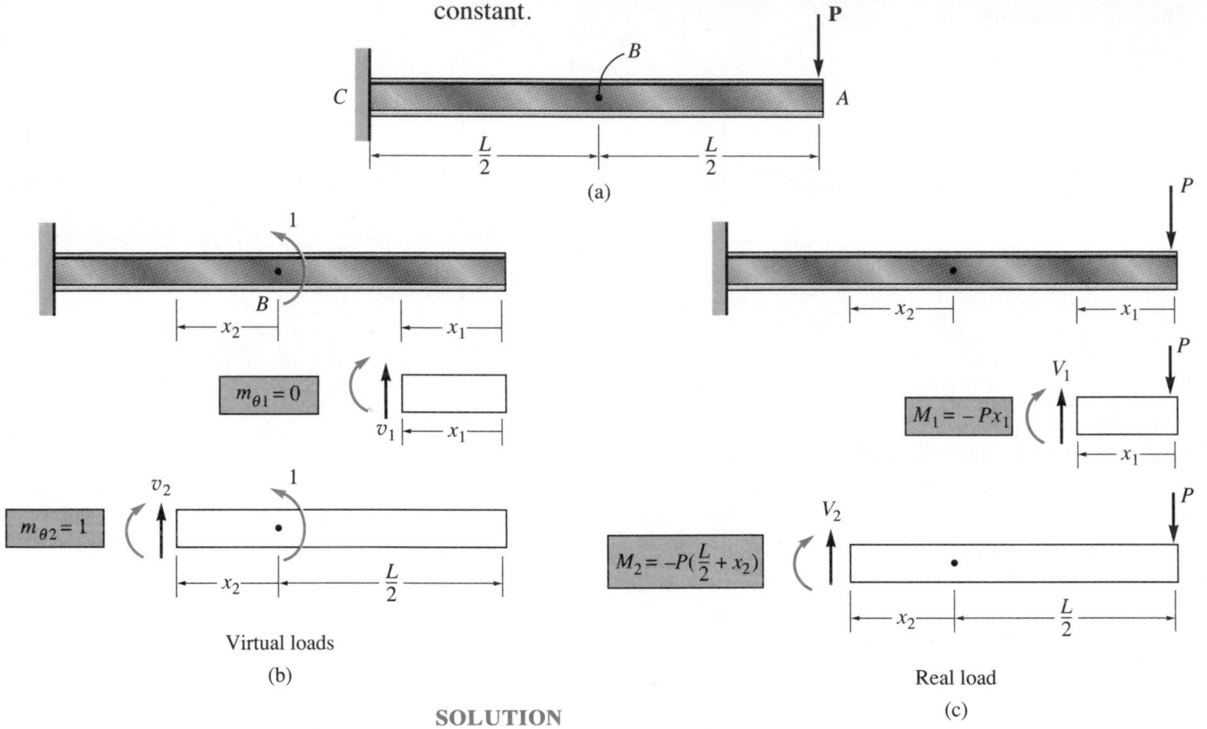

(a)

Virtual loads

(b)

Real load

(c)

Fig. 14–37

SOLUTION

Virtual Moments m_θ.   The slope at B is determined by placing a virtual unit couple moment at B, Fig. 14–37b. Two x coordinates must be selected in order to determine the total virtual strain energy in the beam. Coordinate x_1 accounts for the strain energy within segment AB, and coordinate x_2 accounts for the strain energy in segment BC. The internal moments m_θ within each of these segments are computed using the method of sections as shown in Fig. 14–37b.

Real Moments M.   Using the *same coordinates* x_1 and x_2 (Why?), the internal moments M are computed as shown in Fig. 14–37c.

Virtual-Work Equation.   The slope at B is thus

$$1 \cdot \theta_B = \int \frac{m_\theta M}{EI}\, dx = \int_0^{L/2} \frac{0(-Px_1)\, dx_1}{EI} + \int_0^{L/2} \frac{1\{-P[(L/2) + x_2]\}\, dx_2}{EI}$$

$$\theta_B = -\frac{3PL^2}{8EI} \qquad\qquad Ans.$$

The *negative sign* indicates that θ_B is *opposite* to the direction of the virtual couple moment shown in Fig. 14–37b.

Example 14–16

Determine the displacement of point A of the steel beam shown in Fig. 14–38a. $I = 450$ in^4, $E_{st} = 29(10^3)$ ksi.

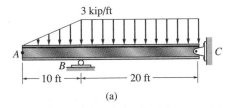

3 kip/ft

10 ft · 20 ft

(a)

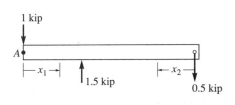

1 kip

1.5 kip

0.5 kip

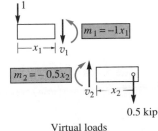

1

$m_1 = -1x_1$

$m_2 = -0.5x_2$

0.5 kip

Virtual loads

(b)

SOLUTION

Virtual Moments m.   The beam is subjected to the virtual unit load at A as shown in Fig. 14–38b. By inspection, two coordinates x_1 and x_2 must be chosen to cover all regions of the beam. For purposes of integration it is simplest to use origins at A and C. Using the method of sections, the internal moments m have been computed in Fig. 14–38b.

Real Moments M.   The reactions on the real beam are computed first. Then, using the *same* x coordinates as those used for m, the internal moments M are determined as shown in Fig. 14–38c.

Virtual-Work Equation.   Applying the equation of virtual work to the beam using the data in Fig. 14–38b and 14–38c, we have

$$1 \text{ kip} \cdot \Delta_A = \int \frac{mM}{EI}\, dx = \int_0^{10} \frac{(-1x_1)(-0.05x_1^3)\, dx_1}{EI} +$$
$$\int_0^{20} \frac{(-0.5x_2)(27.5x_2 - 1.5x_2^2)\, dx_2}{EI}$$

or

$$1 \text{ kip} \cdot \Delta_A = \frac{0.010(10)^5}{EI} - \frac{4.583(20)^3}{EI} + \frac{0.1875(20)^4}{EI}$$

$$\Delta_A = \frac{-5666.7 \text{ kip}^2 \cdot \text{ft}^3}{EI}$$

Substituting in the data for E and I, we get

$$\Delta_A = \frac{-5666.7 \text{ kip} \cdot \text{ft}^3 (12 \text{ in./ft})^3}{[29(10^3) \text{ kip/in}^2]\, 450 \text{ in}^4}$$

$$= -0.750 \text{ in.} \qquad\qquad Ans.$$

The negative sign indicates that point A is displaced upward.

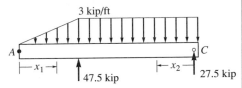

3 kip/ft

47.5 kip · 27.5 kip

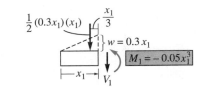

$\frac{1}{2}(0.3x_1)(x_1)$ · $\frac{x_1}{3}$ · $w = 0.3x_1$

$M_1 = -0.05x_1^3$

V_1

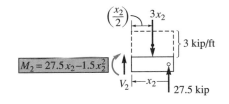

$\left(\frac{x_2}{2}\right)$ · $3x_2$ · 3 kip/ft

$M_2 = 27.5x_2 - 1.5x_2^2$

V_2 · x_2 · 27.5 kip

real loads

(c)

Fig. 14–38

PROBLEMS

14–67. Determine the displacement at point C and the slope at point A. EI is constant.

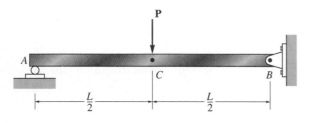

Prob. 14–67

14–70. The steel beam has a moment of inertia of $I = 53.8 \text{ in}^4$. Determine the slope at B and the displacement at C. $E_{st} = 29(10^3)$ ksi.

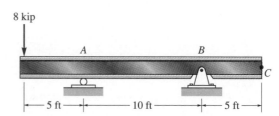

Prob. 14–70

*14–68. Determine the displacement at point C. EI is constant.

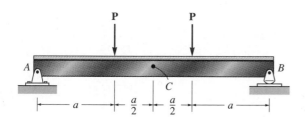

Prob. 14–68

14–71. Determine the displacement of the 1.5-in.-diameter steel shaft at B and its slope at A. $E_{st} = 29(10^3)$ ksi.

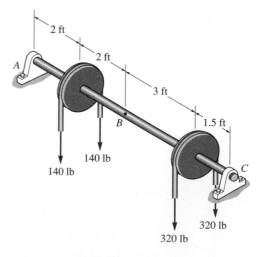

Prob. 14–71

14–69. Determine the displacement at point C and the slope at point A. EI is constant.

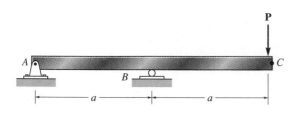

Prob. 14–69

*14–72. Determine the displacement and slope at the pulley B. The steel shaft has a diameter of 30 mm. $E_{st} = 200$ GPa.

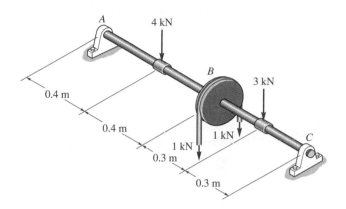

Prob. 14–72

14–73. Determine the displacement at point C and the slope at point A. EI is constant.

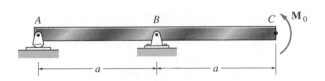

Prob. 14–73

14–74. The steel beam has a moment of inertia of $I = 125(10^6)$ mm⁴. Determine the slope at A and the displacement at D. $E_{st} = 200$ GPa.

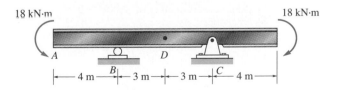

Prob. 14–74

14–75. Determine the displacement at B and the slope at A. The moment of inertia of the center portion DG of the shaft is 2I, whereas the end segments AD and GC have a moment of inertia I. The modulus of elasticity for the material is E.

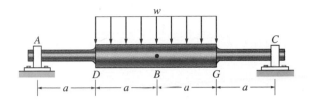

Prob. 14–75

*14–76. Determine the slope and displacement at point C. EI is constant.

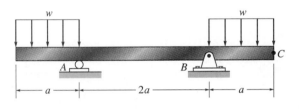

Prob. 14–76

14–77. The beam is made of southern pine for which $E_p = 13$ GPa. Determine the displacement at point A and its slope at C.

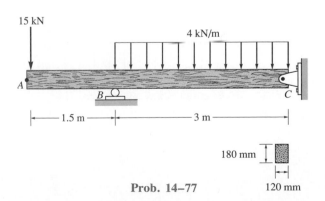

Prob. 14–77

14–78. The steel beam has a moment of inertia of $I = 53.4$ in^4. Determine the slope at A and the displacement at C. $E_{st} = 29(10^3)$ ksi.

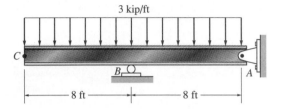

Prob. 14–78

14–79. Determine the displacement of the shaft at C and the slope at A. EI is constant.

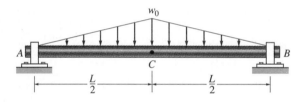

Prob. 14–79

*14–80. The beam is made of oak, for which $E_o = 11$ GPa. Determine the slope and displacement at point A.

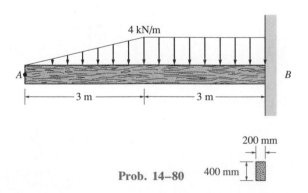

Prob. 14–80

14–81. Beam AB has a square cross section of 100 mm by 100 mm. Bar BC has a diameter of 10 mm. If both members are made of steel for which $E_{st} = 200$ GPa, determine the vertical displacement of point D due to the loading of 15 kN.

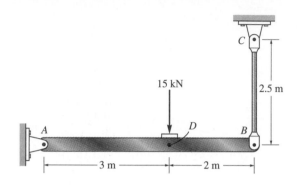

Prob. 14–81

14–82. Bar ABC has a rectangular cross section of 150 mm by 50 mm. Attached rod DB has a diameter of 10 mm. If both members are made of steel for which $E_{st} = 200$ GPa, determine the vertical displacement of point C due to the loading. Consider the effect of bending in ABC and axial force in DB.

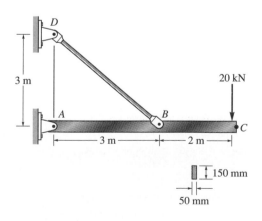

Prob. 14–82

14–83. Bar *ABC* has a square cross section of 2 in. by 2 in. Attached rod *DC* has a diameter of 0.25 in. If both members are made of steel for which $E_{st} = 29(10^3)$ ksi, determine the vertical displacement of point *C* due to the 800-lb loading. Consider the effect of bending in *ABC* and axial force in *DC*.

*14–84. Bar *ABC* has a square cross section of 2 in. by 2 in. Attached rod *DC* has a diameter of 0.25 in. If both members are made of steel for which $E_{st} = 29(10^3)$ ksi, determine the vertical displacement of point *A* due to the 800-lb loading. Consider the effect of bending in *ABC* and axial force in *DC*.

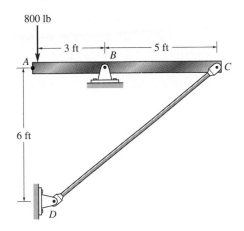

Probs. 14–83/14–84

14–85. Determine the vertical displacement of point *A* on the angle bracket due to the concentrated force **P**. The bracket is fixed-connected to its support. *EI* is constant.

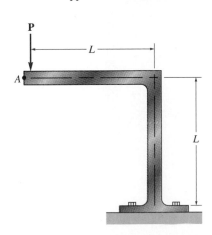

Prob. 14–85

14–86. The L-shaped frame is made from two segments, each of length *L* and flexural stiffness *EI*. If it is subjected to the uniform distributed load, determine the horizontal displacement of the end *C*.

14–87. The L-shaped frame is made from two segments, each of length *L* and flexural stiffness *EI*. If it is subjected to the uniform distributed load, determine the vertical displacement of point *B*.

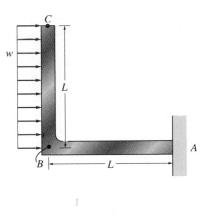

Probs. 14–86/14–87

*14–88. Determine the horizontal and vertical displacements of point *C*. *EI* is constant. There is a fixed support at *A*.

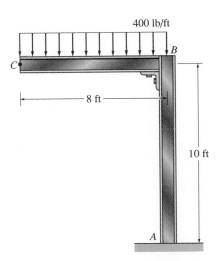

Prob. 14–88

*14.8 Castigliano's Theorem

In 1879 Alberto Castigliano, an Italian railroad engineer, published a book in which he outlined a method for determining the displacement and slope at a point in a body. This method, which is referred to as *Castigliano's second theorem,* applies only to bodies that have constant temperature, rigid supports, and material with linear-elastic behavior. If the displacement at a point is to be determined, the theorem states that the displacement is equal to the first partial derivative of the strain energy in the body with respect to a force acting at the point and in the direction of displacement. In a similar manner, the slope of the tangent at a point in a body is equal to the first partial derivative of the strain energy in the body with respect to a couple moment acting at the point and in the direction of the slope angle.

To derive Castigliano's second theorem, consider a body of any arbitrary shape, which is subjected to a series of n forces $\mathbf{P}_1, \mathbf{P}_2, \ldots, \mathbf{P}_n$, Fig. 14–39. Since the external work done by these forces is equal to the internal strain energy stored in the body, we can apply the conservation of energy, i.e.,

$$U_i = U_e$$

However, the external work is a function of the external loads, $U_e = \Sigma \int P \, dx$, Eq. 14–1, so the internal work is also a function of the external loads. Thus,

$$U_i = U_e = f(P_1, P_2, \ldots, P_n) \qquad (14\text{–}44)$$

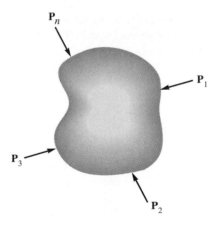

Fig. 14–39

Now, if any one of the forces, say P_i, is increased by a differential amount dP_i, the internal work will also be increased, such that the strain energy becomes

$$U_i + dU_i = U_i + \frac{\partial U_i}{\partial P_i} \, dP_i \qquad (14\text{--}45)$$

This value, however, should not depend on the sequence in which the n forces are applied to the body. For example, we could apply $d\mathbf{P}_i$ to the body *first,* then apply the loads $\mathbf{P}_1, \mathbf{P}_2, \ldots, \mathbf{P}_n$. In this case, $d\mathbf{P}_i$ would cause the body to be displaced a differential amount $d\Delta_i$ in the direction of $d\mathbf{P}_i$. By Eq. 14–2 ($U_e = \frac{1}{2} P_i \Delta_i$), the increment of strain energy would be $\frac{1}{2} \, dP_i \, d\Delta_i$. This quantity, however, is a second-order differential and may be neglected. Further application of the loads $\mathbf{P}_1, \mathbf{P}_2, \ldots, \mathbf{P}_n$ causes $d\mathbf{P}_i$ to move through the displacement Δ_i so that now the strain energy becomes

$$U_i + dU_i = U_i + dP_i \, \Delta_i \qquad (14\text{--}46)$$

Here, as above, U_i is the internal strain energy in the body, caused by the loads $\mathbf{P}_1, \mathbf{P}_2, \ldots, \mathbf{P}_n$, and $dU_i = dP_i \, \Delta_i$ is the *additional* strain energy caused by $d\mathbf{P}_i$.

In summary, then, Eq. 14–45 represents the strain energy in the body determined by first applying the loads $\mathbf{P}_1, \mathbf{P}_2, \ldots, \mathbf{P}_n$, then $d\mathbf{P}_i$; Eq. 14–46 represents the strain energy determined by first applying $d\mathbf{P}_i$ and then the loads $\mathbf{P}_1, \mathbf{P}_2, \ldots, \mathbf{P}_n$. Since these two equations must be equal, we require

$$\Delta_i = \frac{\partial U_i}{\partial P_i} \qquad (14\text{--}47)$$

which proves the theorem; i.e., the displacement Δ_i in the direction of $\mathbf{P}_i$ is equal to the first partial derivative of the strain energy with respect to P_i.

It should be noted that Eq. 14–47 is a statement regarding the body's *compatibility requirements,* since it is a condition related to displacement. Also, the above derivation requires that *only conservative forces* be considered for the analysis. These forces can be applied in any order, and furthermore, they do work that is independent of the path and therefore create no energy loss. As long as the material has linear-elastic behavior, the applied forces will be conservative and the theorem is valid. This is unlike the method of virtual force discussed in the previous sections, which applies to *both* elastic and inelastic material behavior. It should also be mentioned that Castigliano's first theorem is similar to his second theorem; however, it relates the load P_i to the partial derivative of the strain energy with respect to the corresponding displacement, that is, $P_i = \partial U_i / \partial \Delta_i$. The proof is similar to that given above, and like the method of virtual displacement, Castigliano's first theorem applies to both elastic and inelastic material behavior. This theorem is another way of expressing the *equilibrium requirements* for the body; however, it has limited application and therefore it will not be used in this book.

*14.9 Castigliano's Theorem Applied to Trusses

Since a truss member is subjected to an axial load, the strain energy is given by Eq. 14–16, $U_i = N^2L/2AE$. Substituting this equation into Eq. 14–47 and omitting the subscript i, we have

$$\Delta = \frac{\partial}{\partial P} \sum \frac{N^2L}{2AE}$$

It is generally easier to perform the differentiation prior to summation. Also, L, A, and E are constant for a given member, and therefore we may write

$$\Delta = \sum N \left(\frac{\partial N}{\partial P}\right) \frac{L}{AE} \qquad (14\text{--}48)$$

Here

Δ = joint displacement of the truss
P = external force of *variable magnitude* applied to the truss joint in the direction of Δ
N = internal axial force in a member caused by *both* the force P and the loads on the truss
L = length of a member
A = cross-sectional area of a member
E = modulus of elasticity of the material

In order to determine the partial derivative $\partial N/\partial P$, it will be necessary to treat P as a *variable,* not a specific numerical quantity. In other words, each internal axial force N must be expressed as a function of P.

By comparison, Eq. 14–48 is similar to that used for the method of virtual work, Eq. 14–39 ($1 \cdot \Delta = \sum nNL/AE$), except that n is replaced by $\partial N/\partial P$. These terms, n and $\partial N/\partial P$, will, however, be the *same*, since they represent the rate of change of the internal axial force with respect to the load P or, in other words, the axial force per unit load.

PROCEDURE FOR ANALYSIS

The following procedure provides a method that may be used to determine the displacement of any joint on a truss using Castigliano's second theorem.

External Force **P.**   Place a force **P** on the truss at the joint where the desired displacement is to be determined. This force is assumed to have a *variable magnitude* and should be directed along the line of action of the displacement.

Internal Forces N. Determine the force *N* in each member caused by both the real (numerical) loads and the variable force *P*. Assume that tensile forces are positive and compressive forces are negative. Also compute the respective partial derivative $\partial N/\partial P$ for each member. *After N and $\partial N/\partial P$ have been determined, assign P its numerical value if it has actually replaced a real force on the truss. Otherwise, set P equal to zero.*

Castigliano's Second Theorem. Apply Eq. 14–48 to determine the desired displacement Δ. It is important to retain the algebraic signs for corresponding values of *N* and $\partial N/\partial P$ when substituting these terms into the equation. If the resultant sum $\Sigma N(\partial N/\partial P)L/AE$ is positive, Δ is in the same direction as **P**. If a negative value results, Δ is opposite to **P**.
The following examples illustrate this method of solution.

Example 14–17

Determine the horizontal displacement of joint *C* of the steel truss shown in Fig. 14–40a. The cross-sectional area of each member is indicated in the figure. Take $E_{st} = 29(10^3)$ ksi.

SOLUTION
External Force P. Since the horizontal displacement of *C* is to be determined, a horizontal *variable* force **P** is applied to joint *C*, Fig. 14–40b. Later this force will be set equal to the fixed value of 8 kip.

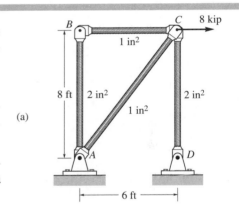

(a)

Internal Forces N. Using the method of joints, the force *N* in each member is computed. The results are shown in Fig. 14–40b. Arranging the data in tabular form, we have

Member	N	$\dfrac{\partial N}{\partial P}$	$N(P = 8 \text{ kip})$	L	$N\left(\dfrac{\partial N}{\partial P}\right)L$
AB	0	0	0	8	0
BC	0	0	0	6	0
AC	1.67P	1.67	13.33	10	222.2
CD	−1.33P	−1.33	−10.67	8	113.8

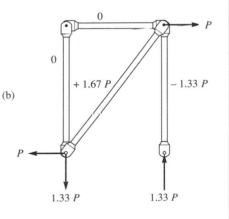

(b)

Castigliano's Second Theorem. Applying Eq. 14–48 gives

$$\Delta_{C_h} = \Sigma N\left(\frac{\partial N}{\partial P}\right)\frac{L}{AE}$$

$$= 0 + 0 + \frac{(222.2 \text{ kip} \cdot \text{ft})(12 \text{ in./ft})}{(1 \text{ in}^2)\, 29(10^3) \text{ kip/in}^2} + \frac{(113.8 \text{ kip} \cdot \text{ft})(12 \text{ in./ft})}{(2 \text{ in}^2)\, 29(10^3) \text{ kip/in}^2}$$

$$= 0.115 \text{ in.} \qquad\qquad Ans.$$

Fig. 14–40

Example 14–18

Determine the vertical displacement of joint C of the steel truss shown in Fig. 14–41a. The cross-sectional area of each member is $A = 400$ mm^2, and $E_{st} = 200$ GPa.

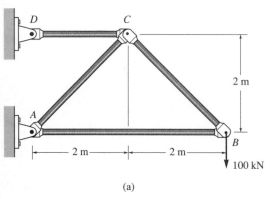

(a)

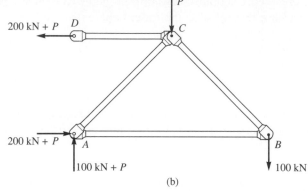

(b)

Fig. 14–41

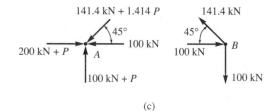

(c)

SOLUTION

External Force P.   A vertical force **P** is applied to the truss at joint C, since this is where the vertical displacement is to be determined, Fig. 14–41b.

Internal Forces N.   The reactions at the truss supports A and D are calculated and the results are shown in Fig. 14–41b. Using the method of joints, the **N** forces in each member are determined, Fig. 14–41c.* For convenience, these results along with the partial derivatives $\partial N/\partial P$ are listed in tabular form. Note that since **P** does not actually exist as a real load on the truss, we require $P = 0$.

Member	N	$\dfrac{\partial N}{\partial P}$	$N(P = 0)$	L	$N\left(\dfrac{\partial N}{\partial P}\right)L$
AB	-100	0	-100	4	0
BC	141.4	0	141.4	2.828	0
AC	$-141.4 - 1.414P$	-1.414	-141.4	2.828	565.7
CD	$200 + P$	1	200	2	400
					Σ 965.7 kN $\cdot$ m

Castigliano's Second Theorem.   Applying Eq. 14–48, we have

$$\Delta_{C_v} = \Sigma N\left(\frac{\partial N}{\partial P}\right)\frac{L}{AE} = \frac{965.7 \text{ kN} \cdot \text{m}}{AE}$$

Substituting the numerical values for A and E, we get

$$\Delta_{C_v} = \frac{965.7 \text{ kN} \cdot \text{m}}{[400(10^{-6}) \text{ m}^2]\ 200(10^6) \text{ kN/m}^2}$$

$$= 0.01207 \text{ m} = 12.1 \text{ mm} \qquad \textit{Ans.}$$

This solution should be compared with that of Example 14–11, using the virtual-work method.

*It may be more convenient to analyze the truss with just the 100-kN load on it, then analyze the truss with the **P** load on it. The results can then be added together to give the **N** forces.

PROBLEMS

14–89.   Solve Prob. 14–54 using Castigliano's theorem.

14–90.   Solve Prob. 14–55 using Castigliano's theorem.

14–91.   Solve Prob. 14–57 using Castigliano's theorem.

*14–92.   Solve Prob. 14–56 using Castigliano's theorem.

14–93.   Solve Prob. 14–58 using Castigliano's theorem.

14–94.   Solve Prob. 14–59 using Castigliano's theorem.

14–95.   Solve Prob. 14–61 using Castigliano's theorem.

*14–96.   Solve Prob. 14–60 using Castigliano's theorem.

14–97.   Solve Prob. 14–62 using Castigliano's theorem.

14–98.   Solve Prob. 14–63 using Castigliano's theorem.

14–99.   Solve Prob. 14–65 using Castigliano's theorem.

*14–100.   Solve Prob. 14–64 using Castigliano's theorem.

14–101.   Solve Prob. 14–66 using Castigliano's theorem.

*14.10 Castigliano's Theorem Applied to Beams

The internal strain energy for a beam is caused by both bending and shear. However, as pointed out in Example 14–7, if the beam is long and slender, the strain energy due to shear can be neglected compared with that of bending. Assuming this to be the case, the internal strain energy for a beam is given by Eq. 14–17 ($U_i = \int M^2\,dx/2EI$). Substituting into Eq. 14–47, $\Delta_i = \partial U_i/\partial P_i$, and omitting the subscript i, we have

$$\Delta = \frac{\partial}{\partial P} \int_0^L \frac{M^2\,dx}{2EI}$$

Rather than squaring the expression for internal moment M, integrating, and then taking the partial derivative, it is generally easier to differentiate prior to integration. Provided E and I are constant, we have

$$\Delta = \int_0^L M\left(\frac{\partial M}{\partial P}\right)\frac{dx}{EI} \qquad (14\text{–}49)$$

where

Δ = displacement of the point caused by the real loads acting on the beam
P = external force of variable magnitude applied to the beam in the direction of Δ
M = internal moment in the beam, expressed as a function of x and caused by both the force P and the loads on the beam
E = modulus of elasticity of the material
I = moment of inertia of cross-sectional area computed about the neutral axis

If the slope of the tangent θ at a point on the elastic curve is to be determined, the partial derivative of the internal moment M with respect to an *external couple moment* M' acting at the point must be computed. For this case,

$$\theta = \int_0^L M\left(\frac{\partial M}{\partial M'}\right)\frac{dx}{EI} \qquad (14\text{–}50)$$

The above equations are similar to those used for the method of virtual work, Eqs. 14–42 and 14–43, except m and m_θ replace $\partial M/\partial P$ and $\partial M/\partial M'$, respectively.

It should also be mentioned that if the loading on a member causes significant strain energy within the member due to axial load, shear, bending moment, and torsional moment, then the effects of all these loadings should be included when applying Castigliano's theorem. To do this we must use the strain-energy functions developed in Sec. 14.2, along with their associated partial derivatives. The result is

$$\Delta = \Sigma N\left(\frac{\partial N}{\partial P}\right)\frac{L}{AE} + \int_0^L f_s V\left(\frac{\partial V}{\partial P}\right)\frac{dx}{GA} + \int_0^L M\left(\frac{\partial M}{\partial P}\right)\frac{dx}{EI} + \int_0^L T\left(\frac{\partial T}{\partial P}\right)\frac{dx}{GJ}$$

(14–51)

The method of applying this general formulation is similar to that used to apply Eqs. 14–49 and 14–50.

PROCEDURE FOR ANALYSIS

The following procedure provides a method that may be used to determine the displacement or slope at a point on the elastic curve of a beam using Castigliano's second theorem.

External Force **P** *or Couple Moment* **M′**.   Place a force **P** on the beam at the point and directed along the line of action of the desired displacement. If the slope of the tangent is to be determined, place a couple moment **M′** at the point. It is assumed that both **P** and **M′** have a variable magnitude.

Internal Moments **M**.   Establish appropriate x coordinates that are valid within regions of the beam where there is no discontinuity of force, distributed load, or couple moment. Calculate the internal moments M as a function of P or M' and the partial derivatives $\partial M/\partial P$ or $\partial M/\partial M'$ for each coordinate x. After M and $\partial M/\partial P$ or $\partial M/\partial M'$ have been determined, assign P or M' its numerical value if it has actually replaced a real force or couple moment. Otherwise, set P or M' equal to zero.

Castigliano's Second Theorem.   Apply Eq. 14–49 or 14–50 to determine the desired displacement Δ or θ. It is important to retain the algebraic signs for corresponding values of M and $\partial M/\partial P$ or $\partial M/\partial M'$. If the resultant sum of all the definite integrals is positive, Δ or θ is in the same direction as **P** or **M′**. If a negative value results, Δ or θ is opposite to **P** or **M′**.
    The following examples illustrate this method of solution.

Example 14–19

Determine the displacement of point B on the beam shown in Fig. 14–42a. EI is constant.

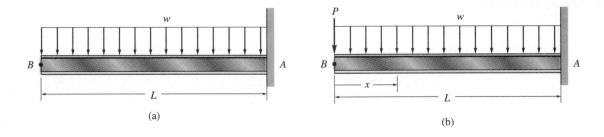

(a)   (b)

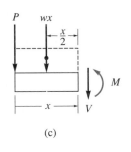

(c)

Fig. 14–42

SOLUTION

External Force $\mathbf{P}$. A vertical force $\mathbf{P}$ is placed on the beam at B as shown in Fig. 14–42b.

Internal Moments M. A single x coordinate is needed for the solution, since there are no discontinuities of loading between A and B. Using the method of sections, Fig. 14–42c, the internal moment and partial derivative are computed as follows:

$$\downarrow^+ \ \Sigma M_{NA} = 0; \qquad M + wx\left(\frac{x}{2}\right) + P(x) = 0$$

$$M = -\frac{wx^2}{2} - Px$$

$$\frac{\partial M}{\partial P} = -x$$

Setting $P = 0$ gives

$$M = \frac{-wx^2}{2} \qquad \text{and} \qquad \frac{\partial M}{\partial P} = -x$$

Castigliano's Second Theorem. Applying Eq. 14–49, we have

$$\Delta_B = \int_0^L M\left(\frac{\partial M}{\partial P}\right)\frac{dx}{EI} = \int_0^L \frac{(-wx^2/2)(-x)\,dx}{EI}$$

$$= \frac{wL^4}{8EI} \qquad\qquad\qquad\qquad Ans.$$

The similarity between this solution and that of the virtual-work method, Example 14–14, should be noted.

Example 14–20

Determine the slope at point B of the beam shown in Fig. 14–43a. EI is constant.

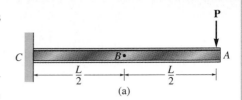

(a)

SOLUTION

External Couple Moment M′. Since the slope at point B is to be determined, an external couple moment $\mathbf{M'}$ is placed on the beam at this point, Fig. 14–43b.

Internal Moments M. Two coordinates, x_1 and x_2, must be used to determine the internal moments within the beam since there is a discontinuity, $\mathbf{M'}$, at B. As shown in Fig. 14–43b, x_1 ranges from A to B and x_2 ranges from B to C. Using the method of sections, Fig. 14–43c, the internal moments and the partial derivatives are computed as follows:

For x_1,

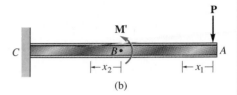

(b)

$$\underset{\curvearrowright}{+}\ \Sigma M_{NA} = 0; \qquad -M_1 - Px_1 = 0$$

$$M_1 = -Px_1$$

$$\frac{\partial M_1}{\partial M'} = 0$$

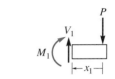

For x_2,

$$\underset{\curvearrowright}{+}\ \Sigma M_{NA} = 0; \qquad -M_2 + M' - P\left(\frac{L}{2} + x_2\right) = 0$$

$$M_2 = M' - P\left(\frac{L}{2} + x_2\right)$$

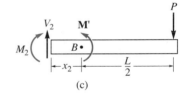

(c)

$$\frac{\partial M_2}{\partial M'} = 1$$

Castigliano's Second Theorem. Setting $M' = 0$ and applying Eq. 14–50, we have

$$\theta_B = \int_0^L M\left(\frac{\partial M}{\partial M'}\right)\frac{dx}{EI}$$

$$= \int_0^{L/2} \frac{(-Px_1)(0)\ dx_1}{EI} + \int_0^{L/2} \frac{-P[(L/2) + x_2](1)\ dx_2}{EI}$$

$$= -\frac{3PL^2}{8EI}$$

Fig. 14–43

The negative sign indicates that θ_B is opposite to the direction of the couple moment $\mathbf{M'}$. Note the similarity between this solution and that of Example 14–15.

Example 14–21

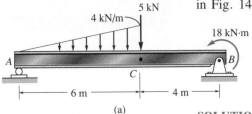

(a)

Determine the vertical displacement of point C of the steel beam shown in Fig. 14–44a. Take $E_{st} = 200$ GPa, $I = 125(10^{-6})$ m^4.

SOLUTION

External Force P.   A vertical force **P** is applied at point C, Fig. 14–44b. Later this force will be set equal to the fixed value of 5 kN.

Internal Moments M.   In this case two x coordinates are needed for the integration, Fig. 14–44b, since the load is discontinuous at C. Using the method of sections, Fig. 14–44c, the internal moments and partial derivatives are computed as follows:

For x_1,

$$\curvearrowright + \ \Sigma M_{NA} = 0; \quad M_1 + \frac{1}{3}x_1^2\left(\frac{x_1}{3}\right) - (9 + 0.4P)x_1 = 0$$

$$M_1 = (9 + 0.4P)x_1 - \frac{1}{9}x_1^3$$

$$\frac{\partial M_1}{\partial P} = 0.4x_1$$

For x_2,

$$\curvearrowright + \ \Sigma M_{NA} = 0; \quad -M_2 + 18 + (3 + 0.6P)x_2 = 0$$

$$M_2 = 18 + (3 + 0.6P)x_2$$

$$\frac{\partial M_2}{\partial P} = 0.6x_2$$

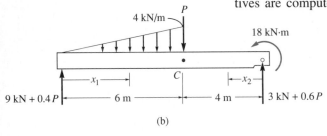

(b)

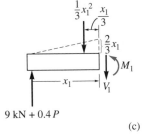

(c)

Fig. 14–44

Castigliano's Second Theorem.   Setting $P = 5$ kN and applying Eq. 14–49, we get

$$\Delta_{C_v} = \int_0^L M\left(\frac{\partial M}{\partial P}\right)\frac{dx}{EI}$$

$$= \int_0^6 \frac{(11x_1 - \frac{1}{9}x_1^3)(0.4x_1)\ dx_1}{EI} + \int_0^4 \frac{(18 + 6x_2)(0.6x_2)\ dx_2}{EI}$$

$$= \frac{410.9\ \text{kN} \cdot \text{m}^3}{[200(10^6)\ \text{kN/m}^2]\ 125(10^{-6})\ \text{m}^4}$$

$$= 0.0164\ \text{m} = 16.4\ \text{mm} \qquad\qquad Ans.$$

PROBLEMS

14–102. Solve Prob. 14–67 using Castigliano's theorem.

14–103. Solve Prob. 14–69 using Castigliano's theorem.

*14–104. Solve Prob. 14–68 using Castigliano's theorem.

14–105. Solve Prob. 14–70 using Castigliano's theorem.

14–106. Solve Prob. 14–71 using Castigliano's theorem.

14–107. Solve Prob. 14–73 using Castigliano's theorem.

*14–108. Solve Prob. 14–72 using Castigliano's theorem.

14–109. Solve Prob. 14–74 using Castigliano's theorem.

14–110. Solve Prob. 14–75 using Castigliano's theorem.

14–111. Solve Prob. 14–77 using Castigliano's theorem.

*14–112. Solve Prob. 14–76 using Castigliano's theorem.

14–113. Solve Prob. 14–78 using Castigliano's theorem.

14–114. Solve Prob. 14–79 using Castigliano's theorem.

14–115. Solve Prob. 14–85 using Castigliano's theorem.

*14–116. Solve Prob. 14–80 using Castigliano's theorem.

14–117. Solve Prob. 14–86 using Castigliano's theorem.

14–118. Solve Prob. 14–88 using Castigliano's theorem.

REVIEW PROBLEMS

14–119. Use the conservation of energy to determine the displacement of point B on the steel beam. $I = 250$ in^4, $E_{st} = 29(10^3)$ ksi.

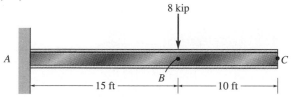

Prob. 14–119

*14–120. The sack of concrete has a weight of 90 lb. If it is dropped from rest at a height of 4 ft onto the center of the steel beam, determine the maximum stress developed in the beam due to the impact. Also, what is the impact factor? The beam's depth is 12.5 in. and $I = 285$ in^4, $E_{st} = 29(10^3)$ ksi, $\sigma_y = 36$ ksi.

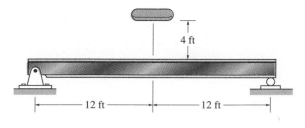

Prob. 14–120

14–121. Determine the displacement at the end C of the steel beam. $E_{st} = 200$ GPa, $I = 70(10^6)$ mm^4. Use the method of virtual work.

14–122. Solve Prob. 14–121 using Castigliano's theorem.

14–123. Determine the slope at point A of the steel beam. $E_{st} = 200$ GPa, $I = 70(10^6)$ mm^4. Use the method of virtual work.

*14–124. Solve Prob. 14–123 using Castigliano's theorem.

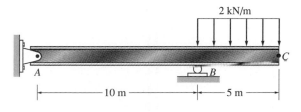

Probs. 14–121/14–122/14–123/14–124

14–125.   Determine the vertical displacement of point A. Each steel member has a cross-sectional area of 400 mm². $E_{st} =$ 200 GPa. Use the method of virtual work.

14–126.   Solve Prob. 14–125 using Castigliano's theorem.

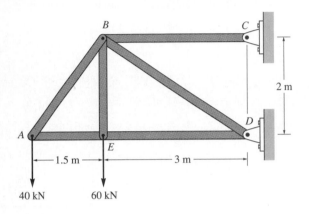

40 kN   60 kN

Probs. 14–125/126

14–127.   The steel bolt is required to absorb the energy of a 2-kg mass that falls $h = 30$ mm. If the bolt has a diameter of 4 mm, determine its required length L so the stress in the bolt does not exceed 150 MPa. $E_{st} = 200$ GPa.

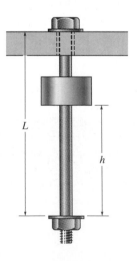

Prob. 14–127

***14–128.**   Determine the displacement of point C and the slope at point B. EI is constant. Use the principle of virtual work.

14–129.   Determine the displacement of point B and the slope at point C. EI is constant. Use Castigliano's theorem.

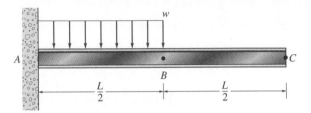

Probs. 14–128/14–129

14–130.   The cantilevered beam is subjected to a couple moment M_0 applied at its end. Use the conservation of energy to determine the slope of the beam at B. EI is constant.

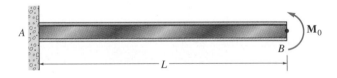

Prob. 14–130

14–131.   Determine the bending strain energy in the beam due to the distributed load. EI is constant.

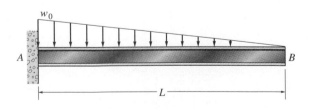

Prob. 14–131

A Geometric Properties of an Area

A.1 Centroid of an Area

The *centroid* of an area refers to the point that defines the geometric center for the area. If the area has an arbitrary shape, as shown in Fig. *A–1a*, the x and y coordinates defining the location of the centroid C are determined using the formulas

$$\bar{x} = \frac{\int_A x \, dA}{\int_A dA} \qquad \bar{y} = \frac{\int_A y \, dA}{\int_A dA} \qquad (A–1)$$

The numerators in these equations are formulations of the "first moment" of the area element dA about the y and the x axis, respectively, Fig. *A–1b*; the denominators represent the total area A of the shape.

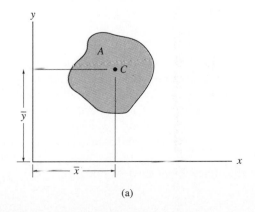

(a)

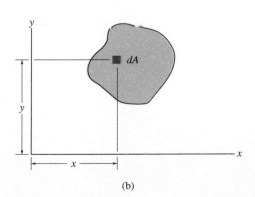

(b)

Fig. A–1

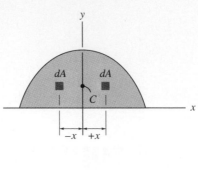

Fig. A–2

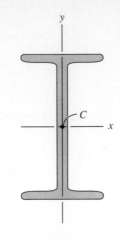

Fig. A–3

It should be noted that the location of the centroid for some areas may be partially or completely specified by using symmetry conditions. In cases where the area has an axis of symmetry, the centroid for the area will lie along that axis. For example, the centroid C for the area shown in Fig. A–2 must lie along the y axis, since for every elemental area dA a distance $+x$ to the right of the y axis, there is an identical element a distance $-x$ to the left. The total moment for all the elements about the axis of symmetry will therefore cancel; that is, $\int x \, dA = 0$ (Eq. A–1), so that $\bar{x} = 0$. In cases where a shape has two axes of symmetry, it follows that the centroid lies at the intersection of these axes, Fig. A–3. Based on the principle of symmetry, or using Eq. A–1, the locations of the centroid for common area shapes are listed on the inside front cover of this book.

Composite Areas.
Often an area can be sectioned or divided into several parts having simpler shapes. Provided the area and location of the centroid of each of these "composite shapes" are known, one can eliminate the need for integration to determine the centroid for the entire area. In this case, equations analogous to Eq. A–1 must be used, except that finite summation signs replace the integrals; i.e.,

$$\bar{x} = \frac{\Sigma \tilde{x} A}{\Sigma A} \qquad \bar{y} = \frac{\Sigma \tilde{y} A}{\Sigma A} \qquad\qquad (A–2)$$

Here $\tilde{x}$ and $\tilde{y}$ represent the *algebraic distances* or x, y coordinates for the centroid of each composite part, and ΣA represents the sum of the areas of the composite parts or simply the *total area*. In particular, if a hole, or a geometric region having no material, is located within a composite part, the hole is considered as an additional composite part having a *negative* area. Also, as discussed above, if the total area is symmetrical about an axis, the centroid of the area lies on the axis.

The following example illustrates application of Eq. A–2.

Example A–1

Locate the centroid C of the cross-sectional area for the T-beam shown in Fig. A–4a.

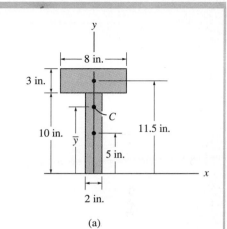

(a)

SOLUTION I

The y axis is placed along the axis of symmetry so that $\bar{x} = 0$, Fig. A–4a. To obtain $\bar{y}$ we will establish the x axis (reference axis) through the base of the area. The area is segmented into two rectangles as shown, and the centroidal location $\tilde{y}$ for each is established. Applying Eq. A–2, we have

$$\bar{y} = \frac{\Sigma \tilde{y} A}{\Sigma A} = \frac{[5 \text{ in.}](10 \text{ in.})(2 \text{ in.}) + [11.5 \text{ in.}](3 \text{ in.})(8 \text{ in.})}{(10 \text{ in.})(2 \text{ in.}) + (3 \text{ in.})(8 \text{ in.})}$$

$$= 8.55 \text{ in.} \hspace{3cm} \textit{Ans.}$$

SOLUTION II

Using the same two segments, the x axis can be located at the top of the area as shown in Fig. A–4b. Here

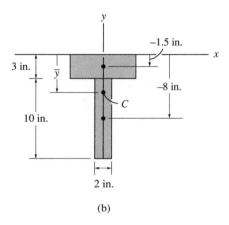

(b)

$$\bar{y} = \frac{\Sigma \tilde{y} A}{\Sigma A} = \frac{[-1.5 \text{ in.}](3 \text{ in.})(8 \text{ in.}) + [-8 \text{ in.}](10 \text{ in.})(2 \text{ in.})}{(3 \text{ in.})(8 \text{ in.}) + (10 \text{ in.})(2 \text{ in.})}$$

$$= -4.45 \text{ in.} \hspace{3cm} \textit{Ans.}$$

The negative sign indicates that C is located *below* the origin, which is to be expected. Also note that from the two answers 8.55 in. + 4.45 in. = 13.0 in., which is the depth of the beam as expected.

SOLUTION III

It is also possible to consider the cross-sectional area to be one large rectangle *less* two small rectangles, Fig. A–4c. Hence we have

$$\bar{y} = \frac{\Sigma \tilde{y} A}{\Sigma A} = \frac{[6.5 \text{ in.}](13 \text{ in.})(8 \text{ in.}) - 2[5 \text{ in.}](10 \text{ in.})(3 \text{ in.})}{(13 \text{ in.})(8 \text{ in.}) - 2(10 \text{ in.})(3 \text{ in.})}$$

$$= 8.55 \text{ in.} \hspace{3cm} \textit{Ans.}$$

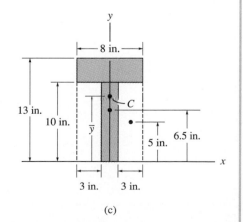

(c)

Fig. A–4

A.2 Moment of Inertia for an Area

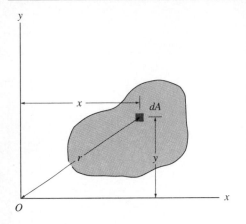

Fig. A–5

When calculating the centroid for an area, we considered the first moment of the area about an axis; i.e., for the computation it was necessary to evaluate an integral of the form $\int x \, dA$. There are some topics in mechanics of materials that require evaluation of an integral of the second moment of an area, that is, $\int x^2 dA$. This integral is referred to as the *moment of inertia* for an area. To show formally how it is defined, consider the area A, shown in Fig. A–5, which lies in the x–y plane. By definition, the moments of inertia of the differential element dA about the x and y axes are $dI_x = y^2 dA$ and $dI_y = x^2 dA$, respectively. For the entire area, the moment of inertia is determined by integration, i.e.,

$$I_x = \int_A y^2 dA$$

$$I_y = \int_A x^2 dA \qquad\qquad (A–3)$$

We can also formulate the second moment of the differential element about the pole O or z axis, Fig. A–5. This is referred to as the *polar moment of inertia*, $dJ_O = r^2 dA$. Here r is the perpendicular distance from the pole (z axis) to the element dA. For the entire area, the polar moment of inertia is

$$J_O = \int_A r^2 dA = I_x + I_y \qquad (A–4)$$

The relationship between J_O and I_x, I_y is possible since $r^2 = x^2 + y^2$, Fig. A–5.

From the above formulations it is seen that I_x, I_y, and J_O will *always* be *positive*, since they involve the product of distance squared and area. Furthermore, the units for moment of inertia involve length raised to the fourth power, e.g., m^4, mm^4, or ft^4, in^4.

Using the above equations, the moments of inertia for some common shapes of areas have been calculated about their *centroidal axes* and are listed on the inside front cover.

Parallel-Axis Theorem for an Area.

If the moment of inertia for an area is known about a centroidal axis, we can determine the moment of inertia of the area about a corresponding parallel axis using the *parallel-axis theorem*. To derive this theorem, consider finding the moment of inertia of the shaded area shown in Fig. A–6 about the x axis. In this case, a differential element dA is located at an arbitrary distance y' from the centroidal x' axis, whereas the *fixed distance* between the parallel x and x' axes is defined as d_y.

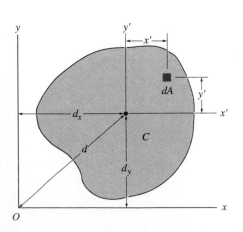

Fig. A–6

Since the moment of inertia of dA about the x axis is $dI_x = (y' + d_y)^2 dA$, then for the entire area,

$$I_x = \int_A (y' + d_y)^2 dA = \int_A y'^2 dA + 2d_y \int_A y' dA + d_y^2 \int_A dA$$

The first term on the right represents the moment of inertia of the area about the x' axis, $\bar{I}_{x'}$. The second term is zero since the x' axis passes through the area's centroid C, that is, $\int y' dA = \bar{y}A = 0$ since $\bar{y} = 0$. The final result is therefore

$$\boxed{I_x = \bar{I}_{x'} + Ad_y^2} \qquad (A\text{--}5)$$

A similar expression can be written for I_y, that is,

$$\boxed{I_y = \bar{I}_{y'} + Ad_x^2} \qquad (A\text{--}6)$$

And finally, for the polar moment of inertia about an axis perpendicular to the x–y plane and passing through the pole O (z axis), Fig. A–6, we have

$$\boxed{J_O = \bar{J}_C + Ad^2} \qquad (A\text{--}7)$$

The form of each of the above equations states that the moment of inertia of an area about an axis is equal to the area's moment of inertia about a parallel axis passing through the "centroid" plus the product of the area and the square of the perpendicular distance between the axes.

Composite Areas.

Many cross-sectional areas consist of a series of connected simpler shapes, such as rectangles, triangles, and semicircles. Provided the moment of inertia of each of these shapes is known, or can be computed about a common axis, then the moment of inertia of the "composite area" can be determined as the *algebraic sum* of the moments of inertia of its composite parts.

In order to properly determine the moment of inertia of such an area about a specified axis, it is first necessary to divide the area into its composite parts and indicate the perpendicular distance from the axis to the parallel centroidal axis for each part. Using the table on the inside back cover of the book, the moment of inertia of each part is computed about the centroidal axis. If this axis does not coincide with the specified axis, the parallel-axis theorem, $I = \bar{I} + Ad^2$, should be used to determine the moment of inertia of the part about the specified axis. The moment of inertia of the entire area about this axis is then determined by summing the results of its composite parts. In particular, if a composite part has a "hole," the moment of inertia for the composite is found by "subtracting" the moment of inertia for the hole from the moment of inertia of the entire area including the hole.

The following examples illustrate application of this method.

Example A-2

Determine the moment of inertia of the cross-sectional area of the T-beam shown in Fig. A–7a about the centroidal x' axis.

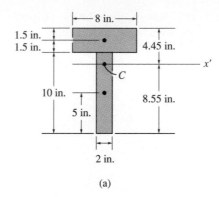

(a)

Fig. A–7

SOLUTION I

The area is segmented into two rectangles as shown in Fig. A–7a, and the distance from the x' axis and each centroidal axis is determined. Using the table on the inside front cover, the moment of inertia of a rectangle about its centroidal axis is $I = \frac{1}{12}bh^3$. Applying the parallel-axis theorem, Eq. A–5, to each rectangle and adding the results, we therefore have

$$I = \Sigma \bar{I}_{x'} + Ad_y^2$$

$$= \left[\frac{1}{12}(2 \text{ in.})(10 \text{ in.})^3 + (2 \text{ in.})(10 \text{ in.})(8.55 \text{ in.} - 5 \text{ in.})^2\right]$$

$$+ \left[\frac{1}{12}(8 \text{ in.})(3 \text{ in.})^3 + (8 \text{ in.})(3 \text{ in.})(4.45 \text{ in.} - 1.5 \text{ in.})^2\right]$$

$$I = 645.6 \text{ in}^4 \qquad\qquad Ans.$$

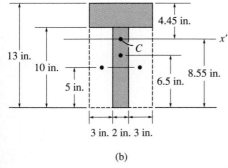

(b)

SOLUTION II

The area can be considered as one large rectangle less two small rectangles, shown dashed in Fig. A–7b. We have

$$I = \Sigma \bar{I}_{x'} + Ad_y^2$$

$$= \left[\frac{1}{12}(8 \text{ in.})(13 \text{ in.})^3 + (8 \text{ in.})(13 \text{ in.})(8.55 \text{ in.} - 6.5 \text{ in.})^2\right]$$

$$-2\left[\frac{1}{12}(3 \text{ in.})(10 \text{ in.})^3 + (3 \text{ in.})(10 \text{ in.})(8.55 \text{ in.} - 5 \text{ in.})^2\right]$$

$$I = 645.6 \text{ in}^4 \qquad\qquad Ans.$$

Example A–3

Compute the moments of inertia of the beam's cross-sectional area shown in Fig. A–8a about the x and y centroidal axes.

SOLUTION

The cross section can be considered as three composite rectangular areas A, B, and D shown in Fig. A–8b. For the calculation, the centroid of each of these rectangles is located in the figure. From the table on the inside front cover, the moment of inertia of a rectangle about its centroidal axis is $I = \frac{1}{12}bh^3$. Hence, using the parallel-axis theorem for rectangles A and D, the computations are as follows:

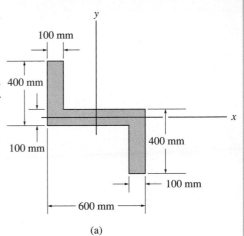

(a)

Rectangle A:

$$I_x = \bar{I}_{x'} + Ad_y^2 = \frac{1}{12}(100)(300)^3 + (100)(300)(200)^2$$

$$= 1.425(10^9) \text{ mm}^4$$

$$I_y = \bar{I}_{y'} + Ad_x^2 = \frac{1}{12}(300)(100)^3 + (100)(300)(250)^2$$

$$= 1.90(10^9) \text{ mm}^4$$

Rectangle B:

$$I_x = \frac{1}{12}(600)(100)^3 = 0.05(10^9) \text{ mm}^4$$

$$I_y = \frac{1}{12}(100)(600)^3 = 1.80(10^9) \text{ mm}^4$$

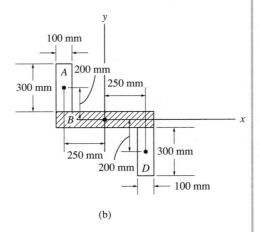

(b)

Fig. A–8

Rectangle D:

$$I_x = \bar{I}_{x'} + Ad_y^2 = \frac{1}{12}(100)(300)^3 + (100)(300)(200)^2$$

$$= 1.425(10^9) \text{ mm}^4$$

$$I_y = \bar{I}_{y'} + Ad_x^2 = \frac{1}{12}(300)(100)^3 + (100)(300)(250)^2$$

$$= 1.90(10^9) \text{ mm}^4$$

The moments of inertia for the entire cross section are thus

$$I_x = 1.425(10^9) + 0.05(10^9) + 1.425(10^9)$$
$$= 2.90(10^9) \text{ mm}^4 \qquad \textit{Ans.}$$
$$I_y = 1.90(10^9) + 1.80(10^9) + 1.90(10^9)$$
$$= 5.60(10^9) \text{ mm}^4 \qquad \textit{Ans.}$$

A.3 Product of Inertia for an Area

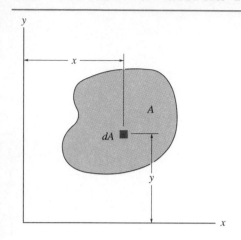

Fig. A–9

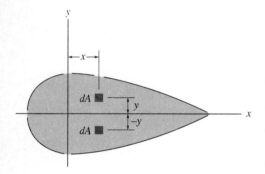

Fig. A–10

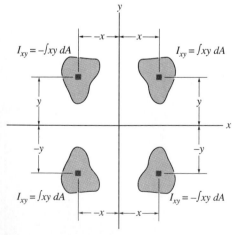

Fig. A–11

In general, the moment of inertia for an area is different for every axis about which it is computed. In some applications of mechanical or structural design it is necessary to know the orientation of those axes that give, respectively, the maximum and minimum moments of inertia for the area. The method for determining this is discussed in Sec. A.4. To use this method, however, one must first compute the product of inertia for the area as well as its moments of inertia for given x, y axes.

The *product of inertia* for the differential element dA in Fig. A–9, which is located at point (x, y), is defined as $dI_{xy} = xy \, dA$. Thus, for the entire area A, the product of inertia is

$$I_{xy} = \int xy \, dA \qquad (A–8)$$

Like the moment of inertia, the product of inertia has units of length raised to the fourth power, e.g., m^4, mm^4 or ft^4, in^4. However, since x or y may be a negative quantity, while the element of area is always positive, the product of inertia may be positive, negative, or zero, depending on the location and orientation of the coordinate axes. For example, the product of inertia I_{xy} for an area will be *zero* if either the x or y axis is an axis of *symmetry* for the area. To show this, consider the shaded area in Fig. A–10, where for every element dA located at point (x, y) there is a corresponding element dA located at $(x, -y)$. Since the products of inertia for these elements are, respectively, $xy \, dA$ and $-xy \, dA$, their algebraic sum or the integration of all the elements of area chosen in this way will cancel each other. Consequently, the product of inertia for the total area becomes zero. It also follows from the definition of I_{xy} that the "sign" of this quantity depends on the quadrant where the area is located. As shown in Fig. A–11, the sign of I_{xy} will change as the area is rotated from one quadrant to the next.

Parallel-Axis Theorem. Consider the shaded area shown in Fig. A–12, where x' and y' represent a set of centroidal axes, and x and y represent a corresponding set of parallel axes. Since the product of inertia of dA with respect to the x and y axes is $dI_{xy} = (x' + d_x)(y' + d_y) \, dA$, then for the entire area,

$$I_{xy} = \int (x' + d_x)(y' + d_y) \, dA$$

$$= \int x'y' \, dA + d_x \int y' \, dA + d_y \int x' \, dA + d_x d_y \int dA$$

The first term on the right represents the product of inertia of the area with respect to the centroidal axis, $\bar{I}_{x'y'}$. The second and third terms are zero since the moments of the area are taken about the centroidal axis. Realizing that the fourth integral represents the total area A, we therefore have as the final result

$$\boxed{I_{xy} = \bar{I}_{x'y'} + A d_x d_y} \qquad (A\text{--}9)$$

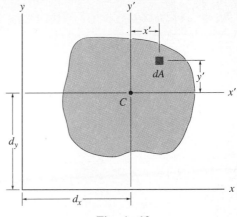

Fig. A–12

The similarity between this equation and the parallel-axis theorem for moments of inertia should be noted. In particular, it is important that the algebraic signs for d_x and d_y be maintained when applying Eq. A–9. As illustrated in the following example, the parallel-axis theorem finds important application in determining the product of inertia of a *composite area* with respect to a set of x, y axes.

Example A–4

Compute the product of inertia of the beam's cross-sectional area, shown in Fig. A–13a, about the x and y centroidal axes.

SOLUTION

As in Example A–3, the cross section can be considered as three composite rectangular areas A, B, and D, Fig. A–13b. The coordinates for the centroid of each of these rectangles are shown in the figure. Due to symmetry, the product of inertia of *each rectangle* is *zero* about a set of x', y' axes that pass through the rectangle's centroid. Hence, application of the parallel-axis theorem to each of the rectangles yields

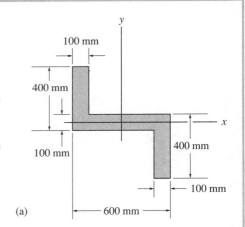

Rectangle A:

$$\begin{aligned} I_{xy} &= \bar{I}_{x'y'} + A d_x d_y \\ &= 0 + (300)(100)(-250)(200) \\ &= -1.50(10^9) \ \text{mm}^4 \end{aligned}$$

Rectangle B:

$$\begin{aligned} I_{xy} &= \bar{I}_{x'y'} + A d_x d_y \\ &= 0 + 0 \\ &= 0 \end{aligned}$$

Rectangle D:

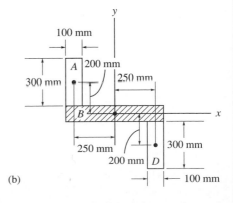

$$\begin{aligned} I_{xy} &= \bar{I}_{x'y'} + A d_x d_y \\ &= 0 + (300)(100)(250)(-200) \\ &= -1.50(10^9) \ \text{mm}^4 \end{aligned}$$

The product of inertia for the entire cross section is thus

$$\begin{aligned} I_{xy} &= [-1.50(10^9)] + 0 + [-1.50(10^9)] \\ &= -3.00(10^9) \ \text{mm}^4 \end{aligned}$$

Ans.

Fig. A–13

A.4 Moments of Inertia for an Area about Inclined Axes

In mechanical or structural design, it is sometimes necessary to calculate the moments and product of inertia $I_{x'}$, $I_{y'}$, and $I_{x'y'}$ for an area with respect to a set of inclined x' and y' axes when the values for θ, I_x, I_y, and I_{xy} are *known*. As shown in Fig. A–14, the coordinates to the area element dA from the two coordinate systems are related by using the *transformation equations*

$$x' = x \cos \theta + y \sin \theta$$
$$y' = y \cos \theta - x \sin \theta$$

Using these equations, the moments and product of inertia of dA about the x' and y' axes become

$$dI_{x'} = y'^2 \, dA = (y \cos \theta - x \sin \theta)^2 \, dA$$
$$dI_{y'} = x'^2 \, dA = (x \cos \theta + y \sin \theta)^2 \, dA$$
$$dI_{x'y'} = x'y' \, dA = (x \cos \theta + y \sin \theta)(y \cos \theta - x \sin \theta) \, dA$$

Expanding each expression and integrating, realizing that $I_x = \int y^2 \, dA$, $I_y = \int x^2 \, dA$, and $I_{xy} = \int xy \, dA$, we obtain

$$I_{x'} = I_x \cos^2 \theta + I_y \sin^2 \theta - 2I_{xy} \sin \theta \cos \theta$$
$$I_{y'} = 1_x \sin^2 \theta + I_y \cos^2 \theta + 2I_{xy} \sin \theta \cos \theta$$
$$I_{x'y'} = I_x \sin \theta \cos \theta - I_y \sin \theta \cos \theta + I_{xy}(\cos^2 \theta - \sin^2 \theta)$$

These equations may be simplified by using the trigonometric identities $\sin 2\theta = 2 \sin \theta \cos \theta$ and $\cos 2\theta = \cos^2 \theta - \sin^2 \theta$, in which case

Fig. A–14

$$I_{x'} = \frac{I_x + I_y}{2} + \frac{I_x - I_y}{2} \cos 2\theta - I_{xy} \sin 2\theta$$

$$I_{y'} = \frac{I_x + I_y}{2} - \frac{I_x - I_y}{2} \cos 2\theta + I_{xy} \sin 2\theta \qquad (A-10)$$

$$I_{x'y'} = \frac{I_x - I_y}{2} \sin 2\theta + I_{xy} \cos 2\theta$$

Note that if the first and second equations are added together, it is seen that the polar moment of inertia about the z axis passing through point O is *independent* of the orientation of the x' and y' axes; i.e.,

$$J_O = I_{x'} + I_{y'} = I_x + I_y$$

Principal Moments of Inertia. From Eq. A–10, it may be seen that $I_{x'}$, $I_{y'}$, and $I_{x'y'}$ depend on the angle of inclination, θ, of the x', y' axes. We will now determine the orientation of the x', y' axes about which the moments of inertia for the area, $I_{x'}$ and $I_{y'}$, are maximum and minimum. This particular set of axes is called the *principal axes* of inertia for the area, and the corresponding moments of inertia with respect to these axes are called the *principal moments of inertia*. In general, there is a set of principal axes for every chosen origin O; however, the area's centroid is the most important location for O.

The angle $\theta = \theta_p$, which defines the orientation of the principal axes for the area, may be found by differentiating the first of Eq. A–10 with respect to θ and setting the result equal to zero. Thus,

$$\frac{dI_{x'}}{d\theta} = -2\left(\frac{I_x - I_y}{2}\right)\sin 2\theta - 2\,I_{xy}\cos 2\theta = 0$$

Therefore, at $\theta = \theta_p$,

$$\tan 2\theta_p = \frac{-I_{xy}}{(I_x - I_y)/2} \qquad (A\text{–}11)$$

This equation has two roots, θ_{p_1} and θ_{p_2}, which are 90° apart and so specify the inclination of each principal axis.

Assuming I_{xy} and $(I_x - I_y)$ are both positive quantities, the sine and cosine of $2\theta_{p_1}$ and $2\theta_{p_2}$ can be obtained from the triangles shown in Fig. A–15, which are based on Eq. A–11. If these trigonometric relations are substituted into the first or second of Eq. A–10 and simplified, the result is

$$I_{\substack{max \\ min}} = \frac{I_x + I_y}{2} \pm \sqrt{\left(\frac{I_x - I_y}{2}\right)^2 + I_{xy}^2} \qquad (A\text{–}12)$$

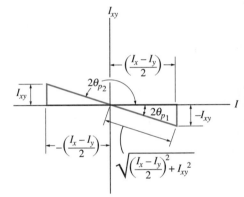

Fig. A–15

Depending on the sign chosen, this result gives the maximum or minimum moment of inertia for the area. Furthermore, if the above trigonometric relations for θ_{p_1} and θ_{p_2} are substituted into the third of Eq. A–10, it may be seen that $I_{x'y'} = 0$; that is, the *product of inertia with respect to the principal axes is zero*. Since it was indicated in Sec. A.3 that the product of inertia is zero with respect to any symmetrical axis, it therefore follows that *any symmetrical axis represents a principal axis of inertia for the area*.

Notice that the equations derived in this section are identical to those for stress and strain transformation developed in Chapters 9 and 10, respectively. The following example illustrates their application.

Determine the principal moments of inertia for the beam's cross-sectional area shown in Fig. A–16 with respect to an axis passing through the centroid C.

SOLUTION

The moments and product of inertia of the cross section with respect to the x, y axes have been computed in Examples A–3 and A–4. The results are

$$I_x = 2.90(10^9) \text{ mm}^4 \qquad I_y = 5.60(10^9) \text{ mm}^4 \qquad I_{xy} = -3.00(10^9) \text{ mm}^4$$

Using Eq. A–11, the angles of inclination of the principal axes x' and y' are

$$\tan 2\theta_p = \frac{-I_{xy}}{(I_x - I_y)/2} = \frac{3.00(10^9)}{[2.90(10^9) - 5.60(10^9)]/2} = -2.22$$

$$2\theta_{p_1} = -65.8° \qquad \text{and} \qquad 2\theta_{p_2} = 114.2°$$

Thus, as shown in Fig. A–16,

$$\theta_{p_1} = 57.1° \qquad \text{and} \qquad \theta_{p_2} = -32.9°$$

The principal moments of inertia with respect to the x' and y' axes are determined by using Eq. A–12. Hence,

$$I_{\substack{\max \\ \min}} = \frac{I_x + I_y}{2} \pm \sqrt{\left(\frac{I_x - I_y}{2}\right)^2 + I_{xy}^2}$$

$$= \frac{2.90(10^9) + 5.60(10^9)}{2} \pm \sqrt{\left[\frac{2.90(10^9) - 5.60(10^9)}{2}\right]^2 + [-3.00(10^9)]^2}$$

$$= 4.25(10^9) \pm 3.29(10^9)$$

or

$$I_{\max} = 7.54(10^9) \text{ mm}^4 \qquad I_{\min} = 0.960(10^9) \text{ mm}^4 \qquad \textit{Ans.}$$

Specifically, the maximum moment of inertia, $I_{\max} = 7.54(10^9) \text{ mm}^4$, occurs with respect to the x' axis (major axis), since *by inspection* most of the cross-sectional area is farthest away from this axis. To prove this, substitute the data with $\theta = 57.1°$ into the first of Eq. A–10.

Fig. A–16

A.5 Mohr's Circle for Moments of Inertia

Equations A–10 through A–12 have a graphical solution that is convenient to use and generally easy to remember. Squaring the first and third of Eq. A–10 and adding, it is found that

$$\left(I_{x'} - \frac{I_x + I_y}{2}\right)^2 + I_{x'y'}^2 = \left(\frac{I_x - I_y}{2}\right)^2 + I_{xy}^2 \qquad (A\text{–}13)$$

In any given problem, $I_{x'}$ and $I_{x'y'}$ are *variables,* and I_x, I_y, and I_{xy} are *known constants.* Thus, Eq. A–13 may be written in compact form as

$$(I_{x'} - a)^2 + I_{x'y'}^2 = R^2$$

When this equation is plotted, the resulting graph represents a *circle* of radius

$$R = \sqrt{\left(\frac{I_x - I_y}{2}\right)^2 + I_{xy}^2}$$

having its center located at point $(a, 0)$, where $a = (I_x + I_y)/2$. The circle so constructed is called *Mohr's circle.* Its application is similar to that used for stress and strain transformation developed in Chapters 9 and 10, respectively.

PROCEDURE FOR ANALYSIS

The main purpose for using Mohr's circle here is to have a convenient means of transforming I_x, I_y, and I_{xy} into the principal moments of inertia. The following procedure provides a method for doing this.

Compute I_x, I_y, I_{xy}.   Establish the x, y axes for the area, with the origin located at the point P of interest, usually the centroid, and determine I_x, I_y, and I_{xy}, Fig. A–17a.

Construct the Circle.   Establish a rectangular coordinate system such that the abscissa represents the moment of inertia I, and the ordinate represents the product of inertia I_{xy}, Fig. A–17b. Determine the center of the circle, C, which is located a distance $(I_x + I_y)/2$ from the origin, and plot the "reference point" A having coordinates (I_x, I_{xy}). By definition, I_x is always positive, whereas I_{xy} will be either positive or negative. Connect the reference point A with the center of the circle, and determine the distance CA by trigonometry. This distance represents the radius of the circle, Fig. A–17b. Finally, draw the circle.

Principal Moments of Inertia.   The points where the circle intersects the abscissa give the values of the principal moments of inertia I_{min} and I_{max}. Notice that the *product of inertia will be zero at these points,* Fig. A–17b.
   To find the direction of the major principal axis, determine by trigonometry the angle $2\theta_{p_1}$, *measured from the radius CA to the positive I axis,* Fig. A–17b. This angle represents twice the angle from the x axis to the axis of maximum moment of inertia I_{max}, Fig. A–17a. Both the angle on the circle, $2\theta_{p_1}$, and the angle on the area, θ_{p_1}, *must be measured in the same sense,* as shown in Fig. A–17. The minor axis is for minimum moment of inertia I_{min}, which is perpendicular to the major axis defining I_{max}.

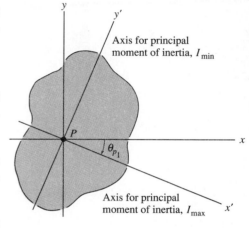

(a)

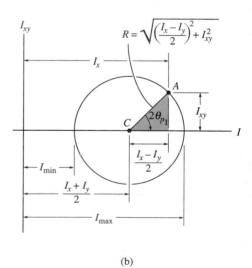

(b)

Fig. A–17

Example A–6

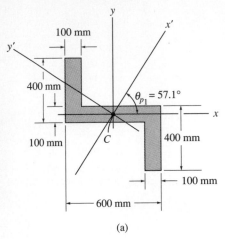

(a)

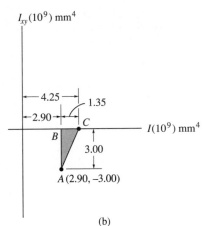

(b)

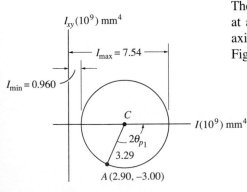

(c)

Fig. A–18

Use Mohr's circle to determine the principal moments of inertia for the beam's cross-sectional area, shown in Fig. A–18a, with respect to axes passing through the centroid C.

SOLUTION

Compute I_x, I_y, I_{xy}. The moments of inertia and the product of inertia have been determined in Examples A–3 and A–4 with respect to the x, y axes shown in Fig. A–18a. The results are $I_x = 2.90(10^9)$ mm⁴, $I_y = 5.60(10^9)$ mm⁴, and $I_{xy} = -3.00(10^9)$ mm⁴.

Construct the Circle. The I and I_{xy} axes are shown in Fig. A–18b. The center of the circle, C, lies at a distance $(I_x + I_y)/2 = (2.90 + 5.60)/2 = 4.25$ from the origin. When the reference point $A(2.90, -3.00)$ is connected to point C, the radius CA is determined from the shaded triangle CBA using the Pythagorean theorem:

$$CA = \sqrt{(1.35)^2 + (-3.00)^2} = 3.29$$

The circle is constructed in Fig. A–18c.

Principal Moments of Inertia. The circle intersects the I axis at points $(7.54, 0)$ and $(0.960, 0)$. Hence,

$$I_{\max} = 7.54(10^9) \text{ mm}^4 \qquad \textit{Ans.}$$
$$I_{\min} = 0.960(10^9) \text{ mm}^4 \qquad \textit{Ans.}$$

As shown in Fig. A–18c, the angle $2\theta_{p_1}$ is determined from the circle by measuring *counterclockwise* from CA to the direction of the *positive I* axis. Hence,

$$2\theta_{p_1} = 180° - \tan^{-1}\left(\frac{|BA|}{|BC|}\right) = 180° - \tan^{-1}\left(\frac{3.00}{1.35}\right) = 114.2°$$

The major principal axis (for $I_{\max} = 7.54(10^9)$ mm⁴) is therefore oriented at an angle $\theta_{p_1} = 57.1°$, measured *counterclockwise,* from the *positive x* axis. The minor axis is perpendicular to this axis. The results are shown in Fig. A–18a.

PROBLEMS

A–1.   Determine the location $\bar{y}$ of the centroid C for the beam's cross-sectional area. The beam is symmetric with respect to the y axis.

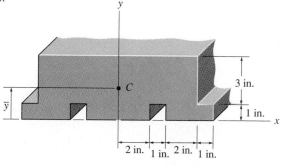

Prob. A–1

A–2.   Determine the centroid $\bar{y}$ for the beam's cross-sectional area, then find $\bar{I}_{x'}$.

A–3.   Determine I_y for the beam having the cross-sectional area shown.

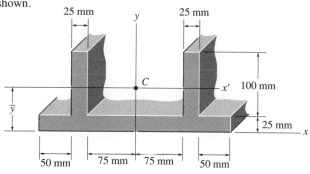

Probs. A–2/A–3

*A–4.   Determine the moments of inertia I_x and I_y of the Z-section. The origin of coordinates is at the centroid C.

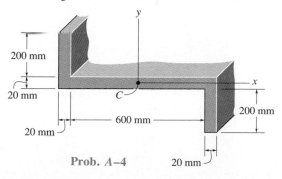

Prob. A–4

A–5.   Determine $\bar{y}$, which locates the centroid, and then find the moments of inertia $\bar{I}_{x'}$ and $\bar{I}_y$ for the T-beam.

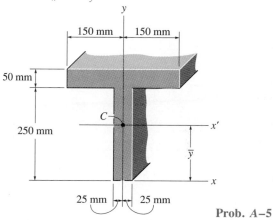

Prob. A–5

A–6.   Determine $\bar{x}$ which locates the centroid C, and then compute the moments of inertia $\bar{I}_{x'}$ and $\bar{I}_{y'}$ for the shaded area.

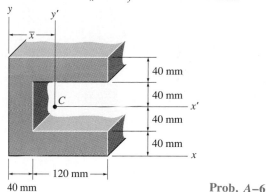

Prob. A–6

A–7.   Determine the location $(\bar{x}, \bar{y})$ of the centroid C of the cross-sectional area for the angle, then find the moments of inertia $\bar{I}_{x'}$ and $\bar{I}_{y'}$.

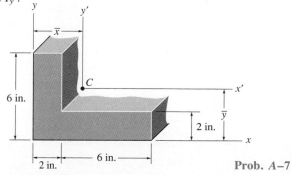

Prob. A–7

*A–8. Determine the location $(\bar{x}, \bar{y})$ of the centroid C of the cross-sectional area for the angle, then find the product of inertia with respect to the x and y axes and with respect to the x' and y' axes.

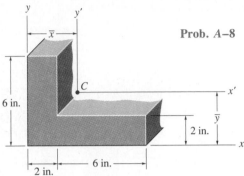

Prob. A–8

A–9. Determine the product of inertia of the cross-sectional area with respect to the x and y axes that have their origin located at the centroid C.

Prob. A–9

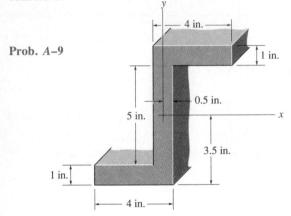

A–10. Determine the product of inertia of the area with respect to the x and y axes.

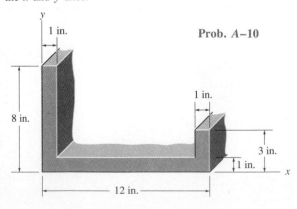

Prob. A–10

A–11. Locate the centroid $(\bar{x}, \bar{y})$ of the channel section and then determine the moments of inertia $\bar{I}_{x'}$ and $\bar{I}_{y'}$.

*A–12. Locate the centroid $(\bar{x}, \bar{y})$ of the channel section and then determine the product of inertia $\bar{I}_{x'y'}$ with respect to the x' and y' axes.

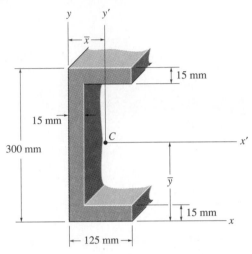

Probs. A–11/A–12

A–13. Locate the position $(\bar{x}, \bar{y})$ for the centroid C of the cross-sectional area and then compute the product of inertia with respect to the x' and y' axes.

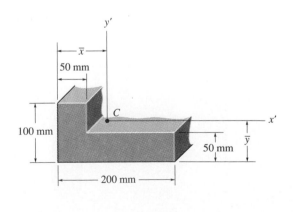

Prob. A–13

A–14. Compute the moments of inertia $I_{x'}$ and $I_{y'}$ of the shaded area.

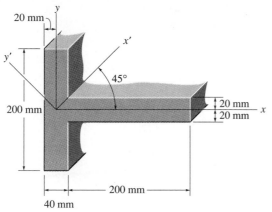

Prob. A–14

A–15. Determine the moments of inertia $I_{x'}$ and $I_{y'}$ and the product of inertia $I_{x'y'}$ for the semicircular area.

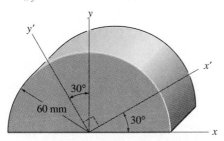

Prob. A–15

*A–16. Determine the moments of inertia $I_{x'}$ and $I_{y'}$ and the product of inertia $I_{x'y'}$ for the rectangular area. The x' and y' axes pass through the centroid C.

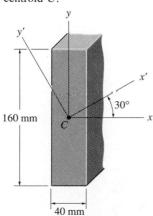

Prob. A–16

A–17. Determine the principal moments of inertia for the angle's cross-sectional area with respect to a set of principal axes that have their origin located at the centroid C. Use the equations developed in Sec. A.4. For the calculation, assume all corners to be square.

A–18. Solve Prob. A–17 using Mohr's circle.

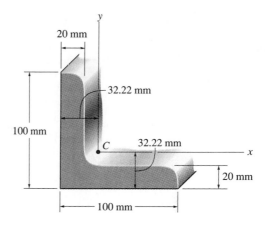

Probs. A–17/A–18

A–19. Determine the principal moments of inertia of the cross-sectional area about the principal axes that have their origin located at the centroid C. Use the equations developed in Sec. A.4. For the calculation, assume all corners to be square.

*A–20. Solve Prob. A–19 using Mohr's circle.

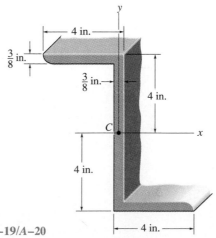

Probs. A–19/A–20

B Properties of Wide-Flange Sections

			Web	Flange		x-x axis			y-y axis		
	Area	Depth	thickness	width	thickness	I	S	r	I	S	r
	A	d	t_w	b	t_f						
Designation	in^2	in.	in.	in.	in.	in^4	in^3	in.	in^4	in^3	in.
W24 × 104	30.6	24.06	0.500	12.750	0.750	3100	258	10.1	259	40.7	2.91
W24 × 94	27.7	24.31	0.515	9.065	0.875	2700	222	9.87	109	24.0	1.98
W24 × 84	24.7	24.10	0.470	9.020	0.770	2370	196	9.79	94.4	20.9	1.95
W24 × 76	22.4	23.92	0.440	8.990	0.680	2100	176	9.69	82.5	18.4	1.92
W24 × 68	20.1	23.73	0.415	8.965	0.585	1830	154	9.55	70.4	15.7	1.87
W24 × 62	18.2	23.74	0.430	7.040	0.590	1550	131	9.23	34.5	9.80	1.38
W24 × 55	16.2	23.57	0.395	7.005	0.505	1350	114	9.11	29.1	8.30	1.34
W18 × 65	19.1	18.35	0.450	7.590	0.750	1070	117	7.49	54.8	14.4	1.69
W18 × 60	17.6	18.24	0.415	7.555	0.695	984	108	7.47	50.1	13.3	1.69
W18 × 65	16.2	18.11	0.390	7.530	0.630	890	98.3	7.41	44.9	11.9	1.67
W18 × 50	14.7	17.99	0.355	7.495	0.570	800	88.9	7.38	40.1	10.7	1.65
W18 × 46	13.5	18.06	0.360	6.060	0.605	712	78.8	7.25	22.5	7.43	1.29
W18 × 40	11.8	17.90	0.315	6.015	0.525	612	68.4	7.21	19.1	6.35	1.27
W18 × 35	10.3	17.70	0.300	6.000	0.425	510	57.6	7.04	15.3	5.12	1.22
W16 × 57	16.8	16.43	0.430	7.120	0.715	758	92.2	6.72	43.1	12.1	1.60
W16 × 50	14.7	16.26	0.380	7.070	0.630	659	81.0	6.68	37.2	10.5	1.59
W16 × 45	13.3	16.13	0.345	7.035	0.565	586	72.7	6.65	32.8	9.34	1.57
W16 × 36	10.6	15.86	0.295	6.985	0.430	448	56.5	6.51	24.5	7.00	1.52
W16 × 31	9.12	15.88	0.275	5.525	0.440	375	47.2	6.41	12.4	4.49	1.17
W16 × 26	7.68	15.69	0.250	5.500	0.345	301	38.4	6.26	9.59	3.49	1.12
W14 × 53	15.6	13.92	0.370	8.060	0.660	541	77.8	5.89	57.7	14.3	1.92
W14 × 43	12.6	13.66	0.305	7.995	0.530	428	62.7	5.82	45.2	11.3	1.89
W14 × 38	11.2	14.10	0.310	6.770	0.515	385	54.6	5.87	26.7	7.88	1.55
W14 × 34	10.0	13.98	0.285	6.745	0.455	340	48.6	5.83	23.3	6.91	1.53
W14 × 30	8.85	13.84	0.270	6.730	0.385	291	42.0	5.73	19.6	5.82	1.49
W14 × 26	7.69	13.91	0.255	5.025	0.420	245	35.3	5.65	8.91	3.54	1.08
W14 × 22	6.49	13.74	0.230	5.000	0.335	199	29.0	5.54	7.00	2.80	1.04

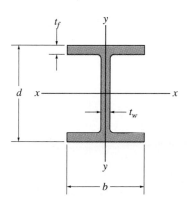

Designation	Area A in²	Depth d in.	Web thickness t_w in.	Flange width b in.	Flange thickness t_f in.	x-x axis I in⁴	x-x axis S in³	x-x axis r in.	y-y axis I in⁴	y-y axis S in³	y-y axis r in.
W12 × 87	25.6	12.53	0.515	12.125	0.810	740	118	5.38	241	39.7	3.07
W12 × 50	14.7	12.19	0.370	8.080	0.640	394	64.7	5.18	56.3	13.9	1.96
W12 × 45	13.2	12.06	0.335	8.045	0.575	350	58.1	5.15	50.0	12.4	1.94
W12 × 26	7.65	12.22	0.230	6.490	0.380	204	33.4	5.17	17.3	5.34	1.51
W12 × 22	6.48	12.31	0.260	4.030	0.425	156	25.4	4.91	4.66	2.31	0.847
W12 × 16	4.71	11.99	0.220	3.990	0.265	103	17.1	4.67	2.82	1.41	0.773
W12 × 14	4.16	11.91	0.200	3.970	0.225	88.6	14.9	4.62	2.36	1.19	0.753
W10 × 100	29.4	11.10	0.680	10.340	1.120	623	112	4.60	207	40.0	2.65
W10 × 54	15.8	10.09	0.370	10.030	0.615	303	60.0	4.37	103	20.6	2.56
W10 × 45	13.3	10.10	0.350	8.020	0.620	248	49.1	4.32	53.4	13.3	2.01
W10 × 30	8.84	10.47	0.300	5.810	0.510	170	32.4	4.38	16.7	5.75	1.37
W10 × 39	11.5	9.92	0.315	7.985	0.530	209	42.1	4.27	45.0	11.3	1.98
W10 × 19	5.62	10.24	0.250	4.020	0.395	96.3	18.8	4.14	4.29	2.14	0.874
W10 × 15	4.41	9.99	0.230	4.000	0.270	68.9	13.8	3.95	2.89	1.45	0.810
W10 × 12	3.54	9.87	0.190	3.960	0.210	53.8	10.9	3.90	2.18	1.10	0.785
W8 × 67	19.7	9.00	0.570	8.280	0.935	272	60.4	3.72	88.6	21.4	2.12
W8 × 58	17.1	8.75	0.510	8.220	0.810	228	52.0	3.65	75.1	18.3	2.10
W8 × 48	14.1	8.50	0.400	8.110	0.685	184	43.3	3.61	60.9	15.0	2.08
W8 × 40	11.7	8.25	0.360	8.070	0.560	146	35.5	3.53	49.1	12.2	2.04
W8 × 31	9.13	8.00	0.285	7.995	0.435	110	27.5	3.47	37.1	9.27	2.02
W8 × 24	7.08	7.93	0.245	6.495	0.400	82.8	20.9	3.42	18.3	5.63	1.61
W8 × 15	4.44	8.11	0.245	4.015	0.315	48.0	11.8	3.29	3.41	1.70	0.876
W6 × 25	7.34	6.38	0.320	6.080	0.455	53.4	16.7	2.70	17.1	5.61	1.52
W6 × 20	5.87	6.20	0.260	6.020	0.365	41.4	13.4	2.66	13.3	4.41	1.50
W6 × 15	4.43	5.99	0.230	5.990	0.260	29.1	9.72	2.56	9.32	3.11	1.46
W6 × 16	4.74	6.28	0.260	4.030	0.405	32.1	10.2	2.60	4.43	2.20	0.966
W6 × 12	3.55	6.03	0.230	4.000	0.280	22.1	7.31	2.49	2.99	1.50	0.918
W6 × 9	2.68	5.90	0.170	3.940	0.215	1.64	5.56	2.47	2.19	1.11	0.905

Answers[*]

Chapter 1

1–1. (a) $F_A = 13.8$ kip, (b) $F_A = 34.9$ kN

1–2. (a) $T_C = 250$ N · m, $T_D = 0$, (b) $T_D = 500$ lb · ft, $T_C = 150$ lb · ft

1–3. (a) $N_a = 500$ lb, $V_a = 0$, (b) $V_b = 250$ lb, $N_b = 433$ lb

1–5. $N_D = 131$ N, $V_D = 175$ N, $M_D = 8.75$ N · m

1–6. $N_D = 1.20$ kip, $V_D = 0.625$ kip, $M_D = 0.769$ kip · ft, $N_E = 2.00$ kip, $V_E = 0$, $M_E = 0$

1–7. $N_D = 1.44$ kN, $V_D = 0$, $M_D = 3.75$ kN · m, $N_E = 2.89$ kN, $V_E = 0$, $M_E = 0$

1–9. $V_C = 60$ N, $N_C = 0$, $M_C = 0.9$ N · m, $V_D = 17.3$ N, $N_D = 10$ N, $M_D = 1.60$ N · m

1–10. (a) $N_C = 3.60$ kip, $V_C = 1.20$ kip, $M_C = 9.60$ kip · ft, (b) $N_C = 1.70$ kip, $V_C = 3.39$ kip, $M_C = 9.60$ kip · ft

1–11. $N_C = 18.2$ N, $V_C = 10.5$ N, $M_C = 9.46$ N · m

1–13. $V_C = 2.94$ kN, $N_C = 2.94$ kN, $M_C = 1.47$ kN · m

1–14. $N_B = 0.4$ kip, $V_B = 0.96$ kip, $M_B = 3.12$ kip · ft

1–15. $(N_B)_x = 0$, $(V_B)_y = 0$, $(V_B)_z = 70.6$ N, $(T_B)_x = 9.42$ N · m, $(M_B)_y = 6.23$ N · m, $(M_B)_z = 0$

1–17. $(V_C)_x = 250$ N, $(N_C)_y = 0$, $(V_C)_z = 240$ N, $(M_C)_x = 108$ N · m, $(T_C)_y = 0$, $(M_C)_z = 138$ N · m

1–18. $V_x = 0$, $N_y = 25$ lb, $V_z = 43.3$ lb, $M_x = 303$ lb · in., $T_y = 130$ lb · in., $M_z = 75$ lb · in.

1–19. $V_B = 0.785wr$, $N_B = 0$, $T_B = 0.0783wr^2$, $M_B = 0.293wr^2$

1–21. $\sigma = 1.82$ MPa

1–22. $\sigma_{40} = 3.98$ MPa, $\sigma_{30} = 7.07$ MPa, $\tau_{avg} = 5.09$ MPa

1–23. $\tau_B = \tau_C = 324$ MPa, $\tau_A = 324$ MPa

1–25. $\bar{x} = 4$ in., $\bar{y} = 4$ in., $\sigma = 9.26$ psi

1–26. $\sigma = 8$ MPa, $\tau_{avg} = 4.62$ MPa

1–27. $\sigma = 1.90$ psi, $\tau_{avg} = 7.08$ psi

1–29. $\sigma_{AB} = 10.7$ ksi (T), $\sigma_{AE} = 8.53$ ksi (C), $\sigma_{ED} = 8.53$ ksi (C), $\sigma_{EB} = 4.80$ ksi (T), $\sigma_{BC} = 23.5$ ksi (T), $\sigma_{BD} = 18.7$ ksi (C)

1–30. $\sigma_B = 151$ kPa, $\sigma_C = 32.5$ kPa, $\sigma_D = 25.5$ kPa

1–31. $\sigma = (238 - 22.6z)$ kPa

1–33. $\theta = 63.6°$

1–34. $\sigma_s = 208$ MPa, $(\tau_{avg})_a = 4.72$ MPa, $(\tau_{avg})_b = 45.5$ MPa

1–35. $\tau_{avg} = 61.3$ MPa

1–37. $d = 1.20$ m

1–38. $\sigma_{a\text{-}a} = 33.9$ psi, $\tau_{a\text{-}a} = 0$, $\sigma_{b\text{-}b} = 8.48$ psi, $\tau_{b\text{-}b} = 14.7$ psi

1–39. $\sigma_{a\text{-}a} = 12.2$ MPa, $\tau_{a\text{-}a} = 0$, $\sigma_{b\text{-}b} = 4.41$ MPa, $\tau_{b\text{-}b} = 5.88$ MPa

1–41. $w = w_1 e^{(w_1{}^2 \gamma/2p)z}$

1–42. $r = r_1 e^{(\pi r_1{}^2 \rho g/2p)z}$

1–43. $d = 1$ in.

1–45. $d_{AB} = 19.9$ mm, $d_{AC} = 19.1$ mm

1–46. $h = 1.5$ in.

1–47. 3 in. $\times$ 3 in., $4\frac{1}{2}$ in. $\times$ $4\frac{1}{2}$ in.

1–49. $d = 144$ mm

1–50. $A_{BC} = 0.577$ in^2, $d_A = 0.743$ in., $d_B = 0.525$ in.

1–51. $P = 5.11$ kN

1–53. $P = 3.26$ kip

1–54. $d_b = 6.11$ mm, $d_w = 15.4$ mm

1–55. $P = 0.491$ kip

1–57. $F.S. = 2.71$, $F.S. = 1.53$

1–58. $w = 0.530$ kip/ft

1–59. $\tau = 23.9$ ksi, $F.S. = 1.05$, $\sigma = 30.6$ ksi, $F.S. = 1.24$

1–61. $P = 561$ kN

1–62. $P = 140$ kN

1–63. $\sigma_{AB} = 417$ psi (C), $\sigma_{BC} = 469$ psi (T), $\sigma_{AC} = 833$ psi (T)

1–65. $N_D = 17.6$ kN, $V_D = 2.61$ kN, $M_D = 5.22$ kN $\cdot$ m, $N_E = 10.1$ kN, $V_E = 4.89$ kN, $M_E = 15.3$ kN $\cdot$ m

1–66. $\tau_{a\text{-}a} = 52.0$ kPa, $\sigma_{a\text{-}a} = 90.0$ kPa

1–67. $\sigma_D = 13.3$ MPa (C), $\sigma_E = 70.7$ MPa (T)

1–69. $N_C = 0$, $V_C = 3.50$ kip, $M_C = 47.5$ kip $\cdot$ ft, $N_D = 0$, $V_D = 0.240$ kip, $M_D = 0.360$ kip $\cdot$ ft

1–70. $\tau_{\text{avg}} = 42.4$ kPa

1–71. $N_B = 0$, $V_{BR} = 22.5$ kN, $M_B = 102$ kN $\cdot$ m, $V_{BL} = 62.1$ kN

1–73. $\tau = 7.50$ MPa

1–74. $N_B = wr\,\theta \cos\theta$, $V_B = wr\,\theta \sin\theta$, $M_B = wr^2\,(\theta \cos\theta - \sin\theta)$

Chapter 2

2–1. $\epsilon = 0.0472$ in./in.

2–2. $\epsilon = 0.167$ in./in.

2–3. $\epsilon_{CE} = 0.00250$ mm/mm, $\epsilon_{BD} = 0.00107$ mm/mm

2–5. $\epsilon_{AB} = 0.0343$

2–6. $\epsilon_{AB} = 0.5\ \Delta L/L$

2–7. $\epsilon_{AB} = \dfrac{\Delta L_1}{L_1} \cos^2\theta + \dfrac{\Delta L_2}{L_2} \sin^2\theta$

2–9. $\epsilon = 2kx$

2–10. $\gamma_{xy} = 0.00880$ rad

2–11. $\epsilon_x = 0.00443$ mm/mm

2–13. $\gamma_{xy} = 0.0502$ rad, $\epsilon_x = -0.03$ in./in., $\epsilon_y = 0.02$ in./in.

2–14. $\gamma_{x'y'} = -0.00599$ rad, $\gamma_{xy} = 0$

2–15. $\epsilon_{BE} = 3.82(10^{-3})$ in./in., $\epsilon_{CF} = 7.31(10^{-3})$ in./in., $\epsilon_{AD} = 0.333(10^{-3})$ in./in.

2–18. $\gamma_{xy} = 0.02$ rad

2–19. $\gamma_{xy} = 0.00125$ rad

2–21. $\epsilon_{AB} = \dfrac{v_B \sin\theta}{L} - \dfrac{u_A \cos\theta}{L}$

Chapter 3

3–1. $E = 286$ GPa, $U_r = 91.6$ kJ/m^3

3–2. $E = 20.0(10^3)$ ksi, $U_r = 25.6$ in. $\cdot$ lb/in^3

3–3. $E_{\text{approx}} = 3.13(10^3)$ ksi

3–5. $U_T = 16.3$ in. $\cdot$ kip/in^3

3–6. $(\sigma_{\text{ult}})_{\text{approx}} = 110$ ksi, $\sigma_R = 93.1$ ksi, $(\sigma_Y)_{\text{approx}} = 55$ ksi, $E_{\text{approx}} = 32.0(10^3)$ ksi

3–7. $E = 28.7(10^3)$ ksi, $\sigma_{pl} = 43$ ksi, $\sigma_{\text{ult}} = 85$ ksi, $U_{r(\text{approx})} = 32.2$ in. $\cdot$ lb/in^3, elastic recovery = 0.00251 in./in., permanent set = 0.0475 in./in.

3–9. $\Delta P = 50.1$ kip

3–10. $\Delta L = 0.304$ in.

3–11. $\epsilon_b = 0.00227$ mm/mm, $\epsilon_s = 0.000884$ mm/mm

3–13. $\epsilon_{DE} = 0.00116$ in./in., $W = 112$ lb, $\epsilon_{BC} = 0.00193$ in./in.

3–14. $A = 0.209$ in^2, $P = 1.62$ kip

3–15. $P = 2.46$ kN

3–17. (a) $\Delta L = -0.611(10^{-3})$ in., (b) $D' = 0.500067$ in.

3–18. $\nu = 0.300$

3–19. $\epsilon_x = 0.0075$ in./in., $\epsilon_y = -0.00375$ in./in., $\gamma_{xy} = 0.0122$ rad

3–21. 8.33 mm

3–22. $\Delta_h = 3.02$ mm

3–23. $\delta = \dfrac{Pa}{2bhG}$

3–25. $\nu = 0.33$, $h' = 2.000176$ in.

3–26. $d_{AB} = 3.54$ mm, $d_{AC} = 3.23$ mm, $l_{AB} = 750.488$ mm

3–27. $x = 1.53$ m, $d'_A = 30.00782$ mm

3–29. $L = 2.792$ in.

3–30. $G = 4.31(10^3)$ ksi

3–31. $E = 28.6(10^3)$ ksi

Chapter 4

4–1. $\sigma_{AB} = 22.2$ ksi (T), $\sigma_{BC} = 41.7$ ksi (C), $\sigma_{CD} = 25.0$ ksi (C), $\Delta_{A/D} = -0.00157$ in.

4–2. $\Delta_{B/C} = -0.0278$ in.

4–3. $\Delta_B = 1.59$ mm, $\Delta_A = 6.14$ mm

4–5. $\Delta_{A/D} = 0.111$ in.

4–6. $\Delta_B = 2.31$ mm, $\Delta_A = 2.64$ mm

4–7. $\Delta_D = 1.17$ mm

4–9. $\Delta_{C_h} = 0.0975$ mm

4–10. $x = 10$ in.

4–11. $\Delta_F = 2.23$ mm

4–13. $\Delta = \dfrac{PL}{AE} + \dfrac{\gamma L^2}{2E}$

4–14. $\sigma_A = \sigma_{max} = 136$ ksi, $\Delta_D = 29.9$ ft

4–15. $F = 1.30$ kip, $\Delta = 8.45$ ft

4–17. $\Delta = \dfrac{PL}{\pi E r_2 r_1}$

4–18. $\Delta = \dfrac{PL}{\pi E r_1 r_2} + \dfrac{\gamma L^2(r_2 + r_1)}{6E(r_2 - r_1)} - \dfrac{\gamma L^2 r_1^2}{3E r_2(r_2 - r_1)}$

4–19. $\Delta = 0.00711$ in.

4–21. $\Delta = 2.37$ mm

4–22. $\sigma_{st} = 48.8$ MPa, $\sigma_{con} = 5.85$ MPa

4–23. $\sigma_{st} = 65.9$ MPa, $\sigma_{con} = 8.24$ MPa

4–25. $\sigma_{st} = 4.05$ ksi, $\sigma_{con} = 0.488$ ksi

4–26. $F_C = 4.80$ kN, $F_A = 11.2$ kN

4–27. $\sigma_{BC} = 14.1$ MPa (T), $\sigma_{AB} = 88.1$ MPa (T), $\sigma_{AD} = 113$ MPa (C)

4–29. $T_{AB} = 1.12$ kip, $T_{AC} = 1.68$ kip

4–30. $P_h = 5.25$ kN

4–31. $P_t = P_b = 116$ kN

4–33. $T_{AB} = T_{CD} = 16.7$ kN, $T_{EF} = 33.3$ kN

4–34. $T_{AB} = 361$ lb, $T_{A'B'} = 289$ lb

4–35. $T_{AB} = T_{AD} = \dfrac{P \cos \theta}{1 + 2 \cos^2 \theta}$, $T_{AC} = \dfrac{P}{1 + 2 \cos^2 \theta}$

4–37. $\sigma_{AB} = \dfrac{7P}{12A}$, $\sigma_{CD} = \dfrac{P}{3A}$, $\sigma_{EF} = \dfrac{P}{12A}$

4–38. $\sigma_{AB} = \sigma_{CD} = \dfrac{3E_1 M_0}{Ad(9E_1 + E_2)}$, $\sigma_{GH} = \sigma_{EF} = \dfrac{E_2 M_0}{Ad(9E_1 + E_2)}$

4–39. $T_B = 45.7$ lb, $T_C = 137$ lb, $T_D = 274$ lb

4–41. $\sigma_A = \sigma_C = 189$ MPa, $\sigma_B = 21.4$ MPa

4–42. $w = 45.9$ kN/m

4–43. $\Delta_B = 0.0489$ in.

4–45. $T_B = 86.6$ lb, $T_C = 195$ lb

4–46. $F = 40.0$ kN

4–47. $F_A = 14.0$ kN

4–49. $F = 67.2$ kip

4–50. $F = 65.7$ kip

4–51. $F = 4.20$ kN

4–53. $F = 16.5$ kip

4–54. 4.23 kN

4–55. 1.85 kN

4–57. $\sigma_b = 29.5$ MPa, $\sigma_s = 40.1$ MPa

4–58. $F_{AD} = 7.51$ kip (T), $F_{AB} = F_{AC} = 4.69$ kip (C)

4–59. $F = \dfrac{\alpha AE(T_B - T_A)}{2}$

4–61. $P = 44.1$ kN

4–62. $\sigma_{max} = 190$ MPa

4–63. $P = 5.05$ kN

4–65. $P = 41.5$ kN, $\Delta = 0.385$ mm

4–66. $w = 0.836$ in.

4–67. $P = 19$ kN, $K = 1.26$

4–69. (a) $\Delta_D = 0.375$ in., (b) $\Delta_D = 6.40$ in.

4–70. $\Delta = 0.952$ in.

4–71. (a) $T_{st} = 444$ N, $T_{al} = 156$ N, (b) $T_{st} = 480$ N, $T_{al} = 240$ N

4–73. (a) $P = 92.8$ kN, (b) $P = 181$ kN

4–74. $d_B = 17.8$ mm

4–75. $\Delta = 8.69$ in., $\sigma = 53.3$ ksi

4–77. (a) $w = 18.7$ kN/m, (b) $w = 21.9$ kN/m

4–78. $\Delta = \dfrac{\gamma^2 L^3}{3c^2}$

4–79. $P = 130$ kip, $F_{BC} = 120$ kip, $F_{AB} = 10$ kip; $P = 150$ kip, $F_{BC} = 120$ kip, $F_{AB} = 30$ kip

4–81. $(\sigma_{AB})_r = 3.75$ ksi (T), $(\sigma_{BC})_r = 3.75$ ksi (T)

4–82. $L = 463.41$ ft

4–83. $P = 56.5$ kN, $\Delta_{B/A} = 0.0918$ mm

4–85. $x = 2.12$ ft, $P = 4.33$ kip

4–86. $F = 2.39$ kip

4–87. (a) $F_{AC} = 62$ N, $F_{BC} = 42$ N, (b) $P = 125$ N

4–89. $\Delta = 0.129$ mm, $h' = 49.9988$ mm, $w' = 59.9986$ mm

4–90. $\sigma_{st} = 102$ MPa, $\sigma_{b} = 50.9$ MPa

4–91. $P = 126$ kN

4–93. $F = 11.8$ kN

4–94. $\sigma_B = 8.90$ ksi, $\Delta_{A/B} = [0.307(10^{-3})y]$ ft

Chapter 5

5–1. (a) $T = 7.95$ kip $\cdot$ in., (b) $T = 6.38$ kip $\cdot$ in.

5–2. $T' = 510$ N $\cdot$ m

5–3. $r' = 0.841r$

5–5. $\tau_{max} = 26.7$ MPa

5–6. $\tau_A = 14.1$ MPa, $\tau_B = 8.06$ MPa

5–7. $\tau_A = 3.45$ ksi, $\tau_B = 2.76$ ksi

5–9. $\tau_C = 3.91$ ksi, $\tau_D = 1.56$ ksi

5–10. $(\tau_{BC})_{max} = 5.79$ ksi, $(\tau_{DE})_{max} = 8.26$ ksi

5–11. $(\tau_{CD})_{max} = 3.62$ ksi, $(\tau_{EF})_{max} = 4.78$ ksi

5–13. $F = 154$ N

5–14. $T' = 125$ N · m, $(\tau_{CD})_{max} = 14.8$ MPa, $(\tau_{AB})_{max} = 9.43$ MPa

5–15. $T_2 = 147$ N · m, $T_1 = 4.74$ N · m

5–17. $T = 71.5$ N · m, $(\tau_s)_{max} = 23.3$ MPa

5–18. $\tau_{max} = 3.18$ MPa, $\tau_{min} = 1.59$ MPa

5–19. $t_0 = 133$ N · m/m, $\tau_A = 255$ kPa, $\tau_B = 141$ kPa

5–21. $\tau_{max} = \dfrac{T}{2\pi r_i^2 h}$

5–22. $\tau_{max}(x) = \dfrac{2TL^3}{\pi[Lr_A + x(r_B - r_A)]^3}$

5–23. $d = 1.25$ in.

5–25. $T = 1.51$ lb · ft, $\tau_{max} = 219$ psi

5–26. $(\tau_{AB})_{max} = 7.90$ ksi, $(\tau_{BC})_{max} = 3.60$ ksi

5–27. $t = 0.174$ in.

5–29. $\omega = 296$ rad/s

5–30. $d = 4.39$ mm

5–31. $(\tau_{AB})_{max} = 1.04$ MPa, $(\tau_{BC})_{max} = 3.11$ MPa

5–33. $\phi_{A/C} = 0.362°$

5–34. $\gamma_{C/2} = \dfrac{Tc}{2JG}$

5–35. $\phi_{B/A} = 0.578°$

5–37. $\phi_{A/D} = 0.879°$

5–38. $d = 2.75$ in.

5–39. $d = 2.50$ in.

5–41. $\tau_{max} = 64.0$ MPa

5–42. $t = 7.53$ mm

5–43. $\omega = 131$ rad/s

5–45. $\phi_A = 1.92°$

5–46. $\phi_B = 3.06°$

5–47. $\phi_A = 3.45°$

5–49. $\phi_B = \dfrac{t_0 L^2}{\pi c^4 G}$

5–50. $\phi = \dfrac{2L(t_0 L + 3T_A)}{3\pi(r_o^4 - r_i^4)G}$, $T_B = \dfrac{1}{2}t_0 L + T_A$

5–51. $\phi_A = \dfrac{7LT}{12\pi r^4 G}$

5–53. $\phi_{A/B} = \dfrac{T}{2a\pi G}(1 - e^{-4aL})$

5–54. $\phi = \dfrac{T}{4\pi h G}\left(\dfrac{1}{r_i^2} - \dfrac{1}{r_o^2}\right)$

5–55. $F = 23.4$ lb

5–57. $d = 42.7$ mm

5–58. $x = \dfrac{17}{32}L$

5–59. $\tau_{max} = \tau_{AC} = 29.3$ ksi

5–61. $d = 403$ mm

5–62. $T_A = 127$ N · m, $T_B = 424$ N · m

5–63. $(\tau_{BC})_{max} = 1.47$ ksi, $(\tau_{BD})_{max} = 1.96$ ksi, $\phi = 0.338°$

5–65. $T_A = 371$ lb · ft, $T_B = 429$ lb · ft

5–66. $T_B = \dfrac{7t_0 L}{12}$, $T_A = \dfrac{3t_0 L}{4}$

5–67. $\left(\dfrac{a}{b}\right)^2$

5–69. $\tau_{max} = 1.33$ ksi, $\phi = 2.45(10)^{-3}$ rad, $\tau_{max} = 1.80$ ksi, $\phi = 2.77(10^{-3})$ rad

5–70. $T = 7.74$ N · m

5–71. $a = 0.350$ in., $\phi = 37.6°$

5–73. $F = 104$ lb

5–74. $\tau_{max} = 2.31$ ksi, $\Delta_F = 0.0303$ in.

5–75. $\tau_A = 2.86$ MPa, $\phi = 0.899°$

5–77. $T_{max} = T_{cir} = 0.282A^{3/2}\tau_Y$, 73.7%, 62.2%

5–78. $T_B = 32$ lb · ft, $T_A = 48$ lb · ft, $\phi_C = 0.0925°$

5–79. $\dfrac{1}{k^2}$

5–81. $\tau_{max} = 702$ kPa

5–82. $T = 5.76$ kN · m, $\phi = 8.83°$

5–83. $(\tau_{avg})_{max} = 2.03$ MPa, $\phi = 0.258°$

5–85. $b = 0.773$ in.

5–86. $\tau_A = \tau_B = 602$ kPa, $\phi = 0.0217°$

5–87. $\tau_A = \tau_B = 357$ kPa

5–89. 1.66

5–90. $\tau_{max} = 3.35$ ksi

5–91. $\tau_A = \tau_B = 50$ kPa

5–93. $\tau_{max} = 7.42$ ksi

5–94. $r = 6$ mm

5–95. 90.2 kW

5–97. $\tau_{max} = 33.7$ MPa

5–98. $T = 8.16$ N · m

5–99. $T_p = 0.105$ N · m
5–101. $T = 193$ lb · ft, $\phi = 17.2°$
5–102. $\rho_Y = 1.16$ in.
5–103. $T = 19.2$ kN · m, $\phi = 24.9°$, $\phi_r = 6.72°$
5–105. $T_Y = 1.26$ kN · m, $\phi_Y = 3.58°$, $\phi = 4.86°$
5–106. $T_Y = 1.18$ kN · m, $\phi_Y = 3.58°$, $\phi = 5.02°$
5–107. $T_p = 13.6$ kN · m, $\phi_r = 12.3°$
5–109. 110 lb · ft.
5–110. $T = 3.27$ kN · m, $\phi = 68.8°$
5–111. $\tau_{max} = \dfrac{19T}{12\pi r^3}$
5–113. $T = 331$ lb · ft
5–114. $t = 0.104$ in.
5–115. $T = 0.0820$ N · m, $\phi = 24.2$ rad
5–117. $(\tau_{BC})_{max} = 9.55$ MPa, $(\tau_{AC})_{max} = 14.3$ MPa
5–118. 35.7%
5–119. At $\rho_Y = 18.7$ mm, $\tau = 1.27$ MPa, at $\rho = 20$ mm, $\tau = 9.15$ MPa
5–121. 25%
5–122. $T = 2.71$ kip · ft, $T_p = 2.79$ kip · ft
5–123. $\phi_C = 0.115°$

Chapter 6
6–1. (a) $(\sigma_z)_{max} = 11.3$ psi, (b) $(\sigma_y)_{max} = 22.7$ psi
6–2. $\sigma_{max} = 6.44$ ksi
6–3. $\sigma_A = 115$ psi (C), $\sigma_B = 68.9$ psi (T)
6–5. $\sigma_{max} = 1.35$ ksi
6–6. $F_R = 5.88$ kN
6–7. $\sigma_B = 3.61$ MPa, $\sigma_C = 1.55$ MPa
6–9. $F_{R_A} = 0$, $F_{R_B} = 1.50$ kN
6–10. $\%\left(\dfrac{M'}{M}\right) = 84.6\%$
6–11. $M = 36.5$ kN · m, $\sigma_{max} = 40$ MPa
6–13. $\sigma_{max} = 635$ kPa
6–14. $\sigma_{max} = 11.8$ ksi
6–15. $\sigma_{max} = 12.2$ ksi
6–17. $F_R = 11.8$ kip
6–18. $\sigma_B = 13.3$ ksi, $\sigma_A = 11.8$ ksi
6–19. $w_2 = 800$ lb/in., $w_1 = 533$ lb/in., $\sigma_{max} = 45.1$ ksi
6–21. $\sigma_{max} = 497$ kPa
6–22. $P = 7.29$ kip
6–23. $P = 5.97$ kip
6–25. $\sigma_{max} = 11.5$ MPa

6–26. $w = 1.65$ kip/ft
6–27. $\sigma_{max} = 66.8$ ksi
6–29. $\sigma_A = 0$, $\sigma_B = 462$ kPa, $\sigma_D = -462$ kPa, $\sigma_E = 0$
6–30. $\sigma_A = -119$ kPa, $\sigma_B = 446$ kPa, $\sigma_D = -446$ kPa, $\sigma_E = 119$ kPa
6–31. $\sigma_{max} = 2.90$ MPa, $\alpha = -66.6°$
6–33. $\bar{y} = 4.83$ in., $\sigma_A = -2.35$ ksi, $\sigma_B = -2.62$ ksi
6–34. $\bar{z} = 36.6$ mm, $\sigma_A = 4.38$ MPa, $\sigma_B = -1.13$ MPa, $\alpha = -87.1°$
6–35. $\sigma_A = 164$ psi, $\alpha = -62.1°$, $\sigma_A = 125$ psi
6–38. $\sigma_A = 58.5$ kPa (C)
6–39. $\sigma_A = 2.60$ MPa (T)
6–41. $(\sigma_{br})_{max} = 3.04$ MPa, $(\sigma_{st})_{max} = 4.65$ MPa, $\sigma_{br} = 1.25$ MPa, $\sigma_{st} = 2.51$ MPa
6–42. $(\sigma_{st})_{max} = 3.70$ MPa, $(\sigma_w)_{max} = 0.179$ MPa
6–43. $M = 14.9$ kN · m
6–45. $(\sigma_{st})_{max} = 1.40$ ksi, $(\sigma_w)_{max} = 77.0$ psi
6–46. $\sigma_{max} = 20.1$ MPa
6–49. $\sigma_{max} = 1.16$ ksi (T)
6–50. $\sigma_A = 792$ kPa (C), $\sigma_B = 1.02$ MPa (T)
6–51. $\sigma_{max} = 3.89$ MPa (T)
6–53. $\sigma_A = 144$ psi (T), $\sigma_B = 106$ psi (C), $\sigma_C = 6.11$ psi (C)
6–54. $\sigma_{max} = 842$ psi (T)
6–55. $\sigma_A = 522$ psi (T), $\sigma_B = 458$ psi (C)
6–57. $\sigma_{max} = 204$ psi (T)
6–58. $\sigma_A = 446$ kPa (T), $\sigma_B = 224$ kPa (C), No
6–59. $r = 8.0$ mm
6–61. $\sigma_{max} = 27.0$ MPa
6–62. $L = 950$ mm
6–63. $\sigma_{max} = 768$ psi
6–65. $M = 24.0$ N · m
6–66. $\sigma_{max} = 125$ MPa
6–67. $M = 14.5$ kip · ft
6–69. $P = 105$ kN
6–70. $\sigma_{max} = 129$ MPa
6–71. $M = 8.25$ kip · ft
6–73. $\sigma_{top} = \sigma_{bottom} = 43.5$ MPa
6–74. $K = 1.57$
6–75. $\sigma_t = \sigma_b = 142$ MPa
6–77. $M_Y = 193$ kip · ft, $M_P = 342$ kip · ft
6–78. $K = 1.78$, $Z = 114$ in^3
6–79. $M_Y = 9.44$ kN · m, $M_P = 17.1$ kN · m
6–81. $K = 1.70$, $Z = 36.0$ in^3

6–82. (a) $P = 66.7$ kN, (b) $P = 100$ kN

6–83. (a) $w = 4.27$ kip/ft, (b) $w = 6.40$ kip/ft

6–85. $M = 73.5$ kip · ft

6–86. $M = 81.7$ kip · ft

6–87. (a) $M = 35.0$ kip · ft, (b) $M_P = 59.8$ kip · ft

6–89. $M = 1.47$ MN · m

6–90. $(\sigma_{max})_c = 1.39$ MPa, $(\sigma_{max})_t = 1.39$ MPa

6–91. $\sigma_{max} = 5.0$ ksi, $F_A = 17.7$ kip, $F_B = 13.7$ kip

6–93. $M = 1.25$ kip · ft

6–94. $(\sigma_{max})_t = 18.4$ ksi, $(\sigma_{max})_c = 19.2$ ksi

6–95. $\sigma_{max} = 2.06$ MPa

6–97. $\sigma_r = 25.1$ ksi

6–98. $M_P = 172$ kip · ft

Chapter 7

7–1. $\tau_A = 1.99$ MPa, $\tau_B = 1.65$ MPa

7–2. $\tau_A = \tau_B = 108$ psi

7–3. (a) $\tau_{max} = 2.29$ ksi, (b) $\tau_{avg} = 2.00$ ksi

7–5. $\tau_{max} = 276$ psi, shear-stress jump $= 156$ psi

7–6. $\tau_{max} = 2.44$ MPa

7–7. $V = 307$ kN

7–9. $\tau_B = 795$ psi, $\tau_C = 596$ psi

7–10. $\tau_{max} = 928$ psi

7–11. $\tau_B = 4.41$ MPa

7–13. $V = 9.96$ kip

7–14. $V = 131$ kN

7–15. $V = 50.3$ kN

7–18. $V = 32.1$ kip

7–19. $\tau_{max} = 4.48$ ksi

7–21. $\tau = \dfrac{2V}{bh^3}(h^2 + 2yh - 8y^2)$

7–22. $a = 1.27$ in.

7–23. $\tau_{max} = 281$ psi

7–25. $\tau_{max} = 1.20$ ksi

7–26. $\tau_{max} = 1.05$ MPa

7–29. $\tau_{avg} = 97.2$ MPa

7–30. $s_t = 1.42$ in., $s_b = 1.69$ in.

7–31. $V = 4.97$ kip, $s_t = 1.14$ in., $s_b = 1.36$ in.

7–33. $F_C = 197$ lb, $F_D = 1.38$ kip, $\tau_{max} = 495$ psi

7–34. $s' = 1.49$ in., $s = 9.88$ in.

7–35. $s = 5.53$ in.

7–37. (a) $V = 4.10$ kip, (b) $V = 749$ lb

7–38. $q_A = 200$ kN/m

7–39. $q_A = 145$ kN/m, $q_B = 50.4$ kN/m, $\tau_{max} = 17.2$ MPa

7–41. (a) 12.6 kN/m, (b) 22.5 kN/m

7–42. $q_A = 0$, $q_B = 1.21$ kN/m, $q_C = 3.78$ kN/m

7–43. $q_{max} = 8.89$ kN/m

7–45. $q_A = 196$ lb/in., $q_B = 452$ lb/in., $q_{max} = 641$ lb/in.

7–46. $q_A = 0$, $q_B = 375$ N/m, $q_C = 0$

7–47. $q_B = 417$ lb/in., $q_A = 0$

7–49. $\tau = \dfrac{V}{\pi R^2 t}(\sqrt{R^2 - y^2})$

7–50. $e = 2.25$ in.

7–51. $e = \dfrac{3(b_2^2 - b_1^2)}{h + 6b_1 + 6b_2}$

7–53. $h = 171$ mm

7–54. $e = 1.07$ in.

7–55. $e = \dfrac{3[h^2 b^2 - (h - 2h_1)^2 b_1^2]}{h^3 + 6bh^2 + 6b_1(h - 2h_1)^2}$

7–57. $e = \dfrac{1.5b^2}{d + 3b}$

7–58. $e = \dfrac{4r}{\pi} - r$

7–59. $e = \dfrac{4r}{2\alpha - \sin 2\alpha}(\sin \alpha - \alpha \cos \alpha)$

7–61. $\tau = \dfrac{4V(c^2 - y^2)}{3\pi c^4}$

7–62. $e = 70.0$ mm

7–63. $V = 317$ lb

7–65. $e = \dfrac{b(6h_1 h^2 + 3h^2 b - 8h_1^3)}{(h + 2h_1)^3 + 6bh^2}$

7–66. 256 lb/in. at top and bottom, 612 lb/in. at center

7–67. $e = \dfrac{b(6h_1 h^2 + 3h^2 b - 8h_1^3)}{2h^3 + 6bh^2 - (h - 2h_1)^3}$

7–69. $V = 28.8$ kip

Chapter 8

8–1. $t = 18.8$ mm

8–2. $\sigma_1 = 209$ psi, $\sigma_2 = 0$

8–3. $p = 3.60$ MPa, $n = 113$

8–5. (a) $\sigma_1 = 127$ MPa, (b) $\sigma_1 = 63.3$ MPa, $\tau_{avg} = 322$ MPa

8–6. $\sigma_h = 432$ psi, $\sigma_b = 8.80$ ksi

8–7. $s = 32.7$ in.

8–9. $\sigma_{max} = 1.07$ MPa

8–10. $(\sigma_{max})_c = 1.12$ ksi, $(\sigma_{max})_t = 1.28$ ksi

8–11. $w = 79.7$ mm

8–13. $(\sigma_{AB})_{max} = 667$ psi, $(\sigma_{CD})_{max} = 40.7$ ksi

8–14. $\theta = 0.286°$

8–15. $\theta = 0.215°$

8–17. $y = 0.750 - 1.50x$

8–18. $\sigma_A = 9.88$ kPa, $\sigma_B = -49.4$ kPa, $\sigma_C = -128$ kPa, $\sigma_D = -69.1$ kPa

8–19. $\sigma_A = -21.3$ psi, $\sigma_B = -12.2$ psi

8–21. $\sigma_A = 0.398$ MPa, $\sigma_B = -1.19$ MPa

8–22. $\sigma_F = -30.7$ MPa, $\tau_F = 0$, $\sigma_E = -1.23$ MPa, $\tau_E = 3.09$ MPa

8–23. $\sigma_A = 444$ kPa, $\tau_A = 130$ kPa

8–25. $\sigma_C = -62.5$ psi, $\tau_C = 162$ psi, $\sigma_B = 5.56$ ksi, $\tau_B = 0$

8–26. $\sigma_A = 15.3$ MPa, $\tau_A = 0$, $\sigma_B = 0$, $\tau_B = 0.637$ MPa

8–27. $\sigma_A = -9.41$ ksi, $\tau_A = 0$, $\sigma_B = 2.69$ ksi, $\tau_B = 0.869$ ksi

8–29. $F = 2.15$ kip

8–30. $\sigma_D = 576$ psi, $\tau_D = 0$, $\sigma_E = 0$, $\tau_E = 9.60$ psi

8–31. $(\sigma_A)_y = 16.2$ ksi, $(\tau_A)_{yx} = 2.84$ ksi, $(\tau_A)_{yz} = 0$

8–33. $\tau_A = 0$, $\sigma_A = -30.2$ ksi, $\sigma_B = 0$, $\tau_B = 0.377$ ksi

8–34. $e_y = \dfrac{r}{4}$

8–35. $(\sigma_{max})_t = 15.8$ ksi, $(\sigma_{max})_c = 9.47$ ksi

8–37. $(\sigma_D)_y = -178$ psi, $(\sigma_E)_y = 9.78$ psi

8–38. $(\sigma_D)_y = -126$ psi, $(\tau_D)_{yx} = 57.2$ psi, $(\sigma_E)_y = -347$ psi, $(\tau_E)_{yz} = 66.4$ psi

8–39. $(\sigma_D)_z = 124$ ksi, $(\tau_D)_{zx} = 62.4$ ksi, $(\sigma_C)_z = 15.6$ ksi, $(\tau_C)_{zy} = 52.4$ ksi

8–41. $\sigma_A = 94.4$ psi, $\sigma_B = -59.0$ psi

8–42. $(\sigma_{max})_t = 12.6$ ksi, $(\sigma_{max})_c = 19.3$ ksi

8–45. $(\sigma_{max})_t = 106$ MPa, $(\sigma_{max})_c = 159$ MPa

8–46. $(\sigma_{max})_t = 228$ MPa, $(\sigma_{max})_c = 168$ MPa

Chapter 9

9–2. $\sigma_{x'} = -388$ psi, $\tau_{x'y'} = 455$ psi

9–3. $\sigma_{x'} = -4.05$ ksi, $\tau_{x'y'} = -0.404$ ksi

9–5. $\sigma_{x'} = -388$ psi, $\tau_{x'y'} = 455$ psi

9–6. $\sigma_{x'} = -4.05$ ksi, $\tau_{x'y'} = -0.404$ ksi

9–7. $\sigma_{x'} = 49.7$ MPa, $\tau_{x'y'} = -34.8$ MPa

9–9. $\sigma_{x'} = -898$ psi, $\tau_{x'y'} = 605$ psi, $\sigma_{y'} = 598$ psi

9–10. $\sigma_{x'} = -19.9$ ksi, $\tau_{x'y'} = 7.70$ ksi, $\sigma_{y'} = 9.89$ ksi

9–11. $\sigma_{x'} = 0.507$ MPa, $\tau_{x'y'} = 0.958$ MPa

9–13. (a) $\sigma_1 = 5.88$ ksi, $\sigma_2 = -10.9$ ksi, (b) $\tau_{max}^{in\text{-}plane} = 8.38$ ksi, $\sigma_{avg} = -2.5$ ksi, $\theta_{p_1} = 53.7°$, $\theta_{p_2} = -36.3°$, $\theta_s = 8.68°$ or $98.7°$

9–14. (a) $\sigma_1 = -0.875$ ksi, $\sigma_2 = -33.1$ ksi, (b) $\tau_{max}^{in\text{-}plane} = 16.1$ ksi, $\sigma_{avg} = -17.0$ ksi, $\theta_{p_1} = 59.9°$, $\theta_{p_2} = -30.1°$, $\theta_s = 14.9°$ or $-75.1°$

9–15. (a) $\sigma_1 = 265$ MPa, $\sigma_2 = -84.9$ MPa, (b) $\tau_{max}^{in\text{-}plane} = 175$ MPa, $\sigma_{avg} = 90.0$ MPa, (c) $\theta_{p_2} = 60.5°$, $\theta_{p_1} = -29.5°$, $\theta_s = 15.5°$ or $-74.5°$

9–17. (a) $\sigma_1 = 4.21$ ksi, $\sigma_2 = -34.2$ ksi, (b) $\tau_{max}^{in\text{-}plane} = 19.2$ ksi, $\sigma_{avg} = -15$ ksi, $\theta_{p_2} = 19.3°$, $\theta_{p_1} = -70.7°$, $\theta_s = 25.7°$ or $64.3°$

9–18. $\tau_{max}^{in\text{-}plane} = 5$ kPa, $\sigma_{avg} = 0$

9–19. $\sigma_1 = 32$ psi, $\sigma_2 = -32$ psi

9–21. $\sigma_1 = 0$, $\sigma_2 = -20$ kPa, $\tau_{max}^{in\text{-}plane} = 10$ kPa

9–22. $\sigma_1 = 167$ kPa, $\sigma_2 = -21.6$ kPa, $\tau_{max}^{in\text{-}plane} = 94.1$ kPa

9–23. $\sigma_1 = 74.6$ kPa, $\sigma_2 = -27.1$ kPa, $\tau_{max}^{in\text{-}plane} = 50.9$ kPa

9–25. point A: $\sigma_1 = 0$, $\sigma_2 = -192$ MPa, point B: $\sigma_1 = 24.0$ MPa, $\sigma_2 = -24.0$ MPa

9–26. point A: $\sigma_1 = 1.50$ ksi, $\sigma_2 = -0.0235$ ksi, point B: $\sigma_1 = 0.0723$ ksi, $\sigma_2 = -0.683$ ksi

9–27. $\sigma_1 = 0$, $\sigma_2 = -85.7$ MPa

9–29. point D: $\sigma_1 = 1.03$ MPa, $\sigma_2 = -0.00919$ MPa, point E: $\sigma_1 = 0.0257$ MPa, $\sigma_2 = -0.570$ MPa

9–30. point A: $\sigma_1 = 111$ MPa, $\sigma_2 = 0$, point B: $\sigma_1 = 2.40$ MPa, $\sigma_2 = -6.68$ MPa

9–31. point A: $\sigma_1 = 0.0378$ MPa, $\sigma_2 = -10.8$ MPa, point B: $\sigma_1 = 42.0$ MPa, $\sigma_2 = -0.0106$ MPa

9–33. point A: $\sigma_1 = 48.8$ ksi, $\sigma_2 = -25.4$ ksi, point B: $\sigma_1 = 34.7$ ksi, $\sigma_2 = -34.7$ ksi

9–34. point A: $\sigma_1 = 5.50$ MPa, $\sigma_2 = -0.611$ MPa, point B: $\sigma_1 = 1.29$ MPa, $\sigma_2 = -1.29$ MPa

9–35. $\sigma_1 = 233$ psi, $\sigma_2 = -774$ psi, $\tau_{max}^{in\text{-}plane} = 503$ psi

9–37. point A: $\sigma_1 = 61.7$ psi, $\sigma_2 = 0$, point B: $\sigma_1 = 0$, $\sigma_2 = -46.3$ psi

9–38. $M = 81.8$ lb · ft, $T = 231$ lb · ft

9–39. $\sigma_{x'} = -388$ psi, $\tau_{x'y'} = 455$ psi

9–41. $\sigma_1 = 265$ MPa, $\sigma_2 = -84.9$ MPa, $\tau_{max}^{in\text{-}plane} = 175$ MPa, $\sigma_{avg} = 90$ MPa, $\theta_p = 29.5°\downarrow$, $\theta_s = 15.5°\nwarrow$

9–42. $\sigma_{x'} = -22.8$ psi, $\sigma_{y'} = -27.2$ psi, $\tau_{x'y'} = -896$ psi

9–43. $\sigma_{x'} = -0.611$ ksi, $\sigma_{y'} = -3.39$ ksi, $\tau_{x'y'} = 7.88$ ksi

9–45. $\sigma_1 = 12.3$ ksi, $\sigma_2 = -17.3$ ksi, $\tau_{max}^{in\text{-}plane} = 14.8$ ksi, $\sigma_{avg} = -2.50$ ksi

9–46. $\sigma_1 = 64.1$ MPa, $\sigma_2 = -14.1$ MPa, $\tau_{max}^{in\text{-}plane} = 39.1$ MPa, $\sigma_{avg} = 25$ MPa

9–47. $\sigma_1 = -5.53$ ksi, $\sigma_2 = -14.5$ ksi, $\tau_{\max_{\text{in-plane}}} = 4.47$ ksi, $\sigma_{\text{avg}} = -10$ ksi

9–50. $\sigma_x = 74.4$ psi, $\sigma_y = -42.4$ psi, $\tau_{xy} = 22.7$ psi

9–53. $\sigma_1 = 0.129$ psi, $\sigma_2 = -121$ psi

9–54. point A: $\sigma_2 = -1.20$ ksi, $\sigma_1 = 0$, point B: $\sigma_1 = 9.88$ psi, $\sigma_2 = -43.1$ psi

9–55. $\sigma_1 = 13.9$ ksi, $\sigma_2 = -9.52$ ksi

9–57. The shear stress is zero for any orientation of the element

9–58. $\sigma_{x'} = 500$ MPa, $\tau_{x'y'} = 167$ MPa

9–61. $\sigma_1 = 75.4$ psi, $\sigma_2 = -95.4$ psi, $\sigma_3 = -120$ psi, $\tau_{\text{abs}_{\max}} = 97.7$ psi

9–62. $\sigma_1 = 150$ MPa, $\sigma_2 = 137$ MPa, $\sigma_3 = -46.8$ MPa, $\tau_{\text{abs}_{\max}} = 98.4$ MPa

9–63. $\sigma_1 = 6.73$ ksi, $\sigma_2 = 0$, $\sigma_3 = -4.23$ ksi, $\tau_{\text{abs}_{\max}} = 5.48$ ksi

9–65. $M = 8.73$ kip $\cdot$ in.

9–66. (a) $\sigma_1 = 264$ MPa, $\sigma_2 = -464$ MPa, $\theta_{p_1} = -37.0°$, $\theta_{p_2} = 53.0°$, (b) $\tau_{\max_{\text{in-plane}}} = 364$ MPa, $\sigma_{\text{avg}} = -100$ MPa, $\theta_s = 7.97°$ or $-82.0°$

9–67. $(\tau_{xy})_{\max} = 40$ MPa, $\sigma_{\text{avg}} = 0$; $(\tau_{yz})_{\max} = 0$, $\sigma_{\text{avg}} = 40$ MPa; $\tau_{xz} = 40$ MPa, $\sigma_{\text{avg}} = 0$

9–69. $\sigma_1 = 3.17$ ksi, $\sigma_2 = -3.86$ ksi, $\tau_{\max_{\text{in-plane}}} = 3.51$ ksi

9–70. $\sigma_1 = 4.00$ ksi, $\sigma_2 = -0.0317$ ksi

9–71. $\sigma_{x'} = 736$ MPa, $\sigma_{y'} = -156$ MPa, $\tau_{x'y'} = 188$ MPa

Chapter 10

10–2. $\epsilon_1 = 283(10^{-6})$, $\epsilon_2 = -133(10^{-6})$, $\gamma_{\max_{\text{in-plane}}} = 417(10^{-6})$, $\epsilon_{\text{avg}} = 75.0(10^{-6})$, $\theta_{p_1} = 84.8°$, $\theta_{p_2} = -5.18°$, $\theta_s = 39.8°$ or $130°$

10–3. $\epsilon_{x'} = -380(10^{-6})$, $\epsilon_{y'} = -130(10^{-6})$, $\gamma_{x'y'} = 1.21(10^{-3})$

10–5. $\epsilon_1 = 1.04(10^{-3})$, $\epsilon_2 = 291(10^{-6})$, $\gamma_{\max_{\text{in-plane}}} = 748(10^{-6})$, $\epsilon_{\text{avg}} = 665(10^{-6})$, $\theta_{p_1} = 30.2°$, $\theta_{p_2} = 120°$, $\theta_s = -14.8°$ or $75.2°$

10–6. $\epsilon_{x'} = 103(10^{-6})$, $\epsilon_{y'} = 46.7(10^{-6})$, $\gamma_{x'y'} = 718(10^{-16})$

10–7. $\epsilon_{x'} = -309(10^{-6})$, $\epsilon_{y'} = -541(10^{-6})$, $\gamma_{x'y'} = -423(10^{-6})$

10–9. $\epsilon_{x'} = -380(10^{-6})$, $\epsilon_{y'} = -130(10^{-6})$, $\gamma_{x'y'} = 1.21(10^{-3})$

10–10. $\epsilon_{x'} = 103(10^{-6})$, $\epsilon_{y'} = 46.7(10^{-6})$, $\gamma_{x'y'} = 718(10^{-6})$

10–11. $\epsilon_1 = 1.04(10^{-3})$, $\epsilon_2 = 291(10^{-6})$, $\gamma_{\max_{\text{in-plane}}} = 748(10^{-6})$, $\epsilon_{\text{avg}} = 665(10^{-6})$, $\theta_s = 14.8°$ ↙, $\theta_{p_1} = 30.2°$ ↖

10–13. $\epsilon_{x'} = -309(10^{-6})$, $\epsilon_{y'} = -541(10^{-6})$, $\gamma_{x'y'} = -423(10^{-6})$

10–14. (a) $\epsilon_1 = 482(10^{-6})$, $\epsilon_2 = 168(10^{-6})$, (b) $\gamma_{\max_{\text{in-plane}}} = 313(10^{-6})$, (c) $\gamma_{\text{abs}_{\max}} = 482(10^{-6})$

10–15. (a) $\epsilon_1 = 773(10^{-6})$, $\epsilon_2 = 76.8(10^{-6})$, (b) $\gamma_{\max_{\text{in-plane}}} = 696(10^{-6})$, (c) $\gamma_{\text{abs}_{\max}} = 773(10^{-6})$

10–17. (a) $\epsilon_1 = 946(10^{-6})$, $\epsilon_2 = 254(10^{-6})$, (b) $\gamma_{\max_{\text{in-plane}}} = 693(10^{-6})$, (c) $\gamma_{\text{abs}_{\max}} = 946(10^{-6})$

10–18. $\gamma_{\text{abs}_{\max}} = 22.4(10^{-6})$

10–19. (a) $\epsilon_1 = 192(10^{-6})$, $\epsilon_2 = -152(10^{-6})$, (b,c) $\gamma_{\text{abs}_{\max}} = \gamma_{\max_{\text{in-plane}}} = 344(10^{-6})$

10–21. (a) $\epsilon_1 = 487(10^{-6})$, $\epsilon_2 = -400(10^{-6})$, (b) $\epsilon_{\text{avg}} = 43.3(10^{-6})$, $\gamma_{\max_{\text{in-plane}}} = 887(10^{-6})$

10–22. $\epsilon_1 = 1.05(10^{-3})$, $\epsilon_2 = -306(10^{-6})$, $\theta_{p_1} = 15.4°$ ↙

10–23. $\epsilon_1 = 889(10^{-6})$, $\epsilon_2 = -539(10^{-6})$, $\theta_{p_1} = 30.6°$ ↙

10–25. (a) $K = 3.33$ ksi, (b) $K = 5.13$ ksi

10–26. $a' = 100.248$ mm, $b' = 100.514$ mm, $t' = 1.989$ mm

10–27. $a' = 99.685$ mm, $b' = 100.750$ mm, $t' = 1.9937$ mm

10–29. $\epsilon_{\max} = 0.903(10^{-3})$, $\epsilon_{\text{int}} = -0.0828(10^{-3})$, $\epsilon_{\min} = -0.710(10^{-3})$

10–30. $E = 14.9(10^3)$ ksi, $e = 0.282(10^{-3})$

10–31. $\nu = 0.291$, $E = 30.7(10^3)$ ksi

10–33. $\epsilon_{\max} = 5.28(10^{-3})$, $\epsilon_{\text{int}} = 1.36(10^{-3})$, $\epsilon_{\min} = -5.92(10^{-3})$

10–35. $\sigma_1 = 10.2$ ksi, $\sigma_2 = 7.38$ ksi, $\sigma_3 = 0$

10–37. $p = 3.43$ MPa, $\tau_{\text{abs}_{\max}} = 85.7$ MPa

10–38. $P = 86.9$ lb

10–39. $\epsilon_{\max} = 250(10^{-6})$, $\epsilon_{\text{int}} = 0$, $\epsilon_{\min} = -250(10^{-6})$

10–41. $\Delta L_{\text{top}} = 1.2(10^{-3})$ in., $\Delta L_{\text{bottom}} = -1.2(10^{-3})$ in.

10–42. $\gamma_{\max_{\text{in-plane}}} = \dfrac{Pr}{2tE}(1 + \nu)$, $\gamma_{\text{abs}_{\max}} = \dfrac{Pr}{tE}(1 + \nu)$

10–43. $\Delta d = 0.800$ mm, $\sigma_{AB} = 315$ MPa

10–45. $p = 0.967$ ksi, $\gamma_{\max_{\text{in-plane}}} = 1.30(10^{-3})$

10–46. $\gamma_{xy} = -160(10^{-6})$, $\epsilon_x = \epsilon_y = 0$, $T = 65.2$ N $\cdot$ m

10–47. $T = 25.2$ kip $\cdot$ ft

10–49. $\dfrac{dV}{V} = \dfrac{Pr}{Et}(2.5 - 2\nu)$

10–50. $t_h = 0.206$ in.

10–51. $(\sigma_x - \sigma_y)^2 + 4\tau_{xy}^2 = \sigma_Y^2$

10–53. No yielding occurs for both cases

10–54. Yielding occurs for both cases

10–55. $\sigma_2 = 37.6$ ksi or 4.36 ksi

10–57. $d = 2.40$ in.

10–58. material fails

10–61. $F.S. = 5.38$

10–62. $M_e = \sqrt{M^2 + \dfrac{3}{4}\,T^2}$

10–63. (a) $\sigma_1 = 10.2$ ksi, (b) $\sigma_1 = 11.6$ ksi

10–65. (a) $F.S. = 1.59$, (b) $F.S. = 1.80$

10–66. $d = 1.88$ in.

10–67. $d = 1.59$ in.

10–69. $\sigma_Y = 18.2$ ksi

10–70. $\sigma_Y = 19.7$ ksi

10–71. OK

10–73. $d = 1.45$ in.

10–74. $\epsilon_{cir} = \dfrac{Pr}{2Et}\,(2-\nu)$, $\epsilon_{long} = \dfrac{Pr}{2Et}\,(1-2\nu)$,

$\Delta d = 3.19$ mm, $\Delta L = 2.25$ mm

10–75. $\Delta V = 0.0168$ m^3

10–77. $T_e = \sqrt{\dfrac{4}{3}\,M^2 + T^2}$

10–78. $\epsilon_1 = 996(10^{-6})$, $\epsilon_2 = 374(10^{-6})$, $\gamma_{\substack{max \\ in\text{-}plane}} = 622(10^{-6})$, $\epsilon_{avg} = 685(10^{-6})$, $\theta_{p_1} = -15.8°$, $\theta_{p_2} = 74.2°$, $\theta_s = 29.2°$ or $119°$

10–79. $\epsilon_1 = 996(10^{-6})$, $\epsilon_2 = 374(10^{-6})$, $\gamma_{\substack{max \\ in\text{-}plane}} = 622(10^{-6})$, $\epsilon_{avg} = 685(10^{-6})$, $\theta_{p_1} = 15.8°\ \downarrow$, $\theta_s = 29.2°\ \nwarrow$

10–81. $\epsilon_x = 2.35(10^{-3})$, $\epsilon_y = -0.972(10^{-3})$, $\epsilon_z = -2.44(10^{-3})$

10–82. $\epsilon_{x'} = -116(10^{-6})$, $\epsilon_{y'} = 466(10^{-6})$, $\gamma_{x'y'} = 393(10^{-6})$

10–83. $\epsilon_{x'} = -116(10^{-6})$, $\epsilon_{y'} = 466(10^{-6})$, $\gamma_{x'y'} = 393(10^{-6})$

10–85. (a) $\epsilon_1 = 862(10^{-6})$, $\epsilon_2 = -782(10^{-6})$, $\theta_{p_2} = 2.01°\ \downarrow$,
(b) $\epsilon_{avg} = 40(10^{-6})$, $\gamma_{\substack{max \\ in\text{-}plane}} = 1.64(10^{-3})$, $\theta_s = 43.0°\ \nwarrow$

Chapter 11

11–1. $x = 0$, $V = P$, $M = 0$; $x = a^+$, $V = 0$, $M = Pa$

11–2. $x = 0$, $V = 4$, $M = 0$; $x = 4^+$, $V = 2$, $M = 16$
$x = 8^+$, $V = 0$, $M = 24$

11–3. $x = 0$, $V = -24$, $M = 0$; $x = 0.25^-$, $V = 7.5$, $M = -6.0$

11–5. $x = 0$, $V = 217$, $M = 0$; $x = 0.2^+$, $V = -233$, $M = 43.3$; $x = 0.6^+$, $V = 66.7$, $M = -50$; $x = 0.9^+$, $V = 150$, $M = -30$

11–6. $\sigma_{max} = 142$ MPa, $V = 15.6$ N,
$M = (15.6x + 100)$ N $\cdot$ m

11–7. $x = 0$, $V = 18$, $M = -75$; $x = 2^+$, $V = 8$,
$M = -39$; $x = 5$, $V = 8$, $M = -15$

11–9. $P = 809$ lb, $C_x = 404$ lb, $C_y = 1.50$ kip; $x = 0$,
$V = -701$, $M = 0$; $x = 8^-$, $V = -701$, $M = -5604$;
$x = 8^+$, $V = 800$, $M = -5200$; $x = 13$, $V = 800$,
$M = -1200$

11–10. $x = 0$, $V = 11.7$, $M = 0$; $x = 4^+$, $V = -3.33$,
$M = 46.7$; $x = 8^-$, $V = -3.33$, $M = 33.3$; $x = 8^+$,
$V = -3.33$, $M = 13.3$

11–11. $x = 0$, $V = 343$, $M = 0$; $x = 18^+$, $V = -557$,
$M = 6171$, $x = 42^+$, $V = 600$, $M = -7200$, $\sigma_{max} = 73.3$ ksi

11–13. $V = 17.7 - 1.5x$, $M = -0.75x^2 + 17.7x - 96.25$

11–14. $x = 0$, $V = 0$, $M = 51200$; $x = 8$, $V = -6400$,
$M = 25600$

11–15. $M_{max} = 281$ lb $\cdot$ ft

11–17. $x = 0$, $V = 0$, $M = 0$; $x = 5^-$, $V = -10$, $M = -25$;
$x = 5^+$, $V = -0.5$, $M = -25$; $x = 10^-$, $V = -0.5$,
$M = -27.5$; $x = 10^+$, $V = -0.5$, $M = 2.5$

11–18. $V = 20 - 2x$, $M = -x^2 + 20x - 166$

11–19. $V = 1050 - 150x$, $M = -75x^2 + 1050x - 3200$

11–21. (a) $x = 0$, $V = 7.5$, $M = 0$; $x = 2.75^+$, $V = -7.5$,
$M = 20.6$; (b) $x = 0$, $V = 7.5$, $M = 0$; $x = 2$, $V = 7.5$, $M = 15$; $x = 2.75$, $V = 0$, $M = 17.8$

11–22. $x = 0$, $V = -2$, $M = 0$; $x = 6^+$, $V = 1.57$, $M = -12$; $(\sigma_{max})_t = 2.87$ ksi (T), $(\sigma_{max})_c = 5.16$ ksi (C)

11–23. For $0 \le x < L/2$, $V = \dfrac{3w_0L}{4} - w_0x$,

$M = \dfrac{-w_0}{2}x^2 + \dfrac{3w_0L}{4}x - \dfrac{7w_0L^2}{24}$;

For $L/2 < x \le L$, $V = \dfrac{w_0(L-x)^2}{L}$, $M = -\dfrac{w_0(L-x)^3}{3L}$

11–25. $x = 0$, $V = -1.625$, $M = 1.5$; $x = 12^+$, $V = 9.0$,
$M = -18$

11–26. $w_0 = 1.2$ kN/m, $x = 0$, $V = 0$, $M = 0$; $x = 0.02$,
$V = 12$, $M = 0.08$; $x = 0.05$, $V = 0$, $M = 0.26$

11–27. $x = 0$, $V = 112.5$, $M = 0$; $x = 4.5$, $V = 0$, $M = 169$

11–29. $x = 0$, $V = w_0L/12$, $M = 0$; $x = 0.630L$, $V = 0$,
$M = 0.0394\,w_0L^2$

11–30. $x = 0$, $V = 21.3$, $M = -128$; $x = 8$, $V = 0$, $M = 0$

11–31. $\sigma_{max} = 581$ psi

11–33. $x = 0$, $V = w_0L/3$, $M = 0$; $x = L/3$, $V = w_0L/6$, $M = 5w_0L^2/54$; $x = L/2$, $V = 0$, $M = 23w_0L^2/216$

11–34. $x = 0$, $V = 2w_0L/\pi$, $M = -w_0L^2/\pi$

11–35. $a = \dfrac{L}{\sqrt{2}}$; $x = 0$, $V = 0.293wL$, $M = 0$;
$x = 0.293L$, $V = 0$, $M = 0.0429wL^2$; $x = 0.707L^-$,

$V = -0.414wL$, $M = -0.0429wL^2$; $x = 0.707L^+$,
$V = 0.293wL$, $M = -0.0429wL^2$

11–37. $b = 211$ mm

11–38. $b = 2.71$ in.

11–39. $b = 3.40$ in.

11–41. $b = 15.5$ in.

11–42. $P = 3.20$ kip, $h = 8.0$ in.

11–43. $W\ 16 \times 31$

11–45. $W\ 24 \times 62$

11–46. $W\ 18 \times 50$

11–47. $w = 1.66$ kip/ft

11–49. $P = 2.90$ kN

11–50. $P = 9.52$ kN

11–51. $P = 6.67$ kip, $h = 7.14$ in.

11–53. $s = 26.8$ mm, $s' = 10.7$ mm

11–54. $s = 1.00$ in., $s' = 0.80$ in., $s'' = 4.00$ in.

11–55. Segment AB: $W\ 16 \times 26$, Segment BC: $W\ 12 \times 14$

11–57. $P = 178$ lb, $s = 12.0$ in.

11–58. $w = \dfrac{w_0}{L} x$

11–59. $d = d_0 \sqrt{\dfrac{x}{L}}$

11–61. $y = \left(\dfrac{4P}{\pi\sigma_{\text{allow}}} x \right)^{\frac{1}{3}}$

11–62. $\sigma_{\max} = \dfrac{3PL}{8b_0h_0^2}$

11–63. $\sigma_{\max} = \dfrac{wL^2}{4b_0h_0^2}$

11–65. $x = 0$, $\sigma_{\max} = 0.566$ ksi

11–66. $x = 40$ mm, $\sigma_{\max} = 4.20$ MPa

11–67. $d = 1\frac{5}{8}$ in.

11–69. $d = 36$ mm

11–70. $d = 1\frac{1}{4}$ in.

11–71. $d = 23$ mm

11–73. $d = 29$ mm

11–74. $\sigma_{\max} = 61.5$ ksi

11–75. $x = 0$, $V = -800$, $M = 0$; $x = 6^-$, $V = -800$,
$M = -4800$; $x = 6^+$, $V = 400$, $M = -1600$

11–77. $x = L$, $\sigma_{\max} = \dfrac{3PL}{t^2b_2}$

11–78. $0 \le x < 3$, $V = 200$, $M = 200x$; $3 < x \le 6$,
$V = 500 - \dfrac{100}{3} x^2$, $M = -\dfrac{100}{9} x^3 + 500x - 600$

11–79. $d = 44$ mm

11–81. $W\ 12 \times 16$

11–82. $P = 13.3$ kN, $s = 60$ mm

Chapter 12

12–1. $\sigma = 3.02$ ksi

12–2. $\sigma = 75.5$ ksi

12–3. $\theta_A = -\dfrac{PL^2}{16EI}$, $v_{\max} = -\dfrac{PL^3}{48EI}$, $v = \dfrac{Px}{48EI}(4x^2 - 3L^2)$

12–5. $\rho = 336$ ft, $v_B = -\dfrac{M_0L^2}{2EI}$

12–6. $\theta_A = \dfrac{Pa(a - L)}{2EI}$, $v_1 = \dfrac{Px_1}{6EI}[x_1^2 + 3a(a - L)]$,
$v_2 = \dfrac{Pa}{6EI}[3x(x - L) + a^2]$,
$v_{\max} = \dfrac{Pa}{24EI}(4a^2 - 3L^2)$

12–7. $v_1 = \dfrac{wb}{6aEI}[x_1^3 - a^2x_1]$,
$v_2 = \dfrac{w}{6EI}[-x_2^3 + (2ab + 3b^2)x_2 - 2(b^3 + ab^2)]$

12–9. $v_1 = \dfrac{1}{EI}(4.44x_1^3 - 640x_1)$ lb $\cdot$ in^3,
$v_2 = \dfrac{1}{EI}(-4.44x_2^3 + 640x_2)$ lb $\cdot$ in^3

12–10. $\theta_{\max} = -\dfrac{M_0L}{EI}$, $v = -\dfrac{M_0x^2}{2EI}$, $v_{\max} = -\dfrac{M_0L^2}{2EI}$

12–11. $\theta_A = -\dfrac{M_0L}{3EI}$, $\theta_B = \dfrac{M_0L}{6EI}$,
$v = \dfrac{M_0}{6EIL}(3Lx^2 - x^3 - 2L^2x)$,
$v_{\max} = \dfrac{-0.0642\ M_0L^2}{EI}$

12–13. $\theta_A = \dfrac{3wa^3}{16EI}$, $v_1 = \dfrac{wx_1}{48EI}[6ax_1^2 - 2x_1^3 - 9a^3]$,
$v_2 = \dfrac{wax_2}{48EI}[2x_2^2 - 7a^2]$, $v_C = -\dfrac{5wa^4}{48EI}$

12–14. $\theta_A = \dfrac{wL^3}{6EI}$, $v_1 = \dfrac{w}{24EI}(-x_1^4 + 8L^3x_1 - 7L^4)$,
$v_2 = \dfrac{wL}{12EI}(L^2x_2 - x_2^3)$, $v_{\max} = \dfrac{7wL^4}{24EI}$

12–15. $v_{\max} = \dfrac{wL^4}{18\sqrt{3}EI}$

12–17. $v_{max} = -\dfrac{18.8 \text{ kip} \cdot \text{ft}^3}{EI}$

12–18. $v = \dfrac{w_0}{120EIL}(-x^5 + 5L^4x - 4L^5)$, $v_A = -\dfrac{w_0L^4}{30EI}$,

$\theta_A = \dfrac{w_0L^3}{24EI}$

12–19. $\theta_A = 0.0611$ rad, $v_A = 3.52$ in.

12–21. $F = 1.375$ N

12–22. $\sigma_{max} = 1.56$ ksi

12–23. $v_B = -\dfrac{PL^3}{2EI_0}$

12–25. $v = \dfrac{1}{EI}[-2.5x^2 + 2<x-4>^3 - \frac{1}{8}<x-4>^4 + 2<x-12>^3 + \frac{1}{8}<x-12>^4 - 24x + 136]$ kip $\cdot$ ft^3

12–26. $\dfrac{dv}{dx} = \dfrac{1}{EI}[2.25x^2 - 0.5x^3 + 5.25<x-5>^2 + 0.5<x-5>^3 - 3.125]$ kN $\cdot$ m^2, $v = \dfrac{1}{EI}[0.75x^3 - 0.125x^4 + 1.75<x-5>^3 + 0.125<x-5>^4 - 3.125x]$ kN $\cdot$ m^3

12–27. $v = \dfrac{1}{EI}[-0.00556x^5 + 12.9<x-9>^3 + 0.00556<x-9>^5 - 256x + 2637]$ kip $\cdot$ ft^3

12–29. $v = \dfrac{1}{EI}\left[-31.5x^2 + \dfrac{8}{3}x^3 - \dfrac{x^4}{12} + \dfrac{1}{12}<x-3>^4 - \dfrac{2}{3}<x-4.5>^3\right]$ kN $\cdot$ m^3, $\theta_B = -0.705°$, $v_B = -51.7$ mm

12–30. $v = \dfrac{1}{EI}[-30x^3 + 46.25<x-12>^3 - 11.7<x-24>^3 + 38700x - 412560]$ lb $\cdot$ in^3

12–31. $v = \dfrac{1}{EI}[-0.25x^4 + 0.208<x-1.5>^3 + 0.25<x-1.5>^4 + 4.625<x-4.5>^3 + 25.1x - 36.4]$ kN $\cdot$ m^3

12–33. $v = \dfrac{1}{EI}[0.417x^3 - 0.333<x-8>^3 - 0.667<x-16>^3 - 169x]$ kip $\cdot$ ft^3

12–34. $v = \dfrac{1}{EI}[-0.05x^3 - 0.000741x^5 + 0.9<x-9>^3 + 0.0333<x-9>^4 + 0.000741<x-9>^5 + 8.91x]$ kip $\cdot$ ft^3, $v_C = -0.765$ in.

12–35. $\theta_B = 0.574°$, $v_C = 0.200$ in.

12–37. $\theta_A = 0.128° \downarrow$, $\theta_B = 0.128° \nwarrow$

12–38. $v_{max} = -3.64$ mm

12–39. $\theta_B = \dfrac{PL^2}{2EI} \downarrow$, $\Delta_B = \dfrac{PL^3}{3EI} \downarrow$

12–41. $\theta_B = \dfrac{Pa^2}{12EI} \nwarrow$, $\Delta_C = \dfrac{Pa^3}{12EI} \downarrow$

12–42. $\theta_B = \dfrac{Pa^2}{4EI} \nwarrow$, $\Delta_C = \dfrac{Pa^3}{4EI} \uparrow$

12–43. $\Delta_{max} = 12.2$ mm $\downarrow$

12–45. $a = \dfrac{L}{3}$

12–46. $\theta_C = \dfrac{4M_0L}{3EI} \downarrow$, $\Delta_C = \dfrac{5M_0L^2}{6EI} \downarrow$

12–47. $\theta_A = \dfrac{M_0a}{2EI} \downarrow$, $\Delta_{max} = \dfrac{5M_0a^2}{8EI} \downarrow$

12–49. $\theta_C = \dfrac{5Pa^2}{2EI} \downarrow$, $\Delta_B = \dfrac{25Pa^3}{6EI} \downarrow$

12–50. $\theta_A = \dfrac{17Pa^2}{12EI} \downarrow$, $\Delta_{max} = \dfrac{481Pa^3}{288EI} \downarrow$

12–51. $\theta_B = \dfrac{7Pa^2}{4EI} \downarrow$, $\Delta_C = \dfrac{9Pa^3}{4EI} \downarrow$

12–53. $P = 6.43$ lb, $\Delta_B = 2.14$ in. $\downarrow$

12–54. $\theta_A = \dfrac{3PL^2}{8EI} \downarrow$, $\Delta_C = \dfrac{PL^3}{6EI} \downarrow$

12–55. $\Delta_{max} = 0.0674$ in. $\downarrow$

12–57. $\theta_C = \dfrac{wa^3}{EI} \downarrow$, $\Delta_B = \dfrac{41wa^4}{24EI} \downarrow$

12–58. $\theta_A = 0.175° \downarrow$, $\theta_B = 0.175° \downarrow$

12–59. $\theta_B = \dfrac{7wa^3}{12EI} \downarrow$, $\Delta_C = \dfrac{25wa^4}{48EI} \downarrow$

12–61. $\Delta_{max} = \dfrac{3wa^4}{8EI} \downarrow$

12–62. $\theta_A = \dfrac{Pa^2}{4EI} \downarrow$, $\Delta_D = \dfrac{Pa^3}{4EI} \uparrow$

12 63. $\Delta_A = 0.933$ in. $\downarrow$

12–65. $\dfrac{x_{max}}{y_{max}} = \dfrac{I_x}{I_y} \tan\theta$, $y_{max} = 0.736$ in., $x_{max} = 3.09$ in.

12–66. $\Delta_A = 0.916$ in.

12–67. $\Delta_C = 1.90$ in. $\downarrow$

12–69. $\Delta_C = \dfrac{2133}{EI} \downarrow$

12–70. $\theta_A = \dfrac{416}{EI} \downarrow$, $\theta_B = \dfrac{437}{EI} \downarrow$

12–71. $\Delta_A = \dfrac{Pa^2(3b + a)}{3EI}$

12–73. $\Delta_A = PL^3\left(\dfrac{1}{12EI} + \dfrac{1}{8JG}\right)$

12–74. $\Delta_G = 5.82$ in. $\downarrow$

12–75. $A_x = 0$, $B_y = \dfrac{5}{16}P$, $A_y = \dfrac{11}{16}P$, $M_A = \dfrac{3PL}{16}$

12–77. $B_x = 0$, $A_y = 394$ lb, $B_y = 3.21$ kip,
$M_B = 122$ kip $\cdot$ in.

12–78. $A_x = 0$, $B_y = \tfrac{1}{10}w_0L$, $A_y = \tfrac{2}{5}w_0L$, $M_A = \tfrac{1}{15}w_0L^2$

12–79. $C_x = 0$, $A_y = 12.0$ kN, $B_y = 40.0$ kN, $C_y = 12.0$ kN

12–81. $T_{AC} = \dfrac{3A_2E_2wL_1{}^4}{8(A_2E_2L_1{}^3 + 3E_1I_1L_2)}$

12–82. $M_A = \dfrac{wL^2}{12}$, $M_B = \dfrac{wL^2}{12}$

12–83. $M_A = \dfrac{w_0L^2}{20}$, $M_B = \dfrac{w_0L^2}{30}$

12–85. $M_A = \dfrac{M_0}{3}$, $M_B = \dfrac{M_0}{3}$

12–86. $A_y = \dfrac{P}{2}$, $M_A = \dfrac{PL}{8}$, $B_y = \dfrac{P}{2}$, $M_B = \dfrac{PL}{8}$

12–87. $A_y = 81.3$ N, $B_y = 138$ N, $C_y = 18.8$ N

12–89. $A_y = \dfrac{4}{3}P$, $B_y = \dfrac{2}{3}P$, $M_A = \dfrac{PL}{3}$

12–90. $a = 0.414L$

12–91. $T_{CD} = \dfrac{5AwL^3}{8(48I + L^2A)}$

12–93. $B_y = \dfrac{7wL}{128}$, $A_y = \dfrac{57wL}{128}$, $M_A = \dfrac{9wL^2}{128}$, $A_x = 0$

12–94. $\Delta_B = 1.50$ mm $\downarrow$

12–95. $M_A = 0.639$ kip $\cdot$ in., $M_B = 1.76$ kip $\cdot$ in.

12–97. $M_A = \dfrac{PL}{8}$, $M_B = \dfrac{3PL}{8}$, $B_y = P$

12–98. $B_y = 634$ lb, $A_y = 243$ lb, $C_y = 76.8$ lb

12–99. $B_y = C_y = \dfrac{11wL}{10}$, $A_y = D_y = \dfrac{2wL}{5}$

12–101. $\Delta_{\max} = \dfrac{3PL^3}{256EI}$ $\downarrow$

12–102. $v_{\max} = \dfrac{3PL^3}{256EI}$ $\downarrow$

12–103. $v = \dfrac{1}{EI}[4.10x^3 - 0.125x^4 + 0.125 <x - 4>^4$
$- 8.33 <x - 7>^3 - 279x]$ kN $\cdot$ m^3

12–105. $F = \dfrac{P}{4}$

12–106. $v = \dfrac{P}{12EI}[-2x^3 + 3 <x - a>^3 - 2 <x - 2a>^3$
$+ 3 <x - 3a>^3 + 15a^2x - 13a^3]$, $v_{\max} = \dfrac{Pa^3}{3EI}$ $\uparrow$

12–107. $\theta_A = 6.54°$ $\downarrow$

12–109. $v_{\max} = \dfrac{0.00652w_0L^4}{EI}$ $\downarrow$

12–110. $T = 2.79$ kN

Chapter 13

13–1. $d = \tfrac{9}{16}$ in.

13–2. $P_{cr} = 272$ kN

13–3. $P_{cr} = 377$ kip

13–5. $P = 17.6$ kip

13–6. $F.S. = 2.19$

13–7. (a) $P_{cr} = 28.4$ kip, (b) $P_{cr} = 58.0$ kip

13–9. $P = 4.23$ kip

13–10. $P = 73.2$ kip

13–11. $P = 5.79$ kip

13–13. $d_{AB} = 2\tfrac{1}{8}$ in., $d_{BC} = 2$ in.

13–14. $F.S. = 2.38$

13–15. $P = 2.42$ kip

13–17. $w = 1.17$ kN/m

13–18. $w = 5.55$ kN/m

13–19. $d = 8.43$ in., $P_{cr} = 245$ kip

13–21. $d = 4.73$ in.

13–22. $d_{st} = 40.2$ mm, $d_{al} = 68.8$ mm

13–23. $P_{cr} = 20.4$ kip

13–25. $P_{cr} = 5.97$ kip

13–26. $T_2 = \dfrac{\pi^2r^2}{4\alpha L^2} + T_1$

13–27. $T_2 = 133°C$

13–29. $P = 31.4$ kN

13–30. $F.S. = 1.12$

13–31. Column fails

13–33. $P = 113$ kip

13–34. $P = 73.5$ kip

13–35. $P = 76.6$ kip

13–37. $P = 88.5$ kip

13–38. $P_{allow} = 5.71$ kN

13–39. $P_{allow} = 15.6$ kN

13–41. $\sigma_{max} = 2.86$ ksi

13–42. $L = 8.34$ m

13–43. $P = 3.20$ MN, $v_{max} = 70.5$ mm

13–45. $P_{allow} = 7.89$ kN

13–46. $M = 44.3$ kN $\cdot$ m

13–47. $M = 60.5$ kN $\cdot$ m

13–49. $\sigma_{max} = 178$ MPa, $v_{max} = 10.8$ mm

13–50. $P_{cr} = 794$ kN

13–51. $P_{cr} = 3.18$ MN

13–53. $E_t = 14.6(10^3)$ ksi

13–54. $0 < L/r < 35.6$, $P/A = 16\ 449/(L/r)^2$; $35.6 < L/r < 70.2$, $P/A = 13$; $70.2 < L/r$, $P/A = 64\ 152/(L/r)^2$

13–55. $L = 8.99$ ft

13–57. $L = 15.1$ ft

13–58. $L = 22.0$ ft

13–59. $W\ 6 \times 12$

13–61. $W\ 6 \times 9$

13–62. $W\ 6 \times 9$

13–63. $L = 1.92$ ft

13–65. $P_{allow} = 299$ kip

13–66. $P_{allow} = 380$ kip

13–67. $b = 0.808$ in.

13–69. $P_{allow} = 57.6$ kip

13–70. $a = 7\frac{1}{2}$ in.

13–71. $d = 14.2$ in.

13–73. $P_{allow} = 27.7$ kip

13–74. $L = 5.03$ ft

13–75. $P_{allow} = 4.22$ kip

13–77. $P = 98.6$ kip

13–78. $P = 48.0$ kip

13–79. $P = 84.7$ kip

13–81. $P = 1.48$ kip

13–82. $P = 25.0$ kip

13–83. $P = 3.91$ kip

13–85. The column is inadequate

13–86. $P = 5.76$ kip

13–87. $P = 9.01$ kip

13–89. $0 < L/r < 49.7$, $P/A = 61.7(10^3)/(L/r)^2$; $49.7 < L/r < 99.3$, $P/A = 25$ ksi, $99.3 < L/r$, $P/A = 247(10^3)/(L/r)^2$

13–90. $F.S. = 4.72$

13–91. $v_{max} = 0.387$ in.

13–93. The column is adequate

13–94. $d = 1.42$ in.

13–95. The column is adequate

13–97. The column is adequate

Chapter 14

14–1. $V = 224$ ft^3

14–3. $\dfrac{\sigma_x}{\sigma_y} = \nu$

14–5. $U_i = \dfrac{5P^2a^3}{6EI}$

14–6. $U_i = \dfrac{M_0^2L}{24EI}$

14–7. $U_i = \dfrac{w^2L^5}{40EI}$

14–9. $U_i = 496$ J

14–10. $U_i = \dfrac{w^2L^3}{20GA}$

14–11. $U_i = \dfrac{0.00111w^2L^5}{EI}$

14–13. $U_i = 0.129$ ft $\cdot$ kip

14–14. $(U_b)_i = 1.78$ kJ, $(U_a)_i = 51.4\ \mu$J

14–15. $(U_b)_i = 62.5$ J, $(U_i)_{sp} = 0.391$ J

14–17. $(U_b)_i = 10.1$ ft $\cdot$ lb, $(U_a)_i = 0.344$ in. $\cdot$ lb

14–18. $U_i = \dfrac{P^2r^3}{GJ}\left(\dfrac{3\pi}{8} - 1\right)$

14–21. $\Delta_D = \dfrac{3.50\ PL}{AE}$

14–22. $\Delta_A = 0.536$ mm

14–23. $\Delta_B = 2.42$ in.

14–25. $\theta_B = \dfrac{M_0L}{12EI}$

14–26. $\theta_A = \dfrac{4M_0a}{3EI}$

14–27. $\theta_C = 1.93^\circ$

14–29. $\Delta_E = 5.46$ in.

14–30. $\theta_E = 3.15^\circ$

14–31. $\Delta = \dfrac{Pr^3\pi}{2}\left(\dfrac{3}{GJ} + \dfrac{1}{EI}\right)$

14–33. (a) 4.52 kJ, (b) 3.31 kJ

14–34. (a) $\sigma_{max} = 45.4$ ksi, (b) $\sigma_{max} = 509$ psi, (c) $\sigma_{max} = 254$ psi

14–35. $h = 1.75$ ft

14–37. $h = 69.6$ mm

14–38. $\sigma_{max} = 98.1$ MPa, $\Delta_{max} = 0.981$ mm

14–39. $\sigma_{max} = 7.75$ ksi

14–41. $\sigma_{max} = 43.6$ ksi

14–42. $\sigma_{max} = 20.3$ ksi

14–43. $\sigma_{max} = 26.0$ ksi

14–45. $h = \dfrac{\sigma_{max}L^2}{3Ec}\left[\dfrac{\sigma_{max}I}{WLc} - 2\right]$

14–46. $\sigma_{max} = 4.49$ ksi

14–47. $\Delta_{max} = 3.95$ mm, $\sigma_{max} = 237$ MPa

14–49. $\Delta_B = 0.226$ in., $\sigma_{max} = 3.22$ ksi

14–50. $\Delta_{max} = 5.41$ in., $\sigma_{max} = 27.6$ ksi

14–51. $n = 16.7$

14–53. $\sigma_{max} = 10.1$ ksi, $\Delta_{beam} = 0.481$ in.

14–54. $\Delta_{B_h} = 0.00191$ in.

14–55. $\Delta_{B_v} = 11.3$ mm

14–57. $\Delta_{C_h} = 0.0482$ in.

14–58. $\Delta_{B_h} = 0.0420$ in.

14–59. $\Delta_{B_v} = 3.38$ mm

14–61. $\Delta_{B_h} = 0.367$ mm

14–62. $\Delta_{C_v} = 0.0375$ mm

14–63. $\Delta_{C_v} = 20.4$ mm

14–65. $\Delta_{C_v} = 0.163$ in.

14–66. $\Delta_{H_v} = 0.156$ in.

14–67. $\Delta_C = \dfrac{PL^3}{48EI}$, $\theta_A = \dfrac{PL^2}{16EI}$

14–69. $\Delta_C = \dfrac{2Pa^3}{3EI}$, $\theta_A = \dfrac{Pa^2}{6EI}$

14–70. $\theta_B = 0.353°$, $\Delta_C = 0.369$ in.

14–71. $\Delta_B = 1.54$ in., $\theta_A = 2.73°$

14–73. $\Delta_C = \dfrac{5M_0 a^2}{6EI}$, $\theta_A = \dfrac{M_0 a}{6EI}$

14–74. $\Delta_D = 3.24$ mm, $\theta_A = 0.289°$

14–75. $\Delta_B = \dfrac{65wa^4}{48EI}$, $\theta_A = \dfrac{7wa^3}{6EI}$

14–77. $\Delta_A = 57.9$ mm, $\theta_C = 0.510°$

14–78. $\Delta_C = 3.43$ in., $\theta_A = 0.341°$

14–79. $\Delta_C = \dfrac{w_0 L^4}{120EI}$, $\theta_A = \dfrac{5w_0 L^3}{192EI}$

14–81. $\Delta_D = 22.5$ mm

14–82. $\Delta_C = 77.4$ mm

14–83. $\Delta_{C_v} = 0.112$ in.

14–85. $\Delta_A = \dfrac{4PL^3}{3EI}$

14–86. $\Delta_C = \dfrac{5wL^4}{8EI}$

14–87. $\Delta_B = \dfrac{wL^4}{4EI}$

14–89. $\Delta_{B_h} = 0.00191$ in.

14–90. $\Delta_{B_v} = 11.3$ mm

14–91. $\Delta_{C_h} = 0.0482$ in.

14–93. $\Delta_{B_h} = 0.0420$ in.

14–94. $\Delta_{B_v} = 3.38$ mm

14–95. $\Delta_{B_h} = 0.367$ mm

14–97. $\Delta_{C_v} = 0.0375$ mm

14–98. $\Delta_{C_v} = 20.4$ mm

14–99. $\Delta_{C_v} = 0.163$ in.

14–101. $\Delta_{H_v} = 0.156$ in.

14–102. $\Delta_C = \dfrac{PL^3}{48EI}$, $\theta_A = \dfrac{PL^2}{16EI}$

14–103. $\Delta_C = \dfrac{2Pa^3}{3EI}$, $\theta_A = \dfrac{Pa^2}{6EI}$

14–105. $\theta_B = 0.353°$, $\Delta_C = 0.369$ in.

14–106. $\Delta_B = 1.54$ in., $\theta_A = 2.73°$

14–107. $\Delta_C = \dfrac{5M_0 a^2}{6EI}$, $\theta_A = \dfrac{M_0 a}{6EI}$

14–109. $\Delta_D = 3.24$ mm, $\theta_A = 0.289°$

14–110. $\Delta_B = \dfrac{65wa^4}{48EI}$, $\theta_A = \dfrac{7wa^3}{6EI}$

14–111. $\Delta_A = 57.9$ mm, $\theta_C = 0.510°$

14–113. $\Delta_C = 3.43$ in., $\theta_A = 0.341°$

14–114. $\Delta_C = \dfrac{w_0 L^4}{120EI}$, $\theta_A = \dfrac{5w_0 L^3}{192EI}$

14–115. $\Delta_A = \dfrac{4PL^3}{3EI}$

14–117. $\Delta_{C_h} = \dfrac{5wL^4}{8EI}$

14–118. $\Delta_{C_v} = \dfrac{1,228,800}{EI}$ lb · ft³, $\Delta_{C_h} = \dfrac{640,000}{EI}$ lb · ft³

14–119. $\Delta_B = 2.15$ in.

14–121. $\Delta_C = 40.9$ mm

14–122. $\Delta_C = 40.9$ mm

14–123. $\theta_A = 0.171°$

14–125. $\Delta_{A_v} = 33.1$ mm

14–126. $\Delta_{A_v} = 33.1$ mm

14–127. $L = 850$ mm

14–129. $\Delta_C = \dfrac{7wL^4}{384EI}$, $\theta_B = \dfrac{wL^3}{48EI}$

14–130. $\theta_B = \dfrac{M_0 L}{EI}$

14–131. $U_i = \dfrac{w_0^2 L^5}{504EI}$

Appendix A

A–1. $\bar{y} = 2$ in.

A–2. $\bar{y} = 37.5$ mm, $\bar{I}_{x'} = 16.3(10^6)$ mm^4

A–3. $I_y = 94.8(10^6)$ mm^4

A–5. $\bar{y} = 207$ mm, $\bar{I}_{x'} = 222(10^6)$ mm^4, $\bar{I}_{y'} = 115(10^6)$ mm^4

A–6. $\bar{x} = 68.0$ mm, $\bar{I}_{x'} = 49.5(10^6)$ mm^4, $\bar{I}_{y'} = 36.9(10^6)$ mm^4

A–7. $\bar{x} = 3$ in., $\bar{y} = 2$ in., $\bar{I}_{x'} = 64$ in^4, $\bar{I}_{y'} = 136$ in^4

A–9. $I_{xy} = 36$ in^4

A–10. $I_{xy} = 97.75$ in^4

A–11. $\bar{x} = 33.9$ mm, $\bar{y} = 150$ mm, $\bar{I}_{x'} = 101(10^6)$ mm^4, $\bar{I}_{y'} = 10.8(10^6)$ mm^4

A–13. $\bar{x} = 85$ mm, $\bar{y} = 35$ mm, $I_{x'y'} = -7.50(10^6)$ mm^4

A–14. $I_{x'} = 85.3(10^6)$ mm^4, $I_{y'} = 85.3(10^6)$ mm^4

A–15. $I_{x'} = 5.09(10^6)$ mm^4, $I_{y'} = 5.09(10^6)$ mm^4, $I_{x'y'} = 0$

A–17. $I_{max} = 4.92(10^6)$ mm^4, $I_{min} = 1.36(10^6)$ mm^4

A–18. $I_{max} = 4.92(10^6)$ mm^4, $I_{min} = 1.36(10^6)$ mm^4

A–19. $I_{max} = 64.1$ in^4, $I_{min} = 5.33$ in^4

Index